D1762168

STELLAR EVOLUTION

STELLAR EVOLUTION

Edited by
Hong-Yee Chiu
and
Amador Muriel

QB
801
S7

The MIT Press

Cambridge, Massachusetts, and London, England

WITHDRAWN
OHIO UNIVERSITY
LIBRARY

Copyright © 1972 by
The Massachusetts Institute of Technology

Printed and bound
in the United States of America.

All rights reserved. No part of this book may be reproduced in any form or by any means, electronic or mechanical, including photocopying, recording, or by any information storage and retrieval system, without permission in writing from the publisher

Library of Congress Cataloging in Publication Data
Main entry under title:

Stellar evolution.

Based on lectures given at the 3d Summer Institute for Astronomy and Astrophysics, held at the State University of New York at Stony Brook, June 18-July 16, 1969.

1. Stars--Evolution--Addresses, essays, lectures. I. Chiu, Hong-Yee, 1932- ed. II. Muriel, Amador, ed. III. Summer Institute for Astronomy and Astrophysics, 3d, State University of New York at Stony Brook, 1969.

QB801.S7 523.8 79-38325
ISBN 0-262-03043-8

WITHDRAWN
LIBRARY

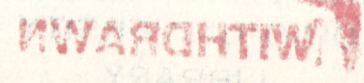

PUBLISHER'S NOTE

The aim of this format is to close the time gap between the preparation of certain works and their publication in book form. A large number of significant though specialized manuscripts make the transition to formal publication either after a considerable delay or not at all. The time and expense of detailed text editing and composition in print may act to prevent publication or so to delay it that currency of content is affected.

The text of this book has been photographed directly from the author's typescript. It is edited to a satisfactory level of completeness and comprehensibility though not necessarily to the standard of consistency of minor editorial detail present in typeset books issued under our imprint.

The MIT Press

PREFACE

Stellar Evolution is based on lectures given at the Third Summer Institute for Astronomy and Astrophysics held at the State University of New York at Stony Brook from June 18 to July 16, 1969. Director of the program was Dr. Hong-Yee Chiu; associate director, Dr. Richard Stothers. The Summer Institute, supported by a grant from the National Aeronautics and Space Administration, was organized by the University's Department of Earth and Space Sciences in cooperation with the Goddard Institute for Space Studies in New York.

Stellar evolution has become one of the most active topics of research in astrophysics in recent years. Attention was first centered around various nuclear processes inside a star and has now expanded to cover processes hitherto regarded as speculative, including neutron stars and gravitational waves. These lectures are presented more or less in the order of the life of a star--from main sequence to white dwarfs, nova, supernova, and neutron stars--beginning with basic principles and developing into problems of current interest. This material should be of interest to workers in the field as well as to physicists who wish to take a quick glance into the problem of stellar evolution.

The editors are grateful to the authors for their respective contributions; to Dr. John Fix and Dr. Richard Sears, Mrs. Miriam Forman, Miss Mary Ann Sweeney, Mr. S. Ridgway, and Mr. J. Warman for their invaluable assistance in the preparation of the manuscripts; to Miss Mary Yastishak for her assistance in the planning of the conference and in coordinating the manuscripts; to Miss Miriam Koral and Miss Mary Marvin for the typing of the manuscripts.

Hong-Yee Chiu
Amador Muriel

INTRODUCTION

The universe around us appears to be shrouded with nothing but stars, and in fact they constitute over 90 percent of the visible universe. Modern stellar evolution theory may well have started when Hipparchus looked with bewilderment at the appearance and disappearance of a nova around 150 B.C., wondering at the validity of the ancient belief that the stars are permanent and perfect. For a long time afterward, every attempt to unravel the mysteries of stellar life ended in vain, for the most important fact about a star--its energy sources--was not understood until 1939. Further, the tool most needed for studying the life of a star (the fast electronic computer) did not come into general use until the 1960s--the same decade as the crucial discovery of neutron stars, which are now identified with pulsars.

The field of stellar evolution is composed of many branches of physics: from the celestial mechanics formulated by Newton to general relativity; from the ancient subject of hydrostatics to the most modern topics of relativistic shock waves; from Boyle's law to Fermi degeneracy; from corpuscles to neutrinos. Each part of this knowledge, standing alone, may appear quite unimpressive. Yet when all this knowledge is put together, a comprehensive picture of the lives of stars emerges.

We believe that a star is born when an interstellar gas cloud condenses under its own gravitation. We know that the density involved is quite low, being about one hydrogen atom per cubic centimeter. We also know that the temperature and magnetic field are small, being in the neighborhood of 3°K and 10^{-5}-10^{-6} gauss, respectively. However, very little is known for certain about the actual processes and the events that lead to the formation of stars. Even more puzzling is the fact that some apparent stars-in-condensation are strong emitters of molecular lines in the radio band (and in some cases, even involve maser mechanism). Skipping this interesting but currently difficult-to-understand episode of a star's life, we come to the next stage: the onset of nuclear burning. This is the topic presented in Chapters 1-3 (Iben, Demarque, Paczyński).

Iben and Demarque's presentations center around the state of a star following partial hydrogen exhaustion to the red giant stage; Paczyński's work reaches even further: from main sequence to white dwarf or carbon ignition. However, the validity of Paczyński's work depends on several factors that are still unknown.

The evolutionary tracks of stars differ according to their masses, as is clear from discussions by Iben and Demarque. It is also known that in massive stars the dominance of the radiation pressure causes instabilities. What are the properties of stars adjacent to this instability region? Immense interest lies in the properties of these stars, for their relatively short lives make them potential progenitors of supernovae and even black holes. Because of the increased importance of radiation pressure and other properties not found in less massive stars, their study requires special treatment, which is discussed in Chapter 4 (Stothers).

The light from some stars varies with a well-regulated period. These are δ-Cepheid and RR Lyrae variables. The periods of δ-Cepheids are long (1-100 days) and those of the RR Lyrae stars are relatively short (2 hours-1 day). They occupy narrow regions in the H-R diagram. Why do these stars pulsate? This has been a puzzling question since their discovery. Now the mechanisms are fairly well understood: In both cases, the pulsations are due to sudden and drastic decreases of opacity when hydrogen or helium go from unionized to ionized states. This is discussed by Baker (Chapter 5), who concentrates on applications to realistic stellar models including nonlinear effects, and by Christy (Chapter 6).

About 10 percent of the stars in the solar neighborhood are white dwarfs, which are dense objects (density $\sim 10^6$ g/cm^3) apparently devoid of nuclear energy source and regarded by many as dead stars. The interpretation of white dwarfs as degenerate stars by Chandrasekhar in the early 1930s marked an important development both in astrophysics and in the then newly developed quantum mechanics. Since Chandrasekhar's time, many more facts about white dwarfs have become known. These are described in Ostriker's article on white dwarfs (Chapter 7).

Some stars are close binaries with periods of days or hours. The evolution of binaries is interesting because the mutual gravitational field of two component stars distorts each other's structure, permitting stellar structure theory to be tested. Furthermore, some close binaries are believed to be progenitors of novae, which show recurrent stellar outbursts with a peak luminosity of around $10^6 L_\odot$ and an ejection of matter of around 10^{-6} $M_\odot$. These problems are discussed in Chapters 8 and 9 (Paczyński, "Close Binaries"; Rose, "Novae").

An even rarer event, and a more intense one, is the outburst of a supernova, occurring at a rate of once every 50-300 years per galaxy. (By contrast, the outburst rate of novae is about 20 per year per galaxy.) During the peak of the outburst, the luminosity of a supernova is about 10^9 $L_\odot$, the mass ejected is a sizable fraction of the total mass of the star, and the total energy emitted during the outburst measures a large part of the entire nuclear energy content of a normal star. It is believed that gravitational collapse is the cause of a supernova. This phenomenon is discussed by Colgate and McKee in Chapter 10.

Under gravitational collapse, the central densities of a supernova can reach 10^{10} g/cm^3 and greater. As early as 1932, Landau, following the earlier success of Chandrasekhar in the explanation of white dwarfs, predicted that under densities even greater than that of white dwarfs, 10^{12} g/cm^3 or greater, electrons will "combine" with protrons to form neutrons, and stellar configurations can exist with a predominant composition of neutrons. The average density of neutron stars is close to that of nuclear matter, around 10^{14} g/cm^3. The radius of neutron stars is not an astronomical number, but a meager 10 km. Although a fairly complete theory of neutron stars was given by Oppenheimer and Volkoff in 1939, and despite the many theories of supernovae which predict the inescapable existence of neutron stars, most astronomers remained skeptical. The theorists were vindicated, however, when pulsars were discovered and identified as neutron stars. In Chapter 11, Cameron gives a comprehensive survey of the properties of neutron stars. In Chapter 12, Chiu gives an account of the properties of pulsars as well as some theoretical interpretations.

Chapters 13 and 14 (Philip, "Photometry of Field Horizontal-Branch Stars," and Strom, "Population I Helium Abundances") describe some observational phenomena that bear on stellar evolution. Special attention must be paid to the helium abundance of the stars. In most stars (including our sun) the helium lines are weak, and the abundance is difficult to determine. Nevertheless, the helium abundance is among the most important input data in stellar evolution calculations.

Chapters 15-17 (Carson, Spiegel, and Truran) summarize our current knowledge of stellar opacities, transport processes, and nuclear reaction rates. The age-old problem of stellar opacity was finally solved by using a computer to evaluate opacities, point by point, taking into account all available facts and knowledge of atomic convection. The best theory available is a phenomenological theory (the mixing length theory). Wherever a transition from convective transport to radiative transport takes place, or vice versa, the mixing length theory is used. However, the uncertainty of the magnitude of the mixing length makes it difficult to ascertain the surface properties of stars that have convective envelopes. Spiegel reviews these problems and suggests a solution. The thermonuclear reaction rate is another important input to stellar structure theory. With improved measurements of nuclear cross sections, our knowledge of the thermonuclear reaction rate is improved. Truran summarizes this process.

In Chapter 18, Rose discusses the instability problems associated with nuclear burning shells. The idea is to investigate how the nuclear star of a planetary nebula can eject a sizable fraction of matter to form a gas shell around the nuclear star, and why a nova erupts. Both types of stars are believed to be very old, presumably having nearly reached the end of their nuclear burning phase, when all available nuclear fuels are located near or at the surface. In Chapter 19, Thorne discusses an old problem that has become increasingly important in view of the discovery of pulsars and the possible detection of gravitational waves. Although for most stars general relativistic theory is not needed, it is almost mandatory for the discussion of neutron stars because of the high density involved. The recent possible discovery of gravitational radiation from our galactic center poses an even more perplexing problem of the energy source and mechanism of emission. The theory of general relativity is usually quite complicated, but Theorne presents here an easy-to-follow account that outlines the most important features of the theory while avoiding too-technical mathematical details.

The problem of magnetic fields in stars is an intriguing one. Normal stars with average field strengths up to 3.4×10^4 gauss have been discovered, using Zeeman line splitting methods. What maintains this field? This question is discussed by Mestel (Chapter 20), who also summarizes current theories of the origin of magnetic fields in stars. A more radical approach is taken by Canuto and Chiu (Chapter 21) on the properties of intense magnetic fields in astrophysics. It is fairly well established that the magnetic fields of neutron stars are as strong as 10^{12}G. In terms of the classical interaction energy $\mu_e H$ (where μ_e is the Bohr magnetron for electron and H is the field strength), it is a large fraction of the electron rest energy. Various properties of intense magnetic fields are discussed and summarized in this chapter.

This book closes with an unusual discussion: What happens when two stars collide? Stellar collision is known to be a very rare phenomenon in our galaxy (the estimated rate is about 10 collisions per 10 billion years per galaxy). However, in very dense star clusters, collision may be more frequent, and in some theories of quasars, stellar collision is taken to be the activity mechanism. Cameron, in Chapter 22, presents a detailed computer computation of a particular stellar collision phenomenon (head-on collision) that, in this case, resulted in stellar coalescence.

Hong-Yee Chiu

1

NORMAL STELLAR EVOLUTION

Icko Iben, Jr.

Massachusetts Institute of Technology
Cambridge, Mass.

I. WHAT DO WE KNOW FROM THE OBSERVATIONS?

The purpose of the theoretical study of the structure and evolution of model stars is to account as well as possible for the observed properties of real stars--their energy output, their color, their pulsation characteristics, and so forth--and to relate their external appearance to their internal structure.

Before going into the physics of matter in the stellar interior and thence into the theory of stellar evolution, I will give a theoretician's oversimplified view of some of the observations which are to be explained.

The Hertzsprung-Russell Diagram

Traditionally, the most readily accessible properties of real stars have been most frequently compared with the corresponding properties of model stars in the Hertzsprung-Russell (H-R) diagram. Individual stars are located in this diagram according to their observed or theoretical magnitude and color. For many years it has been customary among observational astronomers to use the coordinates V and B-V, where V is the visual magnitude and B is the blue magnitude. These quantities are measured photographically with relatively narrow-band filters.

The theoretician traditionally places stars in an H-R diagram with coordinates log L and log T_e, where L is the total luminosity and T_e is the effective surface temperature. Identical data will give similar curves in the H-R diagram regardless of which coordinates are used; nevertheless, the transformation from one system to the other is difficult and draws heavily on the theory of stellar atmospheres.

From the definition of T_e we have a relation between T_e, L, and the stellar radius R,

$$\frac{L}{4\pi R^2} = \sigma T_e^{4} \quad . \tag{1}$$

Consequently, it is possible to draw surfaces of constant R in the H-R diagram as shown in Figure 1. Where stars live in this diagram summarizes the total information we have about most stars.

Observed Properties of Nearby Stars

"Nearby" stars are those whose distances can be measured directly by trigonometric parallax (i.e., stars within about 100 pc). They can be placed absolutely in the H-R diagram. That is, V can be converted into M_V. The majority of nearby stars appear in the H-R diagram in a relatively narrow band, as in Figure 2. This band is known as the main sequence.

Stars with surface temperature below about 8000°K, B-V $\sim$ 0.2, (lower main sequence stars) rotate rather slowly with equatorial velocities on the order of 2 km/sec while stars with T_e larger than 8000°K (upper main sequence stars) are fast rotators with equatorial velocities frequently on the order of 200 km/sec.

For about 100 of the nearby stars which are members of binary systems it is possible to determine masses, and one can construct a mass-luminosity diagram with the appearance shown in Figure 3.

In addition to the main sequence stars, there are giant stars, generally of higher luminosity but lower surface temperature than main sequence stars, and white dwarfs of low luminosity but high surface temperature. These types are approximately located in the H-R diagram in Figure 2.

In order to pick up enough variable stars for a statistically significant statement, it is necessary to explore out beyond several kpc from the sun. One finds that the Cepheids and the RR Lyrae stars lie, in the H-R diagram, within a fairly narrow band of slope opposite in sign to that of the main sequence band. This band is known as the "instability strip."

The RR Lyrae variables are typically 20-80 times as luminous as the sun with periods on the order of a few hours to a day; the δ-Cepheids are 10^3 - 10^4 times as luminous as the sun with periods of a few to roughly a hundred days.

Clusters

It is possible to obtain a great deal of useful information by examining the distribution in the H-R diagram of the stars of a single cluster. One hopes that the stars of a cluster are characterized by the same initial composition. It is also to be hoped that all of the stars of a given cluster are of approximately the same age. In practice, this hope is justified only for clusters that are older than about 5×10^8 yr.

A typical example - the Pleiades - is shown in Figure 4. The slope of the lower portion coincides fairly well with that of the main sequence defined by nearby stars. From the mass-luminosity diagram (Figure 3) we see that more luminous stars are also more massive. On comparing with Figure 4 we infer that, the more massive a star is, the more it departs from the main sequence. We infer further that, the more

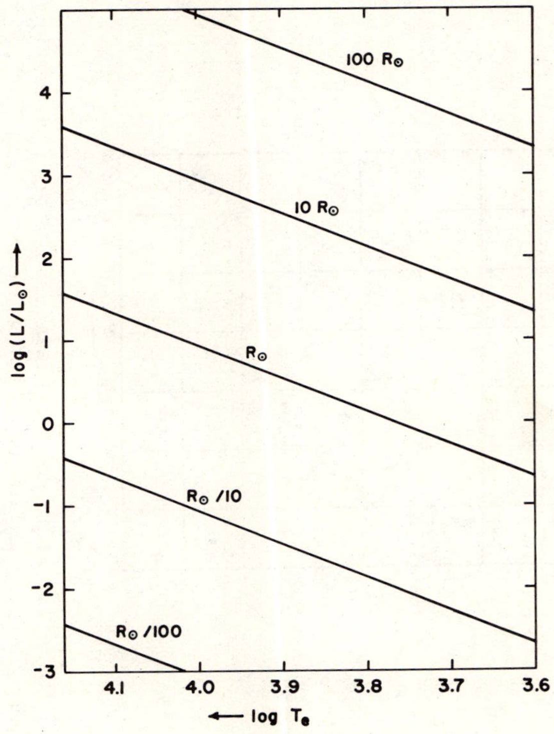

Figure 1. Lines of constant radius in the theoretical H-R diagram. L = stellar luminosity, T_e = stellar surface temperature.

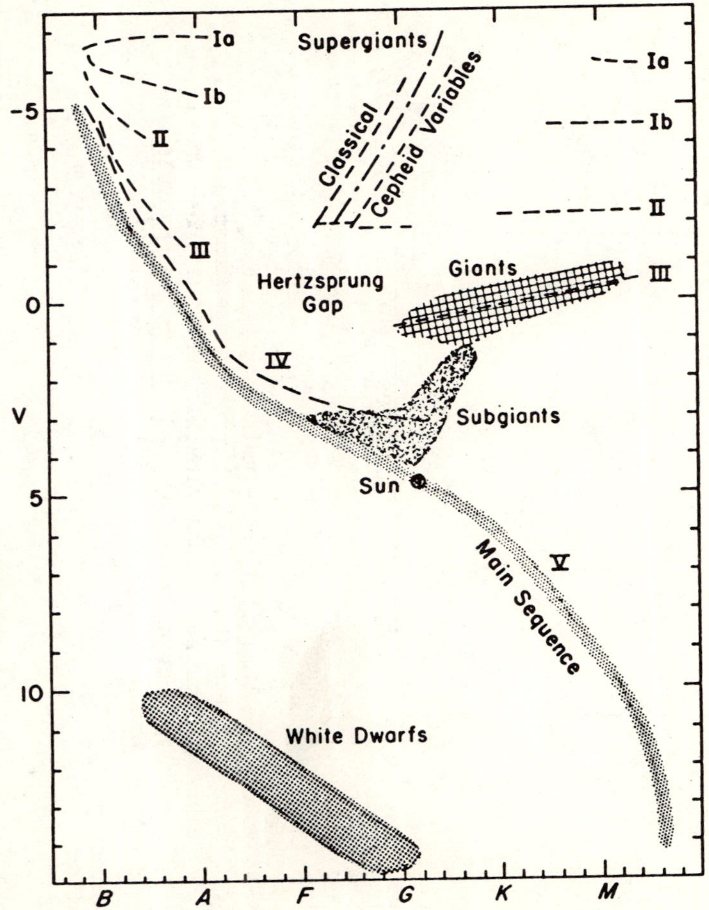

Figure 2. Absolute visual magnitude V as a function of spectral class. From Aller, L.H., Astrophysics: The Atmospheres of the Sun and Stars (Ronald, New York, 1963)

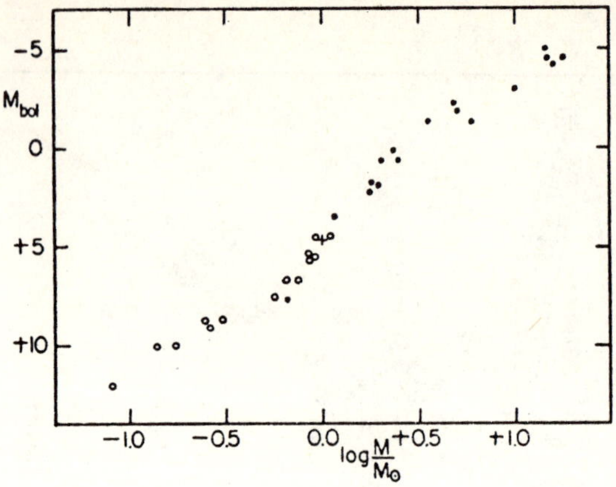

Figure 3. Empirical mass-luminosity relationship for main sequence stars. Dots represent spectroscopic binaries, circles visual binaries, and the cross the sun. From Schwarzschild, M., Structure and Evolution of the Stars (Princeton, 1958).

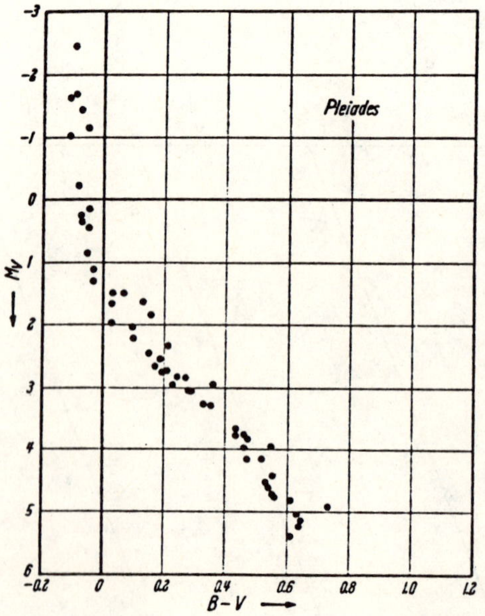

Figure 4. Stars of the Pleiades Cluster located in the Hertzsprung-Russell diagram. From Arp, H. C., Handbuch der Physik, Vol. 51, 1958.

massive a star is, the more rapidly it evolves.

A typical characteristic of stellar distributions in the H-R diagram prepared from observational data is some degree of scatter. This is partially due to parameters, such as rotation rate or the presence of an unresolvable companion, which vary from one star to another within a cluster. It is perhaps due predominantly to inevitable observational uncertainties.

An interesting characteristic found in the distribution in the H-R diagram of stars of several clusters is exhibited by the M67 cluster locus in Figure 5.

Along the main sequence portion of the locus there is a definite gap where very few stars appear. Either stars do not occur in this region, or they evolve through it very rapidly. The significance of this gap will be discussed in some detail later.

A number of cluster loci are represented in a composite diagram in Figure 6. In view of our observation that initially more luminous (hence more massive) stars seem to evolve more rapidly, we may argue that the clusters with a lower turning point are older than those with a higher turning point. A gap appears below turnoff along the loci of several of the older clusters.

All of the clusters shown so far have been from the galactic disk and are relatively young and rich in heavy elements. Stars from the galactic halo are typically old and relatively metal poor. Among these halo stars are the globular clusters - spherical groupings of 10^5 to 10^6 stars. A "typical" example (M3) is shown in Figure 7.

Characteristic of the globular cluster loci is a prominent horizontal branch made up of stars that are about 3.5 magnitudes brighter than turnoff point stars. The apparent gap in the distribution of stars along the horizontal branch is known as the RR Lyrae gap. The gap is actually populated by variable stars. Rather than represent them by blurs or by mean values over a cycle, the RR Lyrae stars have in the past been customarily omitted from globular cluster distributions in the H-R diagram. Hence the misleading usage of the word "gap." Perhaps the words "instability strip" will supplant it.

If we examine a number of globular clusters, we find that the appearance of the horizontal branch varies significantly. In particular, the number of stars occurring to the right or left of the RR Lyrae "gap" or "strip" varies, as does the number of RR Lyrae variables. This is shown in the celebrated study of clusters by Arp as summarized in Figure 8. The magnitude difference between the turnoff point and the RR Lyrae stars is nearly always in the neighborhood of 3.4 or 3.5 magnitudes, regardless of the metallicity of the cluster. See Figure 9.

An Example of How Theory May be Compared with Observations

Once a theory of stellar evolution has been developed we can test it by using the theory to generate a time constant locus in the H-R diagram from model stars of different mass,

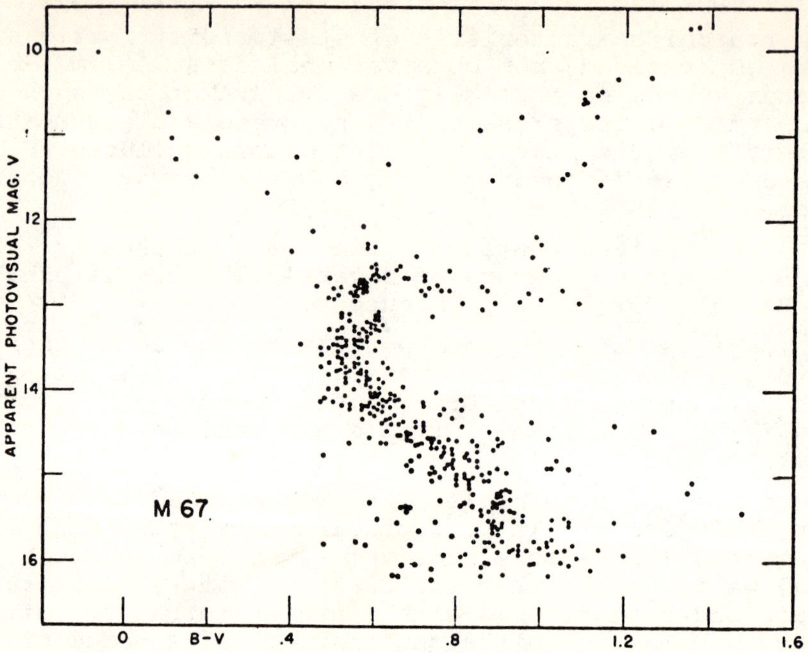

Figure 5. Stars of the cluster M67 located in the H-R diagram. From Johnson, H. L., and Sandage, A., Ap. J., 121, 616 (1955).

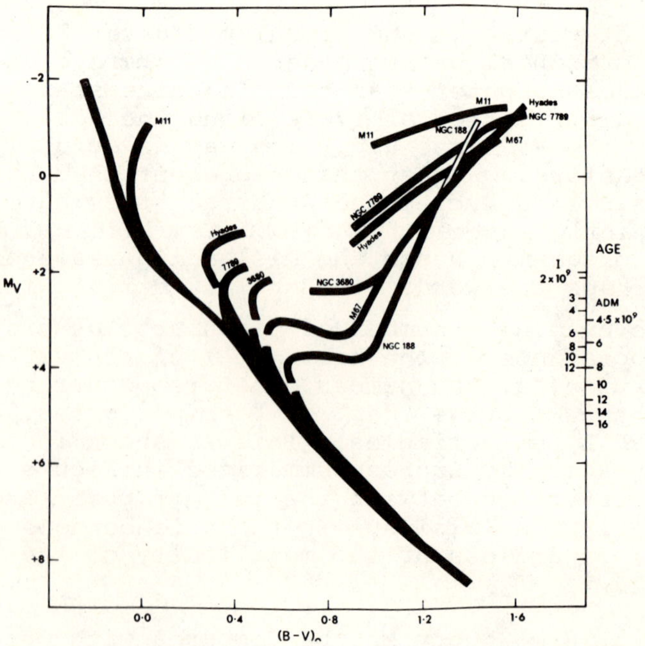

Figure 6. A composite of cluster loci from Sandage (preprint 1969).

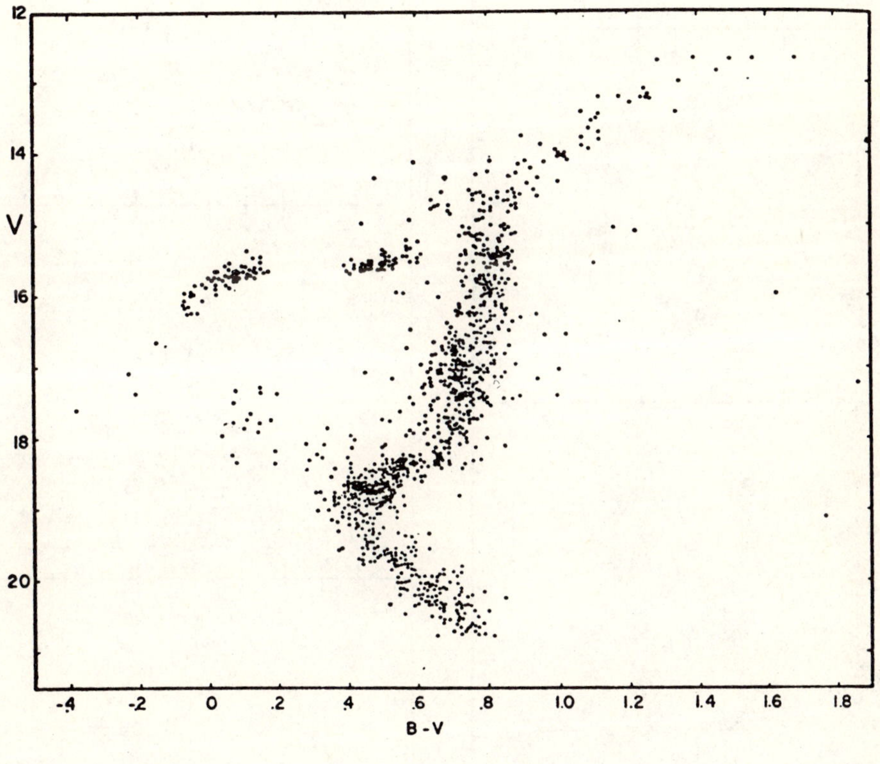

Figure 7. Stars of the globular cluster M3 located in the H-R diagram. From Johnson, H. L., and Sandage, A., Ap. J., 124, 379 (1956).

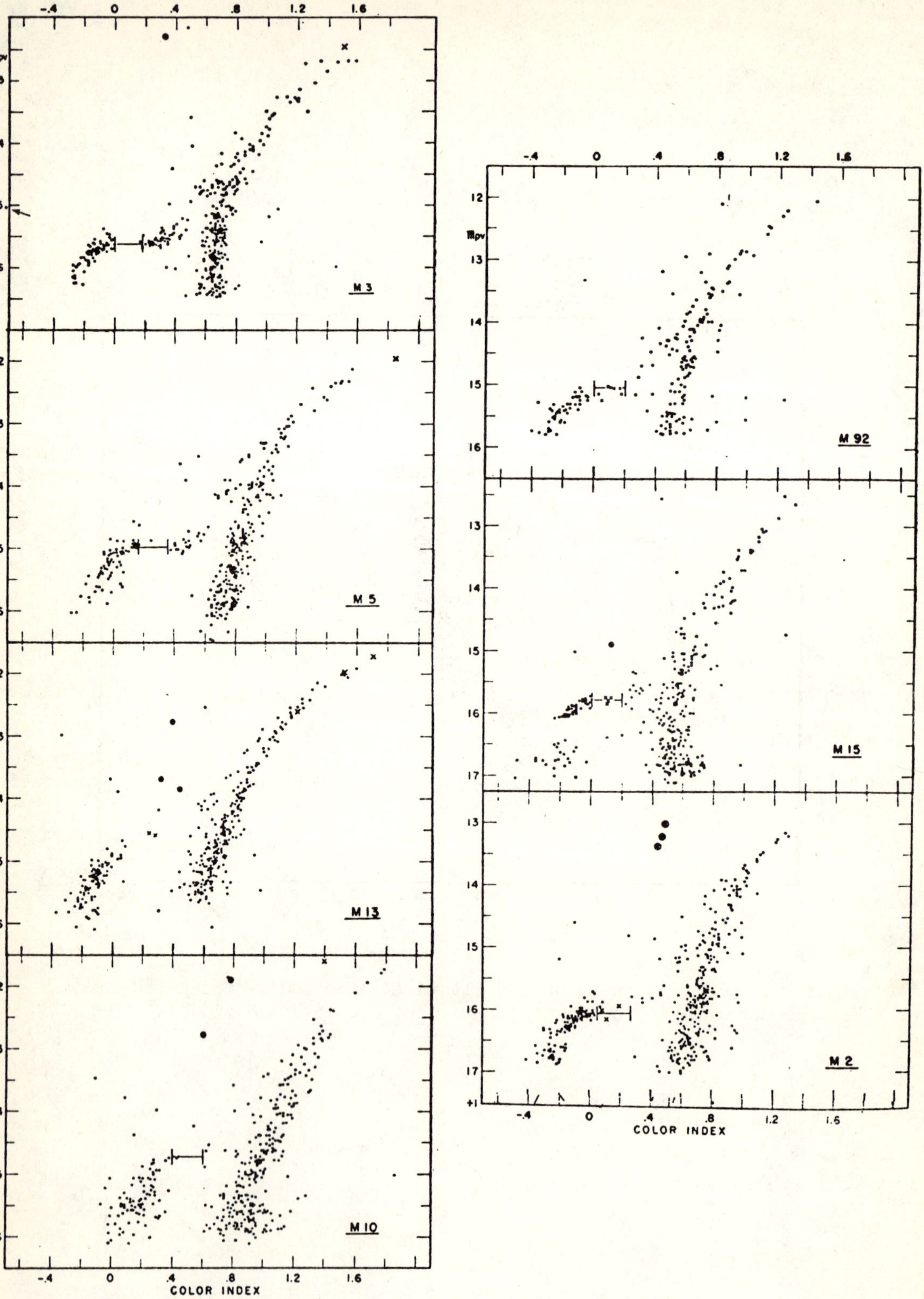

Figure 8. Color-magnitude diagrams for globular clusters. Circled crosses represent cepheids, large crosses long-period or irregular variables, and small crosses RR Lyrae stars. From Arp, A. C., Astronomical Journal, 60, 318 (1955).

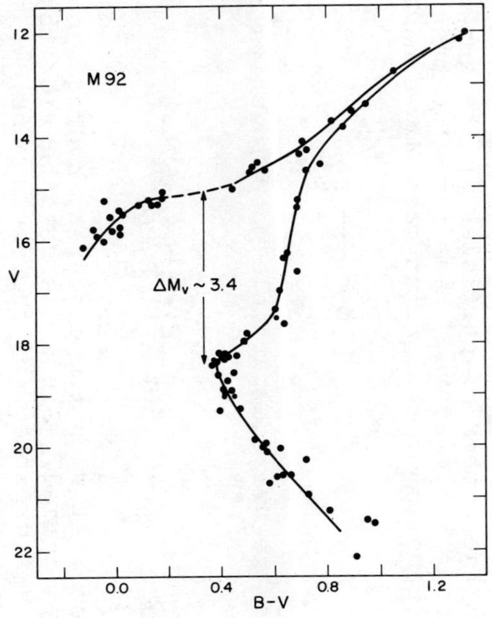

Figure 9. Metal-poor cluster locus as defined by stars in M92. Courtesy of Allan Sandage.

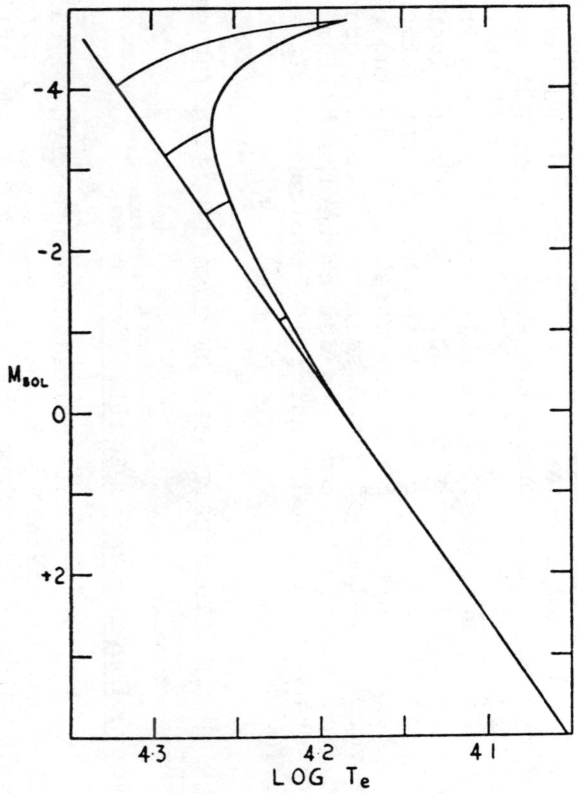

Figure 10. Original main sequence and time constant locus for $t = 4.78 \times 10^7$ yr. Several individual evolutionary tracks are also shown. From Taylor, R., Ap.J., 120, 340 (1954).

but of the same age and initial composition. This locus can then be compared with a locus provided by nature - i.e., a locus defined by the distribution of stars in the H-R diagram for a specific cluster.

In practice, the composition is assigned by choosing values for the parameters (X,Y,Z) which are the abundances by mass of hydrogen, helium and the sum total of all other heavy elements. We then pick a time and compute the evolution of stars of various masses. Hopefully, the resulting theoretical locus in the H-R diagram will coincide with the locus provided by nature for a cluster of the given composition and age and containing stars of various masses. By making computations for various values of Y and seeking a fit with observation one might, for example, attempt to determine the helium concentration in stars of a given cluster. A typical result is shown in Figure 10.

II. ORDER OF MAGNITUDE ESTIMATES OF CONDITIONS IN THE STELLAR INTERIOR

Consequences of Hydrostatic Equilibrium

For the sun we are able to measure directly the radius $R = 7 \times 10^{10}$ cm, the mass $M = 2 \times 10^{33}$ gm, and the luminosity $L = 3.9 \times 10^{33}$ ergs/sec. From the fact that fossils of life exist that are more than 3 billion years old, we infer that the solar luminosity has been reasonably constant over that period of time and that, therefore, the sun must be in hydrostatic and thermal equilibrium.

The assumption of hydrostatic equilibrium implies that the thermal energy content of the sun (which supplies the pressure) is roughly equal to the gravitational binding energy. If the interior is in a gaseous state then

$$NkT \sim \frac{GM^2}{R}, \qquad (2)$$

where N is the number of particles in the sun and T is some mean interior temperature. Measuring the mass of a single particle in units of the proton mass M_p we have

$$N = \frac{M}{\mu M_p} \qquad (3)$$

where μ is the average particle mass in units of M_p. In these units, then,

$$T \sim \left(\frac{\mu M_p}{k}\right)\left(\frac{GM}{R}\right) \sim 10^7 \text{ °K}, \qquad (4)$$

for the sun. At such temperatures, hydrogen and helium are completely ionized. Since elements heavier than helium constitute only a few percent of the sun's mass, we may therefore approximate

$$\mu^{-1} \sim (5X + 3)/4, \qquad (5)$$

where X is the abundance by mass of hydrogen.

We can rediscover the relationship between temperature, mass and radius in a slightly different fashion. In hydrostatic equilibrium the pressure differential across a spherical shell must exactly balance the gravitational force on the matter within that shell. Thus

$$\frac{\partial P}{\partial r} = -\frac{GM(r)}{r^2}\rho(r) \tag{6}$$

where M(r) is the "shell" mass; i.e., the total stellar mass contained between the spherical shell in question and the center. Let us make the following substitutions:

$$\begin{aligned} \frac{\partial P}{\partial r} &\to -\frac{P_c}{R}, \\ M(r) &\to M/2, \\ \rho(r) &\to \rho_c/2, \\ r &\to R/2 \end{aligned} \tag{7}$$

where the subscript c refers to the value of the quantity at the center of the star. We find

$$\frac{P_c}{R} \sim \frac{GM\rho_c}{R^2}. \tag{8}$$

Now if we assume the validity of the ideal gas law,

$$P_c = \frac{\rho_c}{\mu M_p} kT_c \tag{9}$$

and eliminate P_c, we find again that

$$kT_c \sim (\mu M_p)\frac{GM}{R} \tag{10}$$

It is interesting to note in this result that the left hand side is a measure of the thermal energy of a particle at the center of the star and that the right hand side is the gravitational potential energy of a particle at the surface. It turns out that the work required to move a particle from the center to the surface is nearly equal to the work required to carry this particle from the surface to infinity (the potential energy at the surface). Thus, the average kinetic energy of a particle at the center is also of the order of the average potential energy of a particle at the center.

Substituting numbers in the equation for central temperature gives

$$T_c \sim 2 \times 10^7 \frac{\mu \overline{M}}{\overline{R}} \,{}^\circ K. \tag{11}$$

Here the bars indicate that the quantities are expressed in solar units.

The Adequacy of the Ideal Gas Law

We have assumed that the ideal gas law is approximately valid in stellar interiors and that hydrogen and helium are completely ionized. Let us examine how complete the ionization of heavier elements is and then examine several sources of pressure which cause additional deviations from the ideal gas law. We will consider Coulomb forces, the exclusion principle, and radiation pressure.

The mean density of the sun is about 1 gm/cm^3. At this density the average separation between particles is on the order of the first Bohr radius of an electron in an H atom. Clearly, hydrogen will be ionized over most of the solar interior, whatever the temperature.

The thermal energy of a particle is approximately

$$kT \sim 0.865 \text{ keV} \times T_7 \qquad (12)$$

where T_7 denotes temperature in units of 10^7 °K. At the center of the sun, particles have energies on the order of KeV. For comparison, the ionization energy of H is 13.5 eV, of He is 24 eV and of He+ is 54 eV. Clearly, even if no process other than collisional excitation were effective, He as well as H would be completely ionized over much of the solar interior.

The energy required to strip the last electron from an atom is Z^2 times 13.5 ev., where Z is the atomic number. To determine which nuclei are not completely ionized, a good rule of thumb is that the last K-shell electron disappears when kT is on the order of one twentieth to one tenth of the ionization energy. This is because $h\nu$ for an average photon in the Planck distribution is much larger than kT (it is approximately 4kT) and because the relationship between density and temperature that occurs in stars is appropriate.

To determine more carefully the state of ionization of the heavier elements we can use the Saha ionization equation (refer to H. Y. Chiu's book, say), which may be written in the following form

$$\frac{n+}{n}(n_e \lambdabar^3) \sim 2\frac{g+}{g-}\exp\left(-\frac{\chi}{kT}\right) \qquad (13)$$

Here n+ and n are number densities of atoms in different states of ionization (an ion from the set n+ has one fewer electron than one from the set n), n_e is the electron number density, the g's are statistical weights, χ is the ionization energy (the energy to remove one electron from an ion in the set n), and $\lambdabar = \hbar/(2\pi m_e kT)^{1/2}$ is essentially the De Broglie wavelength of an average electron (apart from a small factor).

The results of an application of the Saha equation to the case of N^{14} are shown in Figure 11. Beside each line is the average number of electrons per N^{14} nucleus that holds for all points along the line. We see that, throughout most of all

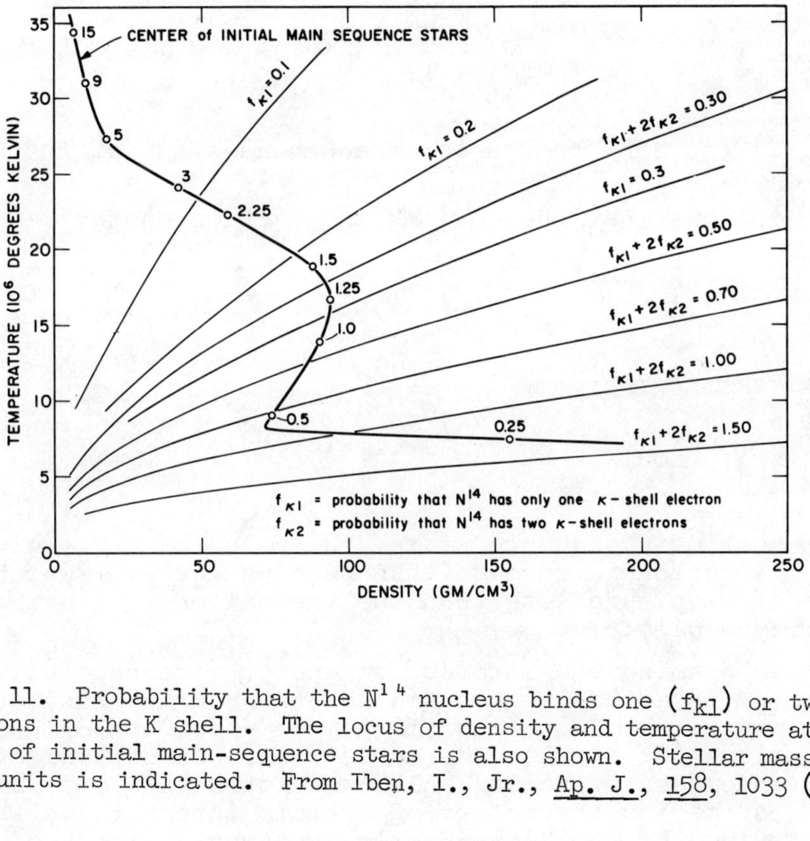

Figure 11. Probability that the N^{14} nucleus binds one (f_{k1}) or two (f_{k2}) electrons in the K shell. The locus of density and temperature at the center of initial main-sequence stars is also shown. Stellar mass in solar units is indicated. From Iben, I., Jr., Ap. J., 158, 1033 (1969).

main sequence stars, ionization is essentially complete.

Since the average density of the sun is about 1 gm/cm^3, and since the central density might be expected to be several orders of magnitude higher, we might anticipate that Coulomb interactions are important. To investigate this possibility we define a mean separation r_o between particles by the following equation:

$$\frac{\rho}{\mu M_p} \frac{4\pi}{3} r_o^3 = 1 \tag{14}$$

This says that the number density multiplied by the volume occupied by each particle is unity. It turns out that

$$r_o \sim \frac{3}{2} a_o \left(\frac{\mu}{\rho}\right)^{1/3} \tag{15}$$

where a_o is the Bohr radius for the ground state of the hydrogen atom $\sim 0.53 \times 10^{-8}$ cm.

It follows that the Coulomb interaction energy

$$E_{coul} \sim -\frac{e^2}{r_o} \tag{16}$$

is given approximately by

$$E_{coul} \sim -\frac{e^2}{a_o} \left(\frac{\rho}{\mu}\right)^{1/3} \sim -18 \text{ ev.} \left(\frac{\rho}{\mu}\right)^{1/3} \tag{17}$$

In most main sequence stars, the value of E_{coul} given by this approximation is no more than perhaps 100 ev., which is an order of magnitude less than the thermal energy, but not necessarily negligible.

If we examine the interaction energy more carefully, we find that it is actually much smaller than our approximation seems to indicate. The reason is simple — our approximation is appropriate for a well ordered lattice, but not for a gas wherein randomness prevails. Although not entirely correct, a way of estimating the effect of Coulomb interactions in a gas is provided by the Debye-Hückle treatment. Think of each charge (whether positive or negative) surrounded by a polarized cloud or Debye sphere; the radius λ_D of the sphere is so chosen that the total charge within it is roughly zero. The average interaction energy between two particles in the Debye sphere multiplied by the number of particles in the sphere will be on the order of kT. Thus,

$$\frac{e^2}{\lambda_D} n_e \, 4\pi \, \lambda_D^3 \sim kT, \tag{18}$$

where the absence of a 3 in the volume factor results from a less heuristic argument. It follows that the average Coulomb energy per particle is given by

$$E_{coul} \sim -\frac{e^2}{\lambda_D} \sim -\sqrt{32\pi n_e a_o^3}\left[\left(\frac{e^2}{2a_o}\right)/kT\right]^{3/2} kT \qquad (19)$$

Upon evaluating this expression at typical points in main sequence stars, one discovers that the Coulomb interaction energy is no more than a percent or a fraction of a percent of the thermal energy.

Under conditions of very high pressure, quantum effects might be expected to manifest themselves and lead to deviations from the ideal gas law. We know that the minimum energy of a particle confined in a box whose linear dimensions are on the order of r_o is

$$E_o \sim \frac{\hbar^2}{2mr_o^2}. \qquad (20)$$

Substituting from an earlier result, we have

$$E_o \sim \frac{\hbar^2}{2ma_o^2}\frac{4}{9}\left(\frac{\rho}{\mu}\right)^{2/3} \qquad (21)$$

Since this is at most only on the order of a percent of the thermal energy, electrons are far from being degenerate and the quantum correction to the ideal gas equation of state is, for most purposes, negligible in main sequence stars.

Finally, let us consider the contribution of "free" photons to the pressure. If the photons are distributed in a black body spectrum, then aT^4, where a is the radiation constant, represents the radiation energy density. If we multiply aT^4 by the average volume associated with a single massive particle, we obtain the "radiation energy per particle." The important result is

$$\frac{E_{Rad}}{kT} \sim aT^4\left(\frac{4\pi}{3} r_o^3\right)\frac{1}{kT}$$

$$= \frac{\pi^2}{15}\left(\frac{kT}{m_e c^2}\right)^3 \frac{1}{\left(\frac{\hbar}{m_e c}\right)^3}\frac{4\pi}{3} r_o^3 \qquad (22)$$

$$\sim 3\left(\frac{kT}{m_e c^2}\right)^3\left(\frac{r_o}{\lambda_C}\right)^3 ,$$

where we have replaced a in terms of other fundamental constants. In these expressions, m_e is the electron mass, c is the velocity of light in vacuum, and λ_C is the Compton wavelength of the electron. In the sun, E_{rad}/kT is on the order of 10^{-3} and the contribution that photons make to the pressure may be neglected.

We have now justified the use of the ideal gas law in the sun and in many other main sequence stars, and have derived expressions that show when corrections to the ideal gas law will be important under varying conditions of temperature and density.

<u>Exercise I</u>. Using the fact that $T \propto \frac{M}{R}$ we have that $T^3 \propto \frac{M^3}{R^3} \propto M^2\rho$, or $\frac{T^3}{\rho} \propto M^2$. Note that the ratio E_{Rad}/kT also goes as T^3/ρ, so $E_{rad}/kT \propto M^2$. Thus, for sufficiently massive stars, radiation pressure becomes important, and even dominant. Examine the case for which $P \sim P_{rad} = aT^4/3$ and find a relation between T_c, M, and R. Also, find how ρ/T^3 depends on M when $E_{rad}/kT >> 1$.

Energy Transport

Let us examine the two types of energy transport that are most important under "normal" stellar conditions.

(i) Convective Transport

An adequate, practical treatment of convective transport does not exist. For this reason, the stellar structure physicist relies on a simple, heuristic "mixing length" model that, if nothing else, has the virtue of being dimensionally correct.

Convection will occur when the local adiabatic temperature gradient is smaller than the radiative temperature gradient. This latter gradient is calculated on the assumption that only photons carry energy outward. We assume that, in a convective region, "bubbles" of matter rise and fall. The matter in a rising bubble is slightly hotter than the matter through which it rises. Consequently, as it rises, it gives off energy to its surroundings until it eventually comes into thermal equilibrium with its surroundings. We call the distance through which a bubble rises before becoming indistinguishable from its surroundings the mixing length, ℓ_{mix}.

The excess energy per unit volume delivered by a bubble to its surroundings may be crudely estimated as the product of a specific heat ρC_p and an average temperature excess

$$< \left| \left(\frac{\partial T}{\partial X}\right)_t - \left(\frac{\partial T}{\partial X}\right)_{ad} \right| > \ell_{mix}, \tag{23}$$

where the temperature gradients are the ambient and adiabatic gradients, respectively, and the carets denote an average over the lifetime of a bubble. Further multiplication by an average bubble speed v_{conv} will then give an estimate of the convective flux. Thus

$$F_{conv} \sim \rho C_p < \left| \left(\frac{\partial T}{\partial X}\right)_t - \left(\frac{\partial T}{\partial X}\right)_{ad} \right| > \ell_{mix}\, v_{conv}. \tag{24}$$

To find the convective speed, remember that the density of the bubble is different from its surroundings, so that it experiences a buoyancy force. The energy gained is approximated by

$$\rho v^2_{conv} \sim \left< \left| \left(\frac{\partial \rho}{\partial X}\right)_t - \left(\frac{\partial \rho}{\partial X}\right)_{ad} \right| \right> \ell_{mix} g \ell_{mix}, \tag{25}$$

where the quantity in carets times ℓ_{mix} is the average density deficiency, and to get a suitable energy we multiply by the acceleration of gravity and mixing length again.

Now, since $P \propto \rho T$ (Eq. of state), we have

$$\begin{aligned} \frac{1}{P}\left(\frac{dP}{dx}\right)_t &= \frac{1}{\rho}\left(\frac{d\rho}{dx}\right)_t + \frac{1}{T}\left(\frac{dT}{dx}\right)_t , \\ \frac{1}{P}\left(\frac{dP}{dx}\right)_{ad} &= \frac{1}{\rho}\left(\frac{d\rho}{dx}\right)_{ad} + \frac{1}{T}\left(\frac{dT}{dx}\right)_{ad} . \end{aligned} \tag{26}$$

Although a bubble is not in thermal equilibrium with its surroundings, it is in mechanical equilibrium. Hence

$$\left| \frac{1}{\rho}\left(\frac{d\rho}{dx}\right)_t - \frac{1}{\rho}\left(\frac{d\rho}{dx}\right)_{ad} \right| = \left| \frac{1}{T}\left(\frac{dT}{dx}\right)_t - \frac{1}{T}\left(\frac{dT}{dx}\right)_{ad} \right| . \tag{27}$$

Substituting, we have

$$F_{conv} \sim \rho C_p \left(\frac{g}{T}\right)^{1/2} T \left[\left< \left| \frac{1}{T}\left(\frac{dT}{dx}\right)_t - \frac{1}{T}\left(\frac{dT}{dx}\right)_{ad} \right| \right> \right]^{3/2} \ell^2_{mix} \tag{28}$$

For the mixing length it has at various times been customary to use not only the density, but the pressure or temperature scale height as well. For example, $\ell_{mix} = H_{pres}$, where

$$\frac{1}{H_{pres}} = \left| \frac{1}{P} \frac{dP}{dx} \right| = \frac{1}{P} g \rho \tag{29}$$

Frequently, however, one simply leaves ℓ_{mix} as a parameter which one varies in attempts to fit observational data.

(ii) Radiative Transport Processes

An approach to radiative transport from a fairly elementary point of view is to consider the diffusion of energy from the center to the surface of a star as a random walk process. This approach is useful because the mean free path of a photon in the stellar interior is much less than the stellar radius.

To find the typical wavelengths occurring in a star, we make use of the fact that the maximum in the distribution of blackbody radiation occurs at a wavelength λ such that

$$(\lambda T)_{max} \sim 0.288 \text{ deg cm}$$
$$\lambda \sim 3\text{Å} \frac{1}{T_7} \qquad (30)$$

Thus, in the deep interior the radiation is largely soft x-rays. The diffusion of energy via such photons can be hampered by a number of processes.

One such process is electron scattering. The cross section is

$$\sigma_{es} \sim \frac{8\pi}{3}\left(\frac{e^2}{m_e c^2}\right)^2 \sim \frac{2}{3} \times 10^{-24} \text{ cm}^2 . \qquad (31)$$

This provides an upper limit to the photon mean free path, ℓ_{ph}:

$$\ell_{ph} < \frac{1}{n_e \sigma_{es}} = \frac{1}{(\rho/\mu_e M_p)} \frac{1}{\sigma_{es}} ,$$
$$\ell_{ph} < \frac{5}{2} \mu_e \frac{1}{\rho} \text{ cm.}, \qquad (32)$$

where ρ is in gm/cm^3 and μ_e is the number of nucleons per electron.

Although ionization is nearly complete in the stellar interior, there is nevertheless some bound-free absorption. An estimate for the cross section at the absorption edge is

$$\sigma_{edge} \sim \left(\frac{a_o}{Z}\right)^2 \sim \frac{1}{4} \times 10^{-16} \frac{1}{Z^2} \text{ cm}^2 . \qquad (33)$$

Believe it or not, this estimate is correct to within a factor of 2.5 (the correct answer is larger). Comparing it with the cross section for electron scattering, it appears that bound-free absorption might be overwhelmingly more important than electron scattering. However, those photons with energies below the threshold energy, χ_k, cannot be absorbed. Further, the cross section for absorption drops off rather steeply with photon energy above threshold:

$$\sigma_{bf} \sim \sigma_{edge}\left(\frac{\chi_k}{h\nu}\right)^{7/2} , \chi_k \sim 13.5 \, Z^2 \text{ eV}. \qquad (34)$$

Finally, the abundance of those ions capable of absorbing photons in the deep stellar interior is quite small. Even in the most metal rich stars, those atoms with one or two K-shell electrons left per nucleus constitute only a fraction of a percent of the total element population. The net result is that bound-free absorption merely competes with the electron scattering process.

Another important process is free-free absorption, also called inverse bremsstrahlung. In the sun it is of an importance comparable to that of electron scattering. Although it can be done, we shall not construct a heuristic expression for its magnitude.

Having identified the important absorption and scattering processes, it remains to show how the temperature gradient is related to the radiative flux through these cross sections. We proceed in analogy with the diffusion approximation approach to conductivity, writing:

$$F_{rad} \sim -\frac{1}{3}\left(\frac{\partial \varepsilon}{\partial X}\right)\ell_{ph}\, v_{ph} \tag{35}$$

where ε is the energy density for photons. Taking $\varepsilon = aT^4$ for a blackbody distribution,

$$F_{rad} \sim -\frac{4ac}{3}\, T^3\, \ell_{ph}\, \frac{\partial T}{\partial X} \tag{36}$$

Now we write symbolically

$$\ell_{ph} = \frac{1}{n\sigma} \tag{37}$$

as a shorthand for a weighted sum over the various scattering processes, and then define an opacity κ,

$$\ell_{ph} = \frac{1}{n\sigma} \equiv \frac{1}{\kappa\rho} \tag{38}$$

which gives the radiative flux

$$F_{rad} = -\frac{4ac}{3}\left(\frac{T^3}{\kappa\rho}\right)\frac{\partial T}{\partial X}\ . \tag{39}$$

<u>Exercise II</u>. Another mode of energy transport in stellar interiors is electron conduction. Determine an order of magnitude expression for electron conductivity in the form $F_{cond} \sim -(?)\frac{\partial T}{\partial X}$. Use the approach that we used in the case of radiative transfer, but with an appropriate distribution for the velocities of electrons. Do two cases: electrons non-degenerate, and electrons non-relativistically degenerate. For an energy density, use the number density of the electrons times $(1/2)mv^2$; for a velocity, the mean velocity in the distribution; and for the mean free path, combine the electron charge and the energy of an average electron to find a suitable length. You will find that, as degeneracy increases, the importance of electron conduction increases rapidly.

(iii) Energy Transport in the Sun

It is of interest to know where in the sun the various transport processes are important. Near the center of the sun energy fluxes may be high enough that convection occurs. Near the surface, opacities become so large that, again, energy is carried out predominantly by convection. Between the central and envelope convective regions, energy is transported by photons. At the higher temperatures near the center, the dominant sources of opacity are free-free absorption and, to a lesser extent, electron scattering. At the lower temperatures and densities further from the core where ionization is less complete, bound-free absorption becomes important. The qualitative aspects of energy flow in the Sun are shown in Figure 12. All main sequence stars are qualitatively much the same. The relative size of the convective regions are, however, very sensitive to the stellar mass. Going up the main sequence from the sun, the more massive the star, the higher are the central temperatures and the larger is the convective core. The larger the stellar mass, the smaller is the convective envelope. Going down the main sequence from the sun, the convective core at first disappears, but below about 1/2 solar mass the convective envelope reaches to the center.

A Theorem Relating Mass and Luminosity

The time for energy produced near the center to "work its way out" by a random walk is

$$t_{ph} \sim \frac{R^2}{c\ell_{ph}} \tag{40}$$

The time for energy to leak out is also related to the total radiation energy content of the star and the luminosity by

$$t_{ph} \sim \frac{aT^4R^3}{L} \tag{41}$$

equating the two expressions for t_{ph} and replacing kT by $\mu M_pGM/R$ and ℓ_{ph} by $1/\kappa\rho \sim R^3/\kappa M$, we obtain

$$L \sim \frac{ca}{k^4} M_p^4 \; G^4 \; \mu^4 \; \frac{M^3}{\kappa} \tag{42}$$

Although it was derived with no reference to the sources of stellar energy, we can conclude that this relationship will be satisfied in nature (it is!) only if the energy sources are very temperature and/or density sensitive. Then the star can adjust slightly to permit the energy sources to supply the demand without upsetting the hydrostatic and radiative equilibria that we assumed to derive the relationship.

Exercise III. Derive a mass-luminosity relationship for a massive star in which $P \sim P_{rad}$ and $\kappa \sim$ const (electron scattering). Use the equation for radiative transport

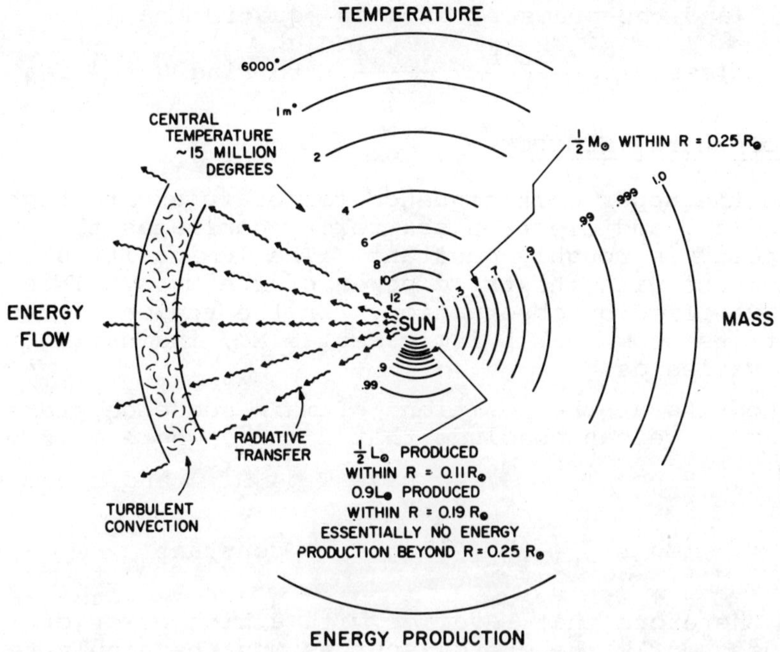

Figure 12. A π^* diagram for the sun showing the distribution of several variables with distance from the center.

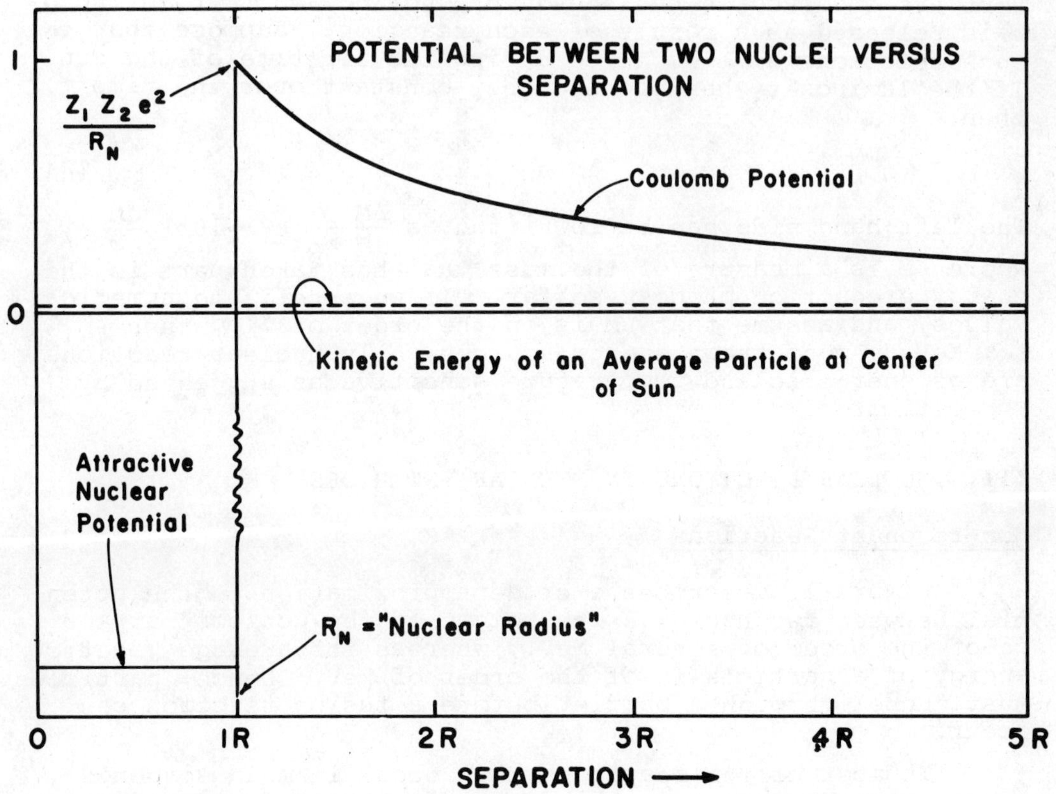

Figure 13. A schematic potential between two nuclei.

and the pressure balance equation

(assuming $\frac{\partial P_{tot}}{\partial x} \sim \frac{\partial P_{rad}}{\partial x}$) to find $L \propto \frac{M}{\kappa}$.

Nature of Energy Sources

On the upper main sequence temperatures are high, densities are low, and electron scattering dominates the opacity so that opacity is roughly constant. The luminosity should therefore increase with the third power of the mass. This prediction is verified by observation. In the region of the sun, observations show that L increases as M^5, indicating that the opacity varies as M^{-2}.

Looking at the positions of main sequence stars in the H-R diagram, we can conclude that $R \propto M^m$, where m is between 1/2 and 1. Since $T_c \propto \frac{M}{R}$, we have $T_c \propto M^{1-m}$ and finally,

$$\log L \sim \left(\frac{3 \text{ or } 5}{1 - m}\right) \log T_c + \text{constant} \tag{43}$$

We find therefore that L varies with a high power of T_c and that consequently the energy sources must be highly temperature sensitive.

We can estimate the order of magnitude of the energies on a particle scale. Assume that binary reactions of some sort are involved in the energy production and that an energy ε is released as a result of each reaction. Suppose that ΔN such reactions have occurred during the lifetime of the sun. If the luminosity has been roughly constant over the time t, then

$$\Delta N \, \varepsilon \sim Lt \tag{44}$$

The left-hand side can be rewritten as $\frac{\Delta M}{M} \frac{M}{M_p} \varepsilon \sim 10^{57} \frac{\Delta M}{M} \varepsilon$, where ΔM is a measure of the mass that has taken part in the energy production process so far. If we substitute numerical values, and assume that ΔM is on the order of .1M, then $\varepsilon \sim$ MeV. Among known energy sources, only nuclear reactions are as energetic and temperature sensitive as suggested by our estimates.

III. NUCLEAR REACTIONS IN STELLAR INTERIORS

Non-resonant Reactions

Figure 13 describes a crude approximation to the potential between two nuclei. The height of the Coulomb barrier is of the order of several MeV., whereas the average kinetic energy of a particle is of the order of keV. Hence, particles must tunnel through a barrier before a fusion reaction can occur.

In most of the reactions that occur in main sequence stars only the s-wave contributes and a typical cross section may be written as

$$\sigma(E) = \frac{S(E)}{E} \exp\left(\frac{-2\pi Z_1 Z_2 e^2}{\hbar v}\right) \tag{45}$$

where S(E) is a slowly varying function of energy E, the exponential is a Coulomb penetration factor, and v is the velocity of interaction. S(E) is called a "center of mass cross section factor." We can also write

$$\sigma v \propto S(E)\, 2\pi\eta \exp(-2\pi\eta) \tag{46}$$

where η is $\frac{Z_1 Z_2 e^2}{\hbar v}$. To find a rate, we integrate over both kinds of particles:

$$\text{Rate} = \iint dn_1 dn_2 \sigma v$$
$$\text{Rate} = n_1 n_2 < \sigma v >, \tag{47}$$

where the carets denote an appropriately weighted average of σv.

Using the Maxwell-Boltzmann distribution, and center of mass coordinates,

$$dn_2 \propto n_2 \left(\frac{1}{2\pi\mu kT}\right)^{3/2} \exp\left(-\frac{E}{kT}\right) E^{1/2}\, dE. \tag{48}$$

The desired average is therefore

$$< \sigma v > \sim \text{const} \left(\frac{1}{kT}\right)^{3/2} \int e^{-E/kT} e^{-2\pi\eta}\, dE. \tag{49}$$

Exercise IV. Evaluate the above integral approximately to show that
$\text{Rate} \propto \rho^2 \frac{1}{T^{2/3}} \exp -\left(\frac{\text{const}}{T^{1/3}}\right)$. The trick is to expand the quantity in the exponent about its maximum and ti extend the integration limits from to + about the position of the maximum. See Fig. 14.

The principle energy producing reactions in the hydrogen burning phase are:

(1) The CN cycle reactions, where the reaction

$$N^{14} + p \rightarrow O^{15} + \gamma \tag{50}$$

is the slowest and hence overall rate determining factor in the cycle.

(2) the pp chains, where

$$p + p \rightarrow d + e^{+} + \nu \tag{51}$$

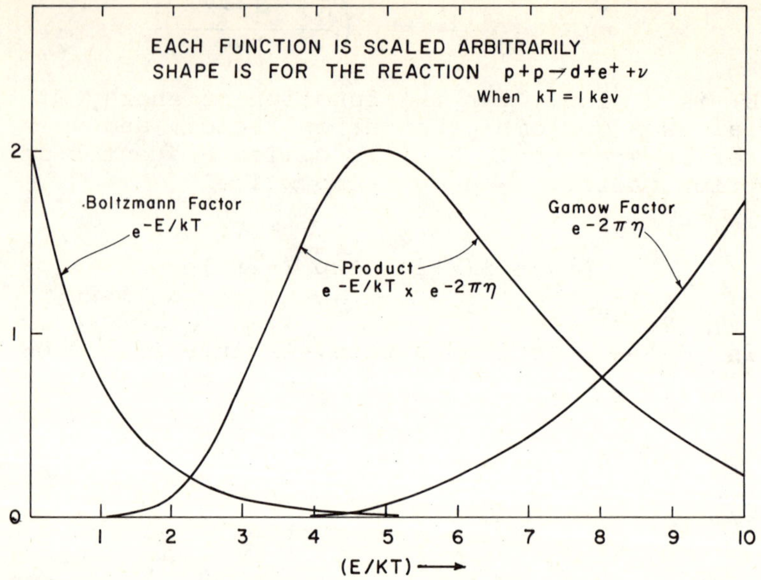

Figure 14. The product of the Coulomb factor $\exp(-2\pi\eta)$ and the Maxwell-Boltzmann factor $\exp(-E/kT)$.

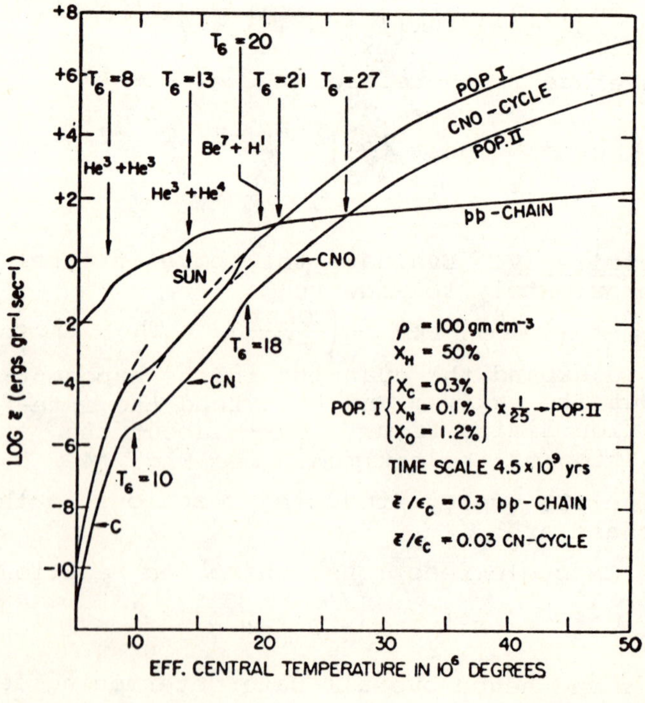

Figure 15. Temperature dependences of the rate determining reactions. From Fowler, W. A., Memoires de la Société Royale des Sciences de Liège, Series 5, Volume III, 1960.

is the slowest and hence major overall rate determining factor.

The temperature dependences of the two rates are shown in Figure 15. Since Bethe and others initiated our quantitative understanding of nuclear reaction rates in stellar interiors in 1938-1939, most of the reactions in the CN cycle and in the pp chains have been investigated in the laboratory.

In order to determine a reaction rate in stars, it is necessary to know the magnitude of the center of mass cross section factor in the neighborhood of the maximum in the integrand for the rate. In almost all cases, laboratory measurements are incapable of reaching the relevant energies and one must be content with an extrapolation from energies that are one or two orders of magnitude larger than required.

Fortunately, the extrapolation from experimental data is in most cases relatively unambiguous. It is also important to realize that a factor of two uncertainty in S(E) does not affect the general features of stellar structure and evolution. This is because of the strong temperature dependence of the reaction rates; a very small shift in temperature can easily compensate for a factor of two adjustment in a cross section factor.

The C-N Cycle

The complete set of reactions in the C-N cycle is:

$N^{14} + p \rightarrow O^{15} + \gamma + 7.4$ meV	$S(E) \sim 3$ keV - barns
$O^{15} \rightarrow N^{15} + \beta^{+} + \nu + 1.7$ meV	$t_{1/2} \sim 3$ min.
$N^{15} + p \rightarrow C^{12} + He^{4} + 5$ meV	$S(E) \sim 10^{5}$ keV - barns
$C^{12} + p \rightarrow N^{13} + \gamma + 2$ meV	$S(E) \sim 1.4$ keV - barns
$N^{13} \rightarrow C^{13} + \beta^{+} + \nu + 1.5$ meV	$t_{1/2} \sim 14$ min
$C^{13} + p \quad N^{14} + \gamma + 7.5$ mev	$S(E) \sim 6$ kev - barns

The reaction rate for two nuclei of mass abundances X_i, X_j can be written

$$R = R_{ij} X_i X_j \text{ gm.}^{-1} \text{ sec.}^{-1} \tag{52}$$

For comparison with beta-decay lifetimes, it is convenient to define the lifetime T_{ij} of a particle i against destruction by a particle j, by

$$\frac{1}{N_i}\left(\frac{dN_i}{dt}\right) = \frac{1}{T_{ij}} = R_{ij} X_j . \tag{53}$$

Typical lifetimes are given in the following table for $T_6 \sim 16$ and $\rho \sim 100$ (X_H = abundance by mass of hydrogen):

Reaction	Lifetime
$C^{12}(p,\gamma)\ N^{13}$	C^{12}: 4×10^5 yr./X_H
$C^{13}(p,\gamma)\ N^{15}$	C^{13}: 10^5 yr./X_H
$N^{14}(p,\gamma)\ O^{15}$	N^{14}: 7×10^7 yr./X_H
$N^{15}(p,\ He^4)\ C^{12}$	N^{15}: 3×10^5 yr./X_H

Note that, as asserted before, the slowest reaction is $N^{14}(p,\gamma)O^{15}$ ($C^{12}(p,\gamma)N^{13}$ has a lower cross section factor but also a lower Coulomb barrier). In fact, all other reactions in the cycle are orders of magnitude faster. The beta-decays are effectively instantaneous. As a result, when an N^{14} nucleus reacts, the resulting nucleus goes through the entire cycle and returns to the N^{14} pool rather quickly. This means that, at any given time, most (by far) of the nuclei involved in this cycle are in the form of N^{14} waiting to react and traverse the cycle.

Since the N^{14} reaction rate governs the whole cycle, the energy production rate can be written as the sum of all the energy released in a cycle times the N^{14} rate. Therefore,

$$\varepsilon_{CN} \cong 25 \text{ meV } R_{N,H}\, X_N\, X_H$$

$$\cong 4 \times 10^{-5} \left(\frac{\rho}{T_6^{2/3}}\right) \exp\left(74.4 - 152.3 T_6^{-1/3}\right) X_N\, X_H \text{ erg/gm-sec,} \tag{54}$$

where X_N is the abundance by mass of nitrogen.

The pp Reaction

In each pp chain, as in the C-N cycle, 4 protons eventually become a helium nucleus. The first reaction in each sequence is

$$p + p \rightarrow d + e^+ + \nu + 1.2 \text{ meV}. \tag{55}$$

This reaction proceeds so slowly that a direct measurement of the cross section in the laboratory is out of the question. Consequently, we must rely upon theory, with only indirect experimental verification (the success of stellar models in accounting for the observations is one such verification). Because of its importance, it is worthwhile to discuss at some length what is involved in the determination of the pp cross section.

The probability per unit time that a reaction will occur with the emission of leptons in specified momentum intervals may be written as

$$dw = \frac{2\pi}{\hbar} \left|\langle H \rangle\right|^2 d\rho_f \tag{56}$$

where < H > is an appropriate matrix element and the density of final momentum states is (for one spin state)

$$d\rho_f = \left(\frac{4\pi p_e^2 dp_e}{h^3}\right) V \left(\frac{4\pi p_\nu^2 dp_\nu}{h^3}\right) V \left(\frac{1}{dE_f}\right). \quad (57)$$

Here V is the volume of a box used for normalization purposes and p_e and p_ν are the momentum of the electron and of the neutrino, respectively. We may convert this expression into

$$d\rho_f = \left(\frac{4\pi V}{h^3}\right)^2 \frac{(m_e c^2)^5}{c^6} (E_f - E_e)^2 (E_e^2 - 1)^{1/2} E_e \, dE_e, \quad (58)$$

where E_f is the final total energy of the electron and neutrino (including the rest mass of the electron) and E_e is the final energy of the electron (including its rest mass). All energies are now in units of $m_e c^2$.

The appropriate matrix element can be written in a rough way as

$$< H > \sim \int (\Psi_d \Psi_e \Psi_\nu)^* G \Psi_{pp} \, d\tau_{nuc} \quad (59)$$

where G is a strength factor. Let us approximate the wave function for the deuteron as a constant out to a finite "deuteron radius," R_d, and as zero beyond this radius. Thus,

$$\Psi_d \sim \begin{Bmatrix} \sqrt{1/V_d}, & r \leq R_d \\ 0 \quad, & r \geq R_d \end{Bmatrix}. \quad (60)$$

Here V_d is the "volume" of the deuteron. The final lepton wave functions may be approximated by plane waves, so that

$$\Psi_e \Psi_\nu \sim \frac{1}{V}. \quad (61)$$

An equally simple approximation to the initial wave function for the two protons can be constructed. If all forces between protons are neglected, then $\Psi_{pp} \sim V^{-1/2}$. Taking only the Coulomb repulsion into account, we can write, for $r \lesssim R_d$,

$$\Psi_{pp} \sim V^{-1/2} \, |2\pi\eta e^{-2\pi\eta}|^{1/2}. \quad (62)$$

To take nuclear forces into account, let us multiply this last estimate by some function $\lambda(r)$ that we may assume takes all other neglected effects into account. Gathering all approximations together, we have

$$\begin{aligned} < H > &\sim \sqrt{\frac{1}{V_d}} \left(\frac{1}{V^{3/2}}\right) G \, |2\pi\eta e^{-2\pi\eta}|^{1/2} \int \lambda(r) \, d\tau_{nuc} \\ &\cong \sqrt{\frac{1}{V_d}} \left(\frac{1}{V^{3/2}}\right) G \, |2\pi\eta e^{-2\pi\eta}|^{1/2} \, \Lambda V_d . \end{aligned} \quad (63)$$

We may regard the quantity Λ as a correction factor that takes on whatever value is necessary to make our expression for $< H >$ exact. We might expect Λ to be on the order of 1 to 100.

Noting that the total probability per unit time, $\int dw$, is equal to the cross section, σ, for the reaction times the flux, v/V, we have

$$\sigma v = 2\pi\eta e^{-2\pi\eta} V_d \Lambda^2 \overline{G}^2 F(E_f) \tag{64}$$

where

$$\overline{G}^2 = \frac{2\pi}{\hbar} G^2 \left(\frac{8\pi}{h^3}\right)^2 \frac{(m_e c^2)^5}{c^6}, \tag{65}$$

and

$$F(E_f) = \int_1^{E_f} (E_f - E_e)^2 (E_e^2 - 1)^{\frac{1}{2}} E_e \, dE_e \tag{66}$$

$$\simeq 0.15.$$

Knowing only the binding energy of the deuteron, we can estimate $V_d \sim \frac{4\pi}{3}$ (few $\times 10^{-13}$ cm)$^3 \sim 3 \times 10^{-38}$ cm^3.

The problem now is to find a reasonable value for G. An estimate can be obtained from a similar decay whose lifetime can be measured accurately. The reaction

$$He^6 \rightarrow Li^6 + e^- + \bar{\nu} \tag{67}$$

has the virtue of involving essentially the same spin sums as does the pp reaction. As before, the decay probability per unit time may be written as

$$w = \frac{2\pi}{\hbar} |< H' >|^2 \rho_f', \tag{68}$$

where

$$< H' > \sim \int \Psi_f^* G \Psi_i \frac{1}{V} d\tau_{nuc} \tag{69}$$

Here Ψ_f and Ψ_i are the final and initial nuclear wave functions respectively. The factor V^{-1} represents the product of the two lepton wave functions. We can write

$$< H' > \sim \frac{G}{V} |M|, \tag{70}$$

Where $|M|$ is a nuclear matrix element. Since He^6 and Li^6 are mirror nuclei, we expect $|M|$ to be on the order of unity, apart from a spin factor that is the same as in the pp case. The integrated density of final states is

$$\rho_f \cong \left(\frac{4\pi V}{h^3}\right)^2 \frac{(m_e c^2)^5}{c^6} F'(E_f') \tag{71}$$

and a straightforward integration gives

$$F'(E_f') \sim 10^3 \tag{72}$$

so that

$$w \sim \bar{G}^2 |M|^2 \times 10^3. \tag{73}$$

From the known half life of He^6 we have that $w \sim 1\ sec^{-1}$. Our final result, then, is that

$$\bar{G}^2 \sim 10^{-3}. \tag{74}$$

Inserting this in our expression for σv and writing $\sigma = S(E)e^{-2\pi\eta}/E$ we obtain

$$S(E) \sim 10^{-24} \Lambda^2 \text{ keV - barns} \tag{75}$$

as the center of mass cross section factor for the pp reaction.

We could say that the cross section is as small as it is because of the weakness of the interaction. In the same spirit, we could say it is as strong as it is because of the diffuseness of the deuteron.

Considering the number of approximations involved, our crude estimate is fairly close to the result of a more careful calculation which gives

$$S(E) \sim (3\tfrac{1}{2} \rightarrow 4) \times 10^{-22} \text{ keV - barns}, \tag{76}$$

and it has the virtue of showing simply what is involved in the most detailed of calculations. The uncertainty in the more careful estimate of $S(E)$ comes from several sources. Uncertainty in the coupling constant may lead to error on the order of 10%, and the integration over the nuclear wave functions leads to uncertainty of 5 - 10%. The latest value is $S(E) = (3.8 \pm 0.7) \times 10^{-22}$ keV barns.

As in the case of the C-N cycle, the uncertainty in cross section is not important so long as we restrict ourselves to stellar structure and evolution; however, if we want to know the neutrino flux and energy distribution, these uncertainties are quite important. We will discuss this later in more detail.

The pp Chains

In addition to the pp reaction itself, the following reactions occur:

$$
\begin{array}{lll}
d + p & \rightarrow He^3 + \gamma + 5.5\ \text{MeV.}; & S(E) = 10^{-4}\ \text{keV - barn} \\
He^3 + He^3 & \rightarrow He^4 + 2p + 13\ \text{MeV.}; & S(E) = 5000\ \text{keV - barn} \\
He^3 + He^4 & \rightarrow Be^7 + \gamma + 1.6\ \text{MeV.}; & S(E) = 0.6\ \text{keV - barn} \\
Be^7 + p & \rightarrow B^8 + \gamma + 0.14\ \text{MeV.}; & S(E) \cong 0.03\ \text{keV - barn} \\
B^8 & \rightarrow Be^{8*} + \beta^+ + \nu + 7.7\ \text{MeV.}; & t_{\frac{1}{2}} \sim \tfrac{1}{2}\ \text{sec.} \\
Be^{8*} & \rightarrow 2He^4 + 3\ \text{MeV.}; & t_{\frac{1}{2}} \sim 10^{-21}\ \text{sec.} \\
Be^7 + e^- & \rightarrow Li^7 + \nu + 0.5\ \text{MeV.}; & t_{\frac{1}{2}} \sim \text{month.} \\
Li^7 + p & \rightarrow 2He^4 + 17.34\ \text{MeV.}; & S(E) = 100\ \text{keV - barn}
\end{array}
\tag{77}
$$

The net energy production rate can be written in terms of the pp rate, since the pp rate is by far the slowest reaction. We find

$$
\varepsilon_{pp} \cong \begin{cases} 13.05\ \text{MeV}\ R_{pp}, & \text{if } He^3 + He^3 \rightarrow He^4 + 2p \\ 19.1\ \text{MeV}\ R_{pp}, & \text{if } B^8 \text{ is formed} \\ 25.7\ \text{MeV}\ R_{pp}, & \text{if } Li^7 \text{ is formed} \end{cases} \tag{78}
$$

where

$$
R_{pp} = (\rho/T_6^{2/3}) \exp(25.44 - 33.81/T_6^{1/3})\ X_H^2 \ \text{reactions/gm-sec} \tag{79}
$$

When the He^4 abundance and the temperature are low, the $He^3 + He^3$ reaction proceeds much more frequently than the $He^3 + He^4$ reactions; the formation at each new He^4 nucleus then requires the prior occurrence of two pp reactions. When the He^4 abundance and the temperature are high enough, the other branches dominate. Then each new He^4 nucleus requires only one prior pp reaction. Hence the difference in the energy produced per pp reaction. Also, the energy lost to neutrinos is different for different branches of the chain.

Exercise V. Define $\alpha \equiv \dfrac{R_{3,4}}{R_{1,1}}$, $\beta \equiv \dfrac{R_{1,14}}{R_{1,1}}$,

$$
\gamma \equiv \frac{R_{1,7}}{R_{e,7}} \quad \text{and} \quad \delta \equiv \alpha(0.957 + 0.457\gamma)(1 + \gamma)^{-1}.
$$

Show that the solar luminosity is given by

$$
L_\odot = 13.10\ \text{MeV} \int R_{1,1}\, dM[1 + \langle\delta\rangle + 1.91\,\langle\beta\rangle]
$$

where $\langle\delta\rangle = \dfrac{\int \delta R_{1,1}\, dM}{\int R_{1,1}\, dM}$, etc.

The subscripts refer to the atomic "weight" of the reactants. Show furthermore that the neutrino flux at the earth from the pp reaction is

$$\phi(pp) = 6.4 \times 10^{10} \, [1 + \langle\delta\rangle + 1.91 \langle\beta\rangle]^{-1} \mathrm{cm}^{-2} \mathrm{secs}^{-1}.$$

Exercise VI. Using a mercury thermometer and your eye, determine the surface temperature of the sun. Do not use your knowledge of our distance from the sun.

A Helium Burning Reaction

The production of C^{12} from helium procedes via the reactions

$$\begin{aligned} He^4 + He^4 &\leftrightarrow Be^8 \text{ (unstable)} \\ Be^8 + He^4 &\rightarrow C^{12} + \gamma + \gamma \, . \end{aligned} \tag{80}$$

One way to find the reaction rate is to first write down the equilibrium density of Be^8 nuclei in terms of the density of He^4 nuclei:

$$\frac{N_4{}^2}{N_8} \sim \left[\frac{2\pi m_{4,8} kT}{\hbar^2}\right]^{3/2} \exp \frac{\Delta\chi}{kT} \, . \tag{81}$$

Here $m_{4,8}$ is the reduced mass of the $He^4 + Be^8$ system and $\Delta\chi$ is the energy difference between $2He^4$ and Be^8. The rate of formation of C^{12} is then given by

$$\frac{dN_{12}}{dt} = n_8 \int dn_4 (\sigma v) \tag{82}$$

where σ is the cross section for the Be^8 alpha capture reaction. Since this reaction proceeds through a resonance, we may write

$$\sigma = \frac{\Gamma_\alpha \Gamma_\gamma}{(E - E_{Res})^2 + (\Gamma_\alpha + \Gamma_\gamma)^2/4} \tag{83}$$

where the Γ's are level widths. On performing the appropriate integrations, we find that

$$(\text{Rate})_{4,4,4} \propto \left[\frac{dn_{12}}{dt}\right] \propto \left[\frac{n_4{}^3}{T^3}\right] \exp(-Q/kT) \tag{84}$$

where Q is the energy difference, between three alpha particles and a C^{12} nucleus in the ground state. When current estimates of level widths are inserted,

$$R_{4,4,4} = (X_4)^3 (\rho^2/T_6{}^3) \exp(52.79 - 4320/T_6) \text{ reactions/gm-sec.} \tag{85}$$

Exercise VII. Supply the steps in the preceding derivations.

IV. METHODS OF SOLUTION

Standard Integration

Before attempting to evolve a model star, we must integrate the equations of stellar structure to find an initial model. At any instant, the equations which must be satisfied at every point are

$$\begin{aligned}
\frac{\partial M(r)}{\partial r} &= 4\pi r^2 \rho \\
\frac{\partial P}{\partial r} &= -\left[\frac{GM(r)}{r^2}\right]\rho \quad , \\
\frac{\partial L}{\partial r} &= 4\pi r^2 \rho\varepsilon(\rho, T, \text{comp.}) \quad , \\
\left(\frac{\partial T}{\partial r}\right)_{\text{rad}} &= -\left[\frac{3}{4ac}\right]\frac{\kappa\rho}{T^3}\left[\frac{L(r)}{4\pi r^2}\right]
\end{aligned} \tag{86}$$

If convection can occur, the last equation is replaced by an expression from the mixing-length treatment.

The boundary conditions are

$$\begin{aligned}
&\text{At} \quad r = 0; \qquad M = 0,\ L = 0 \\
&\text{At} \quad r = R; \qquad T = T_e,\ P = P_e .
\end{aligned} \tag{87}$$

Where T_e and P_e are T and P at the base of a model atmosphere.

We must first decide on a composition distribution (perhaps homogeneous) and guess values for T_c and P_c, L and R. We then integrate the four structure equations, perhaps using M as the independent variable. Beginning at the center, we integrate out part way, but not to the surface. Then, beginning at the surface we integrate inward to the point at which we ceased the outward integrations.

In general, the values found for the variables at the fitting point will differ. Write the discrepancies as

$$\begin{aligned}
(T_2 - T_1) &= Q_T\ (P_c, T_c, R, L) \\
(P_2 - P_1) &= Q_P\ (P_c, T_c, R, L) \\
(R_2 - R_1) &= Q_R\ (P_c, T_c, R, L) \\
(L_2 - L_1) &= Q_L\ (P_c, T_c, R, L)
\end{aligned} \tag{88}$$

Now set $P_c = P_c + \Delta P_c$ and integrate outward to obtain a new set of Q's from which numerical derivatives may be constructed. Thus

$$\begin{aligned}
Q_T(P_c + \Delta P_c, T_c, R, L) &= Q_T + \Delta Q_T \\
&\cong Q_T + \left[\frac{\partial Q_T}{\partial P_c}\right]\Delta P_c
\end{aligned} \tag{89}$$

and we solve for $\frac{\partial Q_T}{\partial P_c}$. Similarly, we find $\frac{\partial Q_P}{\partial P_c}$ $\frac{\partial Q_R}{\partial P_c}$ $\frac{\partial Q_L}{\partial P_c}$. In the same way, we find $\frac{\partial Q_T}{\partial T_c}$, etc.

After we have found all of the necessary partial derivatives numerically, we can estimate what changes we should make in all four variables P_c, T_c, L, and R in order to minimize the fitting discrepancies. Let us use the shorthand notation

$$\begin{aligned} Q_P, T, \text{ etc.} &= Q_i, \quad i = 1,2,3,4 \\ P_c, T_c, \text{ etc.} &= q_j, \quad j = 1,2,3,4 \end{aligned} \tag{90}$$

We solve the four equations

$$Q_i + \sum_{j=1}^{4} \left[\frac{\partial Q_i}{\delta q_j} \right] \delta q_j = 0, \quad i = 1,2,3,4 \tag{91}$$

for the four changes in boundary variables, δq_j. Beginning with new estimates of the boundary variables we integrate inward and outward again. If the fitting discrepancies are still too large, we proceed again as before.

Polytropes

In some situations an adequate initial model can be obtained by solving a simpler set of equations that follow from the use of a temperature independent equation of state.

Without any knowledge of the energy production mechanisms or of radiative transfer processes, it is possible to obtain a model in hydrostatic equilibrium by requiring only that the pressure vary as some function of the density. In particular, it is convenient, as well as traditional, to assume that this relation takes the form $P = K\rho^{1+1/n}$. Since the gravitational field is derivable from a potential, the equation of hydrostatic equilibrium

$$\frac{\partial P}{\partial r} = -g\rho \tag{92}$$

can be written as

$$\frac{\partial P}{\partial r} = \rho \frac{\partial \phi}{\partial r}, \tag{93}$$

Substituting the assumed pressure-density relation, we have

$$\rho\left[\frac{\partial \phi}{\partial r}\right] = K\left(1 + \frac{1}{n}\right)\rho^{1/n}\left[\frac{\partial \rho}{\partial r}\right], \tag{94}$$

Integrating this expression, and choosing the constant of integration to be zero, we have

$$\rho = \left[\frac{1}{K(n+1)}\,\phi\right]^n . \tag{95}$$

Since the gravitational potential must satisfy Poisson's equation,

$$\nabla^2\phi + 4\pi G\rho = 0, \tag{96}$$

we have finally that

$$\nabla^2\phi + 4\pi G\left[\frac{\phi}{K(n+1)}\right]^n = 0 \tag{97}$$

This equation is to be solved subject to the boundary conditions

$$\phi(r)\Big|_{r=0} \neq \infty, \quad \frac{\partial\phi}{\partial r}\Big|_{r=0} = 0 , \tag{98}$$

$$\frac{\partial\phi}{\partial r}\Big|_{r=R} = -\frac{GM}{R} , \tag{99}$$

and

$$\phi(r)\Big|_{r=R} = 0 , \tag{100}$$

where R is the stellar radius.

The solution of the differential equation for ϕ that is subject to the stated boundary conditions (or a star which the solution describes) is called a polytrope of index n. Before carrying out a solution, we make the substitutions

$$\phi(r) = \phi_o u(r), \quad r = \frac{R}{z_s}\, z \tag{101}$$

where ϕ is the potential at the origin, and z_s is chosen so that

$$\left[\frac{\phi_o}{K(n+1)}\right]^n \frac{4\pi G}{\phi_o}\left[\frac{R}{z_s}\right]^2 = 1. \tag{102}$$

With these substitutions, Poisson's equation becomes

$$\frac{\partial^2 u}{\partial z^2} + \frac{2}{z}\left[\frac{\partial u}{\partial z}\right] + u^n = 0. \tag{103}$$

The transformed boundary conditions are

$$\begin{aligned} u(z)\Big|_{z=0} &= 1, & \frac{\partial u}{\partial z}\Big|_{z=0} &= 0 \\ u(z)\Big|_{z=z_s} &= 0, & \frac{\partial u}{\partial z}\Big|_{z=z_s} &\neq 0 . \end{aligned} \tag{104}$$

The value of the derivative $\left.\frac{\partial u}{\partial z}\right|_{z=z_s}$ is found by integrating outward from the origin until arriving at some $z = z_s$ where $u(z)$ vanishes.

If we specify values for M, R, and n, we can relate physical variables to properties of the solution:

$$\phi_o = -\frac{GM}{R}\left[\frac{1}{(z\frac{\partial u}{\partial z})_s}\right],$$

$$\rho_o = -\frac{1}{3}\bar{\rho}\frac{z_s}{(\frac{\partial u}{\partial z})_s}, \tag{105}$$

$$K = \frac{\phi_o}{n+1}\Big/\rho_o^{1/n}$$

Here the subscript s indicates that the enclosed quantity is to be evaluated at $z = z_s$. The mean density is, of course

$$\bar{\rho} = \frac{M}{\frac{4\pi}{3}R^3} \tag{106}$$

We shall shortly make use of the additional identities

$$\frac{P}{\rho} = \frac{\phi}{n+1} \tag{107}$$

and

$$\frac{M(r)}{z^2\frac{\partial u}{\partial z}} = \text{const.} = \frac{M(R)}{(z^2\frac{\partial u}{\partial z})_s}. \tag{108}$$

As an example of the solutions obtainable, let us look at the case n = 1 where

$$\frac{\partial^2 u}{\partial z^2} + \left[\frac{2}{z}\right]\frac{\partial u}{\partial z} + u = 0. \tag{109}$$

With the substitution $u = \frac{\lambda}{z}$ we have

$$\frac{\partial^2 \lambda}{\partial z^2} + \lambda = 0, \tag{110}$$

whose solution will be recognized as $\lambda = \alpha\cos z + \beta\sin z$. Imposing the boundary conditions on u and $\frac{\partial u}{\partial z}$, we are left with

$$\lambda = \sin z, \qquad z_s = \pi$$

or

$$u = \frac{\sin z}{z} . \tag{111}$$

Then,

$$\rho_o = \frac{\pi^2}{3}\bar{\rho} ,$$

$$P_o = \frac{\pi^2}{6}\bar{\rho}\frac{GM}{R} , \tag{112}$$

and

$$\phi_o = \frac{GM}{R} .$$

This last relationship shows that about the same work is done in moving a particle from the center to the surface as is done in moving the particle from the surface to infinity. This result is common to solutions for a wide range of index n. Assuming the ideal gas law and negligible radiation pressure,

$$kT_o = \tfrac{1}{2}(\mu M_p)\frac{GM}{R} . \tag{113}$$

The approximate equality between the kinetic energy of a particle at the center and the work required to carry a particle from the center to the surface exhibited by this last identity is also a characteristic common to polytropes for a large range in n.

One can derive a very illuminating theorem connecting the total gravitational energy of a polytrope with the parameters M and R. Multiply each side of the equation of hydrostatic equilibrium by $4\pi r^3 dr$ and integrate from the origin to the surface of the star

$$\int_0^s 4\pi r^3 \frac{\partial P}{\partial r}\,dr = -\int_0^s \frac{GM(r)}{r}\rho 4\pi r^2 dr. \tag{114}$$

Now integrate the left side by parts to find

$$-3\int_0^s P 4\pi r^2 dr = -\int_0^s \frac{GM(r)}{r}\,dM(r). \tag{115}$$

This can be written

$$3\int_0^s P\,dV + \Omega = 0 \tag{116}$$

where

$$\Omega = -\int_0^s \frac{GM(r)}{r}\,dM(r) \tag{117}$$

is the total gravitation potential energy. Substituting for P gives

$$3 \int_0^s \left[\frac{\phi}{n + 1}\right] dM(r) + \Omega = 0. \quad (118)$$

Integrating the expression for Ω by parts gives

$$\Omega = - \frac{GM^2}{2R} - \frac{G}{2} \int_0^s \left[\frac{M(r)^2}{r^2}\right] dr$$

$$= - \frac{GM^2}{2R} + \tfrac{1}{2} \int_0^s \left[\frac{\partial \phi}{\partial r}\right] M(r)\, dr. \quad (119)$$

One final integration by parts leads to

$$\Omega = - \frac{GM^2}{2R} - \tfrac{1}{2} \int_0^s \phi dM(r). \quad (120)$$

The same integral appears in equation (118); by eliminating it, we find

$$\Omega = - \frac{3}{5 - n} \left[\frac{GM^2}{R}\right]. \quad (121)$$

We have shown not only that the gravitational binding energy $(-\Omega)$ of a star is equal, in order of magnitude, to GM^2/R, but also that this binding energy increases with the degree of central condensation: $\left(\frac{\rho_o}{\bar{\rho}}\right)$ increases with increasing n. Another useful theorem that relates stellar energy sources to polytropic index can be devised.

Exercise VII. Show that, near the origin, one can write

$$u = e^{-z^2/6}\left[1 + \left(\frac{n}{120} - \frac{1}{72}\right) z^4 + \ldots\right]. \quad (122)$$

Exercise VIII. Assume (1) that any variable Q can be approximated by

$$Q = Q_o \left(\frac{\rho}{\rho_o}\right)^{1+k} \left(\frac{T}{T_o}\right)^s , \quad (123)$$

(2) that radiation pressure is negligible, and (3) that μ = const. Then show that

$$\int_0^s Q dM(r) = - \frac{Q_o M}{\left(z^2 \frac{\partial u}{\partial z}\right)_s} \int_0^s u^{2n+nk+s} z^2 dz . \quad (124)$$

Evaluate this approximately by first substituting the approximate expression for u derived in exercise VII and by then integrating to infinity (instead of to z_s). The result is

$$\int_0^s Q dM(r) \cong \frac{Q_o}{2} M_{eff}$$

$$M_{eff} = - \frac{M \left(\frac{\sqrt{\pi}}{2}\right)}{z_s^2 \left(\frac{\partial u}{\partial z}\right)_s} \left[\frac{6}{n(2+k) + s}\right]^{3/2} \left[1 + \frac{3}{8} \frac{(3n - 5)}{n(2+k) + s} + \ldots\right] \tag{125}$$

When will this approximation be reasonable?

As an example of the application of the theorem described in Exercise VIII, suppose that we know the temperature and density dependence of the energy producing nuclear reactions. For the pp chains and the C-N cycle, we have $k = 0$ and $s \sim 4$ or 17-18, respectively. Choosing M, R, L, μ and n, we can find ρ_o and T_o and thence obtain Q_o and finally L (calculated). In general L (calculated) $\neq$ L. By adjusting n to give the required L we determine the suitable n. One can show that for main sequence stars (for which a mean relationship between M and L is known) n is on the order of three. This coincides essentially with Eddington's "standard model."

Exercise IX. Prove that $n \sim 3$ for main sequence stars.

Stellar Evolution by Relaxing Difference Equations

For a polytrope, only one of the equations of stellar structure is necessarily satisfied, i.e., the equation of hydrostatic equilibrium. If we adopt a polytrope as our initial model and then require that it satisfy all the equations, we will find in general that it must undergo considerable internal adjustment before "thermal equilibrium" is achieved. We say that the star relaxes into equilibrium.

To calculate the relaxation process we first divide the star into a series of concentric shells, numbered (let us say) from the origin outward. Instead of writing differential equations, we write difference equations, assigning, at each shell boundary, values for all requisite variables P_i, T_i, L_i, M_i, R_i. For every variable Q we choose shell size in such a way that

$$\left|\frac{\Delta Q_i}{Q_i}\right| < \delta_i, \quad \Delta Q_i \equiv Q_i - Q_{i+1} , \tag{126}$$

where the δ_i, are chosen on the basis of common sense as educated by experience. We then construct the following quantities:

$$A_{1i} = \frac{(P_{i-1} - P_i)R_{i-\frac{1}{2}}^2}{(R_i - R_{i-1})GM_{i-\frac{1}{2}}\rho_{i-\frac{1}{2}}}$$

$$A_{2i} = \frac{4\pi R_{i-\frac{1}{2}}^2 (R_i - R_{i-1})\rho_{i-\frac{1}{2}}}{\Delta M_i}$$

$$A_{3i} = \frac{L_i - L_{i-1}}{\varepsilon_{i-\frac{1}{2}}\ \Delta M_i} \tag{127}$$

$$A_{4i} = \frac{(T_i - T_{i-1})\ P_{i-\frac{1}{2}}}{(P_i - P_{i-1})\ T_{i-\frac{1}{2}}\ V_{i-\frac{1}{2}}}$$

where

$$P_{i-\frac{1}{2}} = \tfrac{1}{2}(P_i + P_{i-1}),\ \text{etc.} \tag{128}$$

In the expression for A_{4i}, we insist that

$$V_{i-\frac{1}{2}} = \begin{cases} \left.\dfrac{d\ \ln T}{d\ \ln P}\right|_{ad} & \text{or} \\[2ex] \left[\dfrac{3\pi}{16}acG\right]^{-1}\left[\dfrac{\kappa PL}{MT^4}\right]_{i-\frac{1}{2}} & \end{cases} \tag{129}$$

For the central sphere, the appropriate A_{ji} are slightly different, since at the origin L = R = 0. The A_{ji} for the central sphere are:

$$A_{12} = \frac{(P_1 - P_2)R_2}{G\ \frac{M_2}{2}\ \rho_1}$$

$$A_{22} = \frac{4\pi}{3} R_2{}^3 \rho_1/M_2$$

$$A_{32} = \frac{L_2}{\varepsilon_1\ M_2} \tag{130}$$

$$A_{42} = (T_1 - T_2)\ \frac{4ac}{3}\ \left(\frac{T^3}{\kappa\rho}\right)_{\frac{1}{2}}\left[\frac{4\pi R_2}{(L_2/2)}\right]$$

It will be recognized that, in equilibrium, all of the A_{ji} must be unity. However, for a typical initial model, this will not be the case. It is our task to find new values of

P_i, T_i, etc., so that the A_{ji} do approach unity.

It is convenient in practice to define

$$C_{ji} = \ln A_{ji} \quad . \tag{131}$$

Take derivatives of all the C_{ji} with respect to $Q_{k,i}$ (= P_i,T_i, etc.) and write

$$C_{ji} + \sum_k \left[\frac{\partial C_{ji}}{\partial \ln Q_{k,i}}\right] \delta \ln Q_{k,i} + \sum_k \left[\frac{\partial C_{ji}}{\partial \ln Q_{k,i-1}}\right] \delta \ln Q_{k,i-1} = 0 \quad . \tag{132}$$

We want to solve these equations simultaneously for the δQ's

At all shells except the first, there are four equations to be solved for eight unknowns. In the central "shell", however, there are only six variables, since $L = R = 0$. So in the central sphere solve for all quantities at the shell interface $i = 2$ in terms of $\delta \ln P_2$ and $\delta \ln T_2$. Thus

$$\begin{aligned}
\delta\ln P_1 &= \alpha(1,1)\ \delta\ln P_2 + \beta(1,1)\ \delta\ln T_2 + \ (1,1) \\
\delta\ln T_1 &= \alpha(2,1)\ \delta\ln P_2 + \beta(2,1)\ \delta\ln T_2 + \ (2,1) \\
\delta\ln L_2 &= \alpha(3,1)\ \delta\ln P_2 + \beta(3,1)\ \delta\ln T_2 + \ (3,1) \\
\delta\ln R_2 &= \alpha(4,1)\ \delta\ln P_2 + \beta(4,1)\ \delta\ln T_2 + \ (4,1)
\end{aligned} \tag{133}$$

In the next shell, we may express $\delta \ln P_2$, $\delta \ln T_2$, $\delta \ln L_3$ and $\delta \ln R_3$ in terms of $\delta \ln P_3$, $\delta \ln T_3$, $\delta \ln L_2$, and $\delta \ln R_2$. But from the first shell we have a relationship between $\delta \ln L_2$, $\delta \ln R_2$ and $\delta \ln P_2$ and $\delta \ln T_2$. We may therefore find the α's and β's and 's in the equations

$$\begin{aligned}
\delta\ln P_2 &= \alpha(1,2)\ \delta\ln P_3 + \beta(1,2)\ \delta\ln T_3 + \ (1,2) \\
\delta\ln T_2 &= \alpha(2,2)\ \delta\ln P_3 + \beta(2,2)\ \delta\ln T_3 + \ (2,2) \\
\delta\ln L_3 &= \alpha(3,2)\ \delta\ln P_3 + \beta(3,2)\ \delta\ln T_3 + \ (3,2) \\
\delta\ln R_3 &= \alpha(4,2)\ \delta\ln P_3 + \beta(4,2)\ \delta\ln T_3 + \ (4,2)
\end{aligned} \tag{134}$$

Continuing in this way we find for the last shell

$$\begin{aligned}
\delta\ln P_{N-1} &= \alpha(1,N-1)\ \delta\ln P_N + \beta(1,N-1)\ \delta\ln T_N + \ (1,N-1) \\
\delta\ln T_{N-1} &= \alpha(2,N-1)\ \delta\ln P_N + \beta(2,N-1)\ \delta\ln T_N + \ (2,N-1) \\
\delta\ln L_N &= \alpha(3,N-1)\ \delta\ln P_N + \beta(3,N-1)\ \delta\ln T_N + \ (3,N-1) \\
\delta\ln R_N &= \alpha(4,N-1)\ \delta\ln P_N + \beta(4,N-1)\ \delta\ln T_N + \ (4,N-1)
\end{aligned} \tag{135}$$

The values of the variables P, T, L, and R at the base of an atmosphere are also known in terms of the variables L_s = surface luminosity and R_s = net stellar radius. Let us denote quantities at the base by P_{NS}, T_{NS}, L_{NS}, and R_{NS}. We can obtain numerical derivatives of these quantities by constructing several atmospheres for different choices of L_s and R_s. We then write

$$\begin{aligned}
\delta \ln P_{NS} &= \alpha(1,N)\ \delta \ln L_s + \beta(1,N)\ \delta \ln R_s \\
\delta \ln T_{NS} &= \alpha(2,N)\ \delta \ln L_s + \beta(2,N)\ \delta \ln R_s \\
\delta \ln L_{NS} &= \alpha(3,N)\ \delta \ln L_s + \beta(3,N)\ \delta \ln R_s \\
\delta \ln R_{NS} &= \alpha(4,N)\ \delta \ln L_s + \beta(4,N)\ \delta \ln R_s
\end{aligned} \tag{136}$$

If we now insist that $\ln P_N + \delta \ln P_N = \delta \ln P_{NS} + \delta \ln P_{NS}$, $\ln T_N + \delta \ln T_N = \ln T_{NS} + \delta \ln T_{NS}$, etc., and eliminate $\delta \ln L_N$ and $\delta \ln R_N$ in terms of $\delta \ln P_N$ and $\delta \ln T_N$, we obtain:

$$\begin{aligned}
&\ln P_N + \delta \ln P_N = \ln P_{NS} + \alpha(1,N)\ \delta \ln L_S + \beta(1,N)\ \delta \ln R_S \\
&\ln T_N + \delta \ln T_N = \ln T_{NS} + \alpha(2,N)\ \delta \ln L_S + \beta(2,N)\ \delta \ln R_S \\
&\ln L_N + \alpha(3,N-1)\ \delta \ln P_N + \beta(3,N-1)\ \delta \ln T_N \\
&\qquad = \ln L_{NS} + \alpha(3,N)\ \delta \ln L_S + \beta(3,N)\ \delta \ln R_S \\
&\ln R_N + \alpha(4,N-1)\ \delta \ln P_N + \beta(4,N-1)\ \delta \ln T_N \\
&\qquad = \ln R_{NS} + \alpha(4,N)\ \delta \ln L_S + \beta(4,N)\ \delta \ln R_S .
\end{aligned} \tag{136'}$$

These last four equations then yield values for $\delta \ln P_N$, $\delta \ln T_N$, $\delta \ln L_S$ and $\delta \ln R_S$. Working back through the star one can then obtain all of the other $\delta \ln Q_{ki}$'s. After obtaining new values of the Q_{ji}, put these values back into the equations for the C_{ji}. Hopefully, the new C_{ji}'s will be closer to zero (i.e., the A_{ji} closer to unity). If so, we can repeat the above procedure until we are satisfied. If the results do not converge, then we must either go back to the initial model and try for improvement there, or we must adjust the time step (not always to smaller values!).

It might be disturbing that time does not seem to appear explicitly in the equations thus far exhibited. Actually, time appears in two distinct fashions.

The energy generation rate can be written as the sum of two terms

$$\varepsilon = \varepsilon_{Nuc} + \varepsilon_{Grav} \tag{137}$$

where ε_{Nuc} is the rate (per gram per second) of nuclear energy production and

$$\varepsilon_{Grav} = -T \frac{dS}{dt} \tag{138}$$

is the rate at which gravitational and thermal energy is being released locally. Here S is entropy per gram. Thus, time enters explicitly into the effective energy production rate through ε_{Grav}.

Wherever nuclear reactions occur, nuclear species are being transmuted, so concentrations are not constant and

$$\frac{dX_i}{dt} \neq 0 \tag{139}$$

This is the second way in which time enters the picture.

One way to incorporate the time dependence into the computational scheme is to write

$$(\varepsilon_{Grav})_{i-\frac{1}{2}} \cong -\frac{1}{2}\left(T_{i-\frac{1}{2}}^{\ n} + T_{i-\frac{1}{2}}^{n+1}\right) \frac{S_{i-\frac{1}{2}}^{n+1} - S_{i-\frac{1}{2}}^{\ n}}{\Delta t} \tag{140}$$

and

$$X_i^{n+1} \cong X_i^{\ n} + \frac{1}{2}\left(\frac{dX_i^{\ n}}{dt} + \frac{dX_i^{n+1}}{dt}\right)\Delta t \tag{141}$$

where $\Delta t = t^{n+1} - t^n$ is the time interval between models n and n+1. In practice, these approximations are adequate as long as Δt is chosen small enough.

A Simple Example--The Gravitational Contraction Phase

One approach to the study of stellar evolution is to first choose an initial model (e.g., by integration and fitting), and to then follow the development of this model in time by means of the relaxation technique. A simple illustration of the procedure is afforded by a spherical homogeneous distribution of matter that is at sufficiently low temperature and density throughout that nuclear energy production can be neglected (assume no deuterium, lithium, or other elements that react at low temperatures).

We know nothing about $\frac{dS}{dt}$ at this stage, so we make the simple assumption that it is a constant throughout the star at any given time, even though this "constant" may depend on time. Then

$$\frac{dS}{dt} = -C(t) \tag{142}$$

After choosing a value for the mass and the central temperature ($T \leq 10^6\,°K$), we vary the parameters P_c, L, R, and C until we obtain (by integration) a model that satisfies the structure equations. It turns out that, for central temperatures as low as $10^6\,°K$, the opacity is very high and energy flow is primarily by convection, with the temperature gradient essentially coinciding with the adiabatic gradient. Consequently, the entropy is constant throughout the model star we have constructed. This justifies our initial assumption that entropy is a function only of time. Since, for low enough stellar masses, radiation pressure is negligible, the model star can be described by a polytrope model of index $n=3/2$ ($P \propto \rho^{5/3}$ under adiabatic conditions). We know therefore that its gravitational potential energy is

$$\Omega = -\frac{6}{7}\frac{GM^2}{R}\ . \tag{143}$$

We can obtain a possibly causally related sequence of similar models simply by varying T_c to other fixed values and iterating on P_c, L, R, and C until we have achieved convergence. If we pick a larger value for central temperature, we find that the new solution has smaller radius and luminosity. A sequence of solutions for varying T_c falls along a line in the H-R diagram known as the"Hayashi track", as shown in Fig. 16. For a long distance along this track, models are completely convective except for a thin outer shell. In order to discover how models along the Hayashi track are causally related, we must first prove a short theorem.

In the earlier result

$$3\int P dV + \Omega = 0 \tag{144}$$

we insert

$$P \cong \frac{2}{3} U_{gas} + \frac{1}{3} U_{rad}\ , \tag{145}$$

where U stands for an energy density (erg/cm^3). We find

$$2\int U_{gas}\, dV + \int U_{rad}\, dV + \Omega = 0. \tag{146}$$

Writing E for an energy density integrated over the entire star, we have

$$E_{total} = E_{gas} + E_{rad} + \Omega = -E_{gas} \tag{147}$$

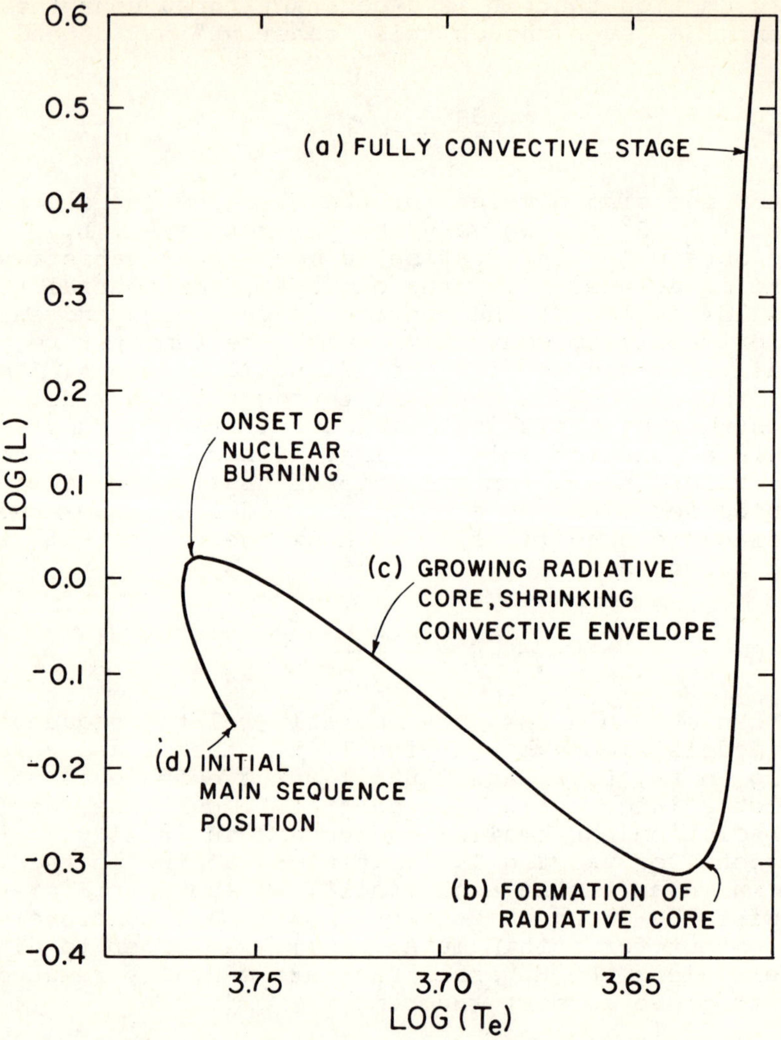

Figure 16. The track of a star in the H-R diagram during the phase of gravitational contraction in hydrostatic equilibrium.

In words: the total binding energy of the star (E_{bind} = $-E_{tot}$) is equal to the total energy content of the star that is in the form of kinetic motions of particles (E_{gas}). Now set $P_{gas} = \beta P$ and write

$$3 \int \frac{P_{gas}}{\beta} dV + \Omega = 0. \tag{148}$$

Define $\overline{\beta}$ in such a way that, with the substitution $P_{gas} = \frac{2}{3} U_{gas}$ we have

$$\frac{2E_{gas}}{\overline{\beta}} + \Omega = 0. \tag{149}$$

Finally,

$$\begin{aligned} E_{tot} &= E_{gas} + E_{rad} + \Omega \\ &= - E_{gas} \\ &= \frac{\overline{\beta}}{2} \Omega \end{aligned} \tag{150}$$

or

$$E_{Binding} = \frac{\overline{\beta}}{2} |\Omega| = E_{gas}. \tag{151}$$

The binding energy can be thought of as the difference between the "gravitational binding energy", $|\Omega|$, and the "excitation energy", $E_{exit} = E_{gas} + E_{rad}$.

Let us now apply our theorem to our contracting models, since $\overline{\beta}$ is close to unity,

$$E_{Binding} \sim \frac{1}{2} |\Omega| = \frac{3}{7} \frac{GM^2}{R} . \tag{152}$$

The rate at which energy is emitted from the star during contraction is just the time derivative of the binding energy!

$$L = \frac{d}{dt} E_{Binding} \sim \frac{d}{dt}(\frac{1}{2}|\Omega|) = - \frac{d}{dt} (\frac{3}{7} \frac{GM^2}{R}) \tag{153}$$

It is obvious now (as it might have been before) that the model with a smaller radius on the Hayashi track follows in time a model with a larger radius. Choosing two models that do not differ much in L, we readily find the actual time separation between the models to be $\Delta t = \Delta E_{Binding}/\overline{L}$, where $\overline{L}$ is the mean luminosity of the two models.

We have just demonstrated that a time sequence of models can be constructed, in some instances, without recourse to the fancy relaxation technique.

At some point, however, our luck runs out.

If we follow a model as it evolves down the Hayashi track, we find that, with rising interior temperatures, the opacity drops until the interior is no longer completely convective; a radiative core develops. Subsequent evolution must be followed by the relaxation method. The resulting track in the H-R diagram is shown in Fig. 16.

The track has been labeled with numbers to indicate qualitatively different stages in the star's evolution. We will discuss briefly some aspects of the development of a star of one solar mass.

a) The star remains on the Hayashi track itself for about 10^6 years. The entire star is in convective, adiabatic equilibrium except for a narrow shell at the surface, where the dominant source of opacity is the H^- ion. This opacity increases with increasing temperature. As the star gives up its gravitational energy, it contracts, and temperatures rise throughout; however, because the surface opacity increases with increasing surface temperature, the luminosity decreases as the star contracts.

b) When temperatures are high enough at the center, the opacity there drops and a radiative core develops. The star remains fairly stationary in the H-R diagram during the development of this core; the binding energy released comes not from a change in radius, but from a redistribution of internal density. The effective polytropic index changes from about 3/2 to about 3. This stage lasts several times 10^6 years.

c) After the radiative core has extended over much of the interior the evolution is determined by the internal opacity instead of by the surface opacity as in earlier stages. During this phase we have roughly $\kappa \propto \rho/T^{3.5}$; but, since $\rho/T^3 \propto 1/M^2$, we have $\kappa \propto \dfrac{1}{M^2}\dfrac{1}{T^{1/2}}$. Gravitational contraction is still the energy source and $T \propto \dfrac{M}{R}$, so as the radius decreases the temperature increases; as a consequence the opacity decreases and, since $L \sim M^3/k$, the luminosity also increases. The star now moves upward and blueward in the H-R diagram toward the main sequence. This period of evolution lasts about 10^7 years.

d) Finally, at the center of the radiative core, nuclear energy production begins. The nuclear energy produced in the core raises the temperature there and acts to brake the contraction rate of the star. Most of the total energy production still comes from contraction. As a result, the luminosity drops because of (rather than in spite of) the additional energy source. In a matter of a few x 10^7 years, the star completes the transition to the main sequence phase when energy production by nuclear sources completely overshadows gravitational energy production. The main sequence hydrogen burning stage persists for about 10^{10} years.

The internal configuration for the four stages described are shown schematically in Fig. 17.

V. Some Results of Evolutionary Calculations

Figs. 18-21 show some of the detailed results of model calculations (with the relaxation technique) of a gravitationally contracting star as it moves down the Hayashi track and across to the main sequence. The two tracks in Fig. 18 describe the evolution of the external characteristics of two stars of one solar mass that differ only in the abundance of elements of low ($\lesssim$ 7.5 eV) ionization potential. The abundance in the star following the dashed track is ten times smaller than that in the star following the solid track. The internal opacity parameter Z is the same in both cases. Since the metals of low ionization potential provide electrons for the H^- ion, reducing their abundance lowers the H^- opacity which dominates near the surface. It is to be emphasized that the opacity law differs only in photospheric layers for the two cases. The prominence of the shift for models in the low temperature Hayashi phase, when most of each star is in adiabatic equilibrium, shows that the tail can, in some instances, really wag the dog.

The time dependences of a number of pertinent quantities are shown in Fig. 19. The curve labeled Q_{RC} describes growth of the radiative core. The quantity $\rho_c/\bar{\rho}$ is a measure of central concentration. Note that as Q_{RC} increases, the central condensation increases from a value appropriate to an $n = 3/2$ polytrope to a value appropriate to an $n \sim 3$ polytrope.

The time dependences of additional internal quantities are shown in Fig. 20. The onset of nuclear reactions can be followed here. The concentration of He^3 begins to rise as it is produced from the pp and pd reactions. As He^3 reaches a peak, C^{12} begins to burn, producing high energy fluxes and initiating a convective core (Q_{CC}). Once the supply of C^{12} has been converted into N^{14}, the energy flux decreases. This is because the N^{14} reaction is slower. We say that the star has reached the main sequence when the energy released by nuclear reactions exceeds by a wide margin the energy released by gravitational contraction. Following the curve L_g/L (= rate of energy release from the gravitational field divided by the total rate of energy efflux from the surface) we see that the star actually expands slightly at one point (when the ratio goes negative). The energy for expansion near the center is supplied by the nuclear sources there.

The variation of luminosity with radius is shown in Fig. 21 for a model just entering the core nuclear burning stage. The steep portion of the luminosity curve occurs in the region of nuclear energy production, while the slow rise in the outer 7/10 of the star is due to gravitational contraction.

The foregoing discussion must be modified in the case of stars more massive than the sun, where the conversion of C^{12} into N^{14} is strong enough to settle the star temporarily

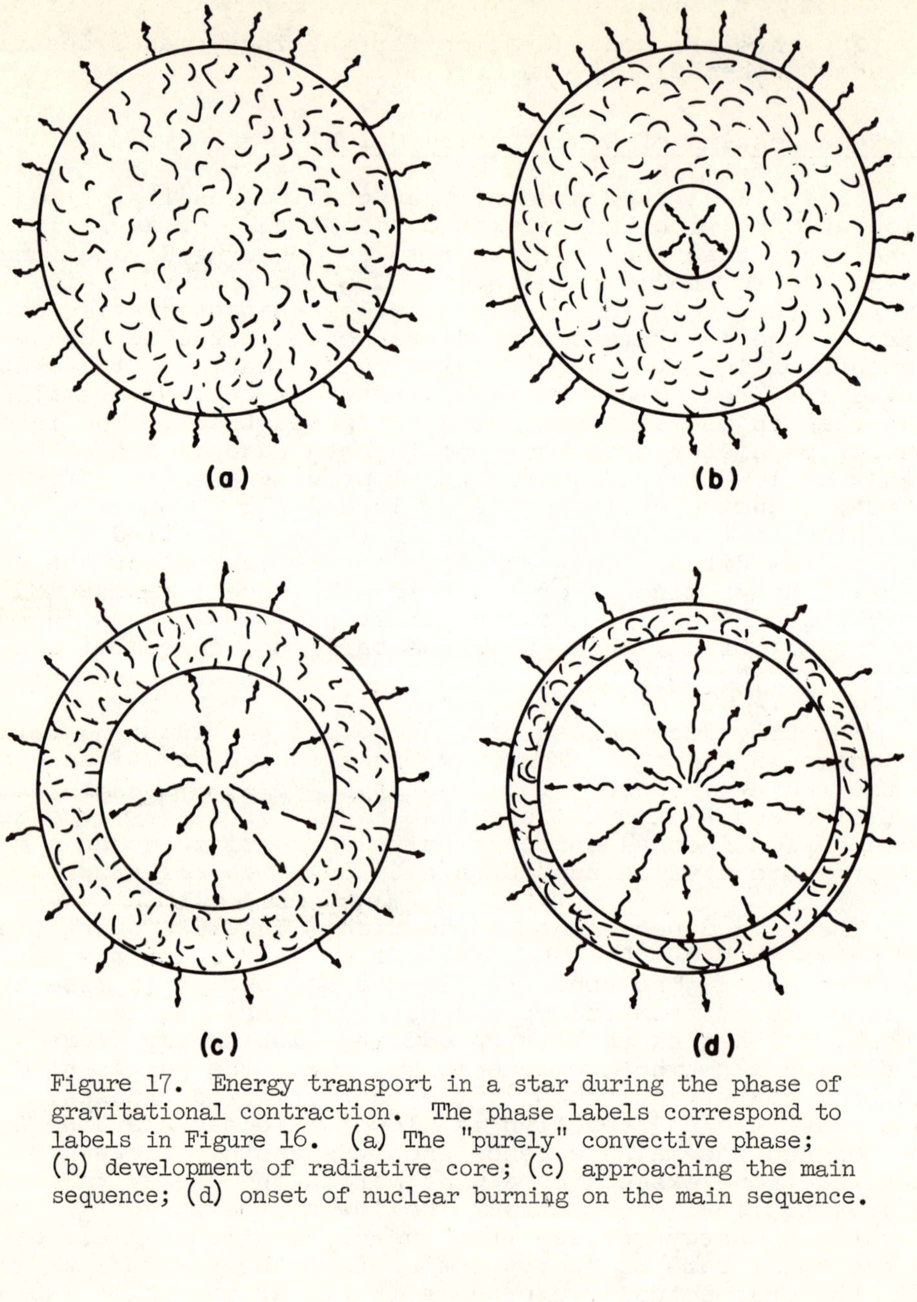

Figure 17. Energy transport in a star during the phase of gravitational contraction. The phase labels correspond to labels in Figure 16. (a) The "purely" convective phase; (b) development of radiative core; (c) approaching the main sequence; (d) onset of nuclear burning on the main sequence.

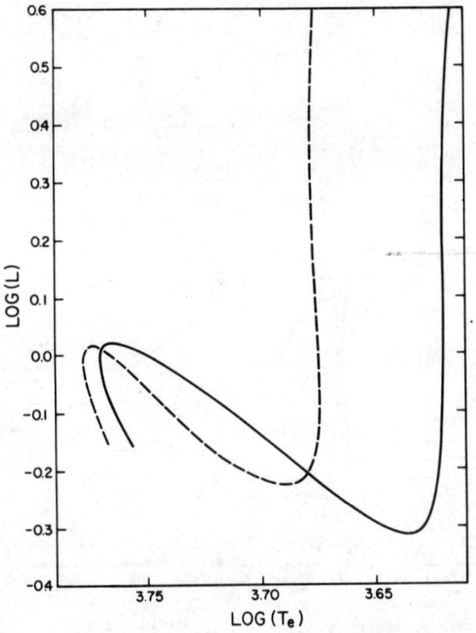

Figure 18. Paths in the theoretical Hertzsprung-Russell diagram for $M = M_\odot$. Luminosity is in units of $L_\odot = 3.86\times10^{33}$ erg/sec, and surface temperature T_e is in units of °K. The solid curve is constructed using a mass fraction of metals with 7.5-eV ionization potential, $X_M = 5.4\times10^{-5}$. The dashed curve is constructed with $X_M = 5.4\times10^{-6}$. From Iben, I., Jr., Ap. J., 141, 993 (1965).

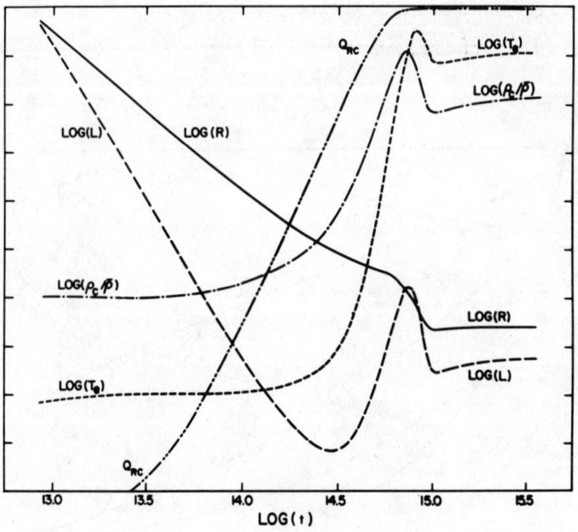

Figure 19. The variation with time t (in sec.) of surface temperature T_e (in units of °K), luminosity L (in units of 3.86×10^{33} erg/sec), stellar radius R (in units of 6.96×10^{10} cm), central over mean density $\rho_c/\bar{\rho}$, and mass fraction in the radiative core Q_{RC} for a stellar model of mass $M = M_\odot$. Maximum and minimum scale limits correspond to $3.58 < \log T_e < 3.78$, $-0.4 < \log L < 0.6$, $-0.4 < \log R < 0.6$, $0.0 < \log (\rho_c/\bar{\rho}) < 2.0$, and $0 < Q_{RC} < 1$. From Iben, I., Jr., Ap. J., 141, 993 (1965).

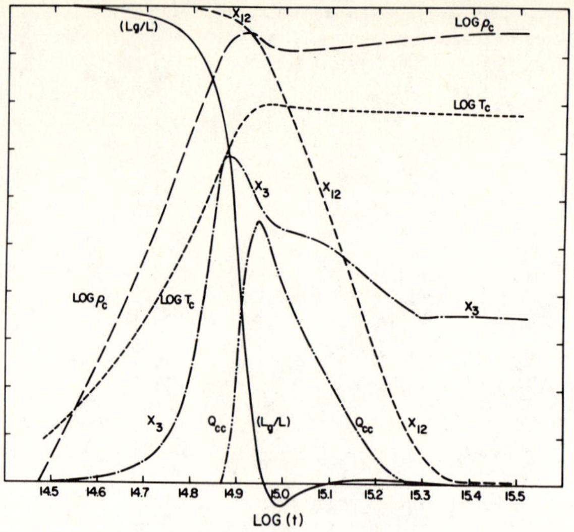

Figure 20. The variation with time of the He^3 abundance by mass (X_3), C^{12} abundance by mass (X_{12}), central temperature (T_e), central density (ρ_c), mass fraction in a convective core ($Q_{cc} = M_{core}/M_{star}$), and the relative contribution of gravitational energy to the total luminosity (L_g/L). Model mass $M = M_\odot$. T_e is in units of 10^6°K, and ρ_c is in units of gm/cm^3. Scale limits for all variables but (L_g/L) are listed in Table 1. Within graph boundaries, L_g/L varies from 1 to 0. From Iben, I., Jr., Ap. J., 141, 993 (1965).

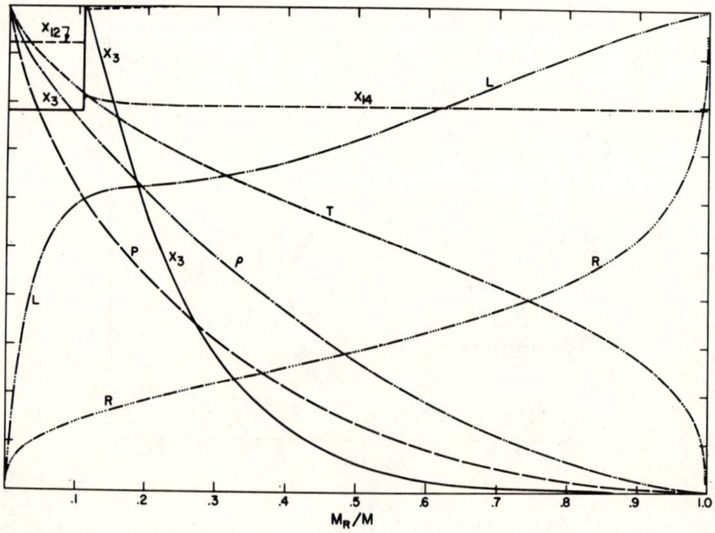

Figure 21. The variation with mass fraction of pressure (P), temperature (T), density (ρ), luminosity (L), radius (R), He^3 abundance by mass (X_3), C^{12} abundance by mass (X_{12}), and N^{14} abundance by mass (X_{14}). Model mass $M = M_\odot$ and $\log(t) = 14.941$. The maximum value of each variable is on the scale of unity. From Iben, I., Jr., Ap. J., 141, 993 (1965).

TABLE 1

Time Scale for Figure 25.

Point	$(M/M_\odot)$								
	15.0	9.0	5.0	3.0	2.25	1.5	1.25	1.0	0.5
1.	6.740×10^{2}	1.443×10^{3}	2.936×10^{4}	3.420×10^{4}	7.862×10^{4}	2.347×10^{5}	4.508×10^{5}	1.189×10^{5}	3.195×10^{5}
2.	3.766×10^{3}	1.473×10^{4}	1.069×10^{5}	2.078×10^{5}	5.940×10^{5}	2.363×10^{6}	3.957×10^{6}	1.058×10^{6}	1.786×10^{6}
3.	9.350×10^{3}	3.645×10^{4}	2.001×10^{5}	7.633×10^{5}	1.883×10^{6}	5.801×10^{6}	8.800×10^{6}	8.910×10^{6}	8.711×10^{6}
4.	2.203×10^{4}	6.987×10^{4}	2.860×10^{5}	1.135×10^{6}	2.505×10^{6}	7.584×10^{6}	1.155×10^{7}	1.821×10^{7}	3.092×10^{7}
5.	2.657×10^{4}	7.922×10^{4}	3.137×10^{5}	1.250×10^{6}	2.818×10^{6}	8.620×10^{6}	1.404×10^{7}	2.529×10^{7}	1.550×10^{8}
6.	3.984×10^{4}	1.019×10^{5}	3.880×10^{5}	1.465×10^{6}	3.319×10^{6}	1.043×10^{7}	1.755×10^{7}	3.418×10^{7}	. . . .
7.	4.585×10^{4}	1.195×10^{5}	4.559×10^{5}	1.741×10^{6}	3.993×10^{6}	1.339×10^{7}	2.796×10^{7}	5.016×10^{7}	. . . .
8.	6.170×10^{4}	1.505×10^{5}	5.759×10^{5}	2.514×10^{6}	5.855×10^{6}	1.821×10^{7}	2.954×10^{7}	. . . .	. . . .

onto a "C^{12} main sequence" until carbon is nearly exhausted at the center. Then, further contraction raises interior temperatures until energy production by the CN cycle settles the star onto the actual initial main sequence. The evolutionary track for such a star is shown in Fig. 22.

In Fig. 23 and Fig. 24 we see a further description of the behavior of the 1.5 $M_\odot$ model. The results are qualitatively similar to those for the 1.0 $M_\odot$ star. The dip in the plot of luminosity against mass fraction in Fig. 24 indicates that energy from the nuclear burning center is being absorbed by expanding material overlying the center.

Evolutionary tracks for still additional models are shown in Fig. 25. In Fig. 26 time constant loci (defined by models of different mass, but same age) are indicated along with a sketch of observational results with which they may be compared. We see a rough agreement between theory and observation in that more massive stars arrive on the main sequence more quickly than less massive stars. The fact that the mean location of cluster stars does not coincide with any single time constant locus suggests that star formation is spread out over a finite interval greater than 10^7 yrs to 10^8 years.

Before describing evolution off the main sequence it is worthwhile commenting on how properties of initial main sequence stars (just beginning to deplete hydrogen in the core) depend on stellar mass and on the choice of initial composition.

The lower curve in Fig. 27 defines the locus of initial main sequence models of a specfic composition ($Y = 0.27$, $Z = 0.02$) in the mass-luminosity diagram. Properties of the initial models are described in parentheses beside a selection of mass points.

For the composition chosen, the energy production rate by the pp reactions equals the energy production rate by the CN cycle in a star of about 1.9 $M_\odot$. Central convection occurs in stars more massive than about 1.0 $M_\odot$, not in less massive stars. Convective envelopes become prominent in stars less massive than about 1.5 $M_\odot$ and convection extends all the way from just below the photosphere to the center in stars less massive than about 0.5 $M_\odot$.

It is interesting that central density is not a monotonic function of stellar mass in the neighborhood of one solar mass, but passes through a relative maximum at some critical mass M_m. Adding <u>or</u> subtracting a fraction of a solar mass to a homogeneous star of mass M_m will cause central regions of that star to expand!

The position of an initial main sequence in the M-L diagram is a function of composition, as is indicated in Fig. 28. The crosses represent the location of several nearby metal-rich stars. Boxes locate stars in several Hyades binaries. Note that, for values of Z thought to be appropriate for metal-rich stars in the solar neighborhood, a helium abundance on the order of 30-40 percent by mass ($Y \sim 0.3$-0.4) seems to be indicated.

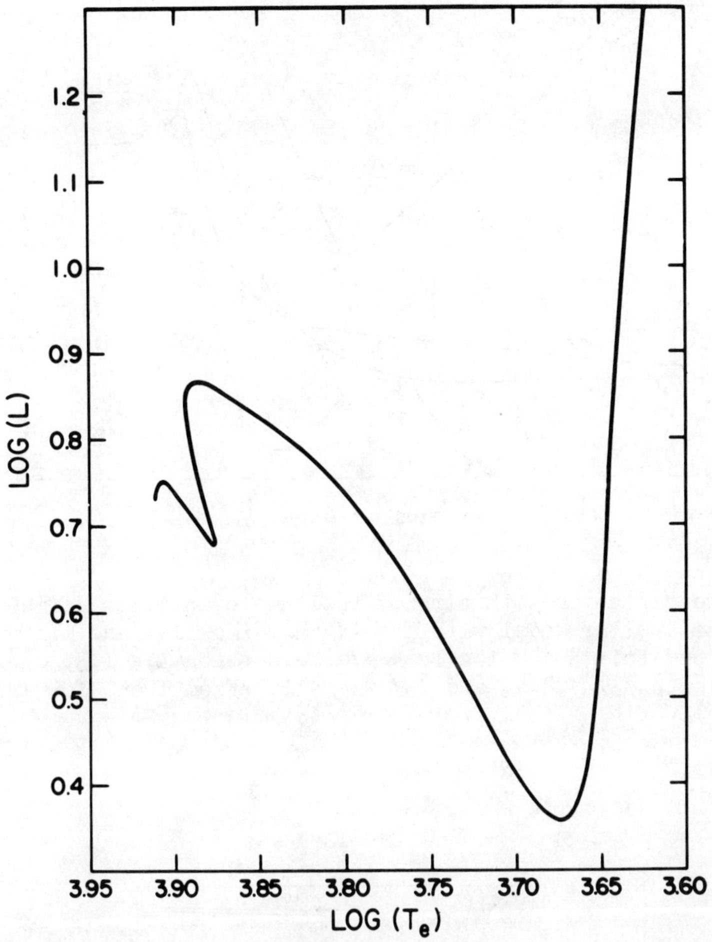

Figure 22. The path in the Hertzsprung-Russell diagram for $M = 1.5\ M_\odot$. Units of luminosity and surface temperature are the same as in Figure 18. From Iben, I., Jr., Ap. J., 141, 993 (1965).

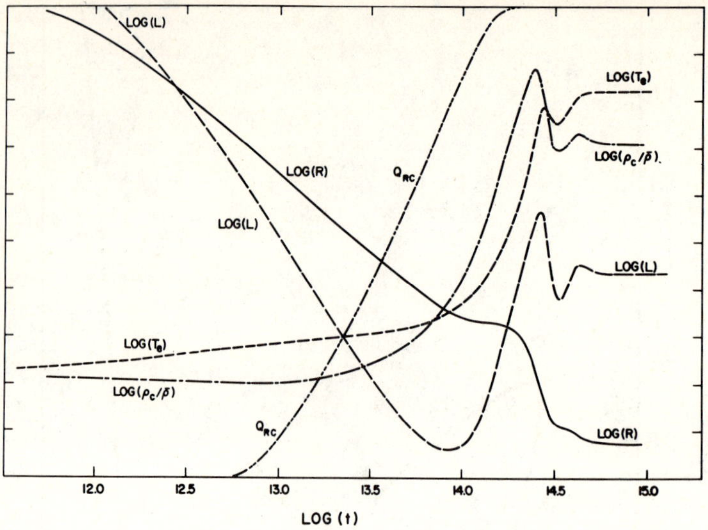

Figure 23. The variation with time of the same quantities defined under Figure 19 for a stellar model with $M = 1.5\ M_\odot$. Maximum and minimum scale limits correspond to: $3.5 < \log T_e < 4.0$, $0.3 < \text{Log } L < 1.3$, $0.0 < \log R < 1.0$, $0.4 < \log(\rho_c/\bar{\rho}) < 2.4$, and $0 < Q_{RC} < 1$. From Iben, I., Jr., Ap. J., 141, 993 (1965).

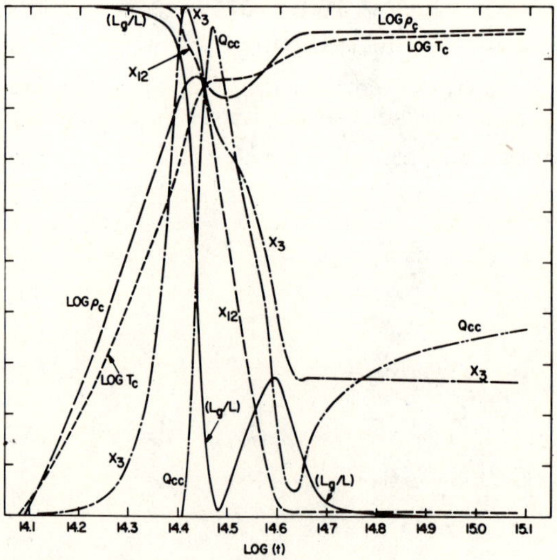

Figure 24. The variation with time of the same quantities defined under Figure 20 for $M = 1.5\ M_\odot$. From Iben, I., Jr., Ap. J., 141, 993 (1965).

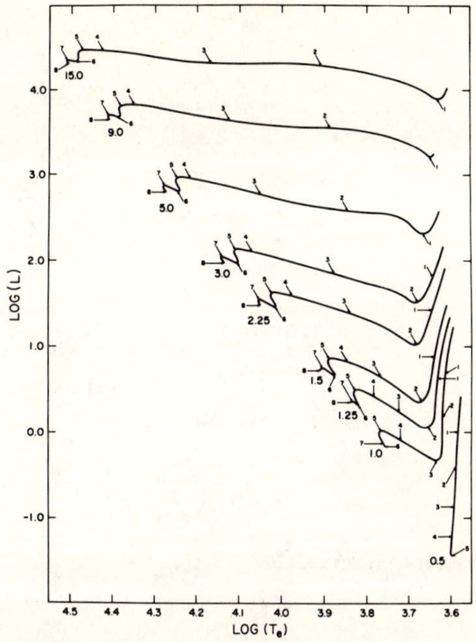

Figure 25. Paths in the H-R diagram for models of mass $(M/M_\odot)$ = 0.5, 1.0, 1.25, 1.5, 2.25, 3.0, 5.0, 9.0 and 15.0. Units of L and T_e are as in Figure 18. From Iben, I., Jr., Ap. J., 141, 993 (1965). Times to reach arrowed points are given in Table 1.

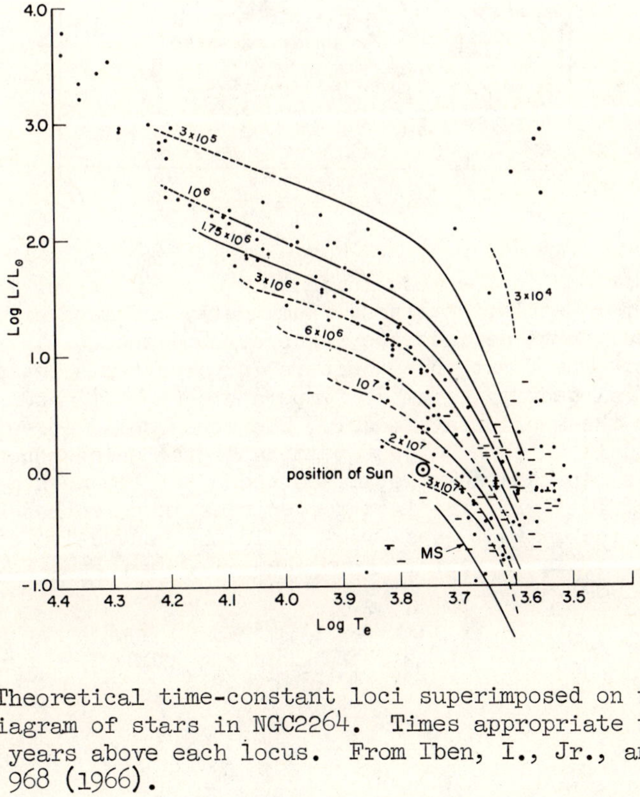

Figure 26. Theoretical time-constant loci superimposed on the position in the H-R diagram of stars in NGC2264. Times appropriate to each locus are given in years above each locus. From Iben, I., Jr., and Talbot, R. J., Ap. J., 144, 968 (1966).

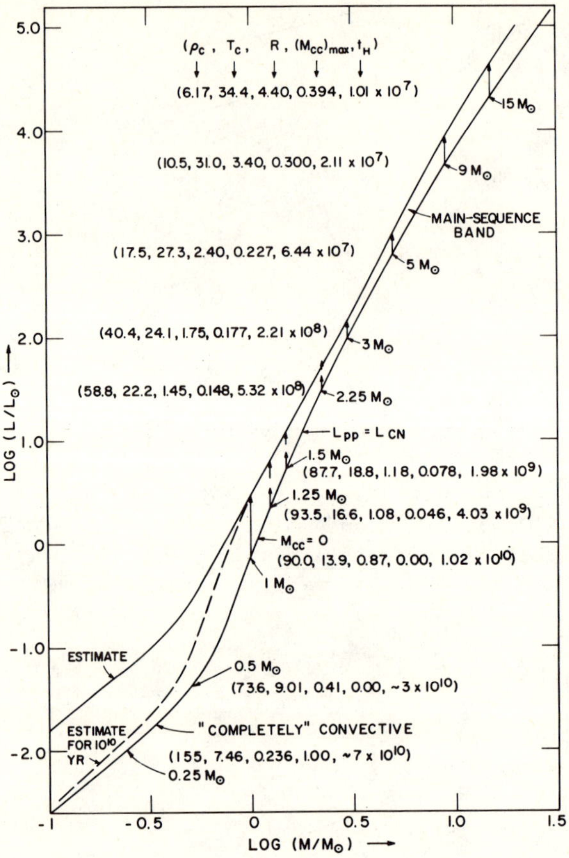

Figure 27. Positions in the mass-luminosity diagram during the main-sequence phase for metal-rich stars. Luminosity and mass are in solar units. The lower curve defines the locus of homogeneous, initial main-sequence models. The first three numbers in parentheses beside each point give central density (g/cm^3), temperature (10^6 °K), and stellar radius ($R_\odot$) for the appropriate model. The upper solid curve represents the locus of models which have just terminated the main-sequence phase. The dashed curve represents an estimated locus for low-mass models which have evolved for 10^{10} yr. The last two entries in parentheses beside each point give the maximum mass fraction in the convective core and the main-sequence lifetime in years. From Iben, I., Jr., Ann. Rev. Ast. & Ap., 5, 571 (1967).

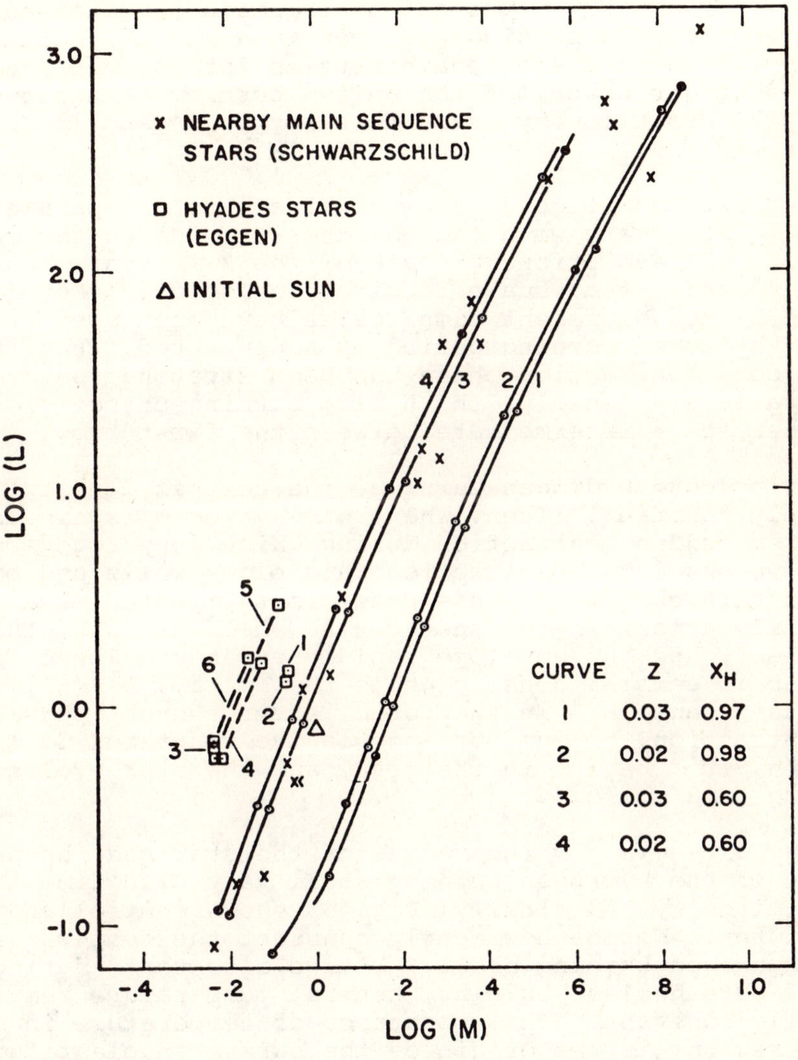

Figure 28. The position of an initial main sequence in the mass-luminosity diagram as a function of composition. The location of Hyades stars is indicated by boxes, and the initial sun is indicated by a triangle. From Iben, I., Jr., Ap. J., 138, 452 (1963).

The helium abundance in Hyades stars seems to be significantly higher than in solar-neighborhood stars.

Once a star has arrived on the main sequence, it remains there, burning hydrogen on a long time scale. When hydrogen has been exhausted from the core, the star undergoes a brief period of overall contraction. If it is sufficiently massive, the star then quickly moves off of the main sequence and over to the giant branch burning hydrogen in a shell. Helium is then ignited and the star again settles into a relatively stable phase that might be called the helium burning main sequence phase. The evolutionary track for a typical case (5 $M_\odot$) is shown in Fig. 29.

The next four figures show how the internal parameters for a 3 $M_\odot$ star vary with time as the star leaves the main sequence. The evolutionary track of the 3 $M_\odot$ star is shown in Fig. 34 and a time scale is given in Table 2. Note in Fig. 30-33 that different time scales have been used; first, the long hydrogen burning period is represented, then the subsequent core contraction phase has been stretched out to show more detail, and finally, the helium burning phase is depicted on another scale intermediate between the first two.

During core hydrogen burning, the central density is relatively constant. Then, when core hydrogen is exhausted, there is a sudden contraction during which a hydrogen burning shell develops. The contraction rate slows while hydrogen burns in a thick shell. This phase is terminated when a thermal instability arises in the shell. The shell narrows, the core "collapses", and the envelope rapidly expands. There is a small dip in central density while the $N^{14}(\alpha,\gamma)F^{18}(\beta^{+},\gamma)O^{18}$ reactions exhaust N^{14} in the core. After a short phase of core contraction, helium burning via the reactions $3\alpha \rightarrow C^{12}$ and $C^{12}(\alpha,\gamma)O^{16}$ begins in the core, and the star evolves again on a long time scale.

Fig. 31 shows how the radius of the star and the spacial location of the hydrogen burning shell vary with time. Fig. 32 shows luminosity and finally, Fig. 33 shows central and shell temperatures. Notice how nearly constant the temperature is in the hydrogen burning shell. Notice also that, during each phase of core nuclear burning, central temperature remains relatively constant. This constancy of temperature in nuclear burning regions is the origin of the phrase "nuclear fuels act as thermostats".

An examination of Fig. 34 shows that, with decreasing stellar mass, the maximum extent to which the evolutionary track extends to the blue during core helium burning decreases until, below about 2.5 $M_\odot$ (for this choice of input physics and composition), the core helium burning phase is confined to the giant branch.

As mass is increased above 2.5 $M_\odot$, the core helium burning phase extends further and further to the blue. For stars more massive than about 3 $M_\odot$, it is possible to define two major phases of core helium burning. This is due to a circumstance

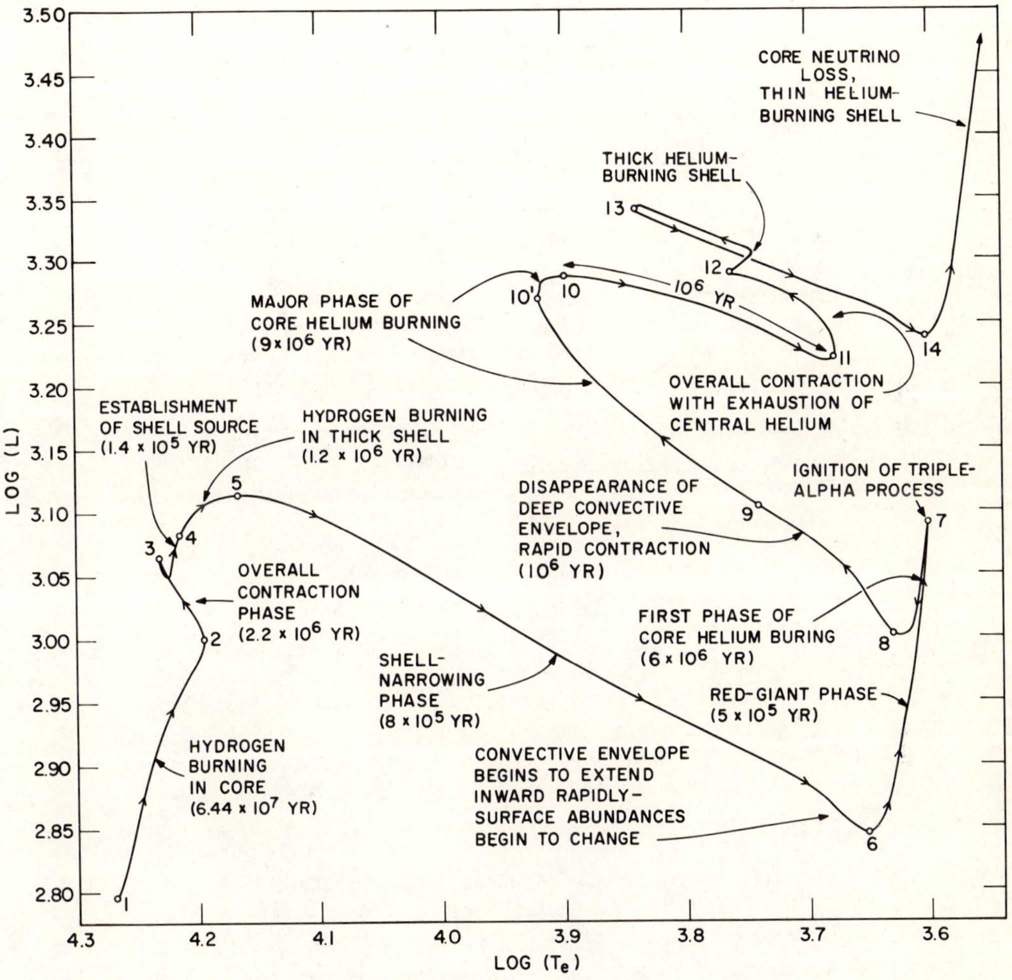

Figure 29. The path of a metal-rich 5 $M_\odot$ star in the Hertzsprung-Russell diagram. Luminosity is in solar units, $L_\odot = 3.86 \times 10^{33}$ erg/sec, and surface temperature T_e is in °K. Traversal times between labeled points are given in years. From Iben, I., Jr., Ann. Rev. of Ast. & Ap., 5, 571 (1967).

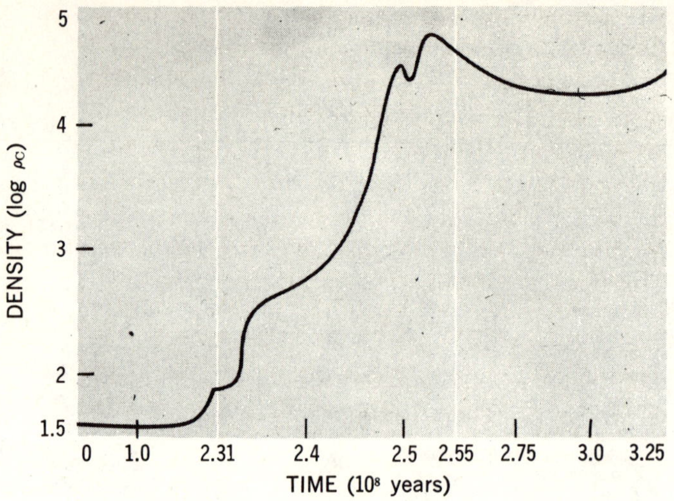

Figure 30. Central density of a 3 $M_\odot$ star as a function of time. From Bethe, H. A., Physics Today, September, 1968, p. 36 ff.

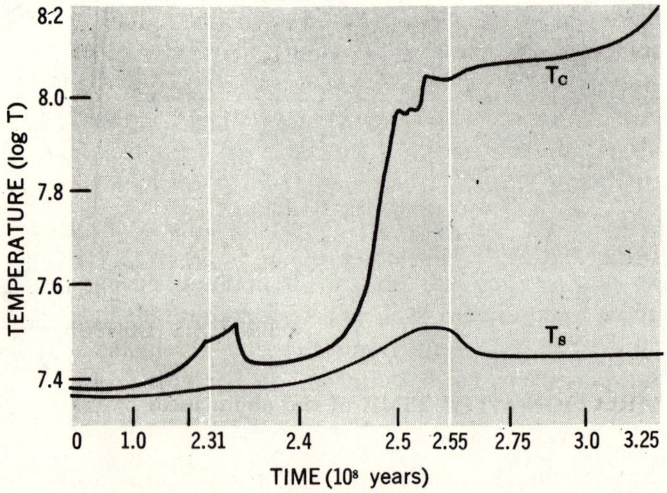

Figure 31. Stellar radius and shell radius of a 3 $M_\odot$ star. From Bethe, H. A., Physics Today, September, 1968, p. 36 ff.

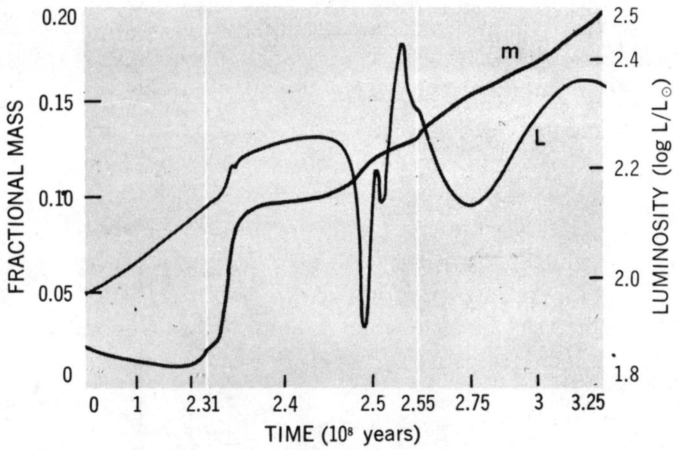

Figure 32. Luminosity of a 3 $M_{\odot}$ star and mass enclosed by the hydrogen burning shell. From Bethe, H. A., Physics Today, September, 1968, p. 36 ff.

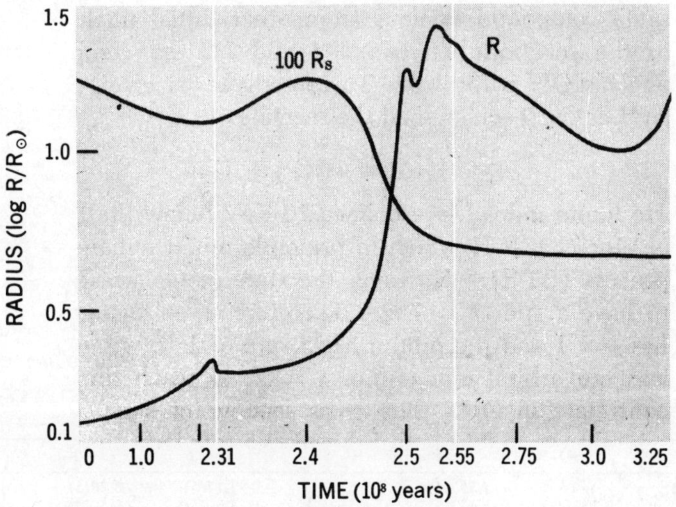

Figure 33. Central and shell temperatures of a 3 $M_{\odot}$ star. From Bethe, H. A., Physics Today, September, 1968, p. 36 ff.

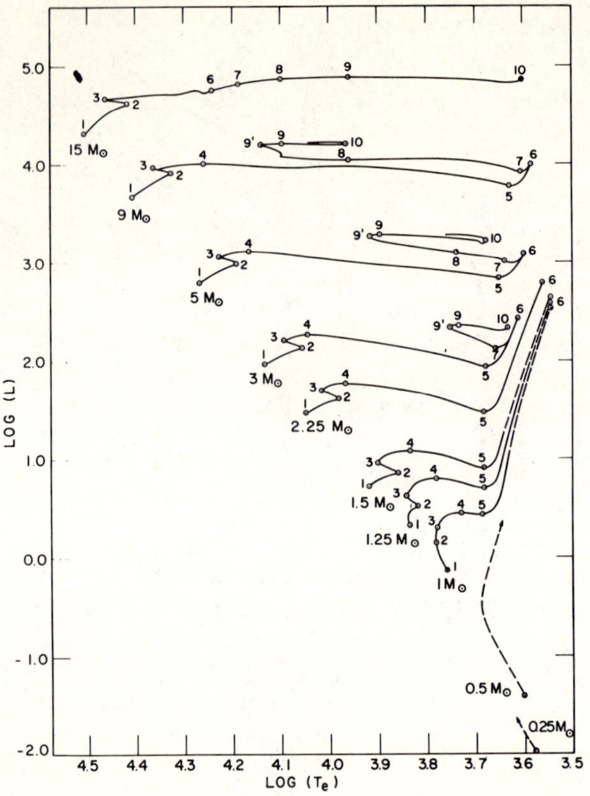

Figure 34. Evolutionary tracks for metal-rich stars leaving the main sequence. From Iben, I., Jr., Ann. Rev. Ast. & Ap., 5, 571 (1967).

TABLE 2

Time (in years) spent by stars of various masses in traversing the intervals between successive labeled points of Fig. 34. The number in parentheses beside each entry indicates the power of 10 to which the entry is to be raised.

Mass ($M_\odot$)	Interval (i–j)								
	(1–2)	(2–3)	(3–4)	(4–5)	(5–6)	(6–7)	(7–8)	(8–9)	(9–10)
15	1.010(7)	2.270(5)		7.55 (4)		7.17(5)	6.20(5)	1.9 (5)	3.5 (4)
9	2.114(7)	6.053(5)	9.113(4)	1.477(5)	6.552(4)	4.90(5)	9.50(4)	3.28(6)	1.55(5)
5	6.547(7)	2.173(6)	1.372(6)	7.532(5)	4.857(5)	6.05(6)	1.14(6)	8.90(6)	9.30(5)
3	2.212(8)	1.042(7)	1.033(7)	4.505(6)	4.238(6)	2.51(7)	4.08(7)		6.00(6)
2.25	4.802(8)	1.647(7)	3.696(7)	1.310(7)	3.829(7)				
1.5	1.553(9)	8.10 (7)	3.490(8)	1.049(8)	≥2 (8)				
1.25	2.803(9)	1.824(8)	1.045(9)	1.463(8)	≥4 (8)				
1.0	7(9)	2(9)	1.20 (9)	1.57 (8)	≥6 (8)				

illustrated in Figs. 35 and 36, where several characteristics of a 5 $M_\odot$ star are given as a function of time during core helium burning. At the red giant tip, a large fraction of the stellar envelope is in convective "equilibrium". As the star descends from the tip, the mass in the convective portion of the envelope decreases gradually and then drops suddenly to very small values as the stellar envelope adjusts on a gravitational (Kelvin) time scale to establish a balance between interior energy production and radiative energy transfer. The rapid contraction of the envelope continues until this adjustment has been achieved. Note, from Fig. 36, that the rate of energy production in the shell is markedly enhanced during the phase of rapid envelope contraction, whereas the rate of core energy production is essentially unaffected.

The fact that the second phase of core helium burning occurs further to the blue as stellar mass increases permits one to understand why it is that Cepheids are confined to a relatively small area in the H-R diagram. The shaded areas in Fig. 37 indicate where stars spend most of their life as core nuclear burners. These regions should define the location of most of the stars out of a statistically large sample of metal rich stars (provided we have chosen the right composition and the right input physics).

Studies of envelope pulsation indicate that instability against self-excited pulsation should occur only for stars in a narrow strip that has a slope opposite to that of the band associated with the second phase of core helium burning. The majority of Cepheids should occur within the area defined by the intersection of the instability strip and the second core helium burning band.

The distribution of Cepheids in number versus luminosity should exhibit a peak coinciding roughly with the luminosity at the center of the intersections. The distribution of Cepheids in M 31, as shown in Fig. 38, and the distribution of the Cepheids in the small Magellanic cloud, as shown in Fig. 39, seem to confirm the theoretical predictions, at least qualitatively.

It should be cautioned that the location of both the instability strip and the second core helium burning band are sensitive to the details of the input physics and are bound to change as input physics is either altered or improved. The strip and band locations are also sensitive to the choice of composition parameters. All other things being equal, the helium burning band moves to lower luminosities and cooler surface temperatures as either the initial helium abundance parameter Y is decreased (keeping the opacity parameter Z fixed) or as Z is increased (keeping Y fixed). In the first case, decreasing envelope Z causes the envelope to expand (in order to achieve pressure balance) and the hydrogen shell luminosity to drop. In the second case, increasing Z has the same net effect, although for a different reason associated with energy flow. Thus, differences in the distributions of Cepheid number versus luminosity are expected to vary from one galactic field to another as a consequence of differences in composition.

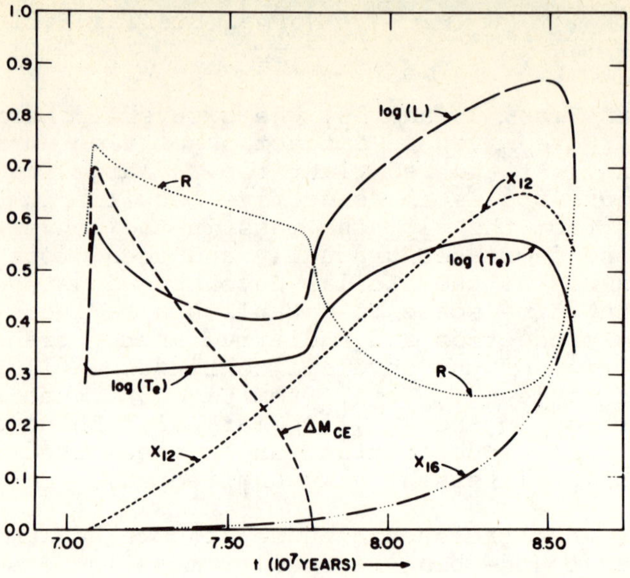

Figure 35. The variation with time of several characteristics of a 5 $M_\odot$ star during core helium burning. Here R = stellar radius, L = luminosity, T_e = surface temperature, ΔM_{CE} = mass of convective envelope, X_{12} = abundance by mass of C^{12} at the stellar center and X_{16} = abundance of O^{16} there. Scale limits for R are $0 < R/R_\odot < 100$. $2.8 \leqslant \log L \leqslant 3.3$; $3.3 < \log T_c < 4.3$; $0 < X_{12}, X_{16} < 1$; $0 < \Delta M_{CE} \leqslant M_\odot$. From Iben, I., Jr., Ap. J., 143, 483 (1966).

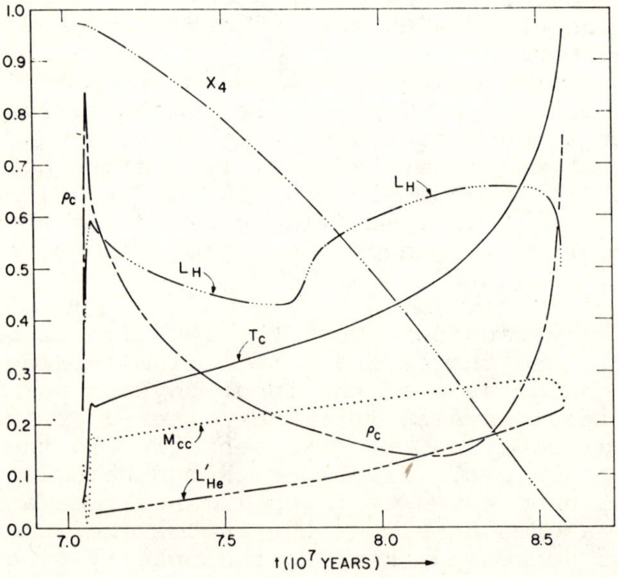

Figure 36. Variation of additional characteristics of a 5 $M_\odot$ star during helium burning. Here T_c = central temperature, P_c = central density, L_{cc} = mass in convective core, X_4 = abundance of helium at the center, L_H = hydrogen burning luminosity, L_{He} = helium burning luminosity. $100 < T_c(10^6 \, ^\circ K) < 200$, $6000 < P_c(\mathrm{erg\,cm^{-3}}) < 16000$, $0 < L_{He}, L_H < 2000 \, L_\odot$, $0.0 < M_{cc} < 0.2 \, M_\odot$. From Iben, I., Jr., Ap. J., 143, 483 (1966).

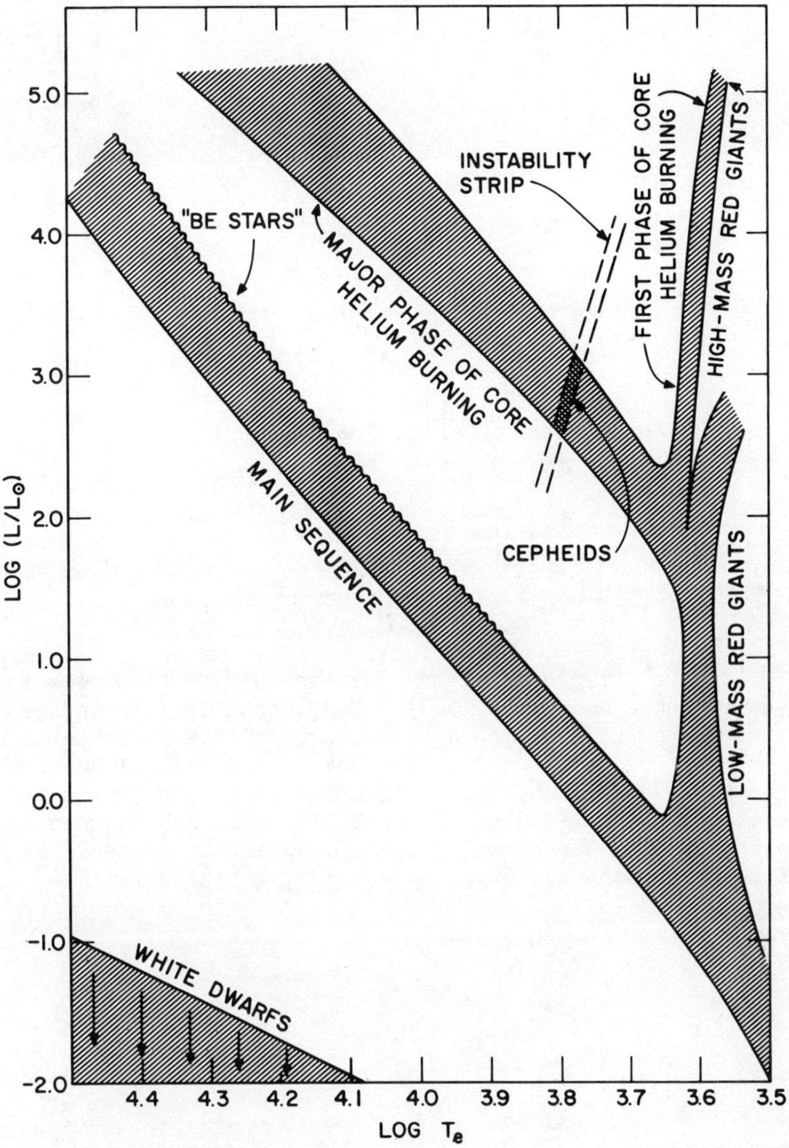

Figure 37. The distribution of metal-rich stars in the H-R diagram expected on the basis of model calculations for a given composition. From Iben, I., Jr., Science, 155, 785 (1967).

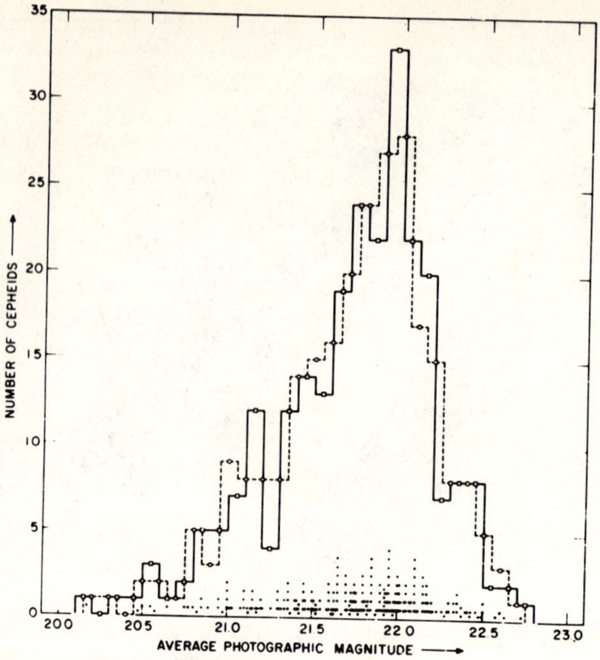

Figure 38. Number versus magnitude relationship for Cepheids in Field III of M31 (after Baade and Swope 1965). Unconnected points indicate the occurrence of a Cepheid at the designated magnitude (arithmetic mean of maximum and minimum magnitudes). For them, the vertical scale has no significance. The solid histogram is obtained by counting Cepheids in 0.1-mag. intervals centered at 20.05, 20.15, etc. The dashed histogram is obtained by counting Cepheids in 0.1-mag. intervals centered at 20.1, 20.2, etc. From Iben, I., Jr., Ap. J., 143, 483 (1966).

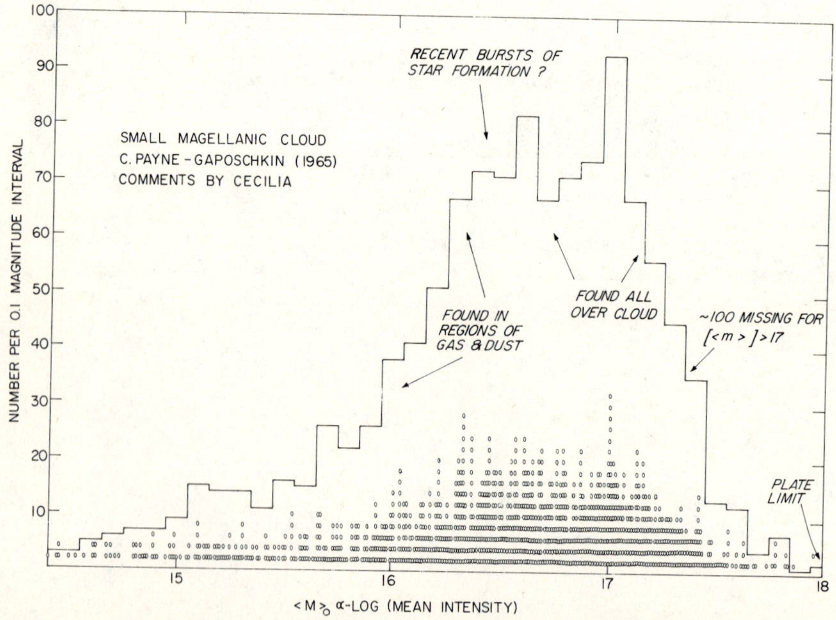

Figure 39. Distribution in number versus magnitude for Cepheids in the small Magellanic cloud. (Work by C. Payne-Gaposchkin.)

Differences in luminosity functions and stellar birth rate functions (see h and X Persei) will also lead to differences, even when composition differences are of minor importance.

In closing, let me remark on two distinctive features characterizing the evolution of low mass stars: 1) the occurrence of a helium "flash" and 2) the occurrence of a well-defined "gap" or phase of rapid evolution separating the phase of core hydrogen burning from a long phase of hydrogen burning in a thick shell.

Following the main sequence phase in stars less massive than about 2.25 $M_{\odot}$, conditions in the hydrogen exhausted core are such that electrons become degenerate before helium burning begins. Under degenerate conditions, electron conduction becomes more efficient, and the helium core becomes more and more isothermal as the star climbs further up the giant branch. This is illustrated by the temperature profiles in Fig. 40 that are appropriate to a 2.25 $M_{\odot}$ star during its climb up the giant branch. As additional helium is added to the core, gravitational energy is converted into thermal energy and then released from the core interior by electron conduction and from the core edge by radiative transfer limited by free-free absorption. As core mass increases, average core temperature continues to rise until, when central temperature approaches a sufficiently high value (in this case, $\sim 7 \times 10^7$ °K), energy production by helium burning (in this case the $N^{14} \rightarrow O^{18}$ reactions) exceeds the rate at which gravitational energy is released. As is illustrated by the luminosity profiles in Fig. 41, the nuclear energy produced remains in the core, raising core temperatures until kT at the center approaches the Fermi energy of the electrons there. Thereafter, any further increase in T means a proportionate increase in pressure which in turn means an expansion of core material. Expansion continues, possibly on a dynamic time scale, until electron degeneracy is essentially removed. The entire phase has been called the helium flash. Following the helium flash, helium burning in the core continues on a long time scale under non-degenerate conditions, just as is the case for the entire core helium burning phase of more massive stars.

The second feature of major interest in low-mass evolution is the prominence of a gap separating two phases of hydrogen burning near the main sequence. Times to reach labeled points along the tracks in Fig. 42 for low mass stars are given in Table 3. Note that the time spent in the phase of hydrogen burning in a thick shell (points 4-7 in Fig. 42) is comparable to the core hydrogen burning lifetime (points 1-3) for both the 1.25 $M_{\odot}$ and 1.5 $M_{\odot}$ stars. Defining τ_{CH} = lifetime in the core hydrogen burning phase, τ_{OC} = lifetime in the phase of overall contraction, and τ_{TS} = lifetime in the thick shell phase, we have $\tau_{TS}/\tau_{CH} = 0.225$ and $\tau_{OC}/\tau_{TS} = 0.234$ for $M = 1.5\ M_{\odot}$ and $\tau_{TS}/\tau_{CH} = 0.372$ and $\tau_{OC}/\tau_{TS} = 0.175$ for $M = 1.25\ M_{\odot}$.

If one proceeds to masses much below about 1.1 $M_{\odot}$ (for the composition and input physics chosen for illustration), a convective core either does not appear or disappears long

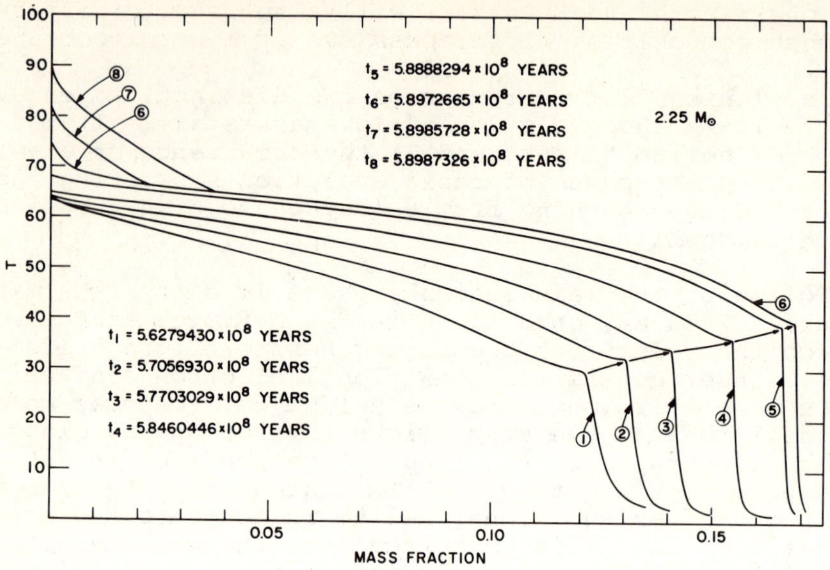

Figure 40. Temperature profiles for a 2.25 $M_\odot$ star during its climb up the giant branch. T is in units of 10^6°K. From Iben, I., Jr., Ap. J., 147, 650 (1967).

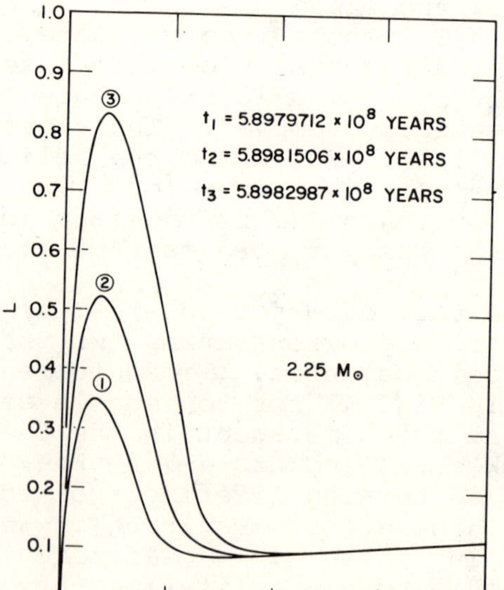

Figure 41. Luminosity profiles near the center of a 2.25 $M_\odot$ star just as the helium flash begins. From Iben, I., Jr., Ap. J., 147, 650 (1967).

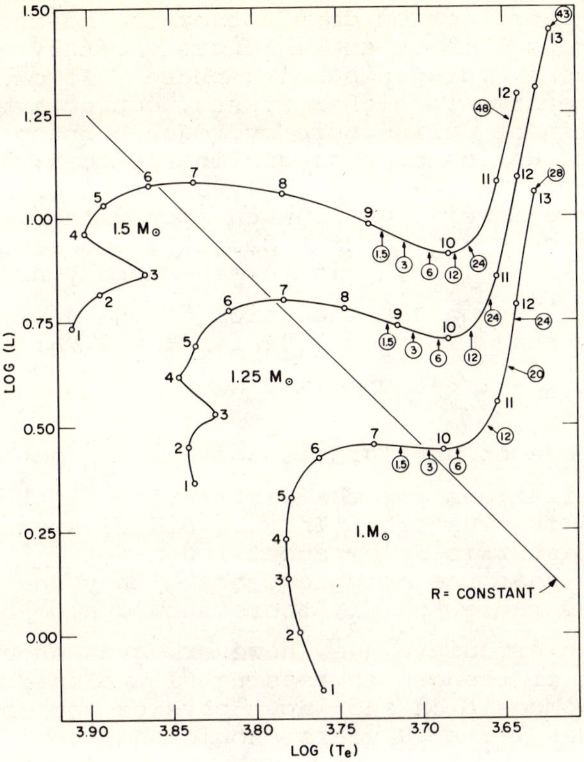

Figure 42. Evolutionary tracks for three low-mass metal-rich stars. From Iben, I., Jr., Ap. J., 147, 624 (1967).

TABLE 3

Evolutionary Lifetimes (10^9 yr) for Figure 42.

Point	1 $M_\odot$	1.25 $M_\odot$	1.50 $M_\odot$
1........	0.05060	0.02954	0.01821
2........	3.8209	1.4220	1.0277
3........	6.7100	2.8320	1.5710
4........	8.1719	3.0144	1.6520
5........	9.2012	3.5524	1.8261
6........	9.9030	3.9213	1.9666
7........	10.195	4.0597	2.0010
8........	....	4.1204	2.0397
9........	....	4.1593	2.0676
10........	10.352	4.2060	2.1059
11........	10.565	4.3427	2.1991
12........	10.750	4.4505	2.2628
13........	10.875	4.5349	....

before hydrogen is exhausted at the center, so that a phase of rapid overall contraction does not occur and the period of hydrogen burning in a thick shell merges smoothly and continuously with the core hydrogen burning phase. If one proceeds in the other direction to higher masses, convective cores become more prominent during core hydrogen burning, but following the exhaustion of hydrogen in the core, less fuel is left to burn before the star evolves into a giant (which occurs in the more massive stars when approximately 10 percent of the star's mass is devoid of hydrogen). Consequently, τ_{TS}/τ_{CH} decreases and τ_{OC}/τ_{TS} increases rapidly as stellar mass is increased beyond the inception of the gap. For example, $\tau_{TS}/\tau_{CH} = 0.0461$ and $\tau_{OC}/\tau_{TS} = 1.02$ for $M = 3\ M_\odot$; $\tau_{TS}/\tau_{CH} = 0.0204$ and $\tau_{OC}/\tau_{TS} = 1.64$ for $M = 5\ M_\odot$.

The dependence on stellar mass of τ_{TS}/τ_{CH} and τ_{OC}/τ_{TS} has obvious implications for the distribution in the H-R diagram of stars in clusters. In very old clusters, where the most massive star is below about 1.0 $M_\odot$ or 1.1 $M_\odot$ there should be no gap near the main sequence. In young clusters, where τ_{OC} is comparable to τ_{TS}, there should also be no gaps. In clusters of intermediate age, however, gaps should appear, the distinctness of the gap increasing with cluster age. If we have chosen composition and input physics correctly, then clusters with ages 1-7 x 10^9 years should show gaps.

The existence of a gap in the distribution of stars in the cluster M67 is well established. The gap is clear in the early data plotted in Fig. 43. Various investigators have attached an age to M 67 in the neighborhood of (4-6) x 10^9 yr.

A gap in the cluster NGC188 is not as well established. It is disturbing that recent attempts to delineate the gap observationally have led to such a clean, unpopulated region. It is clear from our discussion of lifetimes that there should be at least one fifth as many stars _in_ the "gap" as there are stars above it in the phase of thick shell burning. Further, any attempt to make use of a "gap width" in determining composition should, on the same grounds, be viewed with extreme reservation.

VI. The Solar Neutrino Experiment

Although the observation of solar neutrinos is in principle our most _direct_ measurement of conditions in the solar interior, the interpretation of experimental results is difficult. The experiment itself has been discussed elsewhere. We will here restrict ourselves to a description of the theoretical considerations that enter into an interpretation of the observable flux.

The observation, in brief, is that the neutrino flux is significantly lower than predicted by current theory. Since the actual flux is at present not clearly distinguishable from background, we can say only that the "best" theoretical

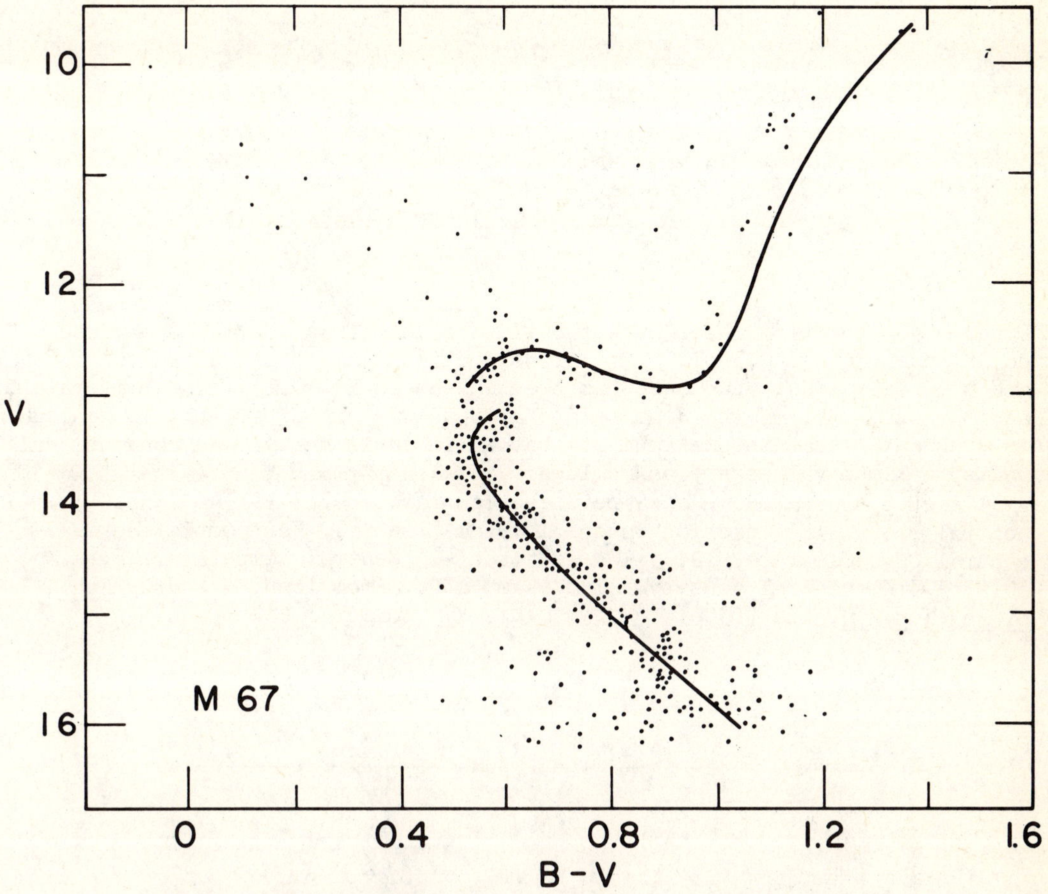

Figure 43. The H-R diagram of M67. From Johnson, H. L., and Sandage, A., Ap. J., 121, 616 (1955).

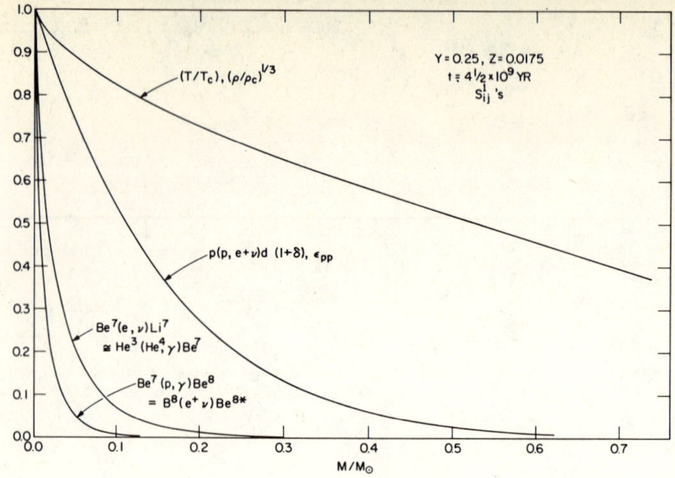

Figure 44. Reaction rates as a function of mass fraction in an unmixed solar model of age 4.50×10^9 yr. Cross section factors S^1_{ij} have been used and the initial He^4 abundance has been chosen as $Y = 0.25$. The temperature and the cube root of the density are also shown. Each variable is scaled in such a way that the maximum of that variable is unity. In conventional units, maximum values are as follows: central p-p reaction rate = 5.93×10^5 gm^{-1} sec^{-1}; central $He^3(He^4, \gamma)Be^7$ reaction rate = 9.19×10^4 gm^{-1} sec^{-1}; central $Be^7(p, \gamma)B^8$ reaction rate = 18.9 gm^{-1} sec^{-1}. For completeness, central CN cycle rate = 2.75×10^4 gm^{-1} sec^{-1}. Central temperature = 15.23×10^6 °K, central density = 148.4 gm cm^{-3}. From Iben, I., Jr., Ann. of Physics, 54, 164 (1969), copyright © Academic Press.

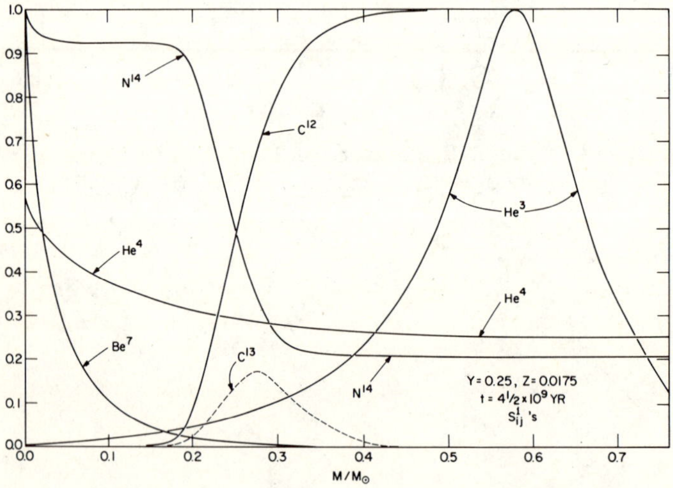

Figure 45. The abundances by mass of He^3, He^4, Be^7, C^{12}, C^{13} and N^{14} as a function of mass fraction in the unmixed solar model partially described in the caption to Figure 44. The He^4 abundance varies from the initial value of $Y = 0.25$ near the surface to 0.57 at the center. All other variables but C^{13} are scaled so that the maximum of that variable is unity. Maximum values are He^3 abundance = 3.06×10^{-3}; Be^7 abundance = 9.41×10^{-12}; C^{12} abundance = 3.14×10^{-3}; and N^{14} abundance = 5.10×10^{-3}. C^{13} is on the same scale as C^{12}. From Iben, I., Jr., Ann. of Physics, 54, 164 (1969), copyright © Academic Press.

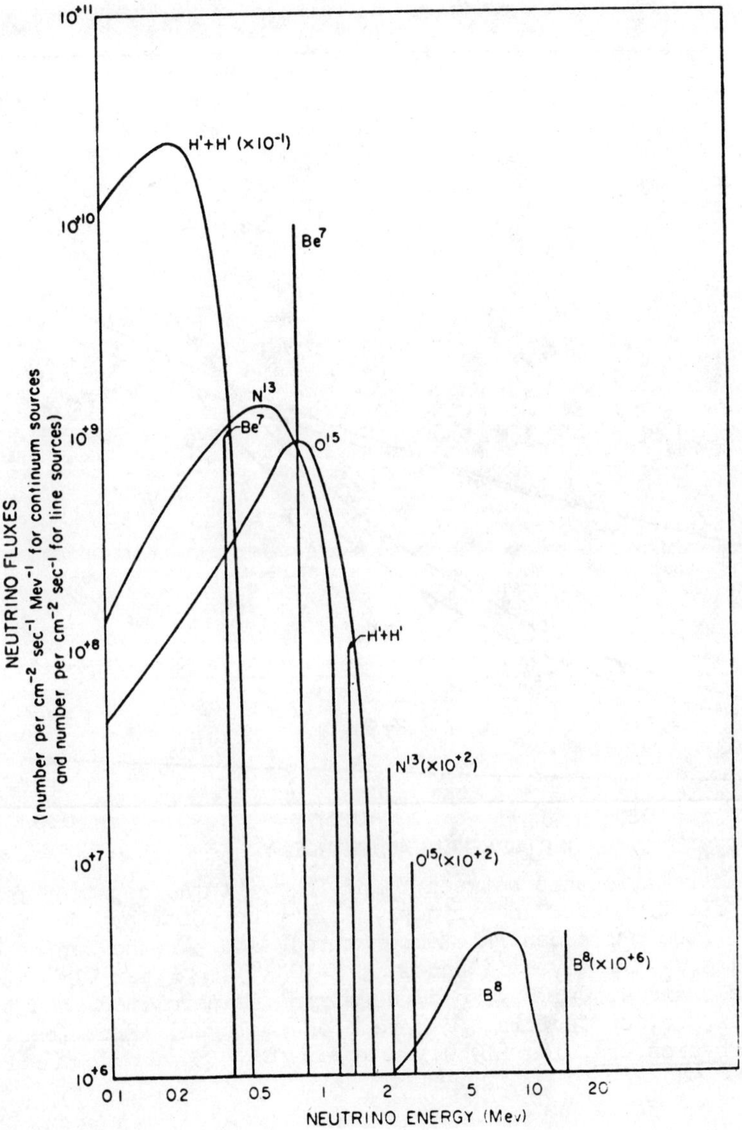

Figure 46. Predicted flux spectrum from an early Sears model of the sun. From Bahcall, J. N., High Energy Physics and Nuclear Structure (North-Holland 1967), page 232.

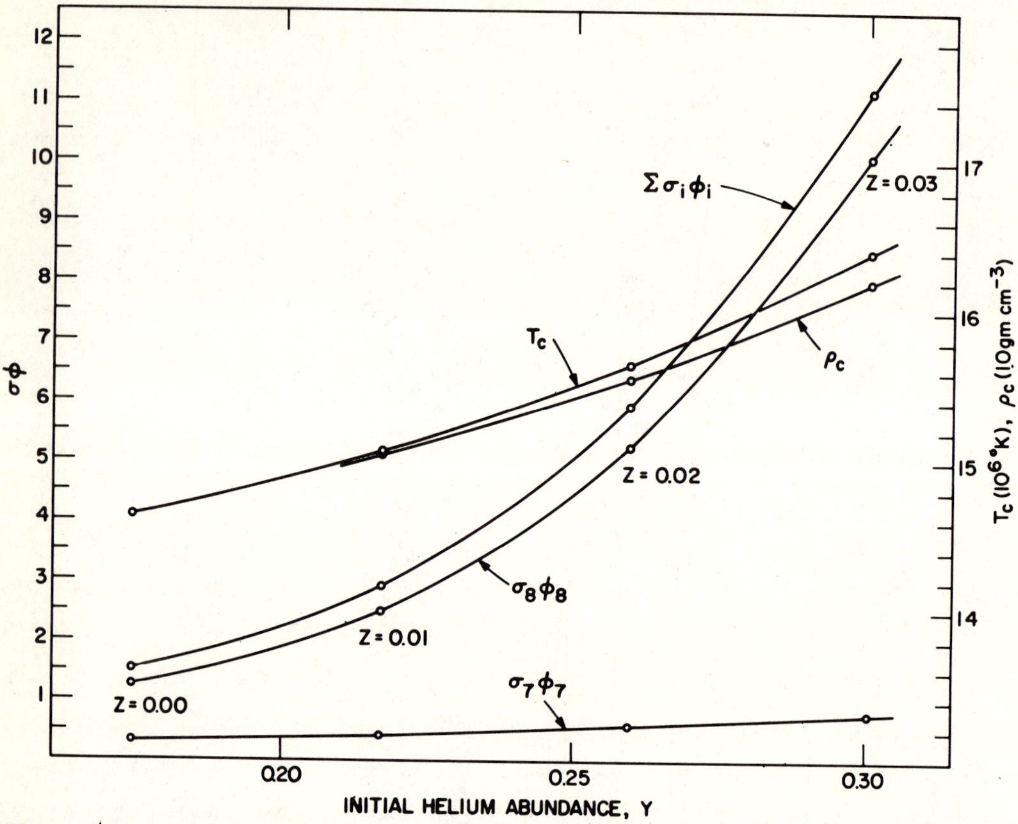

Figure 47. The relationship between Y and $\Sigma\sigma_i\phi_i$ (total counting rate), $\sigma_8\phi_8$ (the contribution of B^8 neutrinos), $\sigma_7\phi_7$ (the contribution of Be^7 neutrinos), T_c (the sun's central temperature), and ρ_c (the sun's central density). All $\sigma_i\phi_i$ are given in units of 3×10^{-36} sec^{-1} per Cl^{37} atom. The solar models are 4.50×10^9 yr old, no mixing is permitted, and the canonical cross section factors S^0_{ij} have been employed. From Iben, I., Jr., <u>Ann. of Physics</u>, 54, 164 (1969), copyright © Academic Press.

values are at least two to four times larger than the actual.

It is worth recalling the names of those who have made major contributions to the study of the sun as a neutrino emitter. Davis, of course, developed the Cl^{37} detection technique. Fowler and Cameron recognized that the cross section for the reaction $He^3 + He^3 \rightarrow Be^7 + \gamma$ as measured by Holmgren and Johnson is sufficiently large that a considerable amount of B^8 might be made via subsequent proton capture on Be^7 and that the neutrinos produced in the $B^8 \rightarrow Be^8 + e^+ + \nu$ decay might be eventually detectable with a Cl^{37} "counter". Bahcall carried through an exhaustive analysis of the cross sections for neutrino absorption on Cl^{37}, discovering that the cross section for the capture of B^8 neutrinos might be an order of magnitude larger than first estimated. Finally, Sears constructed solar models from which neutrino fluxes could be calculated. The calculated flux was thought observable, and Davis set up his detection apparatus.

The predicted counting rate per Cl^{37} atom can be written in the form $\Sigma\sigma_i\phi_i$, where the sum is over the different neutrino sources thought to be present in the sun; ϕ_i is the corresponding flux at the earth and σ_i is the effective absorption cross section on Cl^{37}. To compute the expected counting rate, let us make the following initial assumptions:

1. $L = L_\odot$ after 4.5×10^9 yr.
2. All nuclear and photon absorption cross sections are accurately known and properly incorporated in the available expressions for energy generation rates and opacities.
3. No mixing currents.
4. No rotational effects on structure.
5. Correct neutrino absorption cross sections.

To begin, pick Z and Y to satisfy condition (1). Carrying out the computation of such a model will provide information similar to that plotted in Figs. 44 and 45. Of particular interest to us is the rate of energy producing and neutrino producing reactions as a function of distance from the center. Each reaction rate in Fig. 44 is normalized to unity at the center. Note that the $Be^7(p,\gamma)B^8$ reaction rate is much more temperature sensitive than the $He^3(He^4,\gamma)Be^7$ rate and that this latter rate is in turn more temperature sensitive than the pp reaction rate.

Fig. 45 shows a distribution of abundances built up over the life of the sun for the same model described in Fig. 44. Of particular interest is the distribution of He^4 which builds up in the center. Note also that C^{12} has been converted almost completely into N^{14} over the inner 20 percent of the sun and that a quantity of C^{13} has been made.

Using results from model calculations such as those shown in Figs. 44 and 45, one can calculate neutrino fluxes, obtaining flux distributions similar to those shown in Fig. 46. Although the flux of neutrinos from the pp reaction is very high, the maximum neutrino energy is below threshold for absorption in

Cl^{37}. On the other hand, according to Bahcall's estimates, energetic B^8 neutrinos have an extremely large absorption cross section.

The expected absorption rate can be written in units of the current upper limit as

$$\Sigma\sigma_i\phi_i = (\Sigma\sigma_i\phi_i)_{UL}\left[\frac{\phi(B^8)}{2.2\times10^6} + \frac{\phi(Be^7)}{1.05\times10^{10}} + \frac{\phi(O^{15})}{3.8\times10^9} + \frac{\phi(N^{13})}{1.4\times10^{10}} + \ldots\right], \quad (154)$$

where $\phi(B^8)$, $\phi(Be^7)$, $\phi(O^{15})$, $\phi(N^{13})$, etc. are neutrino fluxes from B^8 decay, Be^7 electron capture, O^{15} decay, N^{13} decay, and so forth.

The dependence of the counting rate on initial Z and Y can be seen in Fig. 47 for a particular choice of input physics. Note that, using current estimates of neutrino absorption cross sections, the dominant contribution is from B^8.

Note further that consistency with the current upper limit on the neutrino flux cannot be achieved without choosing a negative value for the opacity parameter.

Before discussing the effect of changing nuclear reaction cross sections, let us look in more detail at the dependence of neutrino fluxes on the initial helium abundance. In Exercise V you were asked to verify the following result:

$$L_\odot = \varepsilon_{pp}\int r_{11}dM(1 + \langle\delta\rangle + 1.91\langle\beta\rangle). \quad (155)$$

(For definitions of the Greek symbols see Exercise V.) You were also asked to show that the neutrino flux from the pp reaction is

$$\phi_{pp} = 6.44\times10^{10}[1 + \langle\delta\rangle + 1.91\langle\beta\rangle]^{-1}\text{cm}^{-2}\text{sec}^{-1}, \quad (156)$$

The flux of neutrinos from every other source may be related to ϕ_{pp} thusly:

$$\begin{aligned}\phi(Be^7) &= \langle\alpha(1+\gamma)^{-1}\rangle\phi_{pp}\\ \phi(B^8) &= \langle\alpha\gamma(1+\gamma)^{-1}\rangle\phi_{pp}\\ \phi(O^{15}) &\sim \phi(N^{13}) = \langle\beta\rangle\phi_{pp}\ .\end{aligned} \quad (157)$$

Substituting into the sum that gives the total expected counting rate (in terms of the current upper limit) we find

$$\Sigma\sigma_i\phi_i \cong [6.1 < \alpha(1+\gamma)^{-1} > + 29300 < \gamma\,\alpha(1+\gamma)^{-1} > + 21.5 <\beta>][1 + <\delta> + 1.91 <\beta>]^{-1} \tag{158}$$

Since this sum must be less than one we can deduce that

$$(<\beta>) < \frac{1}{20}, \quad (<\alpha>) < \frac{1}{5}, \quad (<\gamma\alpha>) < \frac{1}{29300}, \quad \text{and } (<\gamma>) << 1. \tag{159}$$

Thus the probability of producing a B^8 nucleus is much less than the probability of producing Li^7. Also, the probability that the CN cycle will produce He^4 is no more than 1/20 to 1/10 the probability that the pp chains will produce it. These conclusions are based on the cross sections for the absorption of neutrinos by Cl^{37}, but not on a particular model of the sun.

At temperatures occurring near the solar center we can write $\alpha \propto T^7$, $\gamma \propto T^{13}$. Consequently we can write approximately

$$\Sigma\sigma_i\phi_i \sim A <T^7> + B <T^{20}>, \sim A'T_c{}^7 + B'T_c{}^{20}, \tag{160}$$

demonstrating the strong dependence of a calculated counting rate on the central temperature. This opens the way to understanding why the helium abundance is an important parameter. Recall that central temperature is related to the mean molecular weight, stellar mass, and stellar radius by

$$T_c \propto \mu \frac{M}{R} \tag{161}$$

Since M and R have been fixed, increasing Y means increasing μ, with a resulting increase in T_c and therefore an increase in the calculated counting rate.

Coupling our earlier result that

$$L \propto \frac{\mu^4 M^3}{\kappa(Z)} \tag{162}$$

with the fact that metals contribute strongly to opacity, we discover that if we increase Y, we must also increase Z, if we also fulfill the requirement L = const. = $L_\odot$. This is, then, why the choice of a larger value for Z means a larger calculated value for $\Sigma\sigma_i\phi_i$: a larger Z means a larger Y, and a larger Y means a larger T_c,

It has been customary to assume that the appropriate choice for the opacity parameter Z can be made on the basis of spectroscopic estimates of the abundances of heavy elements at the solar surface. This assumption is a dangerous one to make for several reasons. First, gravitational diffusion can deplete surface abundances of the heavy elements. Some estimates have suggested that many of the elements heavier than helium may have been depleted by a factor of two at the surface of the sun relative to interior abundances. Second, the distribution of elements in the CNO group has been altered considerably in the solar interior relative to the initial distribution. This is not normally taken into account in preparing opacity tables. Finally, even if heavy element abundances were known exactly, the uncertainties in the determination of the opacity are so large (20-50% where heavy elements contribute significantly) that the translation from an abundance parameter Z to the appropriate opacity parameter $\hat{Z}$ (to adjust a calculated opacity to the unknown "correct" opacity) cannot at present be made to better than 50 percent accuracy in many interior regions.

Thus far we have assumed that all cross sections are accurately known. Let us consider the effect of changing these cross sections within "reasonable" ranges about the accepted values. Keep in mind that, in almost every case, the accepted value has been obtained after extensive experimental and theoretical efforts.

1. The temperature dependence of the pp rate is such that

$$L_{\odot} \propto S_{1,1} \int \rho T^4 dM \tag{163}$$

where $S_{1,1}$ is the effective center of mass cross section factor for the pp reaction. For the sun, ρ/T^3 is nearly constant, so

$$L_{\odot} \propto S_{1,1} \langle T^7 \rangle . \tag{164}$$

Consequently, an increase in $S_{1,1}$ requires a drop in interior temperatures. This in turn causes a drop in the fluxes of both B^8 and Be^7 neutrinos.

2. Increasing $S_{3,3}$ causes a drop in $R_{3,4}$ with a consequent drop in ϕ_8, ϕ_7.

3. Increasing $S_{3,4}$ clearly raises ϕ_8 and ϕ_7.

4. Decreasing $S_{1,7}$ of course reduces ϕ_8 but (since, in any case, $\gamma \ll 1$) has a negligible effect on solar structure.

5. Increasing $S_{1,14}$ results in an increase in ϕ_N and ϕ_O and also tends to increase central temperatures (in a non-obvious way) and thereby increases both ϕ_8 and ϕ_7.

In order to gain a feeling for the accuracy of various center of mass cross section estimates, we can examine some of the early experimental data upon which they are based. In Fig. 48 are the results for $Be^7 + p \rightarrow B^8 + \gamma$. The uncertainty in the extrapolated (zero energy) cross section factor is at least 20%.

In Fig. 49 early data for $He^3 + He^4 \rightarrow Be^7 + \gamma$ is plotted, along with a number of curves used at various times for extrapolation. The uncertainty is evident to the eye.

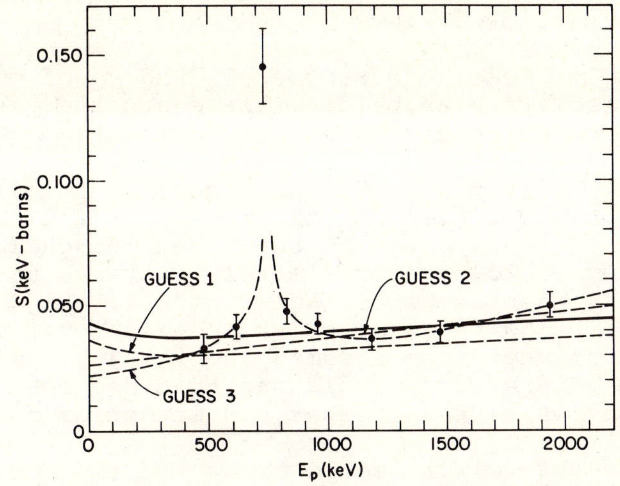

Figure 48. Cross section factor data for the reaction $Be^7 + p \rightarrow B^8 + \gamma$, reproduced from Parker (1966). The solid curve is Parker's choice for a fit to the data. Dashed curves are alternate choices. From Iben, I., Jr., Ann. of Physics, 54, 164 (1969), copyright © Academic Press.

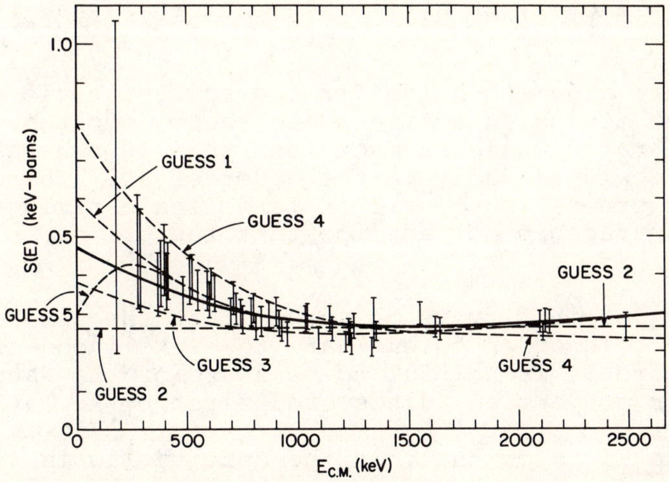

Figure 49. Cross section factor data for the reaction $He^3 + He^4 \rightarrow Be^7 + \gamma$, reproduced from Parker and Kavanagh (1963). The solid curve is the fit suggested by Parker and Kavanagh. Dashed curves are alternate choices. From Iben, I., Jr., Ann. of Physics, 54, 164 (1969), copyright © Academic Press.

The quantitative effect that changes in cross section factors have on the theoretical counting rate is shown in Fig. 50. This plot is similar to Fig. 48, except for the modified cross section factors.

In Fig. 51 we see the effect on T_c and ρ_c of modified cross section factors. As before, these are the results of numerous model calculations.

We have explicitly assumed that there is no mixing in the sun. If there were significant mixing on a time scale short compared to the age of the sun, then the characteristics of the sun today would be approximately the same as if it had just reached the main sequence (age t = "0"). Fig. 52 shows the expected counting rate for solar models of various assumed ages. A model of any given assumed age $t < 4.5 \times 10^9$ yr. is equivalent to a 4.5×10^9 yr. old sun that has been partially mixed. The smaller t, the greater the extent of mixing.

It appears that mixing could bring the expected counting rate down to the observed counting rate; however, there are arguments against this. Only one will be given here. In Fig. 45 is shown the distribution of elements in an unmixed sun. Note that C^{12} has been converted into N^{14} over the inner 20 percent of the sun's mass. If the sun began with no N^{14}, and yet mixed thoroughly on a time scale comparable to the sun's age, one might expect the present surface ratio of C^{12} and N^{14} to be roughly 4. If thorough mixing occurred on a time scale much shorter than 4.5×10^9 yr., then the surface ratio of C^{12} to N^{14} would be much less than 4 since C^{12} would be continuously carried to the center to be converted to N^{14}, which would in turn be mixed out to the surface. Since the observed value is about 4, one may argue that mixing has not occurred continuously on a time scale shorter than the sun's age.

A similar argument holds for the surface ratio of C^{12} to C^{13}. With mixing on a time scale short compared to 4.5×10^9 yr., the ratio would be much less than 40, whereas the observed ratio is actually on the order of 40. These observations (and others) suggest that large scale mixing between center and surface is not an important feature of solar evolution.

Another possibility is to assume that the sun has a convective core. In order to evaluate the likelihood of a small convective core it is helpful to be aware of the Naur-Osterbrock criterion for convection (discovered first by R. Taylor).

<u>Exercise XI</u> Show that, if the opacity law is written as $\kappa = \kappa_o \rho^{\alpha} T^{-\beta}$ and the energy generation law is approximated by $\varepsilon = \varepsilon_o \rho^{1+k} T^s$, ($\kappa_o$, α, β, ε_o, k, s all constants), then a vanishingly small convective core may occur if

$$(\tfrac{3}{2}k + s - 1) > \tfrac{5}{3}(\beta - \tfrac{3}{2}\alpha). \qquad (165)$$

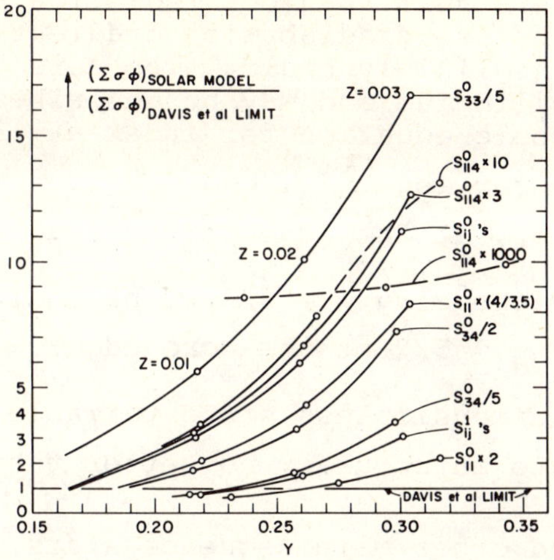

Figure 50. The ratio R of theoretical estimates of $\Sigma\sigma_i\phi_i$ to the upper limit set by Davis, Harmer, and Hoffman. The ratio is given as a function of the assumed initial solar helium abundance Y for various choices of center of mass cross section factors specified relative to a canonical set of cross section factors S^0_{ij}. Each circle designates the ratio for one of three different values of the opacity parameter Z. All results are for an assumed solar age of $4\frac{1}{2}$ billion years. From Iben, I., Jr., <u>Ann. of Physics</u>, <u>54</u>, 164 (1969), copyright © Academic Press.

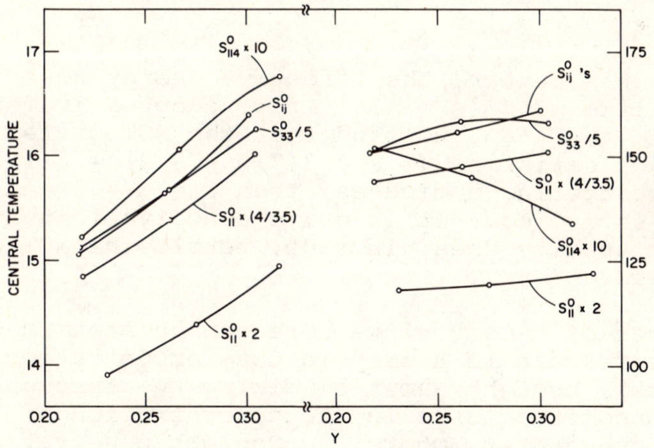

Figure 51. The relationship between central temperature (10^6 °K) and Y and between central density (gm cm^{-3}) and Y as a function of the S_{ij}. The value of that single parameter S_{ij} that differs from the set S^0_{ij} is designated beside each curve. From Iben, I., Jr., <u>Ann. of Physics</u>, <u>54</u>, 164 (1969), copyright © Academic Press.

To derive this result, neglect radiation pressure and assume that the "superadiabatic" gradient in the convective core is negligibly small. The core is then in adiabatic equilibrium and may be described by the solution of the polytropic equation for index n = 3/2. Construct an expression for log V_{rad}, where $V_{rad} = \left(\frac{d\ell nT}{d\ell nP}\right)_{Radiative}$ and insist that, at the edge of the small core, $\frac{dV_{rad}}{dr} < 0$ (by choosing N = 3/2 we have already insured that $V_{rad} = V_{ad} = 5/2$ at the core edge). As long as $\frac{Z^2}{6} << 1$, the dimensionless state variable u need be approximated to an accuracy no better than $u \sim 1 - \frac{Z^2}{6}$.

In the sun and other main sequence stars, all energy generation rates are proportional to only the first power of the density, so the "NOT" criterion says that the inequality (see Exercise XI for a definition of symbols)

$$(s - 1) > \frac{5}{3}(\beta - \frac{3}{2}\alpha) \qquad (166)$$

must be met at the center if a small convective core is to occur. The NOT criterion thus quantifies one's intuitive feeling that convection is relatively more likely if the energy sources are highly temperature sensitive (leading to large fluxes concentrated near the center). A simple interpretation for why a high temperature sensitivity and low density sensitivity for opacity act to suppress convection is not as readily constructed.

At the sun's center, the effective energy generation rate goes roughly like ρT^4, ($s \sim 4$). If we adopt a Kramer's type opacity law, $\kappa \propto \rho T^{-7/2}$, we find that the NOT criterion is just barely not satisfied ($3 \gtrless 5/3(7/2 - 3/2) = 3\ 1/3$). If electron scattering dominates, then $\beta = \alpha = 0$ and the criterion is satisfied. It is quite conceivable that the correct solar opacity does, in fact, permit the NOT criterion to be satisfied.

The effect of a convective core can be seen in Fig. 53. The important feature is a drop in the concentration of Be^7 near the center, brought about by mixing Be^7 throughout the core. The concomitant drop in the concentration of B^8 at the high temperature center, where the Coulomb penetration factor for the reaction $Be^7(p,\gamma)B^8$ is the largest, results in a decrease in the neutrino flux from B^8 decay.

For convective cores of varying mass, the flux behaves as shown in Fig. 54. It is amusing to note that, contrary to traditional superstition, a measurement of the B^8 neutrino flux does not provide a unique determination of the sun's central temperature. In the presence of a convective core, the larger the central temperature, the smaller the B^8 flux. Thus, although the solar neutrino experiment may be a cheap

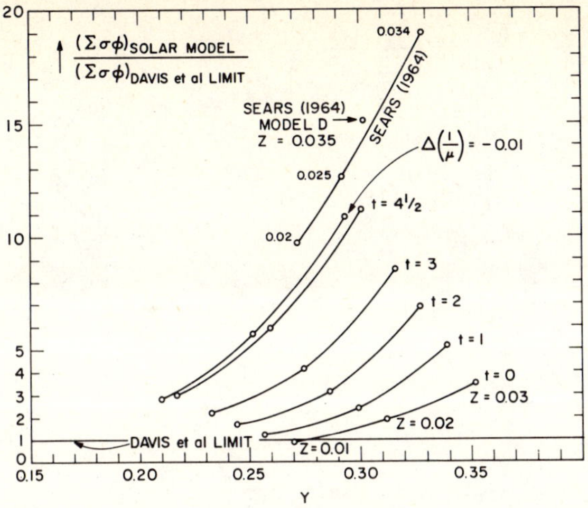

Figure 52. The relationship between the ratio R and the helium abundance Y as a function of (1) assumed solar age, (2) opacity, and (3) equation of state. Curves forming the fifteen point mesh are derived from models made with a "canonical" set of cross section factors S^{0}_{ij}, assuming solar ages of 0, 1, 2, 3, and $4\frac{1}{2}$ billion years. The curve labeled $\Delta(1/\mu)$ = -0.01 is derived from models that differ from those leading to the t = $4\frac{1}{2}$ curve only by a slight modification in the equation of state. From Iben, I., Jr., Ann. of Physics, 54, 164 (1969), copyright © Academic Press.

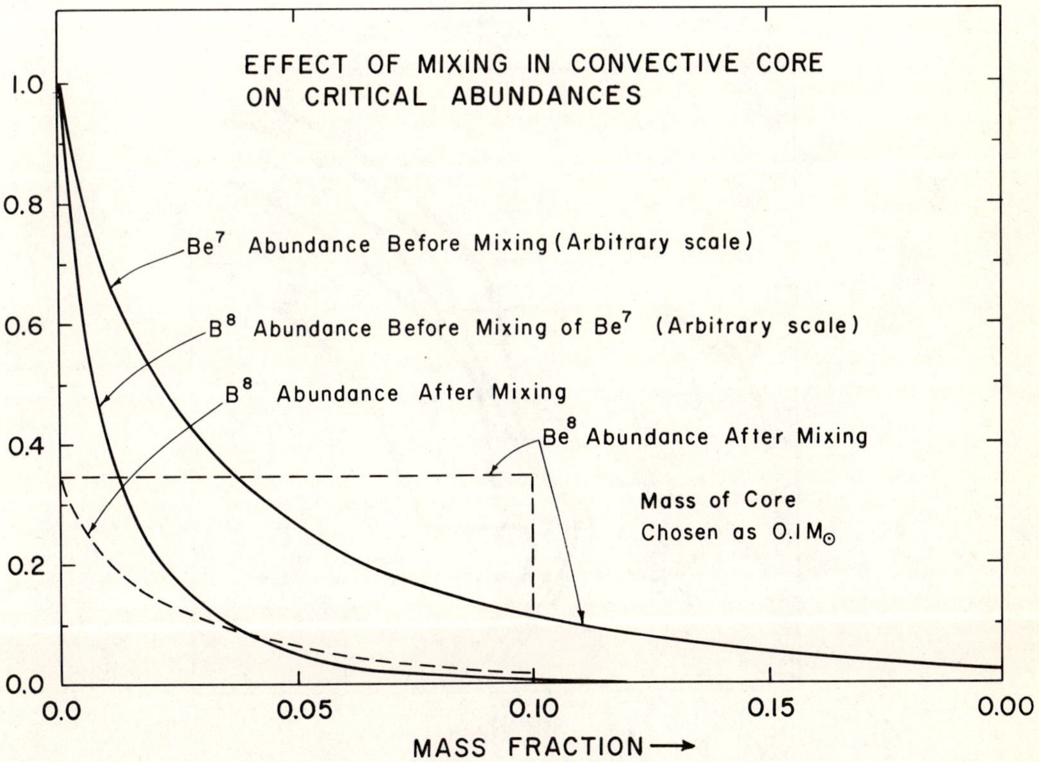

Figure 53. The effect of mixing in a postulated convective core on the abundances of Be^7 and of B^8 near the sun's center.

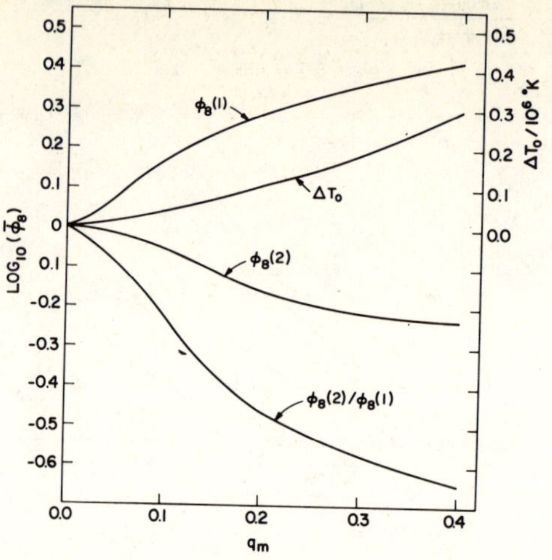

Figure 54. B^8 neutrino flux (logarithmic units) relative to the flux from a standard model. Ordinate is the mass $q_m M_\odot$ in the adiabatic core of models calculated by Shaviv and Salpeter (1968). $\phi_8(1)$ results if the convective mixing time t_m is much larger than t_γ, the lifetime of a Be^7 nucleus in the core. $\phi_8(2)$ results if $t_m \ll t_\gamma$. Shown also is the increment in central temperature in the "adiabatic" series of models of Shaviv and Salpeter. From Iben, I., Jr., Ap. J., 155, L101 (1969).

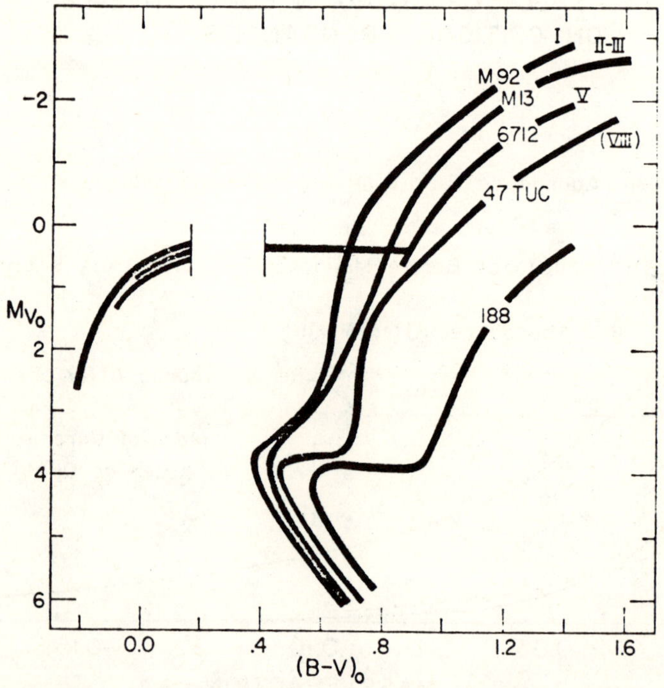

Figure 55. Comparison of color-magnitude diagrams of M92, M13, NGC 6712, 47 Tuc, and NGC 188, corrected for reddening. From Sandage, A., and Smith, L. L., Ap. J., 144, 886 (1966).

thermometer (when measured in terms of cost divided by the magnitude of the temperature to be measured), it has a disadvantage relative to more expensive thermometers that have the virtue of being monotonic sensors of temperature.

The foregoing discussion has not led to any satisfactory explanation of the solar neutrino observation; however, we can at least say that it may be possible to understand the observation if we allow a number of input parameters and cross sections to be shifted within their respective ranges of uncertainty, all in a direction most favorable for a reduction in the calculated B^8 flux.

It is also possible to approach consistency with the Davis result by adhering to current "best estimates" of all of the pertinent nuclear cross sections, but adopting an initial helium abundance for the sun that is near the lower limit obtained by setting the opacity parameter $Z = 0.0$. It is my own personal view that it is much more likely that the "best estimates" of cross sections (cross sections for nuclear reactions in the sun, photon absorption cross sections near the solar center, neutrino absorption cross sections in Cl^{37}) will undergo a revision and that it will not be necessary to accept that the sun's initial helium abundance is significantly less than the helium abundance found for a wide variety of objects in our galaxy other than the sun.

VII. Theoretical Interpretation of Stellar Distributions in Globular Cluster

Important progress has recently been made toward understanding the observed distributions of the metal poor stars in globular clusters in terms of theoretical evolutionary models. As a spin-off from this study it is possible to draw certain conclusions concerning the age and initial composition of these stars.

Before examining the results of evolutionary studies, let us review some of the observational data of importance for comparison with theory. Fig. 55 is a composite of a number of metal poor globular clusters that have been shifted with respect to one another so that their main sequences coincide. Note that the separation between turn-off point and the position of the horizontal branch at the same surface temperature seems to be fairly insensitive to differences in heavy element abundances. This separation is consistently about 3.4-3.5 magnitudes.

In comparison with the metal rich clusters, the sub-giant branch for the metal poor clusters is exceedingly steep. Also, the metal poor clusters exhibit a much more prominent horizontal branch. One further distinction is the absence of a gap along the main sequence defined by metal poor cluster stars.

Other observable parameters that one can hope to understand are the distribution with color of stars along the horizontal branch, the number of stars on the horizontal branch relative to the number of stars on the red giant branch, and, finally, the distribution with luminosity of stars along the red giant branch. The latter quantity is shown in Fig. 56 for the cluster M15.

Let us now turn to the results of theoretical studies of evolution off the main sequence. Fig. 57 shows evolutionary tracks for models of a given mass, but of different initial composition. The luminosity of the initial main sequence is relatively independent of Z. This is because the primary mechanisms of energy production there are the pp chains, and the opacity is almost independent of Z over the interior of the star.

On the other hand, the initial main sequence position is highly Y dependent. The higher Y is, the smaller is the number of particles per gram and therefore the lower is the pressure per gram for a given density and temperature. Consequently, for a given stellar mass, the density in the interior must be higher to generate the pressure required to balance the gravitational force, with resulting higher temperature and higher rate of energy generation. Higher Y thus means higher luminosity, as is verified by the detailed calculations.

As evolution progresses, differences among models become more pronounced and, in particular, Z-related differences begin to show up as the CN cycle becomes a more important energy producer in the heating core.

The letters in Fig. 57 have the following significance. At A the convective core vanishes. At B, $\varepsilon_{CN}(\text{center}) > \varepsilon_{pp}(\text{center})$ for the first time. At C the central hydrogen abundance drops below 0.1% by mass. At D the CN cycle becomes the dominant mode of energy generation for the entire star. At E the density at the center of the hydrogen burning shell reaches a maximum.

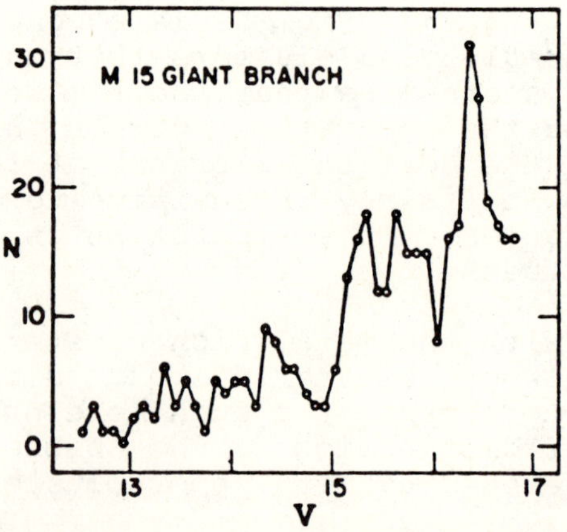

Figure 56. The distribution in number versus magnitude of stars along the giant branch in M15. From Sandage, A. R., Katem, B., and Kristian, J., Ap. J., 153, L129 (1968).

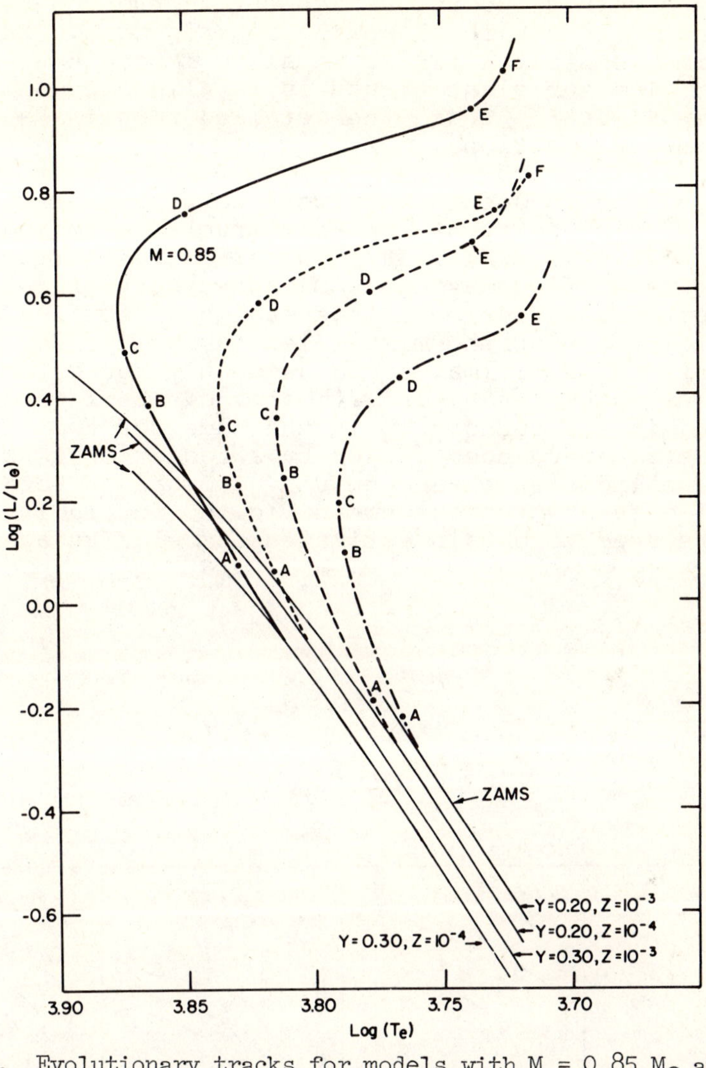

Figure 57. Evolutionary tracks for models with M = 0.85 $M_\odot$ and all initial compositions derivable from Z = 10^{-3}, 10^{-4}, and Y = 0.20, 0.30. The zero-age main sequence is shown for each composition. At the points marked A, the convective core vanishes. At B, energy production at the center by CN-cycle reactions equals that by the p-p chains. At C, the central hydrogen abundance drops below X = 0.001. At D, the CN cycle contributes half the total luminosity. At E, the density in the hydrogen-burning shell reaches a maximum. At F, the CN cycle contributes 90 percent of the total luminosity. From Iben, I., Jr., and Rood, R. T., Ap. J., 159, 605 (1970).

The onset of the CN cycle is an important factor in determining the turnoff point. Since ε_{CN} depends on the initial Z value, the turnoff point likewise depends on Z. Early portions of time constant loci for metal poor models are shown in Fig. 58. The sub-giant branch is in all cases relatively steep, as in Fig. 55. Thus, the detailed results have passed one important observational test.

Figs. 59 and 60 describe turnoff point characteristics versus age of time-constant loci constructed from models with varying compositions. Fig. 59 shows effective surface temperature at turnoff versus age. Because the surface temperature of theoretical models in this region of the H-R diagram depends sensitively on the mixing length, which is not known, T_e at turnoff may legitimately be used only for estimates of relative ages rather than for estimates of absolute ages. For example, if the turnoff T_e is the same for two clusters, but their metal abundances differ by a factor of ten, then their ages will differ by about 50%. Comparison between observational data and the curves in Fig. 60 for luminosity can similarly be used to obtain estimates of relative ages.

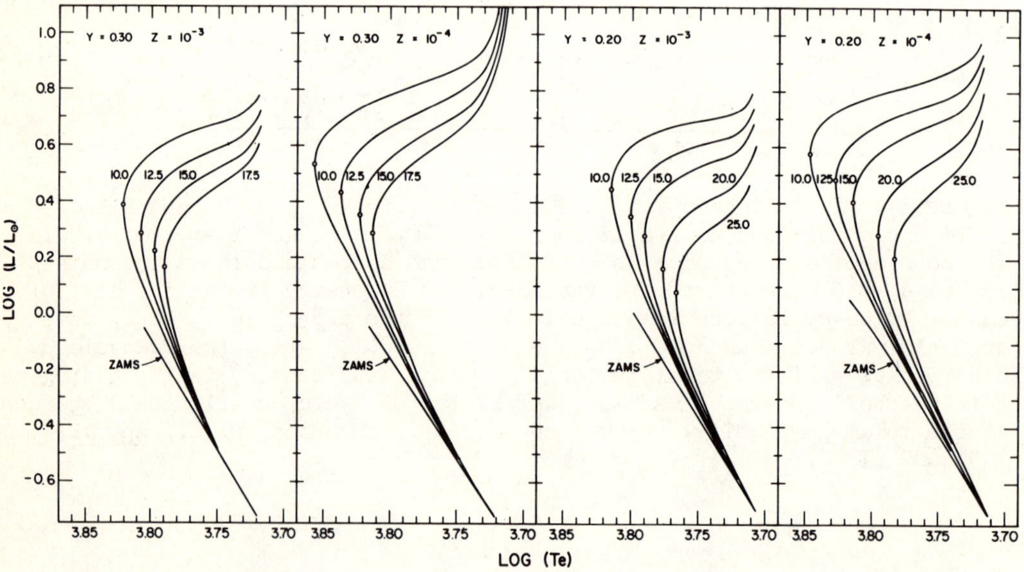

Figure 58. Time-constant loci for four different initial compositions. Age in billions of years is shown beside each curve. Circles denote turnoff points. For reference, the lower part of the zero-age main sequence for each composition is also shown. From Iben, I., Jr., and Rood, R. T., Ap. J., 159, 605 (1970).

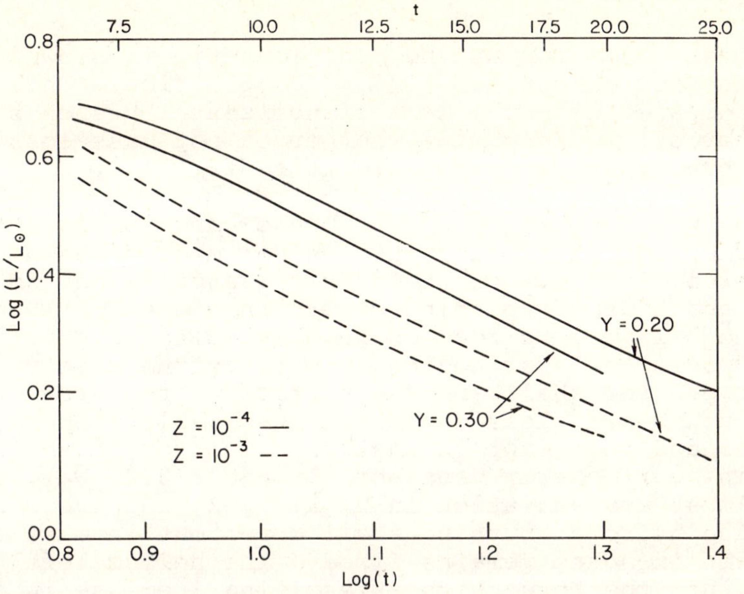

Figure 59. Luminosity at turnoff as a function of age (units 10^9 yr) and composition. From Iben, I., Jr., and Rood, R. T., Ap. J., 159, 605 (1970)

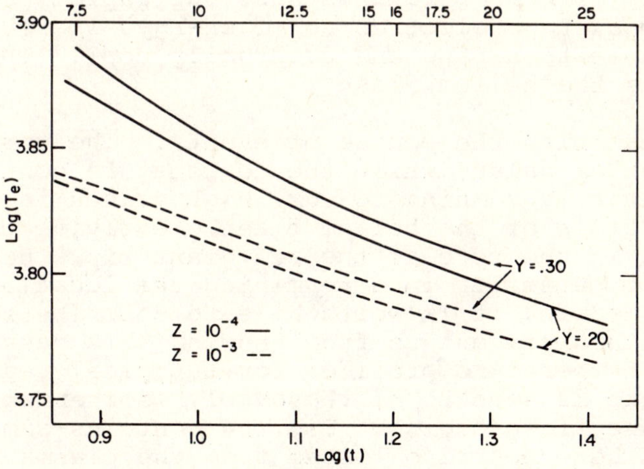

Figure 60. Surface temperature at turnoff as a function of age and composition. From Iben, I., Jr., and Rood, R. T., Ap. J., 159, 605 (1970).

The mass of a star at the red giant tip is shown in Fig. 61 as a function of cluster age and composition. This mass is to be compared with the mass of horizontal branch stars (models thereof) to determine whether or not mass loss along the giant branch is or is not significant.

Let us now move to the giant branch and consider the results of evolutionary calculations for models residing there. As the star evolves off of the main sequence, a helium core develops, and hydrogen burning occurs in a shell. When the mass in the helium core reaches roughly 13 to 20 percent of its total mass, the star begins to move up the giant branch, matter between the shell and the surface of the star continuing to expand and become less dense. A remarkable property of a fully developed red giant is that the diameter of the helium core remains very nearly constant (at about 0.06 $R_{\odot}$), independent of almost any variation in M, X, Y, Z, and even in age. Thus, as the hydrogen burning shell moves out in mass fraction, its position in space remains fixed. The helium left behind is packed into the core, with a resulting increase in core density. Because gravitational energy is being released, the core temperature increases. Eventually the triple-alpha process can begin; however, by this time the core density is 10^5 - 10^6 gm/cc, and the electron Fermi level has become larger than kT by an order of magnitude. The pressure at this stage is thus due largely to degenerate electrons, and is therefore not highly sensitive to temperature. As a result, when helium burning begins, the energy deposited in the core by the nuclear fuel does not force the core to expand and cool. Instead, core temperatures shoot up rapidly until kT approaches the average electron Fermi energy, whereupon the core is then able to expand and eventually cool. This phenomenon is known as the"helium flash".

Fig. 62 exhibits the course of events. The core diameter goes to a constant value, while the temperature continues to increase, eventually running to very high values in the helium flash. The details of the helium flash clearly determine the mass contained in the core at the red giant tip. Because it is an important parameter in determining the location of horizontal branch models, it is worthwhile to describe several uncertainties that prevent us from knowing this mass precisely. Fig. 63 shows temperature profiles for a typical red giant model. The core is seen to be reasonably isothermal near the center until the flash begins. Fig. 64 contains similar information, but here neutrino losses from the plasma neutrino and photo-neutrino processes are included. The neutrino losses are so rapid that they actually keep the center cooler than the surrounding matter, with a resulting temperature inversion. The helium flash does eventually occur, but in a shell instead of at the center. When neutrino losses are considered, the core mass at helium flash is slightly more massive (by roughly 5-10 percent).

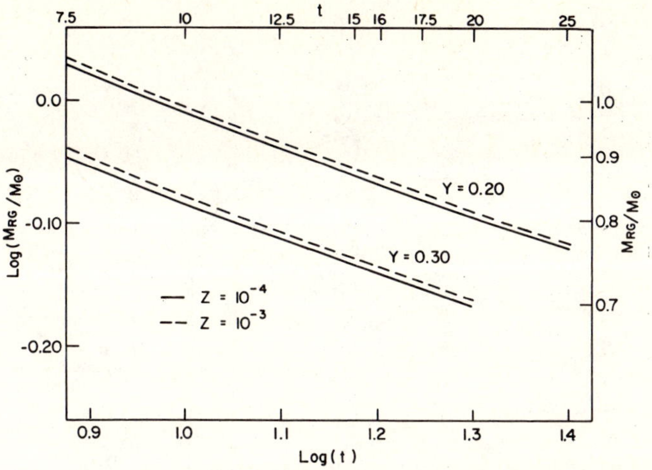

Figure 61. Mass of a star along the red giant branch of a time constant locus as a function of age and composition. From Iben, I., Jr., and Rood, R. T., Ap. J., 159, 605 (1970).

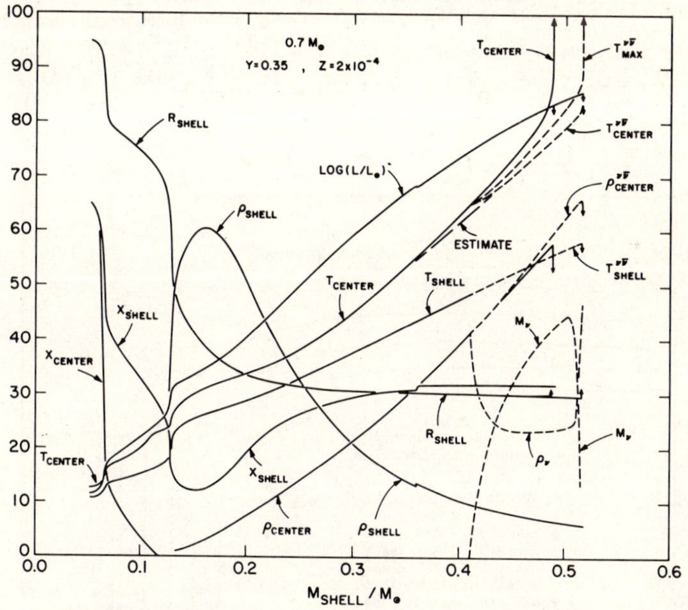

Figure 62. Variation of interior characteristics with core mass for a $0.7\ M_\odot$ metal-poor, helium-rich star ($Y = 0.35$, $Z = 2\times10^{-4}$). The superscript $\nu\bar{\nu}$ signifies that neutrino losses have been included. M_ν represents the mass at which the maximum temperature $T_{max}^{\nu\bar{\nu}}$ occurs and ρ_ν represents the density at this point. Scale limits correspond to $0 \le T_{center}, T_{shell}, T_{max}^{\nu\bar{\nu}} \le 10^8\,{}^\circ K$; $0 \le \rho_{center}, \rho_\nu \le 10^6\ g\ cm^{-3}$; $0 \le \rho_{shell} \le 100\ g\ cm^{-3}$; $0 \le X_{center}, X_{shell} \le 1.0$; $0 \le M_\nu \le 0.5\ M_\odot$; $0 \le R_{shell} \le 0.1\ R_\odot$; $0 \le M_{shell} \le 0.6\ M_\odot$, and $-1 \le \log(L/L_\odot) \le 4$. From Iben, I., Jr., Ap. J., 154, 581 (1968).

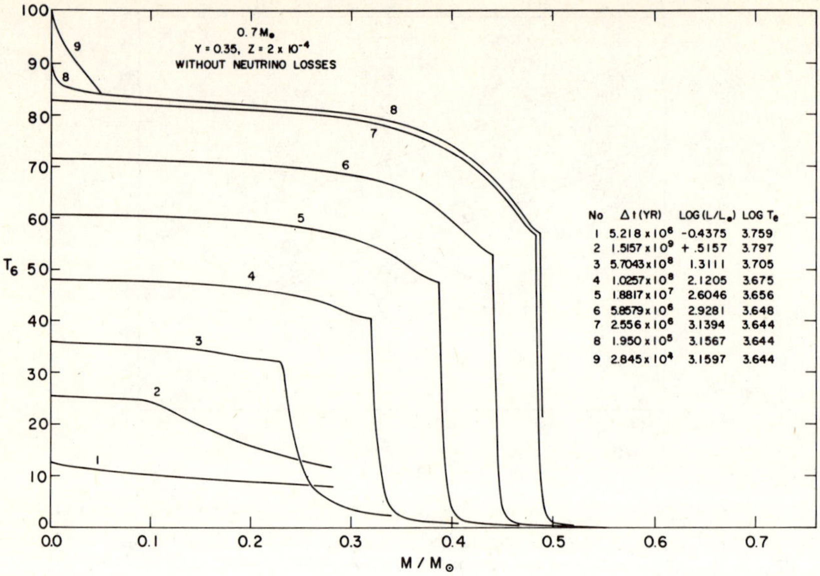

Figure 63. Temperature profiles in a 0.7 $M_\odot$ star characterized by initial $Y = 0.35$, $Z = 2\times10^{-4}$. Neutrino losses by the plasma process are not operating. The ith entry in the column labeled Δt represents the time interval between the ith model and the $(i-1)^{th}$ model. Temperature is in units of 10^6 °K. From Iben, I., Jr., Ap. J., 154, 581 (1968).

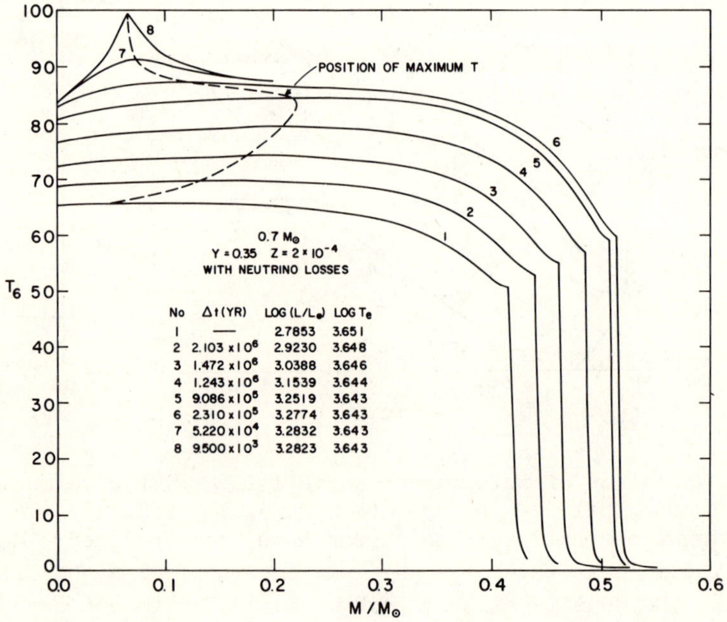

Figure 64. Temperature profiles in a 0.7 $M_\odot$ star with initial $Y = 0.35$, $Z = 2\times10^{-4}$ and with plasma neutrino losses operating. Same notation as in Figure 63. From Iben, I., Jr., Ap. J., 154, 581 (1968).

The opacities in the core are another source of uncertainty. Fig. 65 shows some of the important parameters. The two dominant sources of opacity are electron conduction and the free-free photon absorption process that limits radiative flow. The electron conduction opacity is not shown in the figure, but it decreases very rapidly towards the center, as may be surmised from the behavior indicated for the free-free and total opacities. The crucial region is in the outer 1/3 of the core mass where the two sources of opacity compete. More recent studies of electron conduction opacity have shown that electron-electron collisions are much more important in limiting the mean free path of a conduction electron than had been previously believed. Consequently, the total opacity curve should be raised by a not inconsiderable amount over that shown in Fig. 65. This means that the helium flash will occur for a somewhat lower core mass than computed in this model. The best we can say at the moment is that there is a degree of uncertainty in the determination of core mass at helium flash. Recent work has also suggested that the dynamics of the helium flash may be crucial in these calculations, but we will pretend here that this is not the case.

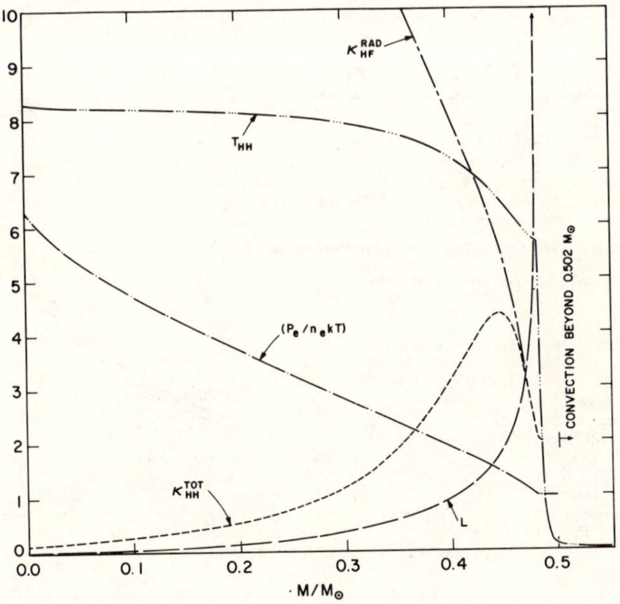

Figure 65. Distribution, in the core of a red-giant star, of luminosity (L), temperature (T), radiative opacity (K_{HF}^{rad}), total opacity (K_{HH}^{tot}), and the degeneracy indicator (P_e/n_ekT). The red giant has a mass $M = 0.7\ M_\odot$, an envelope helium abundance $Y = 0.35$, and a "metal" abundance $Z = 2\times10^{-4}$. Scale limits correspond to $0 \le L < 10\ L_\odot$; $0 \le T < 10^6\,°K$; $0 \le K_{HF}^{rad}$, $K_{HH}^{tot} \le 1\ cm^2\,gm^{-1}$; and $0 \le P_e/n_ekT \le 10$. From Iben, I., Jr., Ap. J., 154, 581 (1968).

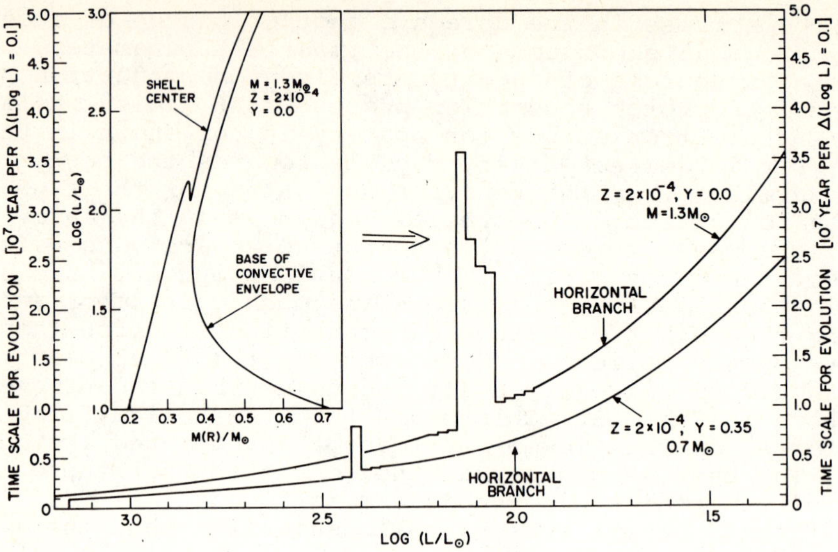

Figure 66. Theoretical giant branch distribution functions for Z = 2×10^{-4} and Y = 0.35. The inset describes the reason for the hump in the Y = 0.0 distribution function. From Iben, I., Jr., Nature, 220, 143 (1968).

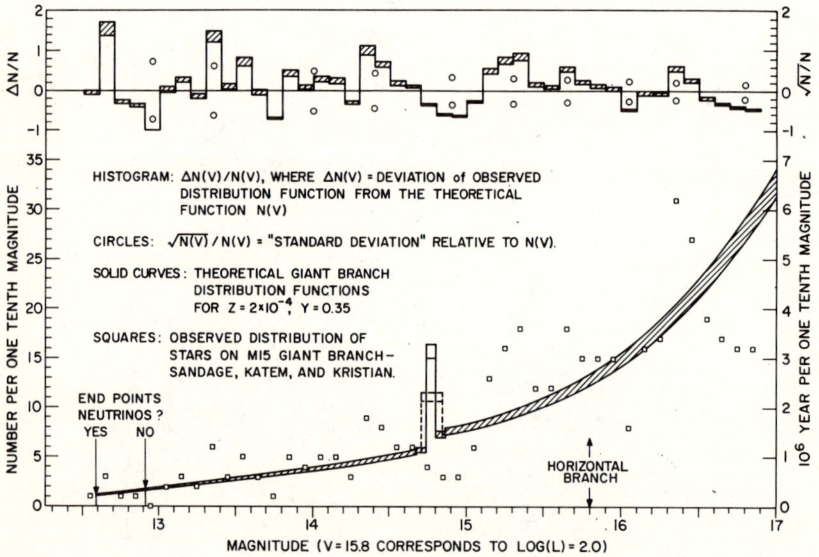

Figure 67. Observed giant branch distribution functions compared with theoretical distribution functions. Each open square represents the number of stars in an interval of 0.1 magnitude along the giant branch in M15, as selected by Sandage, Katem and Kristian (1968). The solid curves bounding the shaded area are theoretical distribution functions for Z = 2×10^{-4}, Y = 0.35. The right-hand scale applies to the upper solid curve. The histogram represents the deviation of the observed distribution from the theoretical distribution functions. Open circles are a measure of the "standard deviation" from the theoretical curves. From Iben, I., Jr., Nature, 220, 143 (1968).

Fig. 66 shows the time spent in a given luminosity interval as a function of luminosity along the giant branch. Except in the region of the humps, dt/d log L is at each point linearly proportional to the hydrogen abundance. This is not surprising, since the amount of fuel available for hydrogen burning is linearly proportional to X. The humps are a consequence of a discontinuity in the hydrogen distribution encountered by the hydrogen-burning shell as it moves outward (in mass fraction). The discontinuity occurs at the point reached earlier by the convective envelope at its maximum extent inward (in mass) during the early giant branch phase. When the shell encounters the discontinuity, the star first dims and then, after the discontinuity has been passed, reascends the giant branch. Thus, the star evolves back and forth three times over a small luminosity interval, producing a hump in the dt/d log L distribution.

In Fig. 67 the observed distribution of stars along the giant branch of the cluster M15 is shown along with a theoretical distribution that has been normalized to match the observations. The deviations between observation and theory are also shown. Except for the humps, the theoretical distributions for all values of X, when approximately normalized, give adequate fits to the observations. The magnitude of the fluctuations in the observed distribution, when compared with the magnitude of humps in the theoretical distributions, permits one to exclude very low values of Y for the stars in M15.

Let us now turn to the horizontal branch. Suppose we know, as a function of initial composition, the mass of the helium core of the star just after it leaves the red giant tip

$$M_{core}(Y, Z) = M^0(Y, Z) + \Delta M(?) \quad . \tag{167}$$

Here ΔM represents the uncertainty in our determination. We want to construct an initial model with a non-degenerate helium core, and an envelope containing the initial abundances. The parameters one can vary are the core mass and the total mass.

To begin, assume we know the age of a cluster, and pick a helium abundance (initial Y) of 0.1. Then we have an idea of the location of the turnoff point and an estimate of the mass of a star on the giant branch, as described earlier. If we choose the mass of a horizontal branch star to be identical with that of a giant branch star (assuming no mass loss), we find that it remains on the giant branch during the core helium burning phase. If, in desperation, we decrease the total mass of the model stars sufficiently (by about 50%), then initial models move down and over to a position on the horizontal branch in the sense that the lower the mass, the bluer the position in the H-R diagram. During subsequent evolution, these models then evolve to the red along tracks that are nearly parallel to (essentially coincide with) the Zero Age Horizontal Branch (ZAHB) defined as the locus of initial core helium burning models of the same (hopefully correctly chosen) core mass but of different (sufficiently low) total mass (Fig. 68).

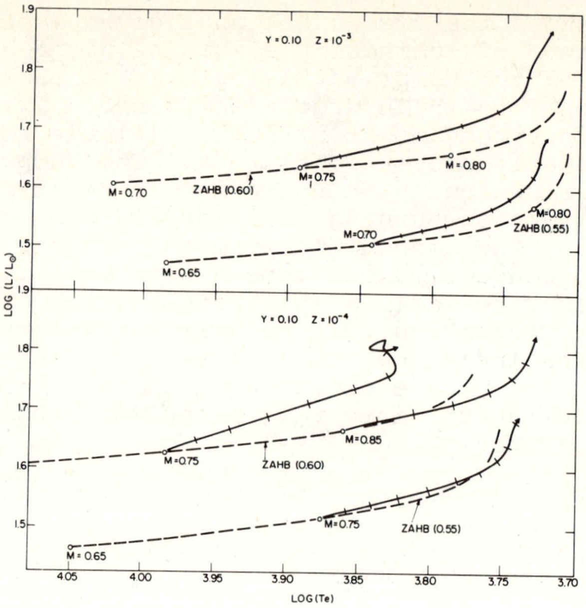

Figure 68. Evolutionary tracks during core helium burning for Y = 0.1. The intervals between adjacent tick marks correspond to 5×10^6 yr. Each dashed line joins zero age horizontal branch models of different total mass but of the same initial helium core mass (indicated in parentheses beside the label ZAHB). All masses are in solar units. From Iben, I., Jr., and Rood, R. T., Ap. J., 159, 605 (1970).

In these models, the hydrogen shell does not burn, so that the horizontal branch luminosity as well as the horizontal branch lifetime are determined by conditions in the hydrogen exhausted core. The conversion of one gram of helium into carbon and oxygen releases only about one tenth as much energy as the conversion of one gram of hydrogen into helium. For Y = 0.1, the ZAHB is roughly 50 times as luminous as the sun which will live roughly 10^{10} yr. as a hydrogen burner. Thus, one might guess that the lifetime of a horizontal branch star for Y = 0.1 might be on the order of 10^7 yr. This is in fact borne out by the calculations. On the other hand, a red giant star (Y = 0.1) requires rougly 10^8 yr. to evolve beyond a luminosity level defined by the mean position of the ZAHB. One might expect then that, if Y = 0.1, there should be 10 times fewer stars on the horizontal branch in a cluster than there are red giants that are more luminous than horizontal branch stars. Detailed calculations show that the factor is closer to 3 than to 10. Observations, however, indicate that the two types are equally frequent, so the theory for Y = 0.1 is at variance with observation.

Let us try again with an initial helium abundance of, say, Y = 0.20. With this choice of helium abundance, only about 25 percent mass loss must be assumed in order to achieve the right color for the horizontal branch models. We find that the initial helium core mass is lower than for Y = 0.1, but that the total luminosity is nevertheless higher because hydrogen burning at the edge of the core contributes significantly. However, the horizontal branch lifetime is determined by the rate at which the helium burns in the core. Because core mass is lower, the core luminosity is also lower and the helium burning lifetime is increased over that found in the Y = 0.1 case. The lifetime of red giant stars beyond luminosities comparable to the Y = 0.20 ZAHB is shorter both because the available fuel is smaller and because the ZAHB is more luminous.

We find that the ratio of horizontal branch stars to red giants of the same or greater luminosity should be ∿ 1/2 if Y ∿ 0.2. This is closer to fact than before. Taking initial Y = 0.30, we find that the predicted ratio of red giants to horizontal branch stars is about 1, as is observed. The hydrogen burning shell contributes somewhat above 50% of the total luminosity, and models evolve to the blue for roughly half the horizontal branch phase along tracks similar to those shown in Fig. 69. Note that the tracks shown in Fig. 69 are very stubby so that it is necessary to assume that there is a spread in the masses of horizontal branch stars on the order of 10-20 percent if we are to account for the observed width

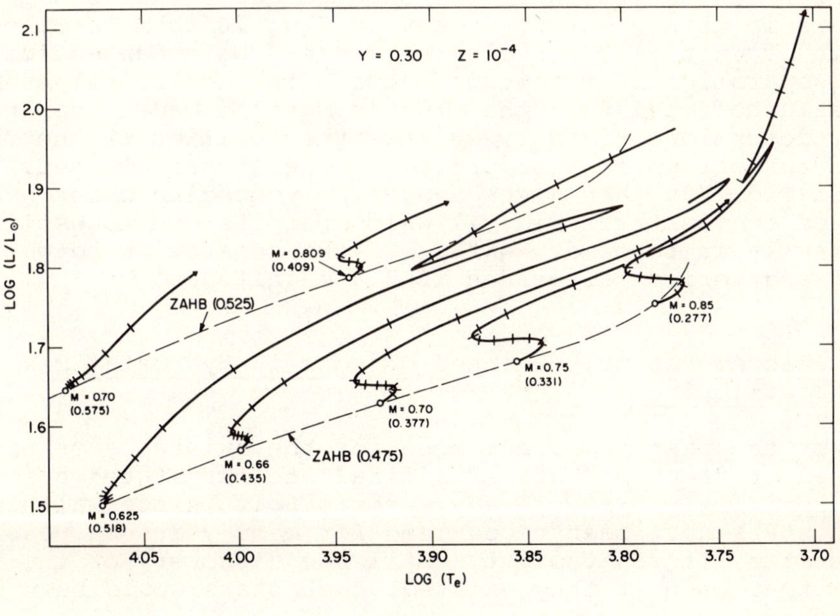

Figure 69. Same as Figure 68 for Y = 0.3. From Iben, I., Jr., and Rood, R. T., Ap. J., 159, 605 (1970).

of the horizontal branch in clusters such as M3 and M15. The stars on the red end of the horizontal branch are of roughly the same mass as their core hydrogen burning progenitors, whereas the blue end of the horizontal branch is populated by stars that are perhaps 20 percent less massive than their progenitors. A fit to the observations thus suggests the existence of a stochastic mass loss mechanism that operates perhaps along the giant branch or perhaps during the helium flash, leading to a loss that averages about 10 percent.

The computed results are summarized in Figs. 70 and 71. The vertical scale is in both figures the ratio of the horizontal branch lifetime to the red giant lifetime beyond the luminosity level of the ZAHB. From observations, we find a value near one.

For each composition, two values of core mass have been chosen in order to estimate the sensitivity of conclusions to several uncertainties. The smaller core masses obtain if the neutrino losses are neglected and the larger ones obtain if these losses are permitted to operate. In every case, modern radiative opacities have been used to derive core masses, but recent improvements in electron conductivity have not been taken into account, so that an additional uncertainty of unknown magnitude in core mass still persists.

The curves in Figs. 70 and 71 are to be interpreted in the following way. Suppose that, for a given cluster, an estimate of N_{hb}/N_{rg} can be made by careful and exhaustive study. Suppose further, that an estimate of the appropriate value of Z can be obtained from spectroscopic analysis of the stars in this cluster. Entry in Fig. 70 then permits an estimate of initial Y for cluster stars. Next, an estimate of the separation in magnitude between the horizontal branch and the turnoff point in the cluster must be made. Having already determined Y and Z, we know the position of the theoretical turnoff point appropriate to the chosen cluster. Entry in Fig. 59, then gives age. The preceding order has been used to construct Fig. 71 which permits one to estimate age directly in terms of N_{hb}/N_{rg} if the separation between turnoff and horizontal branch is 3.4 magnitudes.

VIII. Theorems About Condensed Objects in Hydrostatic Equilibrium

Over the past year, evidence for the existence of highly condensed or "neutron" stars as final stellar states has become surprisingly firm. I say surprisingly, since, although neutron stars have been recognized for many years as possible final states, it was thought, until the discovery of the pulsar, that even if they existed, such stars would be undetectable.

The interpretation of "pulsars" as rotating neutron stars or "rotars" is supported strongly by at least two simple

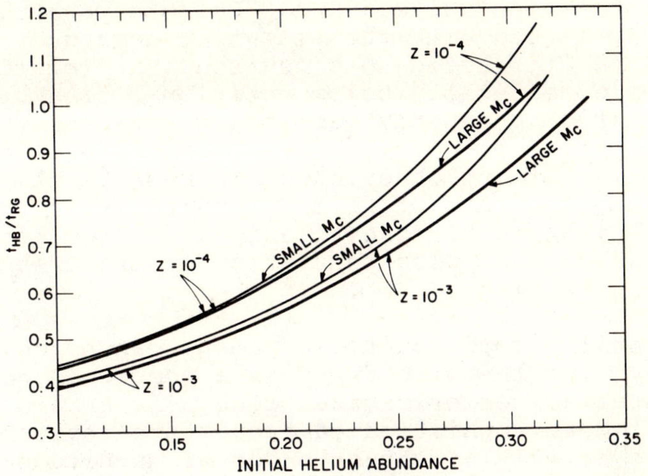

Figure 70. The relationship between initial helium abundance Y and the ratio of horizontal branch lifetime t_{HB} to red giant lifetime t_{RG}. Results are given for two choices of the metal abundance parameter Z and two choices of core mass M_c. From Iben, I., Jr., and Rood, R. T., Nature, 223, 933 (1969)

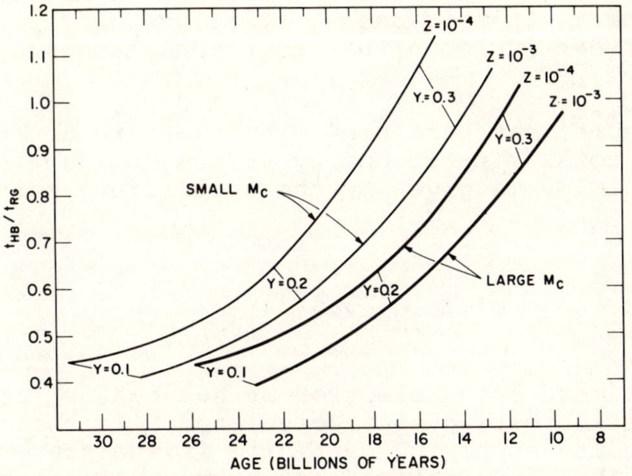

Figure 71. The relationship between t_{HB}/t_{RG} and cluster age for two choices of Z and two choices of M_c. From Iben, I., Jr., and Rood, R. T., Nature, 223, 933 (1969).

arguments: (1) The separation between pulses for at least one pulsar (33 msec) is too short to be attributable to either a pulsating <u>or</u> rotating white dwarf. Pulse separation is too long to be correlated with the pulsation period of a neutron star but can be attributed to a neutron star rotating at an angular speed considerably below rotational breakup speed. (2) the rate at which energy must be injected in the form of relativistic particles into the Crab Nebula in order to maintain the nebulat at its observed brightness in all wavelengths is smaller than the rate at which a neutron star (rotating once every 33 msec) will lose rotational energy if it slows at the rate at which the pulse separation of the Crab pulsar is known to be increasing.

Thus, until a good counter argument can be devised, the neutron star joins the wite dwarf as a member in good standing of the stellar senior citizen club. As an aid to an appreciation of more sophisticated treatments that appear elsewhere, let us briefly construct order of magnitude theorems concerning the radius, maximum mass, maximum internal temperature, pulsation period, and maximum rotation periods of these senior citizens. Although we will achieve only order of magnitude accuracy, the simplicity of the derivations may shed more light on the underlying physics than would be the case had we insisted on greater rigor.

<u>Theorem I</u>. <u>Radii of cold, condensed objects</u>.

Consider a star cool enough that kT is small compared to the kinetic energy of a typical pressure-supplying particle. In a white dwarf, pressure is supplied principally by electrons; whereas, in a neutron star, degenerate nucleons supply pressure as well. In both cases, it is the gravitational attraction between nucleons that must be balanced by the internal pressure.

If we assume that particle velocities are not relativistic, then the total kinetic energy of the dominant pressure-supplying particles is given by the uncertainty principle as, roughly

$$\frac{\hbar^2}{2mr_o{}^2} N, \tag{168}$$

where m is the mass of an electron or nucleon, r_o is the average separation between particles, and N is the total number of pressure-supplying particles in the star. In order of magnitude, this "excitation" energy must equal the gravitational binding energy of the nucleons. Thus

$$\frac{\hbar^2}{mr_o{}^2} N \sim \frac{GM^2}{R} \sim \frac{G(Nm_p)^2}{r_o N^{1/3}} \tag{169}$$

where m_p is the mass of a nucleon and $r_o N^{1/3}$ is roughly equal to the radius of the star. Solving for the average particle separation, we find

$$r_o \sim \left(\frac{\hbar}{mc}\right) \frac{1}{(Gm_p{}^2/\hbar c)} \frac{1}{N^{2/3}} \sim \left(\frac{\hbar}{mc}\right)\left(\frac{N_o}{N}\right)^{2/3} \qquad (170)$$

where we have defined $N_o{}^{2/3} = (Gm_p{}^2/\hbar c)^{-1}$. Physicists tend to feel that further insight has been achieved by calling $\alpha_G = Gm_p{}^2/hc$ the gravitational fine structure constant. I must confess that I do not share this feeling.

Note that, since $N_o \sim 2 \times 10^{57}$ is approximately twice the number of nucleons in the sun, our assumption of non-relativistic degeneracy begins to break down for stars of near solar mass. Since, however, we are only after order of magnitude understanding, this departure from the assumptions is of minor importance.

In the case of white dwarfs, electrons supply the pressure so that $m = m_e$. We have then that the average separation between particles in a white dwarf of near solar mass is just the Compton wavelength of an electron. White dwarf radius ($R \sim r_o N^{1/3}$) is roughly

$$R_{WD} \sim 10^9 \text{cm} \left(\frac{N_o}{N}\right)^{1/3} \sim 10^{-2} R_\odot \left(\frac{N_o}{N}\right)^{1/3} . \qquad (171)$$

In the case of neutron stars, $m \sim m_p$ and, when stellar mass approaches the sun's mass, the separation between particles approaches the Compton wavelength of a nucleon.

The radius of a neutron star of given rest mass is smaller than the radius of a white dwarf of similar mass by a factor that is just the ratio of electron mass to proton mass. Rougly,

$$R_{NS} \sim 10\text{km} \left(\frac{N_o}{N}\right)^{1/3} . \qquad (172)$$

Theorem II. Mass limit for hydrostatically stable, cold stars.

We have noted that, when the average separation between particles approaches the Compton wavelength of the pressure-supplying particles, then the nonrelativistic approximation is no longer valid. The deficiency is easily remedied by using the familiar expression

$$E = \sqrt{(mc^2)^2 + (cp)^2} - mc^2$$

for kinetic energy and replacing p by $\hbar/r_o$. Then

$$Nmc^2 \left\{ \left[1 + \left(\frac{\hbar/mc}{r_o}\right)^2 \right]^{1/2} - 1 \right\} \sim \frac{Gm_p{}^2 N^{5/3}}{r_o} \qquad (173)$$

and we retrieve the previous result in the limit of very large r_o. For small $r_o (< \hbar/mc)$ we find

$$\left(\frac{N}{N_o}\right)^{2/3} \sim 1 - \left(\frac{r_o}{\hbar/mc}\right) + \frac{1}{2}\left(\frac{r_o}{\hbar/mc}\right)^2 - \ldots. \qquad (174)$$

As r_o goes to zero, N approaches N_o so that $N_o \sim \frac{2M_\odot}{m_p}$ represents the maximum number of nucleons that can be assembled into an hydrostatically stable configuration. Note that this maximum number does not contain any statement about whether it is electrons or nucleons that supply the pressure. The reason for this absence is simply that, at relativistic energies, $E \sim cp$, irrespective of particle mass. In practice, of course, the white dwarf mass limit lies somewhat below $N_o m_p$ since, when the electron Fermi energy approaches the energy difference between the neutron and proton, the reaction $p + e^- \rightarrow n + \nu$ turns protons into neutrons.

The fact that N_o is essentially the same for both neutron stars and white dwarfs means of course that these two forms of star arise from two totally different events: in real life, a white dwarf will not of itself turn into a neutron star any more than a stagecoach will turn into a pumpkin. Some external source of pressure is necessary to compress matter at white dwarf densities into matter at neutron star densities. In this connection, note that the gravitational binding energy of a neutron star is roughly 1000 times greater than the gravitational binding energy of a white dwarf of the same mass.

Theorem III. Maximum temperatures in homogeneous stars.

This theorem does not enter into our discussion of pulsars, but it is a pretty result and worthy of presentation. Considering again nonrelativistic energies, but this time admitting a thermal contribution, we have, roughly

$$N \left\{ \frac{\hbar^2}{mr_o{}^2} + kT \right\} \sim \frac{Gm_p{}^2 N}{r_o N^{1/3}} \qquad (175)$$

or

$$\frac{kT}{mc^2} \sim \left(\frac{\hbar/mc}{r_o}\right) \left\{ \left(\frac{N}{N_o}\right)^{2/3} - \left(\frac{\hbar/mc}{r_o}\right) \right\}. \qquad (176)$$

A plot of this relationship in Fig. 72 reveals that, when $N \sim N_o$, the maximum value of kT is, in order of magnitude, equal to the rest mass energy of the dominant pressure-supplying particle and, for smaller values of N, decreases according to the four thirds power of the stellar mass.

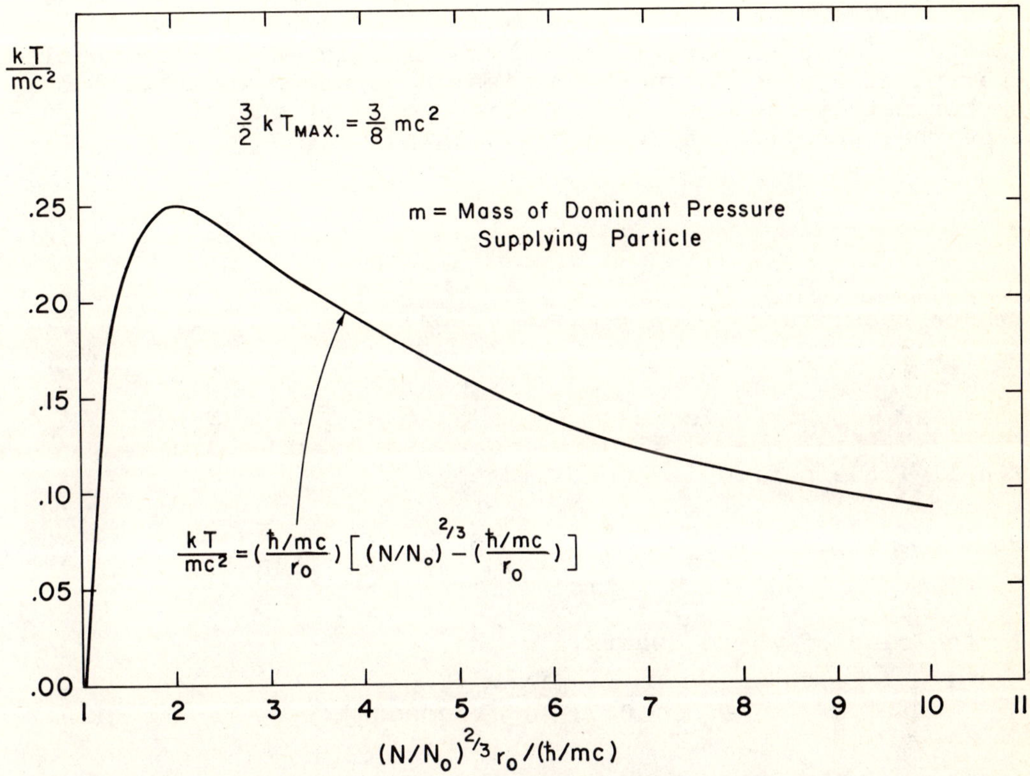

Figure 72. The dependence of the maximum interior temperature (assuming hydrostatic equilibrium) on the average particle separation in the star.

Theorem IV. Pulsation periods.

In first approximation, the fundamental pulsation period of a star is related to the size and internal properties of this star by

$$\text{Period} \sim \frac{\text{Radius}}{\text{Sound Speed in the Interior}} \tag{177}$$

In the cold star approximation, the kinetic energy of a nucleon that participates in the transmission of sonic waves will be approximately equal to the kinetic energy of one of the particles that supplies the pressure. Thus

$$m_p v_s^{\,2} \sim \frac{\hbar^2}{m r_o^{\,2}} \tag{178}$$

or, on substituting for r_o from Theorem I,

$$v_s \sim c\left(\frac{m}{m_p}\right)^{1/2} \left(\frac{N}{N_o}\right)^{2/3} . \tag{179}$$

Finally, the period is approximately

$$P \sim \frac{r_o N^{1/3}}{v_s} \sim \left(\frac{\hbar}{mc^2}\right) \left(\frac{m_p}{m}\right)^{1/2} N_o^{\,1/3} \left(\frac{N_o}{N}\right) . \tag{180}$$

In the case of white dwarfs,

$$P_{WD} \sim 0.7\left(\frac{N_o}{N}\right) \quad \text{seconds} \tag{181}$$

and for neutron stars

$$P_{NS} \sim 10^{-5}\left(\frac{N_o}{N}\right) \text{ seconds} \tag{182}$$

Comparing with the 3.3×10^{-2} second separation between pulses found for the Crab pulsar, we can exclude white dwarf pulsations (in the undamental anyway) as the timing device for pulsars. Further, the pulsation period of a neutron star in the fundamental mode approaches 10^{-2} second only for $N \sim 10^{-3} N_0$, an unlikely possibility on two grounds: (1) $N \sim 10^{-3}N_0$ corresponds roughly to Jupiter's mass and for such a low mass the neutron star configuration is not likely; (2) remnants of this small a mass are unlikely to follow a supernova explosion. Thus neither neutron star nor white dwarf pulsations are likely timing devices for pulsars.

Theorem V. Maximum rotation rates.

In order that a particle at the equator of a rotating star be bound, it is necessary that

$$\frac{v_{surface}^2}{R} < \frac{GM}{R^2} \tag{183}$$

With the usual substitutions, including $P_{rotation} \sim R/v_{surface}$, we have, in order of magnitude, that

$$P_{rot} > \frac{h}{mc^2}\left(\frac{m_p}{m}\right)^{1/2} N_o^{1/3} \left(\frac{N_o}{N}\right) = P_{pulsation} \tag{184}$$

In words, the maximum rotation rate of a star is, in order of magnitude, equal to the lowest pulsational frequency of that star.

Comparing again with the Crab pulsar, we conclude that one pulsar, at least, is not a rotating white dwarf. The remaining possibility -- in the small set of possibilities we have selected -- is a rotating neutron star.

The final numerical argument for the identification of the Crab pulsar as a rotating neutron star proceeds as follows. The rate at which a rotating object converts rotational energy into other forms (in this case, relativistic particles that radiate in strong magnetic fields) may be written as

$$-\frac{dE}{dt} = I\omega^2 \left|\frac{1}{\omega}\frac{d\omega}{dt}\right| = -I\left(\frac{2\pi}{P}\right)^2 \frac{1}{P}\frac{dP}{dt}, \tag{185}$$

where the moment of inertia I is assumed constant and the angular velocity ω is assumed to be approximately constant throughout the object. For the Crab pulsar, period $P \sim 3.3 \times 10^{-2}$ second and $dP/dt \sim 3.65 \times 10^{-8}$ second per <u>day</u>, so that $\frac{1}{P}\frac{dP}{dt} \sim 1.29 \times 10^{-11} \text{sec.}^{-1}$. If we assume that the neutron star is approximately one solar mass, that its radius is roughly 10 km, and that its moment of inertia is $\sim (1/10)MR^2$, then

$$-\frac{dE}{dt} \sim 10^{38} \text{erg/sec.} \tag{186}$$

which is more than ample to account for the known rate at which the Crab nebula is emitting energy in the X-ray, visible, infra-red, and radio bands that have thus far been explored. So, in one stroke, the discovery of the Crab pulsar (tentatively) answers the question, "Do neutron stars exist in nature?", as well as the question, "What is the source of the energy sustaining the light output from the Crab nebula?"

In closing, I would like to emphasize that neither a neutron star nor a white dwarf is the necessary end product of every stellar birth. The mass limits established for cold stars are really only mass limits for cold stars in <u>hydrostatic</u> equilibrium. In principle, there is no reason why the end result of evolution for some stars, at least, might not be a state of "gravitational collapse." The pulses detected by Weber recently might be an indication that this is in fact the final state of some stars.

On the other hand, the fact of the matter may be that stars with $N > N_0$ initially, even though they have no way of knowing that $N > N_0$ will lead them into trouble, nevertheless do get rid of nucleons in excess of N_0 and end life in hydrostatic equilibrium. Evidence in support of this position may be found in the Hyades cluster where a dozen white dwarfs ($M \sim 0.7\ M_\odot$) are found side by side with hydrogen burning stars of mass in excess of 2-3 $M_\odot$. Since all Hyades stars are of roughly the same age and since more massive stars evolve more rapidly than less massive stars, the white dwarfs must be the end products of stars that were initially more massive than (2-3) $M_\odot$.

Acknowledgements

These lecture notes are the result of an experiment to determine whether or not a readable account can be achieved after a minimal modification of the spoken word. The form and content of the notes are relatively faithful reproductions of words delivered during eight 1-1/2 hour lectures. The first draft was typed by S. Ridgway as he listened to lecture tapes. Dick Sears and I then eliminated, as best we could, the most glaring errors of fact and a second draft was typed. Finally, an attempt was made to eliminate some of the more obvious grammatical and logical errors that are characteristic of an (my) oral presentation.

Many thanks to S. Ridgway for his extremely competent work, to R. L. Sears for his wisdom, to H.-Y. Chiu for making these lectures possible, and to the participants in the 1969 Stony Brook Summer Institute for their attentiveness and interest.

Note added in proof: I would like to thank Mary Yastishak who, with fortitude and good cheer, made bearable the painful task of attempting to weed out errors that, despite all precautions, inevitably appear in "final" page proofs.

2

EVOLUTION NEAR THE MAIN SEQUENCE

P. Demarque

Yale University Observatory
New Haven, Connecticut

LECTURE 1. THE MAIN SEQUENCE

The main sequence is where the majority of stars in the solar neighborhood are found. This is because main sequence stars are in the slowest stage of their evolution which is described on the nuclear time-scale.

The basic physics of the main sequence phase of stellar evolution is believed to be well understood, and for this reason it is now possible to consider fine detail on both the theoretical and observational sides. In general, rather than searching for the physical principles responsible for the observed properties of main sequence stars, we can apply our knowledge of stellar interiors as a powerful tool for astronomical research.

Since the physics and mathematics of the various phases of stellar evolution is being described in detail in Iben's lectures, I shall restrict myself to a few comments on the general properties of main sequence stars. Stars are believed by most to form from diffuse interstellar clouds. After an early stage of evolution which is still somewhat uncertain (see e.g., Hunter 1969), they undergo a rapid hydrodynamic collapse (Larson 1969), which is then followed by a period of evolution described on a Kelvin time-scale as they approach the main sequence.

In practice, the main sequence is the locus of chemically homogeneous stars. It is interesting to note that fully mixed stars would follow an evolutionary track from a main-sequence characteristic of one chemical composition to that of another chemical composition. It is clearly not possible to explain the existence of red giants in this way and the realization by Opik (1938), that red giants must be chemically inhomogeneous as a result of unmixed evolution from the main sequence represents one of the most important early discoveries on stellar evolution (Schwarzschild 1958).

The construction of models for main sequence stars involves the integration of the equations of stellar structure for hydrostatic equilibrium and spherical symmetry, which may be written in the following way, using standard notation:

$$\frac{dP}{dM_r} = - \frac{GM_r}{4\pi r^4} \tag{1}$$

$$\frac{dr}{dM_r} = \frac{1}{4\pi r^2 \rho} \tag{2}$$

$$\frac{dL_r}{dM_r} = \varepsilon \tag{3}$$

$$\frac{dT}{dM_r} = - \nabla \frac{GM_r}{4\pi r^4} \frac{T}{P} \tag{4}$$

where

$$\nabla = \frac{d\,ln T}{d\,ln P} .$$

In equation (3), ε need only include the energy generation per gram per second from nuclear sources, and the contribution from the nuclear energy term may be ignored. The gradient ∇ should be either the convective or radiative (including the effects of conduction) gradient, whichever is the smaller, according to the K. Schwarzschild criterion (see Spiegel's notes).

In the deep interior, the adiabatic temperature gradient may be used in convective regions since it is found that the actual temperature gradient exceeds the adiabatic temperature gradient only by a quantity of one part in a million (see M. Wrubel, 1958).

Equations (1) - (4) are solved subject to the boundary conditions:

$$\text{at the center, } M_r = 0;\ r = 0,\ L_r = 0 \tag{5}$$

$$\text{at the surface, } M_r = M;\ P \to 0 \text{ as } T \to T_s \tag{6}$$

For early type stars (hot stars) one can assume that the surface temperature $T_s \to 0$ since the envelope is in radiative equilibrium. For late-type stars, which have an outer convection zone, it is not possible to do so and the structure of the subphotospheric layers must be analyzed with some care.

In addition, one also needs to introduce the details of the physics of the stellar material which are included in the following three relations.

a) the equation of state

$$P = P\ (\rho,\ T,\ \text{chemical composition}) \tag{7}$$

b) the opacity

$$\kappa = \kappa\ (\rho,\ T,\ \text{chemical composition}) \tag{8}$$

c) the nuclear energy generation rates

$$\varepsilon = \varepsilon\ (\rho,\ T,\ \text{chemical composition}) \quad (9)$$

The integration of equations (1) - (4), subject to the boundary condition (5) and (6) and the constitutive equations (7) - (9) must be done numerically, now most frequently performed by the Henyey method, described in Iben's lecture notes.

Stellar masses range from about 50 $M_\odot$ to about 0.80 $M_\odot$. Objects with masses below a critical mass sometimes referred to as Kumar limit (1963), do not become red dwarfs but "black dwarfs", which are unable to provide for their luminosities from nuclear reactions. The reason for the upper end of the mass spectrum is still not clear, although it is known that pulsational instability affects stars with a mass above around 65 $M_\odot$ (Schwarzschild and Harm, 1958).

Stars of high mass derive their energy from hydrogen burning through the CNO cycle which is highly sensitive to the temperature ($\varepsilon \propto T^{\sim 18}$). The main source of opacity is the electron scattering. Such stars exhibit convective cores. Since at their surface temperatures, the hydrogen is ionized, they have envelopes in radiative equilibrium. The size of the convective core is very sensitive to the effect of radiation pressure.

In the vicinity of one solar mass, the central convective region vanishes as the main energy sources become the proton-proton chains. A convective envelope begins to appear due to the presence of a hydrogen ionization zone, and to the effects of molecular dissociation. For stars much less massive than the sun, the outer convective envelope extends to the center and we have fully convective configurations. In these objects the electron gas is partially degenerate. As one goes to very low masses, the effects of electrostatic interactions between charged particles must be included in the equation of state.

The treatment of the hydrogen convection zone involves difficulties which have not yet been fully overcome. The problem one faces is illustrated by K. Schwarzschild's early research on limb darkening on the solar disk in a most beautiful paper (1906). Schwarzschild realized that limb darkening could reveal information on the temperature gradient in the sun's outer layers. He then proceeded to develop the principle of radiative equilibrium and made a comparison of the observed limb darkening with the predictions of adiabatic equilibrium and of radiative equilibrium. His conclusion was that the observed law of limb darkening suggested radiative equilibrium although the existence of some kind of convection was known to exist near the solar surface from the knowledge of granulation. Further work has shown that indeed in the subphotospheric layers, although Schwarzschild's criterion for instability against convection is fulfilled, the actual temperature gradient is sharply superadiabatic and much closer to the radiative gradient. This relatively shallow but strongly superadiabatic region is illustrated in Figure 1. The calculations were made with the mixing length theory, and using a mixing length equal to one pressure scale height in one case and to two pressure scale heights in the other. Our poor

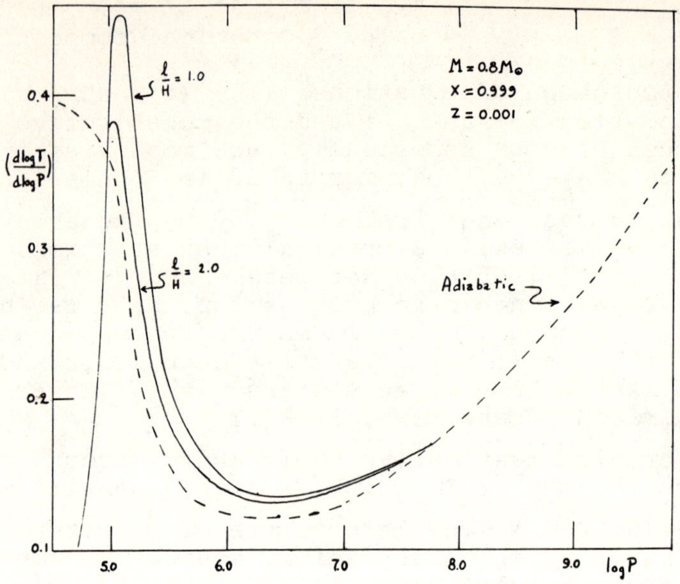

Figure 1. The effect of the choice of ℓ/H_P on the structure of the hydrogen convection zone. For $5.0 < \log P < 8.0$, the value of $(d \log T/d \log P)$ is markedly superadiabatic.

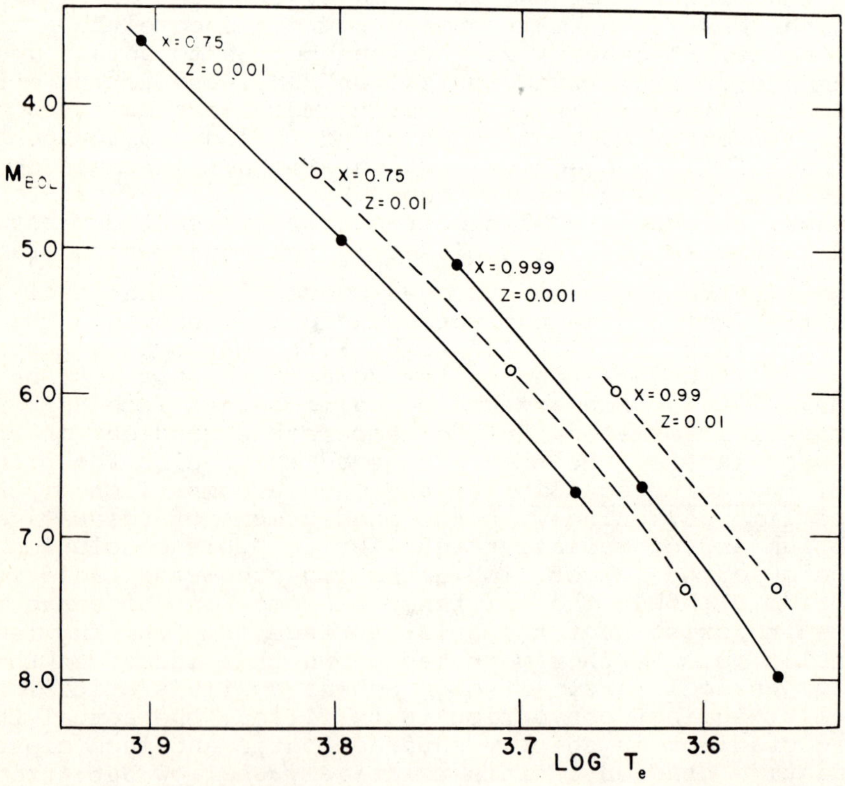

Figure 2. Relative positions of main sequences for different chemical compositions.

understanding of the structure of hydrogen convection zones for late-type stars is particularly serious because of the large uncertainties which it introduces in the calculation of radii of models of late-type stars and therefore of their precise position in the H-R diagram.

One approach is to evaluate the effective mixing length ℓ in the hydrogen convection zone by comparison with observations of a well determined stellar radius, such as that of the sun. Most commonly, one sets ℓ = constant x H_p, where H_p is the local pressure scale height. Some authors prefer the density scale height. Near the main-sequence ℓ/H_p seems approximately between 0.5 and 2.0.

The theoretical models yield two relations which can be compared with observation: a radius-luminosity law which corresponds effectively to the H-R diagram, and a mass-luminosity law. Both of these laws are sensitive functions of the chemical composition and have important astrophysical consequences.

Figure 2 illustrates the effects of chemical composition on the position of main sequence stars of low mass in the H-R diagram. If one denotes by X, Y and Z the mass fractions of hydrogen, helium and the heavier elements respectively so that $X + Y + Z = 1$ it is clear that, at constant mass, the luminosity is a function of chemical composition. If one lowers Z, keeping X fixed, the decrease in opacity leads to a marked increase in luminosity. On the other hand, at fixed Z, the luminosity varies inversely as X because of the strong dependence of luminosity on the mean molecular weight μ. In fact, for stars of low mass, we find that $L \propto \mu^7$ approximately.

The position of the main sequence is affected in such a way that at constant X, a decrease in Z means a lowering of the main sequence position. At constant Z, a decrease in X also means a lowering of the main sequence in the theoretical H-R diagram. This result indicates that for main sequences corresponding to different chemical compositions to be universally superimposed in the theoretical H-R diagram, or ($\log T_{eff} - M_{bol}$) plane, one would require a unique relation between X and Z to hold for all stellar systems. This assumption has often been made in the past for the sake of convenience. It seems unlikely to hold on theoretical grounds and, for that matter, is not confirmed by the most recent observations (Strom and Strom 1967; Cayrel 1968).

Figure 3 shows the effect of chemical composition on the mass luminosity law. It should be pointed out here that in considering masses measured by observation for Population II stars, which are believed to be old, evolutionary corrections must be applied to the models. These corrections may amount to 0.3 magnitude for a GO subdwarf.

Figure 4 illustrates the effect of the choice of the ratio ℓ/H_p on the position of the main sequence. Note that as $\ell \to 0$ we have the radiative surface condition; as $\ell \to \infty$, the convective zone is adiabatic throughout. In the vicinity of the sun, where the hydrogen convection zone is shallow, changes in ℓ/H_p only affect the radii of the model and not their luminosities, i.e. only affect the log T_{eff} variable in the H-R diagram. This relatively fortunate situation does not occur in fully convec-

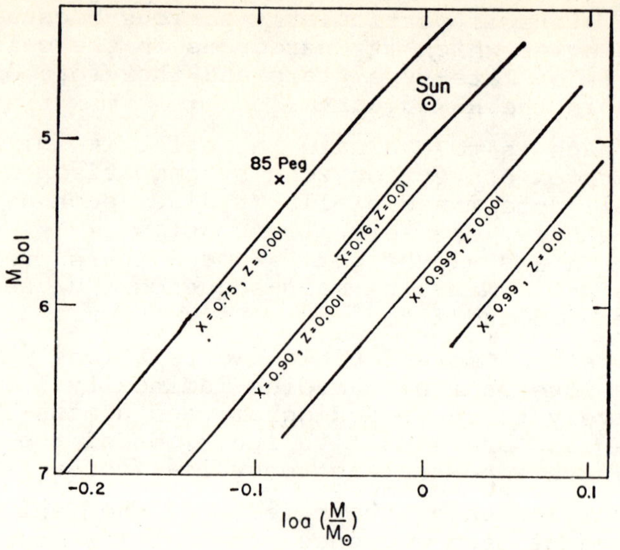

Figure 3. Mass-luminosity laws for different chemical compositions. The position of the sun and the mild subdwarf 85 Peg. are marked on the diagram.

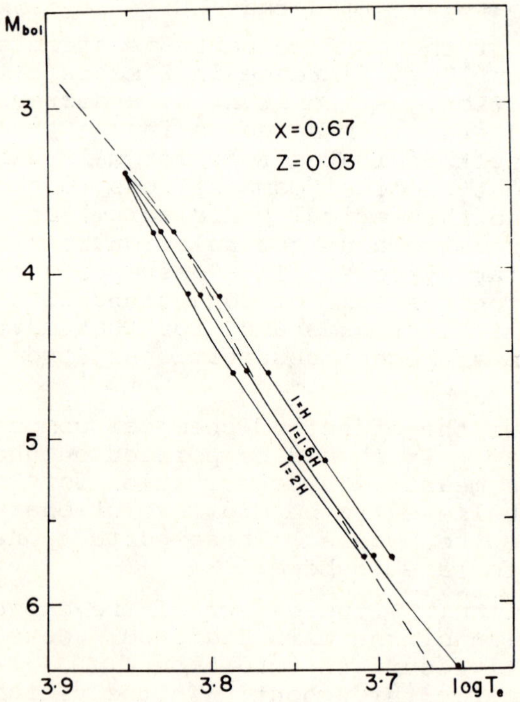

Figure 4. Main sequences for X + 0.67 and Z + 0.03 and several values of the mixing length.

tive stars where the central pressure and temperature are functions of the physical conditions near the surface: in this case the uncertainty in mixing length leads to an uncertainty both in luminosity and radius. Stars near the Hayashi track are thus affected. The radii of models for stars on the faint end of the main sequence which are fully convective, are however, less dependent on ℓ/H_p than giants because their outer convection zones are very nearly in adiabatic equilibrium.

Note that we have discussed only the question of the position of the theoretical H-R diagram. It should be mentioned here that in some cases, large uncertainties in bolometric corrections and in the scale of effective temperature, particularly with regard to the effects of chemical composition on these quantities, make comparison with the observed parameters of stars quite difficult.

The Zero-Age Main-Sequence (ZAMS)

The distance to the Hyades cluster can be estimated by geometrical means by the moving cluster method (van Bueren 1952). The position of the Hyades main sequence can then be determined in the $[(B - V) - M_V]$ plane. It is usually referred to as the ZAMS. Making the assumption that other young star clusters in the field have the same chemical composition as the Hyades (Johnson 1959), and applying approximate evolutionary corrections, it is possible to extend the ZAMS to stars brighter than the brightest main sequence objects in the Hyades. The ZAMS thus used as the standard position for Population I main sequence stars is a very powerful astronomical tool for determining distances of star clusters and variable stars and becomes the basic yardstick for distance determinations in the universe.

For this reason, it is interesting to discuss a well-known problem connected with a discrepancy discussed by Eggen (1963) between the Hyades mass-luminosity laws and the mass-luminosity law for field stars. A discussion of this problem from the point of stellar structure is found in the literature (Demarque 1967) and can be summarized briefly.

Once an estimate of the Hyades distance modulus is obtained from the moving cluster method, it is possible to obtain information about the mass-luminosity law in the cluster from binary system within the cluster in the following way:

a) Kepler's harmonic law yields the total mass of each binary system. We have:

$$m_1 + m_2 = \frac{a^3}{P^2\pi^3} \tag{10}$$

where m_1 and m_2 are the masses of the two components in solar units, a is the semi-major axis of their relative orbit in seconds of arc.

b) By definition, we have also:

$$M = m + 5 + 5 \log \pi \tag{11}$$

If the difference in magnitude between the two components of the binary system is known, and if one assumes that the components are main sequence objects and that the slope of the mass-luminosity law is known, then it is possible to obtain from equation (1) and (2) an empirical mass-luminosity law.

At the same time, the theory of stellar interior provides two relations for homogeneous stellar models:

a) a mass-luminosity relation

$$M_{bol} = M_{bol}\ (M,\ X,\ Z) \qquad (12)$$

which holds for a given pair of composition parameters (X, Z) (Figure 5)

b) a mass radius relation

$$R = R\ (M,\ X,\ Z) \qquad (13)$$

or since the effective temperature

$$T_{eff} = \left[\frac{L}{4\pi R^2 \sigma}\right]^{1/4} \qquad (14)$$

and

$$(M_{bol} - M_{bol}) = 2.5 \log \frac{L}{L} \qquad (15)$$

the position of the main sequence in the H-R diagram:

$$M_{bol} = M_{bol}\ (T_{eff},\ X,\ Z) \qquad (16)$$

for each (X, Y) if one ignores uncertainties in radii due to the pressure of an outer convection zone (Figure 6).

Eggen's data on double stars together with van Bueren's distance modulus gives a mass-luminosity law for the Hyades stars in strong disagreement with that for field stars (Figure 7). After correction for evolution, a Hyades star of one solar mass appeared to be 1.85 magnitudes brighter than the sun. Since the metal abundance of the Hyades does not differ much from that of the sun, this suggests a high helium abundance for the Hyades main sequence in disagreement with van Bueren's distance modulus. The difficulty led Hodge and Wallerstein (1966) to propose that the mass-luminosity law of the Hyades is normal, which results in an upward shift of the main sequence by 0.4 magnitude (Figure 8). Their reasoning unfortunately also leads to a disagreement with the models.

This fundamental problem is not fully solved, although an upward shift of 0.2 magnitude in the position of the Hyades in the H-R diagram may give reasonable agreement between the observations and stellar models (Faulkner 1967; Wallerstein and Hodge 1967; see also Iben 1967). Of course, the last work in measuring the distance to the Hyades should come from the astrometrist, not the theoretical astrophysicist.

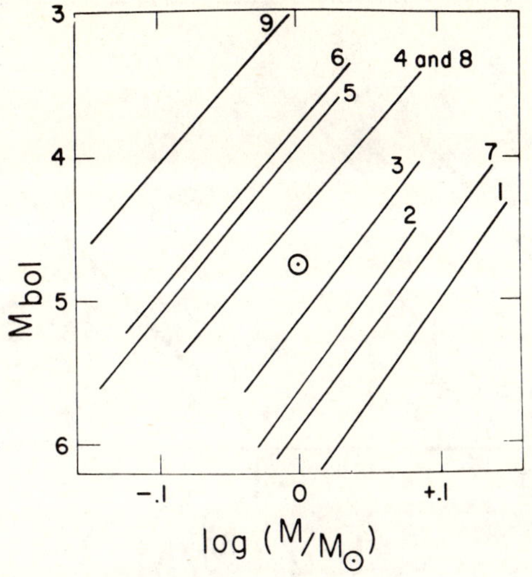

Figure 5. Theoretical mass-luminosity relations. The labels refer to the following compositions: (1) X = 0.67, Z = 0.10; (2) X = 0.67, Z = 0.05; (3) X = 0.67, Z = 0.03; (4) X = 0.67, Z = 0.01; (5) X = 0.67, Z = 0.001; (6) X = 0.67, Z = 0; (7) X = 0.77, Z = 0.03; (8) X = 0.57, Z = 0.03; (9) X = 0.37, Z = 0.03. The present position of the sun is denoted by ⊙.

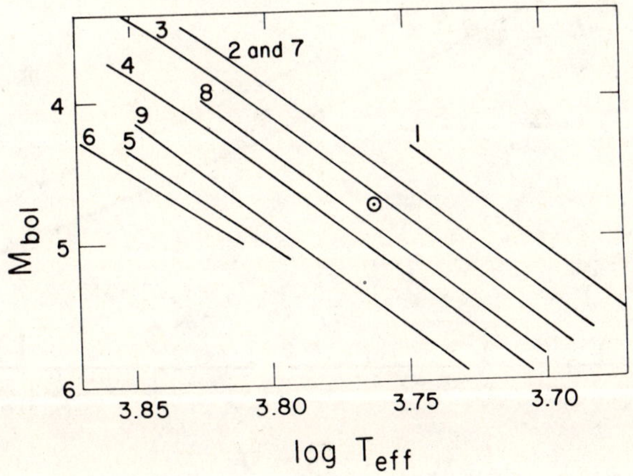

Figure 6. Main sequences in the theoretical H-R diagram. The labels refer to the same composition parameters as in Fig. 5.

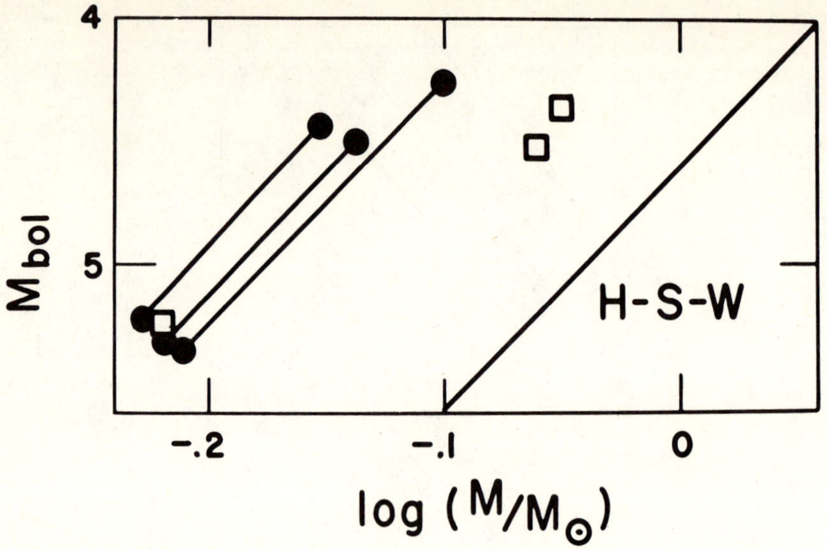

Figure 7. Empirical mass-luminosity diagram for the Hyades binaries listed by Eggen (1963). The circles denote binaries with components of unequal masses. The squares apply to components with nearly equal masses (Iben 1963). The Harris-Strand-Worley relation for nearby stars is represented by a line.

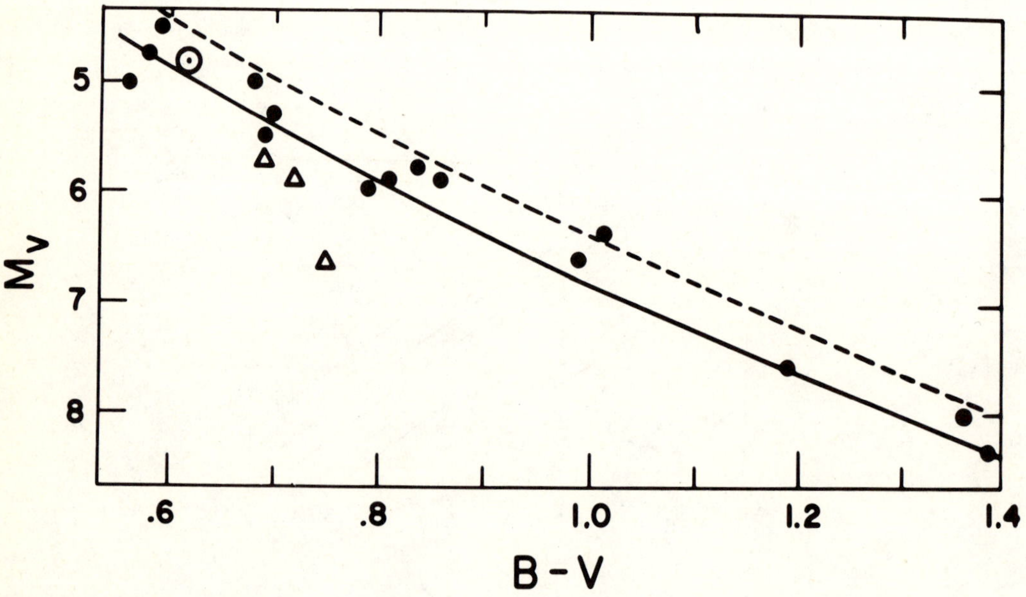

Figure 8. The solid line is the old ZAMS, the dashed line that proposed by Hodge and Wallerstein (1966). The dots are large parallax stars of Population I. The triangles are well-known subdwarfs.

REFERENCES

Bueren, van H.G. 1952, Bull. Astron. Inst. Netherlands, 11, 385.

Cayrel, R. 1968, Astrophys. J., 151, 997.

Demarque, P. 1960, Astrophys. J., 132, 366.

———. 1967, Astrophys. J., 150, 943.

Demarque, P. and Larson, R.B. 1964, Astrophys. J., 140, 544.

Eggen, O.J. 1963, Astrophys. J. Suppl., 8, 125.

Faulkner, J. 1967, Astrophys. J., 147, 617.

Hodge, P. and Wallerstein, G. 1966, Pub. Astron. Soc. Pacific, 78, 411.

Hunter, J.H. Jr. 1969, M.N.R.A.S., 142, 473.

Iben, Icko Jr. 1963, Astrophys. J., 138, 452.

Johnson, H.L. 1963, Photometric Systems in Vol. III of Stars & Stellar Systems, ed. K. Aa. Strand (Chicago: University of Chicago Press).

Kumar, S.S. 1963, Astrophys. J., 137, 1121.

Larson, R.B. 1969, M.N.R.A.S., in press.

Opik, E. 1938, Pub. Obs. Tartu, Vol. 30, Nos. 3 and 4.

Schwarzschild, K. 1966, in Selected Papers on the Transfer of Radiation, ed. D.H. Menzel (New York: Dover Publ.).

Schwarzschild, M. 1958, Structure and Evolution of the Stars (Princeton: Princeton University Press).

Schwarzschild, M. and Härm, R. 1958, Astrophys. J., 128, 348.

Strom, S.E. and Strom K.M. 1967, Astrophys. J., 150, 501.

Wallerstein, G. and Hodge, P. 1967, Astrophys. J., 150, 951.

Wrubel, M. 1958, in Handbuch de Physik, Vol. 51, Stellar Interiors (Berlin: Springer-Verlag).

LECTURE 2

A. Homology Arguments: The mass-luminosity and mass-radius relations

Stellar models generated from a single model by scaling the run of each physical variable with a constant factor are said to be homologous. Homology transformations have been extensively used for two main reasons:

1. Economy in computation. Knowing the constitutive equations (such as (7) - (9) above), it is possible to integrate the equations of stellar structure once and for all for a given set of non-dimensional variables, and thus to construct a whole family of stellar models. This approach turned out to be particularly useful before the development of fast computers.

2. Important properties of families of homologous models can be derived without recourse to numerical integration. In particular the slopes of the mass-luminosity and mass-radius relations can be easily deduced for simple law of opacity and energy generation and the equation of state for a perfect gas.

Let us illustrate this last point by writing the equations of stellar structure in terms of the non-dimensional variable q, f, x, p and t first derived by Schwarzschild (e.g. Schwarzschild 1958). The Schwarzschild variables are chosen in such a way that:

$$M_r = qM \ , \quad L_r = qL \ , \quad r = xR \tag{17}$$

so that

$$P = p\,\frac{GM}{4\pi R^4} \ , \qquad T = t\,\frac{\mu H}{h}\,\frac{GM}{R} \tag{18}$$

For the subsidiary condition (7) - (9) we can write:

$$P = \frac{k}{\mu H}\,\rho T \tag{19}$$

$$\kappa = \kappa_o \rho^{\alpha} T - \beta \tag{20}$$

$$\varepsilon = \varepsilon_o \rho T^{\nu} \tag{21}$$

Equations (1) - (4) then become:

$$\frac{dp}{dq} = -\,\frac{q}{x^4} \tag{22}$$

$$\frac{dx}{dq} = \frac{t}{px^2} \tag{23}$$

and

$$\frac{dt}{dq} = -\,C\,\frac{p^{\alpha}}{t^{3+\beta+\alpha}}\,\frac{f}{x^4} \tag{24}$$

$$\frac{df}{dq} = D\ pt^{\nu-1} \tag{25}$$

where

$$C = \left[\frac{3}{4ac}\left(\frac{k}{GH}\right)^{4+\beta}\left(\frac{1}{4\pi}\right)^{2+\alpha}\right]\left[\frac{\kappa_0}{\mu^{4+\beta}}\right]\left[\frac{L}{R^{3\alpha-\beta}M^{3+\beta-\alpha}}\right] \tag{26}$$

$$D = \left[\frac{1}{4}\left(\frac{GH}{k}\right)^{\nu}\right]\left[\varepsilon_0\mu^{\nu}\right]\left[\frac{M^{\nu+2}}{L\ R^{\nu+3}}\right] \tag{27}$$

Radiative equilibrium has been assumed throughout the star. For a characteristic pair (C,D), we obtain a family of homologous models.

Note that C = C(M, L, R, chemical composition)
and D = D(M, L, R, chemical composition)

so that for any given mass and chemical composition the luminosity and radius of model are uniquely determined.

One finds that:

$$R \propto M^n \ , \quad n = \frac{\nu-\beta+\alpha-1}{\nu-\beta+3\alpha+3} \tag{28}$$

and

$$L \propto M^{\nu+2-n(\nu+3)} \tag{29}$$

If a=1 and β=3.5 (Kramers' Law), ν=4 (p-p chain) then

$$R \propto M^{0.08} \quad \text{and} \quad L \propto M^{5.46} \tag{30}$$

Of course, the proportionality factors can only be obtained through actual integration of the equations of stellar structure subject to the appropriate boundary conditions at the center and at the surface.

B. Evolution Near the Main-Sequence: Stars of low mass

A detailed study of the evolution of stars near the main-sequence is important because it provides what appears to be the most reliable method of determining ages of star clusters. We shall consider here only the evolution of low mass stars (i.e., used in the interpretation of color-magnitude diagrams of old star clusters) since other masses will be treated by other speakers (I. Iben and R. Stothers). With the help of families of evolutionary tracks for stars of slightly different masses, it is possible to construct isochrones in the theoretical H-R diagram (Demarque 1967) which can then in principle be compared to color-magnitude diagrams of star clusters assuming, of course, that the stars within the cluster all have the same chemical composition and are coeval. This last assumption appears to be well justified for old clusters (age $\geq 10^9$ years).

Even though evolution off the main sequence occurs on a nuclear time scale (since it is due to the exhaustion of the supply of hydrogen near the center), one cannot ignore the gravitational energy term in the computations; the gravitational energy release interacts locally with the nuclear energy production and in this way affects the evolution of the stars. One writes:

$$\frac{dL_r}{dM_r} = \varepsilon_N + \varepsilon_G \tag{31}$$

where ε_N is the nuclear energy generation rate and

$$\varepsilon_G = \frac{\partial Q}{\partial \tau}$$

and in most circumstances

$$\varepsilon_G = T\,\frac{\partial S}{\partial \tau} \tag{32}$$

where S is the entropy per unit mass and τ the time.

From our knowledge of ε_G, it is possible to calculate dx/dt for each reaction of interest and therefore to determine the run of chemical composition throughout the star as a function of time. This hydrogen profile controls the structure of the star as it evolves. Since nuclear burning starts during the late stages of pre-main sequence evolution, it may be noted here that when the phase of gravitational contraction ends, the star is already mildly chemically inhomogeneous (since it is not fully mixed even though it may have a small convective core. This effect is small and it is reasonable for most purposes to make the usual assumption that main sequence stars are chemically homogeneous.

It is the chemical inhomogeneity near the center that takes the stars toward the giant branch. The reason for the giant structure seems to be the following: since the core is exhausted of nuclear fuel (the core is not hot enough to burn helium), it contracts and its temperature increases. At the same time, the surface temperature is fixed by the surface physics. The average temperature gradient within the star is determined by the physics of the interior and does not vary very much. It follows that the radius must increase.

The evolution toward the red giant stage follows a pattern with dependence on the presence or absence of a convective core as illustrated by the composition profiles in the adjacent figures.

For stars with masses $\sim 1\ M_\odot$, the high temperature dependence of the CNO cycle, together with the effects of electron scattering opacity produce a central convective region. For the higher masses, the core shrinks continuously as the star evolves off the main-sequence, principally due to a decrease in the opacity as the helium abundance increases. In the lower mass stars with a convective core, the core at first grows due to the progressively increasing relative importance of the CNO cycle over the pp chain as the temperature rises in the core.

When $X < .06$, not enough nuclear energy is produced to keep the star going and the convective core begins to contract while the hydrogen becomes exhausted. During this hydrogen exhaustion phase (HEP), the star adjusts to a new structure, from core burning to shell burning and the temperature just outside the core increases. The HEP is described on a time-scale shorter than the nuclear time-scale. The shell source nuclear burning then stabilizes as the radius of the star expands.

The minimum mass at which a convective core appears depends strongly on the chemical composition and the transition region is of particular interest among Population I stars. (Hallgren 1967; Iben 1967a,b; Demarque 1968; Demarque and Schlesinger 1969; Aizenman, Demarque and Miller 1969).

Old Galactic Clusters (Old Population I)

The color magnitude diagram of M67 exhibits a gap in its stellar distribution just at the turn-off point, at $M_V \simeq 3.3$. This gap can be interpreted as due to the rapid evolutionary phase associated with the HEP described above.

Since at a given luminosity level, the size of the convective core, and consequently the extent of the HEP, is a function of chemical composition, it is of interest to compare the width of the gaps observed in old galactic clusters to that predicted by theoretical tracks for various compositions. The main advantage of such a method of abundance determination is that it is independent of stellar atmospheric factors, and independent of the uncertainties in interstellar reddening and line blanketing effects. The problems associated with the use of the mixing length theory are also removed since the HEP is the result of a phenomenon affecting the deep interior only.

A statistical analysis of the available data for M67 (Eggen and Sandage 1964) reveals the high probability of the existence of a gap in the H-R diagram corresponding to the position of the HEP. Figure 9 shows the luminosity function for M67 in the vicinity of the gap. Figure 10 gives the same information for NGC 188 (Eggen and Sandage 1969). It is instructive to make a comparison of the observations to models obtained for the purpose of calibration of the gap width as a fraction of chemical composition (Aizenman, Demarque and Miller 1969). It is seen that the gap width is a function of both parameters Y and Z. Since the value of Z may be obtained from spectroscopic or photometric evidence, it is then possible to make an estimate for the helium abundance. (Aizenman, Miller and Demarque (1969) suggest the possibility of using the shapes of the evolutionary tracks as a Z indicator as well as a helium indicator. More evolutionary tracks are needed for this purpose.) The results for M67 and NGC 188 are shown in Figure 11. A study of the two old galactic clusters NGC 2360 and NGC 3680

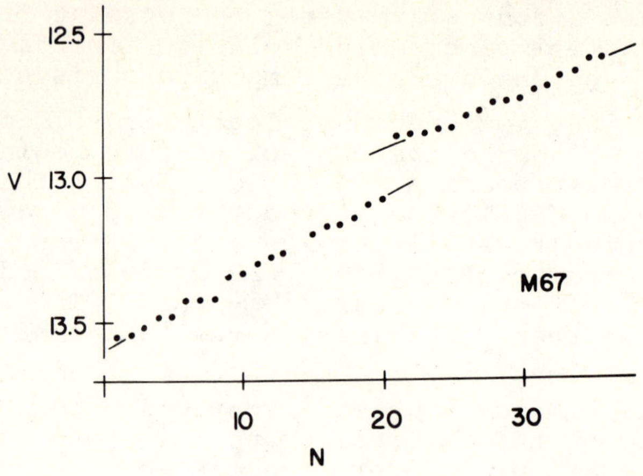

Figure 9. Observational integral distribution of magnitudes of stars for M67. Ordinate is the magnitude V from Sandage and Eggen (1964). Abscissa is the number of stars from the gap region that are fainter than V. Gap occurs between stars 20 and 21.

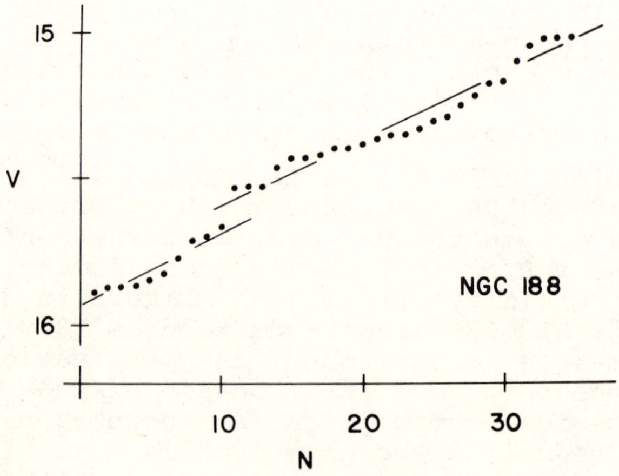

Figure 10. Same as Fig. 9, but for NGC188. Data are from Eggen and Sandage (1969). The gap occurs between stars 10 and 11.

both recently observed by Eggen (1968, 1969) using the same technique raises the interesting possibility that NGC 3680 may be much richer than the sun in heavy elements and perhaps in helium (Demarque and Miller 1969). The existence of such a super metal rich stellar population has recently been suggested on spectroscopic and photometric grounds (Spinrad and Taylor 1967; McClure and van den Bergh 1968; McClure 1969).

Such marked variations in the chemical composition of Population I stars would be reflected in the mass-luminosity and may be the explanation for some of the effects discussed by Eggen (1963) in his studies of mass-luminosity laws.

The Globular Cluster (Population II)

As discussed in the previous lecture, there are still difficulties involved with the interpretation of field main sequence Population II stars or subdwarfs. In studying the globular clusters, the same difficulties arise, compounded by the large distance of the objects, making a detailed study difficult observationally. The importance of the interpretation of globular cluster H-R diagrams, particularly regarding the determination of their ages and helium abundances, is discussed by Iben in his notes (also, Iben and Faulkner 1968; Rood and Iben 1968). It involves not only the consideration of turn offs, but also red giants and horizontal branch stars and is therefore beyond the scope of this discussion. No gaps are observed just above the turn-off point in the Population II color-magnitude diagrams, analogous to the one described in M67. This is in agreement with models for low heavy element abundance ($Z \leq 0.001$) which run predominantly on the proton-proton chain. I shall mention briefly two important topics: (1) the uncertainties in the construction of isochrones near the main sequence; (2) the importance of luminosity functions in the study of stellar evolution.

(1) Figures 12 and 13 illustrate isochrones for composition parameters ($X = 0.999$, $Z = 0.001$) and ($X = 0.75$, $Z = 0.001$) respectively. Both sets of evolutionary tracks were computed with $\ell/H_p = 1$. (Demarque 1967). Figure 14 shows the color-magnitude diagram for M92. The distance modulus was set so that the cluster main sequence superimposes the ZAMS after correction for interstellar reddening and differential line blanketing. The (B-V)-log T_{eff} relation of Population I stars was then assumed.

Uncertainties in the mixing length affect the position of isochrones even more markedly than that of the main sequence (Figure 15). It is interesting to note, however, that the absolute magnitude of the knee (location on track of highest effective temperature) is unaffected by the choice of mixing length, as one would expect since the turn off is a result of composition changes near the center of the star only. The turn off, on the other hand, is very sensitive to small variations in the heavy element abundances as shown recently by Simoda and Iben (1968). The influence of two other factors have also been studied (Demarque, Hartwick and Naylor 1968): the opacity tables and the numerical procedures. Figure 16

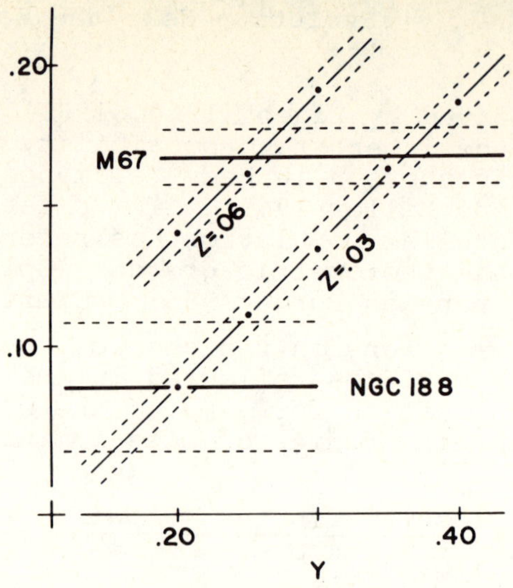

Figure 11. Variation of gap width with composition parameters. Ordinate is the width of the gap in magnitudes, and abscissa is the helium content Y. Sloping lines represent gaps from theoretical tracks and dashed lines bracketing them are estimates of the accuracy of inferring the gap width from the models. Horizontal lines represent the values of the gap width (and error estimates) determined from observational data on the two clusters.

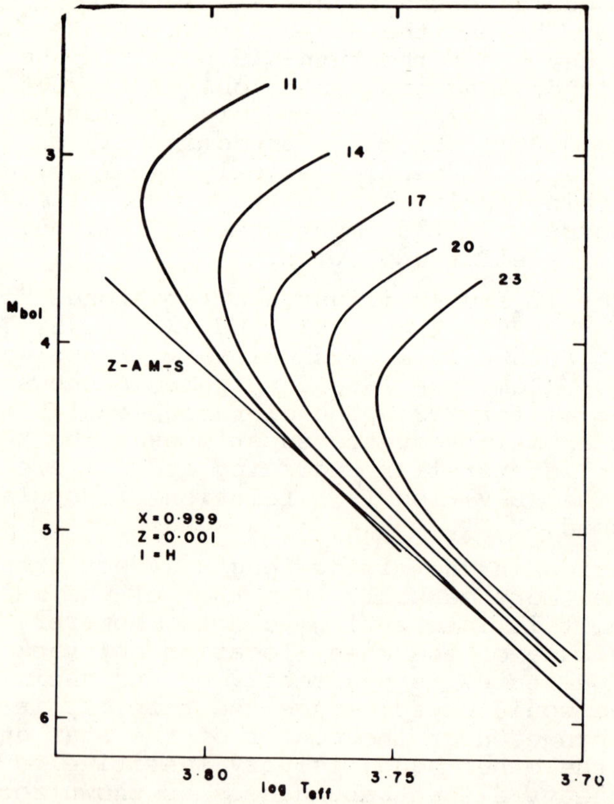

Figure 12. Isochrones for X = 0.999, Z = 0.001 and ℓ/H_p. The zero-age main-sequence is marked. The numbers refer to ages in units of 10^9 years.

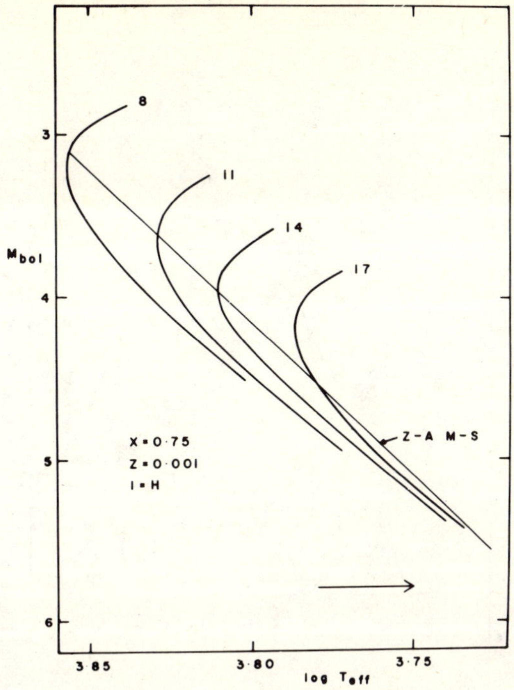

Figure 13. Same as Figure 12, for X = 0.75, Z = 0.001 and $\ell/H_p = 1$.

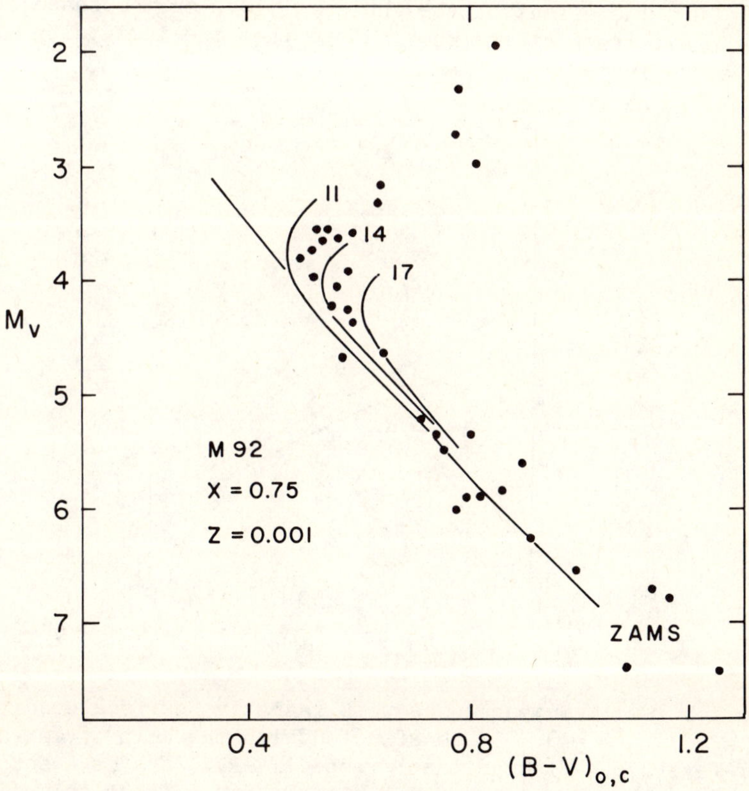

Figure 14. A comparison of isochrones with recent photoelectric observations by Sandage (1969) for M92.

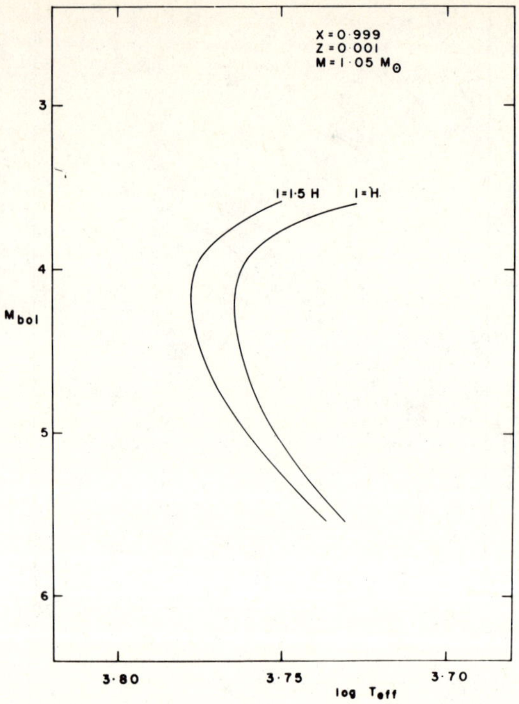

Figure 15. The effect of the choice of ℓ/H_p on the shape of the evolutionary track at the knee. Note that the luminosity at the knee is unaffected.

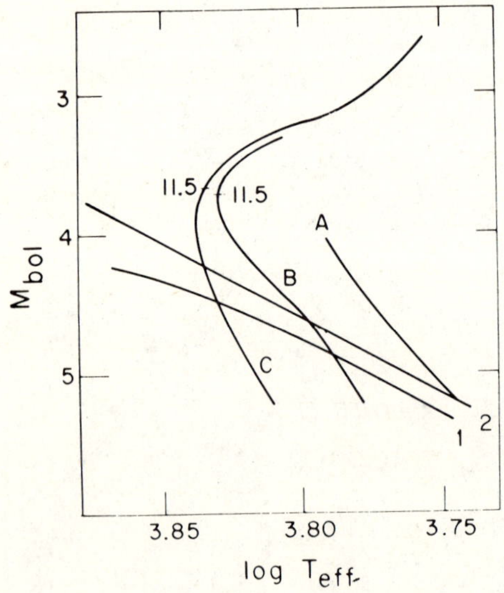

Figure 16. Two main sequences with radiative surface conditions for X = 0/75. Z = 0.001, constructed with the help of Cox and Keller-Meyerot (K-M) opacities, respectively. Tracks A and B show the early evolution of a star of 0.85 $M_\odot$. Also shown is the early evolution of a star with 0.85 $M_\odot$, X = 0.75, Z = 0.001, obtained with two entirely independent programs. The luminosity level corresponding to an age of 11.5 $\times$ 10^9 years is indicated on tracks B and C.

illustrates the comparison models constructed with Keller-Meyerot and Cox opacities respectively. The effect of bound-bound absorption, neglected in the Keller-Meyerot treatment, but included in the Cox models is to increase the opacity at low temperatures and therefore to increase the radii of the models. Figure 16 also indicates that the absolute magnitude of the knee seems to depend little on the method of computation or on the opacity tables used. These results are confirmed by recent more complete calculations (Demarque and Mengel 1969).

(2) The method of semi-empirical evolution pioneered by Sandage (1957) has many possibilities which have been barely exploited. Based primarily on luminosity functions, it is independent of the uncertainties which plague all studies of globular clusters: on the theoretical side, the calculations of radii; on the observational side, the determination of effective temperatures. Simoda and Kimura (1968) who studied M13, and more recently Hartwick (1969) who investigated M92, have pointed out the possibility of using the method for determining helium abundances. Both studies give answers which favor a "high" helium abundance for the two clusters studied (i.e., Y = 0.25-0.35). The development of image tubes will make possible new progress in this promising field.

I am grateful to Dr. B. M. Schlesinger for his assistance in assembling these notes. The author's research described in these lectures was supported by grants GP 8109 and GP 11491 from the National Science Foundation.

References

Aizenman, M.L., Demarque, P. and Miller, R.H. 1969, Astrophys. J., 155, 973.

Demarque, P. 1967, Astrophys. J., 149, 117.

_______. 1968, Astron. J., 73, 669.

Demarque, P., Hartwick, G.D.A. and Naylor, M.D.T. 1968, Astrophys. J., 154, 1143.

Demarque, P. and Mengel, J.G. 1969, in preparation.

Demarque, P. and Miller, R.H. 1969, Astrophys. J., in press.

Demarque, P. and Schlesinger, B. 1969, Astrophys. J., 155, 965.

Eggen, O.J. 1963, Astrophys. J. Suppl., 8, 125.

_______. 1968, Astrophys. J., 152, 83.

_______. 1969, Astrophys. J., 155, 439.

Eggen, O.J. and Sandage, A.R. 1964, Astrophys. J., 140, 130.

_______. 1969, Astrophys. J., in press.

Faulkner, J. and Iben, I. Jr. 1966, Astrophys. J., 144, 995.

Hallgren, L.E. 1967, Astrophys. J., 148, 817.

Hartwick, F.D.A. 1969, preprint.

Iben, I. Jr. 1967a, Astrophys. J., 147, 624.

_______. 1967b, Ann. Rev. Astr. and Astrophys., 5, 571.

Iben, I. Jr. and Faulkner, J. 1968, Astrophys. J., 153, 101.

McClure, R.D. and van den Bergh 1967, Astron. J., 73, 313.

McClure, R.D. 1969, preprint.

Rood, R. and Iben, I. Jr. 1968, Astrophys. J., 154, 215.

Sandage, A.R. 1957, Astrophys. J., 126, 326.

_______. 1969, Astrophys. J., 157, 515.

Schwarzschild, M. 1958, Structure and Evolution of the Stars (Princeton: Princeton University Press).

Simoda, M. and Iben, I. Jr. 1968, Astrophys. J., 152, 509.

Simoda, M. and Kimura, H. 1968, Astrophys. J., 151, 133.

3

Stellar Evolution from Main Sequence to White Dwarf or Carbon Ignition

B. Paczyński

ABSTRACT

Model evolutionary calculations have been made for Population I stars ($X = 0{\cdot}7$, $Z = 0{\cdot}03$) with masses of 0·8, 1·5, 3, 5, 7, 10, and 15 $M_\odot$. Neutrino losses were taken into account. First the computations were made under the assumption that the stars do not lose mass. These computations were started on the main sequence and carried out through the phases of hydrogen and helium exhaustion in the center. Models of 10 and 15 $M_\odot$ ignited carbon in the center nonviolently. Stars of 3, 5, and 7 $M_\odot$ ignited carbon in the center at the density of 3×10^9 g/cm³. This will probably lead to the type of thermonuclear supernova explosion suggested by Arnett. Also, the 1·5 $M_\odot$ model is expected to ignite carbon under similar conditions. When dynamical instability of the red supergiant envelopes was taken into account, the models of 0·8, 1·5, and 3 $M_\odot$ lost their envelopes during the double shell burning phase. The masses of the remaining cores were 0·6, 0·8, and 1·2 $M_\odot$ respectively. These cores evolved through the part of the H — R diagram occupied by the nuclei of planetary nebulae, and were finally cooling down and evolving towards the white dwarf region.

Given our present knowledge of physics and modern computers it is possible to follow stellar evolution from the main sequence to the white dwarf stage or a supernova explosion. In this paper we describe the most essential results of such computations as well as the computer program used. A preliminary report of this work was presented at the 129th Meeting of the American Astronomical Society (Paczyński 1969*b*).

Reprinted courtesy of Acta Astronomica, vol. 20, no. 2, 1970.

The evolutionary computations were performed with a Henyey type code for stars with masses in the range of 0·8 — 15 $M_\odot$. A detailed description of the code, and the code itself are available on request from the Joint Institute for Laboratory Astrophysics.[1] Here we shall describe the most important features of this code.

The model of a star was divided into the interior and the envelope. The outer boundary condition for the interior was obtained from a set of envelope integrations. In most cases the envelope contained 10 per cent of the stellar mass. In some models, however, only $2 \times 10^{-7} M_\odot$ was included in the envelope (model nuclei of planetary nebulae). In others (red supergiants), hydrogen, and even helium shell sources were included in the envelope integrations. The general rule used was that the part of a star that can be adequately described with ordinary differential equations was considered to be the envelope. In the envelope the mass fraction (or radius) was the only independent variable. In most cases these were static envelopes computed with a code described elsewhere (Paczyński 1969*a*). The ionization zones of hydrogen and helium and nonadiabatic convection were treated with this code. The shell sources were included in the envelope integrations when their thickness was considered to be sufficiently small, about 1 per cent of the core mass.

When the shell source is sufficiently narrow one can describe its structure with ordinary differential equations (Eggleton 1967). A time derivative of any physical quantity u can be replaced with its space derivative according to the formula

$$\left(\frac{\partial u}{\partial t}\right)_{M_r} = -\left(\frac{dM_c}{dt}\right) \times \left(\frac{\partial u}{\partial M_r}\right)_t , \tag{1}$$

where $\frac{dM_c}{dt}$ is the rate at which the mass of the core increases, or in other words, the rate of mass flow through the shell source. The thinner the shell source, the more accurate this formula becomes, which makes it possible to obtain the precise distribution of chemical elements within the shell and to include a gravitational energy term in the ordinary differential equations describing the shell. This technique does not allow thermal instabilities in shells to appear.

When static envelopes are used the outer boundary condition for the interior is described by two parameters: the radius and the luminosity of the star. When the shell source is included in the envelope we have a third parameter: the rate of mass flow from the envelope to the core.

---

[1] Requests should be directed to Dr. J. Cox, Joint Institute for Laboratory Astrophysics, University of Colorado, Boulder, Colorado.

Equation of state. Envelope models were always deep enough to cover the hydrogen and helium ionization zones. The matter in the interior was assumed to be completely ionized. A mixture of perfect gas and radiation was taken into account. The electron gas could be arbitrarily degenerate and/or relativistic, but no effects of particle interactions such as crystalization or pair formation were taken into account. All the thermodynamic quantities needed for the evolutionary calculations were stored in tables in the computer memory.

Opacities. Radiative opacities (including lines) were interpolated from tables calculated by Cox and Stewart (1968) for 20 chemical compositions. Conductive opacities calculated by Hubbard and Lampe (1969) and by Canuto (1970) were used. In the course of an evolutionary calculation three opacity tables were stored in the computer memory: one for a hydrogen rich mixture, another for a helium rich mixture, and the third for a carbon and oxygen rich mixture. This sequence corresponded to the evolutionary changes in the chemical composition. In the model envelopes water vapor opacities computed by Auman (1967) were taken into account, and the H_2O abundance was calculated with a code based on Mihalas' (1967) description.

Nuclear reaction rates. Energy generation in the p-p and CNO cycles was calculated with formulae given by Reeves (1965), helium burning rates (nitrogen + helium, triple alpha, carbon + helium) were adopted from the review by Fowler, Caughlan, and Zimmerman (1967), and the carbon burning rate given by Patterson, Winkler, and Zaidins (1969) was used. For helium and carbon burning weak and strong screening corrections were calculated following Salpeter and Van Horn (1969). Reaction rates were stored in the computer memory as a function of temperature and density.

Neutrino emission. Neutrino emission was taken into account throughout the hydrogen-exhausted cores of all model stars. The rate of neutrino emission was calculated with formulae given by Beaudet, Petrosian, and Salpeter (1967) and stored in a table in a computer memory.

As practically all the information about the input physics is stored in tables and calculated by means of linear interpolation in the course of an evolutionary computation, the program is rather fast. On the CDC 6400 computer, it requires 10 msec per zone per iteration and less than one hour of computing time is sufficient to get an evolutionary track from the main sequence to helium exhaustion in the core and formation of a double shell source model. Another hour is required to evolve a small or intermediate mass star to the white dwarf or carbon ignition state.

Our evolutionary calculations were started with homogeneous main sequence models having a metal content $Z = 0.03$, and a hydrogen content $X = 0.70$. The calculations were continued up to the helium flash (owing to the helium + nitrogen reaction) in the core for 0.8 and 1.5 $M_\odot$, and up to carbon ignition for 3, 5, 7, 10, and 15 $M_\odot$. In

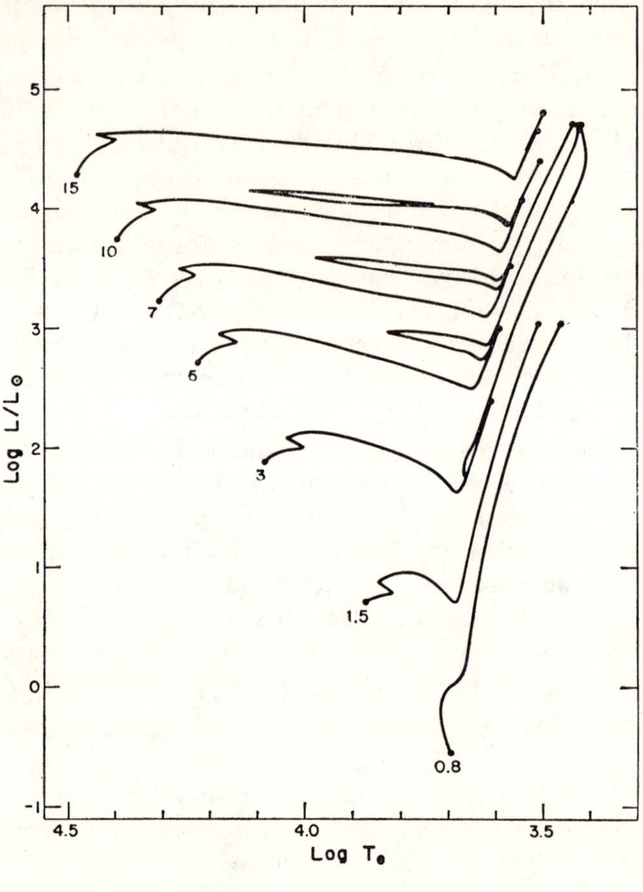

Fig. 1. Evolutionary tracks of Population I ($X = 0.7$, $Z = 0.03$) stars on the H—R diagram. The numbers at the beginning of each track give stellar mass in units of solar mass. Large dots indicate the position of the homogeneous main sequence models and the position of the models at the times of helium and carbon ignition in their cores.

the latter models the helium + nitrogen reaction was neglected. The evolution of those stars on the H—R diagram is shown in Figure 1. Large dots indicate the positions of models on the main sequence, and at the times of helium and carbon ignition. Evolutionary tracks for the 5, 7, and 10 $M_\odot$ models show the familiar loop on the H—R diagram during core helium burning, but the 15 $M_\odot$ model spent all the core helium burning phase in the red giant region. This problem will be discussed in a separate paper. Fortunately the evolution of the stellar cores is hardly affected by the existence or nonexistence of the loops.

The evolution of the centers of stars on the logarithm density — logarithm temperature plane is shown in Figure 2. Also, a line along

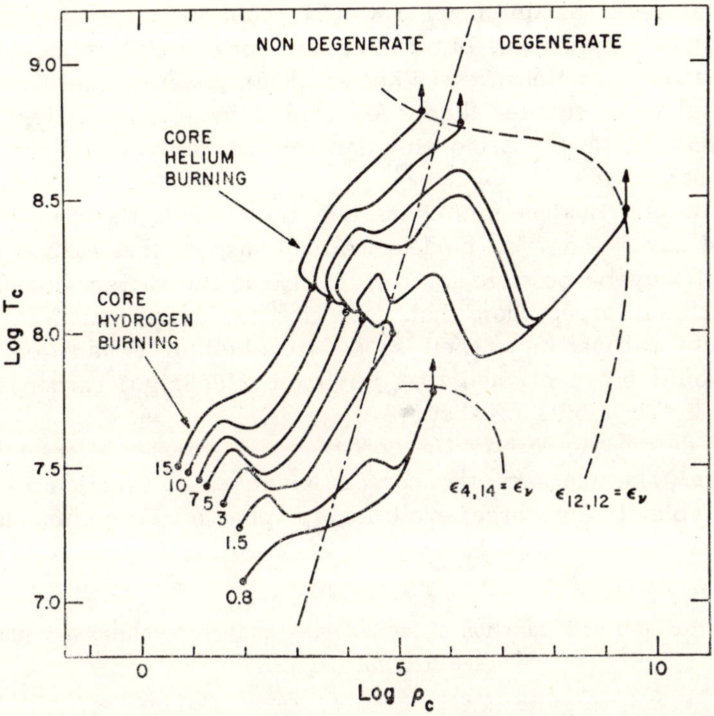

Fig. 2. Evolutionary track of the centers of stellar models in the temperature-density plane. The numbers at the beginning of each track give stellar mass in units of the solar mass. Large dots indicate the position of the centers of the models on the main sequence and at the times of helium and carbon ignition. Along the broken lines neutrino energy losses balance either nitrogen + helium, or carbon burning.

which the carbon burning rate equals neutrino energy losses is shown. Notice that carbon ignition takes place when the center of the star crosses this line. Carbon ignition is nonviolent in models of 10 and 15 $M_\odot$. The cores of the 3, 5, and 7 $M_\odot$ models are smaller and do not contract too rapidly after the helium exhaustion. As a result, neutrino emission cools them down (see also Weigert, 1966), and carbon ignition takes place when the density at the center reaches 3×10^9 g/cm^3, as suggested by Arnett (1968, 1969). This ignition is most likely explosive.

At the density of 3×10^9 g/cm^3 our cores were dynamically stable. However, the effects of crystallization, electron capture, and general relativity were not taken into account. The strong screening correction to the carbon burning rate may also be uncertain. It is not impossible that if all of these effects were properly taken into account, our cores could become dynamically unstable and collapse prior to carbon ignition. In that case carbon burning would probably be unable to cause a thermonuclear explosion, but rather the core would collapse all the way to neutron star densities (Wheeler 1969). In any case a star with a mass equal to or smaller than $7\,M_\odot$ cannot have a nonviolent carbon burning phase if the neutrino emission due to "universal Fermi interaction" exists.

None of our models ignited carbon in a shell. Extrapolating the behavior of our 5 and $7\,M_\odot$ models we may suspect that carbon ignition in the shell may be possible for $8\,M_\odot$. Also if the carbon burning rate were 100 times larger than that used in this study, our $7\,M_\odot$ model would ignite carbon in a shell soon after helium exhaustion in the core. It should be mentioned that Sugimoto (1969) got carbon ignition in a shell in his model of about $6\,M_\odot$.

Let us define the core as the part of a star interior to the hydrogen shell source. The mass of the core is given as a function of stellar mass in Table 1 for three evolutionary phases: 1) helium ignition,

Table 1.

Mass of the core as a function of stellar mass at three evolutionary phases. See text for details.

$M_{star}/M_\odot$	$M_{core}/M_\odot$		
	(1)	(2)	(3)
0·8	0·39	—	—
1·5	0·40	—	—
3·0	0·35	0·51	1·39
5·0	0·56	0·95	1·39
7·0	0·83	1·45 (1·02)	1·39
10·0	1·35	2·32	2·32
15·0	2·54	3·89	3·91

2) helium exhaustion, and 3) carbon ignition. In the case of the $7\,M_\odot$ model the convective envelope penetrated the core after helium

exhaustion and reduced its mass from 1·45 to 1·02 $M_\odot$. The difference in the core mass of the 0·8 and 1·5 $M_\odot$ models at the time of helium flash is probably caused by numerical inaccuracy in the present computations.

Perhaps the most striking feature of Figure 2 is the convergence of the evolutionary tracks for 3, 5, and 7 $M_\odot$ models into a common

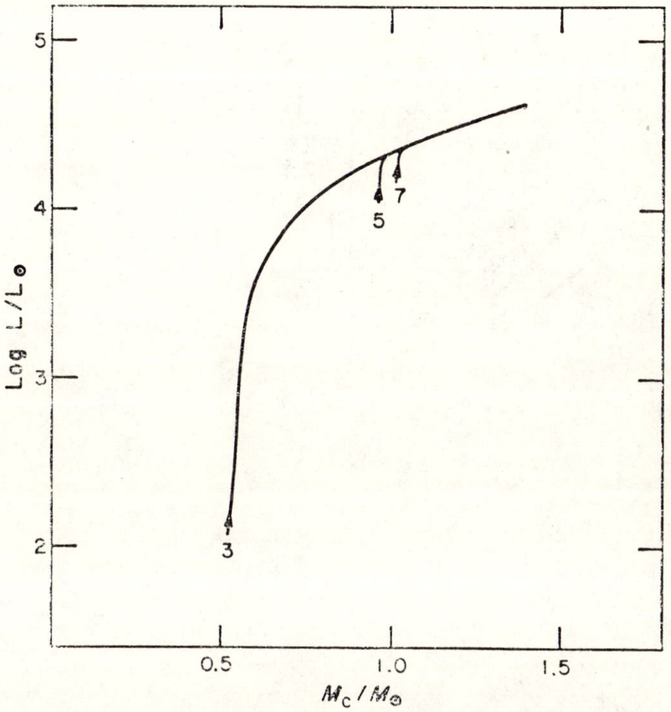

Fig. 3. Variation of the luminosity of 3, 5, and 7 $M_\odot$ models as a function of core mass during the double shell source burning phase.

track after the exhaustion of helium in the core. The same phenomenon may be seen for these models on the mass of the core — luminosity of a star diagram (Fig. 3). Even more surprising, this common mass — luminosity relation can be written as

$$L/L_\odot = 59250 \times M_{core}/M_\odot - 30950 \tag{2}$$

for $0{\cdot}57 < M_{core}/M_\odot < 1{\cdot}39$, with an accuracy better than 2 per cent. In a core of 0·57 $M_\odot$, hydrogen burning in a shell contributes 86 per cent of the total luminosity, helium burning in a shell 13 per cent, and gravitational contraction 1 per cent. In the core of 1·39 $M_\odot$ the corresponding numbers are 80, 10, and 10 per cent.

We expect that the evolutionary tracks of the cores of our 0·8 and 1·5 $M_{\odot}$ models would also merge into the common evolutionary track after the formation of the helium shell source. Therefore, if there is no mass loss during the evolution carbon can be ignited explosively in our 1·5 $M_{\odot}$ model. In general, explosive carbon ignition is possible in a star that can form a degenerate carbon-oxygen core

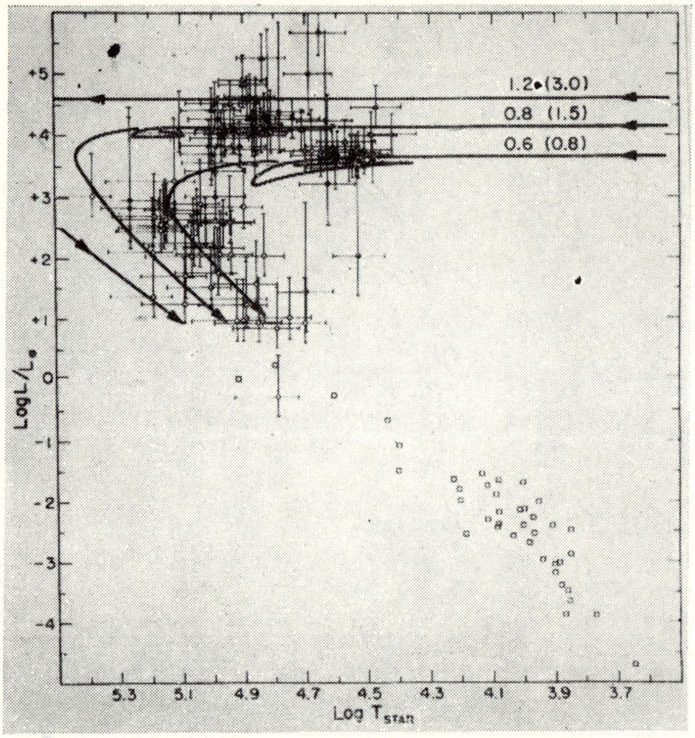

Fig. 4. Evolutionary tracks of the cores of models with initial masses 0·8, 1·5, and 3 $M_{\odot}$ after they have lost their envelopes. Tracks are labelcd with the mass of the core and the initial stellar mass. The positions of the nuclei of planetary nebulae (with error bars) and white dwarfs are shown according to O'Dell (1968). (Courtesy of the International Astronomical Union) .

of 1·39 $M_{\odot}$. In that case we may expect stars in the mass range 1·4 — 7 $M_{\odot}$, or 1·4 — 8 $M_{\odot}$ to ignite carbon explosively. For masses larger than 10 $M_{\odot}$ or perhaps larger than 9 $M_{\odot}$ carbon ignition is not violent. The mass at which transition occurs is somewhere between 7 and 10 $M_{\odot}$.

Our models with two shell sources have very high luminosities, up to $5{\cdot}2 \times 10^4 L_{\odot}$. Envelopes of these models are very extended and may become dynamically unstable (Lucy 1967, Paczyński and Ziółkowski 1968*a*, *b*, Roxburgh 1967). Using Figure 6 of Paczyński and Ziółkowski (1968*b*) to estimate which of our models have dynamically unstable envelopes, we find that instability occurs when the mass of the core is 0·6, 0·8, or 1·2 $M_\odot$ compared to a total mass of 0·8, 1·5, or 3·0 $M_\odot$, respectively. Here we assume that cores of 0·8 and 1·5 $M_\odot$ models follow the mass-luminosity relation given by equation (2). At the onset of dynamical instability the total energy of the envelopes is positive and it is likely that they will be disrupted. It is interesting to see how the remaining core, presumably a nucleus of planetary nebula, will evolve.

The evolution of the cores of 0·6, 0·8, and 1·2 $M_\odot$ (with small hydrogen-rich envelopes attached to them) on the H—R diagram is shown in Figure 4. The positions of nuclei of planetary nebulae and white dwarfs are shown using data from O'Dell (1968). The evolution is almost horizontal as long as there is enough mass left in the envelope to support nuclear burning in the two shell sources. As mass flows from the envelope through the shell sources to the core, the mass of the envelope and the radius of the star decrease, and the effective temperature increases. When the mass of the envelope decreases below a certain limit the shell sources die out and the models cool down to

Table 2.

Evolutionary time scales (in years) for model nuclei of planetary nebulae.

0·6 $M_\odot$			0·8 $M_\odot$			1·2 $M_\odot$		
time	log T_e	log $L/L_\odot$	time	log T_e	log $L/L_\odot$	time	log T_e	log $L/L_\odot$
—4·8 (5)	3·50	3·68	—3·5 (5)	3·50	4·21	—4·0 (5)	3·50	4·60
—1·4 (4)	3·70	3·68	—5·4 (3)	3·70	4·21	—6·0 (3)	3·70	4·60
—4·0 (3)	4·00	3·68	—8·0 (1)	4·00	4·21	—3·0	4·00	4·60
0·0	4·40	3·67	0·0	4·40	4·21	0·0	4·40	4·60
4·0 (3)	4·80	3·53	8·0 (1)	4·80	4·21	0·35	4·80	4·60
1·7 (4)	5·00	3·53	1·4 (2)	5·00	4·20	0·8	5·00	4·60
—	—	—	3·6 (2)	5·20	4·08	1·3	5·20	4·60
—	—	—	—	—	—	3·6	5·60	4·60
—	—	—	—	—	—	1·0 (1)	5·83	4·50
—	—	—	9·0 (2)	5·39	4·00	3·2 (1)	5·79	4·00
1·8 (4)	5·02	3·50	1·5 (3)	5·43	3·50	5·6 (1)	5·70	3·50
3·0 (4)	5·15	3·00	2·4 (3)	5·36	3·00	1·4 (2)	5·59	3·00
3·8 (4)	5·09	2·50	5·2 (3)	5·27	2·50	6·0 (2)	5·46	2·50
7·5 (4)	5·02	2·00	1·7 (4)	5·17	2·00	4·5 (3)	5·35	2·00
1·6 (5)	4·94	1·50	7·5 (4)	5·06	1·50	2·0 (4)	5·23	1·50
4·0 (5)	4·84	1·00	2·3 (5)	4·95	1·00	5·0 (4)	5·11	1·00

the white dwarf region. A similar evolutionary track was recently obtained by Rose and Smith (1970) for $0.85\ M_{\odot}$.

The evolutionary times for our models are given in Table 2. Notice that speed of evolution is extremely sensitive to stellar mass, and is very high for the $1.2\ M_{\odot}$ model when the effective temperature is in the range of $10^4 - 10^5\ ^{\circ}K$. The zero point of the time scale was arbitrarily chosen to correspond to $\log T_{eff} = 4.4$. We do not know how much mass is left in the envelope after the main part of it has been lost. The surface temperature of a model is very sensitive to the amount of mass left, as is the evolutionary time from the moment when the envelope was ejected to the moment when the effective temperature is $10^{4.4}\ ^{\circ}K$. This temperature seems to be the minimum required to excite the nebula formed from the ejected envelope, as no nucleus of a planetary nebula is cooler than this limit. Notice also that the luminosity of our models is completely insensitive to the mass of the envelope, provided there is any envelope left.

Our model cores of 0.6, 0.8, and $1.2\ M_{\odot}$ were taken from our evolutionary calculations for $3\ M_{\odot}$, which were carried out all the way from the main sequence. No attempt was made to choose any parameter so as to fit the observed positions of the nuclei of planetary nebulae with our evolutionary sequences. The structure of our cores is independent of the mechanism which may be responsible for the ejection of the envelope (dynamical instability, stellar wind, thermal instabilities in a helium shell source). In particular we believe that the luminosities of our cores are reliable. However, the relation between the mass of the core, which becomes the nucleus of a planetary nebula, and the total mass of the original star is very uncertain, as this depends on details of the mechanism which disrupts the envelope. If the estimates of dynamical instability made by Paczyński and Ziółkowski (1968*b*) are used in conjunction with relation (2) above, the planetary nebulae may be formed from stars with masses up to $3.5\ M_{\odot}$, and their nuclei may have masses up to $1.39\ M_{\odot}$. The lower mass limit is established by the age of the galaxy, and is likely to be about $0.8\ M_{\odot}$ for the original mass of the star, and $0.6\ M_{\odot}$ for the nucleus of a planetary nebula.

In our calculations of the evolution of model nuclei of planetary nebulae both shell sources were included in the interior which was evolved with the usual Henyey-type technique. As a result, thermal pulses in the helium shell source developed and gave rise to rather large loops on the H—R diagram, which can be seen in Fig. 4 for the 0.6 and $0.8\ M_{\odot}$ models. The $1.2\ M_{\odot}$ model has a loop which falls to the left of the H—R diagram shown in Fig. 4. It took the $0.6\ M_{\odot}$

model 10^3 years to go through the loop. This time was 10^2 years for the $0{\cdot}8\,M_\odot$ model, and 10 years for the $1{\cdot}2\,M_\odot$ model. The position of these loops on the H—R diagram depends on the time when the evolutionary calculations are switched into the ordinary Henyey-type technique, i. e., when the shell sources are included in the interior rather than in the envelope. The loops appear only when the mass of the hydrogen-rich envelope is sufficiently small. If the envelope is too massive, the star remains in the red giant region during the whole thermal pulse of the helium shell (Schwarzschild and Härm 1965).

The present calculations indicate that we may expect fairly rapid changes in those nuclei of planetary nebulae which happen to be seen while developing a thermal pulse in a helium shell source. It is interesting that at least one planetary nebula has a nucleus, FG Saggittae, which has changed very considerably over a period of the last 70 years (Herbig and Boyarchuk 1968).

If our model computations are correct we expect all stars with mass below about $3{\cdot}5\,M_\odot$ to lose their envelopes through dynamical instability, form a planetary nebula and a nucleus which will finally cool down to become a white dwarf. These stars will never burn carbon. The more massive stars never achieve a high enough L/M ratio to have dynamically unstable envelopes. Stars in the mass range $3{\cdot}5 - 7\,M_\odot$ or $3{\cdot}5 - 8\,M_\odot$ are expected to ignite carbon explosively when their central densities reach $3 \times 10^9\,\mathrm{g/cm^3}$. In this case the type of thermonuclear supernova suggested by Arnett would terminate stellar evolution. Because some details of our models are sufficiently uncertain, the possibility remains that the cores of such stars may become dynamically unstable and collapse before carbon ignition. This would lead to the formation of a neutron star. Stars with a mass above 8 or $9\,M_\odot$ ignite carbon nonviolently, and their final evolution is not too clear.

I am very grateful to Dr. A. N. Cox and Dr. J. N. Stewart for sending me their unpublished opacity tables, and to the National Bureau of Standards for providing me with a large amount of computing time on the University of Colorado CDC 6400 computer during my Visiting Fellowship at JILA.

REFERENCES

Arnett, W. D. . . . . 1968 *Nature*, **219**, 1344.
. . . . 1969 *Astroph. and Space Science*, **5**, 180.
Auman, J. Jr. . . . . 1967 *Ap. J. Suppl.*, **14**, 171.

Beaudet, G., Petrosian, V., and Salpeter, E. E. . . . . 1967 *Ap. J.*, **150**, 979.

Canuto, V. . . . . 1970 *Ap. J.*, **159**, 641.

Cox, A. N., and Stewart, J. N. 1969 *Astr. Council, Acad. of Sciences, U.S.S.R. Scientific Informations*, 15.

Eggleton, P. P. . . . . . 1967 *M. N. R. A. S.*, **135**, 243.

Fowler, W.A., Caughlan, G. R., and Zimmerman, B. A. . . 1967 *Ann. Rev. Astr. Ap.*, **5**, 525.

Herbig, G. H., and Boyarchuk, A. A. . . . 1968 *Ap. J.*, **153**, 397.

Hubbard, W. B., and Lampe, M. 1969 *Ap. J.*, **156**, 795.

Lucy, L. B. . . . . 1967 *A. J.*, **72**, 813.

Mihalas, D. . . . . 1967 *Methods in Computational Physics* (New York: Academic Press) Vol. **7**, p. 1.

O'Dell, C. R. . . . . 1968 *Planetary Nebulae, Proceedings of I.A.U. Symposium No. 34* (Dordrecht-Holland: D. Reidel Publ. Co.), p. 369.

Paczyński, B. . . . . . 1969*a* *Acta Astr.*, **19**, 1.

. . . . . 1969*b* *Bull. A. A. S.*, **1**, 256.

Paczyński, B., and Ziółkowski, J. . . . 1968*a* *Planetary Nebulae, Proceedings of I.A.U. Symposium No. 34* (Dordrecht-Holland: D. Reidel Publ. Co.), p. 396.

. . . 1968*b* *Acta Astr.*, **18**, 255.

Patterson, J. R., Winkler, H., and Zaidins, C. S. . . . . 1969 *Ap. J.*, **157**, 367.

Reeves, H. . . . . 1965 *Stellar Structure*, ed., L. H. Aller, and D. B. McLaughlin (Chicago: The University of Chicago Press), p. 156.

Rose, W. K., and Smith, R. L. . 1969 *Ap. J.*, **159**, 903.

Roxburgh, I. W. . . . . . . 1967 *Nature*, **215**, 838.

Selpeter, E. E., and Van Horn, H. M. . . . 1969 *Ap. J.*, **155**, 183.

Schwarzschild, M., and Härm, R. . . . . 1965 *Ap. J.*, **142**, 855.

Sugimoto, D. . . . . 1969 *preprint.*

Weigert, A. . . . . 1966 *Zs. f. Ap.*, **64**, 395.

Wheeler, J. C. . . . . 1969 University of Colorado: *thesis.*

Joint Institute for Laboratory Astrophysics
University of Colorado, Boulder, Colorado, USA
September 1969

B. Paczyński

Visiting Fellow, 1968-69
on leave from Institute of Astronomy
Polish Academy of Sciences, Warsaw

4

STRUCTURE OF MASSIVE MAIN-SEQUENCE STARS

R. Stothers

Institute for Space Studies
New York, N.Y.

I. INTRODUCTION

To define what is meant by a "massive star," it is useful to isolate the physical properties which differentiate stars on the upper main sequence from fainter stars on the lower main sequence. Thus, the following criteria are useful: (1) Population I chemical composition (because present Population II stars in the Galaxy are probably all less massive than $\sim 1.5\ M_\odot$); (2) CNO bi-cycle of nuclear energy generation, which replaces the CN cycle at high temperature; (3) significant contribution of radiation pressure to the equation of state; (4) dominance of electron scattering in the opacity; and (5) developing convective instability just outside the convective core in evolving stellar models. Subject to these criteria, the relevant mass range is, approximately,

$$10 < M/M_\odot < 10^5 .$$

Above $\sim 10^5\ M_\odot$, the effects of general relativity and of electron-positron pair annihilation become appreciable, besides which the mass begins to approach galactic dimensions. In the following, whenever asymptotic limits are evaluated as the mass goes to "infinity," we shall tacitly understand an idealized limit, where complicating effects such as the above effects are absent.

The key factor which determines the evolution of any star is the way in which nuclear fuel is depleted in the interior. Massive stars on the zero-age main sequence (ZAMS) have radiative envelopes and large convective cores; therefore the helium which is manufactured by hydrogen burning near the center becomes quickly mixed throughout a large portion of the star. It is unfortunate, however, that considerable uncertainty exists regarding the subsequent redistribution of this helium outside the core boundary due to the convective instability mentioned under criterion (5) above. For this reason, evolved models beyond the core hydrogen-burning phase are subject to rather large uncertainties, and it seems worthwhile to confine the present review to a theoretical discussion of the core hydrogen-burning models and the associated physical instabilities (ignoring rotation and mass loss). A review of topics omitted here and additional references may be found in an article in an earlier volume of this series (Stothers 1969).

II. BASIC TRENDS

The trend of physical properties of stars with increasing mass is interesting (Stothers, 1966). Quite generally, detailed models of zero-age main-sequence stars show that

$$R \sim M^a, \quad \text{where } a \simeq 1/2. \tag{1}$$

Therefore, the mean density behaves like $\bar{\rho} \sim MR^{-3} \sim M^{-\frac{1}{2}}$. As $M \to \infty$, it follows that $\bar{\rho} \to 0$ and $P \to P_{rad}$. Radiation pressure dominates in massive stars.

At low masses ($\sim 10\ M_\odot$), the mass-luminosity relation can be derived from a simple homology argument; thus

$$L \sim M^3 \quad \text{(low mass)}. \tag{2}$$

In the limit of the highest masses, the radiation pressure force must exactly balance gravity. The balance of forces should be evaluated at the surface of the star, where the flux emerges. The forces are

$$\frac{H}{\ell c} = -\frac{dP_{rad}}{dr} \tag{3}$$

and

$$\frac{GM(r)}{r^2}\rho = -\frac{dP}{dr} \simeq -\frac{dP_{rad}}{dr} \tag{4}$$

where the outward flux $H = L/4\pi r^2$ and the mean free path of a photon $\ell = (\kappa\rho)^{-1}$. Equating the forces at the surface, we obtain

$$L = (4\pi cG/\kappa)M \quad \text{(high mass)}. \tag{5}$$

For electron scattering, $\kappa = 0.2\ (1 + X)$, where X is the hydrogen abundance by mass.

The luminosity-temperature relation on the H-R diagram follows from equation (1) and

$$L \sim M^b, \tag{6}$$

where b = 3 (low mass) and b = 1 (high mass). Since the effective temperature of the surface is given by $\sigma T_e{}^4 = H$, we find

$$T_e \sim M^{\frac{1}{2}} \qquad \text{(low mass)} \tag{7}$$

$$T_e \sim \text{const} \qquad \text{(high mass)} \tag{8}$$

Consequently, the slope of the main sequence on the H-R diagram steepens as the luminosity brightens. This is shown in Figure 1, where we have adopted a zero-age chemical composition given by $X_e = 0.70$, $Z_e = 0.03$, $X_{CNO} = Z_e/2$.

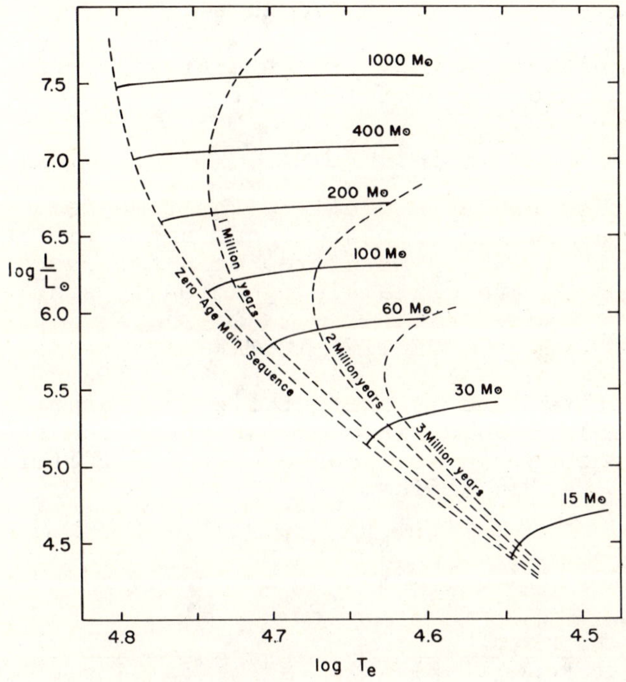

Figure 1. Theoretical H-R diagram for massive stars during core hydrogen burning. Semiconvection is treated according to the LSH prescription. Evolutionary tracks end at $X_c = 0.05$. Dashed curves are loci of constant time.

It may be noted that, for stars with solely electron-scattering opacity and with their energy source confined to the convective core, non-dimensional structural quantities are unchanged for the same value of $M\mu^2$, where μ is the mean molecular weight. Thus, the mass of the radiative envelope on the zero-age main sequence attains a maximum value when the total stellar mass is given by $M/M_{\odot} \approx 60\ \mu^{-2}$.

Three types of stellar instability will be considered in what follows: (1) dynamical, (2) convective, and (3) pulsational. A fourth type, secular instability, is manifested normally as the slow change wrought in the overall structure of the star as the result of gradual hydrogen depletion in the core. This depletion necessitates structural readjustments to accomodate the heating due to nuclear energy release and gravitational contraction in the core and the cooling due to expansion of the envelope. Homology arguments predict that the luminosity and radius will increase during evolution in accordance with (schematically) $L \sim \mu^4$ and $R \sim \mu$, since the average mean molecular weight increases with time.

III. DYNAMICAL INSTABILITY

The first generalized adiabatic exponent is defined as

$$\Gamma_1 = (d \ln P/d \ln \rho)_S \tag{9}$$

evaluated at constant entropy, S. A star is <u>dynamically unstable</u> (Ledoux and Walraven 1958) if

$$\int(\Gamma_1 - 4/3)\, PdV < 0 \ . \tag{10}$$

The dynamical time scale is equal to the free-fall time, $\tau_{ff} \simeq (G\rho)^{-\frac{1}{2}}$. For a mixture of perfect gas and radiation, Γ_1 is always greater than 4/3. Therefore, main-sequence stars of mass up to $\sim 10^5\ M_\odot$ are certainly dynamically stable.

IV. CONVECTIVE INSTABILITY

The convective core is characterized by an adiabatic temperature gradient which is shallower than the radiative temperature gradient, in accordance with the criterion of K. Schwarzschild. Let us define

$$\nabla = d \ln T/d \ln P \tag{11}$$

so that

$$\nabla_{rad} = \frac{\kappa L(r)}{16\pi cGMq(1-\beta)} \ , \qquad \nabla_{ad} = \frac{8-6\beta}{32-24\beta-3\beta^2} \ , \tag{12}$$

where $q = M(r)/M$ and $\beta = P_{gas}/(P_{gas} + P_{rad})$. Then inside the convective core $\nabla_{rad} > \nabla_{ad}$. The core boundary occurs at the point where

$$\nabla_{rad} = \nabla_{ad} \ . \tag{13}$$

This may be written as

$$\begin{aligned} q_f &= \frac{\kappa L}{4\pi cGM} \left(\frac{32-24\beta-3\beta^2}{32-56\beta+24\beta^2}\right)_f \\ &= \frac{\kappa L}{4\pi cGM} \left(1+\beta + \frac{29}{32}\beta^2+\ldots\right)_f \ . \end{aligned} \tag{14}$$

In massive stars, β is considerably less than unity at the core boundary. Since $L\sim M^{1-3}$, equation (14) indicates that q_f increases as the total mass increases. Comparison with equation (5) shows that q_f attains precisely unity in the limit of highest masses.

Consider now the depletion of hydrogen in the convective core. The resulting increase in average mean molecular weight of the star must cause an increase in luminosity according to homology arguments. The convective instability therefore grows outward in mass fraction (if β is small) as indicated by equation (14), for stars more massive than $\sim 10\ M_\odot$.

Fully developed convection, however, cannot occur in the unstable region outside the zero-age core boundary, as the following argument will show. According to the first of equations (12), the function ∇_{rad}/κ must be continuous at the convective-radiative interface, since the luminosity is generated entirely near the center of the star. Continuity of this function on the interior and exterior sides of the interface requires that

$$\left(\frac{\nabla_{rad}}{1+X}\right)_{int} = \left(\frac{\nabla_{rad}}{1+X}\right)_{ext} . \quad (15)$$

Since $X_{int} < X_{ext}$, equation (15) implies that $(\nabla_{rad})_{ext} > (\nabla_{rad})_{int}$ and hence that the exterior side is convective, contrary to our assumption. This contradiction demonstrates that the convective core must actually shrink in mass fraction.

The question then arises as to the nature of the convectively unstable, though chemically inhomogeneous, zone that is left behind by the retreating convective core. Slow mixing, on a time scale of the order of the Kelvin time, must ensue and redistribute the chemical composition until approximate convective neutrality is attained. This process has been called "semiconvection" by Schwarzschild and Härm (1958). They used the equality sign in the usual criterion for convective instability,

$$\nabla_{rad} \geq \nabla_{ad} \quad \text{(SH)} , \quad (16)$$

in order to evaluate the chemical composition in the unstable region. Ledoux (1947) and Sakashita and Hayashi (1961) argued that the μ-gradient should be included in the criterion, since the criterion is fundamentally based on a comparison of density gradients. Thus,

$$\nabla_{rad} \geq \nabla_{ad} + \frac{\beta}{4-3\beta} \frac{d \ln \mu}{d \ln P} \quad \text{(LSH)}. \quad (17)$$

In order to have continuity of the flux, a radiative zone must appear between the semiconvective zone and the convective core, according to the Sakashita-Hayashi prescription.

Figure 2 shows the hydrogen profile which results at the end of core hydrogen burning for each scheme, in a star of $60\ M_\odot$. There is little difference between the two profiles. This is reflected in the similar tracks on the H-R diagram (Figure 3), although the SH scheme produces slightly higher luminosities and a longer lifetime because of the larger hydrogen depletion.

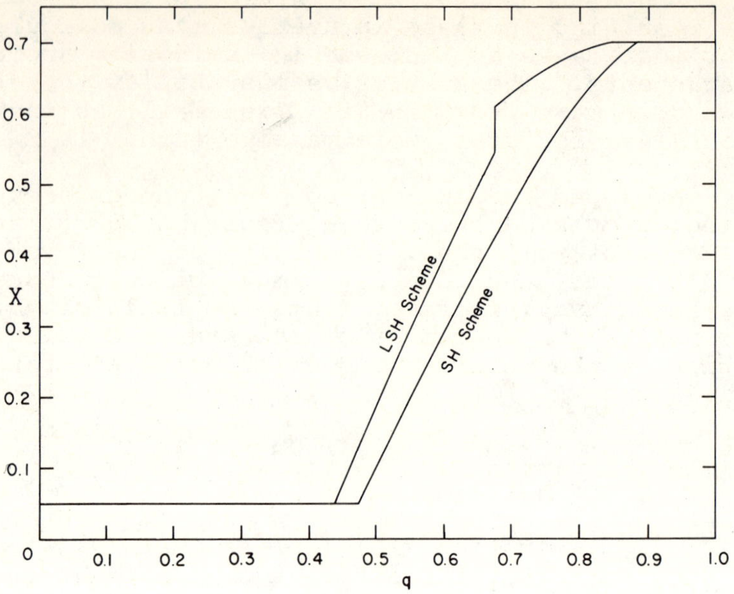

Figure 2. Distribution of hydrogen as a function of mass fraction at the end of core hydrogen burning in a star of 60 $M_{\odot}$. Two schemes of semiconvection are shown.

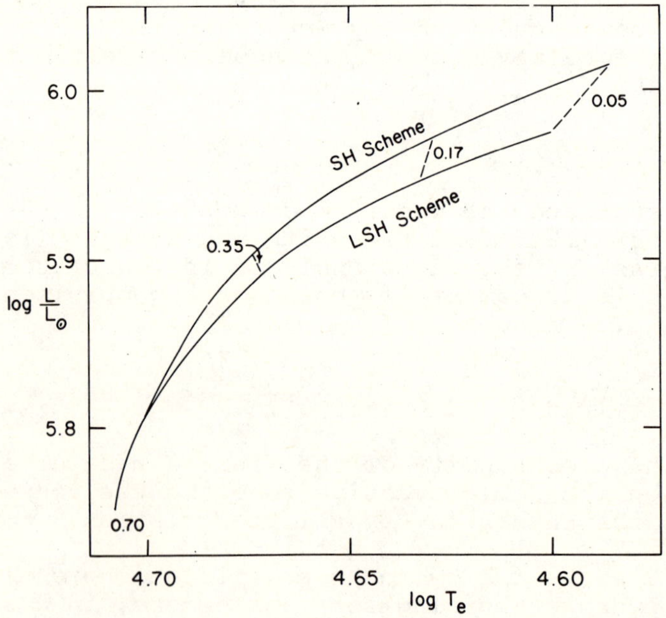

Figure 3. Theoretical H-R diagram for a star of 60 $M_{\odot}$, showing the evolutionary tracks for two schemes of semiconvection. Dashed lines connect points of equal central hydrogen abundance, X_c.

The final arrangement of the zones in stars of mass 15-1000 $M_\odot$ is shown in Figure 4. The zones are denoted: (I) radiative envelope, (II) LSH semiconvective zone, (III) LSH radiative zone, and (IV) convective core. To an excellent approximation, the combination of LSH zones II and III represents the SH semiconvective zone for the same masses. In the LSH scheme, zone II and zones II + IV have a maximum extent in mass fraction at $\sim$ 60 and $\sim$ 200 $M_\odot$, respectively.

Like q_f in the initial homogeneous models (dashed line in Figure 4) the final core boundary at the end of hydrogen burning also approaches an asymptotic limit for very high masses. This can be understood by taking the ratio of initial and final core masses as $M \to \infty$. Then

$$q_f = (1 + X_e)^{-1} . \tag{18}$$

At the present time, it is uncertain which scheme (SH or LSH) should be used in stellar models. Spiegel (1969) has marshalled some laboratory evidence in favor of the original SH scheme. In either case, it is fairly certain that the earliest stages of core hydrogen depletion are characterized by the SH scheme because of overshooting mass motions into the initially thin intermediate zone. As evolution proceeds, the semiconvective zone becomes detached from the shrinking core in all masses below $\sim$ 15 $M_\odot$. Above $\sim$ 15 $M_\odot$, the question of detachment is still open.

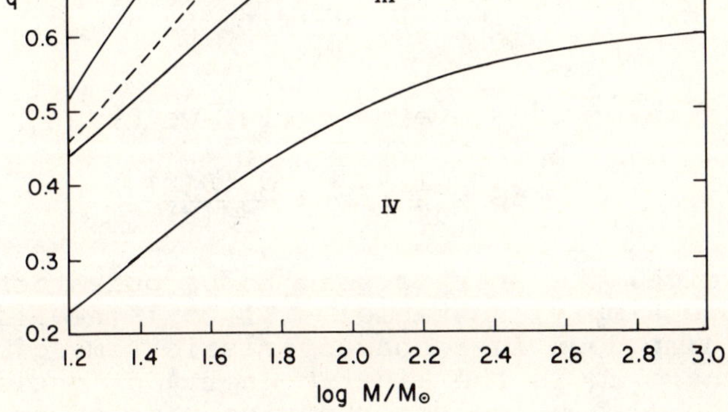

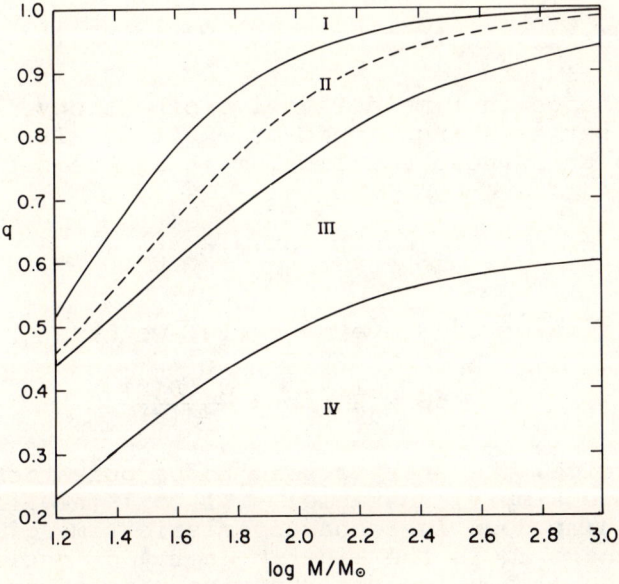

Figure 4. Mass fraction of the zonal interfaces in massive stars (as defined in the text) when X =0.05. Dashed curve denotes the mass fraction of the zero-age convective core. From R. Stothers, Ap.J., 144, 959 (1966)

It is interesting that the lifetime of core hydrogen burning approaches a finite limit, expressed as a combination of fundamental constants, as $M \to \infty$. The lifetime is evaluated from the integral of the rate of hydrogen depletion,

$$L = - E\, d(\bar{X}M)/dt, \tag{19}$$

where $\bar{X} = \int X dq$ and $E = 6 \times 10^{18}$ erg g^{-1} (representing the energy release due to conversion of hydrogen into helium). Since the rate of mass dissipation, $-dM/dt = L/c^2$, due to the radiant energy loss from the surface, is very small ($E/c^2 = 0.007$), we have

$$\begin{aligned} \tau &= \int \frac{EM}{L}\left(1 - \frac{\bar{X}E}{c^2}\right)^{-1} d\bar{X} \to \frac{E\sigma_e}{8\pi c G m_p} X_e (1 + \tfrac{1}{2}X_e) \\ &= 1.4 \times 10^6\, X_e (1 + \tfrac{1}{2}X_e) \text{ yr}, \end{aligned} \tag{20}$$

where σ_e is the Thomson scattering cross-section for free electrons. This formula is valid whether the star evolves inhomogeneously or completely mixed, because of the compensating factors in $\bar{X}$ and κ. It represents a lower limit to the lifetime of any main-sequence star because τ decreases with increasing mass according to equation (19) with $L \sim M^{1-3}$.

V. PULSATIONAL INSTABILITY

The secular stability of a star against growing radial oscillations is measured by the net rate of gain or loss of pulsational energy over a cycle (Ledoux and Walraven 1958):

$$L_p^* = \int_0^M \delta T\, \delta\dot{S}\, dM(r). \tag{21}$$

The secular change in time derivative of entropy, $\delta\dot{S}$, is not zero if the star is being damped or excited, although $\dot{S}$ itself is zero for a star whose equilibrium is stabilized by nuclear reactions:

$$T\dot{S} = \varepsilon - \frac{dL(r)}{dM(r)} = 0 \ . \tag{22}$$

Taking the variation of $\dot{S}$ over a cycle, we have

$$\delta\dot{S} = \frac{1}{T}\left[\delta\varepsilon - \frac{d\ \delta L(r)}{dM(r)}\right] . \tag{23}$$

If L_p^* is positive, the star is said to be pulsationally overstable (or more simply, unstable). If L_p^* is negative, the oscillations damp down in a characteristic time given by $\tau = L_p^*/4E_p$, where E_p is the kinetic energy of oscillation. This expression also holds for the rise time of overstability.

In the so-called quasiadiabatic approximation, one evaluates the pulsational period and pulsational amplitudes by assuming small sinusoidal oscillations and linearizing the equations of motion for the star. The motions are assumed to take place adiabatically (over a cycle) because the pulsational period is far shorter than the Kelvin or photodiffusion time. The pulsational amplitudes thus derived are then used to evaluate $\delta\varepsilon$ and $d\,\delta L(r)/d\,M(r)$.

Nuclear-energized pulsation can arise if the positive contribution from $\delta\varepsilon$ (core) exceeds the negative contribution from $d\,\delta L(r)/dM(r)$ (envelope). The former contribution is evaluated from the nuclear energy-generation rate in the form $\varepsilon = \varepsilon_0 \rho^\lambda T^\nu$ whence

$$\frac{\delta\varepsilon}{\varepsilon} = \lambda \frac{\delta\rho}{\rho} + \nu \frac{\delta T}{T} . \tag{24}$$

Large positive temperature and density exponents in the rate enhance destabilization because then maximum nuclear-energy release occurs at maximum compression, and the stellar core acts like an efficient heat engine. For the CNO bi-cycle, $\lambda = 1$ and $\nu \simeq 15$ while for the triple-alpha reaction, $\lambda = 2$ and $\nu \simeq 18$. These temperature exponents are very high. The contribution from $d\delta L(r)/dM(r)$ is always a damping term in massive main-sequence stars, because the ionization zones of hydrogen and helium lie essentially in the atmosphere and the opacity is due chiefly to electron scattering:

$$\frac{\delta L(r)}{L(r)} = 4 \frac{\delta r}{r} - \alpha \frac{\delta\rho}{\rho} + (4+\eta)\frac{\delta T}{T} + T\frac{d}{dT}\left(\frac{\delta T}{T}\right) \tag{25}$$

for an opacity law of the form $\kappa = \kappa_0 \rho^\alpha T^{-\eta}$. Electron scattering is characterized by $\alpha = 0$ and $\eta = 0$. In order for nuclear sources to overcome the heat leakage in the outer layers of the envelope, the relative amplitudes of pulsation in the core must be large--at least 40 per cent of the surface amplitude in most cases. We can imagine that small oscillations will be set in motion by convective turbulence in the core, but their growth or decay will depend essentially on the run of the amplitudes as a function of position in the star.

To gain insight into the physical mechanism of nuclear-energized pulsation, it will be useful to discuss how large amplitudes in the core can be attained. Consider the limiting case of a <u>homologous displacement</u> of each mass shell, $r' = ar$ ($\delta r/r = \text{const}$), where <u>a</u> is a fixed scale factor. Integrating the equation of hydrostatic equilibrium, we have

$$P = \int_{M(r)}^{M} GM(r)\, dM(r)/4\pi r^4 . \tag{26}$$

The new pressure in the displaced star (required also to be in hydrostatic equilibrium) is

$$P'/P = (r'/r)^{-4} . \tag{27}$$

Similarly, from the equation of mass distribution,

$$\rho = dM(r)/4\pi r^2 dr \ , \tag{28}$$

we find

$$\rho'/\rho = (r'/r)^{-3} \ . \tag{29}$$

Combination of equations (27) and (29) then yields

$$P'/P = (\rho'/\rho)^{4/3} \ . \tag{30}$$

Hence it is sufficient that $\Gamma_1 = 4/3$ throughout the star in order to achieve homologous radial displacements. Dynamical instability ($\overline{\Gamma}_1 \leq 4/3$) arises through the fundamental mode of radial pulsation, and therefore may be considered as a degenerate case of radial pulsation.

It is simple to show from the virial theorem (Chandrasekhar 1939) that a star with $\Gamma_1 = 4/3$ throughout has zero total energy and is therefore essentially unbound (dynamically unstable). The period of oscillation in infinite, or the angular frequency is $\sigma = 2\pi/\mathrm{Per} = 0$. Hence small Γ_1 and small σ favor pulsational instability.

Realistic models of stars must be approached in a more sophisticated way. Let us define a non-dimensional eigenfrequency of the fundamental mode

$$\omega^2 = \sigma^2 R^3/GM \ . \tag{31}$$

It can be shown from the virial theorem (Ledoux and Walraven 1958) that ω^2 must lie between the two extremes given by

$$(3\overline{\Gamma}_1 - 4) \leq \omega^2 \leq (3\overline{\Gamma}_1 - 4)J \ , \tag{32}$$

where

$$J = \int \frac{q}{x}\, dq / \int x^2 dq \tag{33}$$

and $x = r/R$. The quantity J is a non-dimensional ratio of gravitational potential energy to moment of inertia about the center. The left-hand member of inequality (32) is simply the pulsational eigenfrequency of a star of uniform density, which has J = 1 and, like a star with $\Gamma_1 = 4/3$ throughout, $\delta r/r$ = const.

In general, low central condensation of a star favors pulsational instability because J is roughly proportional to $\rho_c/\overline{\rho}$ and therefore ω^2 will be small. Table 1 presents some relevant properties of polytropic models covering the whole range of central condensation (polytropic index $\underline{n}$). In each of these models, the central value of $\delta r/r$ depends not only on $\rho_c/\overline{\rho}$ but also on $\overline{\Gamma}_1$, with smaller values of $\overline{\Gamma}_1$ being re-

TABLE 1

Polytropic Quantities

n	$\rho_c/\bar{\rho}$	ω^2 ($\Gamma_1 = 5/3$)	ZAMS Example	
			$M/M_\odot$	$\bar{\Gamma}_1$
0	1	1.0	....	....
1.5	6	2.7	0	5/3
2.5	23	6.1	100	1.4
3	54	9.3	1 and ∞	5/3 and 4/3
5	∞	∞	....	....

quired to attain the same central value of $\delta r/r$ as $\rho_c/\bar{\rho}$ increases. This requirement is apparent from an examination of the approximate gradient of $\delta r/r$ at the surface

$$\frac{d}{dx}\left(\frac{\delta r}{r}\right) \approx (3 - 4/\bar{\Gamma}_1)(J - 1)\left(\frac{\delta R}{R}\right). \qquad (34)$$

Figure 5 shows several runs of pulsation amplitudes. In massive main-sequence stars, ω^2 is small because of the high radiation pressure (small Γ_1) and the relatively low central condensation.

A large number of detailed stellar models in the mass range 6-1000 $M_\odot$ has been calculated by N. R. Simon and the author, for every extreme combination of hydrogen and helium abundance in the core, envelope, and intermediate zone (Simon and Stothers 1969). Empirically, for the models which have $L_p^* = 0$ (called "critical models"), it is found that the critical eigenfrequency is very narrowly bounded within the range $\omega^2 = 2$-4, corresponding to a range of central condensation $\rho_c/\bar{\rho} = 60$-10. Thus, according to equation (32), radiation pressure "overcompensates" for a high central condensation. Since the right-hand member of inequality (32) can be evaluated from the equilibrium model alone, it is convenient to denote it as ω^2_{max} and to establish its critical range: $\omega^2_{max} = 2.5 - 4.5$. These ranges are general because, like the equilibrium equations, the pulsation equations can be cast into non-dimensional form for massive stars, so that $M\mu^2$ is the defining free parameter.

For chemically homogeneous stars with solely electron-scattering opacity and the CNO bi-cycle of nuclear energy generation, it is found (Schwarzschild and Härm 1959) that the critical eigenfrequency and the critical mass are given by, respectively, $\omega^2 = 3$ and

$$M/M_\odot = 21\ \mu^{-2}. \qquad (35)$$

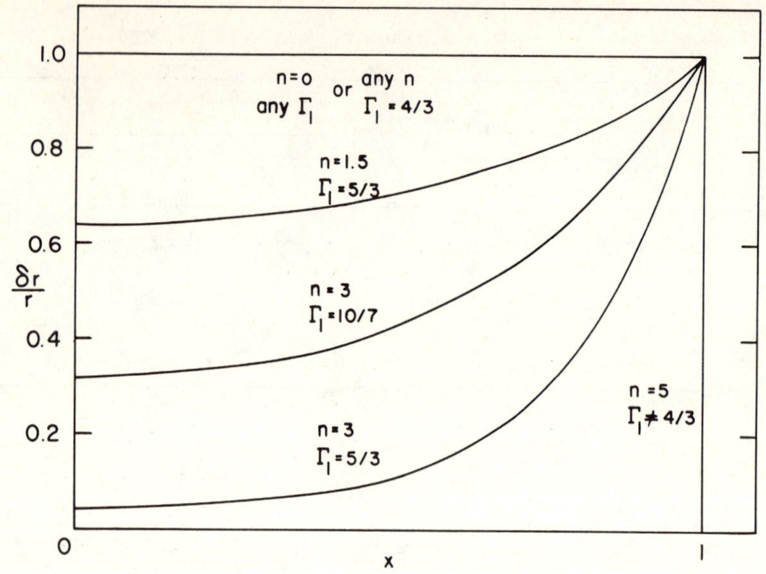

Figure 5. Relative radius amplitude of radial pulsation in the fundamental mode as a function of radius fraction, for polytropes of index n with constant Γ_1. Normalization of $\delta r/r$ at the surface is arbitrary.

For a hydrogen abundance of X = 0.7, the critical mass is $\sim 60\ M_\odot$. During evolution off the zero-age main-sequence, the critical mass increases because the stabilizing effect of growing central condensation must be offset by higher radiation pressure. The critical mass of a pure helium star burning via the triple-alpha process is $\sim 8\ M_\odot$. Models of stars burning core hydrogen or helium and having thin hydrogen-poor envelopes may represent the Wolf-Rayet stars. Hydrogen-burning models with a helium overabundance which is assumed to have been accreted from a stellar companion may represent the β Cephei stars.

It should be noted as a final remark that observed stellar masses range certainly up to 60 $M_\odot$ and, less certainly, up to several hundred solar masses.

REFERENCES

Chandrasekhar, S. 1939, An Introduction to the Study of Stellar Structure (Chicago: University of Chicago Press).

Ledoux, P. 1947, Astrophys. J., 105, 305.

Ledoux, P., and Walraven, Th. 1958, Handbuch der Physik, ed. S. Fluegge, Vol. 51 (Berlin: Springer-Verlag), p. 353.

Sakashita, S., and Hayashi, C. 1961, Prog. Theoret. Phys. (Kyoto), 26, 942.

Schwarzschild, M., and Härm, R. 1958, Astrophys. J., 128, 348.

———. 1959, ibid., 129, 637.

Simon, N. R., and Stothers, R. 1969, Astrophys. J., 156, 377, and subsequent papers.

Spiegel, E. A. 1969, Comments on Astrophysics and Space Physics, 1, 57.

Stothers, R. 1966, Astrophys. J., 144, 959, and earlier papers.

———. 1969, Stellar Astronomy, ed. H. Y. Chiu, J. Remo, and R. Warasila, Vol. 2 (New York: Gordon and Breach), p. 183.

5

STELLAR STABILITY AND STELLAR PULSATION

N. Baker

Columbia University
New York, N.Y.

References

LL Landau, L.D. and Lifschitz, E.M. 1959. Fluid Mechanics (Course of Theoretical Physics, Vol. 6) (Reading, Mass.: Addison-Wesley).

L58 Ledoux, P. 1958. "Stellar Stability" in Handbuch der Physik, Vol. 51 (Berlin-Göttingen-Heidelberg: Springer), pp. 605-688.

LW Ledoux, P. and Walraven, Th. 1958. "Variable Stars" in Handbuch der Physik, Vol. 51 (Berlin-Göttingen-Heidelberg: Springer), pp. 353-604.

L63 Ledoux, P. 1963. "Stellar Stability and Stellar Evolution" in Star Evolution (28th course in International School of Physics "Enrico Fermi", Aug. 1962, Varenna), (New York and London: Academic Press), pp. 394-445.

These basic references will be referred to by the code letters to the left. Readers not familiar with the essentials of fluid mechanics should read parts of chapters I, II, and V in LL. LW is the definitive reference for the essentials of pulsation theory, at least for the linear theory. Most other aspects of stability theory are covered in L58. Although there have been many developments in the last decade, LW and L58 together provide the essentials. L63 is a good, simple introduction to the subject and can be taken as the "textbook" for these lectures. In part we follow closely the development in L63.

Scope of these lectures

Our ambitions are necessarily very limited. In the available time one can give only the most essential introduction to the subject. We shall attempt the following:

1. A definition of the problem to be solved and its general structure.
2. Elaboration of some very simple examples and some of the physical mechanisms involved.
3. A survey of a few of the problems that have been solved or at least attempted. An indication of where these problems, especially those to be discussed by other lecturers, fit into the general framework.

Introduction

In stability theory one normally asks about the stability of an equilibrium configuration against an arbitrary small perturbation. But a star is never strictly in equilibrium, for it evolves, so we have really to consider the stability of a time-dependent, non-steady, process. But for the most part we consider stellar evolution to be a succession of quasi-equilibrium states. Then we can do stability analysis in the usual way, provided that we consider processes which would take place on a time scale much shorter than that of the stellar evolution. A star of luminosity L evolving on a nuclear time scale takes a time $T_N \simeq E_N/L$, where E_N is the total energy available to whatever nuclear process may be going on, to change its composition and hence its structure, significantly. If these sources of nuclear energy were not available, the star would evolve on a gravitational time scale $T_G \simeq |\Omega|/L$ where $\Omega \simeq - GM^2/R$ is the gravitational energy. T_G is really a mmeasure of the time it takes thermal energy to diffuse out of the star--it depends on the nature of the transport processes. The dynamical time scale $T_D \simeq (R/g)^{1/2} \simeq (R^3/GM)^{1/2}$ governs hydrodynamical flows occurring in the absence of strict pressure equilibrium.

Usually, $T_N >> T_G >> T_D$, but this need not be true. If it is true, conventional stability theory should be valid for processes occurring on a time scale T_G or T_D, provided evolution takes place on T_N.

T_G is the time associated with pre-main-sequence evolution, with evolution across the Hertzsprung gap, and with the growth or damping time of pulsations. Pulsation or dynamical collapse takes place in a time $\simeq T_D$.

Stability theory is done much as in plasma physics, except that the confinement here is gravitational, not magnetic. (Magnetic fields may be important in some cases, but these are beyond our scope.) One complication is that "scale effects" are usually important. Scale heights differ greatly from the center of a star to the surface, and usually the scale height is not large compared to the scale of the perturbation ("wave length"). For this reason local analysis (e.g., dispersion relations) is not usually very useful. Apart from this, the main complications are typically compressibility effects and dissipative ones.

Basic equations of stellar hydrodynamics.

Continuity:

$$\frac{d\rho}{dt} + \rho \text{ div } \vec{v} = 0 \tag{1}$$

where $\frac{d}{dt}$ is the "substantial" or Lagrangian time derivative.

Momentum conservation:

$$\rho \frac{d\vec{v}}{dt} = - \text{grad } p - \rho \text{ grad } \Phi + \eta \nabla^2\vec{v} + (\zeta+\tfrac{1}{3}\eta)\text{grad div } \vec{v} \tag{2}$$

where Φ is the gravitational potential, η the usual dynamic viscosity coefficient, ζ is the "second" viscosity, and

$$\nabla^2\Phi = 4\pi G\rho \tag{3}$$

Energy conservation:

$$\rho T \frac{ds}{dt} = \sigma_{ik} \frac{\partial v_i}{\partial x_k} + \rho\varepsilon_N - \text{div } \vec{F} \tag{4}$$

where the summation convention is employed, ε_N is the amount of nuclear energy converted to thermal energy per unit time per unit mass, $\vec{F}$ is the total energy flux density, and

$$\sigma_{ik} = \eta\left(\frac{\partial v_i}{\partial x_k} + \frac{\partial v_k}{\partial x_i} - \frac{2}{3}\delta_{ik}\frac{\partial v_\ell}{\partial x_\ell}\right) + \zeta\delta_{ik}\frac{\partial v_\ell}{\partial x_\ell}. \tag{5}$$

Energy transport:

$$\vec{F} = - K \text{ grad } T. \tag{6}$$

Notes: In the momentum equation we include the "second viscosity", important only in compressible fluids. This contribution might be expected to be at least as important as the usual viscosity. In the energy equation the nuclear energy is treated as a special input term. But in some extreme conditions, nuclear processes are in equilibrium and ε_N depends on the time derivatives of the state variables, and would properly be included on the left-hand side. The transport equation will have the form given regardless of the nature of the transport process (provided a local description is valid). For radiative diffusion, $K = \frac{4acT^3}{3\bar{\kappa}\rho}$, but sometimes conduction (in electron-degenerate regions) or turbulent convection may be the dominant process.

Viscous Effects: If the energy equation is made dimensionless, the dimensionless quantity which enters is the Prandtl number, $\sigma = \nu/\chi$, where $\nu = \eta/\rho$ is the kinematic viscosity and $\chi = \kappa/\rho C_p$ is the thermometric conductivity. For molecular viscosity

$$\nu \simeq (\text{thermal velocity}) \times (\text{molecular mean free path})$$

and for radiative transport,

$$\chi \simeq \text{(speed of light) x (mean free path of photons)}$$

in the absence of very dominant radiation pressure. Thus $\sigma << 1$ everywhere in most stars, even near the surface, where it is greatest (since the opacity is normally greatest there). (Near the surface of the sun, $\sigma \simeq 10^{-8}$.) Radiative viscosity is usually even less important (unless $P_{gas} << P_{rad}$). "Turbulent" or "eddy" viscosity (Reynolds stresses due to small-scale turbulence) may be important, e.g., in pulsating stars with deep convective zones, but this sort of treatment is probably not adequate there.

The Prandtl number should be a good measure of relative size of viscous and thermal effects, provided velocity fluctuations are of the same order as thermal ones, which is usually the case. To be sure, one should evaluate the viscous effects for a given flow. For the lower modes of radial pulsation (LW p. 507), the time scale for (molecular) viscous effects has been estimated at 10^{10} - 10^{13} years, much too long to have any effect. Thus viscous effects can usually be entirely neglected, and we shall do so in what follows.

Equations of state

The nature of the equation of state is often crucial to questions of stability. A well-known example is found in the criterion for instability against thermal convection in a compressible fluid. It is necessary that the actual temperature gradient (typically controlled in a star by the opacity) be greater than that which would exist under adiabatic conditions. The latter, however, depends strongly on the equation of state and when the specific heat of the gas is large (e.g., due to ionization) the adiabatic temperature gradient is depressed, and convection takes place more easily (i.e., the "radiative gradient" need not be so great.) Indeed we shall find that stability which depends on pressure support is always more easily destroyed in a region of high specific heat, because there are internal degrees of freedom which can absorb any energy input; this energy will not then be available to increase the pressure.

It is convenient, and has become quite common in stellar structure work, to describe the equation of state by the "adiabatic exponents":

$$\Gamma_1 = \frac{\rho}{p}\left(\frac{\partial p}{\partial \rho}\right)_s , \quad \frac{\Gamma_2 - 1}{\Gamma_2} = \frac{p}{T}\left(\frac{\partial T}{\partial p}\right)_s , \quad \Gamma_3 - 1 = \frac{\rho}{T}\left(\frac{\partial T}{\partial \rho}\right)_s \qquad (7)$$

where the subscript "s" indicates that the partial derivative is to be taken for an isentropic process. (Only two of the Γ's are independent and they do not entirely replace the specific heats and bulk moduli more conventionally used to define the equation of state.) With this notation the left-hand side of the energy equation may be written

$$\rho T\frac{ds}{dt} = \frac{1}{(\Gamma_3-1)}\left[\frac{dp}{dt} - \frac{\Gamma_1 p}{\rho}\frac{d\rho}{dt}\right] \tag{8}$$

A discussion of the various forms of the equation of state would lead us too far afield. We note, however, that all of the Γ_i are equal to 5/3 for a perfect monatonic gas. For a mixture of such a gas and radiation, the Γ_i lie between 4/3 and 5/3, while for a pure photon gas all the Γ_i are 4/3. In a fully nonrelativistic degenerate gas, $\Gamma_1 = 5/3$; if the gas is relativistic and fully degenerate, $\Gamma_1 = 4/3$. When nuclear processes are in equilibrium, these too can strongly affect the Γ's and at very high densities relativistic effects on the equation of state may influence the Γ's. The effect of a magnetic field providing pressure support is very similar to the effect of radiation pressure.

Linearized equations

Various methods are available for the study of stability and indeed to some extent the definition of stability may depend on the formulation used. It is possible to attempt to generalize the requirement (for a conservative mechanical system) that the potential energy be an absolute minimum. In this connection see L58. A more more commonly used technique is to hypothesize small perturbations about an assumed equilibrium, and see whether or not they will grow.

If the perturbations are sufficiently small, the problem is thus a linear one (although with variable coefficients), and hence much easier to solve. Clearly this only solves the beginning of the problem--it doesn't tell us how the system develops in time after the first perturbation, and it doesn't detect an instability that needs a finite perturbation to set it off. Both of these problems require a nonlinear approach.

We suppose that a dependent variable, say f, is given in equilibrium by $f_o(\underline{r})$. We shall write the equations in the Eulerian form, recalling that $df/dt = \partial f/\partial t + \underline{v}\cdot\nabla f$. The function f_o is then perturbed thus:

$$f(\underline{r},t) = f_o(\underline{r}) + f'(\underline{r},t) \tag{9}$$

and the Eulerian perturbation f' is related to the Lagrangian perturbation δf by

$$\delta f = f' + \underline{\delta r}\cdot\nabla f \tag{10}$$

(Henceforth we drop the subscript o on equilibrium quantities).

The equilibrium equations (the usual problem of stellar structure) are

$$\begin{aligned} \underline{v} &= 0 \\ \text{grad}\, p &= -\rho\ \text{grad}\,\Phi \\ \nabla^2\Phi &= 4\pi G\rho \end{aligned} \tag{11}$$

$$\operatorname{div} \underline{F} = \rho\varepsilon_N \qquad (12)$$
$$\underline{F} = -K \operatorname{grad} T .$$

(We do not consider rotating stars and other equilibrium models in which $\underline{v} \neq 0$.)

The perturbed equations are obtained by taking the first variation of the general equations and then subtracting the equilibrium equations. The velocity,

$$\underline{v} = \frac{d}{dt} \underline{\delta r} \qquad (13)$$

is taken to be small, as are all the perturbations f'. The first order equations are then

$$\frac{\partial\rho'}{\partial t} + \nabla\cdot(\rho\underline{v}) = 0$$

$$\frac{\partial\underline{v}}{\partial t} = - \nabla\Phi' + \frac{\rho'}{\rho^2}\nabla p - \frac{1}{\rho}\nabla P'$$

$$\nabla^2\Phi' = 4\pi G\rho'$$

$$\frac{\partial P'}{\partial t} + \underline{v} \cdot \operatorname{grad} p - \frac{\Gamma_1 p}{\rho}\left(\frac{\partial\rho'}{\partial t} + \underline{v} \cdot \operatorname{grad} \rho\right) \qquad (14)$$
$$= (\Gamma_3-1)\rho\left(\varepsilon_N - \frac{1}{\rho}\nabla\cdot\underline{F}\right)'$$

$$F' = - K' \operatorname{grad} T - K \operatorname{grad} T' .$$

On the right-hand side of the last equation appears the variation of a quantity which is identically zero in equilibrium. Hence its gradient also vanishes (in zero order) and we may interchange the Eulerian and Langrangian variations. It is also convenient to define $L(r) = 4\pi r^2 F(r)$, and $m = \int_0^r 4\pi r^2 \rho dr$.

Then we write

$$(\varepsilon_N - \frac{1}{\rho}\nabla\cdot F)' = \delta\varepsilon_N - \frac{d}{dm} \delta L . \qquad (15)$$

The equation for Φ can be integrated once by first differentiating with respect to time. If the space derivatives of ρ', p' and Φ' are then eliminated, one obtains a third order equation (in time).

$$\frac{\partial^3}{\partial t^3} \delta r = \frac{1}{\rho}\frac{\partial}{\partial r}\left[\frac{\Gamma_1 p}{r^2}\frac{\partial}{\partial r}\left(r^2\frac{\partial\delta r}{\partial t}\right)\right]$$
$$- \frac{4}{\rho r}\frac{\partial p}{\partial r}\frac{\partial}{\partial t}\delta r - \frac{1}{\rho}\frac{\partial}{\partial r}\left[(\Gamma_3-1)\rho\left(\delta\varepsilon_N - \frac{d}{dm}\delta L\right)\right]. \qquad (16)$$

Assuming that we can separate the time (which does not appear explicitly) we write

$$\frac{\delta r}{r} = \xi(r)e^{st} \tag{17}$$

Equation (16) becomes

$$s^3 r\xi = \frac{s}{\rho}\frac{1}{r^3}\left\{\frac{d}{dr}\left(\Gamma_1 p r^4 \frac{d\xi}{dr}\right) + r^3\xi\,\frac{d}{dr}[(3\Gamma_1-4)p]\right\}$$
$$-\frac{1}{\rho}\frac{d}{dr}\left[(\Gamma_3-1)\rho\{\delta\varepsilon_N - \frac{d}{dm}\delta L\}\right] \tag{18}$$

This is to be integrated, subject to the boundary conditions:

$$\delta r = 0 \quad \text{at } r = 0$$
$$\delta p = \delta p(\xi,\frac{d\xi}{dr}) = 0 \quad \text{at } r = R. \tag{19}$$

The analysis to this point is standard. In what follows we use the approach of L63, which is most helpful for a first understanding of the problem. Wherever possible we follow the notation used there, so that further details can easily be found. (Note that there are many typographical errors in the equations in L63.)

In general ξ and s are both complex, though often their imaginary parts are small. We multiply the last equation by $4\pi r^3\rho\,\xi^*\,dr$ and integrate from r=0 to r=R. Following L63 we write the result as

$$s^3 + s(A-B) + C = 0 \tag{20}$$

where

$$A = \frac{1}{J}\int_0^R 4\pi\Gamma_1 p r^4 \left|\frac{d\xi}{dr}\right|^2 dr$$

(21)

$$B = \frac{1}{J}\int_0^R |\xi|^2 \frac{d}{dr}[(3\Gamma_1-4)p]\,4\pi r^3 dr$$

$$C = \frac{1}{J}\int_0^R 4\pi r^3\,\xi^*\,\frac{d}{dr}[(\Gamma_3-1)\rho\{\delta\varepsilon_N - \frac{d\delta L}{dm}\}dr$$

$$J = \int_0^M |\xi|^2 r^2 dm\,.$$

A, B, and J are real. C may be complex: $C = C' + iC''$. Usually $|C'|, |C''| \ll |A|, |B|$ and $|C''| \ll |C'|$. We now look for a solution of the form

$$(s^2 + 2\,Ps + Q)\,(s-s_1) = 0 \tag{22}$$

Equating coefficients of like powers of s, one finds

$$s_1 = -C/Q$$

$$P = s_1/2 = -\ C/2Q \tag{23}$$

$$Q - 2Ps_1 = Q - s_1^{\ 2} = A - B$$

The second and third roots are thus

$$s_{\substack{2\\3}} = -\ P \pm (P^2 - Q)^{\frac{1}{2}} \tag{24}$$

The analysis of these three roots gives a preliminary, simplified look at some of the problems of stellar stability.

Dynamical Stability

Suppose that hydrodynamical changes occur so fast that there is no time for the establishment of thermal equilibrium. In other words, $ds/dt \simeq 0$ and the gas contracts or expands adiabatically.

Then

$$|C| << |A\text{-}B|, \quad P^2 << Q$$

$$s_1 \simeq 0$$

$$s_{\substack{2\\3}} = \pm\ i\ \sqrt{Q} = \pm\ i(A\text{-}B)^{\frac{1}{2}} \tag{25}$$

where A and B must be evaluated with ξ_a corresponding to an adiabatic solution. Write

$$s_{\substack{2\\3}}^{\ 2} = s_a^{\ 2} = -\sigma_a^{\ 2} \tag{26}$$

where σ_a will be a (real) frequency. Then

$$\sigma_a^{\ 2} = \frac{1}{J_a}\left\{\int_0^R 4\pi\Gamma_1 pr^4\left(\frac{d\xi_a}{dr}\right)^2 dr\right.$$

$$\left. - \int_0^R \xi_a^{\ 2}\ \frac{d}{dr}[(3\Gamma_1-4)p]\,4\pi r^3 dr\right\}. \tag{27}$$

This is a well-known result, and also can be obtained from a variational principle.

To solve this, one must of course know ξ_a (as well as the equilibrium quantities) as a function of r. For the fundamental mode of pulsation the second term is normally much greater than the first, and if we assume ξ_a = constant,

$$\sigma_a^2 = \frac{-\int_0^R \frac{d}{dr}[(3\Gamma_1-4)p]4\pi r^3 dr}{\int_0^M r^2 dm} \tag{28}$$

$$= \langle 3\Gamma_1 - 4\rangle \frac{\int_0^M \frac{Gm}{r} dm}{\int_0^M r^2 \, dm}$$

$$= \langle 3\Gamma_1 - 4\rangle \frac{|\Omega|}{I}$$

where Ω is the gravitational potential energy and I the moment of inertia. If $\sigma^2 > 0$, we have undamped pulsations of the star. If $\sigma^2 < 0$, then s_2 and s_3 are both real and one of them is positive. In general, any perturbation will have a component which increases exponentially in time, and is thus unstable. Thus the criterion for dynamical stability is

$$\langle 3\Gamma_1 - 4\rangle > 0. \tag{29}$$

Of course the mean must be taken in the sense of the above equation, but in general if $\Gamma_1 < 4/3$ in a large part of a star there will be dynamical instability.

To see what the time scale is, we put Ω and I in dimensionless form. If we write

$$r = xR, \quad m = qM$$

we find

$$\sigma_a^2 = \langle 3\Gamma_1 - 4\rangle \frac{GM}{R^3} \frac{\int_0^1 \frac{q}{x} dq}{\int_0^1 x^2 dq}$$

$$= \langle 3\Gamma_1 - 4\rangle \frac{4}{3}\pi G\bar{\rho} \frac{\int_0^1 \frac{q}{x} dq}{\int_0^1 x^2 dq} \tag{30}$$

Hence the <u>dynamical time scale</u> is

$$\frac{1}{\sigma_a} \simeq \left(\frac{3}{4\pi G\bar{\rho}}\right)^{1/2}$$

a time which is of the order of one hour for the sun. For cepheids it is the order of days, for RR Lyrae stars, tenths of a day, for white dwarfs, seconds to one minute, and for neutron stars, milliseconds.

The integrals in this equation are invariant to homology transformations, so the dimensionless frequency $\left(\frac{\sigma^2}{G\rho}\right)^{\frac{1}{2}}$ is the same for homologous stars. In fact this result is more general because even the exact second-order equation for ξ_a is invariant to homology transformations. Thus in stars which are homologous to one another, the period of a given mode of radial adiabatic pulsation is inversely proportional to the square root of the mean density (period-mean density relation).

If a star is dynamically stable it will pulsate if given a small perturbation. The question of whether or not it will continue to do so, i.e., whether the oscillations are damped or will grow, depends on the dissipative (non-adiabatic) terms and requires a discussion of at least first-order terms in C. This is the problem of pulsational or vibrational stability and will be discussed later.

Examples of dynamical instability

A number of cases have been studied in detail. The crucial quantity is of course Γ_1, which can be reduced below 4/3 by various processes.

1. Very luminous, very red stars. When the convection and ionization zones occupy a considerable part of a star's mass, as is the case for extreme red giants, Γ_1 may be reduced well below 4/3 by ionization of H and He. This has recently been investigated by Upton, Little and Dworetsky (1968) for protostars, and by Paczynski and Ziolkowski (1967, 1968) for evolved red supergiants. In the latter case this instability may be what initiates mass loss in such stars, perhaps followed by planetary-nebula and white-dwarf stages. It is very difficult to make much progress at present on this problem because we lack a satisfactory time-dependent theory of convection. It should also be remarked that the thermal time scale for these stars may be comparable to the dynamical time scale, in which case it is not satisfactory to ignore non-adiabatic effects in the dynamical problem.

2. Highly evolved stellar cores. If nuclear "burning" proceeds to the stage in which most of the matter is in the form of Fe and other elements near the bottom of the packing-fraction curve, then at higher temperatures the nuclear equilibrium changes to a mixture of these elements and He plus neutrons. This process of "nuclear dissociation" can reduce Γ_1 below 4/3, as pointed out by Hoyle (1946) who suggested the mechanism as a possible "trigger" for supernova explosions. (See the lectures on this subject by Colgate.)

3. Very condensed configurations. Several cases of dynamical instability appear in the theory of very dense, degenerate stars. An example is the instability of

very dense white dwarfs, where the equation of state is affected critically by inverse beta decay of protons (see the book of Harrison, Thorne, Wakano and Wheeler). At the highest densities the Newtonian hydrodynamics we have used is no longer an adequate approximation, and relativistic effects must be included. Relativistic effects in stellar structure are discussed in Thorne's lectures.

4. Convection. This is an example of a local dynamical instability. It would appear naturally in the present type of analysis, if the assumption of spherical symmetry were not made. (See the lectures by Spiegel.)

In most cases of dynamical instability that have been studied, the nonlinear effects are not discussed. It is commonly assumed that drastic consequences follow, usually either mass loss or collapse. But in some cases the motion might be limited by nonlinear effects and then the motion would be a sort of relaxation oscillation. It has often been remarked that nova light curves are very reminiscent of relaxation oscillations.

Physically, the meaning of the dynamical stability criterion is simply that when the hydrostatic equilibrium between pressure and gravitational forces is slightly disturbed, the restoring force must exceed the force which would pull the system farther from equilibrium. In an infinitesimal contraction, for example, the density increases, thereby increasing the gravity. But this will be stable, provided that the pressure increase is more than enough to counteract the increased gravity. The ratio between $\frac{\delta\rho}{\rho}$ and $\frac{\delta P}{P}$ is just Γ_1. The factor 4/3 basically reflects the 3-dimensional geometry.

Secular stability

In general $|s_1| << \frac{|s_2|}{3}$. If a star is dynamically stable, we can examine s_1. Thus one needs to know C and hence the actual form of ε_N and L. An interesting example (L58 and L63) which gives some insight is a homologous contraction. Let us assume radiative transfer throughout and a homologous perturbation. We write

$$\kappa = \kappa_o \rho^{\kappa_\rho} T^{\kappa_T}, \quad \varepsilon = \varepsilon_o \rho^{\varepsilon_\rho} T^{\varepsilon_T} \tag{31}$$

Then it is easily shown (see L63) that s_1 for such a change is given by

$$s_1 = -\frac{C}{Q} = \frac{3(\Gamma_3-1)}{(3\Gamma_1-4)}\left[-(\kappa_T+3\kappa_\rho)-(3\varepsilon_\rho+\varepsilon_T)\right]\frac{L}{(-\Omega)} \tag{32}$$

The time scale is, of course, the gravitational one, and the stability criterion is $s_1 < 0$, or

$$-(\kappa_T+3\kappa_\rho) < 3\varepsilon_\rho+\varepsilon_T. \tag{33}$$

Of course this will never hold everywhere in a star, since there is no energy generation in most of it. But in a typical case it will certainly hold.

One has little information about the form of ξ in such a secular perturbation--surely there is no reason to expect that it will look like ξ for an adiabatic dynamical mode, which is what we used above. It is quite possible for the perturbation to be large locally and small elsewhere. Unless the instability later leads to major changes in the structure of the star, the time scale may then be the thermal time associated with the part of the star where the perturbation takes place, not that of the whole star.

Physically, this type of instability is a strictly thermal phenomenon--it appears even if the dynamics is completely ignored. In the above case of an homologous perturbation, the criterion tells us that an increase in the energy generation rate with T and ρ increases the star's resistance to gravitational collapse. A small contraction accompanied by heating would be unstable only if κ_T or κ_ρ were strongly negative; i.e., if an increase in T or ρ makes the stellar material so transparent that the excess energy flows out so readily as to overcome the effect of ε_ρ and ε_T.

In passing, it may be noted that $s_1 \propto C$ and C is in general complex. Thus, there may be an oscillatory part to the perturbation corresponding to the time scale of C": $T \sim C''/Q$. This has in fact been found by Schwarzschild and Härm (1965) in the instabilities arising from He burning in thin shells.

Secular instability is not normally a violent or disruptive phenomenon and in a sense it may be regarded as a part of "normal" stellar evolution, since different nuclear-burning phases are typically separated by short gravitational-contraction stages. Most computer programs do in fact include gravitational-contraction (T ds/dt) terms and will automatically detect phases of secular instability if the time step is sufficiently small. (This is how the thin-shell instabilities were first recognized.) It is perhaps worth pointing out that if, as is common now among stellar-model builders, a generalized Newton-Raphson iteration scheme is used (the so-called "Henyey-method") it is possible to evaluate the stability eigenvalues with practically no additional effort. In such a scheme one sets out to solve a set of ordinary differential equations of the form

$$\frac{dy_i}{dx} = f_i(y_j) \tag{34}$$

(in stellar structure usually i,j = 1,...,4.) Some form of a difference scheme between steps n and n+1 is introduced, e.g.,

$$\frac{y_i^{n+1} - y_i^{n}}{\Delta x^n} = \frac{f_i(y_j^{n+1})+f_i(y_j^n)}{2} \tag{35}$$

Next one assumes a trial solution $\bar{y}_i$ and writes

$$y_i^n = \bar{y}_i^n + \delta y_i^n . \quad (36)$$

The set of difference equations is expanded to first order in δy_i. For example

$$f_i(y_j^n) = f_i(\bar{y}_j^n) + \sum_j \frac{\partial f_i}{\partial y_j^n}(\bar{y}_j^n)\ \delta y_j^n, \text{ etc.} \quad (37)$$

One then gets a set of linear equations for the y_i

$$\sum_j A_{ij}^n \ y_j^n = \sum_j B_{ij}^n \ y_j^{n+1} + D_i^n \quad (38)$$

where the coefficient matrices A_{ij}, B_{ij} and the inhomogeneous terms D_i are evaluated by using the y_i. The "Henyey method" is a scheme for solving the set of equations of this sort, together with the boundary conditions. The corrections y_i are added to the $\bar{y}_i$ and a new iteration is begun, starting with the "corrected" solution. The scheme "converges" when the y_i become sufficiently small.

Note now that the time-dependent equations of stellar structure may be written in the general form

$$\frac{\partial y_i}{\partial x} = f_i(y_i) + \sum_j \Lambda_{ij}(y_k)\frac{\partial y_j}{\partial t} . \quad (39)$$

If we assume a time dependence $y_i \propto e^{st}$ this becomes

$$\frac{\partial y_i}{\partial x} = f_i(y_j) + s \sum_j \Lambda_{ij}(y_k) y_j . \quad (40)$$

A set of difference equations can be constructed in the same way for this set. Since we want to find the eigenvalue s we must allow it to vary as well: $s = \bar{s} + \ s$. Then we get a set of difference equations of the form

$$\sum_j A_{ij}'^n \ \delta y_j^n = \sum_j B_{ij}'^n \ \delta y_j^{n+1} + C_i^n \delta s + D_i'^n . \quad (41)$$

The A_{ij}', B_{ij}', and D_i' are closely related to the old matrices A_{ij}, B_{ij}, and D_i and are hence easily constructed if we are solving the stellar structure problem. The "trial" solution $\bar{y}_i$ is the converged equilibrium solution, hence the D_i are all zero, if we set $\bar{s} = 0$. Thus the system is homogeneous, and it is of higher order (due to δs). Therefore another condition is needed, which is essentially a normalization condition on δy_i.

Examples of secular instability

1. Helium flash. Helium begins to burn very rapidly in an isothermal degenerate core because the rise in temperature does not result in an expansion (Mestel 1952; Schwarzschild and Härm 1962). Other elements may also start to burn this way (carbon-flash).

2. Schonberg-Chandrasekhar limit. An isothermal nondegenerate core begins to contract gravitationally when it reaches a certain mass. Usually this is looked at as a "hydrostatic incompatibility", in that a certain type of model cannot be made. It can, however, be shown that the onset of core contraction is an example of secular instability (Gabriel and Ledoux 1967).

   (1 and 2 are part of Iben's subject in these lectures.)

3. Instabilities of nuclear burning shells. As an example, He burning in a thin shell gives rise to "thermal pulses" (see Rose's lectures).

4. Mass flow in close binaries. When a star's radius becomes greater than the distance to inner Lagrangian point, mass flow begins, which is a form of secular instability as shown by Morton (1960).

Pulsational or vibrational instability

This is sometimes called overstability. It has the form of oscillations of growing amplitude, which may or may not be limited by non-linear effects.

Consider again the roots $s_{2,3}$ of the cubic, in the case where C is small but not zero. Then to first order in P,

$$s_{2,3} = \pm i\sqrt{A-B} - P = \pm i\sigma_a - \sigma'. \qquad (42)$$

The stability is thus determined by the real part of P, i.e., by C'. The "damping coefficient" is

$$\sigma' = P = -\frac{C}{2(A-B)} = -\frac{C'}{2\sigma_a^{\,2}} \qquad (43)$$

As shown in L63, C' may be written

$$C' = \frac{1}{J}\int_0^R \left(\frac{\delta T}{T}\right)_a \left(\delta E_N - \frac{d\delta L}{dm}\right)_a dm \qquad (44)$$

where the subscript a implies evaluation using the adiabatic eigenfunctions. So long as $\sigma' \ll \sigma_a$, this is not a bad approximation, (called the "quasiadiabatic" approximation.) Of course, we are assuming σ_a to be real.

We saw before that $s_1 = P/Q$ is of order T_G^{-1} for secular stability, and thus the time scale associated with σ', which is also of order Q/P, will be the gravitational one. Of course, there may be important differences because the eigenfunctions used to evaluate P and Q may be very different. Nonetheless, we can say in a general way that pulsations will grow or damp on a thermal time scale, and it is these thermal dissipative effects which stabilize or destabilize.

It is at first sight paradoxical that dissipation can destabilize rather than simply damp. In fact, as Cowling (1957) has pointed out, some sort of basic time asymmetry, like that provided by dissipation, is always needed for overstability. In addition, of course, a restoring force is needed. An example is thermal convection in a rotating medium. The centrifugal and coriolis forces provide the restoring force and viscosity the dissipation, resulting in oscillatory convection under certain circumstances. Another example is semi-convection, discussed in Spiegel's lectures.

The destabilizing influence of thermal dissipation may be seen in the following thought experiment, studied by Moore and Spiegel (1966). Imagine an element of fluid, in a stratified medium, attached by a spring to a fixed point. The buoyant force on the element is opposed by the restoring force of the spring. If there is no thermal dissipation, the element will oscillate about the equilibrium position. Now suppose that there is thermal dissipation--heat leaks from the element to the medium, and <u>vice versa</u>. As the element rises, it cools through heat loss to the surrounding medium. When it comes back down it gains heat, but, because of the finite time for thermal diffusion, the heat content of the element, when it reaches the equilibrium point, will be less than when it started to rise. The result is a net force in the downward direction. A similar effect will take place on the downward excursion, so that the thermal dissipation will gradually increase the amplitude of the oscillations. It is readily shown that the equilibrium position is unstable in the sense of "overstability", provided only that the spring force is strong enough, compared to the bouyancy.

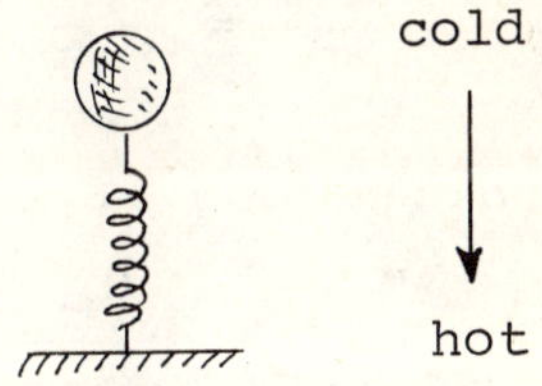

We now discuss further the damping coefficient,

$$\sigma' = -\frac{1}{2\sigma_a^{2}J}\int_0^M \left(\frac{\delta T}{T}\right)_a \left\{\delta\varepsilon_N - \frac{d}{dm}\delta L\right\} dm . \qquad (45)$$

We can write

$$\delta\varepsilon_N = \varepsilon_N\left(\varepsilon_\rho\frac{\delta\rho}{\rho} + \varepsilon_T\frac{\delta T}{T}\right) = \varepsilon_N\left[\frac{\varepsilon_\rho}{\Gamma_3-1} + \varepsilon_T\right]\left(\frac{\delta T}{T}\right). \qquad (46)$$

Since ε_ρ and ε_T are always positive, we can see that nuclear processes always act as destabilizing agents for pulsational instability. In most cases $\delta\varepsilon_N$ will not produce instability in the star as a whole. This is because the pulsations are strongly damped in the outer layers, and the amplitude of radial pulsations is generally much larger in the outer layers than in the core. In massive stars, however, radiation pressure reduces the Γ's and the central condensation, so that the pulsation amplitude does not increase so strongly from center to surface. In this case, the nuclear reactions can make the star pulsationally unstable, as was shown by Schwarzschild and Härm (1959) for stars of mass greater than about 60 $M_{\odot}$.

The origin of pulsational instability

The first order equations given on page 9 in terms of Eulerian perturbations can be rewritten in terms of Lagrangian perturbations, given by

$$f_1 = \frac{\delta f}{f} = \frac{1}{f}(f' + \delta\vec{r}\cdot\nabla f) \tag{47}$$

where f is the zero-order quantity given by the equilibrium model. In terms of these variables, the equations are

$$\frac{\partial p_1}{\partial m} = \frac{1}{4\ rp}\left[\sigma_o^2\ (4r_1 + p_1) - \ddot{r}_1\right] \tag{48}$$

where $\sigma_o^2 = Gm/r^3$,

$$\frac{\partial r_1}{\partial m} = -\frac{1}{4\pi r^3\rho}[3r_1 + \rho_1] \tag{49}$$

$$\frac{dt_1}{dm} = \left(\frac{1}{T}\frac{dT}{dm}\right)\left[L_1 - 4r_1 + \kappa_p p_1 + (\kappa_T - 4)T_1\right]$$

$$\frac{\partial \ell_1}{\partial m} = -\frac{p\delta}{L\rho}\left(\frac{\Gamma_1}{\Gamma_3 - 1}\dot{t}_1 - \dot{p}_1\right).$$

From the equation of state

$$\rho_1 = \alpha p_1 - \delta T_1 \tag{50}$$

where

$$\alpha = \frac{p}{\rho}\left(\frac{d\rho}{dp}\right)_T, \quad \delta = -\frac{T}{\rho}\left(\frac{d\rho}{dT}\right)_p$$

$$\kappa_p = \frac{p}{\kappa}\left(\frac{d\kappa}{dp}\right)_T, \quad \kappa_T = \frac{T}{\kappa}\left(\frac{d\kappa}{dT}\right)_p$$

The Lagrangian independent variable here is the mass m interior to a given elementary mass shell, a quantity which clearly remains constant during a radial pulsation.

The time dependence can then be separated, as in the Eulerian formulation, giving a fourth-order set of linear differential equations.

Since nuclear reactions cannot produce pulsational instability in cepheids, the source of pulsational instability must be found elsewhere. As was suggested by Eddington and later demonstrated by Zhevakin, the source of instability is the ionization zones of the most abundant elements, hydrogen, and helium. In these regions (which are located near the surface of the star where the pulsation amplitude is relatively large), the ionization increases the specific heat, diminishing the temperature fluctuation during pulsation, thereby decreasing the radiative damping and favoring instability. Furthermore, the variation in the opacity is given by

$$\kappa_1 = \kappa_T t_1 + \kappa_\rho \rho_1 . \tag{51}$$

If the pulsation is approximately adiabatic, then

$$T_1 \simeq (\Gamma_3 - 1)\rho_1 \tag{52}$$

and so

$$\kappa_1 \simeq \left(\kappa_T(\Gamma_3 - 1) + \kappa_\rho\right)\rho_1 \tag{53}$$

Thus, since $\kappa_\rho > 0$ and $\Gamma_3 - 1$ is small in the ionization zone, the opacity can increase during compression, even if κ_T is negative.

Thus, the ionization zone will tend to dam up radiation during expansion, driving the instability. Baker (1966) has discussed some of the physical effects underlying cepheid instability in terms of a greatly simplified model. A thorough account of the present state of the subject will be found in Cox (1967).

The ionization zone is only effective in producing pulsational instability under certain circumstances. The most critical factor is the depth of the zone. If the ionization zone is too near the surface, it will be of low density and not sufficiently coupled to the radiation to be effective. If the zone is too deep, the pulsations in the zone will be adiabatic and have little destabilizing effect.

Calculations of the linear non-adiabatic pulsations have been carried out by Cox (1963) and by Baker and Kippenhahn (1962, 1965). They show that the pulsational instability of Cepheids and RR Lyrae stars is due to this process, and that the small range of effective temperatures of these stars is a result of the situation described in the last paragraph. (It seems likely that convective transport is an important factor in the stability of stars of lower effective temperature, but this is not at all well understood.)

Linear calculations will give only the pulsation frequency and the damping coefficient. The limiting amplitude and other effects must be studied with non-linear calculations, which will be described by Christy.

Additional References

Baker, N. 1966, in Stellar Evolution, R.F. Stein and A.G.W. Cameron, eds., (Plenum, New York), p. 333.

Baker, N. and Kippenhahn, R. 1962, Z.f. Astrophys. 54, 114.

———. 1965, Astrophys. J. 142, 868.

Cox, J.P. 1963, Astrophys. J. 138, 487.

———. 1967, in Aerodynamic Phenomena in Stellar Atmospheres, R.N. Thomas, ed., (New York and London: Academic Press), p. 3.

Cowling, T.G. 1957, Magnetohydrodynamics (New York: Interscience Publishers), ch. 4.

Gabriel, M. and Ledoux, P. 1967, Ann. D'Astrophys. 30, 975.

Harrison, B.K., Thorne, K.S., Wakano, N. and Wheeler, J.A. 1965, Gravitation Theory and Gravitational Collapse (Chicago: University of Chicago Press).

Hoyle, F. 1946, M.N.Roy. Astr. Soc. 106, 343; see also Burbidge, E.M., Burbidge, G.R., Fowler, W.A. and Hoyle, F. 1957, Rev. Mod. Phys. 29, 547.

Mestel, L. 1952, M.N. Roy. Astr. Soc. 112, 583.

Moore, D.W. and Spiegel, E.A. 1966, Astrophys. J. 143, 871.

Morton, D. 1960, Astrophys. J. 132, 146.

Paczyński, B. and Ziółkowski, J. 1967, Int. Astron. Union Symposium 34: Planetary Nebulae, p. 396.

———. 1968, Acta Astron. 18, 255.

Schwarzschild, M. and Härm, R. 1959, Astrophys. J. 129, 637.

———. 1962, Astrophys. J. 136, 158.

———. 1965, Astrophys. J. 142, 855.

6

VARIABLE STARS — REALISTIC STAR MODELS

R. F. Christy

California Institute of Technology
Pasadena, California

INTRODUCTION

Variable stars are of interest to different sorts of people from several different points of view. From the point of view of the observer, the addition of the time dimension to the usual ones of position, intensity, and frequency suggests a whole new series of possible observations, while, from the point of view of the theoretician, it provides more complicated things to calculate. Both groups may reasonably hope that matching observed and calculated values of an additional set of parameters will yield increased knowledge of the physical properties of stars. In the case of Cepheid Variables, the observed period yields directly the mean density, while the measurement of colors of such variables allows for the comparison of observed and calculated positions of an instability strip in the HR diagram. The attempt to match observed light and velocity curves, their relative phases and other special features, has given us greater understanding of the Cepheids than of most normal stars.

Figure 1 shows the position of many kinds of variable stars on the HR diagram. This series of talks will be largely confined to a consideration of Cepheid type variables, which can be taken to include δ Scuti and RR Lyrae stars as well as Population I and Population II Cepheids, all of which appear to share a common mechanism.

OBSERVED PROPERTIES OF CEPHEID STARS

Figure 2 shows a light curve of a typical RR Lyrae star. The amplitude of variation is about one magnitude. Notice the rapid rise and slow decay (which are common to all such stars) and the bumps on the declining branch (which are moderately frequent). From light curves in several wavelength regions, color and temperature curves can be prepared. (These stars are generally bluest when brightest), and radial velocity curves can be obtained from a series of spectrograms taken at various points in the period. When plotted in the usual

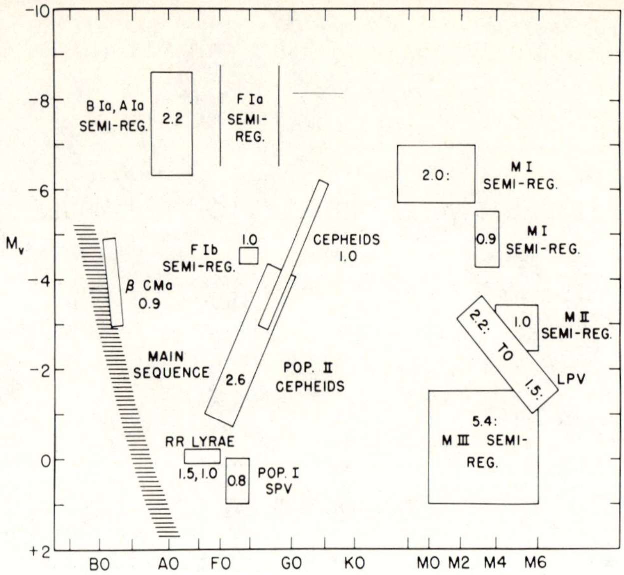

Figure 1

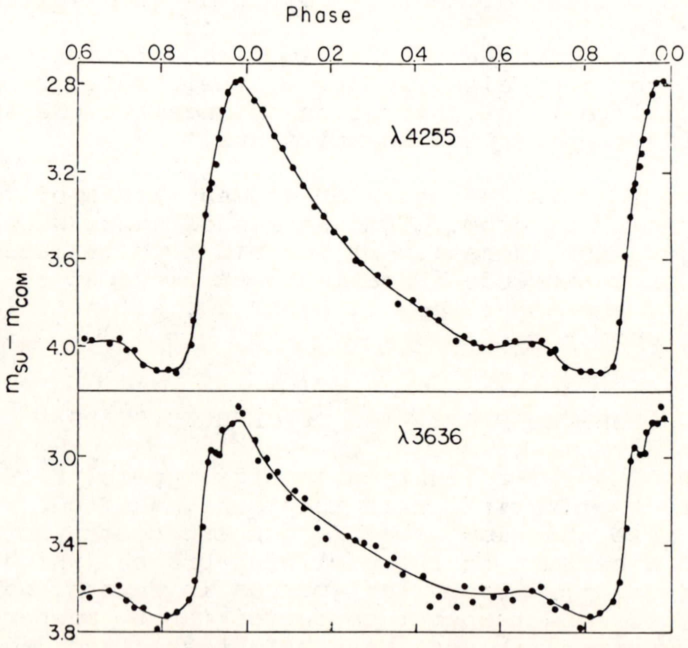

Figure 2

astronomical convention, with positive velocities directed inward in the star, the radial velocity curve is shaped and phased so as to appear much as a "mirror image" of the light curve (see Fig. 7).

METHODS OF CALCULATION - Equations and Limitations

The equations which must be obeyed by a star which is not in hydrostatic equilibrium are

$$M(r) = \int_0^r 4\pi r'^2 \rho(r')dr' \qquad (1)$$

$$V(r) = \frac{1}{\rho(r)} = 4\pi r^2 \frac{dr}{dM} \qquad (2)$$

$$\frac{\partial^2 r}{\partial t^2} = \frac{-GM(r)}{r^2} - 4\pi r^2 \frac{\partial P}{\partial M} \qquad (3)$$

$$L(r) = -(4\pi r^2)^2 \left(\frac{4\sigma}{3}\right) \frac{d(T^4)}{K(V,T)dM} \qquad (4)$$

$$\frac{\partial E}{\partial t} + \frac{P\partial V}{\partial t} + \frac{\partial L}{\partial M} = \varepsilon(V,T) \qquad (5)$$

Notice that pressure and gravitational forces need not balance, the unbalance resulting in an acceleration, nor is the heat flow (which is assumed to be by radiation) necessarily constant, unbalances in luminosity and nuclear energy production being reflected in changes of internal energy of the gas or in work being done.

The general approach will be to take the two time-dependent equations (3 and 5), assume appropriate initial conditions of temperature, density and velocity as a function of position, and integrate the equations forward in time. This approach has been extremely successful for Cepheid type variables, but may not be so for all kinds of variables because of the following time scale problem.

Because the calculation must follow events which move with the speed of sound through the star, time steps must be considered which are small (perhaps 1%) of the vibrational period of the star. Meanwhile, practical considerations of computing time limit us to perhaps at most 100 periods (that is 10^4) steps. It is, therefore, necessary that the phenomenon to be studied develop from the assumed initial conditions in some moderate number of periods, (that is, since Cepheid-type periods are in the hour to week range, the phenomenon must develop in a year or so, which is a very short time from the point of view of the star, through perhaps excessively long from the point of view of the computer) for the method to work. In most cases, 50-100 periods will suffice to develop highly periodic behavior in Cepheid type stars from almost arbitrary initial conditions. Since the number of periods required to develop the phenomenon is of the order of the total amount of energy in the pulsation divided by the energy transferred per period, the relaxation time is likely to vary with the amount of mass involved in the pulsations. This is small in the case of Cepheids, where only the outer envelope has an appreciable pulsation amplitude, but may be much larger for variables, if any, which are energized by nuclear reactions and which must, therefore, have appreciable pulsation amplitudes fairly near their centers, thus involving much more of the mass. The work in progress of Keeley (Caltech) indicates that very luminous red variables also relax quickly. Very massive stars, on the other hand, may not.

Baker (Columbia) and von Sengbush have attempted to develop a non-linear method of calculation which seeks periodic solutions of the equations rather than integrating through time. This method may well be more appropriate for phenomena with longer time scales, but has not yet really been tested.

The linear method, discussed briefly by Baker, can also be applied fruitfully to problems for which the non-linear approach is cumbersome. It will give correctly pulsation periods, whether or not a given mode is unstable, and the rate of growth of the amplitude of an unstable mode. Because the linear method uncovers instabilities far more rapidly than does the non-linear method, it can be used for exploring the effects of varying parameters like mass, luminosity, and effective temperature over a wide range. This has not been done completely even yet for Cepheid type variables.

METHODS OF CALCULATION - Boundary Conditions

The calculations for Cepheid-type variables are greatly simplified by the fact that amplitudes are sensibly zero for a core including a major fraction of the star's mass. One can, therefore, establish an inner boundary to the envelope to be considered on which $U = \dot{r} = 0$ and $L(r) = L_o$, the total luminosity of the star, since all nuclear energy generation will occur in that core.

At the surface, the temperature can be found from the Eddington approximation $T^4 = \frac{3}{4}T_e^4(\tau + \frac{2}{3})$ or $2\sigma T^4(\text{surface}) = L$, is satisfactory to within .1 or so in optical depth from the surface. (So little mass is found above this level that it does very nicely for an effective surface.)

The surface pressure, though taken as zero in many calculations, is better set equal to the radiation pressure resulting from the surface temperature, $P_{surf} = \frac{1}{3}aT^4_{surf}$ while the gas pressure at the surface is taken as going to zero. These approximate boundary conditions, although not of course as good as a proper radiation transport theory treatment, reflect a consistent (gray atmosphere) treatment. The behavior of the extreme outer regions is not important to the calculations except when $\Delta R/R$ is of the order of 0.5 or more, for which case no correct treatment is yet available anyway.

METHODS OF CALCULATION - Numerical Technique

For illustration of the numerical methods, it is desired to integrate numerically the simplest possible sort of time dependent equation for sound, which is of the form:

$$\frac{\partial^2 F(x)}{\partial t^2} = c^2 \frac{\partial^2 F(x)}{\partial x^2} \tag{6}$$

This may be accomplished by defining a discontinuous time variable $t^n = t^o + n\Delta t$, attaching values of $F(x)$ at all t^n, and replacing derivatives by differences. Thus $V = \partial F/\partial t$ can be written as,

$$V(t^{n+\frac{1}{2}}) = V^{n+\frac{1}{2}} = [F^{n+1}(x) - F^n(x)]/\Delta t \tag{7}$$

or

$$F^{n+1}(x) = F^n(x) + \Delta t V^{n+\frac{1}{2}}(x). \tag{8}$$

This can be used to generate values of F at time n + 1 once the values at time n and the V's are known. The V's may be obtained as follows:
First

$$\frac{\partial^2 F^n(x)}{\partial t^2} = \frac{V^{n+\frac{1}{2}}(x) - V^{n-\frac{1}{2}}(x)}{\Delta t} \tag{9}$$

once again approximating the derivative by a difference. Then define a discontinuous space variable:

$$x_I = x_o + I\Delta x$$

then:

$$\frac{\partial^2 F^n(x)}{\partial t^2} = \frac{c^2}{\Delta x^2}[F^n(I+1) - 2F^n(I) + F^n(I-1)] \tag{10}$$

Then, combining (9) and (10) we obtain V at a later time from V at an earlier one, and, returning to (8) can evaluate F(x) at the later time.

$$V^{n+\frac{1}{2}}(I) = V^{n-\frac{1}{2}}(I) + c^2\frac{\Delta t}{\Delta x^2}[F^n(I+1) - 2F^n(I) + F^n(I-1)] \tag{11}$$

Thus, once given F^o and $V^{-\frac{1}{2}}$ everywhere at some one time, the future behavior of the system can be calculated.

In order to guarantee a stable calculation having no numerical oscillations, it is necessary that the parameter $c^2\Delta t^2/\Delta x^2$ be less than one. This is called the Courant condition and implies that the smaller the divisions of space (mass shells or zones) the smaller must be the divisions of time (time steps). It requires that a wave not be able to cross a mass zone in one time step. Since the amount of numerical work goes as the product of the number of space elements and the number of time steps, there is thus a tremendous advantage to be gained by using large x's. These hydrodynamical calculations therefore generally use from 20 to 50 mass zones, compared to as many as 500 used in evolutionary calculations. It is, however, necessary to represent only the outer layers of the star (T less than 10^6°K) with these 20-50 space steps.

It is sometimes possible to make the Courant condition less stringent by artificially varying c, the sound speed (by increasing the inertial but not gravitational mass for example) thus permitting larger time steps and covering a larger fraction of the star's evolution. One may also abandon the above explicit method of calculating the F's and V's at a later time from those at an earlier time for an implicit method not subject to the Courant condition. This is appropriate to phenomena with a long time scale but is less accurate when things happen quickly.

In the case of the heat diffusion equation, which is of the form:

$$\frac{\partial F}{\partial t} = k\frac{\partial^2 F}{\partial x^2} \tag{12}$$

the explicit method of forward differences just outlined yields the following prescription for stepping forward in time:

$$F^{n+1}(I) = F^n(I) + \frac{k\Delta t}{\Delta x^2}[F^n(I + 1) - 2F^n(I) + F^n(I - 1)] \tag{13}$$

In this case, however, the stability condition, $k\Delta t/\Delta x^2 < 1$, may be a very severe restriction. One of the implicit methods which avoids this limitation expresses the F's at a new time in terms of themselves as well as the old F's thusly:

$$F^{n+1}(I) = F^n(I) + \tfrac{1}{2}\frac{k\Delta t}{\Delta x^2}[(F^n(I + 1) - 2F^n(I) + F^n(I - 1)) + + (F^{n+1}(I + 1) - 2F^{n+1}(I) + F^{n+1}(I - 1))] \tag{14}$$

Using this formulation, there is no restriction on the size of the time steps, but it is necessary to solve a set of linear equations relating the new F's to the old ones. These equations are of the form:

$$- \frac{k\Delta t}{2\Delta x^2} F^{n+1}(I + 1) + \left(1 + \frac{k\Delta t}{\Delta x^2}\right) F^{n+1}(I) - \frac{k\Delta t}{2\Delta x^2} F^{n+1}(I - 1) =$$

$$= \frac{k\Delta t}{2\Delta x^2} F^{n}(I + 1) + \left(1 - \frac{k\Delta t}{\Delta x^2}\right) F^{n}(I) + \frac{k\Delta t}{2\Delta x^2} F^{n}(I - 1) \qquad (15)$$

which is just a rearrangement of (14). Solving this system of equations amounts to inverting a matrix which has non-zero elements along only three diagonals. Special numerical techniques for doing this rapidly are available (see Richtmyer).

METHODS OF CALCULATION - Discussion of the Equations Used

The difference equations actually used to calculate the dynamics of the star are:

Mechanics:

$$\Delta M(I - \tfrac{1}{2}) = M(I) - M(I - 1)$$
$$\Delta M(I) = \tfrac{1}{2}\left[\Delta M(I + \tfrac{1}{2}) + \Delta M(I - \tfrac{1}{2})\right] \qquad (16)$$

At time n:

$$V^n(I - \tfrac{1}{2}) = \frac{4}{3}\pi \frac{[(R^n(I))^3 - (R^n(I - 1))^3]}{\Delta M(I - \tfrac{1}{2})} \qquad (17)$$

$$R^{n+1}(I) = R^n(I) + \Delta t^{n+\frac{1}{2}}\, U^{n+\frac{1}{2}}(I) \qquad (18)$$

$$\Delta t^n = \tfrac{1}{2}(\Delta t^{n-\frac{1}{2}} + \Delta I^{n+\frac{1}{2}}) \qquad (19)$$

Viscous Pressure:

$$Q^{n-\frac{1}{2}}(I - \tfrac{1}{2}) = C\frac{[U^{n-\frac{1}{2}}(I) - U^{n-\frac{1}{2}}(I - 1)]^2}{V^n(I - \tfrac{1}{2}) + V^{n-1}(I - \tfrac{1}{2})} \quad \text{if } U(I) - U(I-1) < 0 \qquad (20)$$
$$= 0 \quad \text{if } U(I) - U(I-1) > 0$$

$$U^{n+\frac{1}{2}}(I) = U^{n-\frac{1}{2}}(I) - \Delta t^n \left\{ \frac{GM(I)}{(R^n(I))^2} + \frac{4\pi(R^n(I))^2}{\Delta M(I)}\left[P^n(I + \tfrac{1}{2}) - P^n(I - \tfrac{1}{2}) + Q^{n-\frac{1}{2}}(I + \tfrac{1}{2}) - Q^{n-\frac{1}{2}}(I - \tfrac{1}{2})\right]\right\} \qquad (21)$$

with the boundary conditions:

$$U^{n+\frac{1}{2}}(1) = 0$$

and, defining $P(N + \frac{1}{2}) = -P(N - \frac{1}{2})$ so that $P(N) = 0$

$$U^{n+\frac{1}{2}}(N) = U^{n-\frac{1}{2}}(N) - \Delta t^n \left\{ \frac{GM(N)}{(R^n(N))} - \frac{4\pi(R^n(N))^2}{\Delta M(N - \frac{1}{2})}\left[2P^n(N - \tfrac{1}{2}) + \right.\right.$$

$$\left.\left. + Q^{n-\frac{1}{2}}(N - \tfrac{1}{2})\right]\right\}$$

where $R(N) = R_{star}$, $M(N)$ = mass of star, etc.

Notice that there is no linearization and the correct equation of state has been used but that both time and space variables have only discrete values, while derivatives have been replaced by differences.

The equations include an artificial 'viscous' pressure which solves the following problem: The basic equations permit shock waves — discontinuities in T, p, ρ, and velocity — to develop. It is possible, in principle, to follow such discontinuities in explicit numerical fashion, requiring that matter, momentum, and energy be conserved across them, but this demands excessive amounts of computing time, because the calculation attempts to develop rapid oscillations behind the shock to simulate increasing the entropy behind the shock. If, then, an artificial dissipative force is introduced, no great harm will be done, since the conservation laws are already guaranteed by the equations, but the discontinuity will be spread over several mass elements, thereby greatly reducing the inconvenient numerical oscillation. This artificial dissipation serves as an additional damping mechanism and will inhibit the development of the pulsation if it is too large. For example, if the constant C in the Viscous Pressure Equation is taken to be unity, a wave with $\frac{1}{2}\lambda$ = one mass zone will be critically damped in one oscillation, a wave with half-wave-length equal to 10 elements will be damped in 10 periods, and so forth. It is therefore necessary to vary C in order to determine a value which is large enough to prevent serious oscillations but small enough not to influence the growth rate of the pulsational instability excessively. The choice C = 1 typically reduces the growth rate α in amplitude $\sim e^{\alpha t/\text{period}}$ by order .001.

The difference equations used in the calculation of energy transfer are:

Let $T^4 = W$, which is more nearly linear in M over the envelope

$$W^n(I - \tfrac{1}{2}) = (T^n(I - \tfrac{1}{2}))^4 \tag{22}$$

$$L^n(I) = \left[4\pi(R^n(I))^2\right]^2\left[W^n(I - \tfrac{1}{2}) - W^n(I + \tfrac{1}{2})\right]2F^n(I) \tag{23}$$

where $F^n(I)$ is a suitable approximation to $\frac{4\sigma}{3K\Delta M}$

e.g.,

$$2F(I) = \frac{\frac{4\sigma}{3}\left(\frac{W(I+\frac{1}{2})}{K(I+\frac{1}{2})} + \frac{W(I-\frac{1}{2})}{K(I-\frac{1}{2})}\right)}{\Delta M(I)\,[W(I+\frac{1}{2}) + W(I-\frac{1}{2})]}\,. \tag{24}$$

Then the energy transport equation:

$$E^{n+1}(I+\tfrac{1}{2}) - E^{n}(I+\tfrac{1}{2}) + \left[\tfrac{1}{2}(P^{n}(I+\tfrac{1}{2}) + P^{n+1}(I+\tfrac{1}{2}) + Q^{n+\frac{1}{2}}(I+\tfrac{1}{2})\right] \times \left[V^{n+1}(I+\tfrac{1}{2}) - V^{n}(I+\tfrac{1}{2})\right] \times \Delta M(I+\tfrac{1}{2}) \tag{25}$$

$$= \frac{\Delta t^{n+\frac{1}{2}}}{2}\left[L^{n+1}(I) + L^{n}(I) - L^{n+1}(I+1) - L^{n}(I+1)\right].$$

Notice that it is necessary to define an average opacity within a zone. The way in which this is done will be of some importance in those zones in which the opacity changes by a large factor. It turns out to be most appropriate to average 1/K or the transparency, rather than K directly (see Rev. Mod. Phys. article).

Figure 3 is an approximate representation of the Cox opacities used in this sort of calculation for the temperature and specific volume ($1/\rho$) ranges of interest (for a fixed composition) for a pulsation calculation. Notice at lowest densities the flat, electron scattering opacity at high temperature, the falling Kramer's Law ($K = K_o\rho/T^{7/2}$) opacity at high temperature, and the rapid increase of K with temperature due to the increasing ionization of hydrogen in the low temperature regime. On the average, a change of a factor of 10 in density changes the opacity also by about 10.

Figure 4 shows the distribution of temperature vs. pressure in that part of the star where ionization is taking place. At low T and P is a flat, photospheric region, at high T and P a power law resembling that obtained for an n = 3 polytrope, and in between the ionization zones with a near-discontinuity in temperature which is rather a nuisance in carrying out the calculation because the Courant condition forbids having a large number of mass zones in this region. Since the opacity may change by as much as 100 times across a single mass step in this region, the averaging scheme is of importance. There is a drop in density going through the temperature discontinuity. Vigorous convection in this region is inevitable, but for high temperature Cepheids the region is near the surface and ρ is so low that even the most rapid possible convective heat transport is less than that required. As a result dT/dr is still determined by radiation. In low temperature stars, convection is undoubtedly more important,

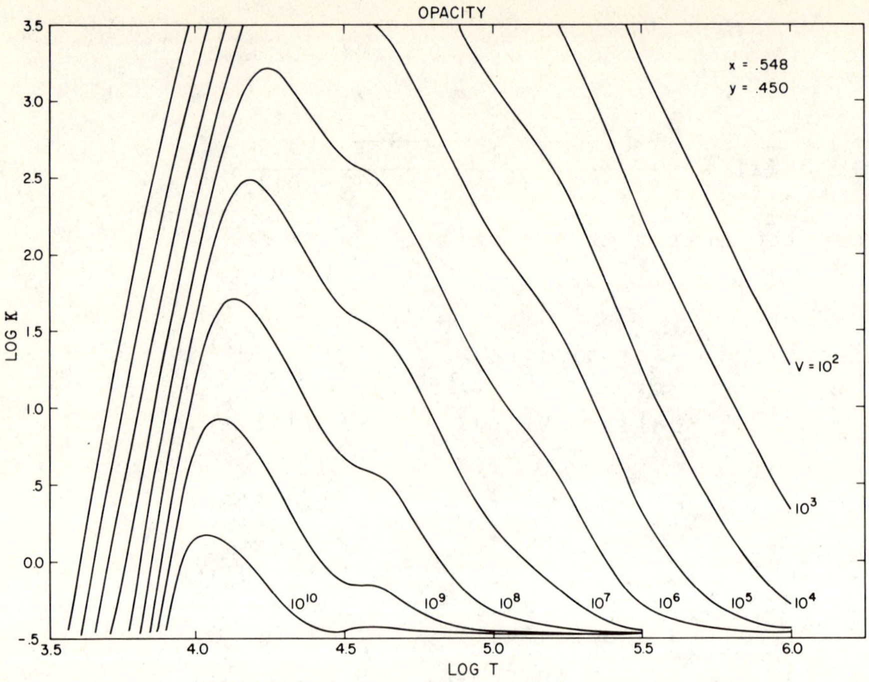

Figure 3

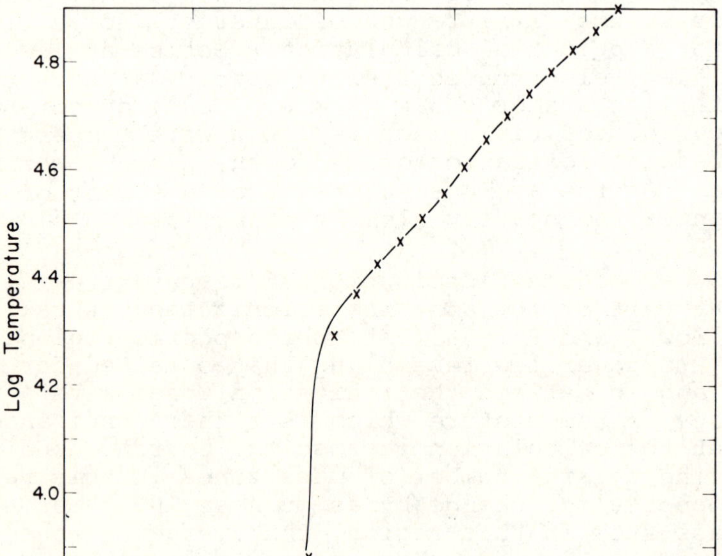

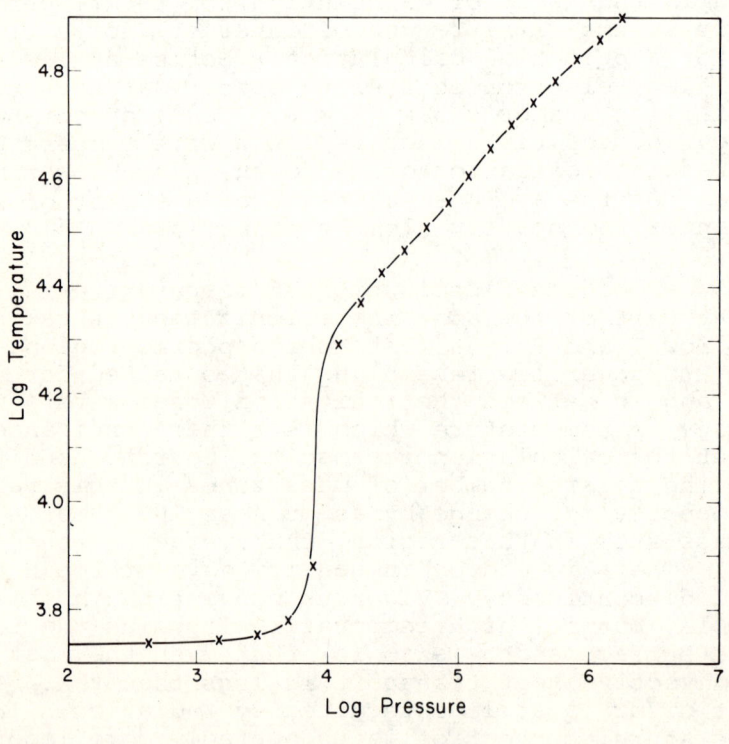

Figure 4

but it has been neglected due to the lack of a good theory for including it.

Figure 5 shows the run of temperature with mass through the star. The mass zones have been chosen very small at the surface and increase exponentially with depth but in such a way that the mass of a given shell is always less than the sum of the masses above it. The solid line shows T vs. M for the static model, and the two dotted lines, the run of T at maximum and minimum light. Notice that the discontinuity moves outward when the star is hottest and inward when it is coolest as would be expected for the ionization zones. The temperature amplitude is of the order of .2 in the log near the stellar surface.

METHODS OF CALCULATION - Initial Conditions

A starting model (T, P, ρ, L and so forth vs. M(r)) may be obtained from the equations tabulated above. All time derivatives are set equal to zero, L_o, T_e, and M chosen, and the equations integrated inward to an inner boundary at about $r = .1R$ and $T \approx 10^6\,°K$. This defines the $F°(x)$ of the discussion following equation (1). The initial velocities U may be taken as essentially zero (equal to machine noise) and allowed to develop in their own way. This is the method of Cox. It is interesting but very slow, since many time steps will be required to develop full pulsation from infinitesimal amplitudes.

An alternate approach is to assume some normal mode and impress its U's on the model. The U's of a stable mode will decrease with time, those of an unstable one increase. It is possible to start with amplitudes about ½ of the final ones to be achieved. This is still in the linear regime and yet considerably speeds up the development of the pulsation. This is the method of Christy.

Figure 6 shows luminosity, surface velocity, and radius as a function of time in a calculation done by the second method. The starting conditions are near but not exactly a normal mode of the star, so that initial behavior is not perfectly periodic, but becomes increasingly so as time goes on. The kinetic energy increases by about 3% per period, so that the pulsation e-folds in about 30 periods and settles into the final configuration in 50 or 60 periods. Using 30 to 40 mass points on an IBM 7094, this requires about one minute per period or 30-60 minutes per model. However it takes many sets of L, M, R, composition, and so forth to understand the many complications of large amplitude pulsation.

METHODS OF CALCULATION - Accuracy Criteria

Accuracy may be judged by varying certain of the parameters of the calculation and examining their effects on the growth rate of pulsation amplitudes. This rate is 3.6% per

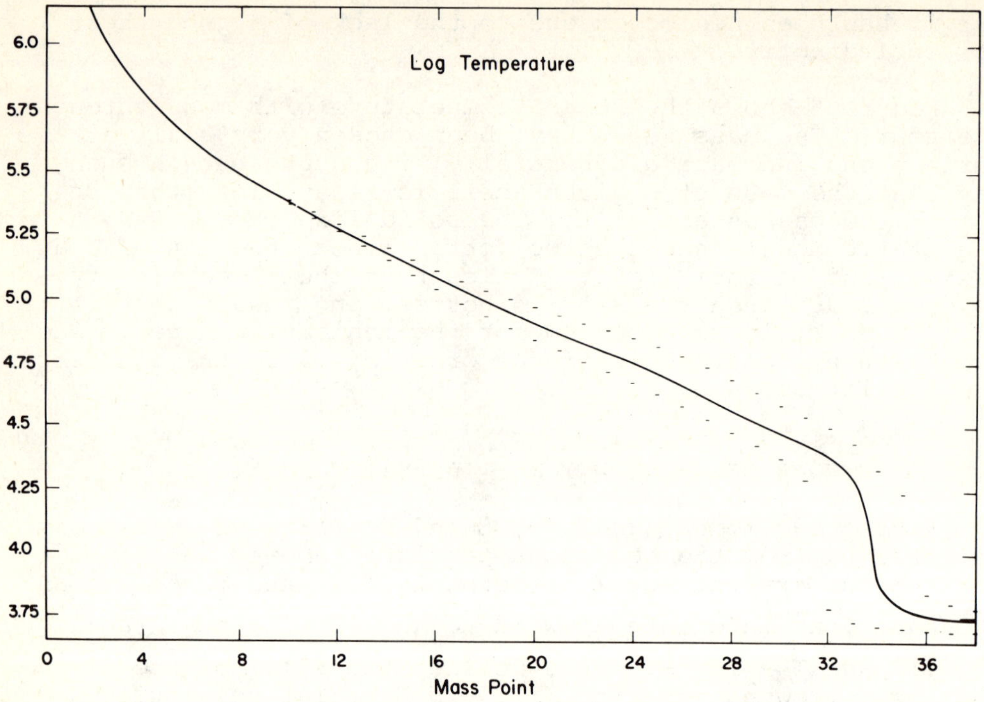

Figure 5

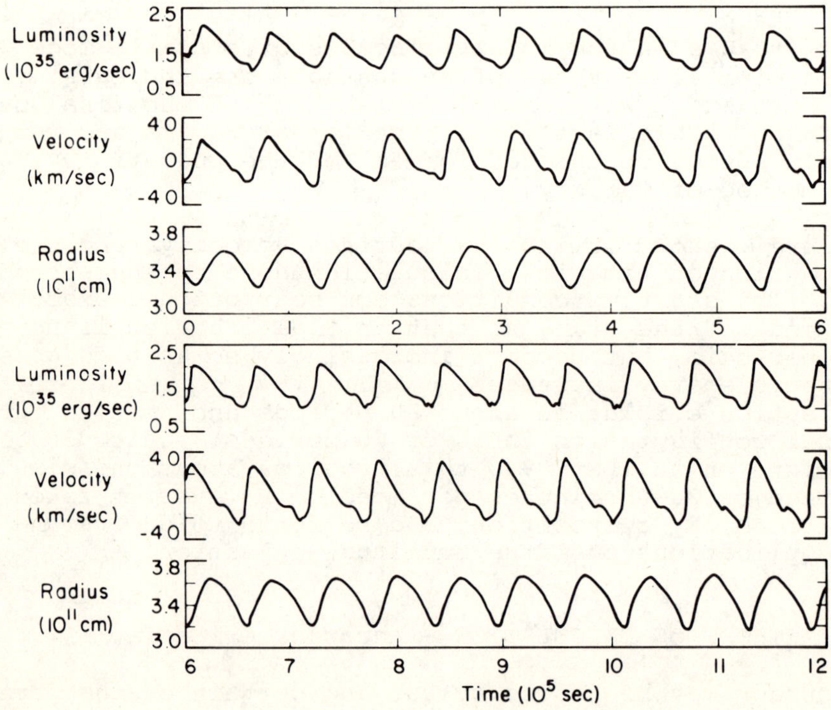

Figure 6

period in energy for the standard case. No significant change was found when the tolerable error in the iteration of the heat equation was increased from 1° to 10° thus saving computing time. Reducing the Courant parameter by reducing the time step by $2^{-\frac{1}{2}}$ had no effect. On the other hand, increasing the viscous parameter C from 1 to 4 reduced the growth rate by 38%, while lowering it to .2 increased the rate by 10% and increased the numerical oscillations to an undesirable degree. Hence the growth rate found with C = 1 is probably good to 10%, a tolerable error. Decreasing the number of mass zones from 38 to 28 increased the period by .5% and reduced the growth rate by 20%. Increasing the number of mass zones to 66 increased the growth rate by a permissible 6%, and the computing time by an impermissible 100%.

Results

Figure 7 shows light and velocity curves for a series of models with $M = .38\ M_{\odot}$, $L = 1.5 \times 10^{35}$, $Y = .45$ and a variety of effective temperatures. The curves have in some cases been smoothed over oscillations which are clearly due to the zone structure. Using a larger number of zones simplifies this smoothing but it is too costly for general use. T_e = 7100° was the hottest model found to be unstable. The apparent instability of the two coolest models is probably incorrect and due to the neglect of convection.

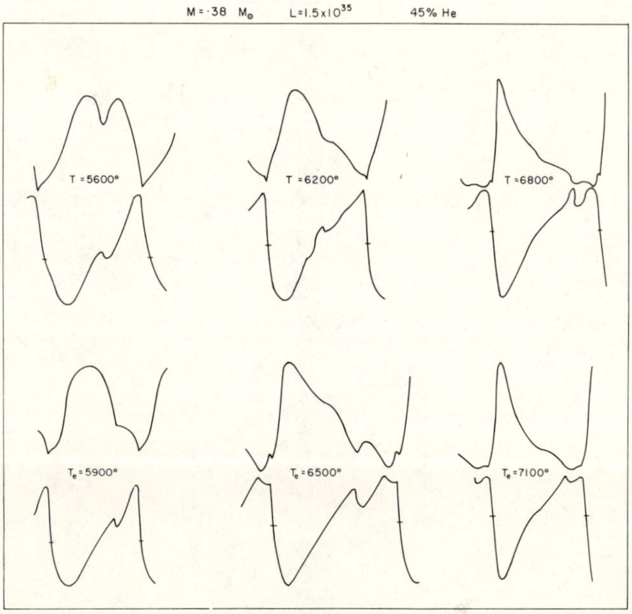

Figure 7

Most of the R.R. Lyrae models calculated had fairly high helium abundance because this puts the instability strip at higher temperature, where numerical difficulities in the calculation are less pronounced.

Advice to the Novice

Everyone who does calculations of this type starts with far too many shells and must eventually drop to 50 or less in order to get any results.

References

General: R. D. Richtmyer, Difference Methods for Initial Value Problems (Interscience Publishers Inc., New York) 1967.

Pulsation and Evolution:

R.F. Christy, Methods in Computational Physics, Vol.7, Academic Press, New York, 1967

____________, Rev. Mod. Phys., 36, 555, 1964.

A.N. Cox, R.R. Brownlee and D.D. Eilers, Ap. J., 144, 1024, 1966.

G.S. Fraley, Astrophys. & Space Sci., 2, 96, 1968.

J. Castor, in press.

Lecture II

PHYSICAL INTERPRETATION AND RESULTS

The results from a systematic program of study of pulsations by the methods of Lecture I can be compared to observations, once suitable mean parameters are defined. The mass is straightforward for both the observed and the calculated star, as is the composition, since these are constant. (The results are found to be more sensitive to helium than to metal abundance.) The luminosity is set equal to the time average and the radius or effective temperature of the static model are approximately equal to the time averages of these parameters in a pulsating mode, and either may be used to characterize the star. In the case of the real star, the mean luminosity is readily defined from the observations, and the average of the maximum and minimum radius is a satisfactory representative value for R. The minimum is somewhat difficult to observe because the star passes quickly through it. (See Fig. 6). There is no really satisfactory definition of the mean effective temperature based on the observations. The temperature corresponding to the B-V color averaged over the period is the parameter most often used.

Given the starting parameters of M, L, R or T_e, Y, and Z, two types of calculations may be carried out. The stability or instability of a given mode may be established in six to eight periods, while the maximum amplitude and full periodicity will be developed in fifty or so periods. It is often possible to accelerate the growth of the instability by numerical techniques.

The Phase Shift

The understanding and deduction of the phase shift has been a central problem of Cepheid variability from the earliest stages (10); even recent non-adiabatic linear calculations (7) have not settled the problem. These calculations found that the phase of the luminosity is delayed after minimum radius by only about 30° in the helium zone and then is delayed by a very large amount in the hydrogen zone giving a total delay of about 180° at the surface. However, all the linear calculations of the behavior of the hydrogen zone show such a large relative amplitude that it is clear that the results of the calculations are no longer meaningful even at amplitudes far smaller than those observed.

The non-linear calculations have followed the behavior of the star, in particular of the surface layers, in the large amplitude region where the phase shift can be compared with observation (1, 2). The results agree with the observation in the shape and phasing of the luminosity and velocity curves. Figure 8 shows a sequence of luminosity curves at different depths. The progressive delay of peak luminosity in the helium zone amounts to about 30° or 1/10 period as in the linear theory. However, the change in the hydrogen zone is better described as a change in shape rather than as a large phase delay. The rising light phase becomes delayed and becomes very abrupt so that mean rising light comes near or

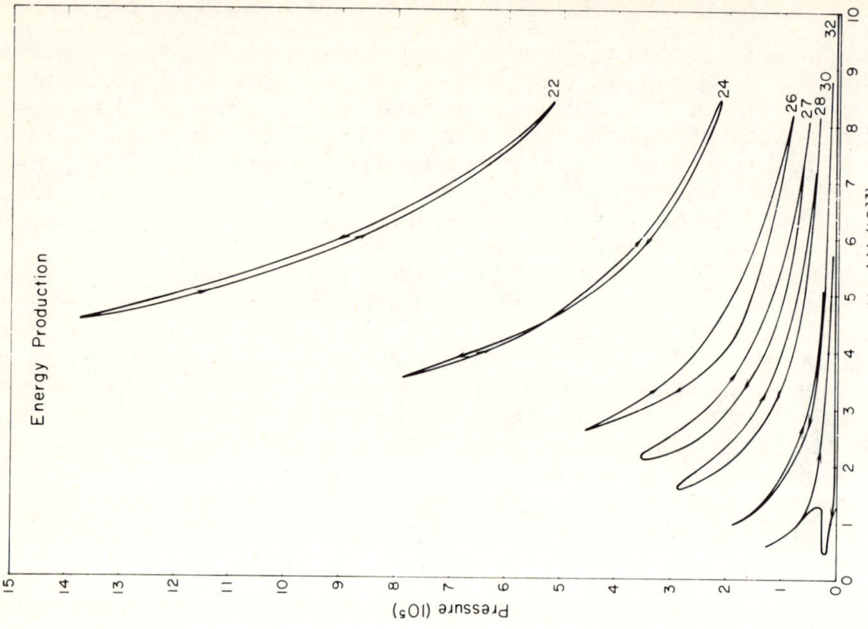

Figure 9

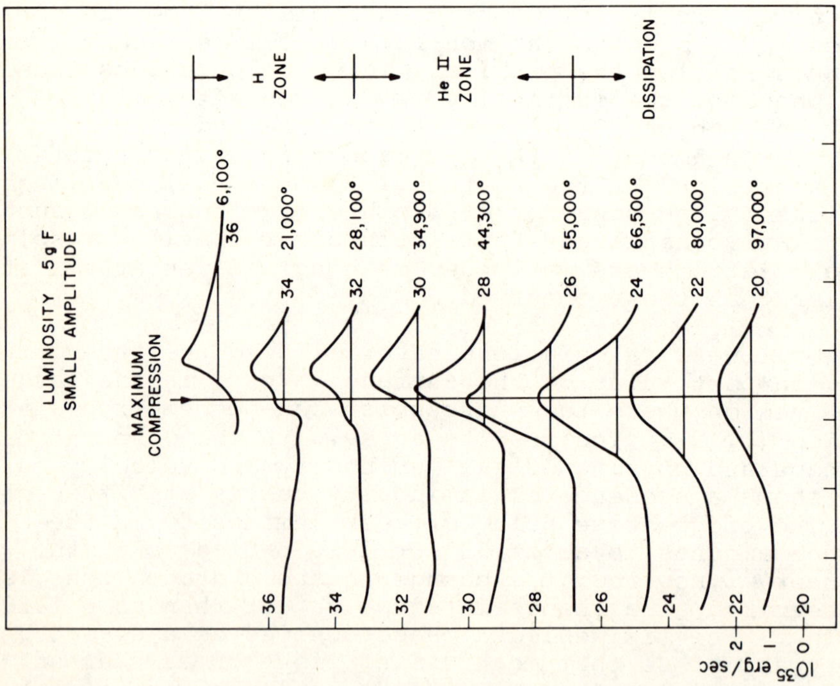

Figure 8

sometimes slightly later than minimum radius. This is the "90° phase lag" but it really is not such a great lag in the center of gravity of the main peak of the luminosity or in the positive phase of the luminosity. The calculated time of mean rising light comes very near minimum radius in most cases, as observed.

The Excitation of Pulsation

In the non-linear calculations, the contribution of each layer to the excitation or dissipation has been found by examining P-V diagrams for each layer during one period. The area enclosed is the work per period, either positive or negative depending on the sense in which it is executed. Figure 9 shows a sampling of such diagrams. Zone 22 shows nearly adiabatic motion but with dissipation; zones 26, 27, 28 show the excitation arising in the helium zone; the phase lag of the temperature rise has now given a considerable area to each loop which is executed with a significant departure from adiabaticity; zone 32 is an example of the behavior in the hydrogen zone where there is a large excitation but no resemblance to adiabatic behavior. These loops in the P-V diagram show how the delay in phase of maximum temperature after maximum compression is so closely related to the excitation of pulsation. The contribution of all zones (for a slightly different case) is shown in Fig. 10 where the separate, but comparable, excitations due to the hydrogen zone and the helium zone are both apparent. At large amplitude, these zones nearly overlap so that there is actually no unambiguous definition of each. The various contributions to the excitation and dissipation are shown in Fig. 11 as a function of amplitude.

Physical Mechanisms

The two mechanisms first recognized as important to pulsation were the nuclear reaction mechanism and the opacity mechanism. The first depends on the rapid increase in nuclear rates with increasing temperature but is ineffective in most stars because of the very small amplitude of pulsation in the core of the star. The second mechanism is dissipative or exciting depending on whether the heat flux is increased or decreased on compression. If $\kappa \propto \rho^r/T^s$, the variation of opacity is:

$$\frac{\delta\kappa}{\kappa} = [r - (\gamma - 1)s]\frac{\delta\rho}{\rho} \tag{26}$$

For the Kramers-type opacity with $r \approx 1$ and $s \approx 3$, this gives a decrease of opacity and an increase of heat flux on compression, unless γ is near 1, and thereby is responsible for the dissipation in much of the envelope. In an ionizing region, this effect may be reversed (4) since γ is near unity (Fig. 12). However, there is another effect of an ionizing region (6), (the γ mechanism) which, even for a constant opacity, leads to excitation if the region is at a critical depth such that the

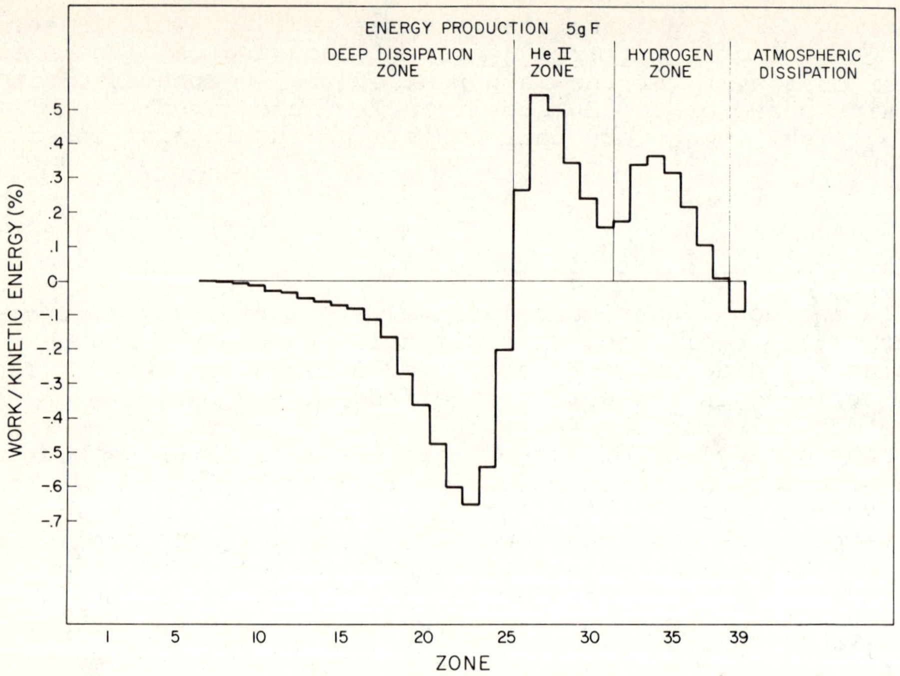

Figure 10

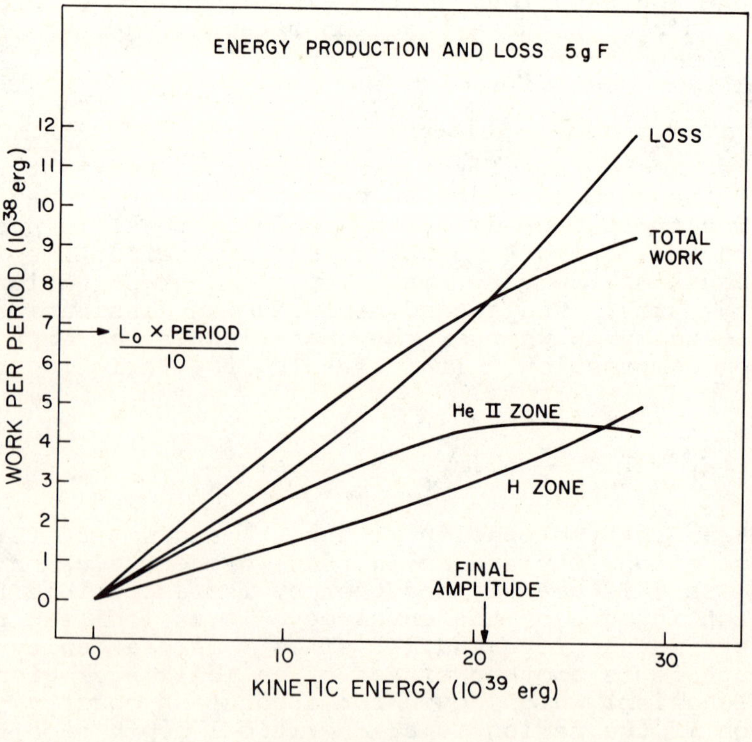

Figure 11

pulsation is nearly adiabatic below the ionizing region whereas above the ionizing region there is too little heat capacity to control the flux. Then, on compression, the ionizing region stays cooler because of the small γ and thus absorbs extra heat from the deeper region at this phase and contributes excitation.

Although the non-linear calculations have clearly demonstrated where the excitation of pulsation arises, they have not automatically yielded an explanation of which mechanism is effective. It is my opinion, however, that the so-called γ mechanism, which depends on the presence of an ionizing region at a critical depth below the surface, is in fact the prime cause of the instability region. It is only that mechanism which localizes the instability to a narrow region in T_e as it is found both observationally and by calculation.

The Maximum Amplitude

In all cases, the non-linear calculations reach a maximum amplitude in the same range as is observed. The velocity amplitude and radius amplitude were similar to those observed as was the luminosity amplitude and the temperature amplitude. It was therefore of interest to explore the effects which control the limiting amplitude. In essentially all cases, the limiting amplitude was approached when the velocity curve began to show a very steep phase at minimum radius. When the duration of the outward acceleration phase had become very short, the amplitude stopped increasing.

In one case (2), a detailed investigation was made of the behavior of the system in its approach to maximum amplitude. It appeared that the rapid acceleration occurred only in and above the helium zone (Fig. 13). It is associated with the rapid expansion as the helium and hydrogen become ionized by the luminosity increase. This expansion leads to the rapid outward acceleration and a rapid increase in pressure and temperature, its effect on the luminosity are shown in Fig. 14 where we see the peak luminosity is no longer delayed in the helium zone. These quantities become so high in the helium zone that the opacity drops and the flux leaks rapidly through the helium region causing the excitation due to this region to level off (Fig. 11). This levelling off of excitation then stops the increase of amplitude as the excitation then is just balanced by the dissipation.

The Instability Region

In view of the origin of the excitation in the hydrogen and helium zones very near the stellar surface, we can readily understand why the instability region is located near $6000°T_{eo}$. It was found that the high T_{eo} boundary of instability is located so that the heat capacity to ionize the unionized matter near the surface is just equal to the total heat flow from the star in 0.2 periods. In other words, there must be

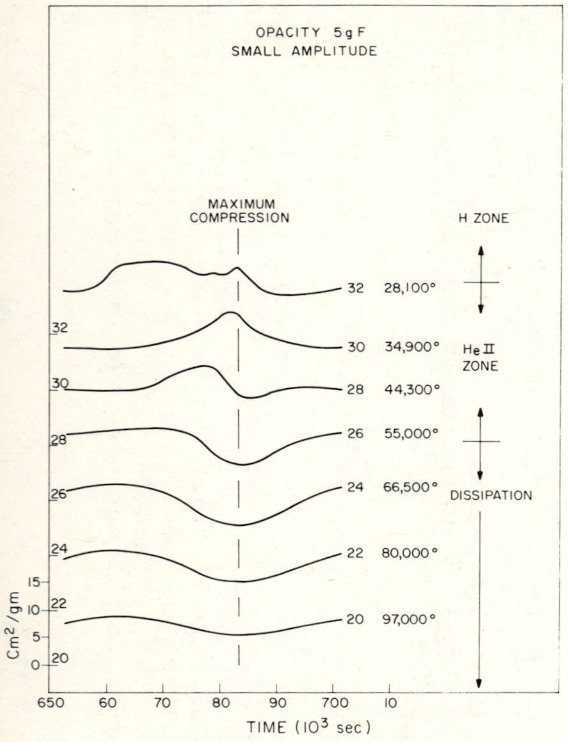

Figure 12

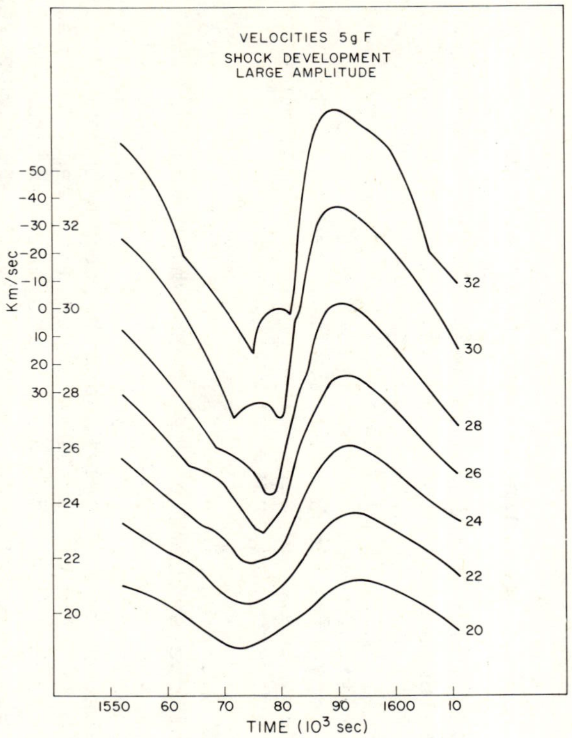

Figure 13

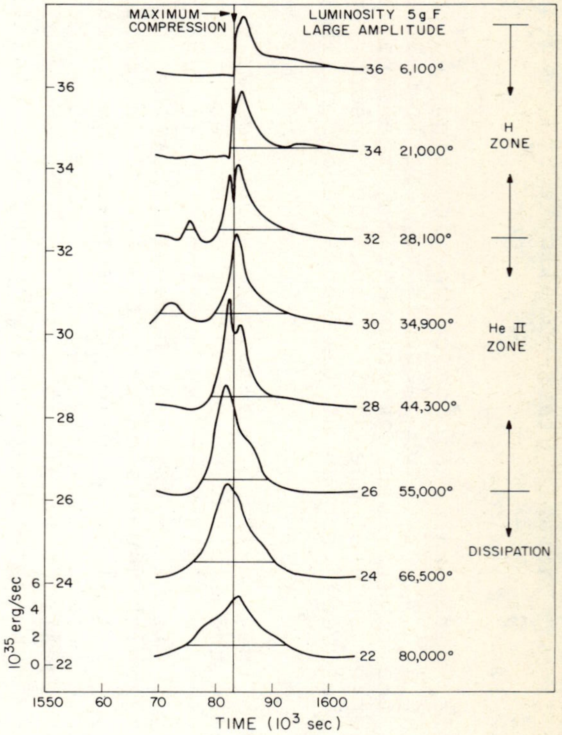

Figure 14

sufficient matter in these regions to be able to give a phase delay of about 0.2 periods, if the excitation is to be great enough to overcome the damping. For higher T_{eo} models, the heat capacity in the driving region is too small for effective action.

On the low T_{eo} side of the instability region, the contribution of the helium zone diminishes again as it becomes so deep that it behaves adiabatically and does not contribute effectively (6, 11). However, the behavior of the hydrogen zone on the low T_{eo} side of the strip is still obscure. Linear, non-adiabatic calculations (7) continue to show significant driving due to hydrogen at lower T_{eo}. The non-linear calculations, on the other hand, have been unable, so far, to give satisfactory results for T_{eo} lower than $\approx$ 5000°K since the coarse zone technique for the treatment of the hydrogen zone breaks down at such low T_{eo} that convection would be prominent. However, it is now generally believed that a time-dependent treatment of convection in the hydrogen zone is required to understand the cessation of pulsational instability at the low T_{eo} boundary.

The boundaries of instability have been shown to depend on T_e and on surface gravity. However, the period is principally dependent on surface gravity and only slightly on mass so that we may expect to be able to define an instability region in a period-T_e plane or a period-color plane. The computed high T_e boundaries of instability were found to fit the relation for Y = 0.3,

$$\log T_e = 3.828 - 0.06 \log P_F(\text{days}), \tag{27}$$

where P_F is the period of the fundamental in days. It was found earlier that the instability strip shifted to higher T_e with increasing helium abundance. This can be represented by

$$\log T_e = 3.798 + 0.1Y - 0.06 \log P_F \tag{28}$$

for the high T_e boundary. The low T_e boundary of instability has been estimated to lie about 0.06 lower in log T_e.

Comparison of Instability Region with Observation

It is found that the position of the instability strip in the HR diagram is dependent upon the mass-luminosity law obeyed by the stars and is therefore different for Population I and Population II stars. Thus, for example, the positions of Magellanic Cloud Cepheids in the HR diagram ought to indicate whether or not they obey a unique mass-luminosity relation. The position of the instability strip in the log g, log T_e plane and in the log Period, log T_e plane (where it is nearly a straight line) is a constant, independent of the M-L law obeyed.

The calculated period-temperature relation may be compared with observation if the effective temperature is derived from B-V in accordance with a relation given by Oke and Kraft:

$$\log T_e = 3.886 - .175 \langle B - V \rangle \qquad (29)$$

This is probably both somewhat uncertain and sensitive to surface gravity and composition but Strom believes that it is adequate for the range 5500-6500 or 7000°K and for log g = 1-2.5.

We then find that the observed variables lie inside a strip of width 0.37 in <B - V> but becoming narrower (≈0.2 in B - V) for RR Lyrae variables (Fig. 15). Except for two or three rare examples (such as Y Oph), the classical Cepheids of our galaxy (12) occupy a strip of width only ≈ 0.2 in B - V and are confined to the red side of the wider instability region. The observations of Gascoigne and Kron (13) and of Dickens (14) in the Magellanic clouds are primarily to be found on the blue side of this instability region. It is also well known (15, 16) that there is a high frequency of short period (1-2 days) Cepheids in the small Magellanic cloud, whereas such variables are almost absent in our galaxy.

We propose the following interpretation of the observed mean colors and frequency of occurrence of various periods (see Fig. 16). The evolutionary calculations of Iben (17), and of Hoffmeister, et al. (18), show a first crossing of the Cepheid instability strip from left to right on the way to the first red giant stage. However, a much slower crossing, during core helium burning, occurs after the first red giant stage when the evolutionary track makes an excursion to higher T_e and returns, after a loop, to the red giant stage. These calculations show that the loop extends further to the blue the higher is the luminosity. In this way these authors have explained a lack of short period Cepheids, when the loop does not intersect the instability region, followed, at higher luminosity, by a peak in the frequency when the loop just reaches the strip and then returns. At high luminosity, the crossing of the strip is complete and is more rapid, corresponding to a lower frequency of Cepheids.

It was noted by Hoffmeister (19) that the loops in the evolutionary track, which are so essential to the Cepheid phase, are prominent for a set of calculations with Z = 0.044 and Y = 0.354 but almost disappear for a set of tracks with Z = 0.021 and Y = 0.24. This implies that the frequency of Cepheids in a given population is very sensitive to either Z or Y or both. I now propose that the Cepheids in our galaxy differ from those in the SMC by a small amount in Z or Y so that the evolutionary tracks show somewhat shorter loops for our galaxy. This view is illustrated schematically in Fig. 16. We would then understand the excess of short period Cepheids in the SMC as due to a greater extent to the blue of the evolutionary tracks at low luminosity compared to the galaxy. At the same time, these tracks would populate the blue side of the instability strip whereas in the galaxy, when the tracks do intersect the instability strip, it would be primarily on the red side and we would thus understand the absence of Cepheids, in the galaxy, on the blue side of the strip. Some of these same differences appear to exist, in a

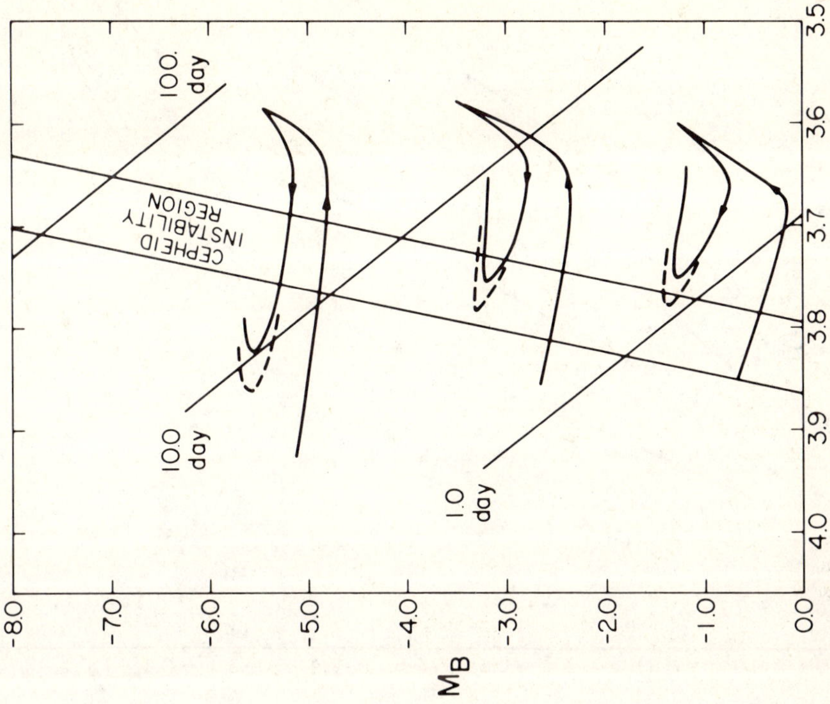

Figure 16

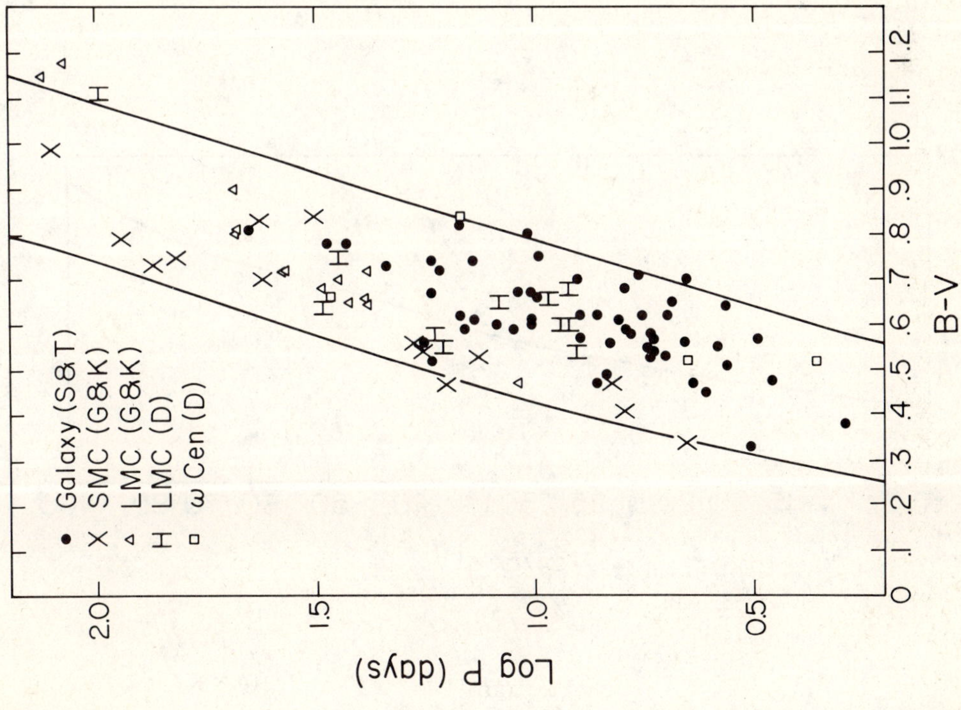

Figure 15

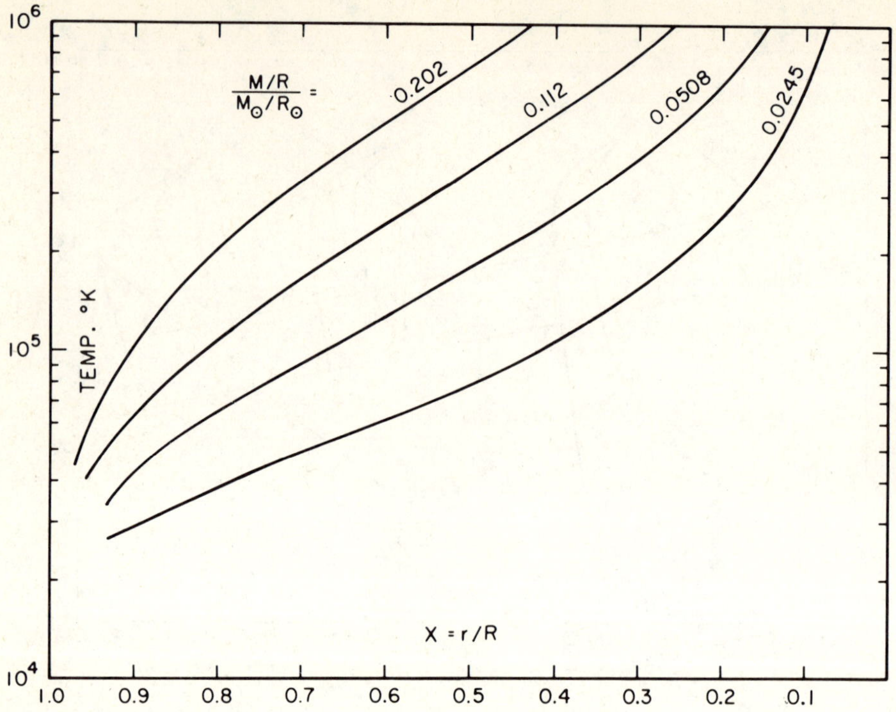

Figure 17

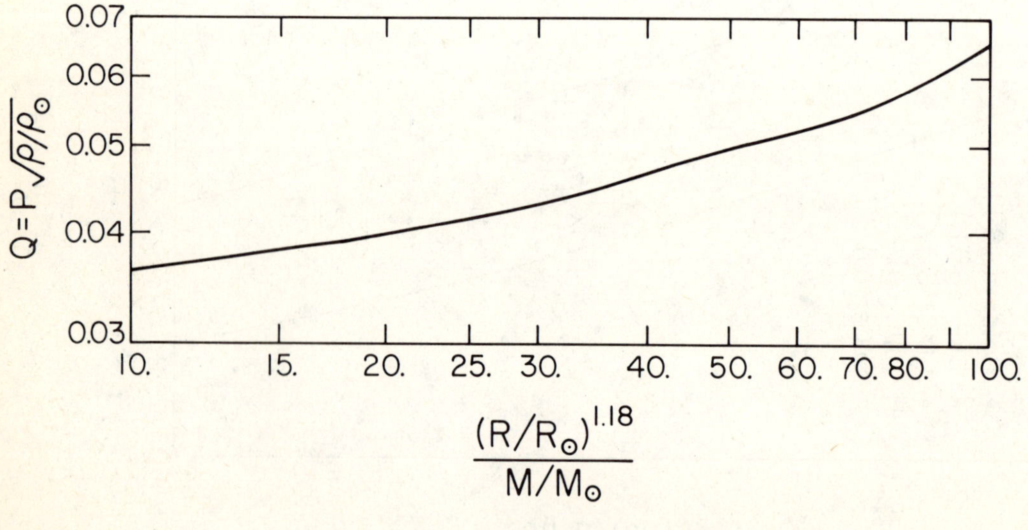

Figure 18

milder form, between Cepheids in the direction of the galactic center and those away from the center which are more like those in the Magellanic clouds. We thus see that the frequency of short period Cepheids is a very sensitive indicator of some, not yet fully understood, aspect of abundances. Further calculation of evolutionary models is needed to clarify the meaning of these evolutionary tracks.

Period-Radius-Mass-Temperature-Luminosity Relations

It was found that many of the differences in behavior of various pulsators could be correlated with the envelope structure. The range of structures studied is shown in Fig. 17 where the upper curve is appropriate to a δ Scuti variable and the lower curve to a much more extreme giant such as a W Virginis variable.

For a very wide range of models, it is found that the quantity $Q = P(\rho/\rho_0)^{\frac{1}{2}}$ when plotted against $(R/R_\odot)^{1.18}/(M/M_\odot)$ gives a single curve which is nearly a straight line in a log-log plot. This is shown in Fig. 18. Replacing density by M and R, this can be used to deduce a period-mass-radius relation of the form

$$P_{days} = .0224(R/R_\odot)^a/(M/M_\odot)^b \qquad (30)$$

where the values of a and b are, respectively, 1.69 and .66 for short period stars (less than 10 days), 1.83 and .78 for long period stars (more than 10 days) and 1.74 and .70 for the average relation.

A useful relationship which applies to individual stars can be derived from equation (30) by replacing R by $L^{\frac{1}{2}}/T_e^2$.

$$P_{days} = CL^{a/2}\ M^{-b}T_e^{-2a} \qquad (31)$$

Notice that some of these relations are given in absolute, others only in relative form. The proportionality constants can always in principle be found from the models and the mass-luminosity relation, but this has not always been done here.

Lecture III

RESULTS OF NON-LINEAR CALCULATIONS

Bumps in Light and Velocity Curves

There is a well-known feature of Cepheid light curves for periods from 7-15 days (the Hertzsprung progression) which is shown in Fig. 19 which is a copy of observational material. Unfortunately, my calculated light curves for $5000° \leq T_e \leq 6000°$, which is the relevant range, show spurious bumps associated with the finite thickness zones sweeping through the ionization region. However, some well-observed velocity curves (see Fig. 20) of Cepheids in this period range also

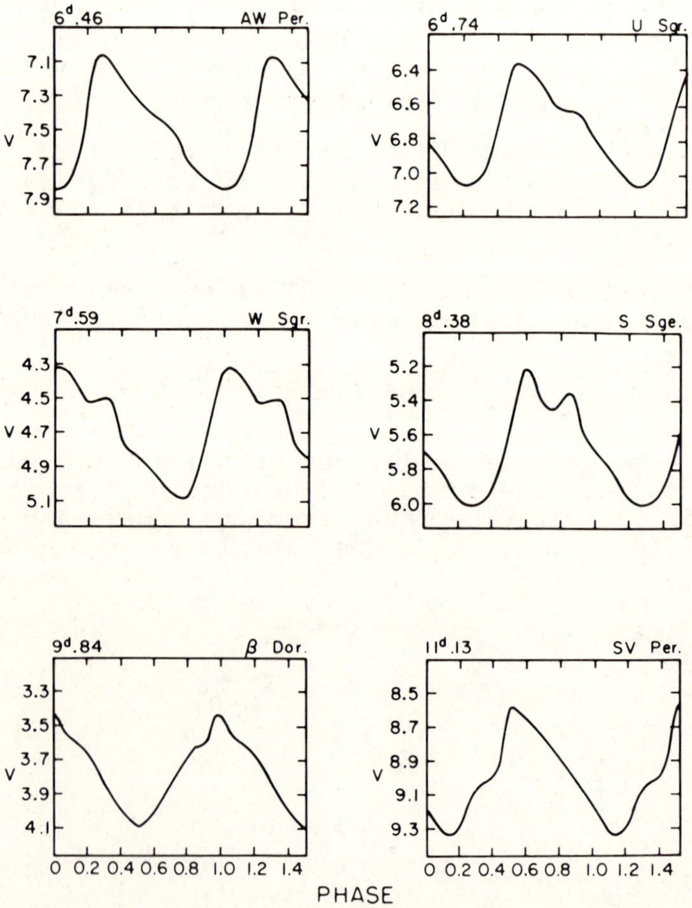

Figure 19

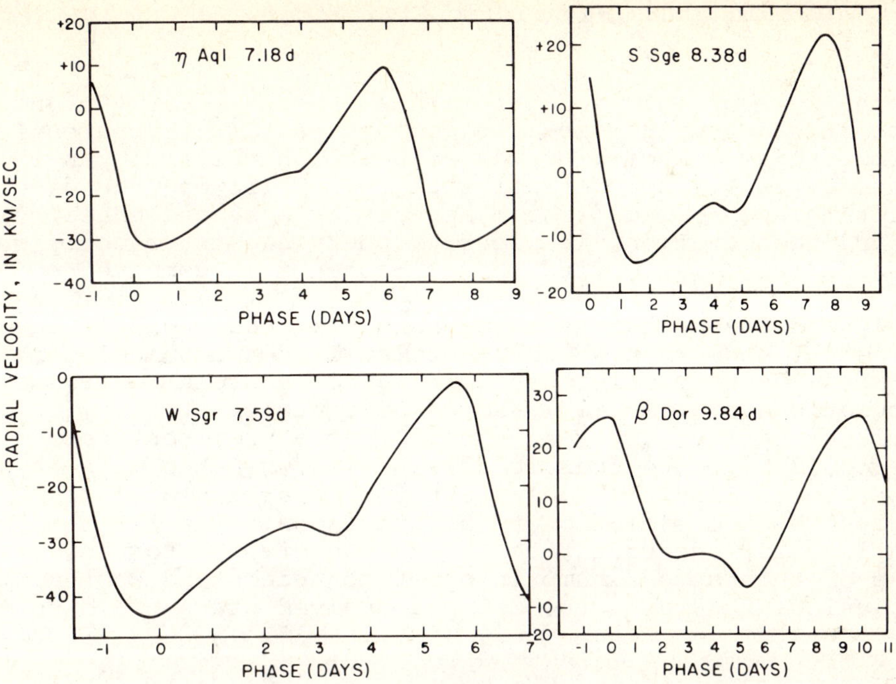

Figure 20

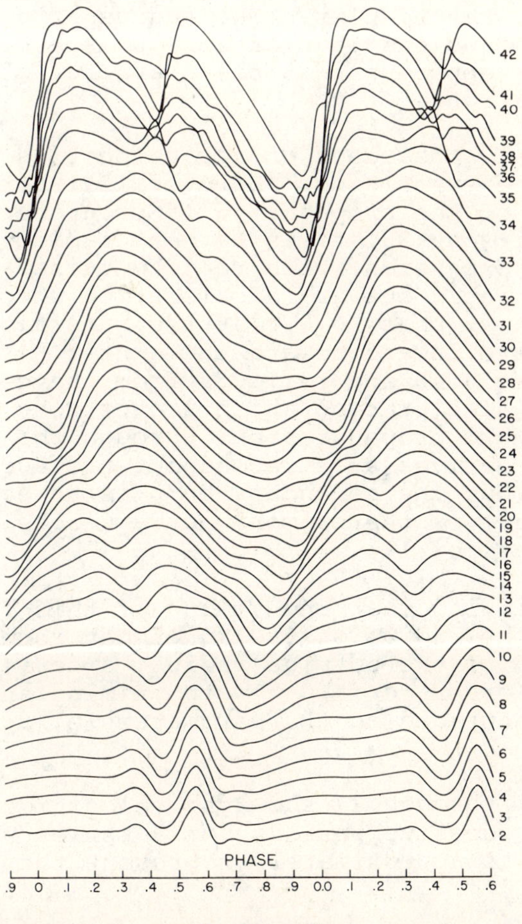

Figure 21

apparently show this feature. This is particularly significant since it has been possible to compute what are, apparently, quite precise velocity curves even though the accompanying light curves may not be satisfactory. Following the discovery that this second bump, which appears as a second phase of rapid outward acceleration, is clearly observable in velocity curves, a search was made to find this phenomenon in computed models.

It was found that this bump did not occur near 10 days period for models in which the relation between luminosity and mass was that which has been found in evolutionary model calculations of Iben (17) and of Hoffmeister _et al_. (18). A further search, however, showed that a prominent second bump could be found near 10 days period when the mass was chosen to be about half as great for a particular luminosity, as given by the evolutionary calculations. The first investigation (20) subsequent to this discovery was to uncover the cause of this second bump in order to understand whether the conclusions regarding the mass of evolved stars could be trusted.

Figure 21 shows the velocities of each zone in a 42-zone model during nearly two periods at maximum amplitude. The actual full amplitude of zone 42 (the outside surface) was 54 km/sec whereas that of zone 2 (the innermost) was 0.014 km/sec. The velocity scale has been progressively shifted and enhanced by a factor reaching 1600 at zone 2 in order to better display the results. The radius of zone 2 is ≈ 0.1 of the outside radius and that of zone 6 is ≈ 0.25 of the outside radius. It is apparent that the character of the motion changes little in this range.

The general appearance of the velocities in the innermost zones was quite surprising. However, I interpret this behavior in terms of the propagation of progressive waves. The helium zone at zone 33 starts to heat rapidly at phase 0.9, causing rapid expansion which rapidly raises the pressure and initiates the rapid outward acceleration which propagates out to the surface at phase 0.0. At the same time it sends a compression wave inward toward the center. This arrives at zone 2 at phase 0.45 and reflects from the central core as a rapid outward acceleration which leaves zone 2 for the outside at phase 0.65 and propagates out through the star to the surface where it arrives as a second outward acceleration at phase 0.45. Thus the secondary bump is an echo (from the center of the star) of the primary outward acceleration about 1.4 periods earlier.

After verifying that the timing of this feature was insensitive to the radius of the innermost zone provided it was less than 0.25 of the outside radius, a series of models was calculated in order to find how the timing and appearance of this feature depended on the parameters M, R and L of the model.

Figures 22 through 27 show the velocities of all zones for several successive periods for a series of models with $L \sim M^3$ chosen to approximately represent the classical

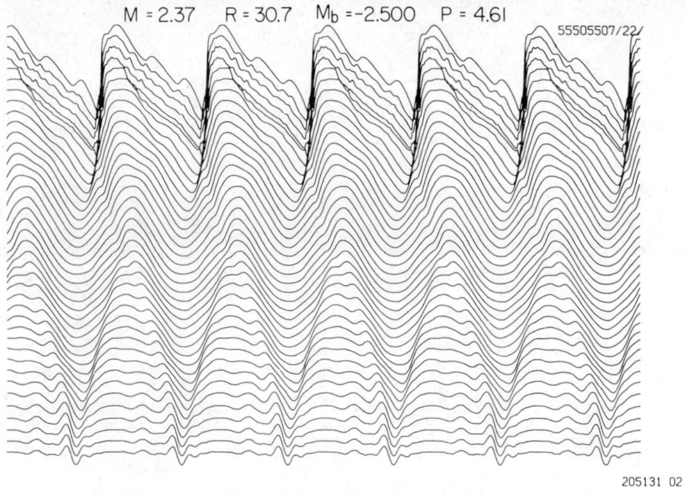

Figure 22

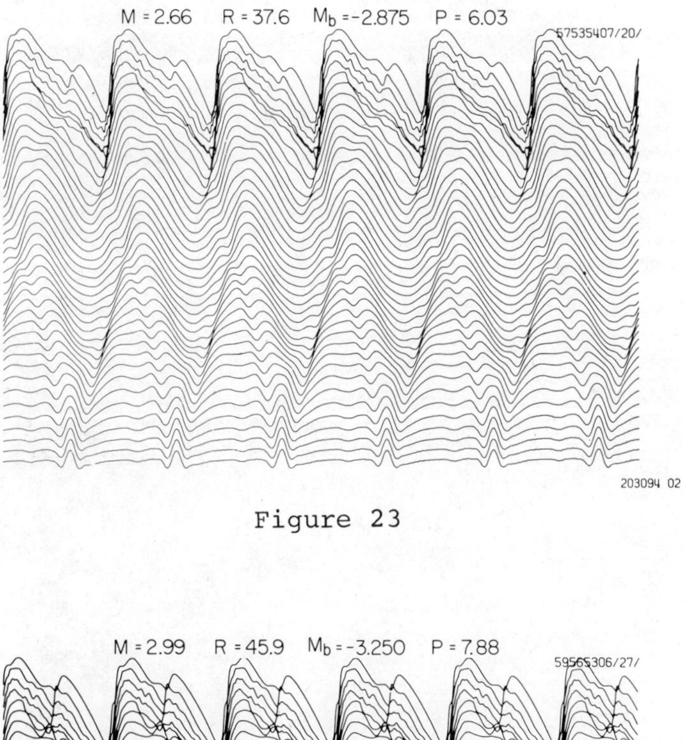

Figure 23

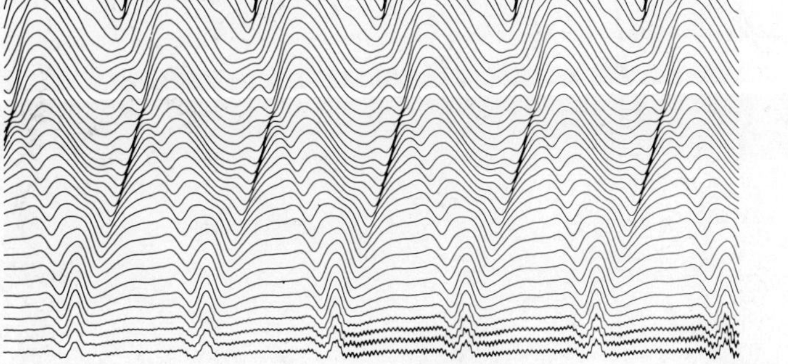

Figure 24

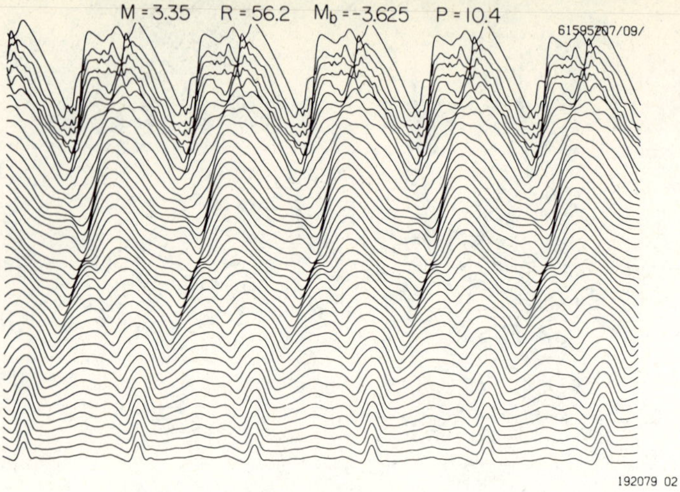

Figure 25

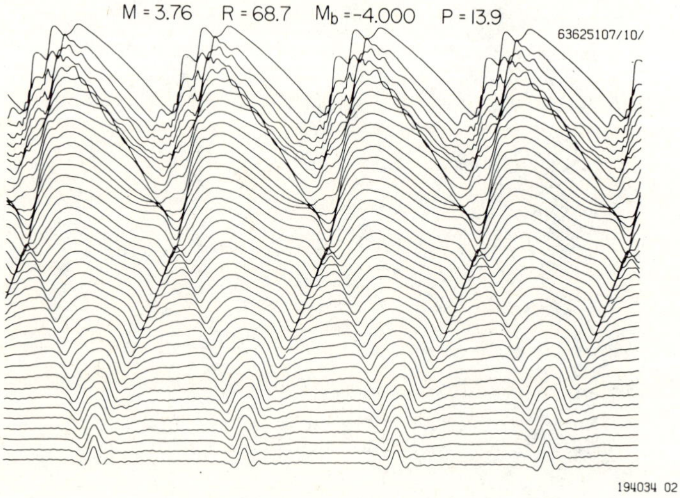

Figure 26

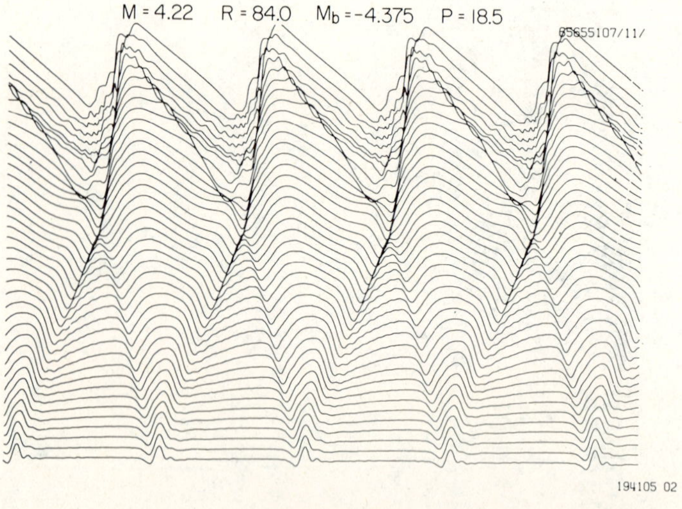

Figure 27

Cepheids. (In one or two cases a numerical instability appeared in the innermost zones but does not seem to have affected the outer part of the envelope).

From a study of a variety of models, it was found that the time delay of the second bump (about 1.4 periods in Fig. 23) correlated well with the structure of the envelope. This time delay was measured from the time the velocity was zero at minimum radius to a time in the middle of the second outward acceleration phase (when the velocity is halfway between its value before the acceleration and after it), including the time the wave was propagating in to the center and out again. This time delay (measured in periods) was found to correlate well with the structure parameter $V_{.83}$ defined in (20) where $V(x) = \frac{\rho(x)GM(x)}{P(x)XR}$ which also correlates with Q. Alternatively, it was found that the phase delay correlates well with the value of $(R/R_{\odot})^{1.18}/M/M_{\odot}$ and we find Delay (periods) = 2.4 - 0.031 $(R/R_{\odot})^{1.18}/M/M_{\odot}$. The most useful correlation was, however, found to be

$$R/R_{\odot} = 4.05 \times \text{delay (days)} \tag{39}$$

which is equivalent to saying that the signal, which travels a distance equal to the stellar diameter, travels at an average speed of about $8R_{\odot}$ per day. This relation shows that the quantity most directly determined by the timing of the second bump is the radius of the star.

It was found that the bump is clearly visible only when the delay, in periods, lies between 1.60 and 1.05, or for $(R/R_{\odot})$/Period between 6.2 and 4.5. At the larger value the bump disappears, apparently as a result of destructive interference between the ingoing and outgoing waves. The bump is strongest for a delay between 1.25 and 1.45 periods, perhaps involving constructive interference, and for a delay less than 1.10 periods and near 1.0 periods the bump appears to reinforce the next principal outward acceleration and results in an enhanced amplitude as is seen for Cepheids of 15 to 20 days period.

We can use the timing of the bump to give the radius; combining this with the Period - Radius - Mass relation and the observed period gives the mass. If we determine the mean T_e from the mean B-V we can also get the mean bolometric luminosity L.

This procedure applied to β Dor, P = 9.84d, gave $M = 3.4M_{\odot}$, $R = 54.5R_{\odot}$ and for $T_e = 6100°$, $L = 3.7 \times 10^3 L_{\odot}$ or $M_b = -4.2$. This value of mass is about half what would be expected for this luminosity. If correct, we infer that the star has lost mass, presumably in the red giant phase, prior to becoming a Cepheid. Forbes (21) and others have shown that the luminosity is not very much affected by mass loss that takes place after the main sequence phase of evolution.

Some preliminary calculations were pursued to test the sensitivity of this method of determining R to the opacity law. It was found that if the Cox and Stewart opacities are consistently too low by a factor of two in the temperature range from 50,000°K to 500,000°K, then the radii could be increased by 12% and the masses by 40%. It would seem to take an increase of a factor of three or four in the opacity to remove the discrepancy in mass.

Overtones

It is possible to use an initial velocity distribution that initiates a nearly pure first overtone mode with only a small contamination of fundamental. In this way it has been possible to explore the region of instability of the first overtone as well as that of the fundamental. Actually this has been attempted only near the region of parameter space where the first overtone is the most rapidly growing mode.

In the course of exploring RR Lyrae models (2), it was found (see Fig. 28) that, for a given luminosity, low mass

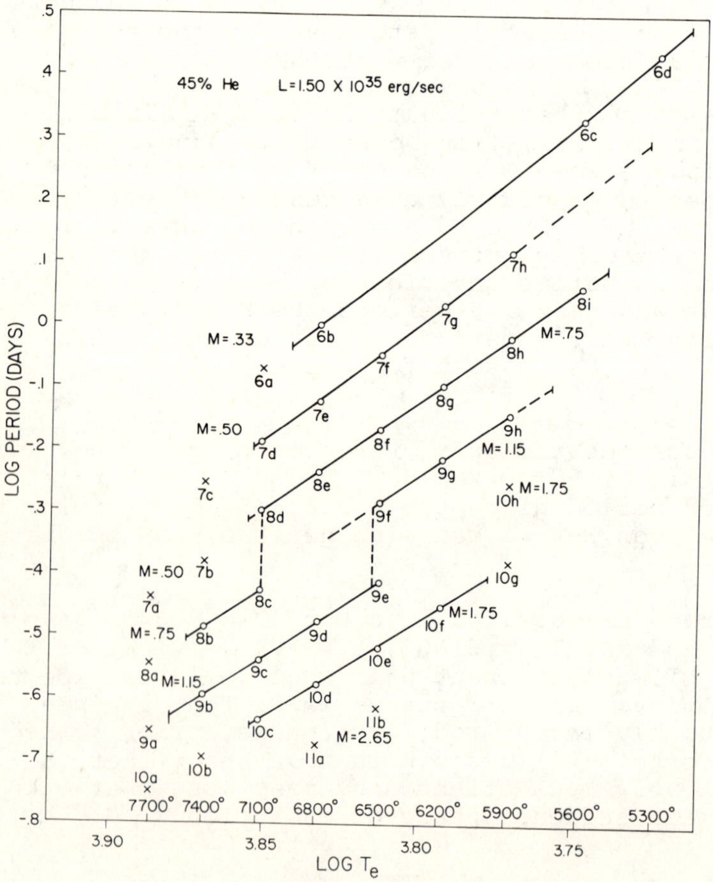

Figure 28

models were unstable only in the fundamental. For higher mass models, the dominant instability at low T_e was in the fundamental whereas at higher T_e it was in the first overtone. For still higher mass models, the entire instability strip was occupied by first overtone pulsators. At that time, no instability in fundamental or first overtone was found at still higher mass; however, recently, Stobie (22) has found second overtone instability in this region. The simplest approximate characterization of the division between fundamental and first overtone instability was that for a given luminosity, there is a shortest period, P_{tr}, at which the fundamental is the dominant mode. At shorter periods, the instability prefers the first overtone.

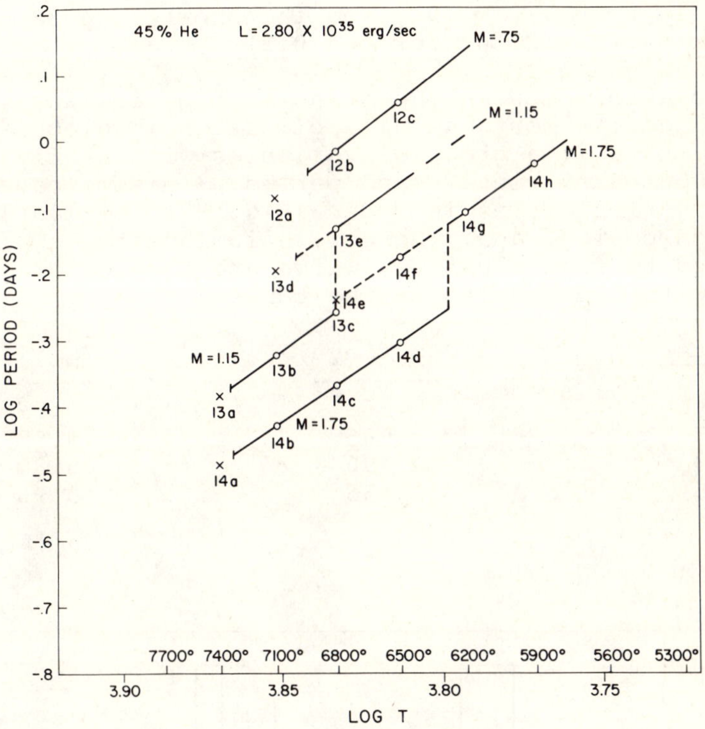

Figure 29

A similar series of models was explored at a higher luminosity (Fig. 29) with the result that again there is a shortest period at which the fundamental is dominant but now P_{tr} is longer. These results led to a relation between P_{tr} and the luminosity which can be written

$$P_{tr} = 0.0625 \, (L/L_{\odot})^{0.57} \qquad (40)$$

This is a line in the period-luminosity diagram (Fig. 31) given by

$$M_b = -0.6 - 4.4 \log P_{tr} \tag{41}$$

such that below and to the right of the line, large amplitude pulsation prefers the fundamental mode, whereas above and to the left of the line, large amplitude pulsation prefers the overtones. I have verified that this relation is approximately independent of He abundance. The P_{tr} has also been explored at periods as long as 6 days at $M_b = -4.0$. Following Stobie's second overtone models, I checked a few examples and surmise that a similar line can be drawn marking the short period boundary of large amplitude first overtone pulsation (Fig. 31). It may well be that a series of lines can be drawn marking regions of instability for higher and higher overtones. We can also approximately separate the modes by the value of the parameter $V_{0.83}$. The shortest period fundamental corresponds approximately to $V_{0.83} = 19.5$ and the shortest period first overtone to $V_{0.83} = 21.0$ (for Y = 0.3).

This relation between the transition period and the luminosity finds its most useful application in globular clusters where there is a group of variables with nearly the same luminosity but differing T_e and period. In some cases there are enough variables, as in ω Cen (Fig. 30) to clearly define a transition period, and, thereby, a luminosity. This has been applied to four clusters, which are particularly rich in variables with the results in the table

Cluster	P_{tr} (days)	M_{bol}
ω Cen	0.565	+0.57
M 15	0.565	+0.57
M 3	0.496	+0.80
M 5	0.455	+0.96

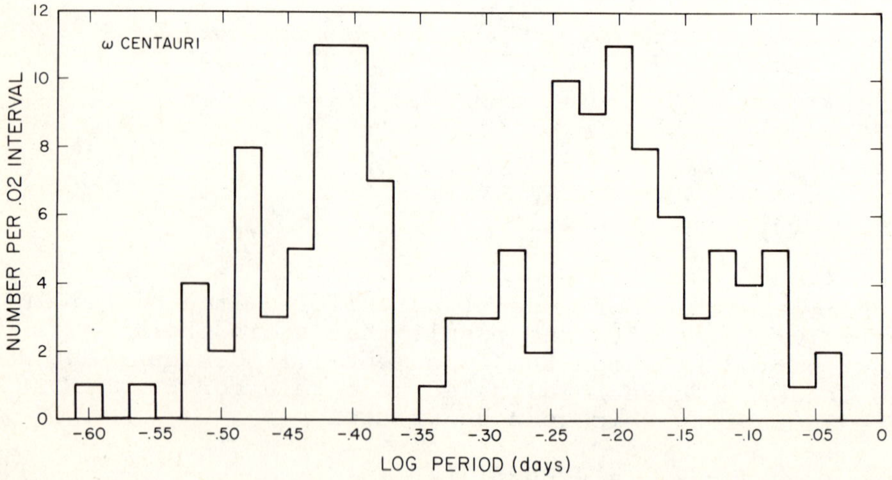

Figure 30

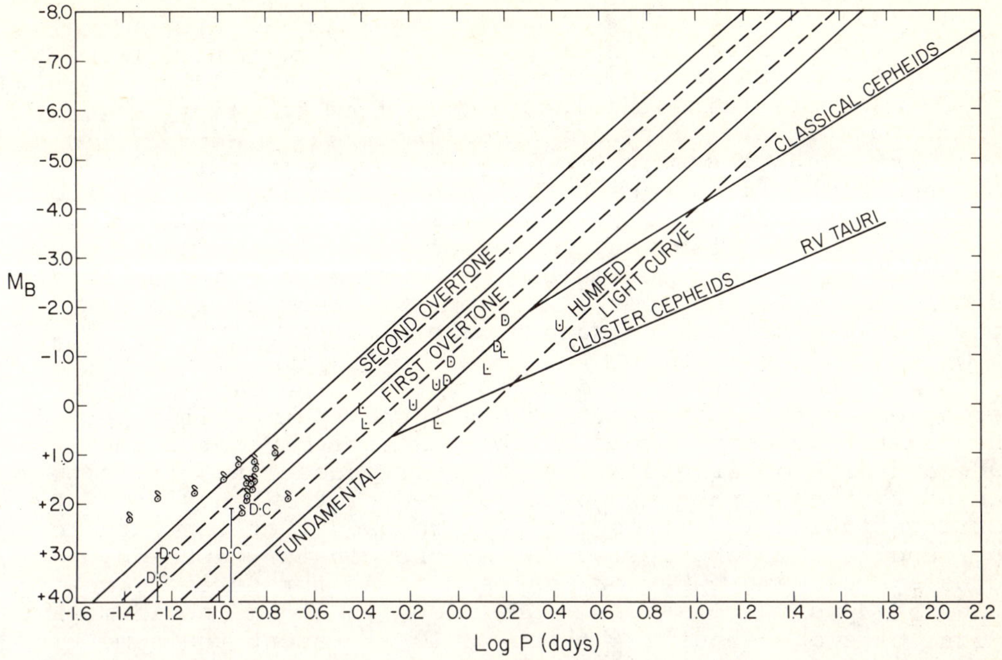

Figure 31

The Period-Luminosity Diagram and Systematics of Cepheid-Type Variables

Much of the systematic behavior of pulsating variables can most easily be represented in a period-luminosity diagram (Fig. 31). By suitable choice of the mass of a star, models which are pulsationally unstable can be found over a wide region in this diagram. Indeed, with the possible variation of T_e because of the finite width of the instability strip, a range of masses can give the same period at fixed luminosity.

The results on the transition period lead to the diagonal lines which separate strips for the second overtone, for the first overtone, and the region to the right which is occupied by the fundamental. The line marked "cluster Cepheids" marks the region which is occupied by variables of fixed mass about $0.5M_\odot$. This line has a period-luminosity relation of the form

$$M_b = +0.05 - 2.1 \log P \tag{42}$$

The line marked "classical Cepheids" marks the region occupied by classical Cepheids with a period-luminosity relation approximately of the form

$$M_b = -1.- 3.0 \log P \tag{43}$$

Finally, a line marked "humped light curve" approximately marks the way in which the appearance of a prominent bump depends systematically on luminosity and period. It is parallel to the line demarking the fundamental from the first overtone and I conclude that lines of this slope ($\Delta M_b = -4.4 \Delta \log P$) are lines of approximate homology as far as pulsation behavior is concerned. We have seen that along this line, R $\sim$ Period so since $L \sim R^2 T_e^4$ and $T_e \sim P^{-0.06}$, we see that $L \sim P^{1.76}$ or $P \sim L^{0.57}$. This line would intersect the cluster Cepheids between 1 and 2 days period and we would expect a prominent second bump also for some of these. No clear example is known but B L Her may be such a case.

Some attempt has been made to fit all the Cepheid-type variables in this diagram. It was found that at long periods, the cluster Cepheid models exhibit excessively large amplitude and develop irregularity in their pulsation. It is believed that this phenomenon is related to the behavior of RV Tauri variables and it is thereby proposed that they are to be found at the long period end of the cluster Cepheids. Actually the same behavior has been found for periods greater than 150 days in classical Cepheid models and we could mark a line of slope about 4.4 which would represent the same kind of behavior for various masses. Observationally, in the Magellanic clouds and in M31, however, there do not seem to be variables between the Population II group, here called cluster Cepheids, and the Population I group, or classical Cepheids.

At short period end, we have explored the location of δ Scuti stars and of dwarf Cepheids. Since the δ Scuti stars are of small amplitude, we cannot be sure that the non-linear aspects are important and we cannot therefore be sure in what mode they are vibrating. If we could determine this, we could then independently determine masses for comparison with estimates from cluster membership. The dwarf Cepheids are apparently less luminous than the δ Scuti stars and would seem to require masses of $\frac{1}{4}M_\odot$ or less. Unfortunately, the luminosities are not quite well enough known to make such a determination certain. Finally, an effort was made to understand the appearance in certain old clusters (23) (Draco, Ursa Minor, Leo II) of short period variables significantly more luminous than RR Lyrae stars. They are shown on the diagram by the letters D, L, U. The only explanation I have found is that these variables are about twice as massive as the bulk of the RR Lyrae stars--a conclusion that is understandable if the RR Lyrae stars have lost mass.

Distance Determinations

The determination of the stellar radius for certain variables which show a prominent second bump, such as is seen in classical Cepheids of period near 10 days, presents a new basis for distance determination. S Nor (24) is perhaps the best example of a cluster variable for calibration purposes. I have estimated the delay of the reflected wave as 1.36 x

9.75 = 12.96 days so $R/R_\odot$ = 53.7. Taking <B - V> = 0.75, we get log T_e = 3.755, log $L/L_\odot$ = 3.428, and M_b = -3.85, whereas Kraft's value is -4.05. This serves to emphasize the fact that the relation (a) leads to consistently smaller radii than is customary and to correspondingly smaller masses.

The best case for application of this method in the Magellanic clouds appears to be HV 2432 (P = 10.92d) (13) in the LMC. It shows a prominent second bump in the light curve and, by comparison with S Sge which is similar and where a velocity curve is known, we deduce a delay of 1.43 periods and $R/R_\odot$ = 63.2. Models of these characteristics have Q = 0.0435 and so we get $M/M_\odot$ = 4.0. The mean <B - V> of this variable is 0.52 and, corrected for a reddening of 0.05 gives $\langle B - V\rangle_o$ = 0.47. Then from (29) we get log T_e = 3.804, log $L/L_\odot$ = 3.766 and M_b = -4.695. Now V = -14.23 so V_o = 14.08 and m_b = 14.06. The modulus is thus 18.75.

No variable measured by Gascoigne and Kron in the SMC shows a similar prominent bump but variables 1400 (P = 6.65d) and 827 (P = 13.47d) closely bracket the range where a prominent bump is anticipated and the distance modulus of the SMC is determined to be 0.4 greater than for the LMC by the requirement that neither of these variables should show the bump.

These moduli are greater by 0.25 than those of Sandage and Tamman. The increase over an earlier report is solely due to an increase in log T_e from a change in (29) by 0.025. This immediately results in an increase in brightness by 0.25 and also in modulus. Alternatively, we can use this method as a device to relate galactic and cloud variables without resorting to absolute calibration. If we adopt Kraft's calibration of S Nor, we would then deduce a modulus of 18.95 for the LMC and 19.35 for SMC. This tentative conclusion emphasizes the need for more careful study of galactic cluster variables and cloud variables which show this bump.

REFERENCES

1. Christy, R.F., Rev. Mod. Phys. 555, 36 (1964).

2. Christy, R.F., Ap. J., 144, 108 (1966).

3. Cox, J.P., Ap. J. 122, 286 (1955).

4. Zhevakin, S.A., Russ. Astron. J. 30, 161 (1953) and Ann. Rev. Astron. and Astrophys. 1, 367 (1963).

5. Baker, N. and Kippenhahn, R., Z. Astrophys. 54, 114 (1962)

6. Cox, J.P., Ap. J. 138, 487 (1963).

7. Baker, N. and Kippenhahn, R., Ap. J. 142, 868 (1965).

8. Christy, R.F., Ap. J. 136, 887 (1962).

9. Cox, A.N., Brownlee, R.R. and Eilers, D.D., Ap. J. 144, 1024 (1966).

10. Eddington, A.S., Observatory 40, 290 (1917).

11. Cox, J.P., Aerodynamic Phenomena in Stellar Atmospheres, R.N. Thomas ed. (Willmer Bros., Birkenhead, England, 1967), p.3.

12. Sandage, A. and Tammann, G.A., Ap. J. 151, 531 (1968).

13. Gascoigne, S.C.B. and Kron, G.E., M.N. 130, 333 (1965).

14. Dickens, R.J., Observatory 86, 18 (1966).

15. Arp, H.C., A.J., 65, 404 (1960).

16. Payne-Gaposchkin, C. and Gaposchkin, S., Smithsonian Contrib. 9, (1966).

17. Iben, I.Jr., Ann. Rev. Astron. and Astrophys. 5, 571 (1967).

18. Hoffmeister, E., Kippenhahn,R. and Weigert, A., Z. Astrophys. 60, 57 (1964).

19. Hoffmeister, E., Z. Astrophys. 65, 164 (1967).

20. Christy, R. F., Q.J.R.A.S. 9, 13 (1968).

21. Forbes, J.E., Ap. J. 153, 495 (1968).

22. Stobie, R.S., Thesis, Cambridge Univ. (1968).

23. Baade, W. and Swope, H.H., A.J. 66, 300 (1961).

24. Feast, M.W., M.N. 136, 141 (1967).

7

WHITE DWARFS

J. P. Ostriker

Princeton University
Princeton, N.J.

I. Introduction

Bessel founded what he called the "astronomy of the invisible" in 1850 when he wrote to Humboldt:

> I adhere to the conviction Procyon and Sirius F form real binary systems, consisting of a visible and invisible star. There is no reason to suppose luminosity an essential quality of cosmical bodies. The visibility of countless stars is no argument against the invisibility of others.

As it turned out, Bessel's statement was truer than he realized, luminosity is not essential for the structure of a white dwarf. Procyon and Sirius have indeed been found to be binary systems. The "invisible star" in each case turned out to have the following characteristics:

STAR	$M/M_\odot$	$L/L_\odot$	$R/R_\odot$	ρ(cgs)
Procyon B	0.7	1.0×10^{-3}	$\simeq10^{-2}$	$\simeq10^{5}$
Sirius B	1.0	3.0×10^{-3}	$\simeq10^{-2}$	$\simeq10^{5}$

How are we to understand the existence of stars having approximately the sun's mass but such low luminosity? They are obviously not on the main sequence. The principal thing to note is the very small radius—about that of a planet—it is this very small radius which allows us to see the stars in the visible region in spite of their very low luminosity. The stars are not burning nuclear fuel or they would be much brighter. The common explanation for such stars is that they have completely exhausted their nuclear fuel—they are dying stars. Alternately, they might be stars which have never turned on. If $M\lesssim0.1\ M_\odot$, stars can contract to the point where degeneracy occurs; but such stars would be red dwarfs rather than white dwarfs, and would have radii intermediate between white dwarf and main sequence values (see crossed line in Figure 8).

Let us consider what happens to a star when its nuclear fuel is used up. For a qualitative idea of the process use the scalar virial theorem. To derive this equation we multiply the equation of hydrostatic equilibrium in its conventional form by $4\pi r^3 \rho$, and integrate over the star. Using integration by parts on the left side and the vanishing of pressure at the surface, we obtain

$$3 \quad P\,\underline{dr} = -\, W \tag{1}$$

where $W = -\int_0^R \rho \frac{GM_r}{r}\,\underline{dr}$. Thus, twice the thermal energy is equal to the negative of the gravitational energy. Assuming the absence of any other forces acting on the star, the total energy is

$$E = W + \frac{1}{\gamma-1} \quad P\,\underline{dr} = -\,\frac{(3\gamma-4)}{3(\gamma-1)}\,|W|, \tag{2}$$

where γ is the correctly defined average ratio of specific heats. For a polytrope, the gravitational energy $W = -\left(\frac{3}{5-n}\right)\frac{GM^2}{R}$. Now imagine the fuel in a star has been used up. Then all the star can do is lose energy. Hence, $\Delta E<0$, so $\Delta R<0$ and the star contracts. Pursuing this line of reasoning further, can we suppose R will go to zero? The star then appears to have an infinite store of energy and should last forever. However such an argument is invalid (even apart from considerations of general relativity) as we shall see presently.

What, then, do we predict for the end point of evolution? We know from evolutionary calculations that massive ($M>4M_\odot$?) main sequence stars will end explosively as supernovae. For low mass stars ($M<1.3M_\odot$?) it is found that burning can terminate at C^{12} or O^{16} after which the stars contract and lose energy. Let us follow these stars to the white dwarf state.

Consider a star of approximately solar mass in the final stages of evolution, let us test the assumption that the pressure can be given by the perfect gas law,

$$P = \frac{k}{\mu H}\,\rho\, T$$

From the virial theorem we see that under these conditions $\bar{T} \propto \frac{M}{R}$, so that the temperature should become very large:

$$\int P\,\underline{dr} = \int \frac{P}{\rho}\,dM = \frac{k}{\mu H}\,\bar{T}M = \left(\frac{1}{5-n}\right)\frac{GM^2}{R} \tag{3}$$

However, $\rho \propto M/R^3$, so the density increases even more steeply with decreasing R than the temperature. We shall see that ultimately the perfect gas law must fail. The typical momentum difference between electrons is

$$\Delta p_e \simeq (M_e kT)^{\frac{1}{2}} \tag{4}$$

and the typical distance is

$$\Delta q_e \simeq \mu_e (M_e/\rho)^{\frac{1}{3}}, \tag{5}$$

where M_e is the mean molecular weight per electron. Using equations (4 & 5) the typical volume occupied by an electron in phase space is

$$(\Delta p_e \; \Delta q_e)^3 \simeq \left[M_e^{\frac{5}{6}} \left(k \; \frac{GM}{R} \; \frac{\mu H}{k} \right)^{\frac{1}{2}} \mu_e \left(\frac{4\pi R^3}{3M} \right)^{\frac{1}{3}} \right]^3$$

$$\approx \left[(M_e^{\frac{5}{6}} \mu_e (G\mu H R_\odot)^{\frac{1}{2}} \left(16 M_\odot \right)^{\frac{1}{6}} \left(\frac{M}{M_\odot} \right)^{\frac{1}{6}} \left(\frac{R}{R_\odot} \right)^{\frac{1}{2}} \right]^3$$

$$\approx \left[6.3 \times 10^{-26} \; \left(\frac{M}{M_\odot} \right)^{\frac{1}{6}} \left(\frac{R}{R_\odot} \right)^{\frac{1}{2}} \text{ gm. cm}^2/\text{sec} \right]^3 \tag{6}$$

using a composition $X = 0$, $Y = 0.9$, $Z = 0.1$.
The constant may be expressed in terms of Planck's constant. Then,

$$\Delta p_e^{\;3} \; \Delta q_e^{\;3} \approx 10^3 h^3 \left(\frac{M}{M_\odot} \right)^{\frac{1}{2}} \left(\frac{R}{R_\odot} \right)^{\frac{3}{2}} \tag{7}$$

When a star of solar mass contracts to a radius $R \simeq 10^{-2} R_\odot$, the volume occupied by an electron in phase space is approximately h^3. As first pointed out by R. H. Fowler in 1926, the Pauli exclusion principle becomes important (the electrons "repel" one another) and Fermi-Dirac statistics must be used at this point. Thus electron degeneracy would appear just at the point in the contraction of stars where the radius was that of the observed white dwarfs. As an aside, we note that neutron degeneracy would appear for ($1\ M_\odot$) stars just at the point where the moment of inertia is that of the recently observed pulsars.

II. The Theory of Zero Temperature Stars

A. The Classical Treatment of Equilibrium

1. The Pressure Density Relation

The classical theoretical treatment of the white dwarf problem involves two assumptions:

(a) The nondegenerate partial pressure of the ions is very small compared to the degenerate electron pressure throughout the main part of the star and hence can be neglected.

(b) The electrons are treated as a zero temperature Fermi gas. The most complete treatment of this can be found in Chandrasekhar's Introduction to Stellar Structure (1939).

According to the Pauli exclusion principle, the maximum number of electrons ΔN in phase space of volume ΔV is

$$\left.\begin{aligned} \Delta N &= \Delta V\ 4\pi p^2 dp\,(2/h^3) \\ \text{or}\quad n(p)\,dp &= 8\pi p^2 dp/h^3 \end{aligned}\right\} \quad \text{for } p \le p_o \tag{8}$$

$$\text{and} \qquad \Delta N = 0 \qquad \text{for } p > p_o$$

where p_o is the maximum momentum and each unit cell of phase space has a volume h^3 and includes two anti-parallel electrons. Integrating from 0 to p_o, the number density of electrons in a completely degenerate electron gas is then related to the maximum momentum by

$$n = \frac{8\pi}{3}\,\frac{p_o^3}{h^3} \tag{9}$$

The pressure integral is given by

$$P = \frac{1}{3}\int_0^{p_o} n(p)\,pv\,dp \tag{10}$$

where pv is the momentum flux and the 1/3 factor arises from integration over a solid angle. At the high densities found in white dwarfs, a large fraction of the electrons may move at velocities comparable to the speed of light; hence, we use the relativistic formula

$$v = \frac{p/m_e}{(1 + p^2/m_e^2c^2)^{\frac{1}{2}}}$$

Insertion of this expression for v into the pressure integral yields

$$P = \frac{\pi m_e^4 c^5}{3h^3}\,f(x) = Af(x) = 6 \times 10^{22}\,f(x)\ \text{dynes/cm}^2 \tag{11}$$

where

$$f(x) = x(2x^2 - 3)(x^2 + 1)^{\frac{1}{2}} + 3\log\,(\sqrt{1 + x^2} + x)$$

and

$$x = p_o/m_e c$$

Writing the electron density in terms of this dimensionless momentum parameter, we obtain

$$n = \frac{8\pi m_e^3 c^3 x^3}{3h^3} = 5.9 \times 10^{29} x^3/\text{cm}^3 \tag{12}$$

The Fermi energy (maximum electron energy) is given by

$$E_F = m_e c^2 (1 + x^2)^{\frac{1}{2}} = (p_o^2 c^2 + m_e^2 c^4)^{\frac{1}{2}} \tag{13}$$

and the specific heat per electron (assuming the temperature is very "small" but not zero) by

$$C_v = \frac{\pi^2 k^2}{mc^2} \frac{(1 + x^2)^{\frac{1}{2}}}{x^2} \text{ T per electron.} \tag{14}$$

In the limit of small x—that is, for non-relativistic electrons—$f(x) \simeq \frac{8}{5} x^5$. In the limit of very large x-relativistic degeneracy-$f(x) \simeq 2x^4$.

The parameter x is related to the physical density of the white dwarf by

$$\rho = Bx^3 = 10^6 \mu_e x^3 \text{ gm/cm}^3. \tag{15}$$

From the asymptotic forms of x, it follows that the relation between P and ρ is such that

$$P = 1.0 \times 10^{13} \left(\frac{\rho}{\mu_e}\right)^{5/3} \text{ gm/sec.} \tag{16}$$

at low densities ($x \ll 1$, $\rho \ll 10^6$ gm/cm^3)

$$P = 1.2 \times 10^{15} \left(\frac{\rho}{\mu_e}\right)^{4/3} \text{ gm/sec.} \tag{17}$$

at high densities ($x \gg 1$, $\rho \gg 10^6$ gm/cm^3). The former is the $P(\rho)$ relation in a polytrope of index $n = 3/2$, the latter, the relation in a polytrope with $n = 3$.

2. The Mass Radius Relation; the Mass Limit

Under the limiting circumstances being considered, the pressure is given in terms of a function of the density, independent of the temperature; thus the hydrostatic problem can be completely separated from the thermodynamic problem. The entire structure of the star can be obtained by combining the equations for hydrostatic equilibrium and mass continuity with appropriate boundary conditions. A one parameter family of white dwarf models is obtained, corresponding to different values of the central density (or, equivalently, different values of the mass or radius).

Figure 2 shows that for white dwarfs unlike main sequence stars the radius decreases with increasing mass. Also, according to Chandrasekhar, an upper limiting mass of $M_3 = 1.44(\frac{2}{\mu_e})^2 M_\odot$ exists for non-rotating stars. As we shall see subsequently, the inclusion of rotation eliminates the existence of a mass limit fixed by equilibrium conditions.

It is one of the great triumphs of theoretical astrophysics that the observed white dwarf stars agree so well with the predictions of this rather simple and straightforward theory.

From Figure 1, we see that the observed masses and radii of three white dwarfs fit fairly closely to the theoretical line, although somewhat above it. The error bars on the observations are large, and there is much dispute as to how big they should really be. But, we can say that the agreement is reasonable. All three have masses less than M_3. However, it is a little suspicious that all the radii are too big.

After noting that the classical theory is almost certainly correct in its general conclusions we will turn to the smaller effects: how will changes in the equation of state, thermal effects, rotation, and magnetic fields alter the classical picture and help us to understand the complexity of the accumulated white dwarf observations.

3. Detailed Comparison of the Theory with Observations

In order to test the validity of the theory the total amount of observational data available must be considered. The three white dwarfs belonging to binary systems with known orbits and masses (Harris, et al., 1963) have already been mentioned:

Star	Mass/$M_\odot$
40 Eri B (α_2Eri B)	0.45
Sirius B (αCMa B)	1.05
Procyon B (αCMi B)	0.65

The mean mass of these binary components is 0.7 $M_\odot$.

From relativity theory the wavelength of light leaving a region of strong gravitational field is redshifted by an amount proportional to M/R. For white dwarfs this redshift is quite large and easily observable. If any independent measure of the

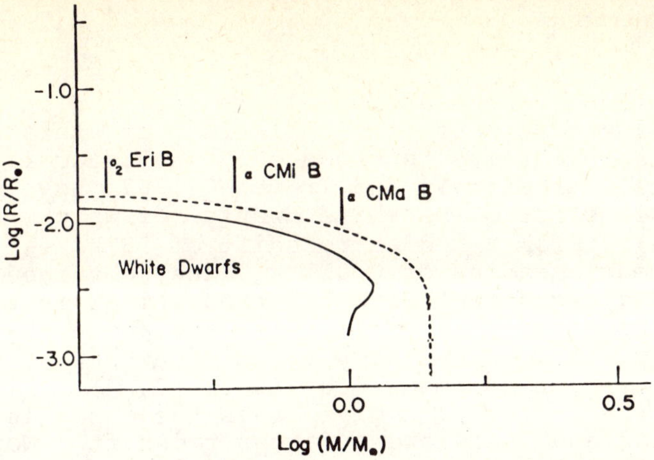

Figure 1. The mass-radius relation for white dwarfs adapted from Auer and Woolf (1965, Fig. 1). The upper dashed curve is for Chandrasekhar's (1938) completely degenerate model with $\mu_e = 2$. The lower curve is for Hamada and Salpeter's (1961) iron white dwarf. The curl near 1.1 $M_\odot$ is due to inverse beta decay. Vertical lines indicate positions of the three white dwarfs with well-determined mass and radius.

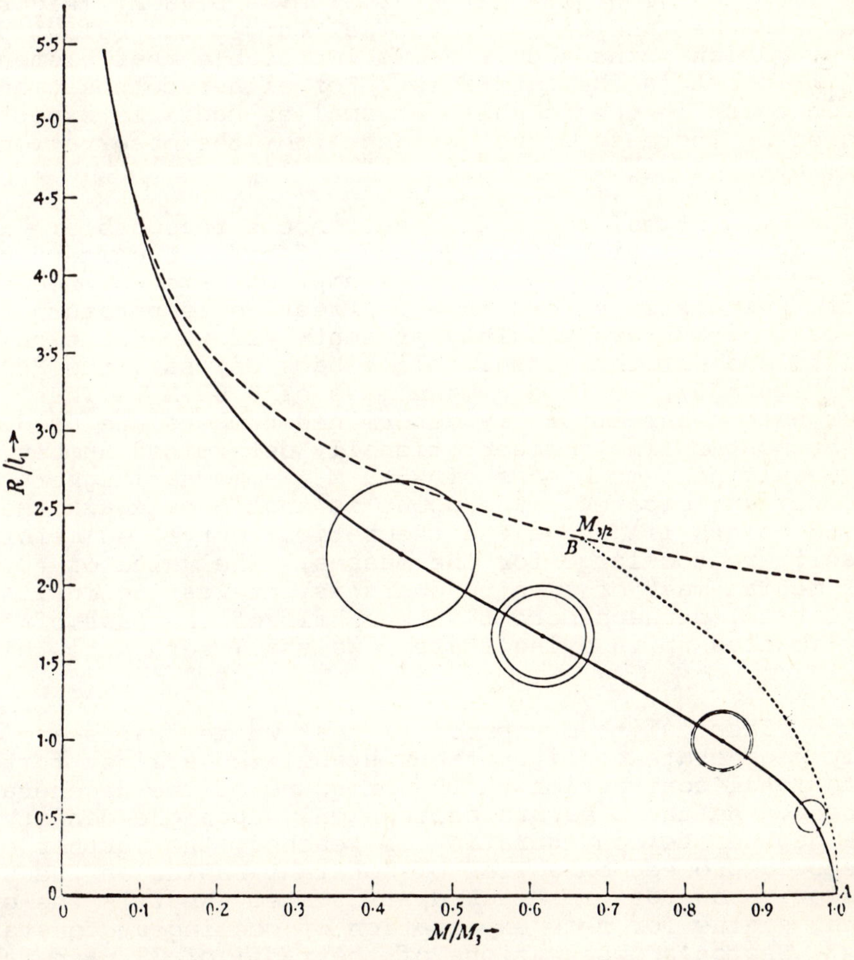

Figure 2. The solid curve represents the mass-radius relation for completely degenerate white dwarfs. The dashed curve, to which the solid curve tends asymptotically as $M \to 0$, is the relation obtained for a polytrope of index $n = 3/2$ by first order approximation for the central density. M_3 is the limiting mass and ℓ_1 is a unit of length $(4\pi m \mu_e H)^{-1}(3h^3/2cG)^{1/2}$. (Chandrasekhar 1939, Fig. 31, p. 428.)

radius is available, such as from luminosity and surface temperature, the mean mass of a collection of stars can be obtained. The luminosity is known only statistically from proper motions. Greenstein and Trimble (1967) have measured Balmer lineshifts for 37 DA white dwarfs (of which 40 Eri B is an example). From a median redshift of 58 km/sec. and a median photometric radius of 0.011 $R_\odot$, they obtained a mass of 0.98 $M_\odot$—in disagreement with the theoretical mass-radius relation. Hamada and Salpeter (1961) have shown that redshifts for a given mass can vary by as much as 40%, depending upon the composition assumed (see Figures 3, 4). Figure 3 shows that an Fe^{56} star of a given mass has a smaller radius than a He^4 star and therefore a larger redshift. Note, however, that, at a given mass, all compositions give a <u>smaller</u> radius than that given by the classical mass-radius relation whereas the observed mean radius (0.011 $R_\odot$) is <u>larger</u> than that anticipated for a 0.98 $M_\odot$ white dwarf on the basis of the Chandrasekhar equation of state. The photometrically derived radius and the observed redshift lie near the theoretical position of a He^4 star of mass 0.78 $M_\odot$ (Figure 5). But both the He^4 and Fe^{56} theoretical curves lie below the observed median point and it seems improbable that elements with $\mu_e<2$ exist in the interiors. For either composition, a steep increase in the redshift at smaller radii is expected. Such a steep increase is not evident from the observations (Figure 6).

The statistical redshift result seems reasonable: it is in good agreement with the measured redshift of white dwarfs in the Hyades where statistical assumptions are not necessary. However, systematic errors in the effective temperature to color calibration are possibly present. If we accept only the redshift, and put the redshift line back on, say, the He^4 curve (Figure 5), we find a mean mass of 0.85 $M_\odot$. Thus whether both observationally determined numbers are used, or only the most reliable observationally determined number and theory, we find a mean mass of $\simeq$0.9 $M_\odot$, somewhat higher than previously anticipated. The point is that once again the observed points lie above the theoretical curve; as before, the radii are too large for the masses. The value of $\simeq$0.9 $M_\odot$ as the median mass of a white dwarf is interesting in itself; for, if true, perhaps more than one half of the mass of our galaxy is tied up in dying stars. We shall return to this point later.

The classical theory predicts that white dwarfs are completely degenerate configurations stabilized against further gravitational contraction by the pressure of the degenerate electrons. As the pressure depends only upon the density, once a degenerate configuration is reached, the further evolution should proceed on a line of constant radius as the stars cool slowly towards the black dwarf state. What is the observational status for this expectation of cooling at constant radius? Seaton's observations of the radii of 37 planetary nebulae central stars show that, indeed, these stars are approaching the pre-white-dwarf stage as the luminosity of the central stars begins to decrease (Figure 7). From his observations he finds a rather short time scale for the pre-white

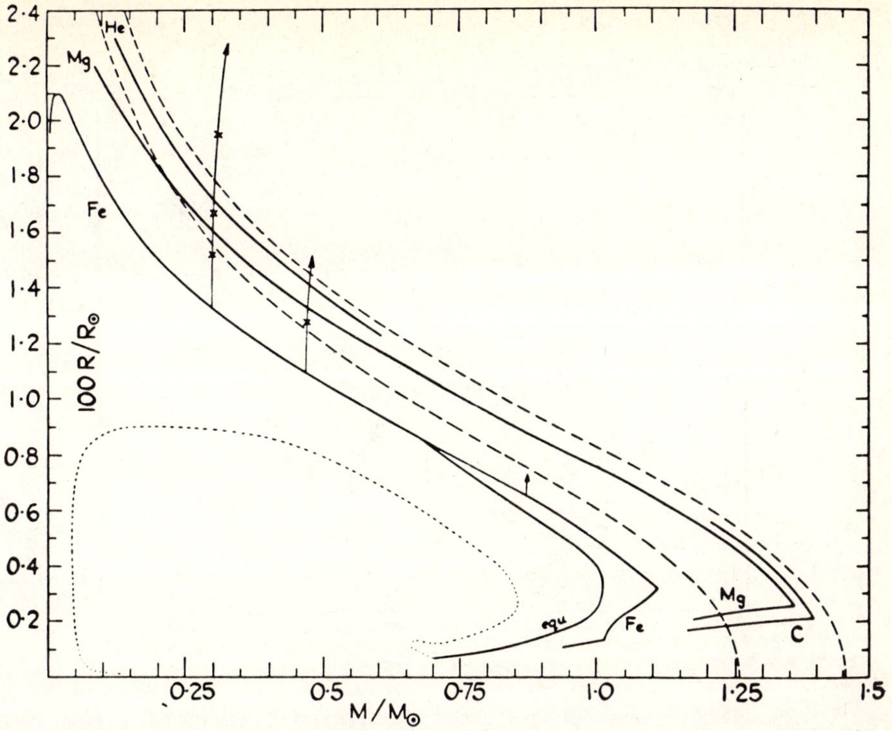

Figure 3. The relation between mass and radius for zero-temperature stars for He^4, C^{12}, Mg^{24}, and Fe^{56}. The curve marked "equ" denotes equilibrium composition at each density. The dashed curves denote the Chandrasekhar models, the upper one for $\mu = 2$ and the lower one for $\mu = 2.15$. The dotted curves denote neutron stars. The vertical arrows denote stars with H^1 in the outer layers. (Hamada and Salpeter 1961, Fig. 1.)

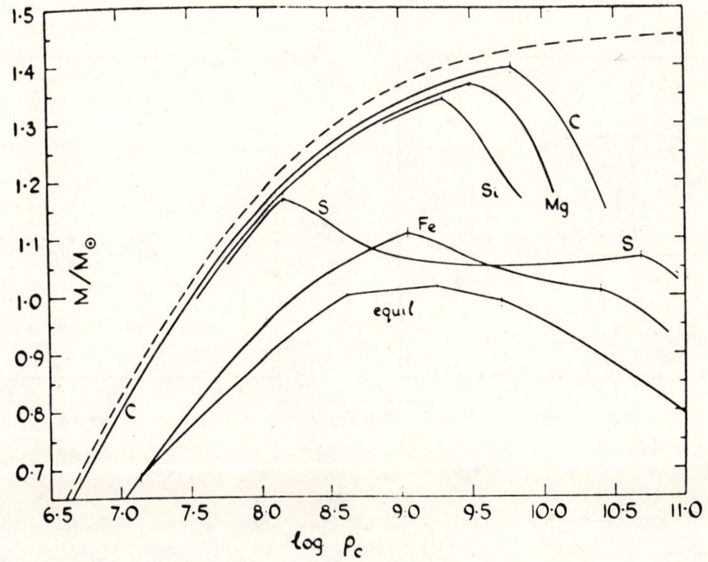

Figure 4. The relation between mass and central density for zero-temperature stars for C^{12}, Mg^{24}, Si^{28}, S^{32}, and Fe^{56} and for equilibrium conditions (Hamada and Salpeter 1961, Fig. 2). The dashed line is derived using the Chandrasekhar equation of state and $\mu_e = 2$.

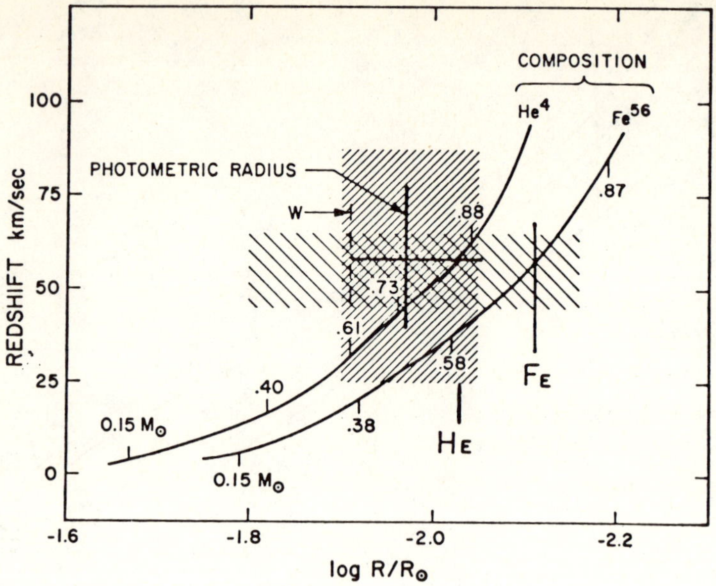

Figure 5. The dependence of the theoretical redshift on mass and radius for He^4 and Fe^{56} (from Hamada and Salpeter 1961). The observed median redshift and radius is shown as a cross with a hatched zone showing the error. The median radius, as given by Weidemann, is the vertical dashed line labeled "W." The two vertical lines marked "He" and "Fe" are the radii given by the redshift and the theoretical mass-radius relation, corresponding to 0.86 $M_\odot$ for He and 0.73 $M_\odot$ for Fe. (Greenstein and Trimble 1967, Fig. 3.)

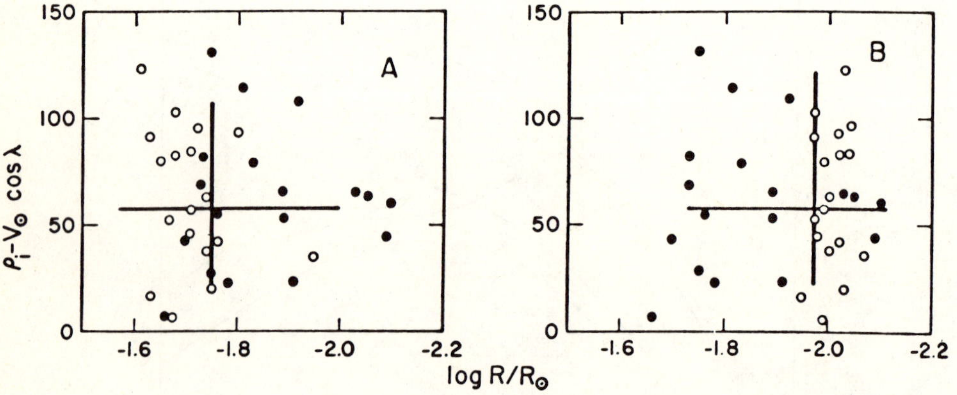

Figure 6. The individual velocities of DA type white dwarfs, corrected for normal solar motion. Photometrically derived radii are indicated by solid dots. Diagram A shows the stars of uncertain radii (open circles) plotted with their luminosities from the upper sequence. Diagram B shows the radii from the lower sequence (open circles). The median photometric radius, for this assumption, and median redshift are shown by crosses. (Greenstein and Trimble 1967, Fig. 2.)

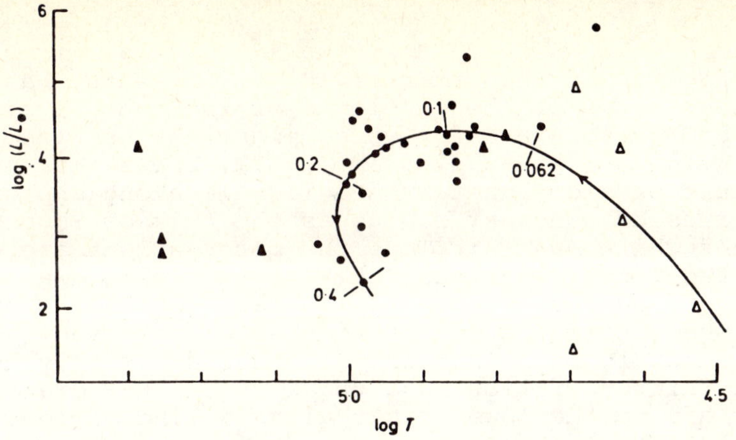

Figure 7. The H-R diagram for the central stars of planetary nebulae. Values of nebular radii, in parsecs, are indicated. ▲ optically thick to HeII and HI; ● optically thick to HeII and thin to HI; △ HeII lines are absent and nebula is optically thick to HI. (Seaton 1966, Fig. 7.) Stars are approaching the white dwarf state of approximately the point labeled "0.4."

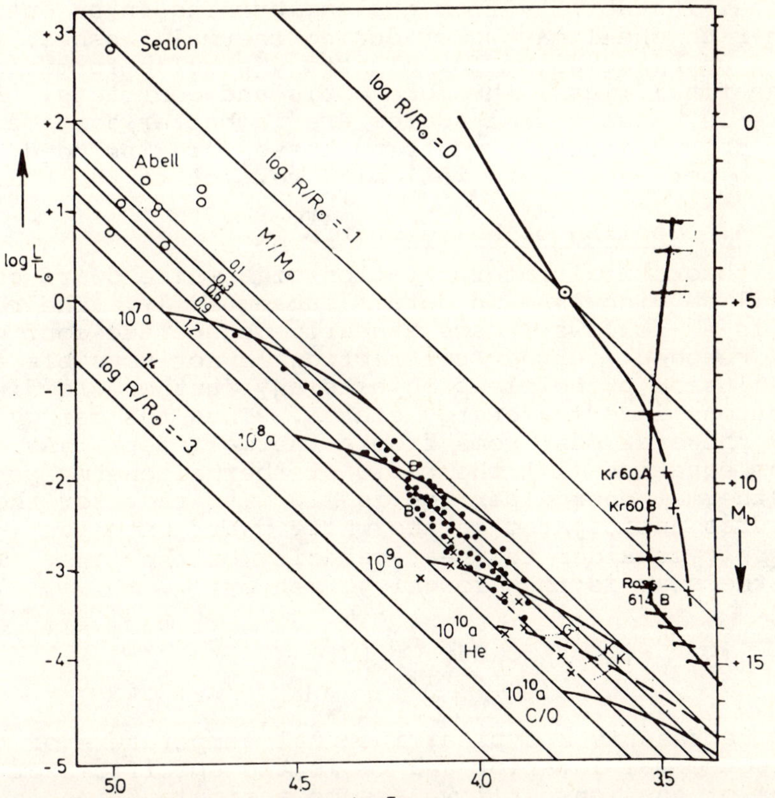

Figure 8. The H-R diagram for white dwarfs. The lines of constant composition are: —— 50% C, 50% O; – – He. Spectral types are ● DA; B DB; x DC, λ4135, and λ4670; G DG; K DK; o central stars of planetary nebulae. (Weidemann 1967, Fig. 1.) Crossed line shows approximate evolutionary path of an 0.05 $M_\odot$ star.

dwarf contraction on the order of $10^4 - 10^5$ years, based on the expansion rates of the envelopes of the central stars. Seaton's observations are for young planetary nebulae. Figure 8, Weidemann's H-R diagram (1967), shows that Abell's central stars of old planetary nebulae lie along approximately the same lines of constant radius as the bulk of Eggen and Greenstein's white dwarfs. Thus the observations are consistent with the picture that the bulk of the white dwarfs were once very hot, central stars of planetary nebulae. Contraction continued until the radius reached $\approx 10^{-2} R_\odot$ and the luminosity fell to $\approx 10\, L_\odot$. Thereafter, the stars simply became cooler and fainter, following a track in the H-R diagram from the upper left to the lower right along a line of constant radius.

Considerable progress has been made recently in beginning to understand the stages of evolution just prior to the white dwarf state. Calculations by Vila (1966) of evolutionary model sequences with energy sources due only to contraction imitate the observed cooling curves on the H-R diagram as seen in Figure 9. However, the calculated cooling times are much too long when compared with Seaton's times based on the estimated envelope age. Rose (1967) proposes an alternate theory. Pulsational instability from nuclear burning in an outer hydrogen-rich shell may occur during thermal flashes; this is suggested as a mechanism for planetary nebulae formation. After the final flash, the core cools and contracts. As seen in Figure 10, Rose's time scales are in better agreement with Seaton's observations if neutrino losses are included, but his results do not agree very well with respect to the radii.

4. Cooling of White Dwarfs

The theoretical interpretation of a white dwarf cooling curve will be discussed in detail subsequently. The rough picture is as follows: once a nearly degenerate configuration has been reached, further contraction is not possible and the Pauli exclusion principle prohibits any further crowding of the electrons into low energy states. Thus the energy liberated by the star must come from a thermal store. As can be seen from equation (14) the ratio of thermal energy per electron to thermal energy per ion is $\approx 10^{-9}$ so that for the temperatures anticipated in the white dwarf interior ($\lesssim 10^7$°K) we need only consider the heat capacity of the ions. The heat loss of the star is thus roughly given by

$$-dQ = -C_V M dT_i = L dt \qquad (18)$$

where T_i represents a typical internal temperature of the nearly isothermal interior and C_V is the specific heat per unit mass of the ions. If we assume a simple Kramer's opacity ($\kappa \propto \kappa_0 \rho T^{-7/2}$) law, the luminosity can be related to the central (i.e. interior) temperature by integrating through to the nondegenerate surface layers, yielding

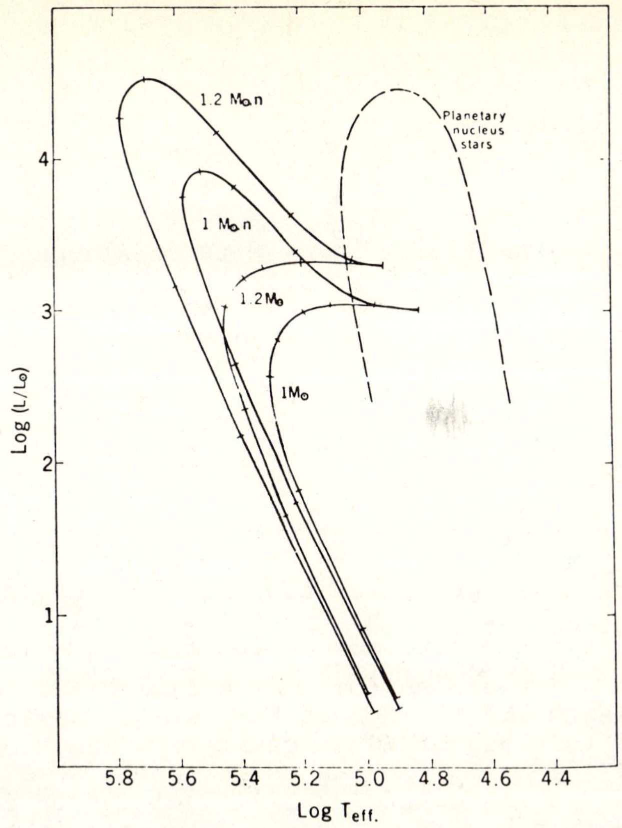

Figure 9. Evolutionary tracks in the H-R diagram. The superscript n indicates a sequence of models with neutrino losses included. The bars correspond to Vila's models listed in his paper. The dashed curve represents the evolution of planetary nuclei as proposed by Seaton. (Vila 1966, Fig. 1.)

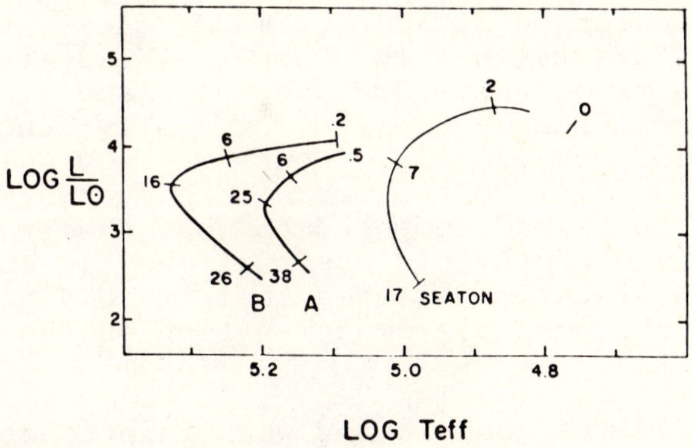

Figure 10. The evolutionary path of a 0.75 $M_\odot$ star after its last relaxation oscillation is compared with Seaton's (1966) results for the observed nuclei of planetary nebulae. Plasma neutrino emission is included in the models described by curve B but not by A. The numbers shown on curves A and B are the time measured in thousands of years. The numbers on the curve labeled "Seaton" are times measured in thousands of years with reference to the point 0.062 pc given in Figure 7 by Seaton (1966). In translating Seaton's estimates of the diameters into units of time, uniform expansion at 20 km/sec was assumed. (Rose 1967, Fig. 2.)

$$L = KMT_i^{7/2} \tag{19}$$

where K is a constant ($\approx 10^{-27}$). With this expression for the luminosity, equation (18) may be written as a differential equation for the central temperature as a function of time. Integration gives (Mestel, 1952; Schwarzschild, 1958)

$$\tau_{cooling} = \frac{2}{5}\frac{C_V T_i M}{L} = \frac{2}{5}\frac{C_V}{K^{2/7}}\left(\frac{L}{M}\right)^{-5/7}$$

In the early work C_V was taken to be 3/2 k per ion (appropriate for a perfect gas). The observations are roughly in agreement with these predicted cooling times so long as we restrict ourselves to hot white dwarfs. For faint white dwarfs ($L \lesssim 10^{-3} L_\odot$) the derived cooling times are much too long, almost 10^{10} years. From Figure 11 it is evident that both the Pleiades and the Hyades have white dwarfs too faint for the predicted ages of these clusters. The Pleiades has one with $L \lesssim 10^{-2.5} L_\odot$, implying an age of about 8×10^8 years for a "typical" white dwarf ($M \approx 1 M_\odot$); yet the predicted age of the Pleiades is 6×10^7 years. Of the ten in the Hyades, two have $M_V \sim 12.5$, yielding an age of about 5×10^9 years. But the Hyades is considered to be only about 4×10^8 years old.

The relatively rapid fading of faint white dwarfs can be expressed in another fashion. From the calculated cooling rates the total number of observed white dwarfs within a given magnitude interval can be predicted. Slower cooling at fainter and fainter magnitudes means that more and more white dwarfs should be observed at fainter magnitudes, even if selection effects are taken into account. Weidemann (1967) evaluates three sets of white dwarf surveys and finds that for $M_V \leq 13$ the empirically derived luminosity function fits theoretical predictions closely. However, Figure 12 shows that above $M_V = 13$ there do not seem to be enough stars at the fainter magnitudes and compared to what theory predicts. Although Eggen and Greenstein's observations lie above the predicted curve at faint magnitudes their result is somewhat uncertain due to the possible contamination of the sample by other types of stars.

Thus, white dwarfs cool faster when faint than is expected. Something is turning them off sooner than is expected and consequently old white dwarfs are black and space may be littered with dead stars. Such a possibility is a very interesting one. If many population I stars are black dwarfs, an even greater percentage of Population II stars are likely to be dwarfs. Most of the "missing matter" of the galaxy may well be locked up in these Population I and II black dwarfs.

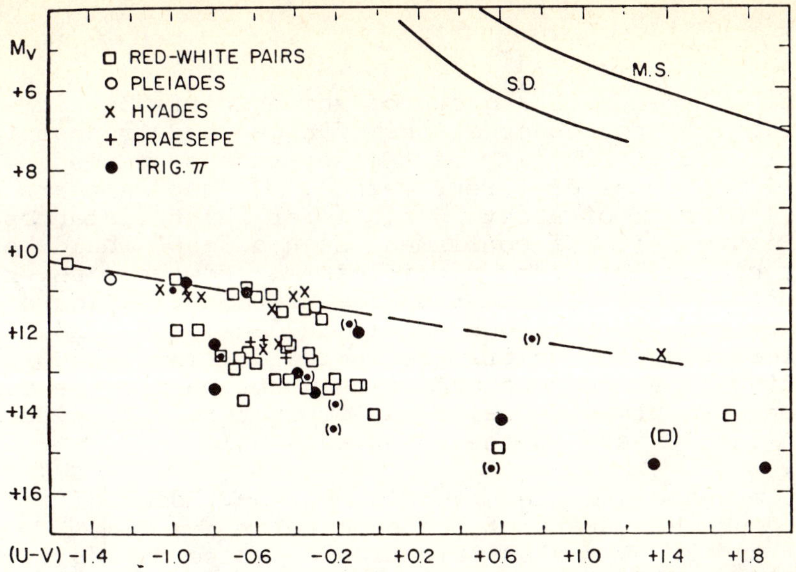

Figure 11. The (M_V, U-V) diagram for white dwarfs (1) of known trigonometric parallaxes, (2) with red-dwarf companions, and (3) which are members of galactic clusters. Part of the normal main sequence and the extreme subdwarf sequence, uncorrected for line-blanketing effects, are also shown. (Eggen and Greenstein 1965, Fig. 4.)

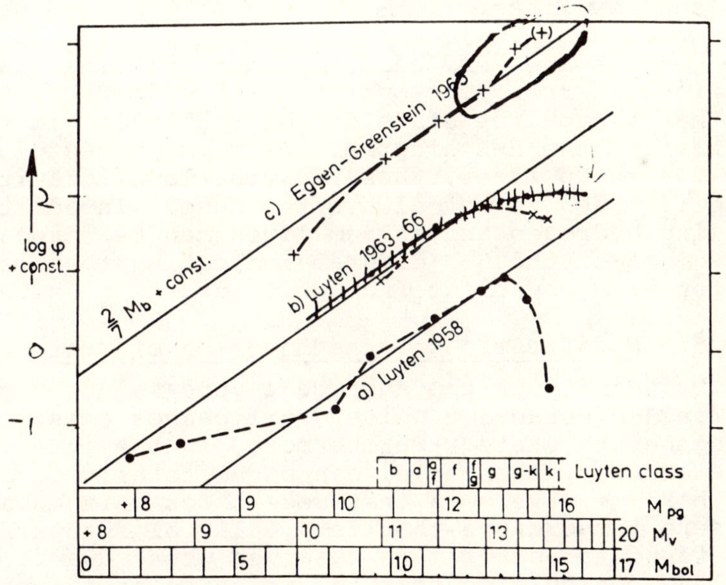

Figure 12. Number distribution of white dwarfs in the solar neighborhood per cubic pc per magnitude interval as a function of various luminosity measures. Weidemann's function log $\varphi(M_b) = 2/7\ M_b$ + constant is given by straight lines. Luyten's 1958 and 1963-1966 surveys of 130 and 1000 stars, respectively, and Eggen and Greenstein's 1965 sample of 167 stars are plotted. (Weidemann 1967, Fig. 2.) Crossed line shows the best fit to observations (1.0 $M_\odot$) for cooling white dwarfs from Ostriker and Axel (1969). The circled portion of line c includes stars which may not be white dwarfs.

The existence of a group of white dwarfs of higher luminosities in the observations of Eggen and Greenstein (1965) has led to the possibility of two separate sequences of different radii and hence different masses. A "gap" appears in their (M_V, U-V) diagram of about 70 stars for which distances are known (Figure 14). If confirmed, such a "gap" would be of considerable evolutionary significance. However, as shown by Weidemann's (1967) H-R diagram of white dwarfs (Figure 13), it disappears when an enlarged sample is considered and the effects of blanketing on the colors are included. The large range in radius at a given temperature is no doubt partly due to observational uncertainties but also may imply that a considerable range in mass exists in the observed sample.

From Eggen and Greenstein's (M_V, U-V) diagram it is also evident that stars of the same color can have different spectral types. The classification of white dwarfs is in some ways roughly parallel to the main sequence spectral classification as shown in Table 1. DA white dwarfs—those with a pure hydrogen spectra—are the most numerous and are found over a wide range in the H-R diagram: $7 \leq M_b \leq 13.5$ and $3.8 \leq \log T_e \leq 4.55$; DF, DG, DM, λ4135 are all very rare. Type DB have helium lines. One of the most interesting classifications, DC, shows no lines and has a range over the whole color spectrum although it predominates at the red end. No spectral scan seems to have been taken for any DC-type star; they may simply be "normal" white dwarfs with extremely broadened shallow lines.

The spectra are very hard to classify: lines can be sharp, diffuse, narrow, very narrow, and so forth. Figure 14 shows that sharp-line stars do not really lie in the expected part of the H-R diagram. We might expect lower pressures for sharp lines, thus implying lower gravity and larger radii. Such is not always the case. In addition, stars showing hydrogen and helium lines can be found in overlapping regions of the H-R diagram indicating that large abundance differences may exist from white dwarf to white dwarf.

5. Modification of the Equation of State

As the density in the surface layers is not high enough for electron degeneracy, a white dwarf cannot exist in a completely degenerate state. Furthermore, in the interior the pressure exerted by the nuclei cannot be totally ignored. In spite of this, we have seen that models for white dwarfs assuming zero temperature—that is, really black dwarf models—have been found to approximate observed white dwarfs closely. Marshak (1940) calculated that the presence of a finite temperature leads to an increase in mass at constant radius of the order of 0.01% of the mass at zero point energy for Sirius B and 40 Eri B.

A more important effect was found by Salpeter (1961) who showed that under expected interior conditions the ions, rather than obeying the perfect gas law, will tend to form a rigid lattice when kT is less than the Coulomb energy of the ion pairs. As a result, corrections must be made to the

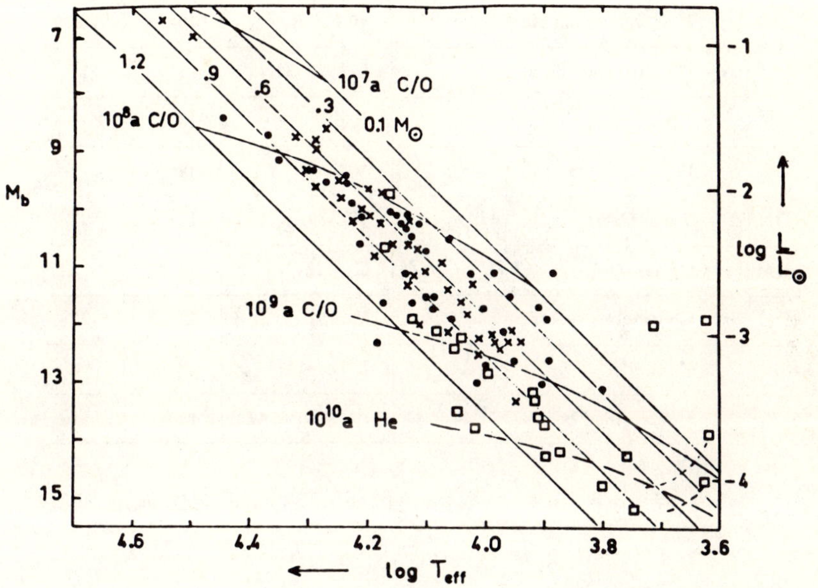

Figure 13. H-R diagram for white dwarfs. • spectral type DA with known distances; x spectral type DA with spectroscopic distances; □ white dwarfs of other spectral types; --- possible shifts for coolest objects. Diagonals are lines of constant radii, with masses labeled for completely degenerate configurations consisting of C to Mg. Lines of constant cooling time (or age) are drawn for C/O and He configurations. (Weidemann 1968, Fig. 1, p. 361.)

TABLE 1

WHITE-DWARF CLASSIFICATION*

Type	Characteristics	Example	M_v	n
D0	He II strong, He I and/or H I	HZ 21	10	5
DA	H I present, no He I	40 Eri B	10–13	136
DB	He I present, no H I	L 1573–31	11	12
DF	Ca II, no H	R 627	13–14	4
DG	Fe I, Ca II, no H	V Ma 2	14	1
DK	Weak Ca II	W 489	15	3
DM	Ca II, Ca I weak, TiO?	G 5–28	14–15	2
DC	Continuous spectra	W 457	12–15	20
λ4670	C_2 bands	L 145–141	13	6
λ4135	Unidentified bands	AC+70°8247	12	1

*The table was compiled from Greenstein's work by Weidemann (1968, Table 1, p. 353).

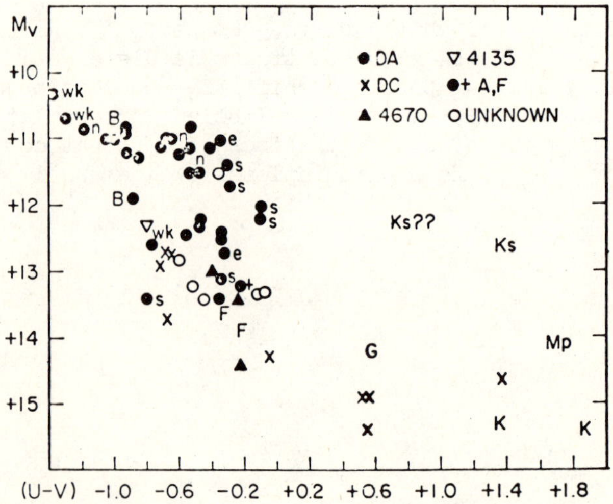

Figure 14. The (M_V, U-V) diagram with spectral types indicated. The doubtful object G14-24 is indicated by Ks?? and αCMaB by the filled circle at M_V = +11.1^m, U-V = -1.1^m. (Eggen and Greenstein 1965, Fig.5.)

energy and pressure of Chandrasekhar's zero temperature models due to the following:

a) The classical Coulomb energy of an ion lattice, assuming a uniformly distributed electron gas;
b) Thomas-Fermi deviations from a uniform electron distribution;
c) Exchange energy and spin-spin interactions between the electrons.

The first effect yields the largest correction.

Using the modified equation of state, Hamada and Salpeter (1961) obtain homogeneous models for different values of μ_e and Z. The computed models (of Figures 3, 4) have smaller radii and larger central densities as compared with Chandrasekhar models of the same mass. At low densities the deviations from Chandrasekhar's equation of state are most marked. At high densities, the important effects are due to inverse beta decay and pycnonuclear reactions. Instead of obtaining a maximum mass of $1.44\left(\frac{2}{\mu_e}\right)^2 M_\odot$ as $\rho_c \to \infty$, a maximum mass for a finite value of ρ_c is obtained. The lowest possible value is 1.01 $M_\odot$ for models with "equilibrium composition"—that composition for which the atomic weight A and the atomic charge Z have the optimum value for the density at each point in the star. The results for C^{12}, Mg^{24}, Si^{28}, S^{32}, and Fe^{56} are shown in Figure 4. The maximum masses in units of $M_\odot$ are 1.396, 1.36, 1.34, 1.17, and 1.11, respectively. For a mass >1.40 $M_\odot$ the star must contract to densities appreciably greater than 10^{10} gm/cm^3 in the absence of mass loss, rotation, or magnetic fields. At these densities inverse beta decay will make the star unstable.

These modifications of the equation of state make worse the discrepancy between theory and observation, since, as already mentioned, the observed radii are larger and the Hamada-Salpeter radii are smaller than those calculated by Chandrasekhar for a white dwarf of given mass. A recent examination of the problem by Hubbard and Wagner (1970) has indicated that for low mass models a careful treatment of thermal effects can remove much of the discrepancy. However, for Sirius B and the Greenstein-Trimble sample we must seek another explanation.

To summarize, the discrepancies found between observation and theory are:

1) The observed radii are apparently somewhat larger than theory predicts for a given mass star.
2) Faint (M_V>13) white dwarfs cool much more rapidly than expected.
3) The classification of white dwarf spectra is not comprehensible on the basis of principles which are successful in classifying main sequence stars (one parameter sequence, small abundance variations, etc.).

What can be done to improve the theory so as to fit observational data better? Possible errors and omissions which should be considered are: rotation and magnetic fields may affect the mass radius relation; the deviation of the specific heat of the ions from that of a perfect gas and the possible convective instability of the outer layers will alter the cooling curves; an understanding of the spectra will require both better observational material (scans over a wide wavelength region) and model atmosphere calculations which include the effects of the strong lines in a detailed fashion.

C. Nonspherical White Dwarfs

1. The Virial Theorem

Now we consider nonspherical white dwarfs. First let us derive the tensor virial theorem, which will also be of use in discussion of stability. Assume an inviscid nonrotating fluid with zero conductivity in which the density and the pressure vanish at the surface. No equilibrium assumption is made. Apart from gas pressure, the only force acting on the medium is that derivable from its own gravitation. Under such conditions the equations of the fluid are:

$$\frac{\partial \rho}{\partial t} + \frac{\partial \rho u_K}{\partial x_K} = 0 \tag{21}$$

$$\frac{\partial^2 V_g}{\partial u_K \partial u_K} = -4\pi G\rho \tag{22}$$

$$\frac{\partial u_i}{\partial t} + u_K \frac{\partial u_i}{\partial x_K} + \frac{1}{\rho}\frac{\partial P}{\partial x_i} \frac{\partial V_g}{\partial x_i} = 0 , \tag{23}$$

where V_g is the gravitational potential, the other symbols have their usual meanings, and summation over repeated indices is assumed.

Multiplying the equation of motion by $x_j dm$ and integrating over the volume V containing the fluid, we obtain

$$\int x_j \frac{\partial u_i}{\partial t} \rho \, \underline{dx} + \int x_j u_K \frac{\partial u_i}{\partial x_K} dm + \int x_j \frac{\partial P}{\partial x_i} \underline{dx}$$

$$- \int x_j \frac{\partial V_g}{\partial x_i} dm = 0 , \tag{24}$$

where $\rho \, \underline{dx} = dm$. Making use of the vanishing of pressure and density at the surface, integration by parts of the first and third terms in the preceding equation yield

$$I = \int x_j \frac{\partial u_i}{\partial t} \rho \underline{dx} = \frac{d}{dt} \int x_j u_i \rho \underline{dx}$$

$$+ \int x_j u_i \frac{\partial (\rho u_K)}{\partial x_K}$$

$$= \frac{d}{dt} \int x_j u_i dm - \int \rho u_K \delta_{jK} u_i \underline{dx}$$

$$- \int x_j u_{i,K} u_K dm$$

$$III = \int x_j \frac{\partial P}{\partial x_i} dx = \int \frac{\partial}{\partial x_i} (x_j P) \underline{dx} - \delta_{ij} \int P\underline{dx}$$

$$= \delta_{ij} \int P\underline{dx} \ . \tag{25}$$

Putting $V_g = \int \frac{dm'}{|x-x'|}$ in the fourth term gives the symmetric gravitational potential energy tensor

$$IV = - \int x_j \frac{\partial V_g}{\partial x_i} dm = - \int dm x_j \frac{\partial}{\partial x_i} \int \frac{dm'}{|\underline{x}-\underline{x}'|}$$

$$= \int\int dm\, dm' \frac{x_j (x_i - x_i')}{|\underline{x}-\underline{x}'|^3}$$

$$= \tfrac{1}{2} \int\int dm\, dm' \frac{(x_i - x_i')(x_j - x_j')}{|x-x'|^3} \ . \tag{26}$$

Using the same notation as in the first term, the second term is

$$II = \int x_j u_K u_{K,i} dm \ .$$

Combining these four terms gives

$$\frac{d}{dt} \int x_j u_i dm - \int \rho u_j u_i \underline{dx} - \delta_{ij} \int P dx$$

$$+ \tfrac{1}{2} \int\int dm\, dm' \frac{(x_i - x_i')(x_j - x_j')}{|x-x'|^3} = 0 \ . \tag{27}$$

With the following definitions the integrated equation of motion simplifies considerably.

$$V_{ji} = \int x_j u_i dm, \text{ the virial tensor} \tag{28a}$$

$$T_{ij} = \tfrac{1}{2} \int u_i u_j dm, \text{ the kinetic energy tensor} \tag{28b}$$

$$\Pi = \int P d\underline{x}, \text{ where } \frac{\Pi}{\gamma - 1} \text{ is the internal heat energy and } \gamma \text{ is the correctly defined average ratio of specific heats} \tag{28c}$$

$$W_{ij} = -\tfrac{1}{2} \int\int dm\, dm' \frac{(x_i - x_i')(x_j - x_j')}{|\underline{x} - \underline{x}'|^3}, \text{ the gravitational potential energy tensor.} \tag{28d}$$

The integrated equation is then expressed in the form

$$\frac{d}{dt} V_{ji} = 2T_{ij} + \delta_{ij}\Pi + W_{ij} \tag{29}$$

All terms on the right hand side are symmetric, V_{ji} is then necessarily symmetric as can be seen by looking at the symmetric and antisymmetric parts:

$$\tfrac{1}{2} \frac{d}{dt} (V_{ji} - V_{ij}) = 0 = \tfrac{1}{2} \frac{d}{dt} \int (x_j u_i - x_i u_j) dm. \tag{30}$$

If the above equation is multiplied by the completely antisymmetric tensor ε_{kij} we find

$$0 = \frac{d}{dt} \int \underline{x} \times \underline{u}\, dm = \dot{\underline{J}}\ ; \tag{31}$$

the total angular momentum of the system is a constant; the proof is due originally to Chandrasekhar (1961).

Considering only the symmetric part of V_{ij} we obtain

$$\tfrac{1}{2} \frac{d}{dt} \int (u_i x_j + u_j x_i) dm = \tfrac{1}{2} \frac{d^2}{dt^2} \int x_i x_j dm$$

$$= \tfrac{1}{2} \frac{d^2}{dt^2} I_{ij}\ , \tag{32}$$

where I_{ij} is the inertia tensor. Equation (29) then becomes

$$\tfrac{1}{2}\dot{I}_{ij} = 2T_{ij} + W_{ij} - 2\mathfrak{M}_{ij} + \delta_{ij}(\Pi + \mathfrak{M}) \,, \qquad (33)$$

where magnetic terms have been included. This is the virial theorem for hydromagnetics. The magnetic energy tensor is defined by

$$\mathfrak{M}_{ij} = \frac{1}{8\pi} \int_V B_i B_j \underline{dx} \,, \qquad (34)$$

where $\mathfrak{M}ii = \mathfrak{M}$ and the magnetic field is integrated over the entire volume v containing the fluid and the field. The surface S is placed at infinity so that surface integrals over the components of $\underline{B}$ vanish. Note that if the surface S is not at infinity or the pressure and density do not vanish at the surface, then the final result (equation 33) will contain surface integrals.

The tensor virial equations have the following important characteristics: they are 1) exact, 2) nonlinear, and 3) valid whether or not the star is in equilibrium. We will use these equations for three somewhat different purposes. First they can be manipulated to show semi-quantitatively the general effect of magnetic and rotational forces. Then, when detailed equilibrium calculations are made for non-spherical stars, the virial equations can be used as a check. And finally, we will determine the stability of the desired equilibrium models by imagining them to be perturbed slightly and then, using a linearized version of the virial tensor equations, we will determine the time development of the perturbations.

Let us apply the foregoing results to white dwarfs. Assume the configuration is at equilibrium in a steady state. Contracting the indices, equation (33) becomes

$$2T + W + \mathfrak{M} + 3\Pi = 0 \,. \qquad (35)$$

If no fluid motions or magnetic fields exist the virial theorem becomes

$$W + 3\int \frac{P}{\rho}\, \rho \underline{dx} = W + 3M\left\langle \frac{P}{\rho} \right\rangle = 0 \,. \qquad (36)$$

For nonrelativistic degeneracy ($P \propto \rho^{5/3}$) the equation becomes

$$-\alpha_{3/2} \frac{GM^2}{R} + \beta_{3/2} \frac{M^{5/2}}{R^2} = 0 \,, \qquad (37)$$

and for relativistic degeneracy ($P \propto \rho^{4/3}$)

$$-\alpha_3 \frac{GM^2}{R} + \beta_3 \frac{M^{4/3}}{R} \approx 0 \,. \qquad (38)$$

The quantities $\alpha_{3/2}$, $\beta_{3/2}$, α_3 and β_3 are positive definite constants.

Now compare the gravitational energy and the internal energy for a star of given mass. For the nonrelativistic case the two energies depend on a different power of the radius. The star can bring the two energies into balance by adjusting its radius. For example, if the gravitational energy were to exceed the internal energy, the star would contract, the decrease in radius increasing the internal energy more than the gravitational energy, the contraction stopping when equilibrium is reached. We see from equation (37) that in the non-relativistic limit (low mass WD*s) the radius relation is $R \propto M^{-1/3}$.

In the relativistic case the energies depend on the same power of the radius. The star cannot bring the two energies into balance by simply adjusting its radius. Assume the star is not quite completely relativistic, so that the internal energy goes as $\frac{1}{R^{1+\varepsilon}}$ rather than $\frac{1}{R}$. A considerable decrease in radius is necessary to bring the two energies into balance. The resulting increase in density brings the radial dependence closer to $\frac{1}{R}$, at which point the star cannot achieve equilibrium if the energies do not balance. Note also that in the relativistic case the two energies depend on different powers of the mass. If the star is not quite relativistically degenerate a moderate increase in mass leads to a large decrease in the radius. A specific mass limit, M_3, exists for which the energies balance; in the notation of equation (38) $M_3 = (\beta_3/G\alpha_3)^{3/2}$. Above this the gravitational energy is always the larger. In a star with a mass smaller than M_3, balance can be achieved by expansion until the density decreases to the point where the star is no longer relativistic.

2. Magnetic White Dwarfs

Consider what happens if forces other than just the star's own gravitation are included. Assume a magnetic field is present. The magnetic energy $\mathfrak{M}$ goes as B^2R^3, but the flux $\Phi \propto BR^2$ must be conserved when the radius changes and therefore $\mathfrak{M} \sim \Phi^2/R$. The virial theorem under conditions of equilibrium is then

$$0 = - \frac{\alpha G M^2}{R} + \frac{\gamma \Phi^2}{R} + 3M < \frac{P}{\rho} > . \tag{39}$$

For both types of degeneracy the effect of the magnetic field is to expand the star slightly.

$$\text{NRD: } 0 = -\alpha_{3/2} \frac{GM^2}{R} + \frac{\gamma_{3/2}\Phi^2}{R} + \beta_{3/2} \frac{M^{5/3}}{R^2} \tag{40}$$

$$\text{RD:}\ 0 = -\alpha_3 \frac{GM^2}{R} + \gamma_3 \frac{\Phi^2}{R} + \beta_3 \frac{M^{4/3}}{R} \tag{40}$$

Speaking very approximately, adding flux, Φ, is equivalent to reducing the gravitational constant G to $G' = G - (\gamma_{3/2}\Phi^2)/(M^2\alpha_{3/2})$. In the relativistic case, it is clear that no finite amount of flux will completely remove the mass limit.

To determine exactly how the presence of the magnetic term affects the mass limit, first consider the case without it: the mass cannot exceed M_3. With the inclusion of a magnetic field, $M = M_3(1 + \delta)$ where one can show $\delta = \frac{3}{2}\frac{M}{|W|}$ to first order in the expansion $1 + \delta$. A magnetic field is thus able to increase the mass limit by only a moderate amount. However, for moderate mass models near the mass limit large changes can therefore be made in the radius by adding reasonable magnetic pressures. Such moderate magnetic pressures could explain the discrepancy between the observed and theoretical radii of white dwarfs; the only question remaining being that of whether the required "moderate" fields will be attained in nature.

Uniformly rotating magnetic models have been numerically constructed by Ostriker and Hartwick (1968) in the following manner. The angular velocity is a function of the distance from the rotation axis so that

$$j = j(\tilde{\omega}) = \Omega(\tilde{\omega})\tilde{\omega}^2 , \tag{41}$$

where $j(\omega)$ is the angular momentum per unit mass. The equation of motion is then

$$\nabla(\tfrac{1}{2}\Omega^2\tilde{\omega}^2) = -\frac{\nabla P}{\rho} + \nabla V_g + \mathcal{L} , \tag{42}$$

where $\mathcal{L}$ is the Lorentz force. The curl of this equation yields

$$\frac{1}{\rho^2}[\nabla\rho \times \nabla P] + \nabla \times \mathcal{L} = 0 ,$$

as the other terms are gradients of scalar quantities. The field configuration was determined by three requirements:

1. Since constant pressure surfaces are parallel to constant density surfaces, $\nabla \times \mathcal{L} = 0$ and the Lorentz force is the gradient of a scalar:

$$\mathcal{L} = \frac{1}{4\pi\rho}(\nabla \times \mathbf{H}) \times \mathbf{H} = \nabla V_M .$$

2. The magnetic field is therefore derivable from poloidal and toroidal "stream functions." Furthermore, the ratio of these stream functions is assumed to be a constant

throughout the star.

3. The pressure and the magnetic field are assumed to vanish at the surface. Under these assumptions we have an eigenvalue problem. Only the lowest mode is considered (no nodes in the radial direction); the eigensolutions are found to have larger toroidal than poloidal parts. The method of calculation will be discussed in the next section.

Results are summarized in Figure 15, where the second model is for a nonrotating star and the other four models are for uniformly rotating stars. Moderate increases in the radius are produced by a relatively small ratio of magnetic to gravitational energy—on the order of a few percent.

If we assume white dwarfs possess fields of the order of 10^{12} to 10^{13} gauss at the center, flux conservation and slow decay of the original fields require that their progenitors should also have moderately high fields of the order of a few percent of the gravitational energy. For homologous contractions, gravitational energy scales in the same way as magnetic energy (as R^{-1}), so it is not expected that magnetic fields will have a larger effect on the structure of a white dwarf than on the main sequence star from which it evolved. Estimating the field strength of the interiors of stars is very difficult. The present degree of agreement between observations and theoretical calculations for stellar interiors probably does not preclude the possibility that the magnetic energy in main sequence stars is as much as a few percent of the gravitational energy—that is, about 10^7 to 10^8 gauss.

Characteristics of the models constructed are presented in Table 2. The mass of model 1 was chosen to represent the white dwarf 40 Eri B, while models 2–5 have a mass close to that of Sirius B. Clearly, the "discrepancy" between theoretical and observed radius can be removed by allowing an interior magnetic field. Model 6 has been included to show that a uniformly rotating white dwarf with an internal magnetic field can have a mass somewhat greater than the Chandrasekhar "limit." Several general effects are evident. With increasing field, the radius increases and the central density decreases. The toroidal field dominates so that in the absence of rotation the stars are slightly prolate. Profiles of the equidensity surfaces of model 5 as presented in Figure 16 show that although the interior is prolate the surface is almost spherical. The magnetic fields of the type considered result in a quadrupole moment in the gravitational potential which affects the ratio of equatorial to polar radius only slightly. The ratio of $T/|W|$ always remains quite small as compared to values reached by nonuniformly rotating models, which will be considered next.

To summarize this work, it is probable that magnetic pressures in white dwarfs have a significant role in determining the observed mass radius relation.

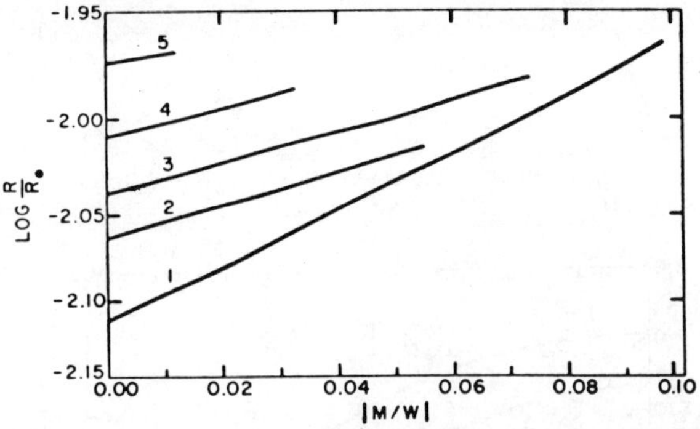

Figure 15. Radius of a white dwarf as a function of magnetic energy and angular momentum. For a 1.05 $M_\odot$ star (representing Sirius B), we plot the radius R versus the ratio of magnetic to gravitational energies $|\mathfrak{M}/W|$; R is the larger of the equatorial and polar radii. The curves labeled 1, 2, 3, 4, and 5 represent uniformly rotating sequences having angular momenta equal to (0,1,2,3,4) × 1.92 × 10^{49} g cm^2 sec^{-1}, respectively. (Ostriker and Hartwick 1968, Fig. 2.)

TABLE 2

MODEL CHARACTERISTICS*

	Model 1	Model 2	Model 3	Model 4	Model 5	Model 6
$M/M_\odot$	0.450	1.05	1.05	1.05	1.05	1.81
J	1.92 (49)	0.00	0.00	5.77 (49)	4.81 (49)	5.77 (49)
R_e	1.33 (9)	5.38 (8)	6.95 (8)	6.83 (8)	7.26 (8)	2.33 (8)
T	6.97 (47)	0.00	0.00	1.17 (49)	6.59 (48)	7.25 (49)
U	1.64 (49)	3.42 (50)	2.44 (50)	2.68 (50)	2.10 (50)	2.93 (51)
$\mathfrak{M}$	4.18 (48)	0.00	3.00 (49)	0.00	2.75 (49)	4.17 (50)
W	−3.63 (49)	−5.29 (50)	−4.23 (50)	−4.53 (50)	−3.87 (50)	−4.14 (51)
R_e/R_p	1.00	1.00	1.00	1.21	1.04	1.03
ρ_c	9.01 (5)	3.74 (7)	2.20 (7)	2.35 (7)	1.73 (7)	1.90 (9)
V_e	9.73 (2)	0.00	0.00	2.73 (3)	1.98 (3)	5.55 (3)
$\Omega^2/4\pi G\rho_c$	7.09 (−3)	0.00	0.00	8.13 (−3)	5.12 (−3)	3.57 (−3)
$\mathfrak{M}_T/\mathfrak{M}_P$	11.7	....	12.0	....	12.6	9.82
H_c	5.03 (11)	0.00	4.07 (12)	0.00	3.35 (12)	9.22 (13)
$H_c^2/8\pi P_c$	0.450	0.00	0.250	0.00	0.239	0.294

*(Ostriker and Hartwick 1968, Table 1.)

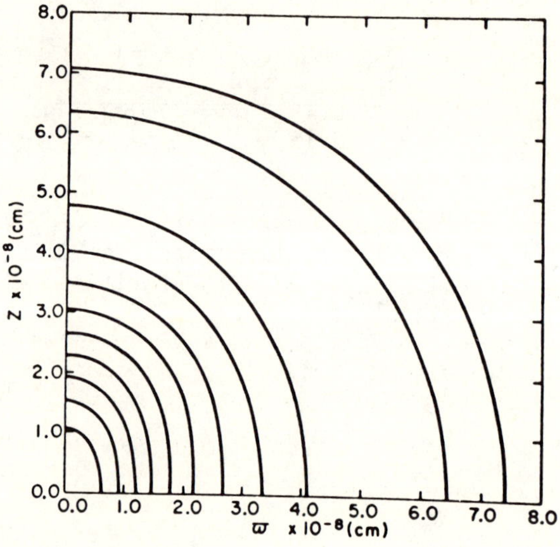

Figure 16. Detailed structure of model 5. A cross-section showing the level surfaces where the density is 0.9, 0.8, 0.7, 0.6, 0.5, 0.4, 0.3, 0.2, 0.1, 0.01, and 0.0 times the central density. The radius in the equatorial plane is indicated by $\bar{\omega}$, that along the axis of rotation by Z. Note the prolate interior and almost spherical exterior surfaces. (Ostriker and Hartwick 1968, Fig. 1.)

3. Rotating White Dwarfs and Elimination of the Equilibrium Mass Limit

Suppose now that no magnetic fields are present but the fluid is in motion. The rotational kinetic energy is

$$T \sim M v_\phi^2 \sim M\Omega^2 R^2 \ . \tag{45}$$

However, just as in the magnetic case, the kinetic energy must be written in terms of conserved quantities

$$T \sim \frac{MJ^2R^2}{I^2} \sim \frac{J^2}{MR^2} \ , \tag{46}$$

with $I \sim MR^2$ and $J = I\Omega$. The kinetic energy therefore has the same dependence on the radius as the internal energy in the nonrelativistic limit and a steeper dependence than in the relativistic limit

$$\text{NRD:}\ 0 = -\alpha_{3/2}\frac{GM^2}{R} + \kappa_{3/2}\frac{J^2}{MR^2} + \beta_{3/2}\frac{M^{5/3}}{R^2} \ , \tag{47}$$

$$\text{RD:}\ 0 = -\alpha_3\frac{GM^2}{R} + \kappa_3\frac{J^2}{MR^2} + \beta_3\frac{M^{4/3}}{R} \ . \tag{48}$$

Consider how the mass limit is affected by inclusion of rotational kinetic energy. As mass is added to the star the negative term (gravitational energy) dominates, but equilibrium can always be achieved by decreasing the radius due to the R^{-2} dependence of the kinetic energy term. We therefore expect that an equilibrium model of any mass and angular momentum can be made provided that the angular momentum J is nonzero. However, such models may or may not be physically reasonable. If, for example, the equilibrium density was found to be 10^{15} gm/cm^3, the equation of state would be invalid and the model nonsensical.

4. SCF Method for Determining the Equilibrium Structure of Distorted Stars

Clearly rotation cannot be treated as a perturbation if significant deviation from the nonrotating model is contemplated. Instead, a scheme is needed which allows for the calculation of severely distorted stars.

The equation of motion in cylindrical coordinates is

$$\Omega^2(\underline{r})\hat{\tilde{\omega}} = -\frac{\nabla P}{\rho} + \nabla V_g \ . \tag{49}$$

Taking the curl of this equation gives

$$\frac{\partial\Omega}{\partial z} = \frac{\partial\Omega}{\partial\phi} = 0; \tag{50}$$

that is, $\Omega=\Omega(\tilde{\omega})$ if $P=P(\rho)$. This result is a general one: for any star with such a pressure density relation in equilibrium with rotation, rotation must be constant on cylinders. As recently shown by Goldreich and Schubert (1967), if secular stability is required we are still forced to this result even if $P\neq P(\rho)$. This result has already been used in the discussion of magnetic white dwarfs.

With the angular velocity a function only of the distance from the rotation axis, a centrifugal potential per unit mass can be defined.

$$V_c(\tilde{\omega}) = \int_0^{\tilde{\omega}} \frac{j^2(\tilde{\omega}')}{\tilde{\omega}'^3} d\tilde{\omega}' \ . \tag{51}$$

Differentiation of V_c yields the centrifugal force. Since the rotational force is derivable from a potential, a total potential

$$V = V_g + V_c + V_M \tag{52}$$

can be defined, if in addition magnetic forces are similarly included. The equation of motion then reduces to

$$\frac{1}{\rho}\nabla P = \nabla V \ . \tag{53}$$

Using Chandrasekhar's parametric pressure density relation for a completely degenerate electron gas

$$P = Af(x)\,,\ \rho = Bx^3 \ , \tag{54}$$

the equation of motion becomes

$$\frac{1}{\rho}\nabla P = \nabla\left\{\frac{8A}{B}\left[1+\left(\frac{\rho}{B}\right)^{2/3}\right]^{\frac{1}{2}}\right\} = \nabla V \ . \tag{55}$$

Integrating this equation and solving for the density, we find

$$\rho(\underline{r}) = B\left[\left\{\frac{B}{8A}\left[V(\underline{r})-V_s\right]+1\right\}^2-1\right]^{3/2} , \tag{56}$$

where V_S is the potential at the surface of the star $[\rho(V=V_S) = 0]$. From equation (56) we see that the density at any point is known explicitly in terms of the potential at that point from equilibrium considerations. On the other hand, the potential is independently derivable from the density via potential theory. For example $V_g(\underline{r})$ can be obtained from $\rho(\underline{r})$ by the linear operator $G\int \frac{\rho(\underline{r})d\underline{r}'}{|r-r'|}$. A similar recipe involving Gegenbauer polynomials exists for the magnetic potential.

The centrifugal potential is obtained as follows. If rotation is confined to cylinders and angular momentum increases outwards, every element can be specified in terms of its angular momentum, or alternatively in terms of the cylinder labeled by mass fraction m on which it is rotating. The angular momentum per unit mass as a function of mass shell, j(m), can then be obtained where the Lagrangian coordinate m is the mass fraction interior to a given cylinder:

$$m(\tilde{\omega}) = \frac{\int_0^{\tilde{\omega}} dM}{\int_0^{R} dM} \; .$$

Given the density, the mass fraction at any point can be obtained; then, from the angular momentum per unit mass, $j_{\tilde{\omega}}(m)$, at any point, the centrifugal potential $V_c(\tilde{\omega})$ can be obtained using equation 51.

The problem of obtaining differentially rotating white dwarf models thus has two parts. Given the density, potential theory determines the potential; given the potential, the equilibrium condition (equation 56) determines the density. No attempt is made to solve the two parts simultaneously, improving initial approximations by further calculations. Instead, each part is solved alternately as exactly as numerical techniques permit. Calculations begin with a guess of the density distribution since the wiggles in a "bad" density distribution tend to be smoothed out under the integral operator, yielding a "good" potential. This "good" potential then yields a "good" density distribution, the new density distribution yields a "better" potential, and so forth. Alterations are continued until self-consistency is obtained. In general, about 20 iterations suffice to give accuracy on the order of 10^{-5}.

As the potential theory part of the problem is not dependent on the star being a white dwarf but the equilibrium part is, models of other stars can be constructed by replacing the algebra of equation (56) by a Henyey code. This is presently being done by Jackson at Princeton.

Using the described self-consistent field (SFC) method, models for differentially rotating white dwarfs have been constructed by specifying the mean molecular weight μ_e (always taken as 2), the total mass M, the total angular momentum J and the distribution j(m) of angular momentum per unit mass. Three series of models were constructed, all with masses in the range $0.5 \leq M/\bar{M} \leq 4.1$. In series A the J for each model is equal to the total angular momentum estimated for a star of the same mass in uniform rotation on the main sequence, using average rotational velocities in Allen (1963) to determine J(M). Series B and C have, respectively, half and twice the total angular momentum for a given mass used in series A.

Each of the resulting twelve models can be specified by a

point in the (M,J) plane (Figure 17). The heavy solid line along the horizontal axis is the locus of points for nonrotating models calculated with Chandrasekhar's equation of state (model 1). The lower hatched curve is the approximate boundary beneath which the models are unstable against inverse beta decay. The small humped region in the lower left corner shows the range for uniformly rotating models (James 1964). As these cannot be made with too much kinetic energy because the effective gravity at the equator will no longer be positive, the mass limit is altered by only a few percent by uniform rotation. The results of the SCF computations clearly indicate that the assumption of a nonzero angular momentum with the resulting differential rotation permits the existence of rapidly rotating stable models for a large range of mass and angular momentum.

Table 3 summarizes the characteristics of twelve models, including the equatorial velocity v_e, the gravitational redshift at the pole, and the ratio of the gyration radius to the equatorial radius R_g/R_e. The kinetic energy T is always less than the internal energy U, even though the kinetic energy is what makes the star possible when $M>M_3$. Thus, the models are supported mainly by the pressure rather than centrifugal force. In all the models, the total energy is always large and negative. Differential rotation is never very extreme: Ω equator/Ω center >0.2 for all models. Note that J and T increase significantly with mass while the equatorial radius itself increases by only a small amount—by a factor of about 3 or 4 over the entire mass range considered. The value of the redshift is rather insensitive to mass when $M \geq 1.3M_\odot$. Thus, unfortunately, it would be difficult to verify the existence of massive models on the basis of gravitational shift especially as extreme rotational broadening will blur line cores. For $M<M_3$, the rotational velocity is essentially zero; for $M>M_3$, the rotational velocity is on the order of 3000 km/sec with radii in the same range as those of nonrotating models below M_3. However, the narrow cores observed in the hydrogen lines of many white dwarf spectra imply that the rapid rotator must be the exceptional case. The members of spectral class DC, which have no observable lines and slightly smaller than normal radii, are the most likely candidates for massive white dwarfs.

Figures 18 and 19 show the detailed structure of models 3 and 6. Both have fairly uniformly rotating high central density regions although the outer regions of model 6 (higher J) are more extended and flattened than those of model 3. In model 6 at 90% of the mass fraction, the angular velocity is about 90% of the central angular velocity Ω_c, i.e., most of the mass rotates as nearly a solid body; yet the mass of this model is more than twice the "mass limit" for nonrotating white dwarfs.

Figure 20 shows the variation of central density, equatorial velocity, equatorial radius, and polar radius with stellar mass for series A. Near the mass limit M_3, rapid changes occur in all quantities. As a result of an apparent coincidence this mass is also roughly the mass on the main sequence for which rotation becomes significant. Thus central

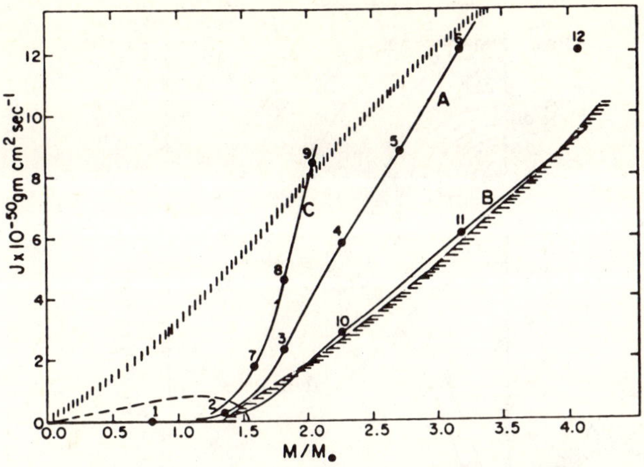

Figure 17. The locations of the twelve selected models, each indicated by a black dot with an identifying number, in the (M,J) plane. The solid lines marked A, B, and C indicate, respectively, the J(M) relations assumed for Series A, B, and C. Other lines are explained in the text. (Ostriker and Bodenheimer 1968, Fig. 1.)

TABLE 3

PROPERTIES OF ROTATING WHITE-DWARF MODELS*

	Model 1	Model 2	Model 3	Model 4	Model 5	Model 6
$M/M_\odot$	0.79	1.36	1.81	2.26	2.72	3.17
J	0.00	2.69 (49)	2.31 (50)	5.77 (50)	8.85 (50)	1.21 (51)
R_e	7.27 (8)	3.04 (8)	6.17 (8)	1.34 (9)	1.67 (9)	1.89 (9)
T	0.00	1.16 (49)	2.01 (50)	2.86 (50)	4.01 (50)	5.37 (50)
U	1.24 (50)	1.38 (51)	1.16 (51)	7.42 (50)	8.17 (50)	9.64 (50)
W	−2.11 (50)	−1.81 (51)	−2.03 (51)	−1.77 (51)	−2.18 (51)	−2.73 (51)
ρ_c	8.81 (6)	4.84 (8)	1.76 (8)	4.07 (7)	3.32 (7)	3.25 (7)
R_e/R_p	1.00	1.05	2.08	3.39	3.96	4.32
V_e	0.00	1.58 (3)	5.01 (3)	4.62 (3)	4.75 (3)	4.90 (3)
Ω_e/Ω_c	0.00	0.476	0.350	0.274	0.254	0.241
$c\Delta\nu/\nu$	4.82 (1)	2.06 (2)	2.46 (2)	2.12 (2)	2.32 (2)	2.58 (2)
g_e	1.99 (8)	1.88 (9)	2.27 (8)	3.71 (6)	1.04 (7)	1.28 (7)
g_p	1.99 (8)	2.12 (9)	2.09 (9)	1.26 (9)	1.23 (9)	1.28 (9)
R_g/R_e	0.424	0.358	0.317	0.270	0.254	0.241

	Model 7	Model 8	Model 9	Model 10	Model 11	Model 12
$M/M_\odot$	1.58	1.81	2.04	2.26	3.17	4.07
J	1.77 (50)	4.62 (50)	8.47 (50)	2.89 (50)	6.06 (50)	1.21 (51)
R_e	6.23 (8)	1.65 (9)	3.27 (9)	4.68 (8)	6.04 (8)	1.00 (9)
T	1.15 (50)	1.44 (50)	1.36 (50)	5.48 (50)	1.28 (51)	1.62 (51)
U	8.12 (50)	3.57 (50)	1.95 (50)	2.81 (51)	4.18 (51)	3.75 (51)
W	−1.50 (51)	−9.01 (50)	−6.58 (50)	−4.76 (51)	−8.21 (51)	−8.79 (51)
ρ_c	1.08 (8)	1.49 (7)	3.82 (6)	8.00 (8)	8.67 (8)	3.37 (8)
R_e/R_p	1.79	3.35	4.67	2.48	3.39	4.19
V_e	4.36 (3)	3.70 (3)	2.62 (3)	6.57 (3)	7.68 (3)	7.18 (3)
Ω_e/Ω_c	0.410	0.284	0.209	0.280	0.239	0.221
$c\Delta\nu/\nu$	1.89 (2)	1.35 (2)	1.00 (2)	4.73 (2)	6.70 (2)	6.19 (2)
g_e	2.45 (8)	4.89 (6)	2.66 (6)	3.71 (8)	1.39 (8)	2.15 (7)
g_p	1.41 (9)	6.26 (8)	2.87 (8)	6.17 (9)	9.29 (9)	5.91 (9)
R_g/R_e	0.339	0.274	0.242	0.281	0.249	0.227

*(Ostriker and Bodenheimer 1968, Tables 1 and 2.)

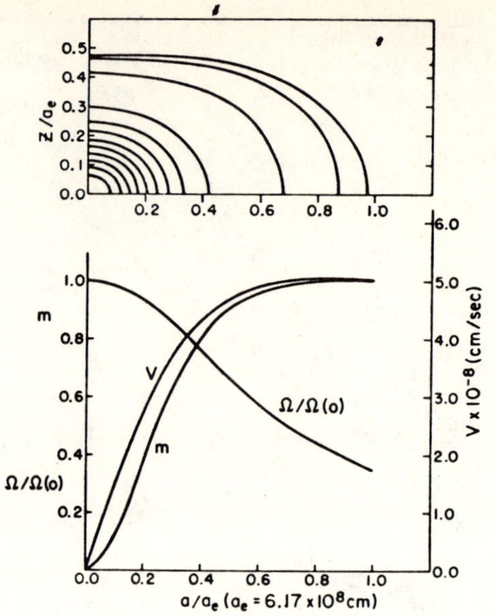

Figure 18. Detailed structure of Model 3. The upper portion shows the radii of the equidensity surfaces of densities 0.9, 0.8, 0.7, 0.6, 0.5, 0.4, 0.3, 0.2, 0.1, 0.01, 0.001, and 0.0001 times the central density. The radius in the equatorial plane is indicated by a; that along the axis of rotation, by Z; while a_e is the total equatorial radius. The lower portion shows the ratio of angular velocity Ω to the central value $\Omega(0)$, the fraction m of the total mass interior to the corresponding cylindrical surface around the axis of rotation, and the circular velocity V, all as a function of a/a_e. (Ostriker and Bodenheimer 1968, Fig. 4.)

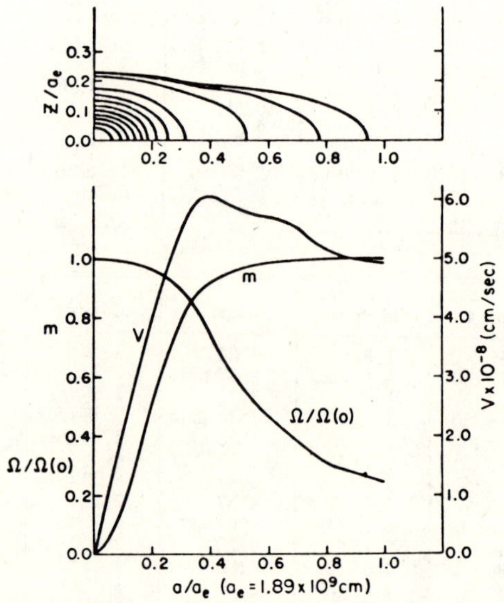

Figure 19. Detailed structure of Model 6. All curves have the same meaning as the corresponding ones in Figure 18. (Ostriker and Bodenheimer 1968, Fig. 5.)

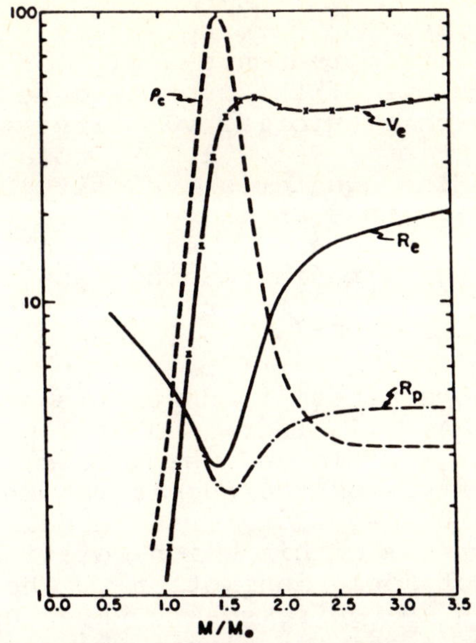

Figure 20. The variation of central density ρ_c, equatorial velocity V_e, equatorial radius R_e, and polar radius R_p as a function of mass along Series A. The unit of ρ_c is 10^7 g/cm^3; of V_e, 100 km/sec; of R_e and R_p, 10^8 cm. (Ostriker and Bodenheimer 1968, Fig. 2).

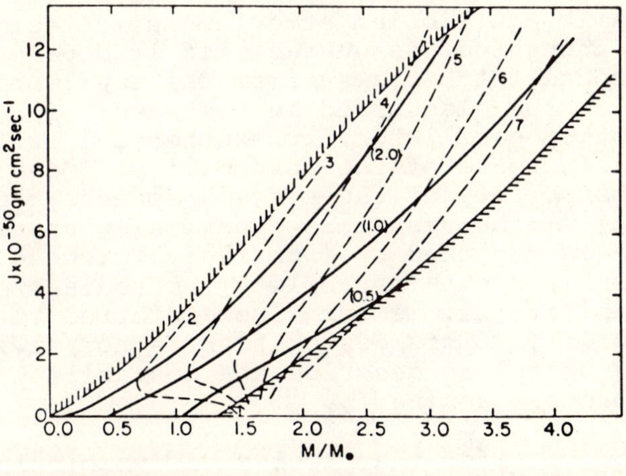

Figure 21. Lines of constant equatorial velocity (dashed curves) and constant equatorial radius (solid curves) in the (M,J) plane. The horizontally and vertically hatched boundaries have the same meaning as those in Figure 17. The velocities are indicated for each curve in units of 10^3 km/sec, while radii are indicated (in parentheses) for each curve in units of 10^9 cm. (Ostriker and Bodenheimer 1968, Fig. 3.)

density increases as $M \to M_3$ for the slowly rotating part of the sequence and then it decreases sharply for higher masses due to the added influence of the increased centrifugal force. The variation of the equatorial velocity with mass is essentially a step function. Below 1.5 $M_\odot$ rotation is not important. Above this the equatorial velocity changes rapidly, leveling off at high mass as increases in the moment of inertia counteract increases in angular momentum. For $M<M_3$ the radius decreases with increasing mass in the expected manner; however, above M_3, the radius increases with mass due to the rapid increase in J along sequence A.

Figure 21 shows lines of constant equatorial radius in the (M,J) plane. As J is increased at low masses, the velocity increases for a while and then drops; at high masses, the velocity decreases as the angular momentum increases. This behavior, which seems at first anomalous, is to be expected in any system within which rotational forces are important (think of Keplerian motion). The changes in the white dwarf radius are quite small. Even up to models of $M=4M_\odot$, the surface areas are almost the same as nonrotating white dwarfs below the Chandrasekhar mass limit.

It has been established that models of white dwarfs with M considerably greater than the Chandrasekhar "limit" can exist in equilibrium if differential rotation or magnetic fields with uniform rotation are assumed; but are such models stable? In order to answer such a question we must first understand what types of instability can occur.

D. Oscillations and Stability of White Dwarfs

1. Classification of Types of Instability

A system is normally thought of as stable if it remains near hydrodynamic equilibrium after an infinitesimal perturbation; it is thought of as unstable if it departs progressively further and further away from the equilibrium state. Instabilities can be classified in two ways: I) according to the kind of motion—oscillatory or exponential or II) according to the rate of increase of the instability. Consider scheme II. If the instability of the system proceeds rapidly on a time scale $(G\bar{\rho})^{-\frac{1}{2}}$, the system is dynamically unstable. If the time scale is governed by a diffusive process like molecular diffusion (viscous instability) or radiative diffusion (thermal instability) it is sometimes referred to as a secular instability. This designation is not always helpful and it is more useful to describe the instability directly in terms of the process acting.

Let us now consider the problem of dynamical stability and classify the motions according to scheme I. Suppose an infinitesimal perturbation is applied to a system in equilibrium. Let the mass element which was at $\underline{x}$ at time t=0 be subsequently at $\underline{x} + \underline{\xi}(\underline{x},t)$ where $\underline{\xi}(\underline{x},t) = \underline{\xi}_o(\underline{x})e^{i\sigma t}$. Solving the perturbed equations of motion for the displacement $\underline{\xi}_o(\underline{x})$ and the frequency σ allows us to classify the system as stable or unstable and, if unstable, to determine the nature of the instability. First of all, if Re $(\sigma)=0$, $\text{Im}(\sigma)<0$, the system is an unstable one; for example, a pencil standing on its tip. The nature of the instability is such that the perturbation grows exponentially with time. The system is stable if $Re(\sigma)\neq 0$, $Im(\sigma)=0$. It oscillates about the perturbed value. An example of such a system is a pencil standing on the flat eraser end—so long as the perturbations are infinitesimal. A round pencil on its side is an example of neutral stability. The perturbation increases linearly with time as $Re(\sigma)=0$ and $Im(\sigma)=0$. The case of overstability, a difficult one to illustrate with pencils, is characterized by oscillations of increasing amplitude: $Re(\sigma)\neq 0$, $Im(\sigma)<0$. A bridge set into oscillation by the wind or by marching soldiers at its natural frequency of vibration is an example of such an instability. These four types of instability are summarized in the table and diagrams on the following page.

In linearized instability theory, it often occurs that many roots (values of σ, eigenvalues) are obtained corresponding to different modes of motion (displacements, eigenfunctions). If any of these is unstable, the object is unstable. Thus, one can prove a system is unstable but not that it is stable. Since, in any continuous system, there is an infinity of degrees of freedom it is always possible that there are modes, which have not been examined, that are unstable.

Another approach is to see if an adjacent equilibrium state of lower energy exists. If no such equilibrium state exists, we say that the system is secularly stable. If such a state does exist, the system is secularly unstable whether or not a mode of motion has been found which can carry the system to the lower state. An example of a secularly unstable system is a pencil on its eraser end. A lower equilibrium position exists—lying on its side. Such a system is dynamically stable when only infinitesimal perturbations are considered, but if the perturbations are large enough the pencil can reach the lower equilibrium position; thus, it is secularly unstable as well.

Consider next the strange (and classic) case of a rapidly rotating frictionless bowl in which a ball is oscillating back and forth. Such a system is dynamically stable, but, strangely enough, secularly unstable; the energy (including centrifugal potential energy) is lower the higher up the side of the bowl the ball can go. If a small amount of friction is introduced between the ball and the bowl the lower energy state can be attained. The ball will oscillate with increasing amplitude and the system is overstable on a diffusive (frictional) time scale. Another example of an overstability driven by a dissipative process is convection in a rapidly rotating star. In the absence of rotation, convection is just a simple instability; rotation tends to stabilize this instability, but when thermal conduction is introduced into the analysis the instability reappears as an overstability. The material oscillates up and down; it goes up, gives off some heat, comes down, gains heat, and so forth. If no diffusive processes were at work (that is, if it could not gain or lose heat) the system would be stable.

Im(σ)	Re(σ) ±	Re(σ) 0
+	damped oscillation (stable)	pure damping (stable)
0	ordinary oscillation (stable)	neutrally stable
−	overstable (increasing amplitudes)	unstable (exponential)

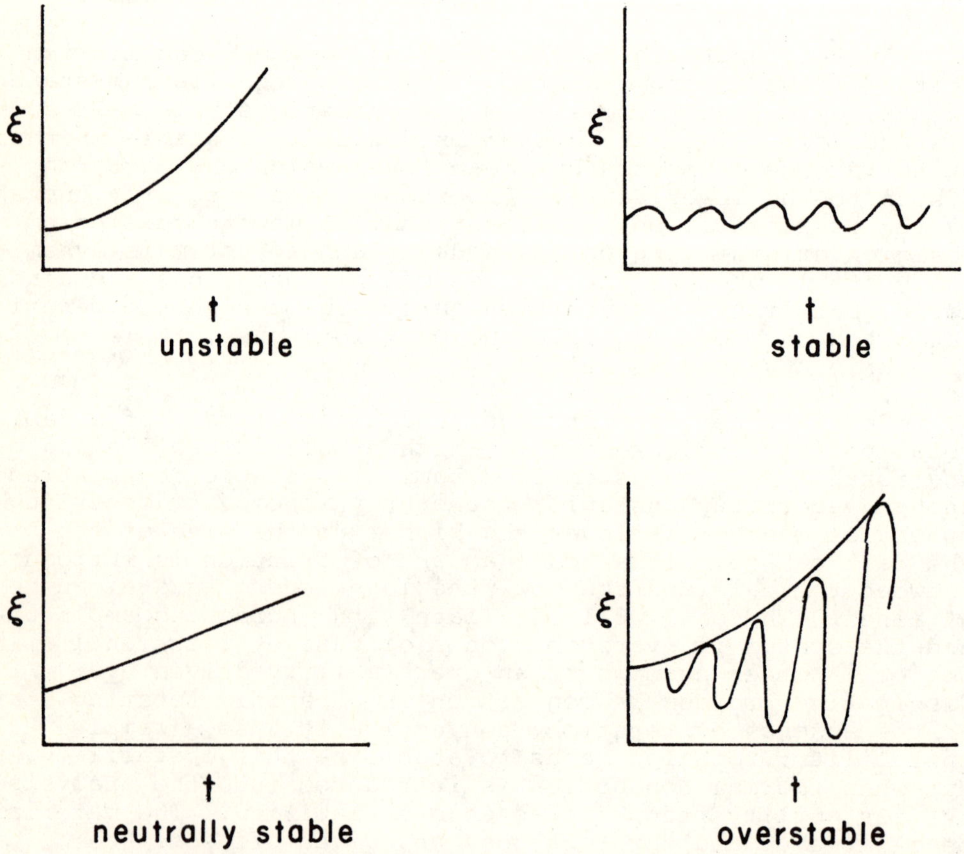

In an important sense all stars are secularly unstable. The possibility of contraction to a lower energy state always exists so long as the Schwarzschild singularity has not been reached and if the star has any way of getting rid of the excess energy. One final and important mixed case, designated vibrational instability, must be described. Here the oscillations are stable on a dynamical time scale but the amplitude of the oscillations grows on a much slower time scale dependent on energy gains and losses. The ordinary variable stars such as the Cepheids are an example of this vibrational instability.

Now we consider the various types of instabilities which can occur in white dwarfs—both for rotating and nonrotating models. Stability analysis will give limits on the allowable mass, angular momentum, and density to supplement the classical limit $M \approx 1.44(\frac{2}{\mu_e})^2 M_\odot$ for spherical models. We begin with nonrotating models.

2. Inverse Beta Decay and Pycnonuclear Reaction, Limits on Density

Nonrotating white dwarf models can be unstable on astronomical time scales against inverse beta decay, in which the electrons in the Fermi sea tumble back into the nucleus. If the energy of the most energetic electron,

$$E_F = MC^2 \left[(1+x^2)^{\frac{1}{2}}-1\right] \propto \rho^{1/3}, \tag{58}$$

exceeds the energy $\Delta E = \frac{1}{2} M_e v^2$ for beta emission by the nucleus (A,Z-1) a spontaneous transformation of (A,Z) into (A,Z-1) occurs:

$$(A,Z) + \beta + \tfrac{1}{2} M_e v^2 \rightarrow (A,Z-1) \tag{59}$$

Such inverse beta decays generally occur in pairs as the odd-odd nucleus usually has a smaller β decay energy than its even-even precursors. Thus the nucleus (A,Z) rapidly undergoes two successive inverse beta decays to (A,Z-2) (Schatzman 1958, Hamada and Salpeter 1961). Beta decay of the (A,Z-1) nucleus into (A,Z) cannot occur since an electron already occupies the energy level into which the emitted electron would fall. If the chemical composition is specified for a homogeneous model, a finite maximum central density is obtained. For a composition of C^{12}, decay into B^{12} with the resulting instability requires $\rho_c \gtrsim 10^{9.7}$ gm/cm^3 and for an oxygen composition $\rho_c \gtrsim 10^{10}$ gm/cm^3. Corresponding to this maximum density for stability is a maximum mass beyond which no homogeneous configuration can exist. These limits are seen on Figures 3 and 4 as points where there is a change in the slope of the mass-radius or mass density relationships. Beyond these points, as can be seen by the following homology arguments, the models are unstable.

From the equations of hydrostatic equilibrium and mass continuity, $\bar{P} \approx GM^2/R^4$ and $\bar{\rho} \approx M/R^3$ imply that

$$\overline{P} = GM^{2/3}\overline{\rho}^{-4/3}. \tag{60}$$

The average ratio of the specific heats is then

$$\overline{\gamma} = \frac{d\ln\overline{P}}{d\ln\overline{\rho}} = \frac{\Delta\overline{P}}{\overline{P}}\,\frac{\overline{\rho}}{\Delta\overline{\rho}} = \frac{4}{3} + \frac{2}{3}\,\frac{\overline{\rho}}{M}\,\frac{\Delta M}{\Delta\overline{\rho}}\,. \tag{61}$$

Assume that the mass of a nonrotating star of a given chemical composition is changed. If the mass increases with the density, $\frac{\Delta M}{\Delta\overline{\rho}}$, is positive and $\overline{\gamma}>4/3$. But if the density is reached where inverse beta decay can occur, Hamada and Salpeter's (1961) curve of mass versus central density turns over (see Figure 4). At this point $\frac{dM}{d\overline{\rho}}<0$ and $\overline{\gamma}>4/3$, so that the star is dynamically unstable. Such calculations have not been checked through in detail since the displacement of the interface must be considered.

Even at zero temperature, nuclear reactions can occur for sufficiently high densities. Such reactions are called "pycnonuclear" and result from the screening of the nuclear Coulomb field by the electrons. A pycnonuclear reaction occurs at $\rho \approx 5\times10^4$ gm/cm^3 for hydrogen. Such a low density limit for hydrogen burning means that a white dwarf of even moderate mass cannot possess hydrogen in the interior and still be in equilibrium. Hence this limit on the density is dependent on equilibrium requirements rather than stability requirements. For heavier nuclei, the pycnonuclear reactions occur at higher densities: He^4 at $\rho\approx8\times10^8$ gm/cm^3 and C^{12} at $\rho\approx6\times10^9$ gm/cm^3. All these densities are lower than the critical densities found for inverse beta decay. Crystallization of the ions (discussed subsequently) will alter these limits.

Thus, if very high densities are reached, thermal equilibrium is precluded by pycnonuclear reactions and mechanical equilibrium by inverse β decay. From these considerations we derive, for carbon or oxygen white dwarfs, the limiting central density of $\approx10^{9.5}$ gm/cm^3; we will see subsequently how this limits the mass and angular momentum stable white dwarfs can have.

3. The Pulsation and Kelvin Modes of Oscillation and the Tensor Virial Method

Consider a completely degenerate zero temperature rotating white dwarf which is stable against local instabilities. Viscosity, meridional circulation, and magnetic fields are neglected. The central density is restricted to values for which inverse beta decay, pycnonuclear reactions, and general relativistic effects are unimportant. The star is axisymmetric and the angular velocity at equilibrium depends only on the distance from the axis of rotation. For differential rotation it is evident that the models will be stable against axisymmetric perturbations since the angular momentum increases outwards and is taken as a constant on cylinders:

$$\frac{\partial j}{\partial \tilde{\omega}} > 0, \frac{\partial j}{\partial z} = 0 \ . \tag{62}$$

It is possible that non-axisymmetric instabilities exist when too much shear occurs, but there has been no theoretical investigation of this point.

Of the adiabatic modes which might become unstable or overstable, two modes are of particular interest:

1) The fundamental modes of oscillation (the p or pressure modes), which in the absence of rotation corresponds to radial oscillations.

2) The non-axisymmetric Kelvin mode (f mode), a relatively incompressible mode which corresponds to surface motions distorting the roughly spheroidal shape into a roughly ellipsoidal form. Unstable gravity modes (g modes) related to convective instability cannot occur in a zero temperature white dwarf since $\bar{\gamma} = \frac{d\ln P}{d\ln \rho}$.

Consider the lowest p and f modes; that is, those modes for which no nodes occur from the center of the star to the surface. Then, in the absence of rotation, the frequencies of oscillation are given by simple formulae due to Ledoux (1945, 1949):

$$\sigma_p^2 \approx (3\bar{\gamma}-4) \frac{|W|}{I} ,$$

$$\sigma_k^2 \approx \frac{4}{5} \frac{|W|}{I} , \tag{63}$$

$$\sigma_f^2 = 0 \ .$$

Such formulae are good to within a few percent for stars that are not very centrally condensed ($n<2.5$). For low mass white dwarfs $\bar{\gamma}=5/3$ and $\sigma_p^2 > \sigma_k^2$; for high masses $\bar{\gamma}=4/3$ and $(\sigma_p^2 \neq \sigma_k^2) \to 0$. In a white dwarf with a mass corresponding to $\bar{\gamma}=8/5$ ($M=0.3\ M_\odot$), a crossover effect occurs: $\sigma_p^2 = \sigma_k^2$. Note also that the radial mode would be dynamically unstable if $\bar{\gamma}<4/3$.

If rotational and general relativistic effects are included, corrections must be applied to Ledoux's formulae. The frequency of oscillation of the p mode is given by

$$\sigma_p^2 = (3\bar{\gamma}-4) \frac{|W|}{I} + c_1^2 (5\bar{\gamma}-3) \frac{T}{I} - c_2^2 \frac{R_s}{R} \frac{|W|}{I} , \tag{64}$$

where c_1 and c_2 are constants of order unity. The second term corrects for the inclusion of rotation, the third, for general relativity effects, where R_s is the Schwarzschild radius. Other modes have similar correction terms. Both terms become very important for very dense white dwarfs (for which $\bar{\gamma} \approx 4/3$), and the first term becomes very small. Although the general relativistic correction is normally small, Chandrasekhar and Tooper (1964) find that in the absence of rotation it can easily make the star unstable, even when $r_s \approx 10^{-2}$ R, since it is cancelling a very small term. However, the general relativistic term only really becomes important for $\rho_c \approx 10^{12}$ gm/cm^3.

From equation (54) it is evident that a white dwarf of the shortest possible period oscillating in the fundamental mode (Ostriker and Tassoul 1968; Faulkner and Gribbin 1968) ($\approx$0.2 sec) occur for models which are dense and rapidly rotating.

Let us employ the virial method developed by Chandrasekhar and Lebovitz (1962) and generalized by Tassoul and Ostriker (1968) to analyze the stability of the pressure and Kelvin modes in nonuniformly rotating, centrally condensed white dwarfs. With the tensor form of the virial equation we have nine second order differential equations and thus nine possible frequencies of oscillation can be determined.

$$\frac{d}{dt} \int_V \rho u_i x_j d\underline{x} = 2T_{ij} + W_{ij} + \delta_{ij}\Pi \ . \tag{65}$$

If the equilibrium configuration is slightly perturbed, the resulting motion can be described by the Lagrangian displacement $\underline{\xi}(\underline{x},t)=\underline{\xi}(\underline{x})e^{i\sigma t}$ of an element of $\underline{x}$. This displacement causes changes in the density, pressure, gravitational energy, and kinetic energy.

As a typical term in the perturbed virial equation, consider

$$\delta\pi = \delta \int P d\underline{x} \ . \tag{66}$$

Assuming that the displacements are adiabatic, the Eulerian variation in pressure, δP, is given by the change in pressure following the motion, ΔP, less the change caused by moving into adjacent regions where the pressure was initially somewhat different:

$$\frac{\Delta P}{P} = \frac{\delta P}{P} + \frac{\underline{\xi}\cdot\nabla P}{P} = \gamma\frac{\Delta\rho}{\rho} \ . \tag{67}$$

But the Lagrangian variation in density is given by

$$\Delta\rho = \delta\rho + \underline{\xi}\cdot\nabla\rho = -\underline{\xi}\cdot\nabla\rho - \rho\nabla\cdot\underline{\xi} + \underline{\xi}\cdot\nabla\rho = -\rho\nabla\cdot\underline{\xi} \ , \tag{68}$$

where $\delta\rho$ is obtained from the equation of continuity. Now we put $\delta P = \Delta P - \underline{\xi}\cdot\nabla P = -\gamma P \nabla\cdot\underline{\xi} - \underline{\xi}\cdot\nabla P$ into equation (66), integrate by parts with the pressure zero at the surface, and find

$$\delta\pi = -\int (\gamma-1) P (\nabla\cdot\underline{\xi}) d\underline{x} \ . \tag{69}$$

The variations in the other terms are obtained in a similar fashion, and the perturbed tensor virial equations become

$$
\begin{aligned}
&-\sigma^2 \int_V \rho \xi_i x_j d\underline{x} - 2i\sigma \int_V \rho Q_{jk} \xi_i x_k d\underline{x} \\
&+ \int_V \rho Q_{jl} Q_{lk} \xi_i x_k d\underline{x} + \int_V \rho Q_{il} Q_{lk} \xi_i x_k d\underline{x} \\
&+ \int_V \rho \frac{\partial Q_{js}}{\partial x_k} Q_{kr} \xi_i x_r x_s d\underline{x} \\
&+ \int_V \rho \frac{\partial Q_{is}}{\partial x_k} Q_{kr} \xi_j x_r x_s d\underline{x} \\
&= -G \int_V \rho \xi_k d\underline{x} \frac{\partial}{\partial x_k} \int_V \rho(\underline{x}') \frac{(x_i - x_i')(x_j - x_j')}{|\underline{x} - \underline{x}'|^3} d\underline{x} \\
&- \delta_{ij} \int_V (\gamma - 1) P \frac{\partial \xi_k}{\partial x_k} d\underline{x} \quad .
\end{aligned}
\tag{70}
$$

This form of the virial equations give nine equations for the characteristic frequencies (eigenvalues) in terms of the equilibrium structure $[\rho(\underline{r}), P(\underline{r}), \Omega(\tilde{\omega})]$ and the eigenfunctions $\underline{\xi}(r)$. $Q_{ij}(\tilde{\omega})$ is the antisymmetric angular velocity tensor (with $\Omega(\tilde{\omega}) = \Omega(\tilde{\omega})\hat{z}$) such that the components of velocity are given by

$$u_i = Q_{ij}(\tilde{\omega}) x_j . \tag{71}$$

The first term in equation (70), containing a second derivative with respect to time, is the inertial term. The second term, containing Q_{jk}, is a coriolis term; the third and fourth terms, containing $Q_{jl}Q_{lk}$ are centrifugal terms. The two terms on the right hand side of equation (70) are the gravitational potential energy of tensor and $(\gamma-1)$ times the internal heat energy.

If the exact form of the Lagrangian displacement $\underline{\xi}$ does not deviate very much from that of a homogeneous, uniformly rotating spheroid, it can be adequately represented by the function

$$\xi_i = L_{i;k} x_k , \tag{72}$$

where $L_{i;k}$ denotes nine constants. If this function is put into equation (70), a set of linear equations is obtained which can be solved to yield the following dispersion relations, for the following types of modes:

1) tesseral oscillations, characterized by a relative shearing of northern and southern hemispheres,

$$\sigma^3-2\langle\Omega\rangle\sigma^2+[\langle\Omega^2\rangle-(L+M)]\sigma+2\langle\Omega\rangle M=0, \qquad \text{(3 roots)} \quad (73)$$

2) sectorial (or toroidal) oscillations which transform the star into a triaxial configuration while preserving the plane of symmetry,

$$\sigma^2-2(S+\langle\Omega\rangle^2-\langle\Omega\rangle^2)\pm2\langle\Omega\rangle(2S+\langle\Omega\rangle^2-2\langle\Omega^2\rangle)^{\frac{1}{2}}=0, \qquad \text{(2 roots)} \quad (74)$$

3) zonal (or pulsational oscillations which preserve the axisymmetry of the star,

$$\sigma^4-(A+C+D)\sigma^2+(AD-BC)=0, \qquad \text{(2 roots)} \quad (75)$$

4) $\sigma^2=0.$ (2 roots)

In these dispersion relations

$$L = W_{13;13}/I_{11}, \qquad M = W_{31;13}/I_{33},$$

$$S = W_{12;12}/I_{11}, \qquad A = 4\langle\Omega\rangle^2-3\langle\Omega^2\rangle+\frac{(W_{33;33}-W_{33;11})}{I_{11}},$$

$$B = W_{33;11}-W_{33;33})/I_{33}, \qquad C = 2(W_{33;11}-P)/I_{11},$$

$$D = (W_{33;33}-P)/I_{33}. \qquad (76)$$

The I_{ii}'s are just moments of inertia and the $W_{ij;k\ell}$'s are gravitational moments (defined by Chandrasekhar and Lebowitz 1962). Provided these various coefficients can be evaluated, the frequencies of oscillation associated with a particular equilibrium configuration can be obtained by solution of trivial algebraic relations (73-75). The gravitational moments are related to the density by quadratic integral operators; the inertial moments are related by linear integral operators. For example,

$$W_{12;12} = \frac{\pi}{4} \iint \rho D_{\tilde\omega}\left[\tilde\omega^4 D_{\tilde\omega}(D_{\tilde\omega}\chi)\right]\tilde\omega d\tilde\omega dz, \qquad (77)$$

where $D_{\tilde\omega}$ is the differential operator $\frac{1}{\tilde\omega}\frac{\partial}{\partial\tilde\omega}$ and $\chi(\underline{x})$ is the superpotential of the gravitational field such that

$$D_{\tilde\omega}\chi(\underline{x}) = -G \int_V \frac{\rho(\underline{x}')}{|\underline{x}-\underline{x}'|}d\underline{x}' \left[1-\frac{\tilde\omega'}{\tilde\omega}\cos(\phi-\phi')\right]. \qquad (78)$$

Ordinarily, the calculations of such integrals would be prohibitive because of the large amount of computer time required. However, the fact that the gravitational moments are quadratic in density means that some four-index array must exist such that multiplication by the densities ρ_{ij} (i,j correspond to selected grid points) expressed as two index arrays and summation over indices gives a good approximation to the integral form. That is, there must exist an array T such that

$$W_{12;12} = \frac{GM^2}{R} \sum_{i=1}^{N} \sum_{j=i}^{N} \sum_{k=1}^{N} \sum_{l=k}^{N} \rho_{ij}\rho_{kl}T_{ijkl} \qquad (79)$$

is an exact relation in the limit $N\to\infty$ and a good approximation for moderate value of N. Once the fixed array T_{ijkl} has been obtained and stored, simple matrix multiplication gives $W_{12;12}$ for any density distribution. The other gravitational moments are obtained in a similar fashion.

4. Modes of Oscillation, Dynamic and Secular Instabilities, Fission and Limits on Angular Momentum

Detailed stability calculations have been made by Ostriker and Tassoul (1969). First let us focus on the relatively simple and previously explored case of non-rotating white dwarfs. As expected, longer periods are associated with lower mass stars. Table 4 shows that the radial (fundamental) mode usually has a period longer than the Kelvin mode. However, for sufficiently low masses the nonradial periods are longer.

Tables 5-7 give the properties and periods of differentially rotating models for two different masses and laws of rotation. As the total angular momentum is increased, the periods of the seven lowest modes change as is seen in Figures 22 and 23. The sectorial modes, labeled P_{-2} and P_2, are represented by solid lines; the tesseral modes, labeled P_{-1}, P_1, and P_0, are represented by the upper and lower dotted lines, respectively. The periods P_Z and P_R increase with increasing angular momentum. That is, rotation increases the pulsation period of a star with given mass. However, for a given <u>central density</u> the greater the angular momentum the smaller the pulsation periods as shown in Figures 22 and 23, where σ and J are given in units of $(4\pi G\rho_c)^{1/2}$ and $M^{5/3}G^{1/2}\rho_c^{\,1/6}$.

From equation (74) it is evident that the period P_{-2} becomes infinite ($\sigma_{-2}=0$) when $\langle\Omega^2\rangle=S$. This point appears at T/W=0.14 and corresponds to the classical bifurcation point. Secular instability is expected beyond this point.

The two sectorial (toroidal) modes become equal when $T/|W|=0.26$. At this point the kinetic energy and the internal energy are of the same order. Beyond this point the models are dynamically overstable, and fission may occur on a dynamical time scale although this last statement cannot be proved on the basis of a linear analysis. The models in Figure 24, which roughly follow the (M,J) relation for main sequence stars, lie below the limit of dynamic overstability. That is, these high mass white dwarfs are dynamically stable. However, some are in the region of a long time scale (10^{10} years)

TABLE 4

FUNDAMENTAL PERIODS OF NON-ROTATING WHITE DWARFS*

$M/M_{\odot}$	$R/R_{\odot}$	ρ_c	P_K	P_R
1.13(−1)	2.60(−2)	5.59(4)	8.29(1)	7.57(1)
2.26(−1)	2.01(−2)	2.51(5)	3.95(1)	3.72(1)
3.40(−1)	1.70(−2)	6.53(5)	2.48(1)	2.43(1)
4.53(−1)	1.49(−2)	1.38(6)	1.73(1)	1.78(1)
5.66(−1)	1.32(−2)	2.65(6)	1.27(1)	1.38(1)
6.79(−1)	1.18(−2)	4.86(6)	9.55	1.10(1)
7.92(−1)	1.05(−2)	8.81(6)	7.24	9.02
9.05(−1)	9.24(−3)	1.62(7)	5.46	7.44
1.02	8.05(−3)	3.14(7)	4.03	6.13
1.13	6.82(−3)	6.69(7)	2.85	4.99
1.24	5.47(−3)	1.74(8)	1.83	3.92
1.30	4.71(−3)	3.23(8)	1.37	3.39

*The calculations are for $\mu_e = 2$. For the other values of μ_e, P should be multiplied by $(2/\mu_e)^{\frac{1}{2}}$. Units are in the cgs system. (Ostriker and Tassoul 1969, Table 1.)

TABLE 5

PROPERTIES AND PERIODS OF DIFFERENTIALLY ROTATING WHITE DWARFS*

(Angular-Momentum Distribution of a Uniformly Rotating Polytrope of Index $n' = \frac{3}{2}$)

$M = 1.13\ M_{\odot}$

J	0	9.62(48)	1.92(49)	4.81(49)	9.62(49)	1.44(50)	1.92(50)	*2.28(50)*	2.41(50)
R_e	4.74(8)	4.77(8)	4.88(8)	5.53(8)	7.38(8)	9.97(8)	1.35(9)	*1.73(9)*	1.88(9)
R_e/R_p	1	1.01	1.02	1.12	1.39	1.75	2.24	*2.82*	2.91
v_e	0	4.32(7)	8.46(7)	1.86(8)	2.80(8)	3.10(8)	3.05(8)	*2.82(8)*	2.75(8)
Ω_c/Ω_e	....	1.60	1.61	1.63	1.72	1.93	2.29	*2.74*	2.90
ρ_c	6.69(7)	6.56(7)	6.21(7)	4.53(7)	2.32(7)	1.25(7)	7.23(6)	*4.97(6)*	4.32(6)
U	4.83(50)	4.79(50)	4.66(50)	3.98(50)	2.82(50)	2.02(50)	1.48(50)	*1.20(50)*	1.11(50)
T	0	5.86(47)	2.25(48)	1.10(49)	2.59(49)	3.51(49)	3.96(49)	*4.06(49)*	4.07(49)
$\lvert W \rvert$	7.18(50)	7.13(50)	7.00(50)	6.31(50)	5.05(50)	4.09(50)	3.37(50)	*2.96(50)*	2.82(50)
P_{-2}	2.85	3.05	3.35	4.94	1.09(1)	2.72(1)	9.75(1)	∞	3.77(2)
P_2	2.85	2.73	2.69	2.88	3.84	5.29	7.24	*9.06*	9.84
P_1	2.85	2.95	3.11	3.86	5.69	7.97	1.06(1)	*1.28(1)*	1.37(1)
P_1	2.85	2.80	2.79	3.02	3.86	5.02	6.47	*7.76*	8.31
P_0	∞	5.21(1)	2.71(1)	1.38(1)	1.16(1)	1.27(1)	1.49(1)	*1.68(1)*	1.75(1)
P_Z	2.85	2.87	2.93	3.31	4.27	5.38	6.58	*7.47*	7.87
P_R	4.99	5.00	5.05	5.39	6.49	8.14	1.04(1)	*1.25(1)*	1.35(1)

*The calculations are for $\mu_e = 2$. For the other values of μ_e, P should be multiplied by $(2/\mu_e)^{\frac{1}{2}}$. Units are in the cgs system. Column in italics pertains to the point of bifurcation. (Ostriker and Tassoul 1969, Table 2.)

TABLE 6

Properties and Periods of Differentially Rotating White Dwarfs*
(Angular-Momentum Distribution of a Uniformly Rotating Polytrope of Index $n' = 0$)
$M = 1.13\ M_\odot$

J	0	7.70(49)	1.54(50)	*2.15(50)*	2.31(50)	3.08(50)	3.85(50)	4.62(50)	*4.94(50)*
R_e	4.74(8)	6.26(8)	9.05(8)	*1.17(9)*	1.24(9)	1.63(9)	2.08(9)	2.59(9)	*2.82(9)*
R_e/R_p	1	1.22	1.61	*1.95*	2.04	2.52	3.03	3.54	*3.75*
v_e	0	1.37(8)	1.89(8)	*2.04(8)*	2.07(8)	2.09(8)	2.06(8)	1.98(8)	*1.94(8)*
Ω_c/Ω_e	....	3.89	3.45	*3.25*	3.22	3.13	3.10	3.08	*3.05*
ρ_c	6.69(7)	2.80(7)	9.65(6)	*4.85(6)*	4.14(6)	2.10(6)	1.16(6)	6.73(5)	*5.43(5)*
U	4.83(50)	3.18(50)	1.83(50)	*1.25(50)*	1.14(50)	7.64(49)	5.35(49)	3.84(49)	*3.37(49)*
T	0	2.13(49)	3.73(49)	*4.20(49)*	4.26(49)	4.30(49)	4.11(49)	3.80(49)	*3.66(49)*
$\|W\|$	7.18(50)	5.45(50)	3.85(50)	*3.05(50)*	2.89(50)	2.27(50)	1.83(50)	1.50(50)	*1.39(50)*
P_{-2}	2.85	8.08	3.71(1)	∞	2.55(2)	6.88(1)	5.51(1)	5.07(1)	*4.43(1)*
P_2	2.85	3.46	5.78	*8.44*	9.25	1.42(1)	2.14(1)	3.29(1)	*4.43(1)*
P_{-1}	2.85	5.01	8.85	*1.27(1)*	1.38(1)	2.00(1)	2.75(1)	3.66(1)	*4.09(1)*
P_1	2.85	3.55	5.40	*7.27*	7.81	1.08(1)	1.45(1)	1.92(1)	*2.14(1)*
P_0	∞	1.20(1)	1.35(1)	*1.68(1)*	1.77(1)	2.35(1)	3.07(1)	3.95(1)	*4.36(1)*
P_Z	2.85	3.92	5.74	*7.29*	7.69	9.82	1.22(1)	1.50(1)	*1.63(1)*
P_R	4.99	6.14	8.76	*1.17(1)*	1.26(1)	1.77(1)	2.42(1)	3.27(1)	*3.69(1)*

*The calculations are for $\mu_e = 2$. For the other values of μ_e, P should be multiplied by $(2/\mu_e)^{\frac{1}{2}}$. Units are in the cgs system. The first and second columns in italics pertain to the models at the point of bifurcation and at the point of dynamical overstability. (Ostriker and Tassoul 1969, Table 3.)

TABLE 7

Properties and Periods of Differentially Rotating White Dwarfs*
(Angular-Momentum Distribution of a Uniformly Rotating Polytrope of Index $n' = 0$)
$M = 2.26\ M_\odot$

J	2.31(50)	*3.40(50)*	3.85(50)	5.39(50)	6.93(50)	8.47(50)	1.00(51)	1.12(51)	*1.15(51)*
R_e	2.51(8)	*4.33(8)*	5.09(8)	7.70(8)	1.04(9)	1.33(9)	1.64(9)	1.90(9)	*1.97(9)*
R_e/R_p	2.02	*2.12*	2.20	2.46	2.77	3.10	3.45	3.71	*3.78*
v_e	5.10(8)	*4.35(8)*	4.19(8)	3.87(8)	3.68(8)	3.52(8)	3.37(8)	3.25(8)	*3.21(8)*
Ω_c/Ω_e	6.23	*5.21*	4.91	4.21	3.81	3.58	3.42	3.32	*3.29*
ρ_c	2.71(9)	*4.26(8)*	2.45(8)	6.17(7)	2.39(7)	1.14(7)	6.12(6)	4.08(6)	*3.68(6)*
U	4.61(51)	*2.17(51)*	1.71(51)	9.13(50)	5.74(50)	3.92(50)	2.81(50)	2.25(50)	*2.13(50)*
T	8.83(50)	*5.45(50)*	4.80(50)	3.62(50)	3.05(50)	2.66(50)	2.37(50)	2.18(50)	*2.14(50)*
$\|W\|$	7.27(51)	*3.95(51)*	3.30(51)	2.10(51)	1.53(51)	1.20(51)	9.71(50)	8.45(50)	*8.15(50)*
P_{-2}	6.56	∞	4.69(1)	2.05(1)	1.87(1)	1.89(1)	1.96(1)	1.93(1)	*1.77(1)*
P_2	4.29(−1)	*1.08*	1.42	2.87	4.79	7.35	1.09(1)	1.50(1)	*1.77(1)*
P_{-1}	6.37(−1)	*1.59*	2.09	4.09	6.51	9.38	1.27(1)	1.55(1)	*1.63(1)*
P_1	3.81(−1)	*9.25(−1)*	1.20	2.27	3.53	5.00	6.71	8.17	*8.60*
P_0	8.89(−1)	*2.11*	2.70	4.96	7.54	1.05(1)	1.38(1)	1.65(1)	*1.73(1)*
P_Z	3.97(−1)	*9.46(−1)*	1.21	2.20	3.23	4.34	5.52	6.45	*6.72*
P_R	1.18	*2.35*	2.86	4.69	6.78	9.29	1.23(1)	1.49(1)	*1.57(1)*

*The calculations are for $\mu_e = 2$. For the other values of μ_e, P should be multiplied by $(2/\mu_e)^{\frac{1}{2}}$. Units are in the cgs system. The first and second columns in italics pertain to the models at the point of bifurcation and at the point of dynamical overstability. (Ostriker and Tassoul 1969, Table 4.)

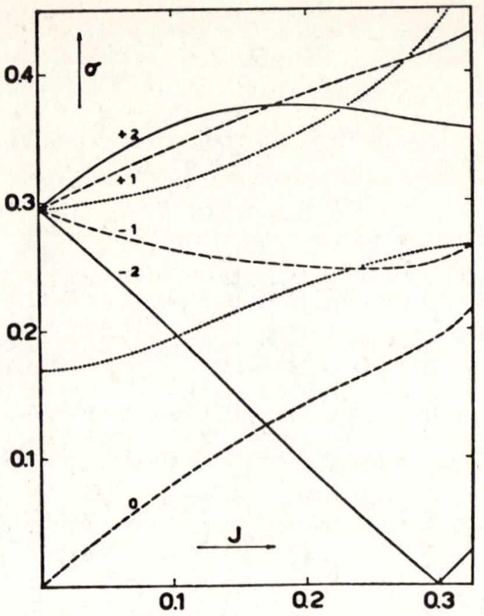

Figure 22. Frequencies of oscillation along a sequence of differentially rotating white dwarfs ($M = 1.13\ M_\odot$). The angular-momentum distribution is that of a uniformly rotating polytrope of index $n' = \frac{3}{2}$. J and σ are given, respectively, in the units $M^{5/3}G^{1/2}\rho_c^{-1/6}$ and $(4\pi G\rho_c)^{1/2}$. Sectorial, tesseral and zonal modes are represented, respectively, by full-line, dashed, and dotted curves. The upper dotted curve represents a zonal Kelvin mode. See text and Table 5 for the labeling of the remaining curves. Note that the frequency of the "radial" mode (lower dotted curve) increases with increasing J; i.e., rotation decreases the pulsation period for given density. Note also the neutral mode (point of bifurcation) at $J \simeq 0.30$. (Ostriker and Tassoul 1969, Fig. 1.)

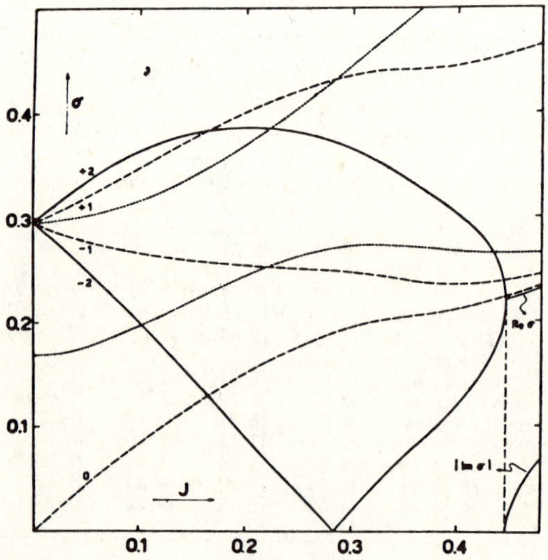

Figure 23. Frequencies of oscillation along a sequence of differentially rotating white dwarfs ($M = 1.13\ M_\odot$). The angular momentum distribution is that of a uniformly rotating, homogeneous spheroid. Units and curve designations as in Fig. 22. Note the neutral mode (bifurcation point) at $J \simeq 0.46$. (Ostriker and Tassoul 1969, Fig. 2.)

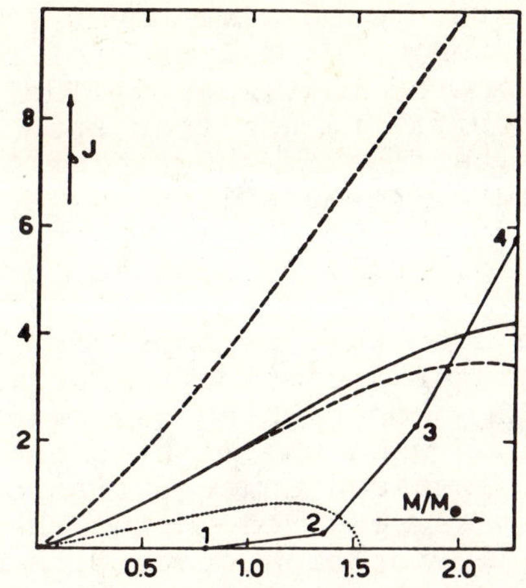

Figure 24. Limits of secular and dynamical overstabilities in the (M,J) plane. J is given in the unit of 10^{50} g cm^2 sec^{-1}. Models labeled 1, 2, 3, and 4 were constructed in Paper I and very roughly define the (M,J) relation for main-sequence stars. White dwarfs having the ($n' = 0$) angular-momentum distribution are overstable above the upper dashed curve if viscosity is negligible, or the lower dashed curve otherwise. The solid curve corresponds to the lower dashed curve when the ($n' = \frac{3}{2}$) angular-momentum distribution is used. Uniformly rotating stars can be constructed within the area defined by the dotted curve. (Ostriker and Tassoul 1969, Fig. 3.)

secular instability driven by viscosity. Of course, the damping mechanism causing secular instability does not have to be viscosity; any other, such as gravitational radiation, could lead to secular instability. Note that both instabilities are almost independent of the total mass and the angular momentum distribution, being determined essentially by the ratio of kinetic gravitational energy.

Physically, it is fairly easy to understand what is going on. If we try to make a non-rotating star in equilibrium, it is spherical because this is the best way of minimizing the gravitational and internal energy. If the star rotates, kinetic energy must also be minimized. For small kinetic energy, the effect is negligible and the star takes on an oblate shape. If the kinetic energy is large, the best way of storing it must be considered—that is, by using the largest moment of inertia. The star chooses a form in which the kinetic energy is minimized without doing too much damage to the minimization of the other forms of energy—an elongated shape is then most advantageous (this occurs at the bifurcation point).

The kinetic energy can become so large that it is more profitable for the star to be a binary system. For a white dwarf, this occurs when $T/|W| \approx 0.2$ and, of course, is also somewhat dependent on the total mass. The angular momentum is then stored as orbital rather than spin angular momentum. On these principles, a fission theory for the formation of binary systems could be re-established. Mathematically, it is still necessary to show that, after the point of dynamical overstability, fission will, in fact, occur.

As an aside we note that perhaps the problem of excess angular momentum in star formation can be resolved in a similar fashion. If a contracting gas cloud has too much angular momentum, then when the kinetic energy builds up to some point the gas cloud can undergo fission. This process is repeated as often as required during the contraction phase, leaving rapidly rotating stars with most of the system's angular momentum stored in orbital form.

Thus to summarize, a white dwarf is (1) stable against implosion if $\rho_c < 3 \times 10^9 \mathrm{gm/cm^3}$; (2) stable against "fission" if $T/|W| < 0.26$ in the absence of viscosity; and (3) stable against "fission" if $T/|W| < 0.13$ with viscosity. The first stability criterion is a rather conservative estimate since the presence of rotation may lead to stable models above this limit. A convenient mnemonic device can be used to summarize the stability analysis for rotating stars in general:

$T/|W|$

$0 \longmapsto$ 1/1 bound system

$0 \longmapsto$ 1/2 equilibrium

$0 \longmapsto$ $\approx$1/4 (0.26) dynamical stability

$0 \longmapsto$ $\approx$1/8 (0.13) secular stability

$0 \mapsto$ $\approx$1/16 equilibrium with uniform rotation (n=3)

The above diagram indicates why the various fissional instabilities were not found before in compressible (n>1) stars. If uniform rotation is insisted upon, not enough kinetic energy can be put in the system to make it secularly or dynamically unstable.

The final stages of stellar evolution can now be speculatively summarized with the aid of the diagram on the following page. Stars with masses above $\approx 5M_\odot$ become supernovae; those with masses below $\approx 5M_\odot$ (some of which become planetary nebulae) have several choices. Stars with masses less than the Chandrasekhar "limit" become black dwarfs if they are slow rotators ($T/|W| < 1/4$) and undergo fission if they are rapid rotators ($T/|W| > 1/4$). Stars above the Chandrasekhar "limit" must rotate. For $T/|W| > 1/4$, these also undergo fission. For small $T/|W|$ equilibrium is very dense and unstable, they become supernovae; for moderate values of $T/|W|$, they become black dwarfs. However, a massive black dwarf can still

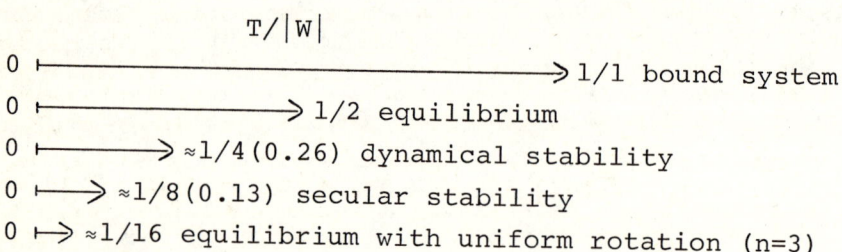

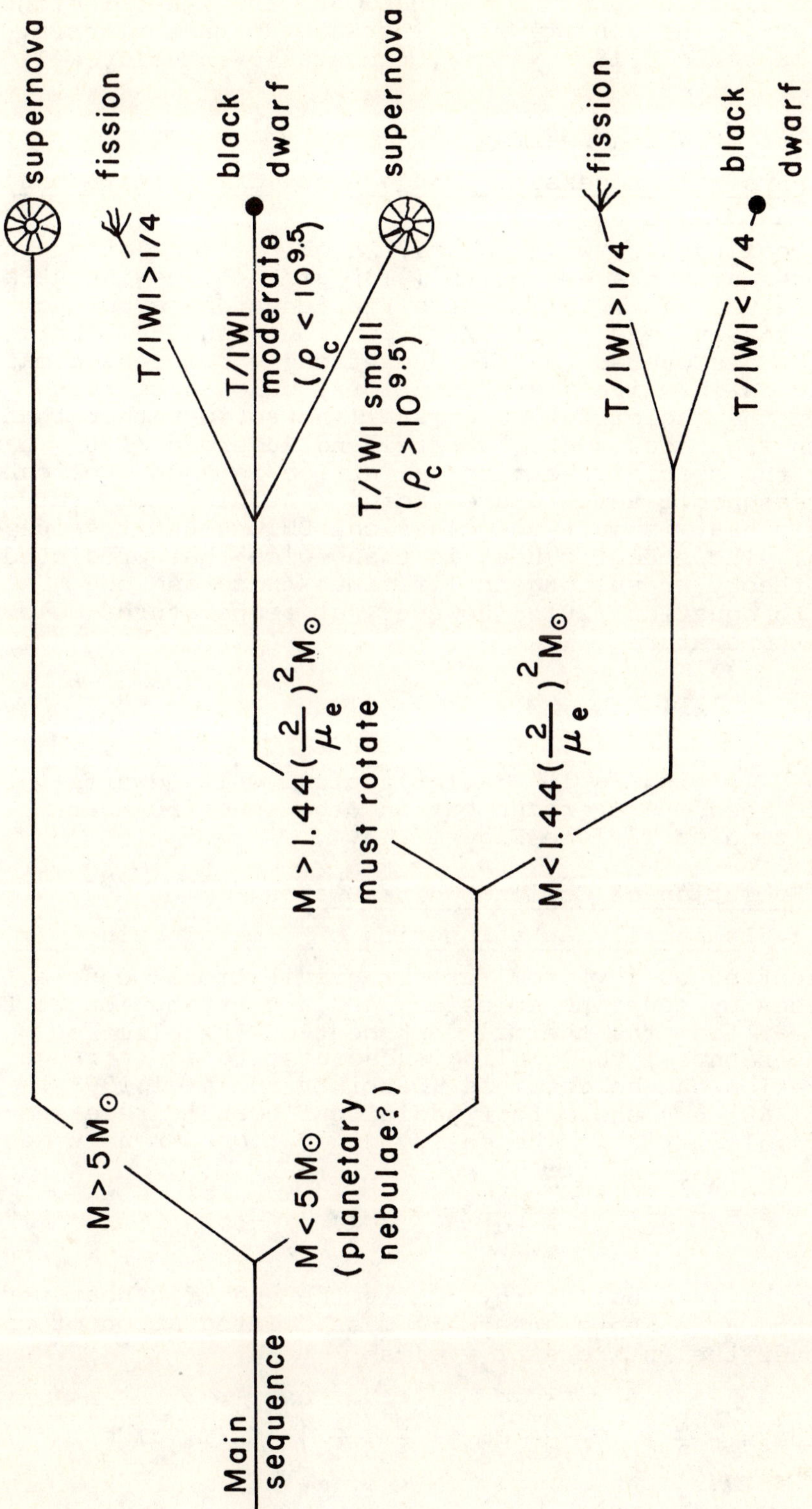
Main sequence
M > 5 M⊙
M < 5 M⊙ (planetary nebulae?)
supernova
M > 1.44 $(\frac{2}{\mu_e})^2$ M⊙
must rotate
M < 1.44 $(\frac{2}{\mu_e})^2$ M⊙
T/|W| > 1/4
fission
T/|W| moderate ($\rho_c < 10^{9.5}$)
black dwarf
T/|W| small ($\rho_c > 10^{9.5}$)
supernova
T/|W| > 1/4
fission
T/|W| < 1/4
black dwarf

eventually become a supernova. The presence of viscous forces will tend to carry angular momentum outwards. The outer regions of the star will expand and the inner regions will contract. The density will increase so that after a viscous time scale (10^{10} years), the star eventually will implode.

III. Cooling of White Dwarfs

A. Physical Processes: Conduction, Crystallization, Convection

The cooling of a white dwarf, which proceeds at approximately constant average density ($\approx 10^5$ gm/km^3), begins at a high temperature ($T_i \approx 10^8$ °K), where the ions form a perfect gas with a thermal energy of 3/2kT per ion. In 1961 Abrikosov and Salpeter discovered that in the strongly compressed matter such as occurs in white dwarf interiors, below a certain temperature the nuclei form a crystalline solid rather than existing in a gaseous state. Mestel and Ruderman (1967) have shown that the specific heat for a lattice is then applicable over all reasonable temperature ranges. At high temperatures, the specific heat is just the classical Dulong-Petit value of 3k per ion. The specific heat is then twice that predicted by classical theory, resulting in lifetimes twice as long according to equation (20). However, at temperatures below the Debye temperature

$$\Theta_D = 4 \times 10^3 \sqrt{\rho} \ °K \quad , \tag{80}$$

where $\mu_e = 2$ (giving $\Theta_D \approx 10^7$ °K for typical white dwarfs), some degrees of freedom are frozen out and the specific heat approaches zero as T^3 (see Figure 26).

B. Evolution of Cooling White Dwarfs

1. Luminosity as a Function of Time; Turning Off

Neglecting energy from further gravitational contraction, energy losses by neutrino emission, and a contribution to the specific heat from the thermal component of the electron gas at low temperatures, the cooling subsequent to short range crystallization can be obtained (Ostriker and Axel 1969). Van Horn (1968) has shown that additional heat is released during crystallization. The energy generation per unit mass is

$$\varepsilon = - \frac{dU}{dt} = - \frac{C_v}{A} \frac{dT}{dt} , \tag{81}$$

where U is the internal energy, C_v is the specific heat per particle, and A is the mean atomic mass. Integration of this equation over the entire star yields

$$L = \int \varepsilon dM = - \frac{M}{A} \int_0^1 C_v \frac{dT}{dt} dm \simeq - \frac{M}{A} \frac{dT_i}{dt} \int_0^1 C_v(T_i, \rho) dm, \tag{82}$$

when the local interior temperature T is replaced by an average interior temperature T_i. The time to cool from some given crystallization temperature T_{cry} is then

$$t(T_i,M) = \frac{M}{A} \int_{T_i}^{T_{cry}} \frac{dT_i}{L(T_i,M)} \int_0^1 C_V(T_i,\rho)\,dm \;. \qquad (83)$$

The luminosity in the preceding equation can then be eliminated if a simple Kramer's-type opacity is assumed:

$$L = KMT_i^{\alpha} \;. \qquad (84)$$

After obtaining $t(T_i)$ for white dwarfs of various masses, inversion of equation (83) and substitution back into equation (84) then yields luminosity as a function of time.

Luminosity-temperature-mass relationships (like equation 84) were obtained from calculated white dwarf models of Vila (1966, 1967), L'Ecuyer (1966), Lee (1950), Rose (1966), Hayashi et al. (1962), and Böhm (1969). Figure 25 shows log L^*/M^* versus log T_i for the different models of cooling to the white dwarf state. The upper part of the curve is the region where Kramer's opacity law holds. At lower central temperatures electron conduction becomes important and Böhm (1969) has shown that convection in the envelope is non-negligible. As the different models had slightly different abundances, it is surprising that the scatter about the mean curve is not larger.

The cooling times for both rotating and nonrotating models are summarized in Table 8. Cooling curves for nonrotating white dwarfs calculated with the Debye approximation in Figure 27 (Ostriker and Axel 1969) show that when the Debye temperature is reached, cooling becomes more rapid than if $C_V = 3/2k$ per ion is assumed. All models then reached low luminosities in moderate times. For Greenstein and Trimble's "typical" mass of $\approx 0.9M_\odot$ the total cooling time is less than 10^{10} years, and proceeds fairly rapidly when the luminosity is below $\approx 10^{-3.5}L_\odot (M_V \approx 13)$. Therefore, white dwarfs fainter than $M_V \approx 13$-14 should be relatively scarce and many stars of moderate mass formed early in the evolution of the Galaxy should now be black dwarfs. The neglect of convection in these calculations means the lifetimes still have been somewhat overestimated and hence even more stars should be invisible. Similar cooling curves for rapid rotators, which are likely to have much higher central densities, show that these cool much faster on a time scale of 10^8 years as compared with 10^{10} years (Figure 28). Hence, there is an even smaller chance of seeing such stars, unless the heat released by internal friction is significant.

The observation of faint white dwarfs in clusters which, by earlier observations, appeared to be much older than the clusters themselves is also explained. The wite dwarf in the Hyades with a luminosity $\approx 10^{-4}L_\odot$ (Eggen and Greenstein 1965) according to old estimates had an age of about 5×10^9

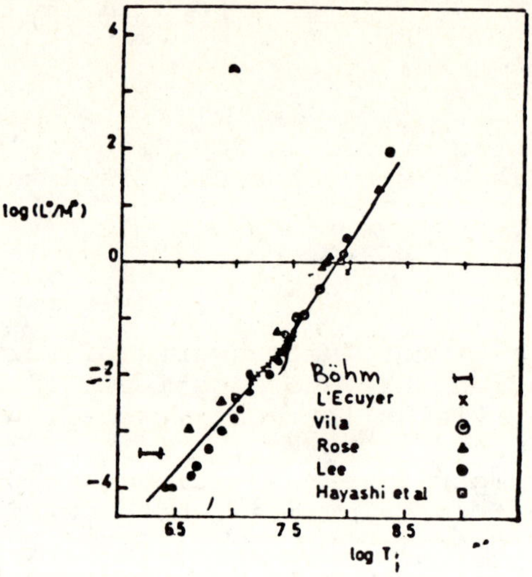

Figure 25. The plotted points represent indicated theoretical calculations; the line shows the approximate relation adopted in this paper. L^* and M^* are the luminosity and mass, respectively, in solar units, and T_i the internal temperature. (Böhm's results have been added to Fig. 2 of Ostriker and Axel 1969.)

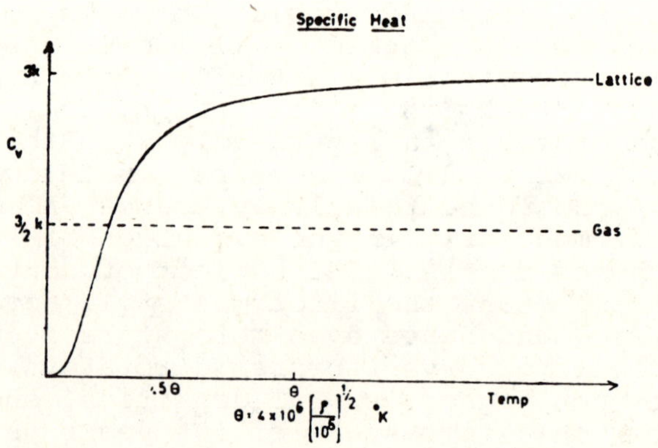

Figure 26. The broken line shows the specific heat per particle as a function of temperature for a monatomic gas, the solid line for a lattice in the Debye approximation. (Ostriker and Axel 1969, Fig. 1.)

TABLE 8

Cooling times of white dwarfs.*

$M/M_\odot$	$J/10^{50}$ (c.g.s.)	log t_c (years)
1.3	0	9.0
1.2	0	9.3
1.0	0	9.7
0.7	0	10.0
0.5	0	>10.0
1.13	6.16×10^{-2}	9.5
1.24	1.32×10^{-1}	8.3
1.52	6.93×10^{-1}	8.6
1.70	1.44	8.8

*(Ostriker and Axel 1969, Table 1.)

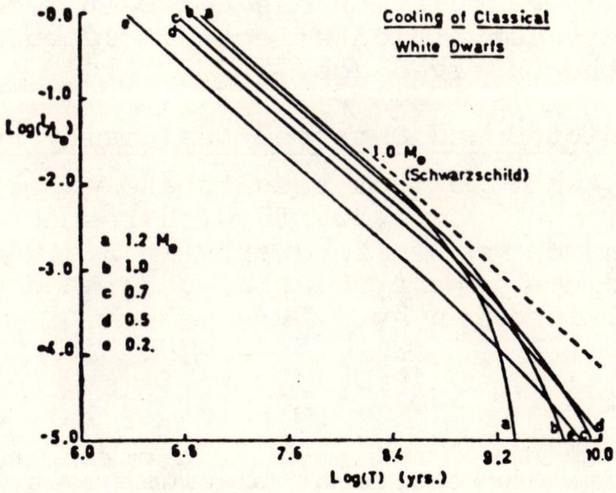

Figure 27. The cooling of classical white dwarfs of a range of masses: luminosity in solar units versus time in years. (Ostriker and Axel 1969, Fig. 3.)

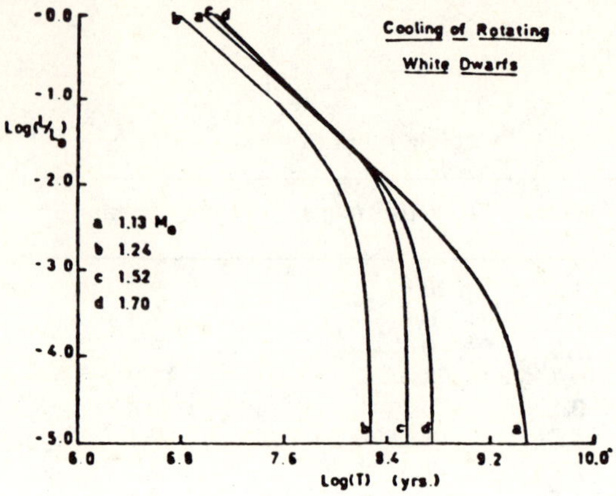

Figure 28. Like Figure 27, but for rapidly rotating models. (Ostriker and Axel 1969, Fig. 4.)

years while the Hyades is believed to be 4×10^8 years old. The new specific heats give an age of about 1.6×10^9 years for a 1.2 $M_\odot$ white dwarf which does not rotate. With rotation, Figure 28 shows that white dwarfs with masses 1.24 $M_\odot$ and 1.52 $M_\odot$ can easily cool down within the required time. Although a detailed calculation has not been made, it seems probable that, were convection effects included, a 1.0 $M_\odot$ star would have the observed age.

2. Predicted and Observed Luminosity Functions

Empirically, what sort of results are expected for the number distribution of white dwarfs with respect to luminosity? In general, the number of white dwarfs in a given magnitude range per unit volume in the solar neighborhood per unit magnitude, $\phi(M_b)$, is given by

$$\text{Log } \phi(M_b) = \text{Log } L - \text{Log } |\dot{L}| + K_1 , \tag{85}$$

where $\dot{L}$ is the time derivative of the luminosity. At high temperatures where Kramer's law holds, $L = K_2 t^{-7/5}$; hence

$$\text{Log } |\dot{L}| = \text{Log } 7/5 - 5/7 \text{ Log } K_2 + 12/7 \text{ Log } L . \tag{86}$$

Therefore,

$$\text{Log } \phi(M_b) = 2/7 \, M_{bol} + \text{constant} , \tag{87}$$

since Log $L = 2/5 \, M_{bol}$ + constant. Weidemann (1967) shows that equation (87), derived on the basis of the older cooling theory, is well satisfied for the brighter white dwarfs (see Figure 12). The break in the curves at $M_V \approx 13$ cannot be explained on the basis of the older theory. It occurs, however, just at the point where the temperature throughout most of the interior of a 1.0 $M_\odot$ star becomes less than the Debye temperature and probably corresponds to the onset of rapid cooling.

3. Space Density of White Dwarfs and the Oort Limit

Weidemann (1967) was able to obtain an estimate of the number of white dwarfs in the solar neighborhood by the following method. Suppose equation (87) can be trusted in some magnitude range. Let us count the number of white dwarfs per unit volume per unit magnitude in this magnitude range. From theory, the rate at which the luminosity changes is known, and hence the rate at which white dwarfs move past a given bolometric magnitude per unit volume in the solar neighborhood per unit time is then calculable. Integrating back for 10^{10} years at the same rate (an underestimate) then gives the number of white dwarfs in the solar neighborhood. Such an idea is an ingenious one. However, Weidemann used a range in M_{bol} of $15^{m}.5$-$16^{m}.5$; whereas it might have been more accurate to apply the method to a magnitude range over which the classical theory gave good results.

The total mass density of the solar neighborhood estimated by Oort (1960) from dynamical studies is $(9.5\pm1) \times 10^{-24}$ gm/cm^3. Of this mass density $(4\pm1) \times 10^{-24}$gm/cm^3 is in stars and $(1.5\pm1) \times 10^{-24}$gm/cm in the interstellar gas (Gliese 1956)*. The unaccounted for mass is thus $(4\pm3) \times 10^{-24}$gm/cm^3. From the Sandage-Luyten sample of 200 stars at $16^{m}.5$ and 103 at $15^{m}.5$, Weidemann obtained the local number density of white dwarfs. If now we take the mean mass to be $\overline{M} = 0.9\ M_\odot$ then the density in white dwarfs is $\approx 3 \times 10^{-24}$gm/cm^3. The white dwarfs are apparently as common as other stars: their total mass is almost as great as that of all others put together. If this is true for Population I stars, the proportion of black dwarfs should be even higher for Population II. Rough estimates indicate the mass of the Galaxy in the form of dead white dwarfs of Population II should be very great, approximately as large as the mass of the Galaxy otherwise accounted for ($\sim 10^{11} M_\odot$). The mass luminosity relation for the Galaxy may then be more comprehensible in comparison to that of other galaxies. Both improved cooling calculations and better white dwarf counts will be required if we are to obtain a realistic picture of the mass distribution in the Galaxy, but it is clear on the basis of available evidence that cold degenerate stars must make a substantial contribution to the mass of both the disc population and the older Population II.

---

*Error bars estimated by J.P.O.

References

Abrikosov, A. A. 1961, Soviet Physics-JETP, 12, 1254.

Allen, C. W. 1963, Astrophysical Quantities (2nd ed.; London: Athlone Press), Chap. X.

Auer, L. H., and Woolf, N. J. 1965, Ap. J., 142, 182.

Böhm, K. H. 1969, in Low Luminosity Stars (New York: Gordon and Breach), p. 393.

Chandrasekhar, S. 1939, Introduction to Stellar Structure (Chicago: Univ. of Chicago Press), Chaps. X and XI.

______. 1961, Hydrodynamic and Hydromagnetic Stability (New York: Oxford Univ. Press), Chap. XIII.

Chandrasekhar, S., and Lebovitz, N. R. 1962, Ap. J., 135, 238.

Eggen, O. J., and Greenstein, J. L. 1965, Ap. J., 141, 83.

Faulkner, J., and Gribbin, J. R. 1968, Nature, 218, 734.

Fowler, R. H. 1926, MNRAS, 87, 114.

Gliese, W. 1956, Zs. f. Ap., 39, 1.

Goldreich, P., and Schubert, G. 1967, Ap. J., 150, 571.

Greenstein, J. L. 1958, Hdb. der Phys., ed. S. Flugge (Berlin: Springer-Verlag), 50, 50, 161.

Greenstein, J. L., and Trimble, V. L. 1967, Ap. J., 149, 283.

Hamada, T., and Salpeter, E. E. 1961, Ap. J., 134, 683.

Harris, D. L., Strand, K. Aa., and Worley, C. E. 1963, in Basic Astronomical Data, ed. K. Aa. Strand (Chicago: Univ. of Chicago Press), Chap. 15.

Hayashi, C., Hoshi, R., and Sugimoto, D. 1962, Prog. Theoret. Phys., Suppl. No. 22.

Hubbard, W. B., and Wagner, R. L. 1970, Ap. J., 159, 93.

James, R. A. 1964, Ap. J., 140, 552.

L'Ecuyer, J. 1966, Ap. J., 146, 845.

Ledoux, P. 1945, Ap. J., 102, 143.

______. 1949, M&M Soc. Roy. Sci. Liège, 9, 1.

Lee, T. D. 1950, Ap. J., 111, 625.

Mestel, L. 1952, MNRAS, 112, 583.

______. 1965, in Stellar Structure, ed. L. Aller and D. B. McLaughlin (Chicago: Univ. of Chicago Press), Chap. V.

Mestel, L., and Ruderman, M. A. 1967, MNRAS, 136, 27.

Oort, J. H. 1960, BAN, 15, 45.

Ostriker, J. P., and Axel, L. 1969, in Low Luminosity Stars (London: Gordon & Breach).

Ostriker, J. P., and Bodenheimer, P. 1968, Ap. J., 151, 1089.

Ostriker, J. P., and Hartwick, F. D. A. 1968, Ap. J., 153, 797.

Ostriker, J. P., and Mark, J. W-K. 1968, Ap. J., 151, 1075.

Ostriker, J. P., and Tassoul, J. L. 1968, Nature, 219, 577.

______. 1969, Ap. J., 155, 987.

Rose, W. K. 1966, Ap. J., 146, 838.

______. 1967, Ap. J., 150, 195.

Schatzman, E. 1958, White Dwarfs (Amsterdam: North-Holland Publ. Co.).

Schwarzschild, M. 1957, Structure and Evolution of the Stars (Princeton: Princeton Univ. Press), Chap. VII.

Seaton, M. J. 1966, MNRAS, 132, 113.

Tassoul, J. L., and Ostriker, J. P. 1968, Ap. J., 154, 613.

Van Horn, H. M. 1968, Ap. J., 151, 227.

Vila, S. C. 1966, Ap. J., 146, 437.

______. 1967, Ap. J., 149, 613.

Weidemann, V. 1967, Zs. f. Ap., 67, 286.

______. 1968, Ann. Rev. Astron. and Astrophys., 6, 351.

8

CLOSE BINARIES

B. Paczyński*

Joint Institute for Laboratory Astrophysics
University of Colorado, Boulder, Colorado

I. Introduction

A number of review articles on problems related to close binaries have been published in recent years. The determination of masses of eclipsing binary stars has been described by Popper (1967). The variation of the elements of orbital motion of close binaries was reviewed by Kruszweski (1966). Plavec (1968b) has given a thorough description of the development of the theory of mass exchange and evolution of close binaries. There is also a short review of the recent advances in this field by Weigert (1969).

We use the abbreviation CBS for a "close binary system". The studies of CBS's are important for the understanding of stellar evolution in general. Our direct information about stellar masses is based on binaries. In those that have undergone a large mass exchange, matter from the deep interior, processed by different nuclear reactions, is now visible on the surface. A large number of stars are binaries, something like 20% or 30%. Some stars are found only among CBS's: U Geminorum stars, novae, and Am stars; in addition, a large fraction of Wolf-Rayet stars are binaries. The theory of stellar evolution can account now for the most essential properties of "normal", well-behaved binaries. Attempts have also been made to explain the evolutionary status of more exotic systems. However, the origin of CBS's has not yet been satisfactorily explained.

It is now well established that the existence of the so-called critical Roche surface is the most essential feature for the evolution of close binaries. Given a binary with a circular orbit, we can place the center of the coordinate system at the center of mass of the CBS and rotate this system with the binary. In such a frame of reference those equipotential surfaces that are close to the center of either com-

**1968-69 Visiting Fellow; on leave from Institute of Astronomy, Polish Academy of Sciences, Warsaw, Poland.*

ponent are almost spherical. The larger the equipotential surface, the more it is distorted. Finally, for a certain value of the potential the surfaces surrounding the two components touch each other at a point between the two stars. This is called the inner Lagrangian point L_1, and the equipotential is called the critical Roche surface. It is composed of two Roche lobes surrounding the two components and touching each other at the point L_1. For larger values of the potential the two stars are enveloped by a common, highly distorted equipotential surface.

It is possible to show that the surface of a star must coincide with an equipotential surface, provided the star rotates with an angular velocity equal to that of orbital motion (synchronous rotation). If a star is significantly smaller than the corresponding Roche lobe, it is almost spherical. The larger the star, the larger the distortion of its surface. If the star increases its radius so much as to overflow the Roche lobe, the matter is free to flow out through the vicinity of the inner Lagrangian point. It may be useful to define here what is understood by the term "close binary system". We shall call the binary "close" if, during the evolution, either component expands so much that if fills up its Roche lobe and mass exchange between the two stars takes place. The stellar radius in the red supergiant phase of evolution may be as large as a few thousand solar radii. Therefore, a binary with a period of orbital motion as long as ten years may still be a close binary.

II. "Normal" Binaries

Kopal (1955) investigated a large number of eclipsing binaries and was able to divide them into three categories:

1. detached systems, with the two components smaller than their Roche lobes;
2. semidetached systems, with one component filling up its Roche lobe;
3. contact systems, with the two components filling up their Roche lobes.

Detached binaries usually have their components close to the main sequence (YY Gem, WW Aur, Y Cyg). If one star appears to be more evolved; it is in most cases the more massive of the two (RS CVn, SZ Psc, 31 Cyg, V380 Cyg).

A typical semidetached binary has a more massive component close to the main sequence with the less massive one filling up its Roche lobe as a subgiant (β Per, V Pup, μ^1 Sco). Massive systems show light curves of the Beta Lyrae type, i.e., variable light outside the eclipses. Low mass binaries in this category show light curves of the Algol type, i.e., light almost constant outside the eclipses. It is very common to see emission lines in the spectra of semidetached binaries. These lines are likely to come from gaseous streams associated with a process of mass exchange. Frequently they indicate the existence of a rotating gaseous ring around the more massive component.

A large number of contact systems are known (W Ursae Majoris stars). All these are composed of relatively low mass stars (total mass below 4 $M_\odot$) close to the main sequence. Many show rapid period changes.

It is easy to account for the evolutionary status of the detached binaries. Apparently not too much interaction occurs between the two components of such a CBS. It is possible to follow the evolution of each component as if it were a single star. The semidetached systems are more difficult to explain. It is commonly believed now that these binaries have experienced a very substantial mass exchange between the two components. I shall spend most of my time describing the theory of these systems. Finally, we have the contact binaries in which mass exchange in both directions should be possible. So far only Lucy (1968a,b) has attempted to build up models of such CBS's.

The most important properties of the subgiant components of semi-detached binaries are shown in Figs. 1 and 2 (after Kopal 1959, p. 484). We see that these stars are above the main sequence, i.e., they look like evolved objects. They are also the less massive components. I would like to point out the systematic difference between components with masses above $1M_\odot$ and below $1M_\odot$. All the massive stars are close to the main sequence. Not a single massive red giant or subgiant fills up its Roche lobe. Among the low mass stars are a large number of red subgiants.

III. Model Computations

Most theoretical calculations related to the evolution of a CBS have been based on the five assumptions listed below.

1. The component of a close binary can be treated as a spherically symmetric star, even if it fills up the Roche lobe. The usual justification for this assumption is as follows. Only the outermost parts of a star are significantly distorted. If we neglect this effect, although a small error is made in the estimated radius, the luminosity should not be affected. The errors involved should not be larger than those in the case of a single star for which rotation is neglected. Most massive single stars are rapid rotators. At the same time most evolutionary calculations for single stars have been done with spherically symmetric models, and it seems that they account adequately for the main observational facts. Benson's (1969) attempt to account for nonsphericity of subgiant components provides a direct indication that this effect is not important. Jackson (1969) comes to the same conclusion.

2. The mean radius of the Roche lobe is defined as the radius of a sphere with the same volume as that of the Roche lobe. With an accuracy of 1% this radius is given by the approximate formula

$$r_1 = A[0.38 + 0.2 \log (M_1/M_2)] ,$$

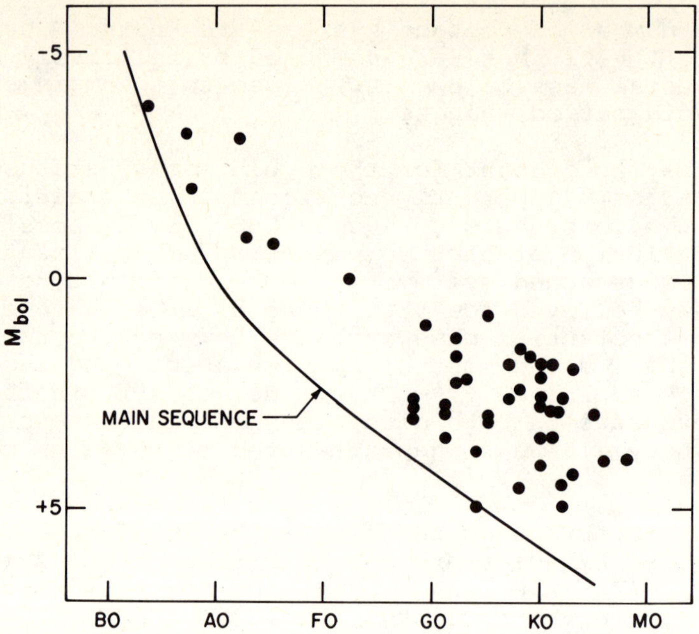

Figure 1. H-R diagram for the subgiant components of elipsing binaries.

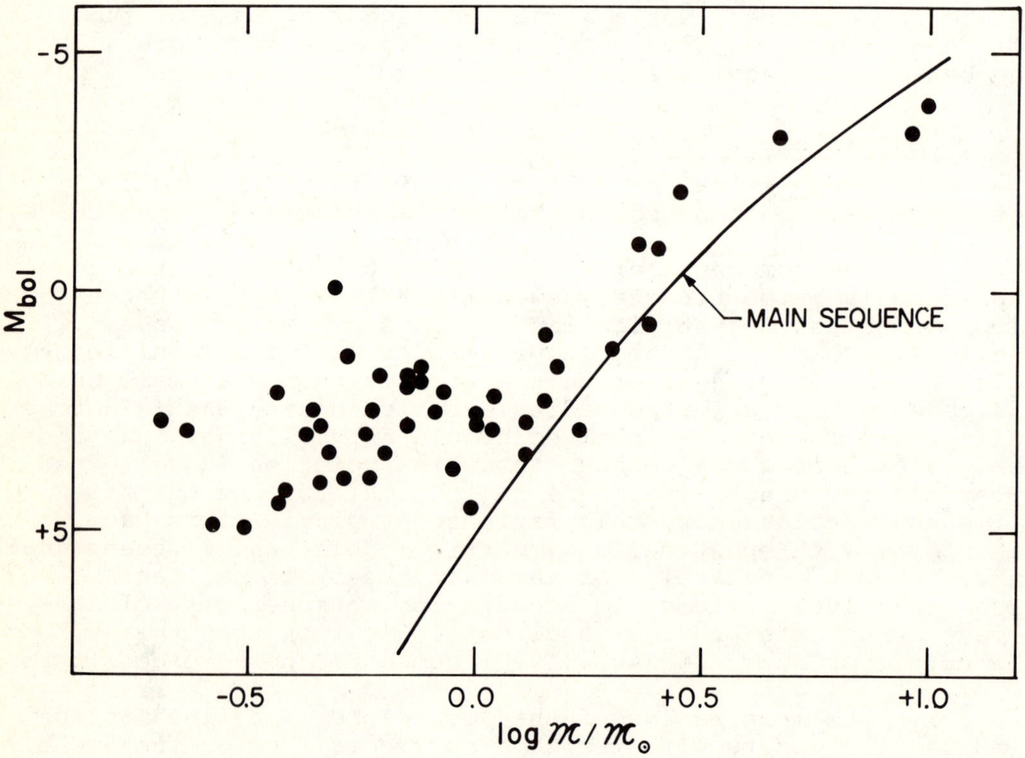

Figure 2. Mass-luminosity relation for the subgiant components of eclipsing binaries.

where A is the separation between the centers of the two components with mass M_1 and M_2 respectively. Thus r_1 is the Roche lobe radius for the star with mass M_1. The definition of equipotential surfaces is possible only for binaries with circular orbits. Therefore the orbit is assumed to be circular. No semidetached binary with a noncircular orbit is known.

3. The radius of a star R_1 cannot be larger than the radius of the corresponding Roche lobe r_1. It is possible to show that a slight excess of R_1 over r_1 is sufficient to cause a violent mass outflow through the vicinity of the inner Lagrangian point. This assumption seems to be well justified, although sometimes it cannot be satisfied. This will be discussed again in connection with stars with convective envelopes.

4. The star is assumed to be in a hydrostatic equilibrium. This is a rather safe assumption which is violated only in a small region very close to the stellar surface. Sometimes the calculations are also done under the assumption that the star is in thermal equilibrium. Some evolutionary phases can be followed very easily if this assumption is made, but in other phases thermal instability is the most essential cause of the mass exchange.

5. The total mass of the binary is

$$M = M_1 + M_2$$

The total angular orbital momentum is

$$T = \left(\frac{G \times A}{M}\right)^{\frac{1}{2}} M_1 M_2 \quad .$$

The total mass M and the orbital angular momentum T are assumed to be conserved during the evolution. This is obviously a simplifying assumption, as we do observe mass loss from many close binaries. Unfortunately, no observational or theoretical estimates of the rate of the mass loss exist. Computational experiments (Paczyński and Ziółkowski 1967) have shown that the main character of the evolution is unaffected by this assumption. If this assumption is accepted, then the separation between the centers of the two components will be inversely proportional to the product of the squares of the two masses:

$$A \sim \frac{1}{M_1^2 M_2^2} = \frac{1}{M_1^2 \ (M - M_1)^2} \quad .$$

Given the total mass M and the orbital angular momentum T the separation A is smallest when $M_1 = M_2$.

Most model calculations are started with the two stars on the main sequence. The masses of the two components and their mutual separation are specified. The component that is originally the more massive is called the primary. In most cases only the evolution of this component is followed in detail. Its evolution is identical with the evolution of a single star of the same mass as long as the stellar radius R_1 is smaller than the radius of the corresponding Roche lobe r_1. When the two radii become equal, mass flow from the primary to the secondary component is possible and the binary nature of the system must be explicitly taken into acount. Morton (1960) has shown that a major mass exchange proceeds on a thermal time scale. I shall here repeat his arguments.

Consider a star evolving up from the main sequence on a mass-radius diagram (Fig. 3). If the star is single it will evolve straight up, from A to B'. If it is a component of a CBS it will fill up the Roche lobe, say at the point B. Consider the simplest case, i.e., the core hydrogen burning phase. The star is in thermal equilibrium, and the evolution proceeds on the nuclear time scale from A to B. When the Roche lobe is filled, mass flow takes place. The result is a change in the mass ratio, in the separation between the two components, and in the radius of the Roche lobe. The component evolving faster is the one initially more massive (the primary). As a result of mass exchange its Roche lobe therefore shrinks (see assumption #5). The U-shaped curve in Fig. 3 shows schematically the changes in the Roche lobe radius with the mass of this component. When the mass ratio is reversed as a result of mass exchange, the radius of the Roche lobe increases.

We must examine now the question of whether this star will be willing to follow the changes in the Roche lobe. If we remove some mass from the surface of the star very rapidly, in a time much shorter than the dynamical (or thermal) time scale of the star, first the star will try to restore hydrostatic equilibrium almost adiabatically, on a dynamical time scale. If its envelope is in radiative equilibrium after this process has been completed, the radius will be smaller than the Roche lobe radius. But the star will not yet be in thermal equilibrium, and it will also try to restore this equilibrium. As a result, the radius will increase somewhat. Calculations show that this equilibrium radius is about the same as the original one, before mass was removed from the surface. But since the new radius of the Roche lobe is smaller than the original one, the thermal equilibrium radius of the star is larger than that of the Roche lobe. Therefore, before thermal equilibrium is restored the star must lose some additional mass. We can calculate the thermal equilibrium radii for models with different amounts of mass removed from the envelope. The chemical composition in the interior does not change. The radii obtained in this way are shown in Fig. 3 by the line B-C-D'. Clearly, in the mass interval M_B to M_C the stellar equilibrium radius is larger than the Roche lobe radius, and therefore an amount of mass (M_B - M_C) must be transferred on a <u>thermal</u> time scale. At point C, $R = r$, and finally the thermal equilibrium model fits its Roche lobe exactly. At this point the rapid mass transfer must come to an end. But hydrogen is burning in the core, and the star has a tendency to expand slowly on a nuclear time scale. As a result, slow mass exchange will occur on a <u>nuclear</u> time scale in order to keep the star within the Roche lobe. The process is now

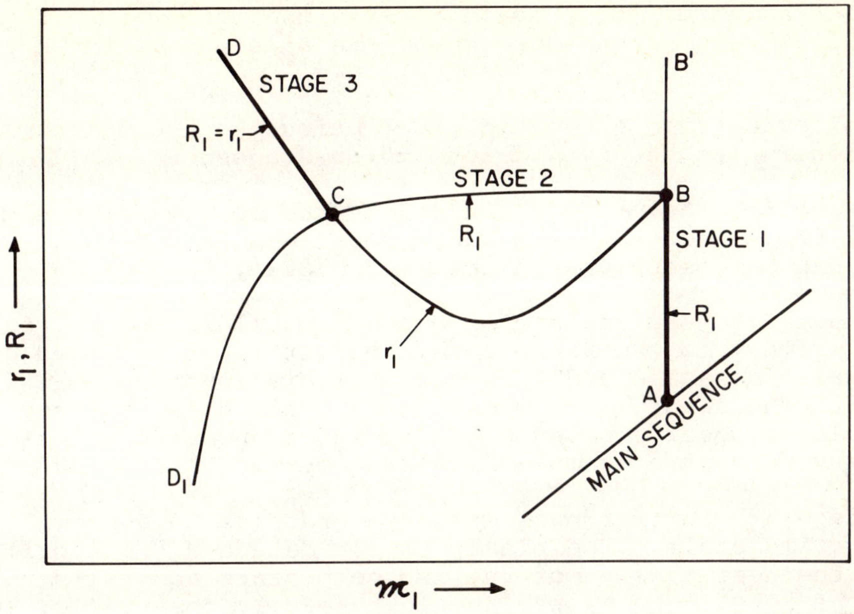

Figure 3. The variation of the thermal equilibrium radius of the primary component R_1 and the radius of its Roche lobe r_1.

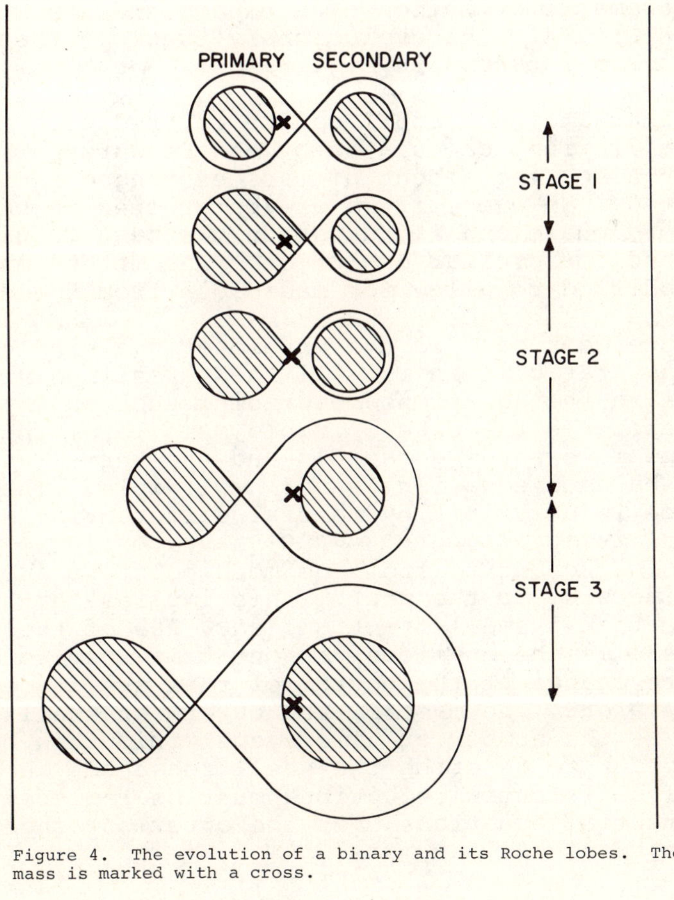

Figure 4. The evolution of a binary and its Roche lobes. The center of mass is marked with a cross.

stable on a thermal time scale because the mass ratio is reversed and the Roche lobe expands as a result of the mass transfer.

Following this schematic description the evolution of a close binary can be divided into three stages:

Stage 1: The primary (initially the more massive component) is smaller than its Roche lobe (A-B in Fig. 3). The binary can be observed as a detached system.

Stage 2: Very rapid mass exchange follows the approach of the primary to the Roche lobe. Matter is transferred on a thermal time scale and the mass ratio is reversed (B-C in Fig. 3). The binary is semidetached. This phase is very difficult to observe because of its short duration. Beta Lyrae may be a rare example of such a system.

Stage 3: Further mass exchange proceeds on a slow, nuclear time scale. The binary is semidetached. The primary is now the less massive of the two components and fills up its Roche lobe. The system should be easily observable, and indeed most semidetached systems are of this kind, i.e., the subgiant filling up its Roche lobe is the less massive star.

The schematic evolution of a binary and its Roche lobe is shown in Fig. 4. The evolutionary track of the primary on the H-R diagram is shown in Fig. 5.

Note that Stages 1 and 3 can be covered with models in thermal equilibrium, calculated by the Schwarzschild (fitting) technique. Also, the amount of mass exchanged in Stage 2 can be calculated. However, if we are interested in the details of this rapid phase and its duration, we have to use the Henyey method for stellar evolution. The method must be slightly modified to allow for mass loss from the star.

The outermost part of the envelope in a single star is described as static; that is, all the terms involving time derivatives in the equations describing such an envelope are equal to zero. In the case of rapid mass loss, the matter in the envelope is expanding rapidly and the term $(T \times \partial s/\partial t)$ in the energy balance equation is not negligible. Therefore, the envelope is described by partial differential equations. In numerical integrations we cannot have too large changes in any physical quantity from one mesh point to another, or from one time step to the next. This implies that we cannot remove more than a small fraction (say, 20% of the mass in the outermost shell (the envelope) in one time step, otherwise, the pressure change at the bottom of this shell would be too large. Therefore, the mass of the outermost shell cannot be too small, as the number of time steps would then be prohibitive. In order to describe the shell reasonably accurately, the partial differential equations must be replaced by ordinary differential equations over the outermost shell, about 3-5% of the total stellar mass.

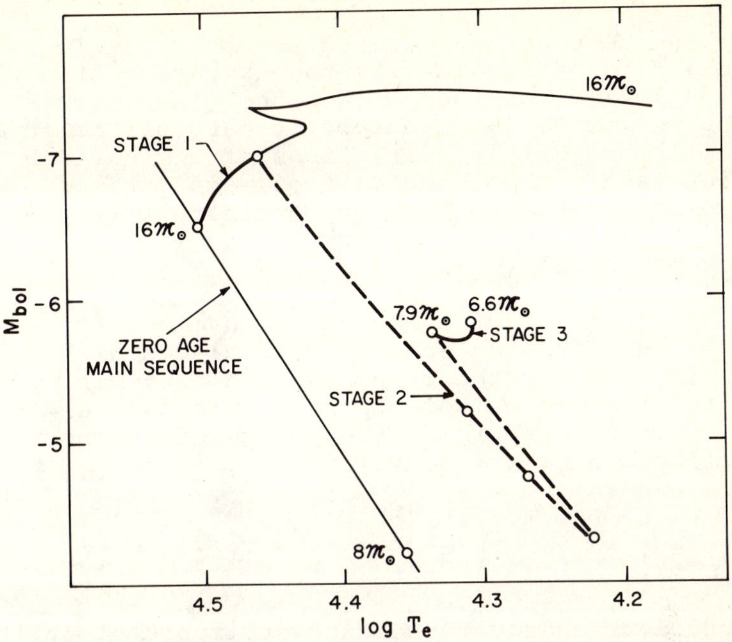

Figure 5. The evolutionary track of the primary component of a massive binary. The Zero age main sequence and the evolutionary track of a single star of 16 $M_{\odot}$ are also shown.

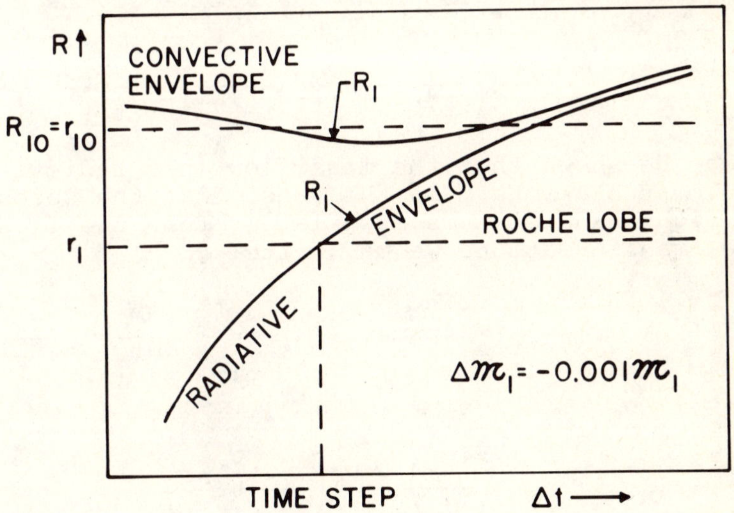

Figure 6. The variation of the radius of the primary component R_1 with the size of the first time step with mass loss Δt.

In the case of a deep convective envelope the temperature gradient is adiabatic with the exception of a thin zone close to the stellar surface. Then the time derivative of the entropy per gram in the adiabatic part of the envelope is given directly by the time variations of the surface boundary conditions and perhaps the rate of mass outflow. Therefore, we can describe the envelope by ordinary differential equations, with the mass fraction M_r or the radius r being the only independent variable.

In the case of a radiative envelope the time derivative of the entropy must somehow be replaced either by the space derivative or by the boundary conditions in order to get ordinary differential equations. This can be done as follows. The equation describing the change of luminosity with mass fraction can be written:

$$\left(\frac{\partial L_r}{\partial M_r}\right)_t = \varepsilon_n - T\left(\frac{\partial s}{\partial t}\right)_{M_r} .$$

The nuclear energy term is never important in the envelope. If the rate of heat loss during quasistatic changes, $-T\left(\frac{\partial s}{\partial t}\right)_{M_r}$, is negligible we say the star is in thermal equilibrium. The Lagrangian time derivative of the entropy can be written as

$$\left(\frac{\partial s}{\partial t}\right)_{M_r} = \left(\frac{\partial s}{\partial t}\right)_r + v\left(\frac{\partial s}{\partial r}\right)_t .$$

It can be shown that the mass flow in a radiative envelope is almost stationary in the sense that the second term on the right-hand side is much larger than the first one. After some transformations we may write:

$$v\left(\frac{\partial s}{\partial r}\right)_t = 4\pi r^2 \rho v\left(\frac{\partial s}{\partial M_r}\right)_t = -\left(\frac{dM_1}{dt}\right)\left(\frac{\partial s}{\partial M_r}\right)_t$$

and

$$\left(\frac{\partial L_r}{\partial M_r}\right)_t = T\left(\frac{dM_1}{dt}\right)\left(\frac{\partial s}{\partial M_r}\right)_t ,$$

where (dM_1/dt) is the rate of mass loss from the star. In this way the time derivative of the entropy is replaced by the space derivative. Then the outermost few percent of the stellar envelope may be described by ordinary differential equations, as in the case of evolution without mass loss.

The surface boundary conditions for a single star with a mass M_1 are described by two parameters: radius R_1 and luminosity L_1. If the star is filling up its Roche lobe, the radius is no longer an independent parameter. It must be equal to the Roche lobe radius, which is a function of the stellar mass M_1. Then the luminosity L_1 and the rate of mass outflow (dM_1/dt) are two parameters that are used to describe the surface boundary conditions.

Making a first time step with mass loss is sometimes difficult because mass loss is so rapid. As a result, the time step is 2 to 4 orders of magnitude shorter than the previous step without mass loss. In order to estimate this first time step the following analysis is helpful (Fig. 6). If we remove a very small amount of mass, say 0.001 M_1 in a time Δt, the rate of mass loss is then $(dM/dt) = -0.001\ M_1/\Delta t$. Let the original stellar radius be $R_{10} = r_{10}$, and the new radius of the Roche lobe be $r_1 < r_{10}$. For a given time step Δt the rate of mass loss is specified, and R_1 and L_1 are the two parameters describing the surface boundary conditions. After calculating the radius of the new model R_1 as a function of the time step Δt, it turns out that if the envelope is in radiative equilibrium the line $R_1(\Delta t)$ always crosses the line $R_1 = r_1$. This intersection gives the value of the time step we need. At this point we can switch completely to the modified Henyey method, with L_1 and (dM_1/dt) as the two surface parameters.

As can be seen from Fig. 6, the line $R_1(\Delta t)$ for convective envelopes in general does not cross the line $R_1 = r_1$, if $r_1 < r_{10}$. If the star filling up the Roche lobe has a convective adiabatic envelope, it is then possible to fit the star into the Roche lobe during the mass loss phase only if the lobe increases. This condition can be satisfied if the star is the less massive of the two components. If it is the more massive one, it is likely to lose mass on a dynamical time scale because the star is larger than its Roche lobe.

A typical distribution of the nuclear and gravitational energy sources in a star losing mass on a thermal time scale is shown in Fig. 7. Notice the very large decrease in L_g close to the surface. This model has a radiative envelope.

The first computations following in detail the phase of rapid mass transfer were reported by Kippenhahn and Weigert (1967), and a few months later by Paczyński (1967a). A large number of papers have been published on this subject within the last three years. The general problems of evolution of the CBS were considered by Giannone, Kohl, and Weigert (1968), van den Heuvel (1969), Paczyński (1967a), Plavec (1967), and Ziółkowski (1969b). Evolutionary calculations for massive binaries were reported by Barbaro et al. (1968), Kippenhahn (1969), Kippenhahn and Weigert (1967), Paczyński (1966, 1967b, c,e), Paczyński and Ziółkowski (1967), and Snezhko (1967); for intermediate mass binaries by Kriz (1969), Plavec (1968a), Plavec et al. (1968, 1969), and Ziółkowski (1969a); and for small mass binaries by Kippenhahn, Kohl and Weigert (1967), Kippenhahn, Thomas and Weigert (1968), and Refsdal and Weigert (1969). The problem of instability of mass outflow from stars with deep convective envelopes was discussed by Paczyński (1965), and Paczyński, Ziółkowski and Zytkow (1968); and shallow convective envelopes were studied by Osaki (1969). The instabilities associated with mass accretion by the secondary component were studied by Giannone and Weigert (1967),

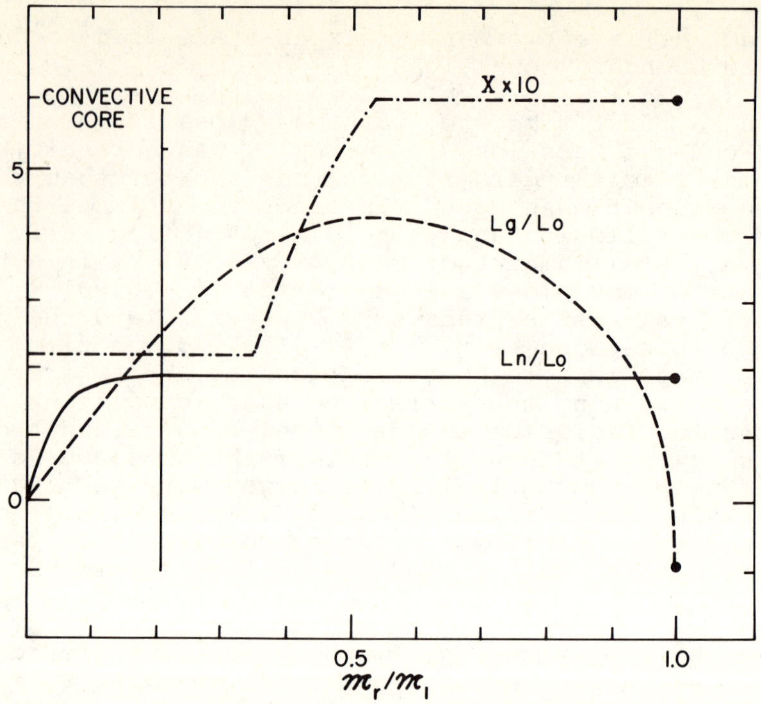

Figure 7. The variation with mass fraction of the gravitational (L_g) and the nuclear (L_n) luminosities, and of the hydrogen content (X) in the primary component during the phase of a rapid mass exchange. L_0 is the surface luminosity of the star.

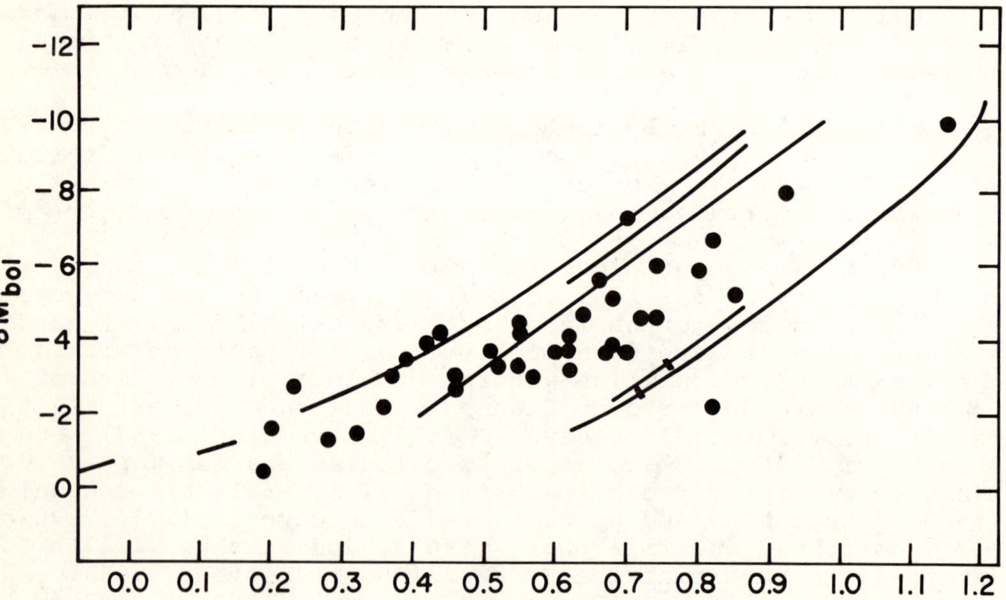

Figure 8. Luminosity excess of the subgiant, $-\delta M_{bo\ell}$, versus mass ratio for low mass semidetached binaries.

Saslaw (1968), and Rose (1968). The tidal interaction was investigated by Dziembowski (1967), and Zahn (1966a,b,c). The possible influence of gravitational radiation on the evolution of a CBS was studied by Paczyński (1967d).

IV. Comparison Between Models and Observations

Evolutionary calculations for binaries with different masses have demonstrated that it is useful to divide the evolutionary patterns into four types. First I shall define case A and case B of evolution following Kippenhahn and Weigert (1967). If the primary (i.e., initially more massive) component of a binary fills up its Roche lobe before hydrogen is exhausted in the core, we have case A. When the Roche lobe is filled up after the hydrogen exhaustion, but before helium ignition in the core, we have case B. If the original mass of the primary is larger than 3 - 4 $M_\odot$ we shall call the system "massive". If the opposite is true, we shall call it a "low mass" binary. It must be emphasized that the mass of the primary after the mass exchange is 2 - 8 times smaller than its original mass. Therefore, the subgiant components of "massive" binaries have masses larger than 1 $M_\odot$, and subgiants in the "low mass" systems are below 1 $M_\odot$.

Case A of evolution is very similar in low and high mass CBS's. After the period of rapid mass exchange the thermal equilibrium of the primary component is restored, and the binary is observable as a semidetached system. The subgiant component is somewhat overluminous for its mass, but is fairly close to the main sequence. It still burns hydrogen in the core.

In a low mass binary, case B leads to a semidetached system with a red subgiant component. This subgiant is highly overluminous and far from the main sequence, burning hydrogen in a shell surrounding a degenerate helium core. In a large mass binary, case B results in a semidetached phase with a thermal time scale. The primary's core is nondegenerate and helium is ignited soon after the beginning of mass exchange. This helium ignition leads to a rapid decrease in the radius of the primary, and a detached system is formed. Therefore it is not possible to get a massive red subgiant (or giant) filling up the Roche lobe in a CBS, a conclusion which compares favorably with the observed distribution of subgiant components on the mass-luminosity and the H-R diagrams (Figs. 1 and 2).

The subgiant components are well known to be overluminous, and this excess in luminosity is correlated with the observed mass ratio of the two stars. The correlation is shown in Fig. 8 and 9 for low and high mass binaries, respectively. A number of evolutionary tracks corresponding to a slow, semidetached phase of evolution are also shown for the model subgiant components. In the case of massive binaries all models burn hydrogen in the core. Among the low mass binaries, models with a small luminosity excess burn hydrogen in the core, and those which are highly overluminous burn hydrogen in the shell. The luminosity excess is defined as the difference in bolometric magnitude between the subgiant and a main sequence star

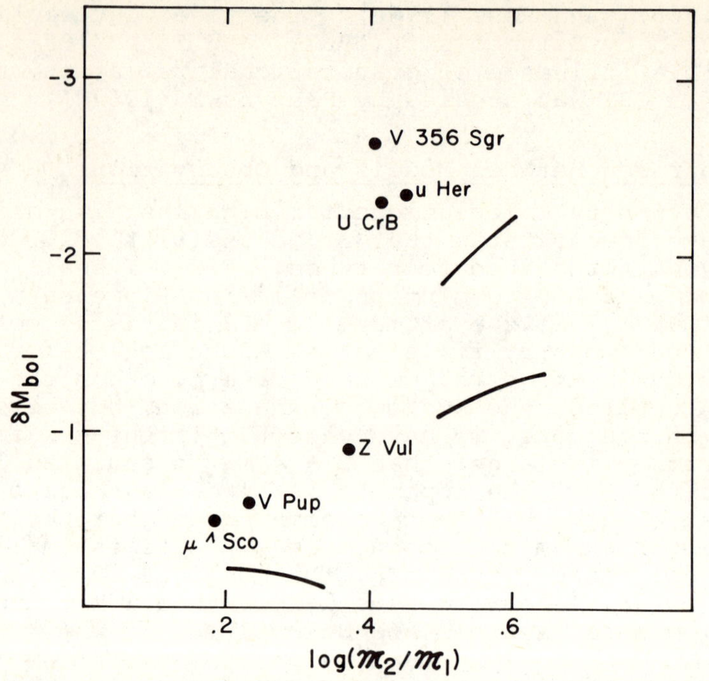

Figure 9. Luminosity of the subgiant, $-\delta M_{bol}$, versus mass ratio for massive semidetached binaries.

with the same mass. The mass ratio is in the sense: secondary/primary (subgiant). The theoretical models seem to agree reasonably well with binaries and can account for the observed correlations.

V. Instabilities

A large number of instabilities appear in a close binary as a result of mass exchange. First consider the primary, the component losing mass. If it has a radiative envelope, the first phase of mass outflow is due to a thermal instability. This thermal instability was first discussed by Morton (1960) and has been described in some detail in this lecture. In the case for which the primary has a deep convective envelope while filling up its Roche lobe, a dynamical instability may set in (Paczyński 1965; Paczyński, Ziółkowski and Zytkow 1968). The transfer of a large fraction of stellar matter on a time scale of a few hours or days may take place. This problem has not been adequately investigated so far.

Osaki (1969) studied the problem of stability of mass outflow from a primary with a shallow convective zone. He assumed that the shear turbulence due to the mass outflow increases the effective mixing length in the convective envelope. In some of his models Osaki found instabilities which led to relaxation oscillations on a time scale $\sim$ 15 days, roughly the thermal time scale for a shallow convective envelope. The light output from Osaki's models is variable, with an amplitude $\sim$ 1.5 magnitudes, and the models have light curves similar to those shown by U Geminorum or SS Cygni. Osaki has suggested that the eruptions of U Gem-type stars are caused by the instability of mass outflow coupled with turbulent convection. It is worth noting that in the case of U Geminorum itself the component that fills its Roche lobe is the one that erupts (Krzeminski 1965).

Consider now the primary component of a very massive binary, evolving according to case B. The process of mass transfer is terminated with helium ignition in the core. What is left of the primary is almost a pure helium star with a thin envelope containing some hydrogen (Kippenhahn and Weigert 1967; Kippenhahn 1969; Barbaro _et al_. 1968). Paczynski has suggested (1967a,e) that this star _may_ be indentified with a Wolf-Rayet component of some massive binaries. About 50% of W-R stars are known to be members of CBS's, and no other systems are known which could be identified with the massive binaries resulting from the mass exchange following case B. If the primary after mass transfer has a mass larger than $\sim 8\ M_{\odot}$ it may be vibrationally unstable (Boury and Ledoux 1965). Unfortunately, Van der Borght (1969) and Stothers (1969) have found that a hydrogen-rich envelope of very small mass may easily stabilize such a star.

The secondary component of a CBS is accreting matter, and this may also lead to new phenomena. Van den Heuvel (1968) has suggested that Am and Ap stars were formerly secondaries in CBS's which have accreted substantial amounts of mass.

More spectacular effects may be expected when the primary becomes a white dwarf as a result of large mass outflow to the secondary (Kippenhahn, Kohl and Weigert 1967). The thin hydrogen shell that is left makes a prominent thermal flash (Kippenhahn, Thomas and Weigert 1968), but finally the star cools down as a white dwarf. In the meantime the secondary exhausts the hydrogen in its core, expands, and a second mass transfer takes place in the opposite direction, from the secondary to the primary. This situation has been studied by Giannone and Weigert (1967), Saslaw (1968), and Rose (1968). As a result of the accretion of hydrogen-rich matter onto the surface of the cool white dwarf, thermally unstable hydrogen burning in the shell takes place. According to Rose (1968) even vibrational instability may set in at the peak of the thermal flash. He would like to associate this phenomenon with a nova explosion.

So far only spherically symmetric models of white dwarfs accreting matter have been studied theoretically. At the same time mass falling down from the inner Lagrangian point onto the surface of the star is well known to have so much angular momentum that it tends to create a rapidly rotating ring or disc (Kruszewski 1967). In fact, such rings or discs are commonly observed in close binaries. Only as a result of tidal interaction may the disc lose angular momentum, and the matter will finally settle down onto the stellar surface. But large mass accretion will also cause the surface layers of the star to rotate rapidly. Again tidal interaction seems to be the most promising mechanism in slowing down these layers. However, the tides may not be very efficient in stars as small as white dwarfs. The rapid rotation of these stars may well be of primary importance in studies of their stability.

As so many instabilities seem to be present in CBS's, it may be a good idea to call the reader's attention to a review of the observational data relating to novae and U Geminorum-type stars by Kraft (1963) and a review of theoretical work on novae explosions by Schatzman (1965).

Finally I would like to mention a suggestion made by Shklovsky (1967) that mass accretion in close binaries might lead to a strong emission of X-rays, and presumably account for some of the observed X-ray sources. No detailed model of this kind has been published so far. Finally, McCrea (1964) has suggested that the so-called "blue stragglers" in star clusters are secondaries of CBS that have experienced a large scale mass exchange.

References

Barbaro, G., Giannone, P., Giannuzzi, M. A., and Summa, C. 1968, Colloquium: "On Mass Loss from Stars," Trieste (Italy).

Benson, H. E. 1969, private communication.

Boury, A., and Ledoux, P. 1965, Ann. d'ap., 28, 353.

Dziembowski, W. 1967, Comm. Obs. Roy. Belgique, Uccle, B17, 105.

Giannone, P., Kohl, K., and Weigert, A. 1968, Z. Astrophys., 68, 107.

Giannone, P., and Weigert, A. 1967, Z. Astrophys., 67, 41.

Jackson, S. 1969, preprint.

van den Heuvel, E.P.J. 1968, Bull. Astr. Inst. Neth., 19, 326.

———. 1969, preprint.

Kippenhahn, R. 1969, Astron. Astrophys., 3, 83.

Kippenhahn, R., Kohl, K., and Weigert, A. 1967, Z. Astrophys., 66, 58.

Kippenhahn, R., Thomas, H.-C., and Weigert, A. 1968, Z. Astrophys., 69, 265.

Kippenhahn, R., and Weigert, A. 1967, Z. Astrophys., 65, 251.

Kopal, Z. 1955, Ann. d'ap., 18, 379.

———. 1959, Close Binary Systems (New York: J. Wiley and Sons).

Kriz, S. 1969, Bull. Astr. Inst. Czech., 20, 127.

Kraft, R. P. 1963, Adv. Astr. Astrophys., 2, 43.

Kruszewski, A. 1966, Adv. Astr. Astrophys., 4, 233.

———. 1967, Acta Astr., 17, 297.

Krzeminski, W. 1965, Ap. J., 142, 1051.

Lucy, L. 1968a, Ap. J., 151, 1123.

———. 1968b, ibid., 153, 877.

McCrea, W. H. 1964, Mon. Not. R. Astr. Soc., 128, 147.

Morton, D. C. 1960, Ap. J., 132, 146.

Osaki, Y. 1969, preprint.

Paczyński, B. 1965, Acta Astr., 15, 89.

———. 1966, ibid., 16, 231.

———. 1967a, Comm. Obs. Roy. Belgique, Uccle, B17, 111.

———. 1967b, Acta Astr., 17, 1.

———. 1967c, ibid., 17, 193.

———. 1967d, ibid., 17, 287.

———. 1967e, ibid., 17, 355.

Paczyński, B., and Ziółkowski, J. 1967, Acta Astr., 17, 7.

Paczyński, B., Ziółkowski, J., and Zytkow, A. 1968, Colloquium: "On Mass Loss from Stars," Trieste (Italy).

Plavec, M. 1967, Comm. Obs. Roy. Belgique, Uccle, B17, 83.

———. 1968a, Astrophys. Space Sci., 1, 239.

———. 1968b, Adv. Astr. Astrophys., 6, 201.

Plavec, M., Kriz, S., Harmanec, P., and Horn, J. 1968, Bull. Astr. Inst. Czech., 19, 24.

Plavec, M., Kriz, S., and Horn, J. 1969, Bull. Astr. Inst. Czech, 19, 24.

Popper, D. 1967, Ann. Rev. Astr. Astrophys., 5, 85.

Refsdal, S., and Weigert, A. 1969, Astr. Astrophys., 1, 167.

Rose, W. K. 1968, Ap. J., 152, 245.

Saslaw, W. C. 1968, Mon. Not. R. Astr. Soc., 138, 337.

Schatzman, E. 1965, Stars and Stellar Systems, Vol. VIII (Chicago: University of Chicago Press), p. 327.

Shklovsky, I. S. 1967, Ap. J., 148, L1.

Snezhko, L. I. 1967, Perem. Zv., 16, 253.

Stothers, R. 1969, preprint.

Van der Borght, R. 1969, preprint.

Weigert, A. 1969, Astr. Gesellshaft Mitt., 25, 19.

Zahn, J.-P. 1966a, Ann. d'ap., 29, 313.

———. 1966b, ibid., 29, 489.

———. 1966c, ibid., 29, 565.

Ziółkowski, J. 1969a, Astrophys. Space Sci., 3, 14.

———. 1969b, private communication.

9

NOVAE

W. K. Rose

Massachusetts Institute of Technology
Cambridge, Mass.

I. Observations

Novae are explosive variable stars. They are observed to brighten suddenly by about 11 magnitudes and then decline. Before and after an outburst novae are hot subdwarfs (i.e., ultraviolet dwarfs) that are generally members of binary systems. At the observed maximum of an outburst, M_v is between -6 to -9, a value that is much less than that observed for a supernova outburst ($M_v \simeq -18$). Although the initial rise to near maximum light is very rapid ($\sim$ 1 day in some instances), the decline is much slower (years or, in extreme cases, decades). Most novae show variability during decline and even after they appear to have returned to their prenova state. There is good evidence that at the time of a nova outburst, a shell of matter is ejected from the star. This rapidly expanding shell produces the effect of an enlarged photosphere. Some novae are observed to recur, and it is generally assumed that most novae would repeat if they could be observed for a sufficiently long interval of time. The characteristics of a nova outburst are not sharply defined. It is clear that each nova outburst has its own peculiarities. For this reason, theories for novae must explain the broad outlines of the phenomenon and still be flexible enough to allow for the variations among novae. Moreover, certain types of variables that are generally classified as nova-like (e.g., U Gem stars, Z Cam stars, η Carina) may be caused by unrelated physical processes.

A schematic light curve for a nova outburst that is due to McLaughlin (1960) is shown in Figure 1. It should be emphasized that the light curves of novae show great variety. However, if they are appropriately normalized, similarities do appear. A rapid rise that is followed by a much slower decline is observed for all novae. McLaughlin (1960) has summarized some of the principal stages of the nova outburst. These are outlined in Figure 1.

The initial rise of a nova to about 2 magnitudes below maximum is quite rapid. At this point a premaximum halt which lasts for hours to days is observed for most novae. This pre-

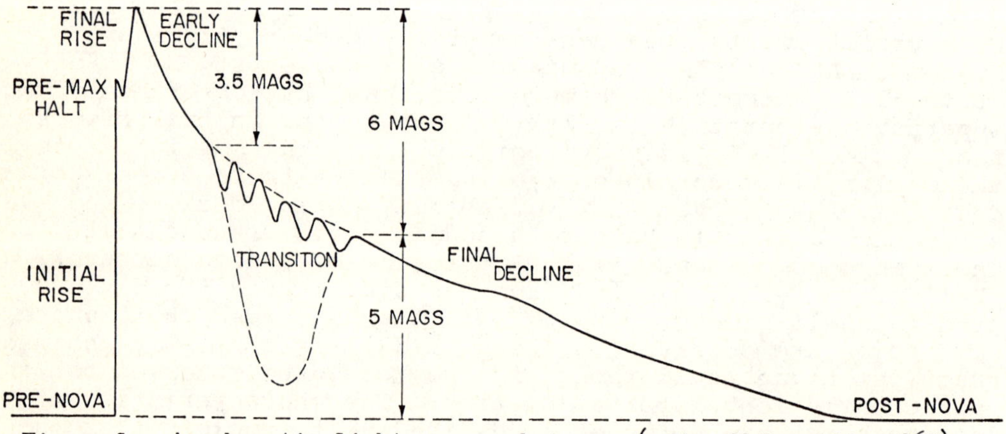

Figure 1. A schematic light curve of a nova (after McLaughlin 1960). The time scale has been magnified during the early stages of the outburst. The magnitudes shown in the figure are visual magnitudes.

maximum halt has been interpreted as an opacity effect. The nova then brightens to maximum and declines. Some novae show strong variations in brightness during their decline while others fade more smoothly. Slow novae sometimes show a deep light minimum during decline (e.g., DQ Her). The final decline is ended when the star appears to return to its prenova state. Some fast novae (e.g., GK Per) do not indicate a premaximum halt. Moreover, the light fluctuations are generally more pronounced for fast novae as compared to novae of moderate or slow rate of decline. The rate of decline from maximum (defined as $\sim$ 3 mag. below maximum) is more rapid for novae with the brightest maxima. The energy released in a nova outburst is typically 10^{44} - 10^{45} erg.

The RT Serpentis stars that make up about 5% of the novae in our galaxy are examples of very slow novae. Although RT Serpentis type novae generally rise to maximum fairly rapidly, their decline is very slow ($\sim$ 15 years for RT Ser).

Although useful information can be obtained from visual brightness measurements, spectroscopy affords a means of obtaining much more detailed information about the physical processes that take place during a nova outburst. The spectral development is found to be related to the amount of decline from maximum light as well as with the observed velocities of absorption. At least five distinct types of spectra, which are associated with the various stages of the nova light curve (see Figure 1), can usually be identified during a nova outburst. The expected line shape from an expanding shell of gas is an absorption feature displaced toward the blue and an emission peak broadened about the true wavelength.

The premaximum spectrum consists of moderately diffuse absorption without strong emission lines (exception: DQ Her). The absorption spectrum usually becomes stronger and always changes to a later spectral class (i.e., A or F supergiant) as the nova rises to maximum. The principal spectrum, which is characterized by strong, symmetrically broadened emission lines (H, Ca II, Na I) and shortward displaced absorption lines, appears immediately after maximum. The principal absorption spectrum has a greater displacement toward the blue than the premaximum spectrum. It is generally assumed that the mass shell is ejected just before the appearance of the principal spectrum. Emission lines (N III, He II, [0 III]) that require relatively high excitation appear while lines requiring lower excitation weaken during the development of the principal spectrum.

The diffuse enhanced spectrum (most prominent for slow novae) appears before the principal spectrum has lost much strength. Its blue displacement is twice the shortward displacement of the principal spectrum. The diffuse enhanced spectrum shows an initially large spread in velocities that tends to sharpen and become multiple as a function of time, indicating some coalescence into clouds or separate shells. It provides evidence for continued ejection of mass after the main outburst. This later-ejected mass is believed responsible for the observed acceleration of the shell of matter ejected during the main outburst.

The Orion spectrum appears next, when the total luminosity is about 3 magnitudes below maximum, and is accompanied by a large increase in the color temperature. This spectrum shows featureless emission and strongly displaced absorption lines that do not sharpen with time. Because large fluctuations in the intensity of these lines accompany fluctuations in the light, it is likely that this spectrum arises from a region close to the star and does not develop into detached clouds.

When the luminosity is about 4 magnitudes below maximum, nebula lines such as [O III], [He III] become visible and then strengthen relative to permitted lines. At $\sim$7 magnitudes below maximum the emission lines are similar to those of planetary nebulae except for the much greater widths. The postnova star is generally variable (sometimes the variability is periodic, e.g., DQ Her).

Novae show a marked concentration toward the galactic equator and center (b = 6°). The most reliable means of estimating the distance to novae depends on measuring the doppler shift and angular rate of expansion of the principal ejected mass shell. This distance determination and therefore the determination of the brightness of a nova outburst is made somewhat uncertain because the outburst is not generally spherically symmetric. However, most novae appear to show symmetry about an axis. The observation of novae in external galaxies such as M31 (Arp 1956) is perhaps the most accurate way of determining their absolute brightness.

It is possible to estimate the mass ejected in a nova outburst. To do this one first estimates the electron temperature in the shell by measuring the relative intensity of forbidden lines (usually [O III]). Since the emission of, say, $H_\delta = f(T_{eff}) \cdot N_e^2 \cdot \text{Volume}$, one can estimate the mass of the shell ($N_e M_p \cdot$Volume) if one can estimate the distance and measure the energy radiated in H_δ (see Pottasch 1959, 1967 for details). The estimates of ejected mass are $10^{-4} - 10^{-6} M_\odot$ and the observed kinetic energy associated with this mass is $10^{43} - 10^{44}$ ergs.

The connection between novae and binary membership is established (Struve 1955; Huang 1956, 1957; Kraft 1963, 1964). Although it is not possible to prove that all novae are members of binary systems, it would appear that the vast majority of novae are members of binary systems such that one component is an ultraviolet dwarf and the other a red star. The observations suggest that mass is flowing from the red to the blue star (Kraft 1964).

A plausible theory of the nova outburst must explain not only the basic observations of the nova outburst itself but also must explain why close binary systems in which one component is a hot white dwarf and the other a red star are generally associated with the occurrence of novae. It should be pointed out that the fact that the blue star is hot and usually variable is very significant statistically since the *a priori* probability of finding a hot white dwarf as compared to a normal white dwarf (lifetime $\sim 10^9$ years) is small due to the relatively short cooling times of hot white dwarfs (lifetime $\sim 10^5 - 10^6$ years). It is therefore likely that excitation of the blue star (keeping it hotter than usual for its age) plays an important role in causing the nova outburst.

The following is a summary of the key observations that a theory of novae must explain.

1. Occurrence in a blue star that is a member of a red-blue binary system.
2. Variable mass, but with many occuring in $< 1M_\odot$ stars (e.g., DQ Her $\lesssim .5M_\odot$, TCrB $\sim 2.1M_\odot$)
3. Recurrence
4. Energy radiated $\sim 10^{44} - 10^{45}$ ergs.
5. Kinetic energy of ejected shell $\sim 10^{43} - 10^{44}$ ergs.
6. Mass of ejected shell $\sim 10^{-4} - 10^{-6}M_\odot$.
7. Sudden rise to maximum followed by more gradual decline and continued activity long after outburst.
8. Axial symmetry but not spherically symmetry is usually associated with the ejected mass.
9. Widely variable decay times.

II. Theory

In discussing the origin of the nova outburst we will take the point of view that nuclear energy generation (H burning or in some instances, He burning) is the principal energy source for the outburst. More specifically, we will assume that mass accretion onto the surface of a hot white dwarf is a key physical process responsible for the nova outburst. A model for the origin of the nova outburst that accepts the above point of view has recently been discussed (Rose 1967b, 1968). These calculations indicate that if a white dwarf accretes small amounts of hydrogen-rich mass, hydrogen burning will increase until the nondegenerate hydrogen-burning shell becomes thermally unstable (see Vol. 2, Instability Problem in Nuclear Burning Shells for discussion of thermal instability). Pulsational instability, which is presumed responsible for the nova outburst itself, is found to develop as a consequence of the thermal instability. Other relevant theoretical discussions have been presented. Schatzman (1953) suggested the possible connection between pulsational instability and the nova outburst. Mestel (1952; see also Saslaw, 1968) discussed the onset of hydrogen burning under degenerate conditions in a white dwarf that had accreted hydrogen-rich matter. Giannone and Weigert (1967) independently showed that mass accretion onto the surface of a white dwarf could lead to thermal instability. Additional theories concerning the origin of novae have been summarized by Schatzman (1965).

The initial model for the calculations described by Rose (1968) is a hot, hydrogen-exhausted $0.75M_\odot$ star that is contracting toward the white dwarf state. Beginning with the initial model, evolutionary sequences were computed for two different mass accretion rates ($7 \times 10^{-8}M_\odot$/yr and $1 \times 10^{-5}M_\odot$/yr).

Mass accretion rates as low as $10^{-9} M_{\odot}/yr$ would be sufficient to lead to physically similar results. The calculations indicate that mass accretion leads to hydrogen burning in a thin shell. This hydrogen burning is dominated by the temperature sensitive CN cycle. The reaction rate used in the calculations has been given by Caughlan and Fowler (1962). An equilibrium distribution was assumed to have been established among the light nuclei that are necessary to complete the CN cycle.

It should be pointed out that the kinetic energy per gram of the infalling matter (neglecting collisions and assuming spherically symmetric infall) is of order

$$\frac{GM}{R} \sim \frac{6.7 \times 10^{-8} \times 2 \times 10^{33}}{10^{9}} \sim 10^{17} \text{ erg/gm} \tag{1}$$

This energy is quite significant but still much less than the nuclear energy content of hydrogen or helium. For an accretion rate as high as $10^{-5} M_{\odot}/yr$ the rate of kinetic energy transferred to the surface of the hot white dwarf could exceed the initial luminosity by a factor of 50. Some of this kinetic energy could be radiated away at X-ray wavelengths. For mass accretion rates as high as $10^{-5} M_{\odot}/yr$ an accurate treatment of the outer boundary layer of the star might have an interesting effect on the solution. However, no attempt to include this additional energy source was included in the calculations.

The final contraction phase before a star reaches the white dwarf state is not uniform. Most of the contractional energy release takes place in the outer layers. This is a consequence of the fully degenerate state of the hydrogen-exhausted interior and the semidegenerate and nondegenerate state of the helium and hydrogen that constitutes the outer layers. For the computed models helium burning has ceased and hydrogen burning takes place at a density of $\sim 2 \times 10^{2} gm/cm^{3}$. The effective temperature and visual magnitude of the initial model are consistent with those observed for old novae. Although the choice of the initial model was somewhat arbitrary, no attempt to optimize conditions necessary for instability was undertaken.

The calculations that are described below show that as hydrogen-rich matter is added to the envelope of the star the energy generation due to hydrogen burning increases. After the burning has reached $\sim 150\ L_{\odot}$ the hydrogen shell source becomes thermally unstable and therefore the burning continues to rise independently of mass accretion. The e-folding time for the increase in hydrogen burning is given approximately by the local Kelvin time of the hydrogen-burning shell,

$$\tau = \frac{3p/2\rho}{\varepsilon_{CN}} , \tag{2}$$

where the variables are evaluated at the mass point with the highest nuclear energy generation per unit mass. This characteristic nuclear rise time is as short as 2 days for some of the computed models. The calculations show that the very high rate of nuclear energy generation resulting from the thermal instability causes the star to become pulsationally unstable. The basic behavior of these unstable hydrogen shell-burning models is similar to those of helium shell-burning stars discussed previously (Rose 1966, 1967a). However, in the present case mass accretion is necessary to initiate the instability. In addition, the Kelvin times that characterize the hydrogen-rich envelope above the hydrogen-burning shell are much shorter than the Kelvin times that characterize the envelopes of unstable helium shell-burning stars. This is true primarily because the mass of the hydrogen-rich envelope is $\sim 10^{-4}M_\odot$, whereas the masses of the helium envelopes of the unstable helium shell-burning models were $\sim 0.1M_\odot$. The observed rapid relaxation of postnovae stars ($\sim$ 10 years) suggests that only a thin layer of mass has been involved in the outburst.

As a star approaches the white dwarf state the radial mode of adiabatic oscillation becomes nearly constant (i.e., $\frac{\delta p}{p}$ = constant throughout). For more condensed stars the relative amplitudes of the radial mode decreases sharply interior to the surface layers. If the relative amplitude of the fundamental pulsational mode is high at the position of the hydrogen-burning shell, it is relatively easy to excite pulsational instability by means of nuclear energy generation if the nuclear source is temperature sensitive (e.g., $\varepsilon_{CN} \propto T^{15}$, $\varepsilon_{3\alpha} \propto T^{30}$). That this is the case can be readily seen from an examination of the stability integral

$$\frac{dW}{dt} = \frac{\delta T}{T}(\delta\varepsilon - \frac{d\delta L}{dM_r})dM_r \qquad (3)$$

where $\delta\varepsilon \sim \nu\varepsilon\frac{\delta T}{T}$ and ν is the temperature sensitivity of the nuclear burning source. If the above integral is positive, pulsations will grow in amplitude. If it is negative, damping will take place. The nuclear term makes a positive contribution while the radiative loss term which is more important near the surface makes a negative contribution unless extensive ionization zones are present. The very short pulsation periods and the reasonably homologous character of the modes implies that it is possible to store a significant amount of energy in the radial mode of oscillation ($\sim 10^{44} - 10^{45}$ erg for the computed models).

Figure 2 shows L, L_N and R as functions of time for the evolutionary sequence such that the mass accretion rate is $7 \times 10^{-8}M_\odot$/yr. The rise time for the hydrogen burning is given approximately by equation 2. The increased nuclear energy generation due to the thermal instability not only causes the star to be pulsationally unstable; it also causes the outer envelope to expand. The maximum rate of expansion is primarily dependent on the peak amplitude of the nuclear energy generation. It remains insignificant until the rate of nuclear energy generation is near its maximum value. The expansion velocities for the computed models never exceed about 5×10^3 cm/sec and therefore are always subsonic.

Figure 3 shows the fundamental radial mode for two models that were selected near the peak nuclear energy generation.

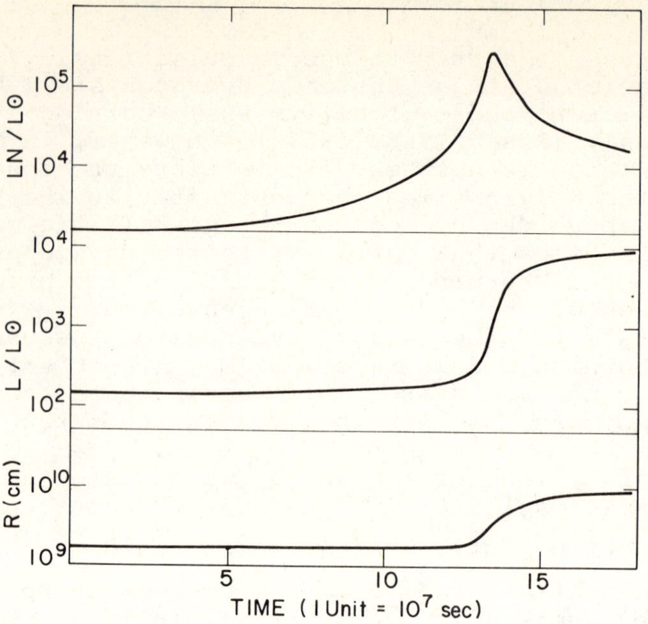

Figure 2. The pulsation amplitude in radial velocity, V, is shown as a function of radius for the fundamental mode computed for two models (1 and 2 shown on the curve). Model 2 is four days later than Model 1. The radial velocities have been normalized so that the velocity for Model 1 is unity at the surface, and the total pulsational energy of Model 2 has been set equal to that of Model 1.

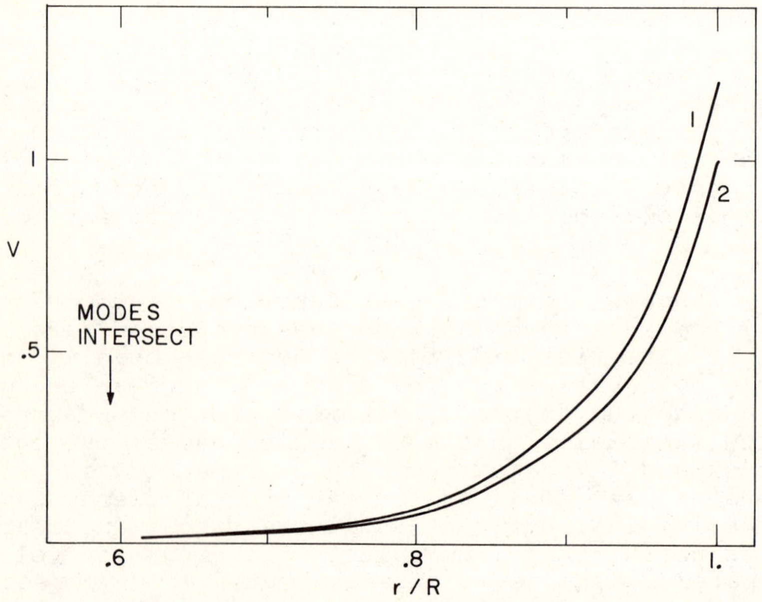

Figure 3. The nuclear energy generation, L_N; the surface luminosity, L; and the radius are shown as a function of time.

The Figure shows that the amplitudes of the modes at the surface are expected to increase significantly in 4 days and therefore illustrates how the expansion of the outer layers of a pulsationally unstable star helps transport mechanical energy to the vicinity of the surface layers where the dissipative processes are the greatest.

The model for a nova outburst that this discussion attempts to make plausible divides the nova light curve shown schematically in Figure 1 into two stages. It is assumed that the radiation and mass ejection that has taken place between the initial rise and the transition region is due to the dissipation of pulsational energy. The second stage includes the time after the transition region and continues until the star has returned to its prenova configuration. It is assumed that the thermal instability is the principal energy source for the radiation emitted during this stage. However, short periodic fluctuations and periodic variations are to be associated with the continued decay of pulsations. The observations of Arp (1956) indicate that the amount of energy that is released in the first stage is $\simeq 3 \times 10^{44}$ erg. This characteristic energy is very nearly the amount of energy that, according to the linear theory, might be stored in pulsations for the computed models.

The question of what determines the limiting amplitude of pulsations is a very important one. On the basis of a linear pulsation theory such as has been used in this paper it is not possible to determine the limiting amplitude. However, non-linear calculations have been carried out for pulsationally unstable models of Cepheids and RR Lyrae stars (for a review of recent results see Christy 1966a). Although the physical causes for the limit to the growth of pulsations in RR Lyrae stars are not fully understood, Christy's calculations (1966b) indicate that the leveling off of the excitation in the He^+ layer is the principal cause. For the models that have been studied in this paper, the amplitude of the adiabatic modes of oscillation are small in the region where the nuclear energy generation has taken place ($\simeq 0.03 \times$ surface amplitude), and the opacity is primarily electron scattering. Therefore, one would not expect non-linear terms to significantly affect the excitation of pulsations until the amplitudes at the surface became quite large.

It is of interest to try to estimate the shortest time scale for an outburst that may be possible by means of a mechanism such as have been discussed in the present paper. If the above model for a nova is correct, nuclear energy generation in a thin, nondegenerate shell by means of the CN cycle is responsible for the nova outburst. For temperatures in excess of 10^8 °K the fast CN cycle (Caughlan and Fowler 1965) replaces the slow CN cycle (Caughlan and Fowler 1962) as the appropriate set of reactions. For the calculated models it is still correct to use the slow cycle. However, if the rate of energy generation were to be increased by about a factor of 10 to $\sim 5 \times 10^{13}$ ergs gm^{-1} sec^{-1}, the slow CN cycle

$$C^{12}(p,\gamma)N^{13}(p,\gamma)O^{14}(\beta^+\nu,\ 100sec)N^{14}$$
$$N^{14}(p,\gamma)O^{15}(\beta^+\nu,\ 200sec)N^{15}(p,\alpha)C^{12} \qquad (4)$$

is limited by the β decay lifetime of O^{14} and O^{15}. It should be pointed out that the accumulation of an overabundance of He^3 could also provide a temperature sensitive nuclear source.

Figure 4 shows the temperature distribution of the outer layers of the initial $.75M_\odot$ model used in carrying out the calculations described above. The mass of the hydrogen-rich layer is about $10^{-4}M_\odot$. The Figure also shows the temperature distribution of a $1.395M_\odot$ pure oxygen star that is contracting toward the white dwarf state. The effective temperature of this star is very high ($\sim$ 8 x 10^5 °K and its luminosity $\sim 10^4L_\odot$. Although its luminosity is quite high the amount of energy radiated in the visual region of the spectrum is $\lesssim 1L_\odot$ because of the high temperature. The extremely steep temperature gradient near the surface of the model (the mass exterior to the temperature at which hydrogen or helium burning would commence is $\sim 10^{-9} - 10^{-10}M_\odot$) indicates that if mass accretion were taking place onto the surface of the star, very high rates of nuclear energy generation would result. This is the case because the burning shell would be very thin and also because the temperatures inside the shell would be $> 10^8$ °K. Under these principal conditions the CN cycle may allow higher rates of nuclear energy generation and moreover unstable helium shell-burning can take place.

Shock waves are expected to play a role in any explosive event such as a nova outburst. The possibility that the nova outburst can be explained as due to a strong shock wave emanating from the interior of a star has been discussed (e.g., Hazlehurst, 1962). Although there are observational grounds for believing that most novae do not originate in this manner, some nova outbursts may be caused by strong shock waves. Moreover, the theory of strong shock waves (i.e., blast waves) is of interest because some relatively simple solutions can be found.

If energy is injected into a gas in a time scale that is much shorter than the time for sound to travel a scale height, a blast wave will be formed. If the blast wave travels into a region of uniform density ρ_o a self-similar solution that depends only on ρ_o and the injected energy E can be found (Sedov, 1959). This solution remains valid so long as the pressure, P_1, behind the shock is much greater than the pressure, P_o, in the undisturbed region ahead of it. The only dimensional combination of E and ρ_o that contains only length and time is $E/\rho_o = [cm^5\ sec^{-2}]$. This implies that the dimensionless variable is

$$\xi = r\left(\frac{\rho_o}{Et^2}\right)^{1/5} \qquad (5)$$

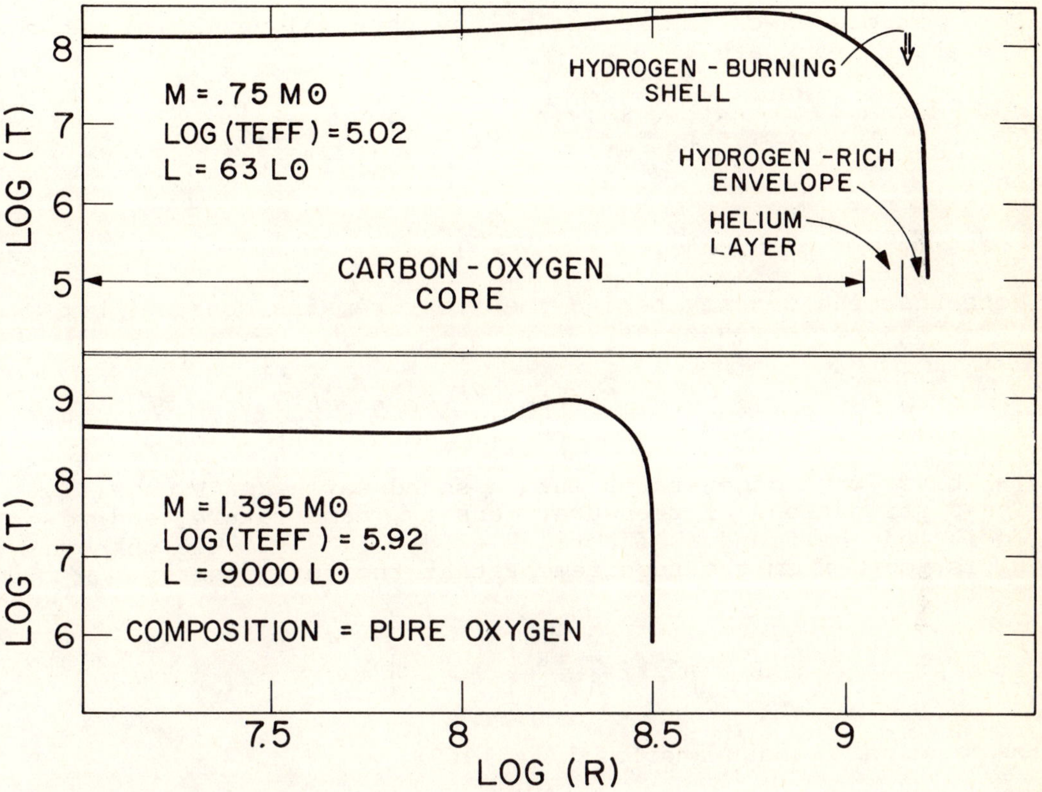

Figure 4. The temperature (T) is shown as a function of radius (R) for a 0.75 $M_\odot$ model and a 1.395 $M_\odot$ model. The 1.395 $M_\odot$ model has a very high effective temperature (T_{eff}) and a very steep rise in temperature near the surface.

and the motion of the wave front is governed by

$$R = \xi_0 \left(\frac{E}{\rho_0}\right)^{1/5} t^{2/5} \tag{6}$$

The propagation velocity of the shock wave is

$$D = \frac{dr}{dt} = \frac{2}{5}\frac{R}{t} = \xi_0 \frac{2}{5}\left(\frac{E}{\rho_0}\right)^{1/5} t^{-3/5} \tag{7}$$

For a strong shock the values of the physical variables behind the shock front are

$$\rho_1 = \rho_0 \frac{\gamma + 1}{\gamma - 1}; \qquad P_1 = \frac{2}{\gamma + 1}\rho_0 D^2$$

$$u_1 = \frac{2}{\gamma + 1} D \tag{8}$$

Note that the density behind the front remains constant but the pressure decreases as

$$P_1 \propto \rho_0 D^2 \sim \rho_0 \left(\frac{E}{\rho_0}\right)^{2/5} t^{-6/5} \sim \frac{E}{R^3} \tag{9}$$

The blast wave degenerates into a sound wave when $P_1 \to P_0$. The distributions of pressure, density, temperature, and velocity can be found as functions of $\xi = \xi_0\ r/R$. The unknown ξ_0 is found from the requirement that the total energy

$$E = \int_0^R 4\pi r^2 \left(\frac{P}{\gamma - 1} + \frac{\rho\mu^2}{2}\right) dr \tag{10}$$

be constant (Landau and Lifshitz 1959).

In the case of stars the density in the outer layers is a steeply decreasing function of radius. It can be shown that the velocity of propagation of a shock wave increases as it propagates outward (Hazlehurst 1962). Sparks (1969a) has pointed out that as the material expands outward from the star an observer should see further into the envelope and on the basis of his calculations has concluded that the velocity that he sees should decrease with time. The observations of most novae indicate that the velocities during the premaximum stage are usually slower than those at maximum. On the other hand, certain recurrent novae such as T Cr B and Rs Oph, which show a very rapid rise to maximum and appear not to show a premaximum halt, may represent novae for which a strong shock wave is the primary mechanism for mass loss. Temperatures in excess of 10^8°K are required if a thermal instability is to lead to a blast wave.

The passage of a strong wave through an envelope takes place in a short time ($\sim$ seconds). The hydrodynamical readjustment should take place in a time scale much shorter than that required for significant radiative losses. It would be difficult to explain the occurrence of post-maximum shells that can arise days or even weeks after an outburst if a strong shock wave is assumed to be the cause of the outburst.

It is generally believed that the rise to maximum of a nova is associated with the development of an extended photosphere. Various mechanisms have been suggested to explain the pre-maximum halt. Pottasch (1959) attributes it to the arrival of energy emitted by the star itself at the surface of the ejected shell. Sobolev (1969) has shown that the proper explanation of the nova light curve may lie in the currect treatment of the non-stationary transfer of energy between the star and the ejected shell. In this theory, a pre-maximum halt occurs because at some time after bolometric maximum the increase in the radius of the star has a greater influence on the visual magnitude than the decrease in the effective temperature.

An expanding photosphere will eventually become optically thin. Because an optically thin gas cloud can radiate much more effectively than an optically thick cloud of comparable mean temperature, it is likely that the initial rise of a nova outburst is due to an expanding photosphere for which an energy source sufficient to prevent it from cooling rapidly is available. To show that an optically thin body can radiate more effectively than a comparable optically thick body we recall that the effective temperature (i.e., brightness temperature) need not be as high as the average temperature.

$$\text{Flux} = \sigma T_{\text{eff}}^4 \sim \frac{\lambda c U}{4L} \sim \frac{\lambda}{L}\, \sigma T^4 \tag{11}$$

where λ = mean free path of photon, L = characteristic length, U = the radiant energy density and T = the average temperature. Equation 11 shows that if the photon mean free path is sufficiently short (i.e., $[\lambda/L]^{1/4} \ll 1$), $T_{eff} \ll T$. On the other hand, if $\lambda \sim L$, the effective temperature can approach the average temperature. The temperature of an expanding photosphere must eventually decrease. Sparks (1969b) has recently pointed out that because $\kappa \sim \kappa_0 \rho^n T^m$ ($n \sim 1$, $m \sim -3.5$) the decrease in temperature during expansion may offset the decrease in density in such a manner as to offer a possible explanation for the premaximum halt. The observations of novae suggest that the final rise to maximum occurs when the temperature in the outer layers falls below that necessary to ionize hydrogen and consequently the opacity is greatly reduced.

In our discussion of pulsational instability, we considered only the radial modes of oscillation. However, the observational evidence indicates that most nova outbursts are not spherically symmetric. Moreover, many nova outbursts appear to have axial symmetry. Photographs of Nova Herculis illustrate this point. It is well known that a star has nonradial as well as radial modes of oscillation. The resonant excitation of nonradial modes by radial pulsations is one means by which these nonradial modes may become excited.

The character of the spectrum of nonradial modes is more complicated than the radial spectrum. Cowling (1941) showed that in addition to a high frequency branch of nonradial modes (p modes), a stable gas sphere will also have a branch whose periods increase indefinitely as the number of radial modes increases (g modes). The motion is primarily radial and the pressure variations large for high frequency oscillations. On the other hand, the motions are primarily horizontal for low frequency oscillations. Modes of intermediate frequency (Kelvin modes) also exist if the angular quantum number (ℓ) ≥ 2. For $\ell = 1$, the Kelvin mode (or f mode in Cowling's terminology) corresponds to a uniform displacement of the gas sphere.

The equations governing the adiabatic oscillations of a compressible gas sphere are

<u>The equation of motion:</u>

$$\frac{d\vec{v}}{dt} = -\frac{1}{\rho}\nabla P - \nabla\Phi \tag{12}$$

<u>The equation of continuity:</u>

$$\frac{d\rho}{dt} = -\rho \operatorname{div} \vec{v} \tag{13}$$

<u>The adiabatic condition:</u>

$$\frac{d}{dt}(P\rho^{-\gamma}) = 0 \tag{14}$$

<u>Poisson's Equation:</u>

$$\nabla^2\Phi = -4\pi G\rho \tag{15}$$

If the above equations are linearized, it can be shown (Pekeris 1938) that the displacement satisfies a fourth order linear differential equation. The solutions of equations governing nonradial oscillation must be solved numerically and for this reason it is convenient to reduce the system of equations to two coupled second order linear differential equations (Hurley, Roberts, and Wright 1966). Since in our present discussion we wish to emphasize the resonant excitation of non-radial modes we will follow a discussion due to Vandakurov (1966). We assume that a star is pulsating with small amplitude in a radial mode such that the radial position of a mass shell is

$$r = a(1 + \varepsilon(a)\sin\omega t), \quad \varepsilon << 1 \tag{16}$$

and from the equation of radial pulsation

$$P = P_o\left[1 - \gamma(3\varepsilon(a) + a\frac{\partial\varepsilon}{\partial a})\sin\omega t\right] \tag{17}$$

If the above expansions are substituted into the equations of non-radial oscillation the problem is reduced to that of finding the solution of a perturbed equation where the solutions for the unperturbed equation are known. The solutions for the equations of nonradial oscillation can be written

$$\psi_k^i\, e^{j\omega_{k\ell}t} \qquad i = 1, 2, 3 \tag{18}$$

The index i denotes an unknown function in the coupled equations and ℓ the angular momentum number. To solve for the solution to the perturbed equation, one assumes that the solution can be written

$$\psi^i(a,t) = \sum_k A_k^i(t)\psi_k^i\, e^{j\omega_{k\ell}t} \tag{19}$$

where $A_k^2 = A_k^1 + \varepsilon B_k$ and $A_k^3 = A_k^1 + \varepsilon C_k$. The above expansions are substituted into the equations of nonradial oscillation and it is assumed that the time variations of A_k^i are slow as compared to the radial pulsation period. Vandakurov (1966) proceeds to show that if the resonant condition

$$|\omega_{k\ell} + \omega_{k'\ell} - \omega| < \varepsilon\omega D_{kk'\ell} \qquad \ell \geq 1 \tag{20}$$

where $D_{kk'\ell}$ (a constant of order unity) is satisfied, the amplitudes $A_k(t)$ grows exponentially

$$A_k(t) = \text{constant x} \exp\left[-\frac{j}{2}(\omega_{k\ell} + \omega_{k'\ell} - \omega)t + h_{kk'}t\right]$$

$$h_{kk'} = \sqrt{\frac{1}{\tau^2_{kk'}} - \frac{1}{4}(\omega_{k\ell} + \omega_{k'\ell} - \omega)^2} \tag{21}$$

$\tau_{kk'}$ represents the shortest time scale for the growth of $A_k(t)$ and is of order $1/\varepsilon\omega$. Since the fundamental radial pulsation period is expected to be $\leq$ 1 minute, the coupling to a nonradial mode takes place in less than 1 hour. If the mechanical energy that is initially stored in a radial mode were transferred to nonradial modes a net transfer of energy toward the surface of the star would result. This is the case because the amplitudes of the nonradial modes have angular dependence and are more strongly concentrated toward the surface than the fundamental radial mode.

Recently (Harper and Rose 1970) have studied the non-radial oscillations of models for hot white dwarfs such as may be responsible for the nova outburst.

REFERENCES

Arp, H. 1956, A.J., 61, 15.

Cauglan, G., and Fowler, W. 1962, Ap.J., 136, 453.

———. 1965, ibid., 70, 670.

Christy, R. F. 1966a, Ann. Rev. Astr. and Ap., 4, 353.

———. 1966b, Ap.J., 144, 108.

Cowling, T. 1941, M.N.R.A.S., 101, 367.

Giannone, P., and Weigert, A. 1967, Zeit. fur Astro., 67, 41.

Harper, R., and Rose, W. K. 1970, in preparation.

Hazlehurst 1962, Advances in Astronomy and Astrophysics, ed. Z. Kopal, Vol. 2 (New York: Academic Press).

Huang, S.-S. 1956, A.J., 61, 49.

———. 1957, Occasional Notes Roy. Soc., 19, 161.

Hurley, W., Roberts, P. H., and Wright, K. 1966, Ap.J., 143, 539.

Kraft, R. 1963, Advances in Astronomy and Astrophysics, ed. Z. Kopal, Vol. 2 (New York: Academic Press).

———. 1964, Ap.J., 139, 457.

Landau, L., and Lifshitz, F. 1959, Fluid Mechanics (Reading, Mass.: Addison Wesley).

McLaughlin, D. 1960, Stellar Atmospheres, ed. J. Greenstein (Chicago: University of Chicago Press).

Mestel, L. 1952, M.N.R.A.S., 112, 583.

Pekeris, C. 1938, Ap.J., 88, 189.

Pottasch, S. 1959, Ann. d'Ap., 22, 297.

———. 1967, Bull. Astr. Inst. Neth., 19, 227.

Rose, W. K. 1966, Ap.J., 146, 838.

———. 1967a, ibid., 150, 193.

Rose, W. K. 1967b, A.J., 72, 824.

———. 1968, Ap.J., 152, 245.

Saslaw, W. 1968, M.N.R.A.S., 138, 337.

Schatzman, E. 1965, Stellar Structure, ed. H. Aller and D. McLaughlin (Chicago: University of Chicago Press).

Sedov, L. 1959, Similarity and Dimensional Methods in Mechanics (New York: Academic Press).

Sobolev, V. V. 1969, Vistas in Astronomy, Vol. 11 (London: Pergamon Press).

Sparks, W. 1969a, Ap.J., 156, 569.

———. 1969b, Bull. Amer. Astron. Soc., 1, 205.

Struve, O. 1955, Sky and Tel., 14, 275.

Vandakurov, Yu. V. 1967, Sov. Ast., 10, 810.

10

EARLY SUPERNOVA LUMINOSITY

Stirling A. Colgate and Chester McKee
New Mexico Institute of Mining and Technology, Socorro
Received October 17, 1968; revised January 20, 1969

ABSTRACT

The diffusion of radiant energy from spherical expanding matter has been analytically and numerically calculated for masses and velocities of model supernova outbursts. The agreement with observation is satisfactory. The production of a large mass fraction of the radioactive isotope ^{56}Ni, which has been predicted from calculations of supernova nucleosynthesis, appears to be critical for the formation of the observed light curves. The radioactive energy from 0.25 $M_\odot$ of ^{56}Ni by the decay process $^{56}\mathrm{Ni} \rightarrow {}^{56}\mathrm{Co}$ (6ᵈ01, Ec, 1.72 MeV of γ-rays per decay) supplies the radiant energy, 10^{49} ergs, during the "diffusive release" phase (5–20 days) of expansion near maximum. The subsequent decay process, $^{56}\mathrm{Co} \rightarrow {}^{56}\mathrm{Fe}$ (77 days, Ec, 3.59 MeV of γ-rays), in conjunction with progressive γ-ray transparency of the expanding matter, gives rise to the long-time exponential light decay of 35–65 days. The velocity distribution with the best fit (Type I supernovae) gives $\langle V^2 \rangle = (1.6 \times 10^9 \text{ cm sec}^{-1})^2$. For a pure thermonuclear supernova requiring a minimum of 1.4 $M_\odot$ of ^{12}C, the implied kinetic energy is 3 times the maximum available from ^{12}C burning, and 6 times that available from the model calculated by Hansen. This implies the possibility that some supernovae originated from neutron stars.

I. INTRODUCTION

The early optical emission from supernova has been both quantitatively and qualitatively a mystery. The rise time of 1 week has been interpreted by Poveda (1964) as indicating ejection of 0.01 $M_\odot$, while line emission and Doppler shifts indicate ejection of 0.1–10 $M_\odot$ (Minkowski 1964, 1968). The spectrum is totally unexplained (Zwicky 1965), and the ejection velocity of 10^8 cm sec^{-1} observed for old remnants (Poveda and Woltjer 1968) is at variance with that of 2×10^9 cm sec^{-1}, interpreted for the newest remnant, Tycho's star (Minkowski 1964). Theory has been no more helpful, as Schatzman (1965) admits. The supernova calculations of Colgate and White (1966) depended on the release of the relatively large binding energy ($\sim 10^{53}$ ergs) of the dynamically formed neutron star, a fraction of which (10–30 percent) appears as kinetic energy of ejected matter. However, the predicted optical energy was small, and an attempt to explain this by assuming the presence of a large (2 $M_\odot$) amount of radioactive (allowed β-decay) matter was even then inadequate to explain the observed luminosity.

The problem is that the heat energy of the expanding matter is converted to kinetic energy by adiabatic expansion long before it is liberated as radiant heat. The later the liberation of heat energy, the smaller the subsequent adiabatic expansion before release. Cameron and Arnett (1967) attempted an explanation based on late burning of deuterium, but the neutron release and subsequent heavy-element production were inconsistent with *r*-process abundances. Instead, and as a consequence, we have been stimulated to perform the present calculation by the results of nucleosynthesis calculations; namely, James Truran recommended that we take a look at the consequences of silicon-burning (Truran, Arnett, and Cameron 1967). This process leads to the α-particle isotope ^{56}Ni as the predominant end product of thermonuclear processes. (^{56}Ni is the minimum in the packing-fraction curve for α-particle nuclei.) The decay scheme of ^{56}Ni (Lederer *et al.* 1967), shown in Figure 1, is $^{56}\mathrm{Ni}(6^{\mathrm{d}}01) \rightarrow {}^{56}\mathrm{Co}(77^{\mathrm{d}}) \rightarrow {}^{56}\mathrm{Fe}$. The release of the relatively large energy (1.72 MeV per ^{56}Ni decay) during the optimum and critical time period of $\sim$6 days, in which diffusive release of the radiant heat energy occurs, produces the optical luminosity. As a consequence, we have taken the material-expansion curves from previous hydrodynamic calculations and appended onto them

Reprinted courtesy of The Astrophysical Journal, vol. 157, August 1969, copyright © The University of Chicago Press.

the β-decay energy of the expected 0.25 $M_\odot$ of ^{56}Ni and ^{56}Co. From these we have reproduced with a numerical diffusion program the observed time and luminosity behavior of early light curves of supernovae. We have started this work with an analytic solution using one set of simplifying assumptions. This solution illuminated the important features of the problem, as well as satisfying the necessity for obtaining a suitable test problem to simulate and test the extreme conditions required of the numerical program. A phenomenological approach to the problem of supernova light curves is developed first.

II. VELOCITY OF THE EJECTED MATTER

A supernova is presumed to be the explosion of a self-gravitating sphere of gas—a star. Furthermore, the star is commonly presumed to be approaching the end of nucleosynthesis. As a consequence, the matter density must be large (10^5–10^{10} g cm^{-3}) in order that the gravitational field can contain in equilibrium the temperature associated

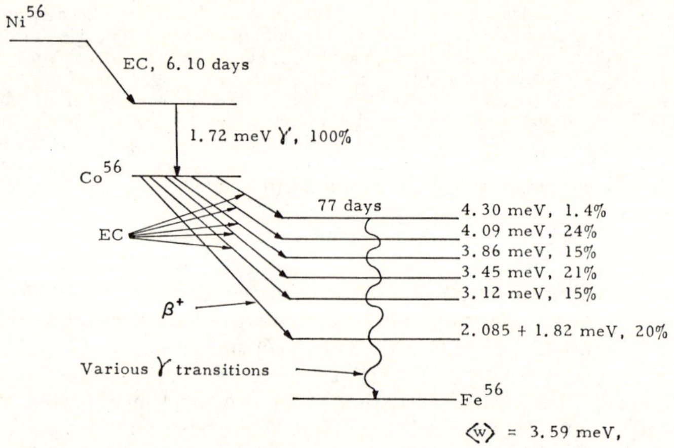

FIG. 1.—^{56}Ni → ^{56}Co → ^{56}Fe decay scheme (*Nuclear Data Sheets*)

with the final stages of nuclear burning. Regardless whether the explosion is thermonuclear, triggered by collapse (Burbidge *et al.* 1957; Rakavy and Shaviv 1967; Fraley 1968; Hansen and Wheeler 1968; Arnett 1968*a,b*), or whether the binding energy of the resulting neutron star is transported to the mantle by neutrinos (Colgate and White 1966; Arnett 1967; Schwartz 1967), the explosion is nearly the same; namely, roughly 1 solar mass is ejected with a mean velocity corresponding to the gravitational binding energy just before explosion and with a velocity distribution depending on the relative location of the mass fraction in question. The simplest velocity distribution is that of a uniform spherical expansion (density independent of radius), $V = V_1 r/r_0$, where r_0 is the outer boundary, $V_1 t$; or

$$V = V_1(1 - F)^{1/3} \tag{1}$$

where F is the external mass fraction,

$$F = \frac{1}{M_{\text{ej}}} \int_r^\infty 4\pi r^2 \rho dr \, , \tag{2}$$

where M_{ej} is the ejected mass. Figure 2 shows where the velocity distribution of equation (1) is matched at the mass fraction, $F = 0.42$, to a velocity distribution (eq.[3]) of the external layers resulting from the speedup of the shock wave progressing in the density gradient of the outer layers of the star. The hydrodynamic results of Colgate and White for supernovae of 1.5 $M_\odot$ and 10 $M_\odot$ are included for comparison. The external velocity distribution is

$$V = V_2 F^{-1/4} . \quad (3)$$

For the case of a supernova produced by the formation of a neutron star, the constants $V_1 = 3 \times 10^9$ cm sec^{-1} and $V_2 = 2 \times 10^9$ cm sec^{-1} result in an average kinetic energy of 3.7×10^{18} ergs g^{-1}. This energy is roughly 4.2 times that achievable from helium burning and 7.4 times that possible from carbon burning, if gravitational binding is excluded. The curves for the velocity distribution when both V_1 and V_2 are reduced by $\frac{1}{2}$ and $\frac{1}{4}$ are shown, as well as the velocity distribution resulting from the ^{12}C thermonuclear hydrodynamic calculation of Hansen and Wheeler (1968). As would be expected, the form of the velocity distribution is approximately independent of the mechanism of explosion; the velocity scale depends only upon the total energy.

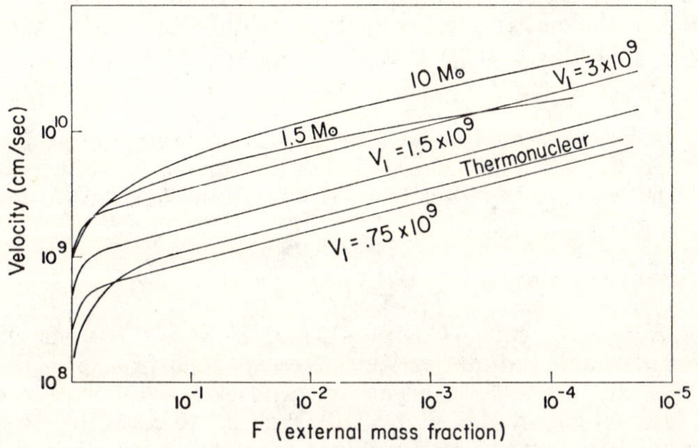

FIG. 2.—Velocity versus mass fraction for a series of analytic curves corresponding to equations (1) and (3). For comparison, the velocity distributions of the neutron-star supernova of Colgate and White (1966) and of the thermonuclear supernova of Hansen and Wheeler (1968) are added.

III. DIFFUSIVE RELEASE OF RADIANT ENERGY

It is assumed that the velocity distributions of equations (1) and (3) are first created by heating the gas by thermonuclear reactions, neutrino deposition, or a shock wave, and that subsequent adiabatic expansion converts the heat into kinetic energy of ejection. It is further assumed that the initial heat source has decayed to insignificance long before the time (several days) that expansion has proceeded to the point of bolometric maximum. The residual energy associated with each mass element is that remaining after adiabatic expansion from the initial heating, as well as that introduced by other energy sources such as radioactivity or collision with the interstellar medium. This latter luminosity due to collisional heating has been discussed in connection with the far greater luminosity of quasi-stellar sources (Colgate and Cameron 1963; Colgate 1967) and in connection with Seyfert galaxies (Colgate 1968), but it is unlikely to be the principal cause of the average supernova luminosity. The minimum circumstellar density required to give the observed peak luminosity is 10^4 H atoms cm^{-3} at $r = 10^{17}$ cm, which is typical of planetary nebulae. Then the luminosity becomes $L \leq 4\pi r^2 \rho_0 V^{3/2}$ ergs sec^{-1}. However, the rise time of the light emission cannot be much less than $\frac{1}{3} rV - 1 \geq 6$ months, compared with the observed 1 week. Although some supernovae might fit this model, the requirement that such a nebula be conditional for all supernovae appears too restrictive. Instead, we will investigate internal sources of energy such as radioactivity.

The energy densities associated with the velocity distributions of equations (1) and (3) are large enough that a major fraction of the internal energy is in the form of radiation rather than particle energy. As a consequence, the diffusive flow of radiation proceeds as a separate "gas," and the specific heat of the matter becomes insignificant. Under these conditions, a diffusive wave of radiation will penetrate a plane-parallel distance ∂ in a time t according to the error-function solution,

$$\partial = (Dt)^{1/2}\,, \tag{4}$$

where the diffusion coefficient $D = \frac{1}{3}c/K\rho$ (where K = opacity, ρ = density). During the expansion of the matter, each mass element will follow an adiabatic law modified by any energy sources until a diffusive wave penetrates from the surface to the mass element in question. This diffusive wave will "release" the internal energy presumably at peak luminosity when the diffusive wave has penetrated to a depth $\partial \simeq \frac{1}{3}r_0$ corresponding to one-half the ejected mass. Equating the diffusion time obtained from equation (4) with $\partial = \frac{1}{3}r_0$ to the expansion time r_0/V_1 gives

$$r_0/\lambda = 3c/V_1\,, \tag{5}$$

where λ = the radiation mean free path = $(K\rho)^{-1}$. In other words, the star will be relatively thick ($3c/V_1 \simeq 30$ mean free paths) at the time of the diffusive release of the bulk of the internal energy. If we let $\rho = 3M/(4\pi r_0^3)$ in equation (5) and neglect the density gradient, we obtain

$$t = \left(\frac{KM}{4\pi c V_1}\right)^{1/2} \text{ sec}\,. \tag{6}$$

If $M = 1\ M_\odot$, $K = 0.2$ g cm^{-2} (Compton), and $V_1 = 3 \times 10^9$ cm sec^{-1}, then $t = 6 \times 10^5$ sec (7 days), and the matter will have expanded from $\rho = 10^{9\pm3}$ g cm^{-3} to $\rho \simeq 2.5 \times 10^{-13}$ g cm^{-3}. The large expansion ratio of $4 \times 10^{21\pm3}$ reduces the largest possible initial internal energy, U_0, of 4×10^{18} ergs g^{-1} to $2.6 \times 10^{11\pm1}$ ergs g^{-1} ($U = U_0\,\rho^{-1/3}$), if an adiabatic expansion is assumed with $\gamma = \frac{4}{3}$ for a radiation-dominated gas. The total internal energy that can be radiated is then only $UM_\odot = 5 \times 10^{44\pm1}$ ergs, which is $10^{-3.3\pm1}$ of that estimated for the average supernova, 10^{49} ergs, of either type (Minkowski 1964). The fact that part of the adiabat may proceed with $\gamma = \frac{5}{3}$ only further reduces the available radiant energy. Two possibilities remain: Either a source of radioactive energy exists in the expanding matter which gives heat energy late in time, or the shock-heated matter undergoes less expansion, owing to an initial expanded stellar structure. This latter possibility was investigated by Colgate and White, and the assumption of a red-giant structure ($r \simeq 1.5 \times 10^{13}$ cm) as calculated by Hofmeister, Kippenhahn, and Weigert (1964) and the relatively large shock-energy input of 10^{52} ergs resulted in a bolometric luminosity one-sixtieth of that observed. As a consequence, one is forced to consider either a more expanded initial structure ($r \simeq 10^{14}$ cm) or late sources of radioactive energy.

IV. ANALYTICAL SOLUTION WITH ALLOWED β-DECAY

a) Method

The most general radioactive energy source is one composed of an ensemble of β-decays. (The α-particle decay of heavy nuclei or spontaneous-fission nuclei are presumed too low in abundance, despite their high decay energies, to contribute significantly to the radioactive energy.) It was further assumed that the ensemble of β-decay nuclei would be on the neutron-rich side of the stability curve analogous to fission-product nuclei, but of lower isotopic mass. Such a statistical ensemble of radioactive decays leads to the well-known rate of decay proportional to $t^{-1.2}$, as observed for fission products. (An elementary derivation is given in Colgate and White 1966.)

The rate of β-decay energy release then becomes $S = S_0 t^{-1.4}$, and the constant, $S_0 = 5 \times 10^{16}$ ergs g^{-1} sec^{-1}, was chosen as if two allowed β-decays occurred in series per fifty nucleons.

Finally, the velocity trajectory of each mass fraction is presumed to remain constant after the initial expansion; the total β-decay energy is small compared with the ejection kinetic energy, and the velocity distribution of equation (3) was assumed as an adequate approximation for the whole star. Since the initial radius is small compared with the radii of interest, the radius of the expanding matter becomes

$$r = V_2 t F^{-a}, \tag{7}$$

where the Lagrange coordinate, F, is given by equation (2), and a is of the order $\frac{1}{4}$. Then, if one equates $\partial F/\partial r$ from equations (2) and (7), the density becomes

$$\rho = \rho_0 \frac{F^{3a+1}}{t^3}, \tag{8}$$

where $\rho_0 = M/(4\pi a\, V_2{}^3)$.

The energy equation in spherical geometry, including diffusive transport, an energy source, and adiabatic expansion with $\gamma = \frac{4}{3}$, is

$$\frac{du}{dt} - \frac{4}{3}\frac{u}{\rho}\frac{d\rho}{dt} = \frac{\rho}{r^2}\frac{\partial}{\partial r}\left(r^2 D \frac{\partial u}{\partial r}\right) + \rho S_0, \tag{9}$$

where u is the specific energy density, the diffusion coefficient $D = \frac{1}{3}c/\kappa\rho$, and S_0 is an energy source or sink.

Transforming to Lagrangian coordinates by equations (7) and (8), we obtain the diffusion equation

$$\frac{\partial u}{\partial t} + \frac{4u}{t} = D' F^{3a+1} \frac{\partial}{\partial F}\left[F^{-4a}\frac{\partial u}{\partial F}\right] + \rho_0 S_0 F^{3a+1} t^{\beta_0 - 1}, \tag{10}$$

where $D' = 4\pi c V_2/3aK\rho$ and the β-decay source is $S = S_0 t^{\beta_0+2}$ and β_0 is presumably -3.4. The terms in equation (10) are respectively the net change of energy following a volume element, the adiabatic work, the net balance in diffusive transport, the energy generated due to β-decay.

Use of the method of similarity transformation (Morgan 1952) obtains the transformation

$$u = \rho_0 S_0 \sigma^{2(3a+1)/(a+1)} t^{\nu} G(z); \quad z = \frac{F^{a+1}}{2\sigma^2 t^2}; \quad \sigma = \frac{a+1}{2}\sqrt{D'}; \quad \nu = \frac{2(3a+1)}{a+1} + \beta_0, \tag{11}$$

and the invariant solution,

$$\begin{aligned} G(z) = {} & AM\left(-\frac{5a+3}{a+1} - \frac{\beta_0}{2}, -\frac{3a}{a+1}; -z\right) + Bz^{(4a+1)/(a+1)} \\ & \times M\left(-\frac{a+2}{a+1} - \frac{\beta_0}{2}, \frac{5a+2}{a+1}; -z\right) \\ & - 2^{2a/(a+1)} \sum_{n=0}^{\infty} \frac{(-1)^n[(2a+3)/(a+1) + \frac{1}{2}\beta_0]_n\, z^{n+(4a+2)/(a+1)}}{n!\,[n + 1/(a+1)][(4a+2)/(a+1)]_{n+1}} \\ & \times M\left(-\frac{a+2}{a+1} - \frac{\beta_0}{2}, n + \frac{5a+3}{a+1}; -z\right), \end{aligned} \tag{12}$$

where the solution is obtained by the method of Laplace transforms (Erdelyi *et al.* 1954), and where $M(a, b; z)$ is the Kummer function (Jahnke, Emde, and Losch 1960) and $G(z)$ satisfies the ordinary differential equation,

$$zG'' + \left(-\frac{3\alpha}{\alpha+1} + z\right)G' - \left(\frac{5\alpha+3}{\alpha+1} + \frac{\beta_0}{2}\right)G = -2^{2\alpha/(\alpha+1)}z^{(3\alpha+1)/(\alpha+1)} . \quad (13)$$

To determine the proper boundary conditions to impose, we observe that, in the region of large optical depth,

$$G(z) = B'z^{(5\alpha+3)/(\alpha+1)+\beta_0/2} + \frac{2^{2\alpha/(\alpha+1)}z^{(3\alpha+1)/(\alpha+1)}}{2+\frac{1}{2}\beta_0} \quad (14)$$

satisfies equation (10) if the diffusion term is zero.

Then, if $t > 0$ and if we assume no energy input other than $S = S_0\, t^{\beta_0+2}$, we can impose the following boundary conditions:

i) $u(0,t) = 0$ implies $G(0) = 0$;

$$\text{ii) } u(F,t) = \frac{\rho_0 S_0 F^{3\alpha+1}}{\beta_0+4}\, t^{\beta_0} \text{ as } F \to 1 \text{ implies } G(z) = \frac{2^{2\alpha/(\alpha+1)}z^{(3\alpha+1)/(\alpha+1)}}{2+\frac{1}{2}\beta_0} \text{ for } z \gg 1 . \quad (15)$$

Under these boundary conditions the solution is most easily found by using the Runge-Kutta technique to integrate equation (13) numerically. The numerical solution was started with $z = 10^{-4}$ near the outer boundary of the star by using the analytic solutions (12) with $A = 0$, as required by condition (15i), and $B = 12$. As the inner boundary is approached, one evaluates B' in equation (14). If one chooses a value of B, one can estimate graphically and obtain $B \approx 4.2$ as the value which satisfies the boundary condition (15ii). A check performed with $B = 4.19725$ resulted in a value for B' of $\sim 10^{-6}$, the more accurate value being obtained by subtracting out the undesired solution.

If, in addition to the β-decay source, we consider the energy input due to the shock wave produced in the gravitational-collapse model, we can obtain a solution by noting that the density distribution before the shocked layers expand is $\rho = \rho_i\, F^{-\lambda}$, where $\lambda \simeq 1$. This approximates the polytropic ($n = 3$) distribution where $\lambda = \frac{4}{3}$, but is modified by the small redistribution occurring in the outer layers during the initial collapse phase. In addition, the internal energy of the fluid immediately behind the shock front is roughly one-half the kinetic energy imparted to the fluid by the shock.

These considerations enable us to solve for the adiabatic behavior of the shock-deposited energy; namely,

$$U_s = U_0 \frac{F^{2\alpha+(4-\lambda)/3}}{t^4} , \quad (16)$$

with

$$U_0 = \frac{\rho_0 V_2^2}{8}\left(\frac{\rho_0}{\rho_i}\right)^{1/3} .$$

If we use equation (16) as the boundary conditions in place of condition (15ii) as $F \to 1$ and the second function in equation (12) as the solution, we note that it satisfies the homogeneous part of equation (10) and the boundary conditions (15i) and (16) if we choose

$$\frac{\beta}{2} = \frac{4-\lambda}{3(\alpha+1)} - 3 .$$

This yields

$$U_s = ct^{-2(\lambda+2)/3(\alpha+1)}z^{(4\alpha+1)/(\alpha+1)}M\left(\frac{6\alpha+\lambda-1}{3(\alpha+1)}, \frac{5\alpha+2}{\alpha+1}; -z\right), \quad (17)$$

with

$$c = U_0(2\sigma^2)^{2a/(a+1)+(4-\lambda)/3(a+1)} \frac{\Gamma[(3a+1)/(a+1)+\frac{1}{3}(4-\lambda)/(a+1)]}{\Gamma[(5a+2)/(a+1)]},$$

as the solution for the shock-deposited energy.

The final solution is then the computer solution plus equation (17). We then obtain the luminosity from the relation,

$$L = -4\pi r^2 \frac{c}{3K\rho} \frac{\partial u}{\partial r}, \quad (18)$$

while the surface is defined by

$$F_s = \left(\frac{4a+2}{3a\rho_0 V_2 K}\right)^{1/(2a+1)} t^{2/(2a+1)} \quad (19)$$

which is obtained by evaluating the integral,

$$\int_r^\infty K\rho dr = 2/3 . \quad (20)$$

b) Results

Figures 3 and 4 show the energy density for 10 $M_\odot$ and 2 $M_\odot$ obtained from the analytic solution as a function of mass fraction and time. The mass fractions smaller than 10^{-4} correspond to relativistic velocities, so the solution is nonphysical for $F \leq 10^{-4}$. The energy density early in time is determined by the shock deposition. Later in time, as adiabatic expansion takes place, the β-decay adds a larger fraction of the residual energy density. The energy density is diffusively released from the mass fractions marked by the solid line $z = 0.6$ in Figures 3 and 4. This region of diffusive release occurs inside the surface (*dashed line*) when the surface velocity is less than c, the limiting velocity for a diffusion wave. The peak luminosity occurs when the diffusive-release condition approaches the center of the star, $F = 1$, at 20 and 7 days, respectively.

Figure 5 shows the luminosity, surface radius, surface velocity, and surface temperature as a function of time for both the 2 $M_\odot$ and 10 $M_\odot$ cases. Even the extreme and unrealistic case of 10 $M_\odot$ of β-radioactive material shows a peak luminosity one-third of that observed. If the expansion velocity were reduced, the luminosity would be further decreased. One is forced to conclude that another energy source exists whose decay is slower than the allowed β-spectrum. (Although a "statistical" initial distribution of neutron-rich nuclei with allowed β-spectra was used as the energy source, $S = S_0 t^{-1.4}$, at any given time t the energy production is predominantly determined by nuclei with decay constant t^{-1}; hence the generalization of excluding the allowed β-spectrum of neutron-rich neuclei.)

V. COMPUTATIONAL MODEL

The numerical diffusion hydrodynamics program written by Richard White is given in the Appendix. The test of the ability of the numerical program to reproduce the analytical solution is also shown in Figure 5, where the curve is analytic and the points are computed. The close agreement gives confidence in the ability of the computational program to calculate the more general problem.

a) Velocity Distribution

The velocity distribution of a 1.5 $M_\odot$ supernova with a presumed 50 percent ejected mass (0.75 $M_\odot$) was chosen as typical of a Type I supernova. As discussed following equations (1) and (3), the inner 0.42 mass fraction expands uniformly, and the outer 0.58 mass fraction follows the law, $V = V_2 F^{-1/4}$. Figure 2 shows the distribution used compared with the original explosion calculation of Colgate and White. The reduction

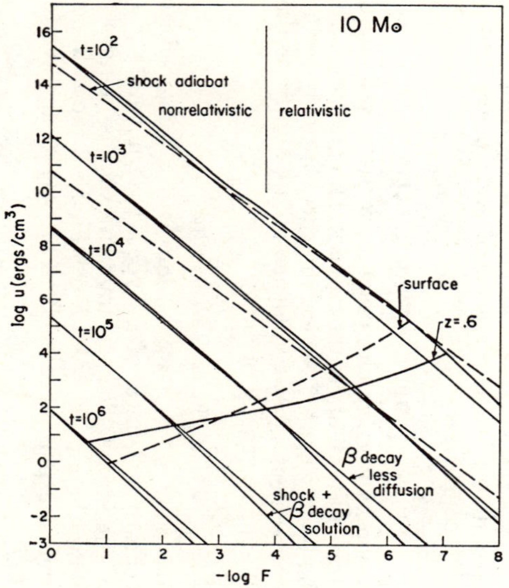

FIG. 3

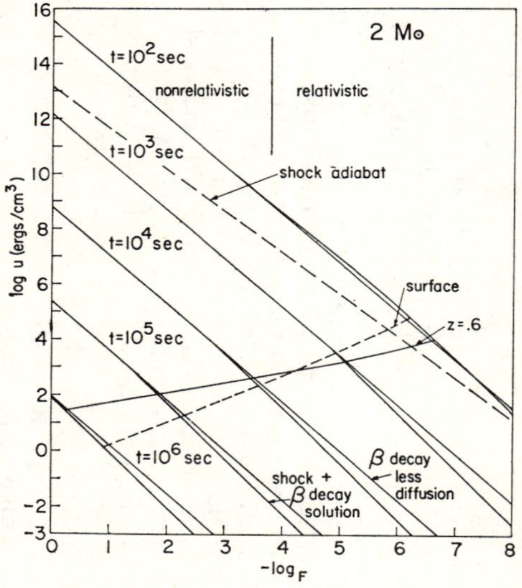

FIG. 4

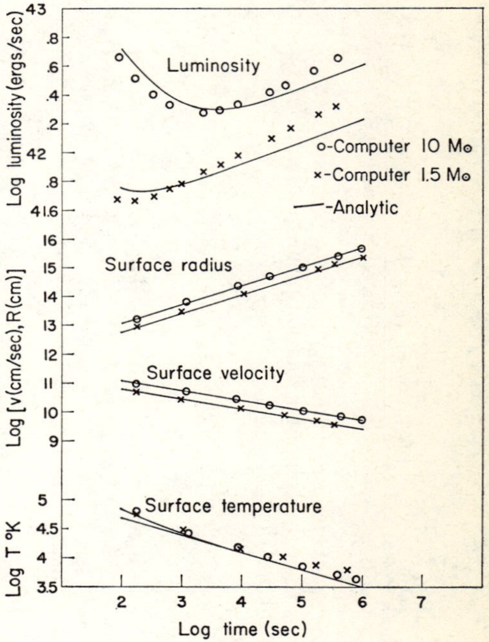

FIG. 5

FIG. 3.—Energy density of supernova ejecta 10 $M\odot$ as a function of mass fraction and time for the analytic solution.

FIG. 4.—Energy density of supernova ejecta 2 $M\odot$ as a function of mass fraction and time for the analytic solution.

FIG. 5.—Comparison of the analytic solution of 2 $M\odot$ and 10 $M\odot$ supernova luminosity versus time with the computational program.

in velocities by a factor of 2 for each distribution was done empirically in an attempt to correlate the resulting light curve with that observed. The same velocity distributions were used for the 10 $M_\odot$ calculations. The quantity $V_1 = 3$, 1.5, and 0.75×10^9 cm sec^{-1}, and $V_2 = 2$, 1, and 0.5×10^9 cm sec^{-1}.

b) Energy Source from the β-Decay $^{56}Ni \rightarrow {}^{56}Co$

The results of nucleosynthesis calculations of silicon burning by Bodansky, Clayton, and Fowler (1968*a*,*b*) simplified and clarified the earlier papers of Fowler and Hoyle (1964), Truran, Cameron, and Gilbert (1966), and Truran, Arnett, and Cameron (1967) on the abundances to be expected, and it now appears likely that silicon-burning shells in supernovae are responsible for most of the nucleosynthesis in the range of atomic weight $28 \leq A \leq 57$. The most striking feature of this work is the prediction of the element abundances that is in good agreement with the observed abundances. In particular, the mass peak of ^{56}Fe stands out a factor of 10 above the surrounding isotopes, each of which is reproduced from the calculations with surprisingly good accuracy. The unique feature of the ^{56}Fe peak for the present work is that its origin depends on the formation of the most tightly bound α-particle nucleus ^{56}Ni, which subsequently decays to ^{56}Co and then to ^{56}Fe according to the decay scheme shown in Figure 1. Although adjacent nuclei are also formed that are radioactive, their observed cosmic abundance ($\leq$10 percent ^{56}Fe) would indicate that they contribute no more than 10–20 percent to the total β-decay energy of the ^{56}Ni chain. Although we will neglect these other radioactive nuclei in this treatment of the light emission by supernovae, they are by no means unimportant for the possible detection of supernova remnants by the γ-rays resulting from such decays. Indeed, Clayton, Colgate, and Fishman (1969) predict that supernovae out to several megaparsecs should be observable by balloon-borne detectors and that the time behavior of the rising portion of these curves should give a unique determination of the velocity-distribution constants, V_1 and V_2, in equations (1) and (3).

The results of silicon-burning calculations that give the best fit to observed relative abundances predict roughly one-third ^{56}Ni formation and two-thirds ^{28}Si, other nuclei making up a small fraction of the total. Considerations of nucleosynthesis conditions occurring in the strong-shock limit for $V_1 = 3 \times 10^9$ cm sec^{-1} and $V_2 = 2 \times 10^9$ cm sec^{-1} would indicate (from calculations of Truran [1967]) almost 100 percent ^{56}Ni synthesis, in which case the observed abundance of silicon would have to emerge from other less violent events. As a first estimate, the ejected matter was chosen to be one-third ^{56}Ni and two-thirds ^{28}Si. In the case of the 1.5 $M_\odot$ supernova with 0.75 $M_\odot$ ejected, this corresponds to 0.25 $M_\odot$ of ^{56}Ni, which is almost twice that chosen by Clayton, Colgate, and Fishman (0.14 $M_\odot$) as being consistent with galactic abundances and supernova frequency. The resulting light curves, on the other hand, show a bolometric luminosity about equal to that observed, so the first calculations of γ-ray detectability have been performed with what is most likely one-half the mass of ^{56}Ni required.

The decay scheme in Figure 1 indicates that only the 1.72-MeV γ-ray of ^{56}Ni will be deposited as heat, the remaining 0.5 MeV escaping as the electron-capture neutrino. During the "thick" phase of the expansion ($\rho r \gg K_\gamma^{-1}$, $t \lesssim 10$ days) the γ-rays as well as the positrons from ^{56}Co will be deposited, giving a deposited energy of 3.59 MeV per decay. The remainder will be lost as neutrino emission.

If the decay of ^{56}Ni is given by $N_1 = N_0\, e^{-\lambda_1 t}$, then the number of ^{56}Co, N_2, with decay constant λ_2, is

$$N_2 = \frac{N_0 \lambda_1}{\lambda_2 - \lambda_1} \left(e^{-\lambda_1 t} - e^{-\lambda_2 t}\right) \tag{21}$$

and the energy rate from ^{56}Ni and ^{56}Co becomes

$$S_1 = N_0\lambda_1\epsilon_1 e^{-\lambda_1 t}\,;$$
$$S_2 = \lambda_2 N_2 \epsilon_2 e^{-\lambda_2 t}, \tag{22}$$

whose sum for the decay constants of ^{56}Ni and ^{56}Co, if one-third ^{56}Ni per gram is chosen, is

$$S_0 = 1.15 \times 10^{10} e^{-\lambda_1 t} + 0.252 \times 10^{10} e^{-\lambda_2 t} \text{ ergs g}^{-1}\text{ sec}^{-1}\,. \tag{23}$$

The case of 0.75 $M_\odot$ ejected gives a peak energy rate of 2×10^{43} ergs sec^{-1}, which is twice the desired luminosity.

VI. OPACITY

The expected temperature of the matter during the time of interest, 10^3–10^6 sec, in the region of diffusive release ($z \simeq 0.6$), can be estimated from the analytic solution to be $5 \times 10^{3°} \leq T \leq 5 \times 10^{4°}$ K, and densities can be estimated to be $10^{-11} \leq \rho \leq 10^{-15}$ g cm^{-3}. This is just the region of temperature where line opacities may become important. For this reason, a computer run was requested of the program for opacity calculations of Cox and Stewart (1965) for the elements Fe, Cu, He, and H in the above density and temperature range. These are displayed in Figures 6–9 and demonstrate the unique properties of iron. The number of lines in the iron spectrum is so large (500–1000) that, even at the low densities in question the lines contribute significantly ($\times 5$) to the total opacity at $T \simeq 10^{4°}$ K. Since this is approximately the surface temperature, the question arises whether a mixture of all other elements might have the same effect of line "blocking" and give rise to an opacity significantly larger than their individual sum. The final result indicates that the heat flow or total luminosity is determined by the diffusive release of the internal energy from the "bulk" of the mass at considerably higher temperature ($\times 5$) and density ($\times 10$) than the surface. The surface condition, as in any star, adjusts to this total luminosity. However, since we did not know this beforehand, we had to test the significance of this possibility, and so two separate opacities were used. One was iron, increased by the ratio 56/37 which corresponds to a mean molecular weight of 37 (i.e., to one-third ^{56}Ni + two-thirds ^{28}Si), and the other was copper, again with a representative molecular weight of 37. Presumably a mean molecular weight of 37 is the best estimate of the result of silicon burning. The opacity of the inner region, the bulk of the mass, is determined at these low densities entirely by the number of free electrons and hence almost solely by the mean molecular weight. This, in turn, will determine the total luminosity, so the assumption of $\langle\mu\rangle = 37$ is an upper limit to the molecular weight and results in a *lower* limit to the number of free electrons and hence opacity. Consequently the calculated luminosities are upper limits, and any addition of light elements will increase the opacity and will broaden and lower the light curves.

The initial calculations comparing Fe and Cu opacities (Fig. 10) were performed by inadvertently using an erroneous heat-source term—the ^{56}Co → ^{56}Fe decay (S_2) multiplied by 10, which increases the total energy input late in time during maximum by a factor of 3.2. The calculated luminosity covers a range of 10^4, overlapping and large compared with this factor of 3.2, and over this entire range the two opacities give the same result. Thus, fortunately, the resulting light curves are not sensitive to the additional line opacity of iron, since this is purely a surface effect and does not determine the heat flow in the interior. The line opacity of iron did lead to a calculational instability and a large investment in computer time, so it was not repeated. The effect of the line opacity is to hold back the heat flow; however, the line opacity has the odd property of decreasing with increasing temperature (Fig. 6), so that the heat held back tends to "burn out" the lines, allowing the radiation to escape and cooling and closing the

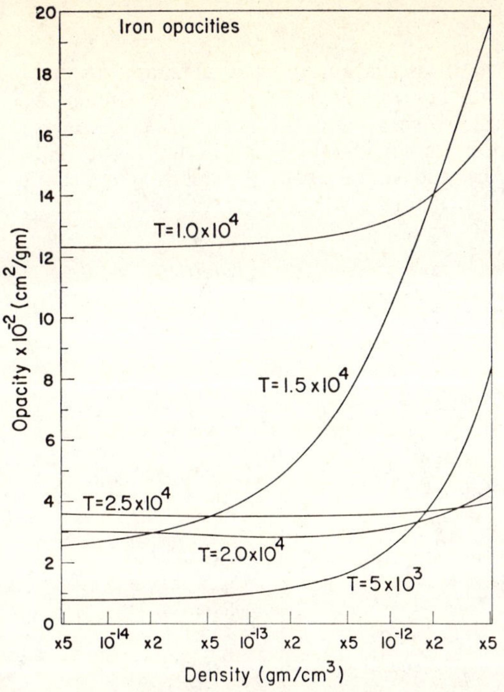

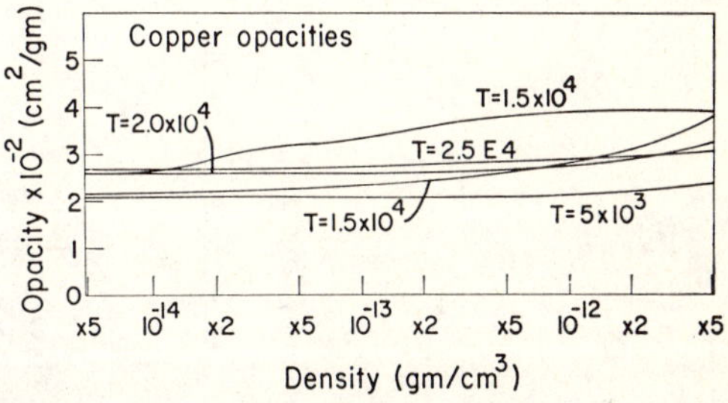

FIG. 6.—Opacity of Fe as a function of temperature and density (Cox and Stewart 1965).
FIG. 7.—Opacity of Cu as a function of temperature and density (Cox and Stewart 1965).

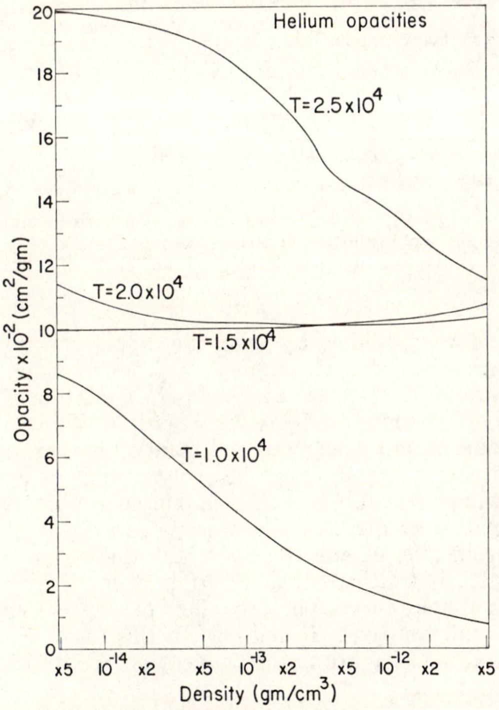

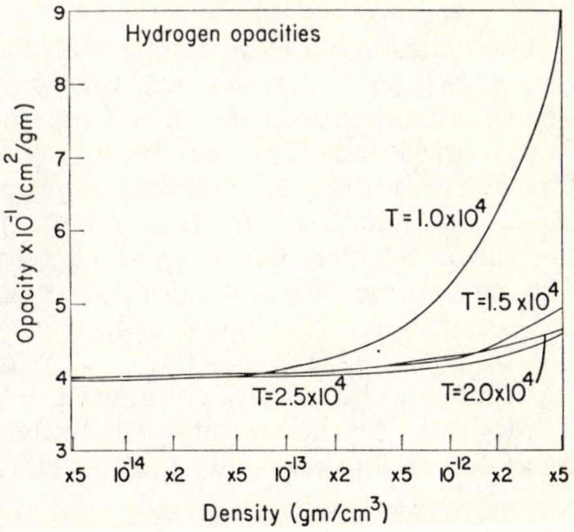

FIG. 8.—Opacity of He as a function of temperature and density (Cox and Stewart 1965).
FIG. 9.—Opacity of H as a function of temperature and density (Cox and Stewart 1965).

"shutter" behind. Numerical errors resulted in an oscillation and caused computational difficulty, but a small smoothing of the opacity law removed the instability—much as would be expected by the addition of light elements—and the result then becomes independent of the line contribution. In order for an instability to occur in the real case, a phase delay must be introduced as by hydrodynamics and an oscillation like that of a Cepheid variable will occur. The period becomes heat capacity/radiation = $3R/KcaT^3 \simeq 1$ sec.

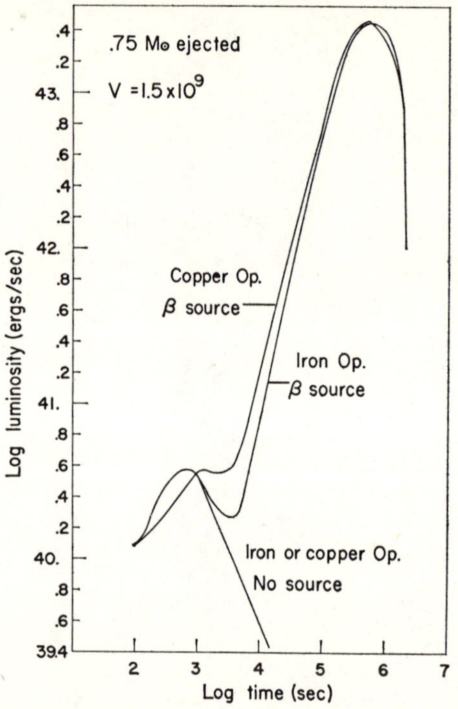

FIG. 10.—Log luminosity of 1.5 $M_\odot$ supernova (0.75 $M_\odot$ ejected) versus log time for both Fe and Cu opacity $V_1 = 1.5 \times 10^9$ cm sec^{-1}, $V_2 = 10^9$ cm sec^{-1}, and an erroneous source $S = S_1e^{-\lambda_1 t} + 10S_2e^{-\lambda_2 t}$ (see eq. [23] in text). The similarity between the two curves indicates that the "line" contribution to the opacity is effectively "burned" out by the heat flux.

The main body of the results was therefore calculated by using the copper opacity ($\langle\mu\rangle = 37$) for the region emitting β-decay in order to simplify the computations. The external mantle of the 10 $M_\odot$ supernova (7.25 $M_\odot$) was assumed to be a mixture of 75 percent H and 25 percent He by number. The inner β-decay region of 0.75 $M_\odot$ was the same as for the 1.5 $M_\odot$ case. The justification for an "external mantle" is that in the collapse model of supernovae only the inner 1–2 $M_\odot$ need be evolved to Si or Fe to result in collapse. The external matter will presumably be lighter in molecular weight. The assumption of the highest-opacity matter at these low densities is equivalent to the assumption of the lowest molecular weight, i.e., H and He. Even with this assumption of *highest*-opacity matter, the "mantle" is transparent at the time of luminosity maximum and hence does not affect the final light curve. The only effect of this matter is to partake in the velocity-distribution function (eqs. [1] and [3]). In this form, the equations of shock velocity and hence final velocity are not dependent on the initial distribution of the mantle, so it applies equally well to an initial model of a red giant or a polytrope of index 3.

The results of the numerical calculations are shown in Figures 11–15 for the conditions given in Table 1.

Figures 11 and 12, the visual luminosity of several typical Type I supernovae (NGC 4621 and NGC 4636 [Minkowski 1964]) are indicated. The long time tail of the bolometric luminosity is indicated by a dashed curve normalized to the diffusion calculations at time t_0 for each curve. The calculation of this tail depends on the condition of γ-ray absorption transparency discussed in the next section. The diffusion calculation becomes invalid at time t_0 because the emissivity of the matter external to the surface is not—and, by the choice of boundary condition usual in diffusion theory, cannot be—included in the total emission. As a consequence, when the β-decay mass fraction external to the surface becomes a major fraction of the total β-decay matter (this fraction $\simeq$

TABLE 1

Conditions for Calculations

Figure	Ejected Mass, $M\odot$	Opacity	t_0 (days)	$V_1 \times 10^{-9}$ (cm sec^{-1})	$V_2 \times 10^{-9}$ (cm sec^{-1})	Inner Ejected Matter		Outer Ejected Matter
						$M\odot$(^{56}Ni)	$M\odot$(^{28}Si)	$M\odot$(He+H)
11, 12, 13....	0.75	Cu	5.8	3	2	0.25	0.5	0
11, 12, 13....	0.75	Cu	11	1.5	1	.25	.5	0
11, 12, 13....	0.75	Cu	22	0.75	0.5	.25	.5	0
14, 15........	8	Cu, H, He	22	3	2	.25	.5	7.25
14, 15........	8	Cu, H, He	44	1.5	1	.25	.5	7.25
14, 15........	8	Cu, H, He	88	0.75	0.5	0.25	0.5	7.25

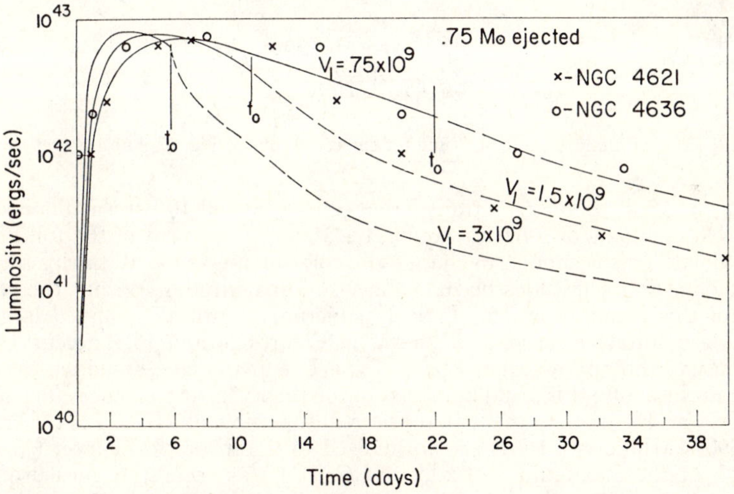

Fig. 11.—Log of luminosity versus time for the three velocity distributions of 0.75 $M\odot$ ejected from an initial 1.5 $M\odot$ star, $V_1 = 0.75, 1.5,$ and 3×10^9 cm sec^{-1}; $V_2 = 0.5, 1,$ and 2×10^9 cm sec^{-1}. The 0.75 $M\odot$ ejected mass was assumed to be 0.25 $M\odot$ ^{56}Ni and 0.5 $M\odot$ of ^{28}Si with an opacity corresponding to copper. Two observed Type II supernovae are included (cross, NGC 4621; open circle, NGC 4636); t_0 and the dashed curves are derived from the condition of γ-ray transparency.

$\frac{1}{2}$ at t_0), then the diffusion calculation loses validity and the transparency calculation of the following section becomes a better approximation. The bolometric correction to the theoretical curves is highly uncertain late in time because of near transparency. However, near maximum, where the temperature (Figs. 13 and 15) is about 6000°–7000°, the bolometric correction should be small. On the other hand, the observed colors of supernovae at this stage, as well as later, indicate a very much higher temperature of 10000°–30000° (Minkowski 1964; Zwicky 1965; Arp 1961). If the temperature were as high as 30000°, the total bolometric luminosity would have to be 30–60 times greater than calculated, and the β-decay energy source would be entirely inadequate. Instead, we propose that the temperatures are indeed as low as calculated and that the color is shifted toward the blue by the "fluorescence" of the very much larger projected area of the low-density, transparent, high-velocity material (see eq. [1]) external to the sur-

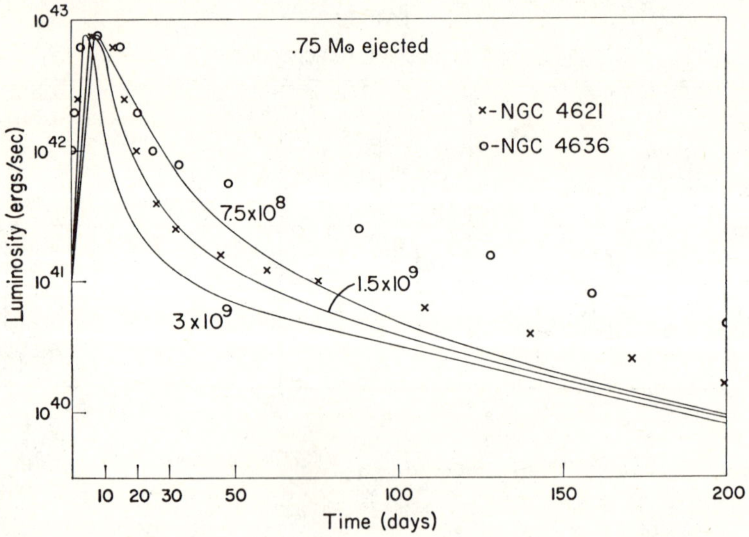

FIG. 12.—Log of luminosity versus time for the conditions of Fig. 12, but a longer time

face. After $t = t_0$, half the mass is transparent, and a typical 10 percent fluorescence of the γ recoil electron and/or β-decay would give almost 10 percent of the total emission in the blue-violet. This is enough to change the color temperature. With this interpretation, the bolometric magnitudes become the visual magnitudes. Within the relatively large error of this assumption, the Type II supernova, with an assumed larger mass 10 $M_\odot$, makes qualitative sense, as is shown in Figures 14 and 15. The curves do not exhibit the usual "hump" associated with Type II supernovae—a sudden decrease at 20–40 days. Presumably, this could be added to the theory as fluorescence efficiency after transparency. In Figure 15, the two observed supernovae, NGC 7331 (Arp 1961) and NGC 5907 (Minkowski 1964) are normalized to the theoretical curve, $V_1 = 1.5 \times 10^9$ cm sec^{-1}, at the maximum. Actually, NGC 7331 was accurately measured photometrically, and the best estimate of the observed bolometric peak was at 1.2×10^{43} ergs sec^{-1}. The bolometric correction assumed for the observed value was small (0.5 mag), so the theoretical curves would require an increase by a factor of 1.4, or to 0.35 $M_\odot$ of ^{56}Ni, to give agreement. The difference is within the error of the many assumptions involved in the model and calculations.

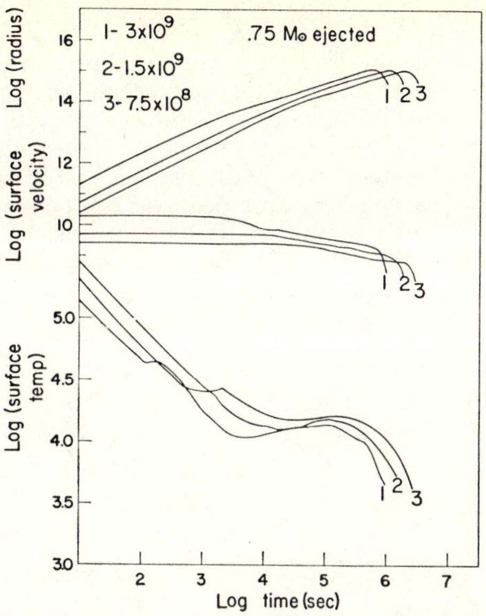

FIG. 13.—Surface temperature and surface velocity and radius versus time for the conditions of Fig. 12.

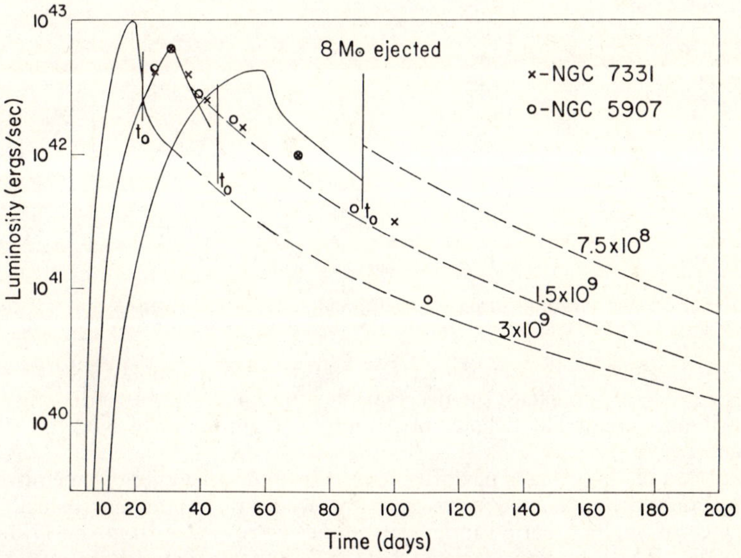

FIG. 14.—Log luminosity versus time for the three velocity distributions of 8 $M\odot$ ejected from an initial 10 $M\odot$ star. $V_1 = 0.75, 1.5$, and 3×10^9 cm sec^{-1}; $V_2 = 0.5, 1$, and 2×10^9 cm sec^{-1}. The inner 0.75 $M\odot$ ejected mass was assumed to be 0.25 $M\odot$ ^{56}Ni and 0.5 $M\odot$ ^{28}Si with an opacity corresponding to that of copper. The outer ejected 7.25 $M\odot$ was assumed to be 75 percent H and 25 percent He by number with the corresponding opacity. Two observed Type I supernovae are included (*cross*, NGC 7331; *open circle* NGC 5907); t_0 and the dashed curves are derived from the condition of γ-ray transparency.

VII. LONGER-TIME LUMINOSITY

Finzi (1965) has advanced the idea that the long-time light curve depends on the decay of the residual vibrational energy of the neutron star. This excludes the possibility of thermonuclear supernovae. Burbidge *et al.* (1957) suggest the energy source ^{254}Cf, but the variability of the observed decay constant and the large mass of transuranic elements required casts some doubt on this mechanism. Morrison and Sartori (1966) have advanced the idea that the long-time decay of the optical luminosity is due to the fluorescence of helium in the interstellar medium excited by photons with energies of approximately 51 eV in the far-ultraviolet. The difficulty lies in the very large bolometric luminosity required to give the flux of ultraviolet photons. The energy

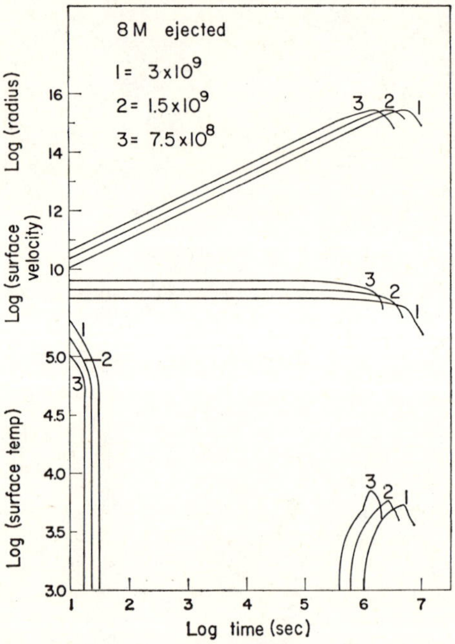

FIG. 15—Surface temperature, surface velocity, and surface radius versus time for the conditions of Fig. 14.

emitted in the slow-decay component (50–100 days) is variously 5×10^{48}–10^{49} ergs. The energy necessary to excite the necessary helium flourescence, under the assumption of the optimum interstellar density of helium, is a minimum of 5×10^{50}–10^{51} ergs of radiant energy.

At no period of the process of diffusive release could this energy be emitted without vastly changing the presumed physics of the whole phenomenon. Instead, we would rather associate the long-period luminosity decay as resulting from the 77-day β-decay of $^{56}Co \rightarrow {}^{56}Fe$. The changing efficiency for the self-absorption of the γ-ray fraction and the reduced efficiency for optical emission by line excitation and bremsstrahlung both conspire to result in a light curve that decays more rapidly that the basic 77-day period.

The mean γ-ray absorption coefficient for the ^{56}Co spectrum in the Si-Co-Fe mixture. is 0.04 cm^2 g^{-1}, so that 50 percent of the γ-ray energy escapes at a time when

$$\int_0^\infty \rho dr = 1/K \simeq 25 \text{ g cm}^{-2} . \qquad (24)$$

At $T \simeq 10000°$, the optical opacity for copper (Ni + Si) is about the same as the γ-ray opacity, so that the optical transparency time t_0 is the same as the γ-ray transparency time. The subsequent fractional γ-ray deposition will decrease approximately as the thickness of the stellar matter $\langle \rho r \rangle \sim r^{-2} \sim (t/t_0)^{-2}$. This deposited energy first appears as recoil-electron kinetic energy, which, in turn, ionizes and excites atomic transitions. The effective recoil-electron, range, including multiple scattering in the Fe-Si mixture for a mean recoil energy of ~0.3–0.5 MeV, is approximately $\langle \lambda_{recoil} \rangle \sim 0.05$ g cm^{-2}. This corrseponds to 500 times the γ-ray opacity, so the transparency condition for the recoil electrons is reached only at a time greater than 22 t_0, a time considerably longer than most observations. With the presence of many elements besides iron, the number of emission lines is so great that a major fraction of the excitation energy should appear in the optical spectrum. The lines will be broadened, owing to both Doppler spread and the very large self-absorption. The optical luminosity for $t_0 \leq t \leq 20\, t_0$ should then approximately follow the deposition of radioactive energy. In addition to the γ-ray emission from electron capture, $^{56}Co \rightarrow {}^{56}Fe$, there exists 20 percent β^+-emission (Fig. 1), where the positron has a mean kinetic energy of 0.75 MeV. The fractional energy, $f = (0.2 \times 0.75)/3.59 = 0.04$, deposited by these positrons (exclusive of the annihilation quanta) will not be subjected to the transparency attenuation as are the γ-rays. As a consequence, the late-time luminosity should be:

$$L = S_2 f + \left(\frac{t}{t_0}\right)^{-2} [S_1 + (1 - f)S_2]$$
$$= 0.015 \times 10^{43} e^{-\lambda_2 t} + \left(\frac{t}{t_0}\right)^{-2} (1.72 \times 10^{43} e^{-\lambda_1 t} + 0.362 \times 10^{43} e^{-\lambda_2 t}) \text{ ergs sec}^{-1}, \quad (25)$$

for 0.75 $M_\odot$ ejected of one-third ^{56}Ni and two-thirds ^{28}Si.

In the case of the 1.5 $M_\odot$ curves (Figs. 11 and 12), the luminosity is matched to the diffusion calculation at time t_0 by reducing the transparency luminosity by a factor of 0.7. This is a partial correction for the fraction of the β-decay matter external to the surface. For the 10 $M_\odot$ curves (Fig. 14) no correction was made owing to the uncertainty in the model of the ^{56}Ni core and H-He shell.

VIII. DISCUSSION

The calculated light curves that best fit the observations are those for $V_1 \simeq 10^9$ cm sec^{-1}. The average kinetic energy is $\frac{1}{2} \langle V^2 \rangle = 0.41\, V^2_1$ for the velocity distributions used, so that the best fit for the case of the small supernovae is therefore 4.1×10^{17} ergs g^{-1}. On the other hand, the kinetic energy resulting from the complete thermonuclear burning of 1.42 $M_\odot$ of ^{12}C (Hansen and Wheeler 1968) is 2.8×10^{17} ergs g^{-1}. One-half the fusion energy released is balanced by the initial net binding energy of the presupernova star. Other models may have substantially less binding energy. Finally, the theoretical kinetic energy should be increased because of the larger ejected mass (1.42 $M_\odot$) compared with that of the calculated model, 0.75 $M_\odot$. From equation (6), if t is kept constant (constant width of the light curve), then $V_1 = 2 \times 10^9$ cm sec^{-1} if the mass ejected is 1.5 $M_\odot$. Then $\frac{1}{2} \langle V^2 \rangle = 1.6 \times 10^{18}$ ergs g^{-1}. This is 6 times larger than that predicted from thermonuclear energy alone for the Hansen model and 3 times larger than the absolute upper limit of thermonuclear energy. Roughly the same discrepancy in energy applies to the 10 $M_\odot$ supernova because of the large inert envelope. The observed velocity of supernova envelopes ranges from 10^8 cm sec^{-1} for the older remnants with ages of 10^4 years (Poveda and Woltjer 1968), to the maximum velocity of 2×10^9 cm sec^{-1} observed for the most recent supernova in our Galaxy, Tycho's star, reported by Minkowski (1968).

At a late stage of expansion the light emission from the transparent remnant gases depends upon the second or higher power of the electron density, so it necessarily strongly emphasizes the mass fractions of the slower-moving interior. The range of observed velocities is of the order of what would be expected, and the maximum observed velocity of 2×10^9 cm sec^{-1} qualitatively agrees with the mass average velocity $\langle V \rangle = 0.82 V_1 = 1.6 \times 10^9$ cm sec^{-1} for the ejected mass of 1.5 $M_\odot$.

Although no absolute conclusions can be made, the qualitative agreement between observed and predicted light curves substantiates the concept of the radioactive source of the luminosity and at least is consistent with both the thermonuclear and neutron-star origins of supernovae. To the best of the authors' present knowledge, no cobalt has been recognized in supernova spectra.

We are deeply indebted to Richard White for first writing the diffusion hydrodynamic program. The suggestion of James Truran concerning the importance of ^{56}Ni was the key to the problem. The opacity calculations of Arthur Cox and John Stewart of the Los Alamos Scientific Laboratory added greatly to the significance of the results. Professor Rudolph Minkowski has supplied many of the experimental curves, and Henry Lazarus has aided in the computations. This work was supported by the Astronomy Section, National Science Foundation, NSF grant GP-6810. W. D. Arnett has greatly contributed to the readability.

APPENDIX

The difference equations for the hydrodynamic part of the program have been given by Colgate and White (1966); therefore, we give only the difference equations for the section concerning diffusion. White's method is similar to that of Cox, Brownlee, and Eilers (1966) in that a system of n linearly coupled equations is used to solve for the temperatures at $n + 1$; however, differences exist in the methods of evaluating the radiation boundary condition, the diffusion coefficient, and the method of coupling the hydrodynamics to the radiation flow.

The equation to be solved is

$$\rho \epsilon_T \frac{\partial T}{\partial t} - \frac{1}{R^2} \frac{\partial}{\partial R} \left(R^2 D \frac{\partial T}{\partial R} \right) = 0 , \tag{A1}$$

where T is the temperature, ϵ is the internal energy per unit mass, $\epsilon_T = (\partial \epsilon / \partial T)_\rho$, and ρ is the density.

Using the notation of Colgate and White, we difference equation (A1) in the following manner:

$$\begin{aligned}(\rho \epsilon_T R^2 \Delta R)_{j-0.5}^{n+0.5} \frac{T_{j-0.5}^{n+1} - T_{j-0.5}^{n}}{\Delta t^{n+0.5}} - \Bigg[\left(\frac{R^2 D}{\Delta R} \right)_j^{n+0.5} (T_{j+0.5}^{n+0.5} - T_{j-0.5}^{n+0.5}) \\ - \left(\frac{R^2 D}{\Delta R} \right)_{j-1.0}^{n+0.5} (T_{j-0.5}^{n+0.5} - T_{j-1.5}^{n+0.5}) \Bigg] = 0 ,\end{aligned} \tag{A2}$$

where the superscripts denote time and the subscripts denote zone numbers. We define

$$a_j^{n+0.5} = 0.5 \left(\frac{R^2 D}{\Delta R} \right)_j^{n+0.5} \Delta t^{n+0.5} \tag{A3}$$

and

$$b_{j-0.5}^{n+0.5} = (\rho \epsilon_T R^2 \Delta R)_{j-0.5}^{n+0.5} . \tag{A4}$$

When equations (A3) and (A4) are substituted into equation (A2), we obtain

$$\begin{aligned}b_{j-0.5}^{n+0.5} (T_{j-0.5}^{n+1.0} - T_{j-0.5}^{n}) - 2a_j^{n+0.5} (T_{j+0.5}^{n+0.5} - T_{j-0.5}^{n+0.5}) \\ + 2a_{j-1}^{n+0.5} (T_{j-0.5}^{n+0.5} - T_{j-1.5}^{n+0.5}) = 0 .\end{aligned} \tag{A5}$$

Letting

$$T_{j-0.5}^{n+0.5} = 0.5(T_{j-0.5}^{n+1} - T_{j-0.5}^{n})$$

and substituting into equation (A5) yield

$$-a_{j-1.0}^{n+0.5}T_{j-1.5}^{n+1.0} + (b_{j-0.5}^{n+0.5} + a_{j-1.0}^{n+0.5} + a_{j}^{n+0.5})T_{j-0.5}^{n+1.0} - a_{j}^{n+0.5}T_{j+0.5}^{n+1.0} = C_{j-0.5}\,; \tag{A6}$$

$$-a_{j-1.0}^{n+0.5}(T_{j-0.5}^{n} - T_{j-1.5}^{n}) + a_{j}^{n+0.5}(T_{j+0.5}^{n} - T_{j-0.5}^{n}) + b_{j-0.5}^{n+0.5}T_{j-0.5}^{n} = C_{j-0.5}\,. \tag{A7}$$

Equation (A6) is a set of linear simultaneous equations linear in T^{n+1} and can be solved by Gaussian elimination, if one uses the following relations in equations (A3) and (A4)

$$(R_{j}^{n+0.5})^{2} = \tfrac{1}{3}[(R_{j}^{n})^{2} + R_{j}^{n}R_{j}^{n+1} + (R_{j}^{n+1})^{2}]\,; \tag{A8}$$

$$(R^{2}\Delta R)_{j-0.5}^{n+0.5} = \frac{\Delta m_{j-0.5}}{4\pi\rho_{j-0.5}^{n+0.5}}\,; \tag{A9}$$

$$\Delta R_{j}^{n+0.5} = 0.5(\Delta R_{j+0.5}^{n+0.5} + \Delta R_{j-0.5}^{n+0.5})\,; \tag{A10}$$

$$D_{j}^{n+0.5} = \left(\frac{4ac}{3}\frac{T^{3}}{K\rho}\right)_{j}^{n+0.5} = \frac{ac}{3}(T_{j-0.5}^{*n+0.5} + T_{j+0.5}^{*n+0.5})[(T_{j-0.5}^{*n+0.5})^{2} + (T_{j+0.5}^{*n+0.5})^{2}] \times \frac{\Delta R_{j+0.5}^{n+0.5} + \Delta R_{j-0.5}^{n+0.5}}{K_{j+}^{n+0.5}(\rho\Delta R)_{j+0.5}^{n+0.5} + K_{j-}^{n+0.5}(\rho\Delta R)_{j-0.5}^{n+0.5}}\,, \tag{A11}$$

where

$$T_{j+0.5}^{*n+0.5} = T_{j+0.5}^{n} + 1/2\,\frac{\Delta t^{n+0.5}}{\Delta t^{n-0.5}}(T_{j+0.5}^{n} - T_{j+0.5}^{n-1})\,, \tag{A12}$$

$$K_{j+}^{n+0.5} = K(\langle T\rangle_{j}^{n+0.5}, \rho_{j+0.5}^{n+0.5})\,, \qquad K_{j-}^{n+0.5} = K(\langle T\rangle_{j}^{n+0.5}, \rho_{j-0.5}^{n+0.5})\,, \tag{A13}$$

$$\langle T\rangle_{j}^{n+0.5} = 0.5(T_{j+0.5}^{*n+0.5} + T_{j-0.5}^{*n+0.5})\,, \tag{A14}$$

and

$$4(T_{j}^{n+0.5})^{3} = \left(\frac{T_{j+0.5}^{4} - T_{j-0.5}^{4}}{T_{j+0.5} - T_{j-0.5}}\right)^{n+0.5} \tag{A15}$$

We therefore average neither the opacity nor the mean free path but rather the optical depth since it is this quantity which determines the absorption.

BOUNDARY CONDITIONS

To solve equation (A1), we furnish

$$T_{J+0.5}^{n+1}$$

by imposing the Milne condition on the outer boundary. This approximation treats the star as a blackbody radiating into space from a surface two-thirds of a mean free path in from the actual surface with an effective temperature given by

$$T_{\text{effective}}^{4} = 2T_{\text{surface}}^{4}\,. \tag{A16}$$

If we consider the outermost zone whose boundaries are R_J and R_{J-1}, then the effective temperature can be determined by linear interpolation

$$\frac{T_{e}^{4} - T_{s}^{4}}{2\lambda/3} = \frac{T_{J-0.5}^{4} - T_{s}^{4}}{\tfrac{1}{2}\Delta R_{J-0.5}}\,, \tag{A17}$$

where λ is the mean free path. The use of equation (A16) in equation (A17) yields

$$T_e^4 = \frac{(8\lambda/3\Delta R_{J-0.5})T_{J-0.5}^4}{1 + 4\lambda/3\Delta R_{J-0.5}} . \tag{A18}$$

Therefore, the flux F which crosses the outermost zone boundary J is

$$F = \frac{ac}{4} T_e^4 = \frac{(2ac\lambda/3\Delta R_{J-0.5})T_{J-0.5}^4}{1 + 4\lambda/3\Delta R_{J-0.5}} , \tag{A19}$$

whereas the flux computed by the difference equation is

$$F = -D_J \left(\frac{T_{J+0.5} - T_{J-0.5}}{0.5\Delta R_{J-0.5}} \right) . \tag{A20}$$

Equating equations (A19) and (A20), we find

$$D_J^{n+0.5} = \frac{\frac{1}{3}ac(T^{*n+0.5}_{J-0.5})^3}{(K\rho\Delta R)_{J-0.5}^{n+0.5} + \frac{4}{3}} \Delta R_{J-0.5} , \tag{A21}$$

where we have set

$$T_{J+0.5}^{n+1} = 0 , \quad 1/\lambda = (K\rho)_{J-0.5}^{n+0.5} \tag{A22}$$

and

$$K_{J-0.5}^{n+0.5} = K(T^{*n+0.5}_{J-0.5}, \rho_{J-0.5}^{n+0.5}) . \tag{A23}$$

For the inner boundary, we chose

$$T_{+0.5}^{n+1} = 0 , \quad D_{+1.0}^{n+0.5} = 0 , \tag{A24}$$

which is the condition of zero flux.

COUPLING OF HYDRODYNAMICS TO DIFFUSION

The equations for advancing temperature of Colgate and White (1966) allows one to compute

$$(\delta T_{j-0.5}^{n+0.5})_{\text{Hydro}} = T_{j-0.5}^{n+1} - T_{j-0.5}^{n} \tag{A25}$$

due to hydrodynamics above. Using one-half this increment, we advance the temperature,

$$\langle T \rangle = T^n + 0.5\delta T_{\text{Hydro}} , \tag{A26}$$

to obtain a mean temperature increase due to hydrodynamics alone. Then, if O defines the operation of the diffusion equation upon this mean temperature, the new temperature for the time step Δt^{n+1} is

$$T^{n+1} = O(T^n + 0.5\delta T_{\text{Hydro}}) + 0.5\delta T_{\text{Hydro}} . \tag{A27}$$

This procedure allows second-order accuracy if the operator is applied only to the temperatures which occur explicitly in equation (A7), but the diffusion coefficient and ϵ_T are assigned their mean extrapolated values from n to $n + 1$.

TIME STEPS

These controls are the same as those of Colgate and White with the exception that the energy is inhibited from changing more that 0.4 percent per cycle in any given zone.

SURFACE TEMPERATURE, RADIUS, VELOCITY, AND LUMINOSITY

To evaluate these surface quantities, we evaluate

$$\int_r^\infty K\rho dr = \tfrac{2}{3} \tag{A28}$$

to the nearest zone boundary denoted by j^*.

We then have

$$\text{Surface radius} = R_{j^*}^{n+0.5}\,; \quad \text{Surface velocity} = U_{j^*}^{n+0.5}\,; \tag{A29}$$

$$\text{Surface temperature} = \left[-\frac{4D_{j^*}^{n+0.5}}{ac}\left(\frac{T_{j^*+0.5}^{n+0.5} - T_{j^*-0.5}^{n+0.5}}{\Delta R_{j^*}^{n+0.5}}\right)\right]^{0\ 25} \tag{A30}$$

and

$$\text{Luminosity} = 4\pi(R_{j^*}^{n+0\ 5})^2 \frac{ac}{4} \times (\text{surface temperature})^4\,. \tag{A31}$$

REFERENCES

Arnett, W. D. 1967, *Canadian J. Phys.*, **44**, 2553.
———. 1968*a*, *Nature*, **219**, 1344.
———. 1968*b*, *Ap. J.*, **153**, 341.
Arp, H. C. 1961, *Ap. J.*, **133**, 883.
Bodansky, D., Clayton, D. D., and Fowler, W. A. 1968*a*, *Phys. Rev. Letters*, **20**, 161.
———. 1968*b*, *Ap. J. Suppl.*, **16**, 299.
Burbidge, E. M., Burbidge, G. R., Fowler, W. A., and Hoyle, F. 1957, *Rev. Mod. Phys.*, **29**, 547.
Cameron, A. G. W., and Arnett, W. D. 1967, *A.J.*, **72**, 292.
Clayton, D. D., Colgate, S. A., and Fishman, G. J. 1969, *Ap. J.*, **155**, 75.
Colgate, S. A. 1967, *Ap. J.*, **150**, 163.
———. 1968, in *Conf. Seyfert Galaxies and Related Objects* (Tucson, Arizona, February 1968 (to be published).
Colgate, S. A., and Cameron, A. G. W. 1963, *Nature*, **200**, 870.
Colgate, S. A., and White, R. H. 1966, *Ap. J.*, **143**, 626.
Cox, A. N., Brownlee, R., and Eilers, D. 1966, *Ap. J.*, **144**, 1024.
Cox, A. N., and Stewart, J. N. 1965, *Ap. J.*, *Suppl.*, **11**, 22.
Erdelyi, A., Magnus, W., Oberhettinger, F., Tricomi, F. G. 1954, *Tables of Integral Transforms*, Vol. **1** (New York: McGraw-Hill Book Co.).
Finzi, A. 1965, *Phys. Rev. Letters*, **15**, 599.
Fowler, W. A., and Hoyle, F. 1964, *Ap. J. Suppl.*, **9**, 201 (No. 91).
Fraley, G. S. 1968, *Ap. and Space Sci.*, **2**, 96.
Hansen, C. J., and Wheeler, J. C. 1968, *Ap. and Space Sci.* (in press).
Hofmeister, E., Kippenhahn, R., and Weigert, A. 1964, *Zs. f. Astr.*, **60**, 57.
Jahnke, E., Emde, F., and Losch, F. 1960, *Tables of Higher Functions* (New York: McGraw-Hill Book Co.).
Lederer, C. M., Hollander, J. M., and Perlman, I. 1967, *Table of Isotopes* (6th ed.; New York: John Wiley & Sons).
Minkowski, R. L. 1964, *Ann. Rev. Astr. and Ap.*, **2**, 247.
———. 1968, in *Stars and Stellar Systems*, Vol. **7**, eds. B. M. Middlehurst and L. H. Aller, chap. 11 (Chicago: University of Chicago Press).
Morgan, A. J. A. 1952, *Quart. J. Math.* (Oxford), **2**, 250.
Morrison, P., and Sartori, L. 1966, *Phys. Rev. Letters*, **16**, 414.
Poveda, A. 1964, *Ann. d'ap.*, **27**, 522.
Poveda, A., and Woltjer, L. 1968, *A.J.*, **73**, 65.
Rakavy, G., and Shaviv, G. 1967, *Ap. J.*, **148**, 803.
Schatzman, E. 1965, in *Stars and Stellar Systems*, Vol. **8**, eds. L. H. Aller and D. B. McLaughlin (Chicago: University of Chicago Press), p. 327.
Schwartz, R. A. 1967, *Ann. Phys.*, **43**, 42.
Truran, J. W. 1967, paper presented at Supernova Conference GISS, New York, November 1967.
Truran, J. W., Arnett, W. D., and Cameron, A. G. W. 1967, *Canadian J. Phys.*, **45**, 2315.
Truran, J. W., Cameron, A. G. W., and Gilbert, A. 1966, *Canadian J. Phys.*, **44**, 563.
Zwicky, F. 1965, in *Stars and Stellar Systems*, Vol. **8**, eds. L. H. Aller and D. B. McLaughlin (Chicago: University of Chicago Press), p. 397.

11

NEUTRON STARS

A.G.W. Cameron

Yeshiva University and
Institute for Space Studies
New York, N.Y.

Interest in neutron stars has increased dramatically during the past two years as a result of observations of pulsars. Since a great deal of work is still in progress in this field, and since some aspects of the subject have been scarcely touched upon yet, I will not present a lot of mathematical detail, but will try to emphasize the underlying physical concepts. I will begin by discussing the equation of state suitable for a neutron star configuration and the composition we might expect. Then I will present the results of model computations and explore some of the properties that might be associated with neutron stars such as pulsations, rotations, magnetic fields, etc.

Equation of State and Composition

We can begin by considering a white dwarf in which the pressure is provided by a degenerate electron gas. As we consider progressively larger masses we must obtain larger central densities to provide the pressure required to support the overlying material. As the density increases the electrons become highly relativistic and the exponent Γ in the expression $\Gamma = \frac{d\ell nP}{d\ell n\rho}$ tends to the value 4/3. A star composed of such material will become unstable and will tend towards collapse. Such considerations lead us to an upper limit for the mass of a stable white dwarf of about 1.3 - $1.4M_\odot$. The exact value of the upper limit will depend slightly on the composition through the mean molecular weight per electron of the material. Recent work by Arnett has suggested that C^{12} will burn explosively in a flash if at all. If we accept this result then we would expect a white dwarf to be composed of largely C^{12} and O^{16} as a result of helium burning. The corresponding molecular weight per electron would be about 2, and the limiting mass about $1.44M_\odot$. A number of more detailed considerations will lower this value by a few percent.

When a star which has a mass exceeding the limit discussed above enters the condensed white dwarf configuration, collapse will necessarily begin, and will continue until some new source of pressure appears. Under these circumstances the electron Fermi level will become very large. Just how large it may become can be seen in Fig. 1 which has been constructed for the case $\mu_e = 2$. The maximum density obtainable in a white dwarf is about 10^{9-10}gm/cm^3. The corresponding electron Fermi level is several MeV. Under these conditions electron capture by heavier nuclei is possible, but not by C^{12} and O^{16}. When the density rises to about 10^{10}gm/cm^3, electron capture by C^{12} or O^{16} can begin, removing electrons from the core, lowering the pressure, and initiating collapse; however, it turns out that such a configuration would already be unstable due to general relativistic effects. When a star collapses to nuclear densities or near nuclear densities the nuclei will become neutron enriched and then disintegrate into free neutrons, and the predominant source of pressure will be a highly degenerate neutron gas.

To see this, suppose that collapse raises densities to 10^{11} or 10^{12} gm/cm^3 without removing the electrons. Then one can see from Fig. 1 that the Fermi level for the electrons is about 30MeV. Now consider a nucleus in its ground state, and an isobar at a somewhat higher mass. Imagine these nuclei immersed in a degenerate electron gas. As the Fermi level rises near the value of the energy difference between the isobars, beta decay from the higher to the lower will still be possible, but it will be less and less probable, because the electron phase space will be filled to such an extent that the decaying nucleus cannot emit its normal beta spectrum. If the Fermi level is higher than the energy difference between the isobars, then no beta decay at all will occur, since it will be energetically impossible to emit an electron into the unoccupied phase space. However, there will then be a supply of electrons with sufficient energy to force inverse beta decay, i.e., electron capture, and under the new conditions beta decay will be impossible. Thus a nucleus which is normally unstable can become stable under these circumstances.

If this is a nucleus of an odd mass number we will have a progression. Electron capture will continue step by step converting protons into neutrons as the electron Fermi level rises. When the Fermi level reaches about 25 - 30 MeV the series of electron captures will have arrived at nuclei in which the last neutron is no longer bound. Eventually the nuclei lose their identities completely, leaving a degenerate neutron gas.

For nuclei of even mass number these considerations take a slightly different twist. Because of the familiar odd-even effect the isobaric energy levels are grouped in pairs, with the two members of each pair closely spaced, but a fairly wide separation between pairs. Even mass number nuclei will be stable against both beta decay and electron capture. As a result, even mass number nuclei will not be available for neutrino energy dissipation through the URCA process (see

H.Y. Chiu, this volume). Of course, for a sufficiently high Fermi level the even mass number nuclei will also be converted to free neutrons.

Specifying an equation of state for a neutron star model is complicated by the fact that the density varies by about 15 orders of magnitude from the surface to the center. Our task is simplified by the assumption that throughout most of the interior we have "cold" material where kT is much less than the Fermi levels of the constituent particles. Under these conditions the pressure will come from two sources; the kinetic pressure of the degenerate particles, and the nuclear force between the free nucleons. In order to know the equation of state we must know the composition. One approach is to determine the equilibrium composition for each density by minimizing the total thermodynamic potential. At low densities the equilibrium composition is Fe^{56}. For higher density the equilibrium composition is dominated by more neutron-rich isobars, and for much higher density the composition is dominated by more massive nuclei. Eventually, as discussed above, nuclei will be produced which have unbound neutrons; however, the neutron Fermi level will then become high enough that some "unbound" neutrons will not be emitted, owing to lack of available phase space, but will remain associated with the nucleus. If we neglect nuclear forces among the nucleons which are not associated with particular nuclei, then we can imagine the neutron Fermi level gradually increasing until it is comparable to the kinetic energy of a nucleon in a nucleus. Then the distinction between the nuclear interiors and the degenerate nucleon gas disappears. Taking nuclear forces into account, the energy/particle in the interstitial medium between the nuclei is lower. Again the nuclei disappear rather abruptly, but at a considerably lower density than in the case mentioned above.

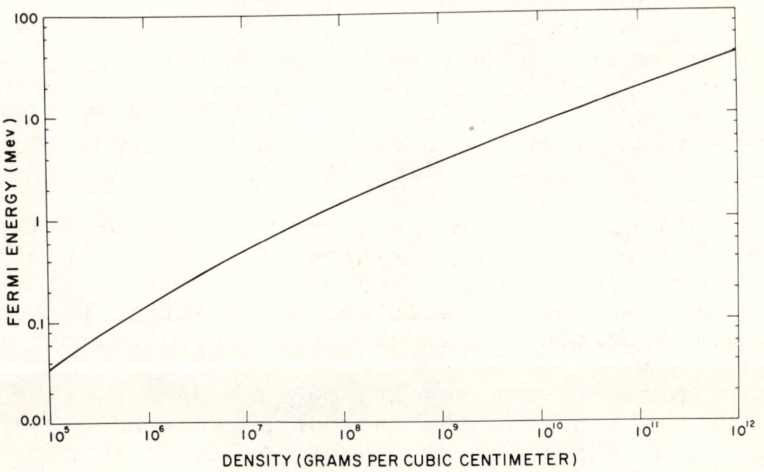

Fig. 1 Fermi energy of the electrons in a medium with a mean molecular weight per electron of two.

Carrying out the computation of composition and equation of state leads to the kind of results shown in Table 1.

Table 1. Composition of Cold Material

Density gm/cm^3	Number densities (10^{30} cm^{-3})					Γ
	neutrons	protons	nuclei	electrons	muons	
3.03×10^{11}	3.51×10^{1}	0.0	1.47×10^{3}	5.79×10^{4}	0.0	0.938
5.81×10^{11}	1.31×10^{5}	0.0	1.68×10^{3}	6.75×10^{4}	0.0	0.402
1.19×10^{12}	4.56×10^{5}	0.0	1.90×10^{3}	7.82×10^{4}	0.0	0.627
3.31×10^{12}	1.64×10^{6}	0.0	2.30×10^{3}	9.83×10^{4}	0.0	0.895
1.03×10^{13}	5.69×10^{6}	0.0	2.82×10^{3}	1.26×10^{5}	0.0	0.854
4.00×10^{13}	2.32×10^{7}	7.77×10^{-4}	3.67×10^{3}	1.77×10^{5}	0.0	2.05
5.12×10^{13}	2.97×10^{7}	5.26×10^{4}	3.11×10^{3}	2.11×10^{5}	0.0	2.50
1.01×10^{14}	5.93×10^{7}	9.81×10^{5}	0.0	9.81×10^{5}	0.0	2.92
2.19×10^{14}	1.25×10^{8}	5.38×10^{6}	0.0	5.37×10^{6}	1.16×10^{4}	2.95
5.03×10^{14}	2.55×10^{8}	3.88×10^{7}	0.0	2.54×10^{7}	1.34×10^{7}	2.89
1.00×10^{15}	4.14×10^{8}	1.33×10^{8}	0.0	7.57×10^{7}	5.74×10^{7}	2.87
5.35×10^{15}	9.78×10^{8}	6.77×10^{8}	0.0	3.54×10^{8}	3.23×10^{8}	2.83

The A and Z columns specify the most abundant nucleus at each density. The Γ is a general relativistic quantity, $\frac{P+U}{U}\frac{\partial U}{\partial P}$. In computing these values a particular assumption about the nuclear force was employed; in particular, the V_α potential of Levinger and Simmons.

At densities of 5×10^{13}gm/cm^3 the electron Fermi energy is about 35 MeV. At higher densities it will be energetically favorable to produce negative muons instead of raising the Fermi level much higher. A number of critical densities are:

neutron threshold	- 2.85×10^{11}gm/cm^3
proton threshold	- 4.0×10^{13}
nuclear breakup	- 6.0×10^{13}
muon threshold	- 2.2×10^{14}

For somewhat greater densities certain hyperons may be produced. The first such reaction will be $N + N \rightarrow P + \Sigma^-$. The models discussed below do not include hyperons; however, such computations are presently in progress and it is already clear that the inclusion of hyperons will not greatly effect the results.

From Table 1 it may be seen that Γ is substantially greater than 4/3 for the range of densities of interest in constructing neutron stars. The large values of Γ indicate a stiff equation of state, and as a result we can expect that the interior of a neutron star model will have relatively uniform density.

Fig. 2 presents the results of some early work for a non-interacting gas. We can see the concentrations of various particles, and in particular the manner in which muons and hyperons come into the picture. If one takes nuclear forces into account, muons and hyperons come in at lower densities than indicated in this figure.

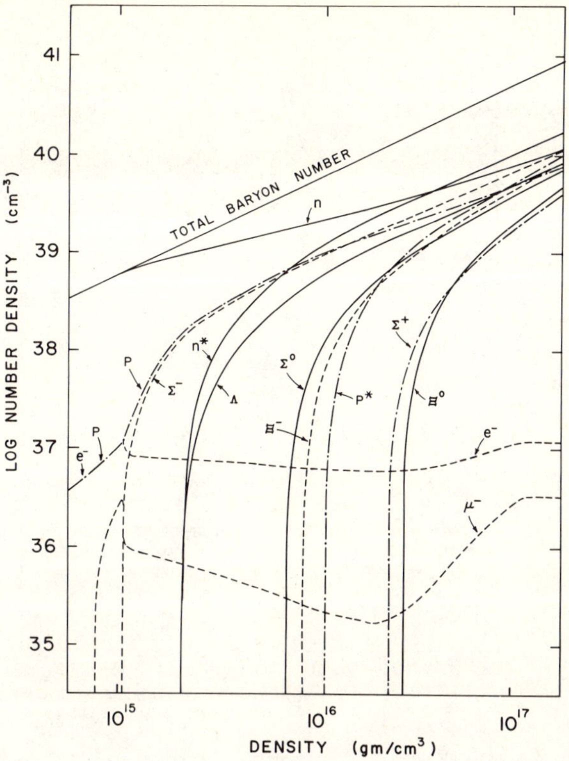

Fig. 2 Composition of dense degenerate matter neglecting nuclear interactions between baryons (calculated by S. Tsuruta).

Fig. 3 shows the Levinger-Simmons potentials V_β and V_γ used in early calculations by Sachiko Tsuruta. The three Levinger-Simmons potentials V_α, V_β, and V_γ have recently been used in computations of the properties of infinite and finite nuclear matter by R. Weiss, and he has found that the V_β potential is much too soft, whereas the other two are fairly reasonable. Nevertheless, models constructed with the pair V_β and V_γ are of special interest because they represent respectively very soft and fairly hard potentials.

Cold Neutron Star Models

The early work on neutron stars by Oppenheimer and Volkoff considered equilibrium configurations of stars consisting of a pure noninteracting neutron gas. The results of such computations, carried out with modern computers, are shown in Fig. 4. The diagram plots gravitational mass and proper mass against central density. Remember that gravitational mass is the mass sensed by a distant test particle, while proper mass is the mass of the individual particles dispersed to infinity. The first region of positive slope represents dynamically stable neutron star models, but the second region does not (see K. Thorne, Vol. 2).

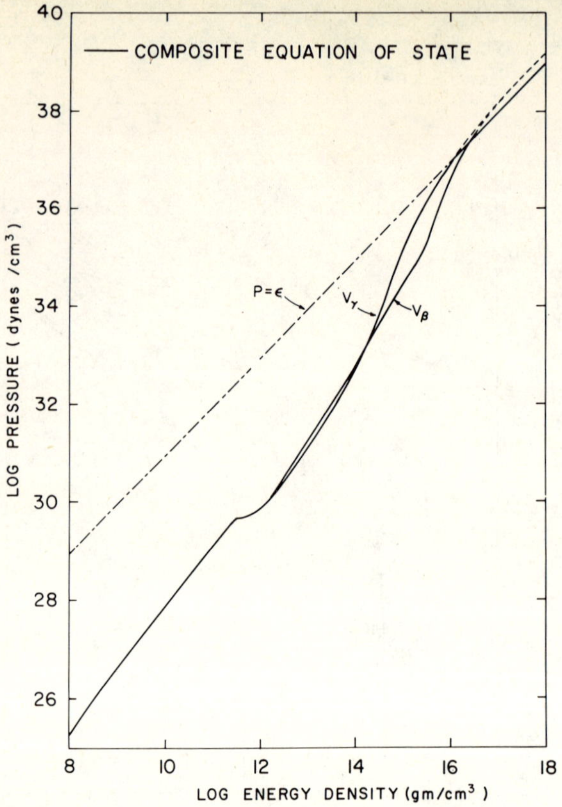

Fig. 3 The V_β and V_γ equations of state used in the neutron star model calculations of Tsuruta and Cameron.

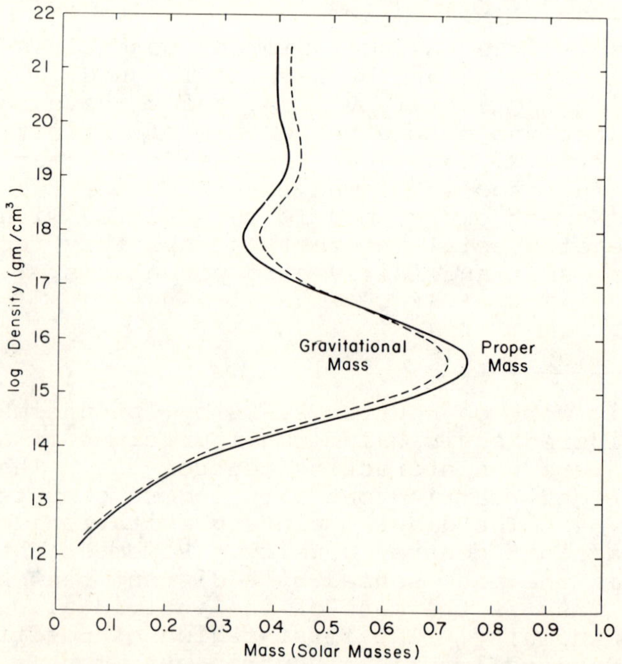

Fig. 4 The mass of a neutron star model, with noninteracting neutrons, as a function of central density.

The results of more realistic models are shown in Fig. 5. These models were computed by Tsuruta using the Levinger-Simmons potentials mentioned earlier. The choice of nuclear force clearly makes a quantitative difference, but the results are qualitatively the same. This is generally true for the many interactions considered by various investigators. Note that the curves become double for sufficiently high density. This is because a hard core nuclear potential predicts sound velocities exceeding the speed of light for high density. To avoid this, one can require that the pressure level not exceed 1/3 the proper energy density (for a non-interacting gas) or never exceed the proper energy density (limit for an interacting gas). Both alternatives are indicated in this figure, with no significant effect on the stable models.

Some properties of the models described above are shown in the figures below. Fig. 6 shows the stellar radius for various central densities. The stable models of interest have central densities on the order of 10^{15}gm/cm^3 and corresponding to radii of 12 or 13 km.

In Fig. 7 we see the plot of density against radius for several stable models. Although this data is for V_β, the soft potential, we still see the remarkably uniform densities in the central regions.

The results of recent computations with the V_α potential are shown in Fig. 8. Compared with some early work, the stable region goes to surprisingly high masses, about $2.5M_\odot$. A neutron star near the peak of the mass range would have a radius only about 40% larger than its gravitational radius. The small intermediate peak is probably real, but it definitely does not correspond to a region of dynamical stability, since both the fundamental and first overtone vibrational periods are imaginary in that region.

Pulsation periods for white dwarfs and neutron stars are shown in Fig. 9. The white dwarf curve here is for cold catalyzed matter (Fe^{56} or some neutron-enriched isobar), but using a more realistic composition would only lower the minimum period to perhaps 2 sec.

A Particular Model

The foregoing discussion has outlined the gross properties of the class of objects known as neutron stars. Now I would like to present in somewhat greater detail the properties of a particular neutron star model. The model we will consider is in the intermediate range of stable models, having a central density of $10^{14.5}$ gm/cm^3. In Table 2 the characteristics of this model are presented in a form useful for discussing the properties of the stellar matter.

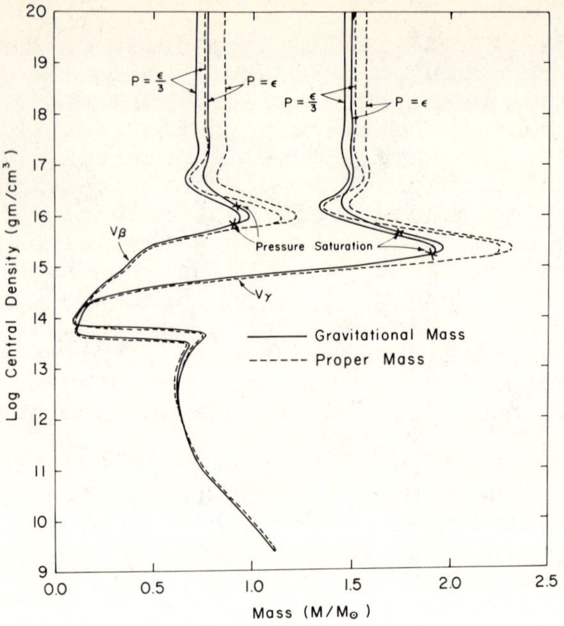

Fig. 5 Masses of neutron star models calculated by S. Tsuruta with V_β and V_γ potentials. The crosses mark the points at which the central pressure exceeds one-third and one times the proper energy density. Beyond the lower density cross two curves are generated, one with the pressure limited to one-third the proper energy density, and the other with no limitation. At the second cross the pressure is limited not to exceed the proper energy density.

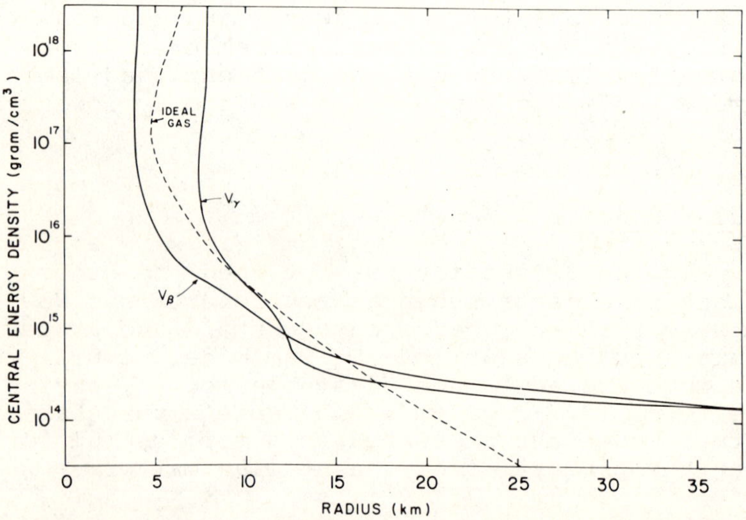

Fig. 6 Radii of neutron star models calculated with V_β and V_γ potentials and also for a noninteracting ("ideal") neutron gas.

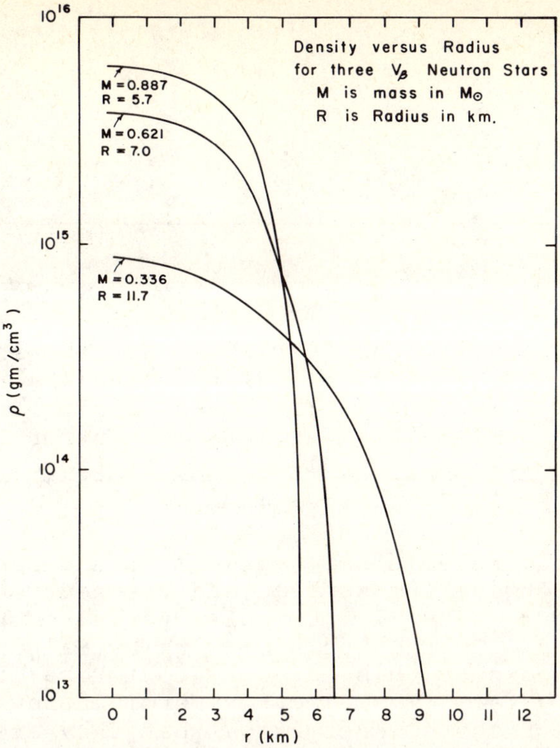

Fig. 7 Internal density distribution for three V_β neutron star models.

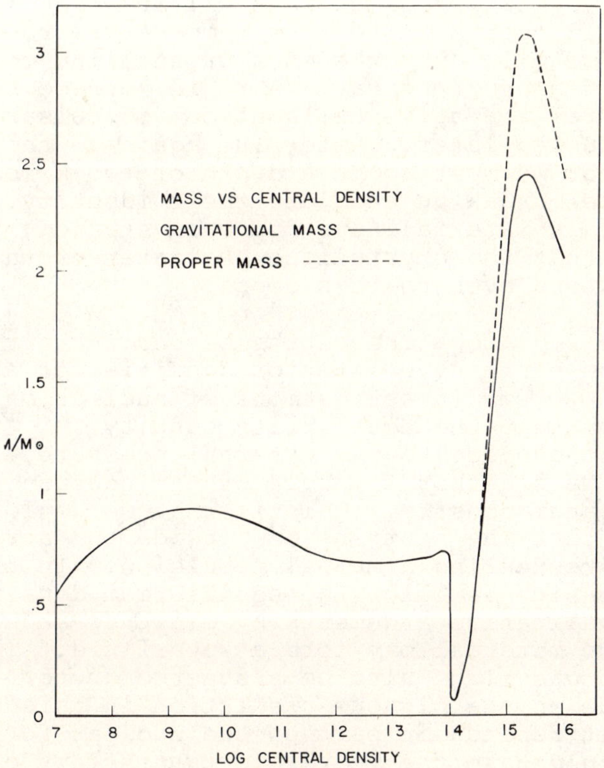

Fig. 8 Models of white dwarfs and neutron stars calculated with the more complicated equation of state due to Langer, Rosen, Cohen, and Cameron, but without inclusion of hyperons.

Table 2

ρ	M(r)	R	
$3.16\text{x}10^{14}$gm/cm^3	0	0	
$2.2\text{x}10^{14}$	$5.01\text{x}10^{32}$gm	7.72km	μ^- threshold
$6.0\text{x}10^{12}$	$1.15\text{x}10^{33}$	11.52	nuclear breakup
$4.0\text{x}10^{13}$	$1.16\text{x}10^{33}$	11.70	proton threshold
$2.8\text{x}10^{4}$	$1.17\text{x}10^{33}$	12.18	neutron threshold
$2.0\text{x}10^{10}$	$1.17\text{x}10^{33}$	13.68	surface

The above table reveals some very important aspects of neutron stars. Between the surface and the neutron threshold we see that there is about 1.5 km. of material composed of electrons and ions. On the basis of work by Salpeter and van Horn we would expect this material to form a Coulomb (Wigner-Seitz) crystal. The electron Fermi energy is so high, and the electrons so energetic, that they are really a smooth charge distribution. Then the repulsion between the ions will set up a lattice, since this will be a configuration of lower energy than a disordered array. The freezing temperature for a lattice of this kind is, for a white dwarf, about 10^8°K, whereas for a neutron star it will range up to 10^9°K in higher density regions. As will be seen later, a neutron star will cool down to this temperature range rather quickly, and so we will expect such a crystalline form for the surface layer of a neutron star. A rigid surface shell is assumed in several tentative explanations of pulsar behavior, as will be discussed later. Referring again to the table above we see that we must go to a depth of two kilometers before the nuclei break up and lose their identity. The properties of this material have not been studied, but it seems possible that the crystalline character of the surface layers might extend even to this depth.

What can we say about the depths beyond nuclear breakup? Stable nuclei on earth have interior densities of about $4\text{x}10^{14}$gm/cm^3; however, investigations of nuclear matter by Weiss have shown that the equilibrium density for a pure neutron gas, not subject to any external pressure, is about $2\text{x}10^{14}$gm/cm^3. We see that the bulk of the neutron star model is at greater density. This is what we would expect, since it means that the neutrons are inside the attractive potential and are feeling a mutual repulsion. In the region in which the density is below the equilibrium density of a neutron gas, qualitative arguments by Ginzburg and others suggest that the material may form a superfluid. That means that the proton gas might also be a superconductor. Theory does not provide an unambiguous prediction in this case, and in particular it depends in part on the nuclear force. Superconductivity would have the very important effect of excluding magnetic fields from some regions of the core.

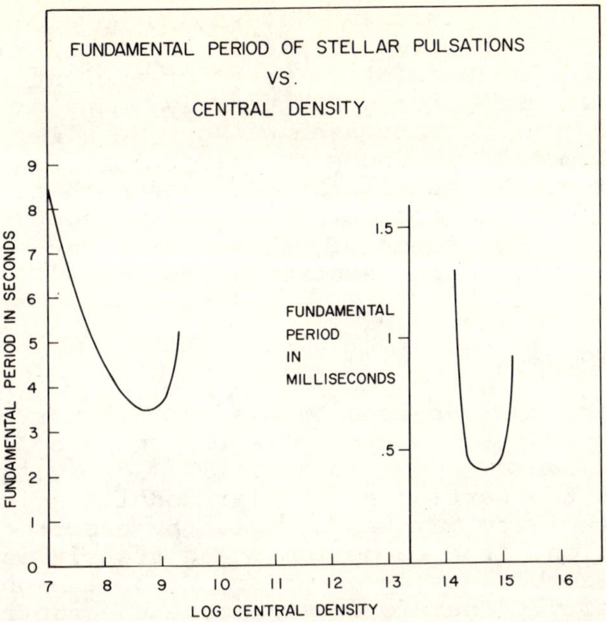

Fig. 9 Fundamental periods of white dwarf and neutron star models shown in Fig. 8.

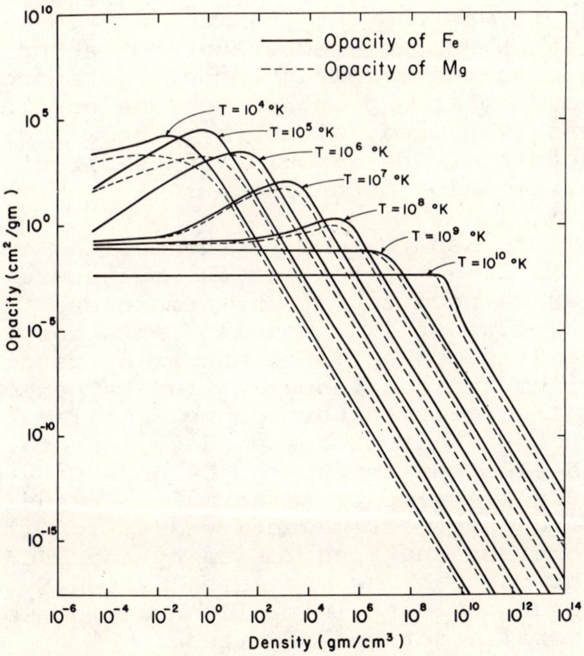

Fig. 10 Opacity of a gas composed of iron or magnesium.

From the above discussion it should be clear that the clarification of the properties of very dense matter is only beginning, and there is still need for a great deal of fundamental work in this area. Fortunately, many properties of neutron stars can be discussed without detailed knowledge of the internal material, and some predictions can be made on the basis of purely qualitative guesses such as those described above. We will now procede to consider the possible disposition of the various forms of energy which may be initially stored in a newly formed neutron star.

Thermal Energy

It appears that any reasonable evolutionary track ending in a neutron star state must pass through a collapse phase from about $10^9 gm/cm^3$ down to about 10^{14} or $10^{15} gm/cm^3$. Depending on the particular stellar model considered, the initial instability may be initiated by electron capture, pair production, iron to helium phase transition, or some other process, but collapse appears to be the only route to extreme density. Therefore we presume the neutron star to be formed violently, and tentatively associate it with a supernova event. This possibility is discussed in some detail elsewhere in this volume (see S. Colgate). For our purposes we assume that the neutron star is formed with large amounts of energy in thermal, vibrational, rotational, and magnetic form.

We will begin by considering the fate of the thermal energy. Several years ago, when the Crab nebula X-ray source was discovered, it was thought that the source might be a neutron star with a surface temperature of about $5 \times 10^7 °K$ radiating as a blackbody. Measurements during occultation by the moon proved that a point source was not involved, and the proposal was abandoned, although we know today that a measurable fraction of the X-radiation is directly associated with the pulsar emission mechanism.

Our problem is then to compute the cooling time for a neutron star. To find the rate of thermal energy loss of a hot neutron star we need to fit an "atmosphere" as in the case of more familiar stellar models. An atmosphere is constructed in a fairly conventional manner by requiring hydrostatic equilibrium and constancy of flux. The computations of opacity are also carried through by familiar techniques. Fig. 10 shows opacity curves computed by Tsuruta. In high density regions electron conduction becomes very efficient and these regions are almost isothermal. In addition to computing cooling by electromagnetic radiation it is also necessary to consider neutrino losses by the URCA, plasma-neutrino, and bremsstrahlung processes (see H.Y. Chiu in this volume). It turns out that initially neutrino losses are dominant by several orders of magnitude.

In the process of computing an atmosphere for a neutron star model we add to our store of information regarding its physical properties. In Fig. 11 the density is plotted against distance from the surface. Note that the density increases by six orders of magnitude over a radial distance of only one meter.

Fig. 12 presents the result of computations of cooling time. The entire procedure was carried out for two surface compositions--once for iron and once for manganese. The composition apparently makes little difference. The crosses indicate the time at which photon energy loss first excedes neutrino energy loss. Clearly the thermal energy is lost rapidly on an astronomical time scale.

I am aware of one consideration which could affect these conclusions rather significantly. If there is a very strong magnetic field, then the surface opacity can be reduced drastically. If the electrons are quantized in the perpendicular case, and the photons do not have sufficient energy to raise them to higher levels, then Compton scattering and the bound-free opacity would be virtually eliminated. The surface temperature would be higher by about a factor of 100, and the rate of energy loss accelerated by a factor of 10^8. Then we would be thinking in terms of a cooling time of about 30 yr.

Vibrational Energy

Although there is no evidence that a significant amount of vibrational energy is imparted to a neutron star at formation, we can still consider the damping time of such vibration if it existed. Because of the vibration the pressure at a point in the star will oscillate. Consequently, the Fermi energy will also oscillate. This is an ideal situation for the loss of energy through URCA process. As the Fermi level rises, energetic electrons force electron capture with the release of a neutrino. As the Fermi level falls, beta decay can occur, with the release of an antineutrino, and so on. The process is somewhat more complicated than the URCA process at constant pressure, and it has not been computed in complete detail, but fairly good results have nevertheless been obtained. Fig. 13 shows the behavior of total vibrational energy with time. It appears that significant energy would remain after perhaps 10^3 yr., but this is only one damping process, and stronger ones have turned up more recently. It turns out that shock deposition of energy at the surface gives the vibrational mode of a half-life of only about 30 yr.

Still another important damping mechanism has been predicted, and if it occurs it would be more important than any of those described above. It is now thought that Σ^- hyperons may appear at densities of $3 \times 10^{14} gm/cm^3$. From Table 2 it may be seen that such densities occur near the center of an intermediate neutron star model. Now consider the reaction $N + N \rightleftarrows P + \Sigma^-$. In a vibrating neutron star the neutron Fermi level will oscillate. As it rises, the reaction will be driven to the right. As it falls the reaction will be

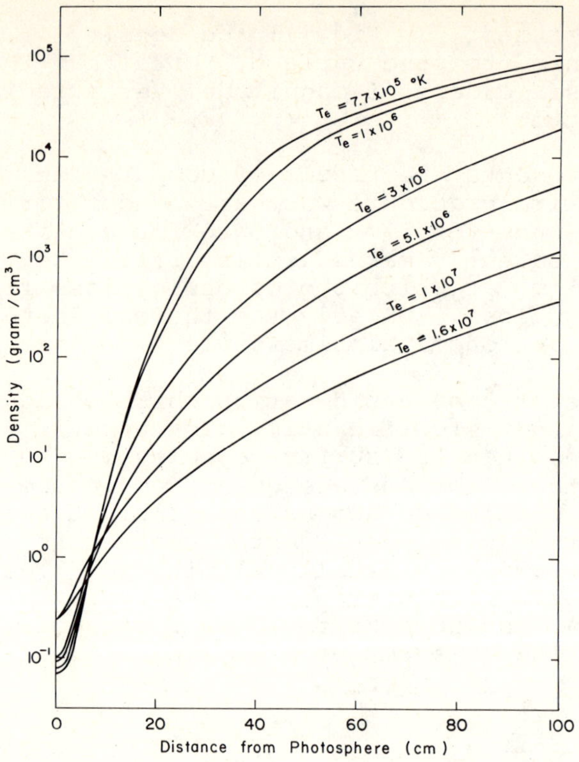

Fig. 11 Density distribution in the outer part of a neutron star atmosphere, as a function of surface temperature.

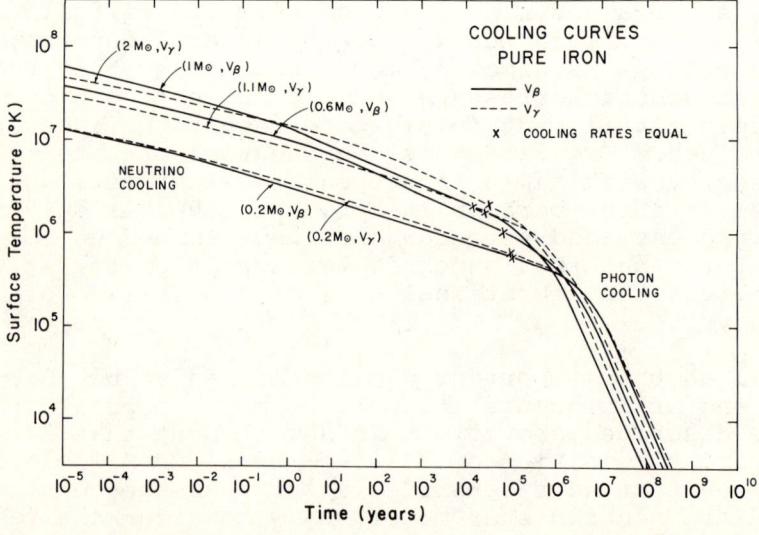

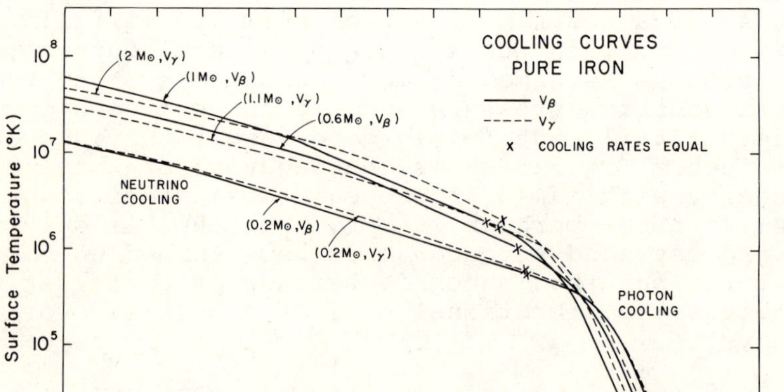

Fig. 12 Surface temperature during the cooling history of V_β and V_γ neutron star models.

allowed to procede to the left (i.e., the Σ^- will be able to decay). Since this reaction does not conserve strangeness, it is a weak process, and relatively slow. Consequently there will be a phase lag between the vibration and the reaction, resulting in a hysteresis loop and dissipation of vibrational energy. Computations indicate that this process would damp the vibrations on a time scale of seconds.

In summary, it appears fairly certain that vibrational processes in neutron stars will disappear so quickly that they will not be of observational interest.

Rotational Energy

From the observed rotation of normal stars we can predict with confidence that newly formed neutron stars will have high rate of rotation. It is to be expected that the newly formed neutron star will be rotationally unstable at the equator. In normal stars in which the density is much higher near the center, the equatorial material would stretch out into a sort of cusp which would then shed mass. Since a neutron star is of almost uniform density, its behavior will be more typical of a sphere of incompressible fluid. The effects of rotation on a sphere of incompressible fluid have been analyzed many years ago. The body will extend into a Jacobi ellipsoid; i.e., a triaxial deformation with unequal axes. It has been estimated that the dividing line between the equatorial cusp and the Jacobi ellipsoid is a polytropic index of about 1.0. A convenient way of assigning polytropic indices to dense stars is by finding the ratio of central density to mean density. For Tsuruta's models, the polytropic character is examined in Fig. 14. The consequences of a triaxial deformation are clear. The object will have a time-varying gravitational quadrupole moment and will emit gravitational radiation until the rotational velocity and the triaxial deformation is reduced. Of course this discussion is highly speculative because the dynamics of neutron star formation are highly uncertain. Nevertheless, it is still possible to discuss the fate of rotational energy. We have good evidence that neutron stars find some mechanism for conversion of their rotational energy into pulsed X-ray, radio and optical radiation and energetic electrons. Before continuing, I would like to make a few comments on magnetic fields, and then we can consider these two forms of energy together in the light of pulsar observations.

Magnetic Energy

Magnetic fields of various strength are known to exist in most stars. It is generally assumed that flux conservation can be assumed during the formation of a neutron star. This leads one to expect magnetic fields of perhaps 10^{12} to 10^{13} gauss. An upper limit to the field in a neutron star can be found by simply equating magnetic energy to gravitational binding energy, giving the value of 10^{16} gauss, but this is certainly unlikely. The question of the decay of

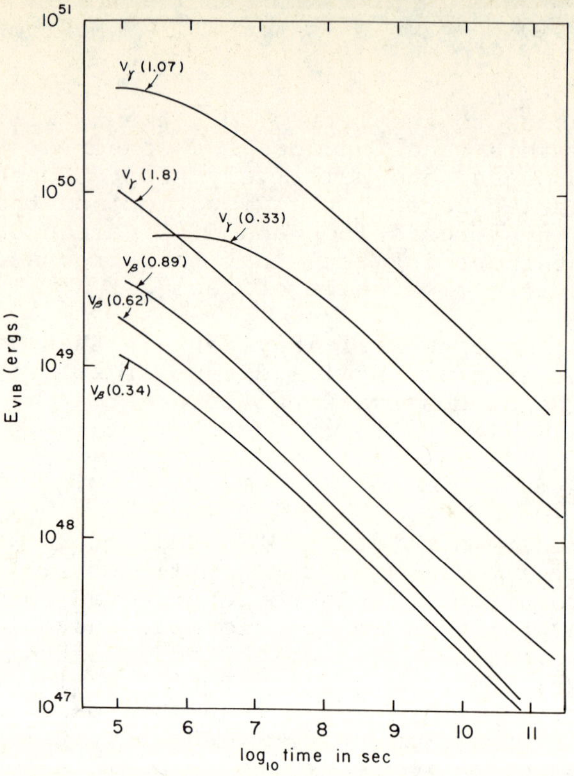

Fig. 13 Total vibrational energy as a function of time for various neutron star models when damping is solely by neutrino-anti-neutrino emission.

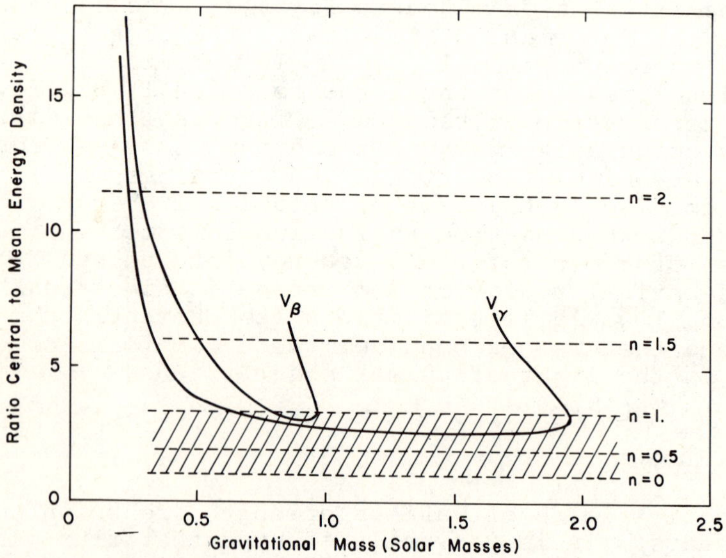

Fig. 14 The ratio of central to mean densities for V_β and V_γ neutron star models. The polytropic indices with various ratios are also shown. For polytropic indices of one or less (shaded region), distortion to form Jacobi ellipsoids at large rotation rates is probable.

the magnetic field is at present unsettled. It clearly depends on the interior electrical conductivity. If the conductivity is high, decay will be slow. Existing calculations of the conductivity of a degenerate electron gas give much too small values for the conductivity in a neutron star interior, since there the protons are also degenerate, and there is a large reduction in the phase space available for proton recoil in the scattering processes. On the other hand, if the interior is superconducting, magnetic fields will be excluded and the situation will be complex.

Combined Effects

One effect of rotating a magnetic neutron star will be the acceleration of particles. In the exterior region there will be an electric field parallel to the magnetic field, and ions will be lifted off the surface. Some workers, e.g., Gold, hope to explain the cosmic rays through such a process; however, computations by Rosen have shown that the surface of a neutron star will definitely be composed of iron regardless of the initial composition, and if Gold's idea is correct we would expect most cosmic rays to be iron. The observed cosmic ray spectrum is not consistent with this prediction.

I do not want to consider the pulsar emission mechanism here, but observations of periods and rates of change of periods already allow us to say quite a bit about the underlying properties of the star. The important thing is the rate of loss of rotational energy. Suppose we set ourselves to the problem to the problem of what kind of energy loss rate will be associated with a particular mass and angular velocity. The computation of angular momentum by Newtonian mechanics is not adequate for this problem. Cohen has derived a general relativistic expression for angular momentum,

$$J = \frac{8\pi}{3}\int_0^R (p + \rho/c^2) r^4 B A^{-1}(\omega - \Omega)\,dr$$

$$A = \sqrt{-g_{00}}\,, \quad B = \sqrt{g_{11}} \tag{1}$$

where ω is the angular velocity of the star and Ω is the angular velocity of the inertial reference frame dragged along (see K. Throne). The rotational energy is given by a more complicated expression,

$$E_{rot} = \frac{4\pi}{3}\int_0^R (p + \rho/c^2) r^4 B A^{-1}(\omega - \Omega)^2 dr +$$

$$+ \int_0^R r^4 \Omega_r^2 c^2 (12ABG)^{-1} dr + GJ^2/R^3c^2 \tag{2}$$

$$\Omega = \frac{2GJ}{r^3 c^2}$$

where the last term is the contribution to the rotational energy of the exterior gravitational field which is also being dragged. These integrals must be carried out by computer. This has been done for the V_α equation of state. Fig. 15 shows the degree of inertial frame drag. Table 3 shows some properties of several models. Since it is not initially clear how to define moment of inertia in general relativity, the last two columns show two possible definitions of moment of inertia.

Table 3. Rotational Properties of Some Neutron Star Models

$\log \rho_c$	mass	radius	J/ω	$2E_{rot}/\omega^2$
15.3	$4.87\text{x}10^{33}$ gm.	11.87 km.	$3.25\text{x}10^{45}$	$5.46\text{x}10^{45}$
15.0	$4.28\text{x}10^{33}$	13.39	$3.19\text{x}10^{45}$	$3.94\text{x}10^{45}$
14.8	$3.11\text{x}10^{33}$	13.91	$2.18\text{x}10^{45}$	$2.36\text{x}10^{45}$
14.6	$1.73\text{x}10^{33}$	13.79	$9.91\text{x}10^{44}$	$1.02\text{x}10^{45}$
14.4	$0.75\text{x}10^{33}$	13.80	$3.14\text{x}10^{44}$	$3.16\text{x}10^{44}$
14.2	$0.28\text{x}10^{33}$	17.72	$7.89\text{x}10^{43}$	$7.90\text{x}10^{43}$

Now we can consider the pulsar associated with the Crab nebula and try to determine what mass neutron star might be compatible with the observations. If we take the observed period and rate of slowing down, and insert them into the expressions above for various models, we can predict the total energy and energy output of those models. The result is shown in Table 4.

Table 4. Possible Combinations of Rotational Properties of Crab Pulsar

$\log \rho_c$	$M/M_\odot$	E_{rot}	$-\dot{E}_{rot}$
15.3	2.44	$9.8\text{x}10^{49}$ erg.	$2.52\text{x}10^{39}$ erg/sec.
15.2	2.41	$9.3\text{x}10^{49}$	$2.35\text{x}10^{39}$
15.0	2.14	$7.2\text{x}10^{49}$	$1.84\text{x}10^{39}$
14.8	1.56	$4.25\text{x}10^{49}$	$1.08\text{x}10^{39}$
14.6	0.86	$1.82\text{x}10^{49}$	$4.64\text{x}10^{38}$
14.4	0.38	$5.9\text{x}10^{48}$	$1.50\text{x}10^{38}$
14.2	0.14	$1.42\text{x}10^{48}$	$3.62\text{x}10^{37}$

For comparison with this table we need a figure for the total observed energy output of the Crab pulsar. Now ever since the X-radiation from that region was discovered it has been thought that the source was synchrotron radiation from electrons trapped in magnetic fields. The half-life for an electron to lose its energy by this process in the fields thought to exist in the Crab nebula (about 10^{-3}gauss) is about 1 year. Clearly the supply of electrons must be

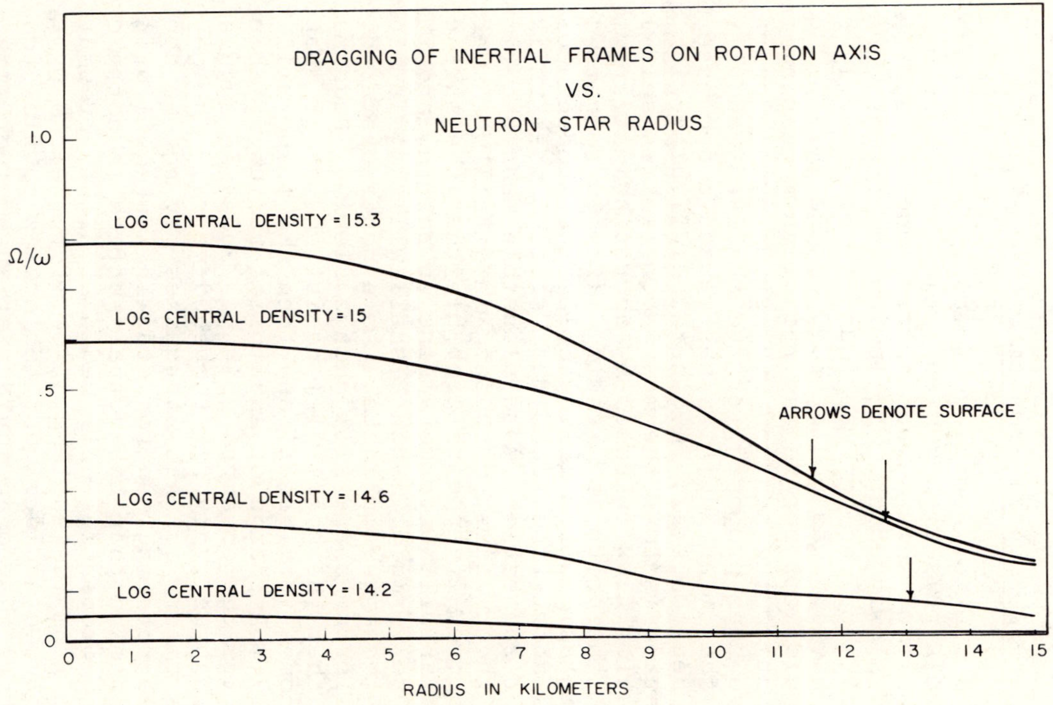

Fig. 15 The rate of dragging of inertial frames (as a fraction of rotational velocity) for selected neutron star models shown in Fig. 8.

continually renewed. Without inquiring into the details, we can reasonably believe (after Wheeler and others) that the neutron star is the source of the total energy output of the Crab nebula. With this assumption, we can set a lower limit for the energy loss of the Crab neutron star at 7×10^{37} - 1.5×10^{38} ergs/sec. Referring to Table 4 we see that we can tentatively assign a lower limit to the mass of the Crab pulsar of $0.4 M_{\odot}$. Of course, if we believe that much of the rotational energy is lost in forms which do not lead to synchrotron radiation, we must consider higher masses.

The Vela Pulsar - A Special Problem

During the week of February 24 to March 3 this year PSR 0833-45 (the Vela pulsar) underwent some kind of spinup process. The result of this spinup corresponded to approximately three weeks of slowing down. Following the change, the rate of loss of rotational energy was found to be about 1% higher. The radiation pattern was undisturbed by the spinup, so presumably we should look not to the radiation mechanism for an explanation, but to the neutron star structure. Several explanations have been proposed for the observed behavior. One suggestion is that mass addition has caused a decrease in radius of the neutron star. A general difficulty with this sort of suggestion is that the mass accretion is self-limiting. The high radiation flux released by accreting matter will severely limit the accretion itself. (See S. Colgate, this volume.) Quite aside from this problem the sudden accretion of matter on a short time scale must surely be an unlikely event. Another suggestion, more plausible than the first, is the starquake theory first discussed in detail by Ruderman. As discussed above, it appears probable that the surface shell of a neutron star will be crystalline in nature, and this shell will be very rigid. Such a shell will have a high shear modulus against changes in shape. As a neutron star slows down its equilibrium shape will change slowly. Ruderman's suggestion is that the shell will resist change of shape until the stress becomes very great, and then the shape will change discontinuously. A change of equatorial radius of about 1cm. would account for the observed spinup. Ruderman suggests that this may happen every few thousand years. The major objection to this theory is that starquakes may not be necessary to maintain the rotating neutron star in its equilibrium shape. Keep in mind that the actual surface is not a crystal. The crystalline properties only begin at higher pressure some centimeters in from the surface. Now if the earth were to slow appreciably we would expect the oceans to flow toward the poles to maintain the ocean surface at an equipotential under the combined effects of rotation and gravity. Likewise on the surface of the neutron star we should expect a flow of nondegenerate material from equatorial regions of higher potential to polar regions of lower potential. At the same time, adjustments would take place on the inner surface of the crystalline layer. The lowered pressure at the equator due to the flow of material away would then allow the formation of ions at a slightly greater depth, and the crystal would grow slightly inward. At the poles this process would be reversed. The net effect would be to maintain the thickness of the crystalline mantle in its new shape. Now if there is a strong magnetic field at the surface, the flow of the ionized material from equator to poles would be seriously hindered; however, I believe that the resculpturing of the interior of the mantle should prevent the buildup of large stresses.

The final proposal which I will present is the possibility of nonrigid rotation. Just as the electrical conductivity for neutron star matter is not known, so also the viscosity is not known. If the interior is of low viscosity, and especially if it is a superconductor and there is no magnetic field linkage, we would certainly expect differential rotation. We may very well have the core spinning faster than the surface layers. Of course the surface layers are continually slowing down, so we must ask ourselves if the surface layers and the interior will ever exchange much angular momentum or if the rotational differential will continue to grow. There is an old result that large scale instability will set in if

$$\frac{dJ}{d\bar{\omega}} + \delta < 0$$

where J is the angular momentum per unit mass, $\bar{\omega}$ is the distance from the rotational axis, and δ expresses in some way the effect of composition gradient. The neutron star does have a composition gradient. Now suppose the rotational differential slowly increases until the above inequality is just barely satisfied. Then the instability will result in large scale mixing. The immediate effect of the mixing will be to reduce the composition gradient and reduce δ. Weak interactions will eventually reestablish the composition gradient, but with a time lag. Thus for a short time the above inequality will be strongly satisfied, and very rapid angular momentum transfer should occur. Qualitatively, we predict that surface effects will be rapid and macroscopic. Needless to say, the detailed computation of this effect cannot be carried out until we know much more about the properties of the interior.

Final Comment

It should be clear by now that the study of neutron stars is a meeting place for many of the major fields of physics: general relativity, particle physics, solid state physics, cryogenics, hydrodynamics, radiative transport, and others. The predicted properties of the neutron stars seem to be verified in the large, by the pulsar observations. On the other hand, the detailed properties will only be found through extensive theoretical study. Because of the many fields involved, clearly beyond the grasp of an individual physicist, this area appears to require cooperative investigation by many physicists of differing background.

I am indebted to Mr. S. Ridgway for preparing the initial draft of these lecture notes. This research has been supported in part by the U.S. Atomic Energy Commission, the National Science Foundation, and the National Aeronautics and Space Administration.

Bibliography

For an extensive bibliography, see article on Neutron Stars by A.G.W. Cameron, in preparation for Annual Reviews of Astronomy and Astrophysics. The figures and data given here have been taken from the following papers:

Tsuruta, S., and A.G.W. Cameron, 1966, Can. J. Phys., 44, 1895.

_______. 1966, Can. J. Phys., 44, 1863.

_______. 1966, Nature, 211, 356.

Hansen, C. J., 1966, Nature, 211, 1069.

Langer, W.D. and A.G.W. Cameron, 1969, Astrophys. and Space Sci., in press.

Langer, W.D., L.C. Rosen, J.M. Cohen, and A.G.W. Cameron, 1969, Astrophys. and Space Sci., in press.

Cohen, J.M., W.D. Langer, L.C. Rosen, and A.G.W. Cameron, 1969, Astrophys. and Space Sci., in press.

Cohen, J.M. and A.G.W. Cameron, 1969, Nature, in press.

12

A REVIEW OF THEORIES OF PULSARS

HONG-YEE CHIU
Institute for Space Studies
Goddard Space Flight Center, NASA
Department of Earth and Space Sciences
State University of New York at Stony Brook

1. Introduction

The nature of pulsars is perhaps one of the most perplexing astrophysical problems of this century. The discovery of pulsars was accompanied by initial disbelief, astonishment, and a rush of theories to explain their enigmatic behavior. Nearly every telescope operating in radio frequencies was turned to a pulsar at least once, to listen to its mesmerizing beeps. Within one year of the initial announcement of the discovery (Hewish et al. 1968), over 300 related papers have appeared in literature, one-third of them dealing with theoretical interpretations (Maran and Cameron 1969). It is the purpose of this paper to discuss, on the basis of observed data and current knowledge of physics, the present theoretical status of pulsars. A complete literature survey is included in the review article by Maran and Cameron (1969).

This paper is divided into ten main sections:

1. Introduction.
2. Geometrical Models.
3. Collapsed Bodies—Neutron Stars.
4. Properties of Intense Magnetic Fields.
5. Physical Emission Mechanisms.
6. Gold-Goldreich-Julian Theory (Magnetosphere of Neutron Stars).
7. Pacini-Ostriker-Gunn Theory (Magnetic Dipole Radiation from Neutron Stars).
8. Ginzburg-Zaitsev Theory (Pulsating and Rotating Neutron Stars).
9. Chiu-Canuto Theory (Intense Magnetic Fields and Radio and Optical Pulses).
10. Summary and the Future of the Pulsar Problem.

Reprinted courtesy of the Publications of the Astronomical Society of the Pacific, vol. 82, no. 486, May 1970.

2. Geometrical Models

Pulsars are observed as pulsed radio sources of unknown dimension and identity, and with one exception (NP 0532), no optical identification has been made. It is necessary to extract their physical nature from observations. Many of the arguments used in constructing the preliminary model are geometrical and independent of the detailed mechanisms that produce these pulses. In particular, the prevailing model of a pulsar—a rotating magnetic neutron star—may be credited to T. Gold (1969).

The pulses from pulsars are observed as short bursts of radio signals occurring in well-defined time sequences with unpredictable amplitudes. The duration of the pulses at half-width ranges from 1 msec to 0.1 sec. Some sub-millisecond ($\sim$ 100 μsec) structures of pulses have been detected. The dimension of the emitting region must be smaller than the smallest pulse-structure time scale multiplied by the velocity of light. In the case of the shortest observed time scale—100 μsec—the maximum dimension of the emission region is deduced to be no greater than 30 km, about the size of a neutron star (Drake 1968).

Interstellar space is populated with electrons which affect the transmission of electromagnetic waves in much the same way as our ionosphere does. The dispersion relation of electromagnetic waves in the interstellar medium is given by the well-known relation

$$\omega^2 = \omega_p^2 + k^2c^2 \quad , \tag{2.1}$$

where ω is the frequency, k is the wave number of the radiation, and ω_p, the plasma frequency, is given by

$$\omega_p = \sqrt{(4\pi N_e e^2)/m} = 5.6 \times 10^4 \sqrt{N_e} \text{ Hz} \quad . \tag{2.2}$$

N_e is the electron number density. The velocity of propagation of a wave, v_p, is therefore

$$v_p = kc^2/\sqrt{(kc^2 + \omega_p^2)} \approx (1 - \omega_p^2/2\omega^2)c \quad (\omega_p << \omega) \tag{2.3}$$

which is smaller than c and is dependent on the frequency ω. The velocity difference between two waves of angular frequency ω_1 and ω_2 is

$$\Delta v_p/c = \omega_p^2(1/\omega_2^2 - 1/\omega_1^2) \simeq 7 \times 10^{-8} N_e(1/\nu_8^2 - 1/\langle\nu_8\rangle^2) \tag{2.4}$$

where $\nu_8 = (\omega_1/2\pi) \times 10^{-8}$ and $\langle\nu_8\rangle = (\omega_2/2\pi) \times 10^{-8}$ are the natural frequencies in units of 10^8 Hz. As a result of this velocity difference, radio pulses of the same time origin from the pulsar will arrive at the earth at different times. The time difference Δt is given by

$$\Delta t = l \Delta v_p / c^2 \quad , \tag{2.5}$$

where l is the distance. For example, if $\nu_8 = 1$, $\langle\nu_8\rangle = 10$, $l/c = 1000$ years, $N_e \approx 0.01$, we find that $\Delta t = 60$ seconds. (Because of this delay, pulses will be completely smeared when the bandwidth of the receiver $\Delta\nu_8$ exceeds $(\Delta\nu_8)_{max}$, where

$$\frac{(\Delta\nu_8)_{max}}{\nu_8} = \frac{10^8 P \nu_8^{\,2}}{2N_e l/c} \quad ; P = \text{period.}) \tag{2.6}$$

When transmission delay is taken into account, radio observations show that all radio emissions from a pulsar appear to originate from the same spatial region after allowing for some variations in pulse width and structure according to the frequency. This transmission delay is absent in optical and X-ray observations, and simultaneous X-ray and optical observations of NP 0532 show that the time of arrival of X-ray and optical pulses essentially agree with each other within an observational accuracy of 10^{-3} sec (Bradt et al. 1969). These coincidences in the arrival times of pulses indicate that all pulsar radiation originates from a common emission region and hence most likely is emitted by the same process. As discussed previously, the emission region cannot be greater than 30 km.

The period of many pulsars has been observed to be increasing at a rate in terms of $\Delta P/P$ between 10^{-12} to 10^{-16}, where P is the period and ΔP is the change of P per period. This is shown in Figure 1. One pulsar (the Vela pulsar) shows a temporary decrease in period of $10^{-6}\,P$ in one week, and resumes increasing its period afterwards with a slightly different rate of increase. The regular lengthening and almost erratic shortening of pulsar periods are two important clues to the pulsing mechanism. Two models have been suggested, the pulsating model and the rotating model. In the pulsating model the star is in a state of pulsation, and radio emission takes place during certain definite phases of pulsation. However, the pulsation period in all linearized theories of stellar pulsation is constant, and in nonlinearized theories the periods tend to decrease

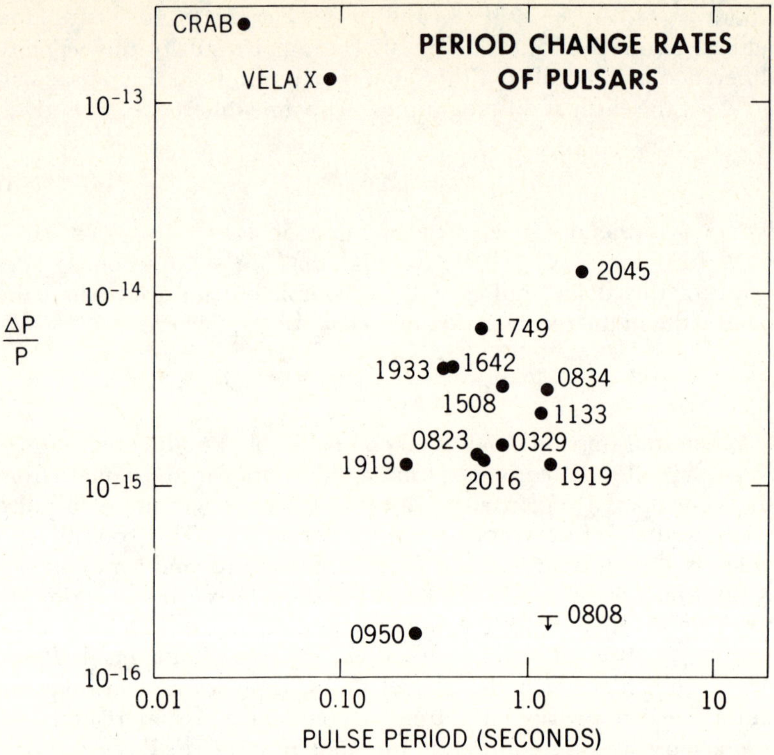

FIG. 1 — Period change rates of pulsars. (Courtesy S. Maran).

as the amplitude of pulsation becomes damped. Thus the pulsating theory can be excluded on observational grounds. Further, most stable neutron stars and white dwarfs do not have periods between 1 msec and 2 sec which is the range of the period of nearly all observed pulsars. (Although, near the instability point, the period of a neutron star can have any value from 0.1 msec to ∞, (Cocke and Cohen 1968), the range of mass associated with this range of period is too small—of the order of 0.01 $\mathfrak{M}_\odot$—to be of any real interest.)

The rotating model pictures the observed pulsed emission as a result of a narrow beam of emission corotating with the star. This narrow beam is produced by some anisotropic emission mechanism, and the emission can take place either from the surface or from some region away from the star. These points will be discussed in more detail later.

The rotating model is supported by the simple fact that the period of a rotating body can be lengthened by losing its rotational energy. This is consistent with observations. Likewise, the period of a rotating body can be decreased by contraction, and the change of period δP due to contraction is proportional to the contracted radius δR. ($\delta P/P \approx 2\, \delta R/R$.) Thus, the strange behavior of the Vela pulsar is easily understandable in terms of a rotating model but nearly impossible in the pulsating model.

After identifying the pulsar as a rotating star whose period of rotation is the observed pulse period, which ranges from 0.03 sec to 3.7 sec, all types of previously observed stars, including white dwarfs, can be immediately excluded as possible candidates. The clue is the observed period. Because ejection of matter will take place when the rotational velocity at the equator of a rotating star exceeds the escape velocity at the surface, a minimum period of stable rotation exists, and this minimum period for normal stars is of the order of a few hours; even in the case of white dwarfs it is no less than 5 sec. The only surviving candidate is the neutron star, which has been speculated to exist for more than three decades, but which has never been identified.

As will be discussed in the next section, a neutron star without a magnetic field can only radiate isotropically as a blackbody, and the radiation is chiefly in the spectral range of 10 Å to 500 Å (Chiu and Salpeter 1964; Tsuruta and Cameron 1966). The pulsed emission from pulsars thus excludes nonmagnetic neutron stars. As we will see later, in the presence of a magnetic field, radiation can be emitted anisotropically. The required field strength for anisotropical radiation in most current theories is greater than 10^{10} gauss, and a favored value is 10^{12}–10^{13} gauss.

As was first pointed out by Pacini (1967), and later by Gold (1969) independently, the rotational energy of a neutron star is very large, being 10^{46} ergs for a rotational period of approximately 1 sec. This rotational energy may be the source of the various activities in the Crab nebula and may be the energy source of pulsars. This view is fairly well accepted at present. The remaining problem is to find a mechanism to couple the rotational energy to an emission mechanism to account for the observed pulse radiations.

In conclusion, by using very simple arguments, a geometrical model for pulsars can be constructed and this geometrical model is the first step towards an understanding of the pulsar problem. This geometrical model pictures the pulsar as a rotating neutron star with a magnetic field of the order of 10^{10} gauss *or greater.* The pulsed radiation is a result of a fan-shaped or a pencil beam co-rotating with the star. A possible energy source for the pulsar is the rotational energy of the neutron star.

3. Neutron Stars

A. *History and Literature (Incomplete).*

Neutron stars were first proposed by L. Landau (1932). Baade and Zwicky (1934) and Zwicky (1939) suggested that neutron stars may be formed at the core of stars depleted of energy sources (supernovae). Theoretical structures of neutron stars were studied extensively in the 1930's by Tolman (1934*a,b*) and Oppenheimer and Volkoff (1939). The formation of neutron stars from supernova collapse has been studied by Chiu (1964), Schwartz (1967), Arnett (1966), Colgate and White (1966), Colgate (1970), and Cameron (1970). X-ray emission from nonmagnetic neutron stars has been studied by Chiu (1964), Chiu and Salpeter (1964), and Tsuruta and Cameron (1966). Rotating models have been considered by J. Hartle (1967) and Hartle and Thorne (1968). The energy transport in the presence of a strong gravitational field, and the pulsational properties have been considered in detail by K. S. Thorne (1967). The equations of state of neutron star matter have been considered by Salpeter (1960) and Levinger and Simmons (1961), and their effects on neutron star structures by Tsuruta and Cameron (1966) and by Tsuruta (1965), who also computed the cooling curves for various neutron star models. A possible existence of magnetic fields as high as 10^{13} gauss has been suggested by Woltjer (1964). The possible existence of a crystalline state in neutron star matter has been suggested and studied by Ruderman (1968).

B. *Formation.*

A normal star, with an interior temperature in the range $1\text{–}3 \times 10^7$° K, uses hydrogen as nuclear fuel and the end product of hydrogen burning is helium. After hydrogen exhaustion in the core, the star moves into the red-giant phase and helium burning commences at a temperature of $10^{8\circ}$ K, when the star reaches the tip of the red-giant phase. The burning of helium results in carbon or oxygen (or both). After helium exhaustion, carbon burning will take place when the temperature rises to $6 \times 10^{8\circ}$ K. The burning of oxygen requires a higher temperature, about $2 \times 10^{9\circ}$ K. At a temperature of $4 \times 10^{9\circ}$ K, elements will come into equilibrium (Burbidge et al. 1957) and all nuclear energy sources are exhausted. From this, stellar evolution proceeds on gravitational energy alone.

Above a temperature of $5 \times 10^{8\circ}$ K, neutrino processes become an important mechanism for dissipating stellar energy. The two important neutrino processes are (Chiu 1966):

(A) The pair annihilation process $e^- + e^+ \rightarrow \nu + \bar{\nu}$. (3.1)
(B) The photoneutrino process $\gamma + e^- \rightarrow e^- + \nu + \bar{\nu}$. (3.2)

The energy loss rate via these two processes increases rapidly with temperature ($\propto T^9$) and the neutrino luminosity exceeds the optical luminosity of the brightest stable stars in the Galaxy when the temperature rises above $6 \times 10^{8\circ}$ K. At a temperature of $5\text{–}7 \times 10^{9\circ}$ K, the rate of neutrino emission increases sufficiently to cause the star to contract by a factor of two in a matter of seconds. Hydrostatic equilibrium of the star cannot be maintained and within seconds, stellar collapse takes place (Chiu 1964).

A different mechanism for collapse has been proposed by Fowler and Hoyle (1964). As the temperature approaches $7 \times 10^{9\circ}$ K, the equilibrium configuration of elements shifts from iron to helium, and the photodisintegration of Fe^{56} into He^4 is proposed as a trigger mechanism for collapse.

However, it is immaterial which mechanism causes collapse. The events following collapse depend critically on neutrino dynamics (the production and absorption of neutrinos) and the hydrodynamics of shock waves (production and propagation), and very little on the nuclear behavior of the star (Schwartz 1967; Arnett 1966; Colgate and White 1966; Colgate 1970; and Cameron 1970). They are virtually independent of the mechanism for collapse. During collapse, the annihilation process and the processes

$$e^- + p \rightarrow n + \nu \tag{3.3}$$

$$e^+ + n \rightarrow p + \bar{\nu} \tag{3.4}$$

continue to dominate as the most important neutrino production processes. Finally, as the density of the collapsing core approaches the nuclear density ($\sim 10^{14}$g/cm^3), the mean free path of neutrinos becomes less than the dimensions of the star. Absorption of neutrinos via the inverse processes to (3.3) and (3.4) in the outer layer of the star then becomes important. In this way the gravitational energy released in the core can be transferred to the outer layer of the star via neutrino absorption processes (Chiu 1964). The absorption of neutrinos strongly enhances the strength of the outgoing shock wave produced by collapse. This enhanced shock wave can eject matter into space to form the observed supernova remnants. The remaining core then forms a neutron star.

The gravitational binding energy of a neutron star is in the neighborhood of 50 MeV per particle and prior to collapse is about 1 MeV per particle. If the efficiency of neutrino energy deposition to the shock wave is 100 percent, then up to 50 $\mathfrak{M}_\odot$ of the envelope may be ejected, assuming that the mass of the neutron star is of the order of 1 $\mathfrak{M}_\odot$. Invariably there is some neutrino loss. Further, if the star is too massive the core temperature may become so high that muon neutrinos are produced (Chiu 1964). Since muon neutrinos cannot be absorbed by electrons, larger amounts of neutrino fluxes can escape the star, and the efficiency of neutrino energy deposition will further suffer.

In fact, according to Schwartz (1967), the maximum ejectable mass is about 10 $\mathfrak{M}_\odot$. However, another calculation by Arnett (1966) showed that the maximum ejectable mass is in the neighborhood of 5 $\mathfrak{M}_\odot$. This range of discrepancy may well represent the dispersion of ejectable mass under realistic conditions. In any case, neutron stars are indeed formed in models for supernova collapse.

C. Structural Characteristics.

In the laboratory, a neutron is unstable and decays into an electron, a proton, and an antineutrino, in about 1115 seconds. The maximum energy of the electron from decay is 1.8 MeV. However, the decay of a neutron in dense matter is inhibited by degeneracy. Theoretically, if all electron states are occupied up to a Fermi energy of 1.8 MeV, free neutrons can exist and are stable against spontaneous decay. However, as neutrons can still participate in nuclear reactions, the required Fermi energy is much higher.

When reactions and stability of nuclei are taken into account, free neutrons can exist when the density exceeds 10^{11} g/cm^3 (Salpeter 1961). A small fraction of matter still exists in the form of protons and electrons and it is the Fermi energy of the residual electrons (about 1%) that keeps neutrons from spontaneous decay. Accordingly, the minimum mass for a stable configuration of a neutron star is the one that gives the density of 10^{11} g/cm^3. The value of the minimum mass is approximately 0.14 $\mathfrak{M}_\odot$ (Landau 1932; Oppenheimer and Volkoff 1939; Tsuruta 1965; and Tsuruta and Cameron 1966).

According to the theory of zero-temperature stars, all dynamically stable configurations require the central density (and the average density) to increase with mass. As the mass departs from the minimum stable configuration the density rapidly increases to 10^{15} g/cm^3. This value of density is typical for all stable configurations. Likewise the radius of a neutron star is approximately 10 km (to within a factor of two). This radius (10 km) is comparable to the gravitational radius or Schwarzschild radius of the star, R_s, which is $2G\mathfrak{M}/c^2 = 2.6(\mathfrak{M}/\mathfrak{M}_\odot)$ km. (The gravitational radius is the radius at which the classical gravitational energy of a mass configuration ($\sim 2G\mathfrak{M}^2/R$) equals the rest energy of the star, $\mathfrak{M}c^2$.) Therefore the corrections to the hydrostatic equilibrium equation due to general relativistic effects become important. The relativistic equation of hydrostatic equilibrium is (Tolman 1934*a*,*b*; Oppenheimer and Volkoff 1939)

$$\frac{dP}{dr} = -G\frac{[\mathfrak{M}(r) + 4\pi r^3 P/c^2](\rho + P/c^2)}{r^2[1 - 2G\mathfrak{M}(r)/rc^2]} , \tag{3.5}$$

where P is the pressure, $\mathfrak{M}(r)$ the shell mass, ρ the density, and r the distance from the center. Comparing this equation with the classical equation of hydrostatic equilibrium

$$\frac{dP}{dr} = -G\frac{\mathfrak{M}(r)\,\rho}{r^2} , \tag{3.6}$$

we find that the shell mass is effectively increased by the contribution from the pressure $4\pi r^3 P/c^2$, so that the pressure appears to have a weight. The density is likewise increased by the mass equivalent of the stress energy, and the radius is effectively decreased by a factor $[1 - 2G\mathfrak{M}(r)/rc^2]^{1/2} < 1$. Thus, for a given mass, the general relativistic hydrostatic equilibrium condition requires a higher pressure gradient to keep a star in equilibrium, as compared to that of a star under the condition of classical hydrostatic equilibrium. As the mass of the star increases, the pressure eventually contributes so much to the shell mass $\mathfrak{M}(r)$ that the resultant hydrostatic equilibrium configuration becomes unstable against even the smallest perturbation. A small perturbation of the star will cause an increase in pressure which results in a stronger gravitational field. A larger pressure gradient is needed to counter-balance this increased gravitational field. As a result a larger gravitational field is produced by the added pressure. A greater pressure gradient is then needed, and so on. A runaway situation eventually occurs, and the star contracts indefinitely towards its gravitational radius whereupon it becomes invisible.

Thus, due to general relativistic effects, dynamical instability will commence when the mass exceeds a critical mass $\mathfrak{M}_{cr}$ whose value is 0.76 $\mathfrak{M}_{\odot}$ if the equation of state of the neutron gas is taken to be that of an ideal Fermi gas. Unfortunately, at a density of 10^{15} g/cm^3 nuclear interactions sufficiently alter the equation of state of a neutron gas so that the equation of state of an ideal Fermi gas cannot apply. The effect of nuclear interactions in a neutron gas has been considered by Salpeter (1960), and by Levinger and Simmons (1961). It appears however, that the critical mass is a sensitive function of the nuclear interaction. Levinger and Simmons constructed two nuclear potentials based on the same set of experimental data but taking two extreme error limits. Tsuruta and Cameron (1966), and Tsuruta (1965) found that these two potentials gave two entirely different critical masses, 1.8 and 3 $\mathfrak{M}_{\odot}$ respectively, which may be taken to be the range of the actual critical mass.

The question also arises as to whether this critical mass can be avoided in some types of equation of state. However, Misner and Zapolsky (1964) showed that the existence of the critical mass is a general relativistic effect and is independent of the equation of state.

Composition, temperature, and density distributions in neutron stars. Near the surface of a neutron star where the density drops below 10^{11} g/cm^3, neutron gas becomes unstable and elements can be formed. Just below the neutron-unstable density of 10^{11} g/cm^3 the stable elements are neutron-rich elements such as $_{39}Y^{120}$. As the density decreases, fewer neutron-rich elements can exist; finally when the density is below 10^7 g/cm^3 the composition is essentially stable nuclides as on earth. Since a neutron star is the remnant core of a supernova explosion, its surface composition is likely to be that of the equilibrium composition, namely Fe^{56}.

Within seconds of its formation, a neutron star very likely possesses an interior temperature of $10^{12\circ}$ K, which is the maximum temperature achieved in collapse (Chiu 1964; Arnett 1966; Schwartz 1967; Colgate and White 1966; Colgate 1970; Cameron 1970). However, this temperature rapidly decreases as neutrino processes remove the thermal energy of the star. The most important neutrino processes are (Chiu 1966):

(A) Plasma process: $\hbar\omega \rightarrow \nu + \bar{\nu}$ (3.7)

(B) Modified URCA process: $n + n \rightarrow p + n + e^- + \nu$. (3.8)

Within one year, the interior temperature will drop to somewhere between $5 \times 10^{8\circ}$ K and $10^{9\circ}$ K. Afterwards the optical emission from the surface takes over as the main energy dissipation mechanism (Chiu and Salpeter 1964; Tsuruta and Cameron 1966).

At a given temperature the electrons are degenerate if the Fermi energy exceeds kT, and the corresponding density for a temperature of a few times $10^{8\circ}$ K is 10^6 g/cm^3. The neutron gas is in any case highly degenerate. The thermal conductivity of degenerate matter is exceedingly high, hence a temperature gradient exists only in the nondegenerate envelope near the surface of the star. The thickness of this nondegenerate envelope is found to be about one meter over which the temperature drops by a factor of 100 (assuming an interior temperature ranging from 10^8 to $10^{9\circ}$ K). Thus, the surface temperature of a neutron star is likely to be a few times $10^{6\circ}$ K within a few hundred thousand years of its formation (Chiu and Salpeter 1964; Tsuruta and Cameron 1966). With a luminosity $\sim L_\odot$, a neutron star (without a magnetic field) will radiate in the range of 10–500 Å depending on the surface temperature. The density scale height in the nondegenerate layer ranges from 0.1 to 1 cm.

Several years ago, celestial X-ray sources of roughly the correct spectrum and with the flux to be expected from neutron stars at a distance of a few thousand light years were discovered. Enthusiasm arose in identifying these X-ray sources as neutron stars. Subsequently, however, these X-ray sources were definitely identified as not being neutron stars, and the enthusiasm dwindled (Friedman 1967; Giacconi, Gursky, and Van Speybroeck 1968). This discrepancy is due to the fact that magnetic fields can severely alter the emission characteristics of a neutron star.

Ruderman (1969) recently advanced a very interesting theory regarding the interior of neutron stars. According to this theory, below a temperature of a few times $10^{8\circ}$ K, crystallization can take place in the interior of neutron stars. Using this theory he was able to explain the sudden change in the period of the Vela pulsar as due to a resettlement of the structure of the star (starquakes). We will discuss this point later.

To summarize, a typical neutron star has a diameter of ten kilometers, and a mass of the order of one solar mass. From the center to about nine kilometers, the compositon is predominantly free neutrons with about one percent protons and electrons, and these electrons keep the neutrons from decaying. (In massive neutron stars hyperons may even exist as stable particles at the core.) In the next kilometer the composition varies from highly beta-unstable neutron-rich elements like ${}_{39}Y^{120}$ at the bottom to common elements near the top. The most likely composition at the surface is Fe^{56}. Only in the last meter of the surface does matter become nondegenerate, and a substantial temperature gradient appears. The mass of the nondegenerate envelope is about 10^{19} g (a higher value is attained in younger neutron stars and vice versa). The surface temperature is about 1/100 of that in the interior. Without a magnetic field a neutron star can only radiate as a blackbody, predominantly in the soft X-ray region.

D. Rotation, Magnetic Fields, and Pulsations.

During the formation of a neutron star the parent star collapses by a factor of 10^5 (from an initial radius of about 10^{11} cm—solar radius—to a terminal radius of 10^6 cm). Because of the size of this contraction factor, those parameters normally regarded as perturbations to stellar structure may become important, and among these are magnetic fields and rotations.

Rotation. The moment of inertia I of a body is of the order of $\mathfrak{M}R^2$ where $\mathfrak{M}$ is the mass and R is the radius. The angular momentum $I\Omega$, where Ω is the rotational frequency, is a conserved quantity. Therefore Ω increases as R^{-2}. A contraction by a factor of 10^5 will cause the rotational angular frequency to increase by a factor of 10^{10}.

The period of rotation of many stars, P, is in the neighborhood of one day, for which $\Omega = 2\pi/P \cong 6 \times 10^{-5}$ sec^{-1}. According to the angular-momentum conservation condition, after the collapse the period of rotation can be as short as 0.01 msec. However, there is a minimum rotational period at which the centrifugal force per unit mass at the equator equals the surface gravity. Matter will be ejected at the equator if the rotational period is less than this minimum period. The centrifugal force f_c is v_e^2/R where v_e is the velocity at the equator which can be written as $R\Omega$, where R is the stellar radius. Therefore we have

$$f_c = v_e^2/R \lesssim G\mathfrak{M}/R^2 \quad , \tag{3.9}$$

and the maximum value of Ω is

$$\begin{aligned} \Omega_{max} &\approx \surd(G\mathfrak{M}/R^3) = 10^4 \surd((\mathfrak{M}/\mathfrak{M}_\odot)(10\ \text{km}/R^3)) \quad \text{sec}^{-1}, \\ P_{max} &= 2\pi/\Omega_{max} \quad . \end{aligned} \tag{3.10}$$

For a radius of 10 km, this limiting rotational period is about 1.6 msec >> 0.01 msec. Hence a newly-formed neutron star most likely will rotate with this limiting period.

The rotational energy of a neutron star is roughly

$$\mathfrak{M}R^2\Omega^2 = 7 \times 10^{46} (\mathfrak{M}/\mathfrak{M}_\odot)(R/10\ \text{km})^2 P^{-2} \ \text{ergs} \quad , \tag{3.11}$$

where P is the period in seconds (Morton 1964; Pacini 1967; Gold 1969).

Magnetic fields. Sky surveys of stars in our Galaxy show a large number of stars possessing magnetic fields greater than 500 gauss, with the maximum observed field in the neighborhood of 3.4×10^4 gauss. Stellar magnetic fields are believed to originate from electric currents in the form of drifting charges inside a star. Because of the large dimension involved the decay time of such a current is usually very long—of the order of the lifetime of a star or longer.

The magnetic field of a current carrying plasma follows the flux conservation law, which states that the field strength is proportional to R^{-2} where R is the dimension of the plasma. Therefore, during collapse the magnetic field of a star can also increase as R^{-2}, just as in the case of rotation. With a contraction factor of 10^5, the field of a neutron star may be 10^{10} times greater than that of the parent star. Assuming an initial magnetic field of 1000 gauss, after collapse the neutron star may acquire a field as high as 10^{13} gauss (Woltjer 1964). The associated magnetic field energy is $10^{43}(H/10^{13})^2(R/10\text{ km})^3$ ergs.

The properties of plasma in such an intense field are quite different from those in weaker fields usually encountered in plasma physics. This will be discussed in a separate section.

Pulsations. Light from many stars has been observed to vary regularly and this variation has been interpreted as due to pulsation. In order for a star to pulsate, some driving mechanism must exist. In the case of cepheids and RR Lyrae stars the ionization of hydrogen and helium plays a dominant role in maintaining pulsation. In the case of neutron stars no mechanism for maintaining pulsation is known to exist. Thus, if a neutron star pulsates, the pulsation must be a free oscillation, and the energy of pulsation can be derived from collapse.

K. S. Thorne (1967) studied the properties of pulsations in neutron stars quite extensively. He found that the fundamental periods of stable neutron stars generally are less than 1 msec.

Any free oscillation is subject to dissipation, and the most important dissipation mechanism in a neutron star is the emission of gravitational waves (Wheeler 1966). A radially pulsating nonrotating neutron star cannot emit gravitational waves, but if rotation is taken into account, the small distortion introduced by rotation will cause the emission of gravitational waves. The time for dissipation is rather short—for a freshly formed neutron star with a rapid rotation, the half-life is less than 1 sec (Wheeler 1966).

Cocke and Cohen (1968) found that near the critical mass the period can become substantially longer, but unfortunately such mass configurations cannot sustain large amplitude oscillations. Further, the mass range for pulsational periods greater than 1 msec is rather narrow, being only 0.01 $\mathfrak{M}_\odot$. Thus, generally speaking, no neutron stars can possess pulsational periods greater than 1 msec.

E. General Relativity and the Structure of Neutron Stars.

The effect of general relativity on stellar structure has been exhaustively studied by Thorne (1967), Hartle (1967), and Hartle and

Thorne (1968). It is fair to say that, apart from the existence of the critical mass, the general relativistic effect on stellar structure is masked by uncertainties in the equation of state for a neutron gas.

4. Intense Magnetic Fields

In a magnetic field as intense as 10^{13} gauss, classical theory of electricity and magnetism breaks down and various quantum effects become important (Erber 1966; Chiu and Canuto 1968*a,b*). In this section we will discuss various such effects which might be relevant to the pulsar problem.

A. *Motion of an Electron in a Magnetic Field.*

In a magnetic field which is along the z direction, a classical electron will describe a helical orbit whose equations are as follows:

$$\begin{aligned} x &= x_0 + R_L \cos(\omega_L + \alpha) \quad , \\ y &= y_0 + R_L \sin(\omega_L + \alpha) \quad , \quad \alpha = \text{phase angle} \quad , \\ z &= z_0 + v_z t \quad , \quad (x_0, y_0, z_0) = \text{initial position} \quad . \end{aligned} \tag{4.1}$$

Equations (4.1) describe a circular motion of radius R_L and frequency ω_L in the plane perpendicular to the field ($\perp$ plane) and a rectilinear motion in the direction parallel to the field (// direction). The Larmor radius R_L and the Larmor frequency ω_L are given by

$$R_L = v_\perp / \omega_L = c\, p_\perp / eH = \gamma v c m / eH \tag{4.2}$$

$$\omega_L = ecH/E \quad , \quad E = mc^2(1 - (v/c)^2)^{-1/2}, \quad \gamma = E/mc^2 \tag{4.3}$$

$$p_\perp = \perp \text{ momentum of the electron.}$$

The magnetic field thus confines the $\perp$ motion.

This classical description of electron motion is valid when the radius of the orbit R_L is large compared to the de Broglie wavelength of the electron, λ_B, which in the nonrelativistic case is given by

$$\lambda_B = 1.22 \times 10^{-7} / \sqrt{E_k(eV)} \text{ cm} \tag{4.4}$$

where $E_k(eV)$ is the kinetic energy of the electron. In a field of 1 gauss, the Larmor orbit is 1 cm if the energy of the electron is 1 eV. In a field of 10^7 gauss, however, the Larmor radius is 10^{-7} cm which is comparable to the de Broglie wavelength. In the case of an atom, quantization arises because the orbit radius is comparable to the de Broglie wavelength. Likewise quantization of electron orbits in a magnetic field takes place when the Larmor radius is comparable to the electron wavelength.

In the case of an atom, the force of the nucleus is isotropic and quantization affects all components of momenta, but in the case of a magnetic field the force is in the $\perp$ plane only. As a result only the $\perp$ motions are quantized. The rule of quantization is

$$p_x^2 + p_y^2 \rightarrow 2(H/H_q)\, n\, (mc)^2 \quad , \qquad (4.5)$$

where n is an integer taking values from 0 to ∞ and $H_q = m^2c^3/e\hbar = 4.414 \times 10^{13}$ gauss. The total energy E of the electron in a magnetic field thus takes the following form (Rabi 1928):

$$E = mc^2[1 + (p_z/mc)^2 + 2n(H/H_q)]^{1/2} \quad , \quad n = 0,1,2, \ldots \infty \qquad (4.6)$$

The energy levels defined by the principal quantum number n are known as Landau levels. The lowest level is the one with $n = 0$ (i.e., there is no $\perp$ motion and the electron moves along the magnetic field). The next higher level is the one with $n = 1$ (an orbit with a classical radius of $(mc^2/eH)\sqrt{2(H/H_q)}$, and so on. This is illustrated in Figure 2.

The nonrelativistic expression for equation (4.6) is

$$E - mc^2 = p_z^2/2m + n(H/H_q)\, mc^2 \qquad (4.7)$$

which gives a value of $E - mc^2$ to within 50 percent for H up to 10^{13} gauss and p_z up to mc.

Because of this quantization an electron behaves as a one-dimensional particle with only one degree of freedom in the // direction as long as the energy imparted to the $\perp$ motion of the electron is small compared to the energy gap between two neighboring magnetic states of the same value of p_z. The value of this energy gap is 11.6 $(H/10^{12}G)$ keV in the nonrealativistic regime.

Impossibility of pair creation. It has often been claimed that pair creation can take place in a magnetic field. The argument is as follows: The classical electron spin energy in a magnetic field is $\mu_B H$ where $\mu_B = e\hbar/(2mc)$ is the Bohr magneton. When $H = 2H_q = 8.8 \times 10^{13}$ gauss the classical spin energy is mc^2. Conceivably a pair of electrons can be created in a magnetic field with their spins

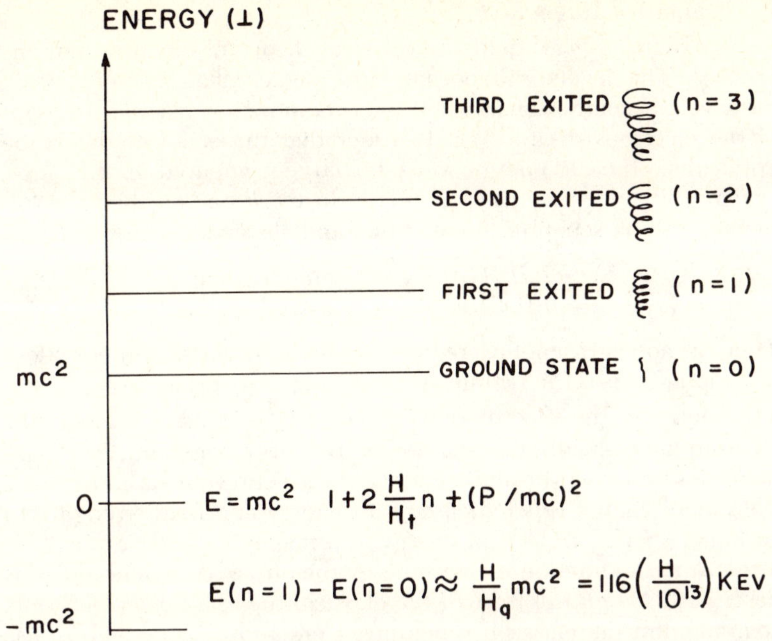

FIG. 2 — Landau orbits of electrons in a magnetic field.

oriented at appropriate directions. The energy needed for creating such pairs comes from the magnetic field.

As a result of the quantized energy levels (eq. 4.6) the rest energy of the electron never decreases below mc^2, and the separation between the positive and negative energy states of the electron is still 2 mc^2. To create a pair, an external energy of at least 2 mc^2 must be supplied. We therefore conclude that pair creation cannot be taking place at the expense of the magnetic field energy.

Jancovici (1969) has investigated this problem with the anomalous magnetic moment of the electron properly taken into account. He found that the separation energy between positive and negative energy states of the electron remains nearly constant ($\sim$ 2 mc^2) and pair creation will not take place at any field strength.

B. Regimes of Magnetic Fields.

Clearly, in a weak field the classical theory of electron motions applies. The levels will become so crowded that discreteness of the levels becomes unimportant, and the one-dimensional behavior of the electrons disappears. In order that the classical theory be applicable, the $\perp$ energy must be large compared to the level spacing. If we equate the $\perp$ energy to the // energy then the following criteria separate the quantum and classical regimes (Fig. 3):

$$\begin{aligned} &(\gamma - 1) \sim 2(H/H_q) \qquad && \text{quantum regime} \\ &(\gamma - 1) >> 2\,(H/H_q) && \text{classical regime.} \end{aligned} \tag{4.8}$$

There is, however, another regime to which the classical theory does not apply: This is the regime of strong coupling (Erber 1966).

In deriving the electron orbits (eq. 4.1) it is assumed that the electron loses a negligible fraction of its energy per orbit. In a magnetic field an electron loses energy via synchrotron radiation, and the rate of energy loss increases with energy and field strength. In certain regimes of electron energy and magnetic field, an electron loses a large fraction of its energy before a complete orbit is described. When this occurs, the classical theory of electron orbits (eq. 4.1) and the classical synchrotron radiation rate, which is also calculated assuming negligible energy loss per orbit, fail to apply. Two corrections must be taken into account: (1) The exact electron wave functions must be used in evaluating the quantity of interest. (2) The emitted radiation exerts a reaction on the electron, and this radiation reaction must be taken into account. While the first point has been considered by Klepikov (1952) and elaborated by Erber (1966), the second point has been considered by Shen (1970).

The regime of strong coupling can be obtained from the criterion that the lifetime of an electron against synchroton radiation should be shorter than or equal to the Larmor period $P_L = 2\pi/\omega_L$. The lifetime of an electron against synchrotron radiation is

$$\tau_s = \left[\frac{2}{3}\frac{e^2}{\hbar c}\frac{mc^2}{\hbar}\gamma\left(\frac{H}{H_q}\right)^2\right]^{-1} . \tag{4.9}$$

The Larmor period is given by

$$P_L = 2\pi\gamma(H/H_q)^{-1}(\hbar/mc^2) . \tag{4.10}$$

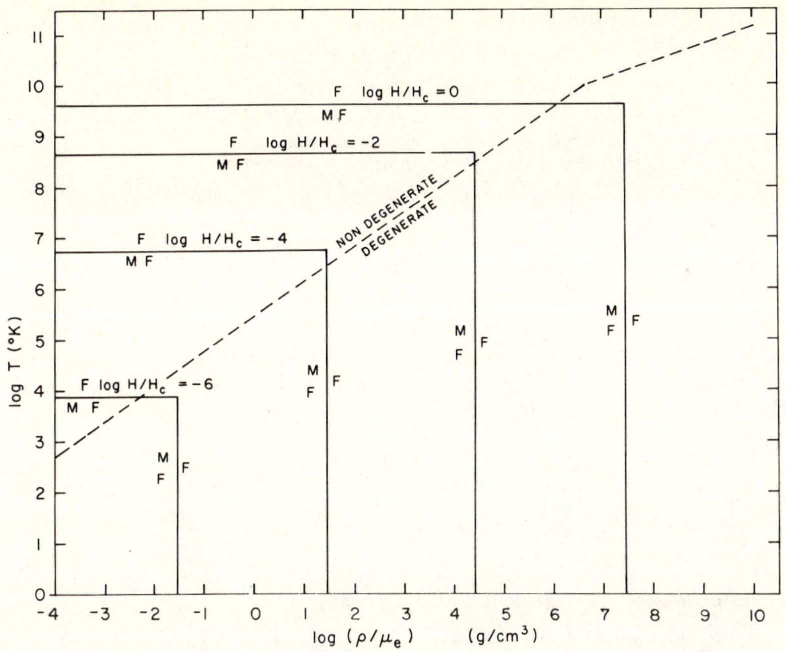

FIG. 3 — Regimes of magnetic fields. M—magnetized regime. F—Fermi gas regime. $H_c = m^2c^3/e\hbar = 4.414 \times 10^{13}$ gauss.

Therefore the strong coupling regime is given by the condition that

$$(e^2/\hbar c)\gamma^2(H/H_q) >> 1 \quad , \quad \text{or} \quad \gamma^2 H >> 6 \times 10^{15} \quad . \quad (4.11)$$

In reality, when $(e^2/\hbar c)\gamma^2(H/H_q) \approx 0.01$, the strong coupling effect already becomes noticeable (Erber 1966).

As a numerical example, consider an electron of energy 10^{14} eV or $\gamma = 10^8$. The classical theory of synchrotron radiation breaks down even when the field is only a modest 1 gauss, the corresponding electron lifetime is 1 sec. In a field of 10^5 gauss, the classical theory breaks down when the electron energy is only 10^{12} eV ($\gamma = 10^6$) and the electron lifetime is less than 10^{-8} sec.

Summaries of the three regimes of magnetic fields. Classical regime: Medium field, medium energy such that

a. weak coupling; $\frac{e^2}{\hbar c}\gamma^2\frac{H}{H_q} << 1$, or $\gamma^2 H << 6 \times 10^{15}$ (4.12)

b. Landau level spacing not important,

$$H/H_q << (\gamma - 1) \quad . \tag{4.13}$$

Strong coupling regime: High field and high energy such that

$$\frac{e^2}{\hbar c}\gamma^2\frac{H}{H_q} \rightarrow 1 \quad . \tag{4.14}$$

Quantum regime: a. The energy of the electron is comparable to the Landau level spacing

$$\begin{aligned} &H/H_q \sim (\gamma - 1) \quad , \\ \text{or } &(E - mc^2) \sim 0.5(H/H_q) \quad \text{MeV} \quad , \end{aligned} \tag{4.15}$$

b. $H/H_q \gtrsim 1$

C. *Radiation Mechanisms and Processes.*

The most important radiation process in a magnetic plasma is the synchrotron process. If the density is high, the bremsstrahlung process may become important.

(*i*). *Synchrotron radiation.* We will only summarize the results.

Classical regime. This problem has been extensively summarized by Jackson (1962). The energy loss rate is

$$\frac{dE}{dt} = \frac{2}{3}\left(\frac{v}{c}\right)^2 \frac{e^2}{\hbar c}\frac{mc^2}{\hbar}\gamma^2\left(\frac{H}{H_q}\right)^2 mc^2 \quad , \tag{4.16}$$

which can be integrated to give the energy as a function of time:

$$E = \frac{E_0}{1 + \frac{2}{3}\frac{e^2}{\hbar c}\frac{mc^2}{\hbar}\gamma\left(\frac{H}{H_q}\right)^2 t} \quad . \tag{4.17}$$

The lifetime, defined as the time for an electron to lose half of its energy (for a classical electron, $\tau_s^{(c)}$) is given by

$$\tau_s^{(c)} = \left[\frac{2}{3}\frac{e^2}{\hbar c}\frac{mc^2}{\hbar}\gamma\left(\frac{H}{H_q}\right)^2\right]^{-1} = 5.1 \times 10^7(\gamma H^2)^{-1} \text{ sec.} \tag{4.18}$$

Very little energy is radiated above a critical frequency ω_c given by

$$\omega_c = 3\gamma^3\omega_L \cong \frac{1}{\hbar} \cdot 116\,\gamma^2\left(\frac{H}{10^{13}}\right) \text{ keV} \quad . \qquad (4.19)$$

Strong coupling regime (Erber 1966, Klepikov 1952). Neglecting radiation reaction, the rate of emission resembles, but does not coincide with, the classical result. For the regime $\gamma(H/H_q) >> 1$, the energy loss rate is similar to the classical case with $h\nu/E$ replaced by $1 + h\nu/E$ where $h\nu$ is the energy of the photon and E the electron energy. An expression that is valid for $E >> h\nu$ is

$$\frac{dE}{dt} = 0.517\frac{e^2}{\hbar c}\, mc^{-2}\frac{mc^2}{\hbar}\left(\frac{\gamma H}{H_q}\right)^{3/2}\left(\frac{\hbar\omega_c}{\gamma mc^2}\right)^{1/3}\left(1 - \frac{2}{3}\frac{\hbar\omega_c}{\gamma mc^2}\right) \quad . \qquad (4.20)$$

When $\hbar\omega \sim \hbar\omega_c$, equation (4.20) gives a much larger rate than this. An upper limit of the lifetime $\tau_s^{(u)}$ (for a classical electron in the strong coupling regime) can be obtained by assuming the validity of (4.20) even at $\omega \sim \omega_c$. The result is:

$$\tau_s < \tau_s^{(u)} \cong 10^{-19}\left(\frac{H_q}{H}\right)^{2/3} \text{ sec} \qquad (\gamma^2 H >> 10^{15}) \quad . \qquad (4.21)$$

Equation (4.21) shows that $\tau_s^{(u)}$ is independent of E. In reality it is expected that τ_s will decrease with increasing E.

Radiation damping.

C. S. Shen (1970) has extended the classical calculation of synchrotron radiation to include radiation damping (the deceleration that causes the spiraling of the electrons is included). The result for energy loss is

$$\frac{dE}{dt} = -\frac{r^2\omega_L{}^2 \sin^2\phi}{\omega_0}\left[1 - 6\frac{R^2}{r}\sin^2\phi\right]mc^2 \qquad (4.22)$$

where

$$\omega_0 = 3mc^3/2e^2 = 1.8 \times 10^{23}\ \text{sec}^{-1}, \quad R = \gamma^2\,\omega_H/\omega_0$$

and ϕ is the angle between the instantaneous velocity **V** and the field **H**. This energy loss rate should be used whenever $\gamma^2 H \gtrsim 10^{15}$.

Quantum regime. In this case the synchrotron radiation is a line of an energy equal to the difference in energy between the two states. For the $n = 1$ to $n = 0$ transition the energy of the photon is 116 (H/H_q) keV and the lifetime (in the quantum regime, $\tau_s^{(q)}$) is given by Chiu and Fassio-Canuto (1970)

$$\tau_s^{(q)} = \left[\frac{1}{2}\,\frac{e^2}{\hbar c}\,\frac{mc^2}{\hbar}\left(\frac{H}{H_q}\right)^2\right]^{-1} = 3.5\times 10^{-19}\left(\frac{H_q}{H}\right)^2 \text{ sec.} \qquad (4.23)$$

(ii) Continuum radiation. The synchrotron radiation gives a continuum only in the classical case or when the emitting region has large-scale magnetic-field inhomogeneities. The true continuum process in a magnetic field is the electron bremsstrahlung process (Canuto and Chiu 1970).

For large quantum numbers the cross section is the same as for the case of the absence of a magnetic field. At low quantum numbers, it is possible to de-excite a level via Coulomb collision, followed by the emission of a photon:

$$e^-(n,p_z) + (Z,A) \rightarrow e^-(n',p_z') + (Z,A) + \gamma \quad . \qquad (4.24)$$

In so doing, the electron changes from an initial state (n,p_z) to a final state (n', p_z'), and as long as energy is conserved there is no restriction on n, n', p_z, and p_z' (e.g., a bremsstrahlung transition from a state n to a state $n' > n$ is allowed as long as energy is conserved). The transition probability for $n = 0 \rightarrow n = 0$ transition in the nonrelativistic regime and in the frequency range $\hbar\omega << \epsilon$ is (Canuto and Chiu 1970):

$$W = W_0(p\omega)^{-1} \ln(\gamma_0\Lambda)^{-1} \quad ; \qquad (4.25)$$

where

$$\begin{aligned} W_0 &= Z^2\alpha^3\pi^2 N_i \lambda\!\!\!^{-}_c{}^3 hc^2 (H/H_q)^{-2} \\ &= 2.09\times 10^{-37}\, Z_1 N_i (H/H_q)^{-2} \ (\text{erg cm}^2\ \text{sec}^{-1}) \end{aligned} \qquad (4.26)$$

$\alpha = e^2/hc$, Z = nuclear charge, N_i = ion density, $\lambda\!\!\!^{-}_c = \hbar/mc$, $\gamma_0 = 1.78102\ldots$ (Euler's constant), $\omega = 2\pi\nu$, $\Lambda = (h\omega)^2/[4\epsilon mc^2 (H/H_q)] << 1$

$$\epsilon = \text{electron energy} = p^2/2m \quad . \qquad (4.27)$$

The transition probability (eq. 4.25) is proportional to $1/p$ which in the classical expression is proportional to p. This is due to the fact that in a strong field the motion perpendicular to the field is restrained and the electron moves kinematically as a one-dimensional particle. For a one-dimensional particle the density of final states is $dp/dE \propto p^{-1}$. For a three-dimensional particle the expression for the density of state is $d^3p/dE \propto p$. As we will see later this makes it possible to have a negative absorption coefficient.

D. The Equations of State (Canuto and Chiu 1968*a*,*b*).

In a classical plasma the gas pressure is isotropic and, for conditions usually encountered in astrophysics, the noninteracting gas model applies. For low density the pressure is accurately given by that of an ideal gas, namely, $P = NkT$ where N is particle density. This is because in the classical approximation an electron can have arbitrary values for the $\perp$ momentum. Under conditions of thermodynamic equilibrium there is an equipartition of all momentum states. The only anisotropy that can be caused by the presence of a magnetic field is the magnetic stress, which is $B^2/8\pi$ in the $//$ direction and is $-B^2/8\pi$ in the $\perp$ direction.

In the quantized version the values of $p_x^2 + p_y^2$ are quantized and anisotropy can exist. As an extreme example, consider a gas in which all electrons are in the ground state (with $n = 0$). The expression for the electron energy is then

$$E = mc^2[1 + (p_z/mc)^2]^{1/2} \quad , \qquad E - mc^2 \approx p_z^2/2m \quad , \tag{4.28}$$

and there are no components of p_x and p_y respectively. Since a pressure is caused by an exchange of particle momentum with a given surface (wall), the absence of a certain component of the momentum implies the absence of pressure in the corresponding direction. (This conclusion is substantiated by a detailed calculation, as will be shown later).

The equation of state of an electron gas in strong quantizing fields has been extensively studied by Canuto and Chiu (1968*a*,*b*). The expressions for the equation of state are fairly complicated compared to the field-free case, and they will not be discussed here. We will only discuss the qualitative behavior of the two most interesting cases, the degenerate case and the nondegenerate case.

In the degenerate case all electron states are occupied up to a Fermi energy ϵ_F. In terms of the principal quantum number n, all states are occupied up to and including the state n such that

$$[1 + 2(H/H_q)n]^{1/2} - 1 \leqslant \epsilon_F/mc^2 \leqslant [1 + 2(H/H_q)(n + 1)]^{1/2} - 1 \quad . \tag{4.29}$$

The Fermi momentum $p_F^{(n)}$ of each Landau level is different and is determined by the condition

$$[1 + 2(H/H_q)n + (p_F^{(n)}/mc)^2]^{1/2} = \epsilon_F/mc^2 \quad . \tag{4.30}$$

As ϵ_F increases, the inequality sign on the right-hand side of equation (4.29) eventually changes direction and a new state $n + 1$ is added to the system. This is the same as adding another discrete function to the expression for the equation of state. Consequently the derivatives of thermodynamic quantities (such as the pressure P, the energy density E, and the number density N) are discontinuous although physical quantities in the examples given are still continuous. This behavior is shown in Figure 4.

In the nondegenerate case all states are occupied to some extent; but the probability for the occupation of a particular state n is proportional to $\exp\{-[E(n,p_z) - E(0,0)]/kT\}$, and, in particular, if $[E(n,0) - E(0,0)] >> kT$, the occupational number of this state n is exceedingly small. Since in the nonrelativistic limit $[E(n,0) - E(0,0)] \approx (H/H_q)nmc^2$, when $(H/H_q)mc^2 >> kT$, the population in all states but the ground state is small. As a result, the $\perp$ pressure is also negligible compared to the // pressure. In this case the expressions for the $\perp$ pressure and the // pressure are as follows:

$$P_{//} = N_e kT$$

$$P_{\perp} = \left(\frac{H}{10^{13}}\right)\frac{72}{T_6^{3/2}}\exp\left(-\frac{H}{10^{13}}\,\frac{1.337 \times 10^3}{T_6}\right)\cdot P_{//}\ ,\quad T_6 = T/10^6 \quad . \tag{4.31}$$

As an example, at $H = 10^{13}$ gauss, $T_6 = 1$, we find that $P_{\perp} \cong e^{-1000}\,P_{//}$.

Because of their larger mass, the ions remain classical until the field is above 10^{19} gauss. Therefore the ion pressure in both the // and the $\perp$ directions are the same and to a high degree of accuracy $P_{ion} = N_{ion}\ kT$. At $H > 10^{10}$ gauss, $T_6 \approx 1$, the $\perp$ pressure is essentially due to that of the ions, while the // is mainly due to electrons (assuming the composition to be Fe^{56}). This means that the molecular weight (the number of proton masses per gas particle) is also anisotropic. As the scale height is proportional to (molecular weight)$^{-1}$, in neutron stars the scale height along the field line is roughly two times that which is $\perp$ to the field.

E. The Propagation of Electromagnetic Waves in a Magnetized Medium.

The propagation of electromagnetic radiation is affected by the presence of a plasma and a magnetic field. The discussion of the

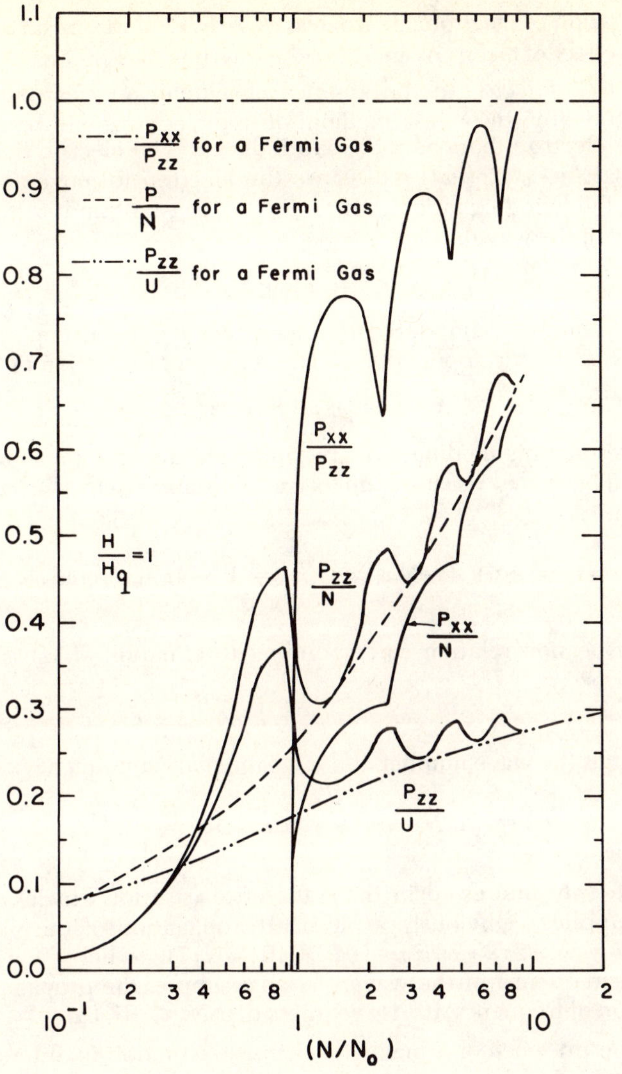

FIG. 4 — Functional dependence of $P_\perp/P_{||}$, $P_{||}/N$, $P_\perp/N$, and $P_{||}/U$ or N for the degenerate case with $H/H_q = 1$. The corresponding functions for a Fermi gas are also shown for comparison.

propagation of electromagnetic waves is divided into two parts, one on the effect of electrons and the other on the effect of ions.

Electron effects. In the absence of a field, an electromagnetic wave traveling through a medium of free electrons will be affected by the electron motions the wave induces. The electric field component $\mathbf{E} = \mathbf{E}_0 \exp(i\omega t)$ will cause the electron to move according to the following equation (where $\mathbf{E}_0$ is a constant and ω is the frequency of the wave):

$$m\ddot{\mathbf{r}} = -e\mathbf{E}_0 e^{i\omega t} \quad . \tag{4.32}$$

In this equation nonrelativistic mechanics has been imposed (this is justified in our applications). The solution of equation (4.32) is

$$\mathbf{r} = +e/m\,\omega^2\,\mathbf{E}_0\,e^{i\omega t} \quad . \tag{4.33}$$

As a result, this medium will be polarized and the polarization is $N_e\mathbf{r}e$. This gives rise to a dielectric constant ϵ which is different from unity:

$$\epsilon\mathbf{E} = \mathbf{E} + 4\pi e N_e \mathbf{r} \quad ; \qquad \epsilon = 1 - 4\pi N_e e^2/m\omega^2 \quad . \tag{4.34}$$

The dispersion relation corresponding to equation (4.34) is (transversal case)

$$\omega^2 = \omega_p{}^2 + k^2c^2 \quad , \tag{4.35}$$

where k is the wave number and ω_p is the plasma frequency

$$\omega_p = [4\pi N_e e^2/m]^{1/2} = 5.6 \times 10^4 \sqrt{N_e} \text{ Hz} \quad . \tag{4.36}$$

We are only interested in the transversal case for which equation (4.35) applies. Obviously a physical propagation of the wave requires $\nu = \omega/2\pi > \omega_p/2\pi = 0.9 \times 10^4 \sqrt{N_e}$ Hz where ν is the natural frequency of the wave. As an example, the propagation of radiation of frequency 10^8 Hz requires that $N_e \leq 10^8 \text{ cm}^{-3}$.

In the presence of a mag ietic field the situation is different. An electric field perpendicular to the magnetic field can only introduce a drifting velocity $v_D = c|\mathbf{E}|/|\mathbf{H}|$ which (a) is small under normal circumstances and (b) does not cause charge separation (and hence no polarization, to the first order). This can substantially reduce the value of ω_p.

The dielectric constant in a magnetic field has been studied widely, both with and without quantum effects. As the frequency of radiation is small compared with the Larmor frequency of the ions and the electrons, as well as the plasma frequency of the electrons and the ions, we can use the dielectric constant computed for a cold ionic neutral plasma in a magnetic field. It turns out that quantum mechanical calculations (Kelly 1964) and classical calculations coincide in this limit.

The dielectric tensor is (assuming that the magnetic field is in the z direction (Stix 1962)):

$$\epsilon_{\alpha\beta} = \begin{pmatrix} S & -iD & 0 \\ iD & S & 0 \\ 0 & 0 & P \end{pmatrix} \tag{4.37}$$

where

$$\begin{aligned} R &= 1 - \frac{\omega_p^2}{\omega^2} \cdot \frac{\omega}{\omega - \omega_H} - \frac{\omega_I^2}{\omega^2} \cdot \frac{\omega}{\omega + \Omega_H} \\ L &= 1 - \frac{\omega_p^2}{\omega^2} \cdot \frac{\omega}{\omega + \omega_H} - \frac{\omega_I^2}{\omega^2} \cdot \frac{\omega}{\omega - \Omega_H} \end{aligned} \tag{4.38}$$

$$P = 1 - \left(\frac{\omega_p}{\omega}\right)^2 - \left(\frac{\omega_I}{\omega}\right)^2, \quad 2D = R - L, \quad 2S = R + L \tag{4.39}$$

$$\begin{aligned} \omega_p^2 &= 4\pi e^2 N_e/m \qquad \omega_I^2 = 4\pi e^2 Z N_i/m \\ \omega_H &= eH/mc \qquad \Omega_H = Z\,\omega_H m/M_i \ . \end{aligned} \tag{4.40}$$

ω_p and ω_I are plasma frequencies of the electron gas and the ion gas, respectively, and ω_H and Ω_H the corresponding Larmor frequencies.

Generally speaking, the photon propagation in a doubly refractive medium such as equation (4.37) can be analyzed into two modes, the ordinary and the extraordinary mode of propagation. The propagation of these two modes can be easily studied for the direction $\theta = 0$ and $\theta = \frac{1}{2}\pi$, and can be studied numerically at other angles.

Along the direction $\theta = 0$ these two modes are right-handed (R) and left-handed (L) circularly polarized, and the refractive indices for these two modes are: (o = ordinary, x = extraordinary)

$$n_x^2 = R \tag{4.41}$$

$$n_o^2 = L \tag{4.42}$$

Now we will substitute into these equations quantities pertinent to a neutron star: $H = 10^{12}$G, $N_e = 10^{24}$. We find $\omega_p \simeq 10^{16}$, $\omega_l \simeq 10^{14}$, $\omega_H \simeq 10^{19}$, $\Omega_H \simeq 10^{16}$. If we choose $\omega = 10^9$ Hz, we find that $E_1{}^{-2} = E_2{}^{-2} = 1$. For R and L we can expand equations (4.38) and (4.39) into power series in ω/ω_H and ω/Ω_H. For a charge neutral plasma, for which equation (4.40) is valid, we find that the first nonvanishing term in the expansion gives:

$$R = 1 + \frac{\omega_p^2}{\omega_H^2} + \frac{\omega_l^2}{\Omega_H^2} \simeq 1 \tag{4.43}$$

$$L = 1 - \frac{\omega_p^2}{\omega_H^2} - \frac{\omega_l^2}{\Omega_H^2} \simeq 1 \tag{4.44}$$

and hence the propagation of electromagnetic radiation is possible.

In the perpendicular direction there is only one mode of propagation; this is the extraordinary mode. The refractive index for the ordinary mode is simply $1 - [(\omega_l/\omega)^2 + (\omega_p/\omega)^2]$, and propagation of this mode is forbidden. For the extraordinary mode the refractive index is:

$$n_x^2 = RL/S \approx 1 \tag{4.45}$$

and propagation is the same as in vacuum. This mode, whose direction is perpendicular to both the wave vector k and the magnetic field H, has 100 percent linear polarization.

F. Electron Scattering Opacity in an Intense Magnetic Field.

Canuto (1970) investigated the electron scattering opacity (the *only* source of continuum opacity other than the free-free transition) and he found that the opacity is strongly suppressed by the pressure of a field in the forward angle. This is shown in Figure 5. This scattering opacity can give rise to a strong beaming effect.

G. The Existence of Permanent Magnetism in Condensed Bodies.

Electrons moving in circular orbits can give rise to a magnetic moment, and, in addition, the spin also contributes to the magnetic

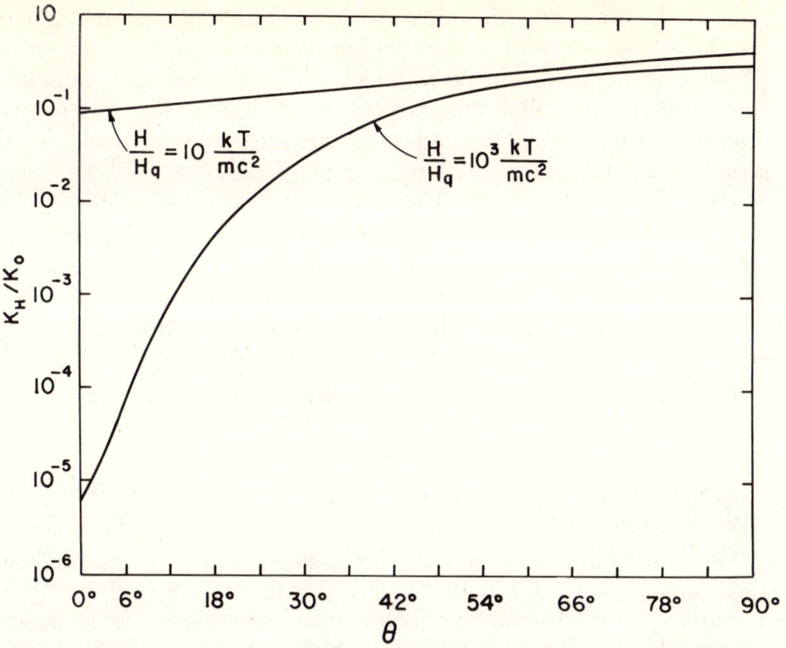

FIG. 5—The angular dependence of the electron scattering opacity K_H with respect to the direction of the field, K_0 is the classical value without a magnetic field, θ is the angle between the photon and the field.

moment. These moments give rise to a macroscopic magnetization $\overline{M}$ such that the magnetic induction B is given by

$$B = H + 4\pi\overline{M} \tag{4.46}$$

where $\overline{M}$ is the sum of moments of all electrons in their respective Landau orbits (Canuto and Chiu 1968*c*). However, as each electron senses the magnetic induction B and not H, $\overline{M}$ is a function of B and not of H, i.e., $\overline{M} = \overline{M}(B)$. The question thus arises whether there exist solutions for equation (4.46) in the case $H = 0$. If solutions exist and are stable against spontaneous decay then an electron gas can have a permanent magnetism.

Such solutions have been found and have been shown to be stable by Lee et al. (1969) and by Canuto et al. (1970). This type of permanent magnetism is different from the ordinary type of ferromagnetism. In the ordinary type of ferromagnetism the permanent magnetic moment arises because of interactions of electrons in atomic orbits of ferrous metals. The permanent magnetic moment discussed here arises from the moments of electrons in their respective Landau orbits. This type of magnetization is tentatively called *Lofer,* extracted from the longer name *La*ndau *o*rbital *fer*romagnetism.

It has been shown that the Lofer state is stable in that the probability for a magnetized system to make a transition into a nonmagnetic state is negligible provided the temperature is less than a critical temperature which is in the neighborhood of $10^{8\text{-}9\circ}$ K for neutron star matter.

For a given density, there exists a maximum Lofer field B_{ML}. It is found that

$$B_{ML} \propto N_e^{2/3} \propto R^{-2} \quad , \tag{4.47}$$

which bears a close resemblance to the flux conservation law discussed earlier.

The Lofer field is strong only in collapsed bodies of high density. In ordinary matter the field strength due to Lofer is of the order of 1 gauss or less. In the case of neutron stars the value of B_{ML} is approximately 10^{13} gauss. In white dwarfs the corresponding Lofer field is of the order of 10^7 gauss.

It is thus highly possible that permanent magnetism is a feature of all collapsed bodies, including neutron stars (and possibly, even present in the case of Schwarzschild singularities).

5. Physical Emission Mechanisms

Basic considerations. In the rotating magnetic neutron-star model, it is necessary to have a radiation mechanism which can emit a sharp pencil or fan-shaped beam of radiation corotating with the star. The first step in searching for such a radiation mechanism is to identify and locate the emission region—whether it is on the surface or away from the star.

Before discussion, it is necessary to bear the following observational facts in mind:

(i) The emission region must not have a physical dimension greater than 30 km.

(ii) The radio flux (based on the estimated distance, the observed flux, and the maximum dimension of the emitting region) from the surface of the emitting region is in excess of 10^{14} ergs cm^{-2} sec^{-1} and in certain cases as high as 10^{17} ergs cm^{-2} sec^{-1}. In the optical emission the energy flux is about 10^{20} ergs cm^{-2} sec^{-1}. The brightness temperature associated with these fluxes is typically of the order of 10^{24}° K.

(iii) The arrivals of individual pulses have been timed to coincide with the calculated arrival times (on the assumption of constant period) to an accuracy of 0.01 times the pulse width.

The average of many pulses over a long-time baseline gives a period accurate to nine decimal places in some cases. From accurate determinations of pulsar periods the slowing down rate has also been measured to six decimal places. This means that the phase of the emission with respect to rotation is constant to a great accuracy. If the phase slips, then a similar accuracy must be maintained in the slowing rate.

The implication of (ii) is that a coherent rather than a random process must operate as a radiation mechanism. A random process cannot give rise to a brightness temperature greater than the kinetic temperature of the emitting particle. In order to produce a brightness temperature of 10^{21}° K by a random process the required electron energy must exceed 10^{17} eV, which is a clear absurdity. A coherent process—such as a laser process, which will be discussed below—can give rise to a very high brightness temperature without requiring uncomfortably high energy electrons. However, a laboratory laser operating on the transition between two atomic states can only give rise to sharply defined lines. A different laser process must operate in the case of pulsars.

The implication of (iii) is that the emitting region must be very tightly bound to the star; otherwise a high accuracy in both the determined period and the time of arrival cannot be achieved. The observed accuracy in the determination of the time of arrival of individual pulses limits the wobbling of the emitting region from a central position by no more than a few kilometers in the case of the Crab nebula pulsar NP 0532.

Location of the emitting region. There are two possibilities: The emitting region may be located either on the surface of the star or away from the star. The respective advantages and difficulties of these two possibilities will now be discussed.

(a) Emission away from the neutron star. This is an extension of the emission theory of the most frequently encountered radio sources. Most radio sources are tenuous nebulae and the emission region extends over large parts of space. The emission process is believed to be the classical synchrotron radiation from energetic electrons (of energy ranging from $10^{9\text{-}14}$ eV) moving in magnetic fields of strength of the order of 10^{-5} gauss. Certain stars, notably the sun, are also radio emitters and the emitting region is located outside the surface of the main body. For the sun the emission of radio waves occurs predominantly outside the surface of the sun in the coronal region.

If we apply the mechanism of radio emission for ordinary nebulae to pulsars, there are a number of difficulties which must be overcome to make this mechanism attractive.

(i) The radio flux per unit area per second from radio nebulae is usually very small and the brightness temperature hardly exceeds $10^{9\circ}$ K. The lack of correlation among particles in a tenuous medium is another difficulty for collective phenomena.

(ii) The repetition rate of pulses is very regular although the amplitudes of pulses are not. If the pulses are produced in a region far away from the surface and energy is transmitted to this region via some type of plasma link, then the width of this plasma link must be smaller than 30 km. It is difficult to see how dynamical stability can be maintained if this plasma link is located, say tens of thousands of miles away, as is required in some theories.

(iii) Because of the intense gravitational field near a neutron star, it is necessary to accelerate particles to high energy to overcome the surface gravity of the neutron star. Acceleration mechanisms have been proposed by Gold (1969) and by Goldreich and Julian (1969). However, the synchrotron energy loss rate for electrons in intense magnetic fields (which are present near the surface of the star) is so large that even for a relatively low energy electron of one MeV, the lifetime is of the order of 10^{-19} sec if the field is a few times 10^{13} gauss. In order to accelerate a particle to relativistic energy with a net gain in energy, the energy supply rate is nearly 10^{15} eV/cm.

(iv) Even if acceleration were possible, these electrons will then continuously emit radiation while traveling from the surface to whatever region is chosen for emission. The total dimension of the radiation zone will have to be considerably greater than 30 km, unless some mechanism is found to restrain the energy loss of the electrons until arrival at the prescribed emission region.

At present there are no solutions to these difficulties. In conclusion, the synchrotron radiation loss is the most severe difficulty if the emission region is required to be outside the star.

(b) Surface emission. This alternative has not been taken seriously until recently, as no celestial object is known to emit intense radio emission from its surface. Since there is an immense difficulty in developing a theory to account for the radio and optical pulses from pulsars via an emission-at-a-distance theory, this alternative, however improbable at first glance, must be taken seriously.

Two difficulties must be overcome.

(i) Plasma-frequency restriction. As we have seen, the plasma-frequency restriction is removed if an intense magnetic field is present. In fact, the emitted radiation is beamed along the magnetic field lines, so that anisotropic emission is achieved. (See Section 4*E*.)

(ii) Self-absorption. In a dense emitting medium, self-absorption limits emission to a layer of one photon mean free path. The resultant radiation flux cannot exceed that given by a blackbody at the same kinetic temperature as the medium.

Self-absorption is a feature encountered predominantly in cases of thermodynamic equilibrium. We will show, in (c) below, that self-absorption disappears when there is a population inversion—i.e., when the level temperature of the medium is negative.

(iii) Another problem is the existence of a coherent-radiation mechanism. This amounts to the same as to require a mechanism to maintain population inversion. This will be discussed in Section 9.

(c) Coherent emission. Theory of lasers. The usual theory of lasers is developed on the basis of the following photon dispersion relation which applies to vacuum only:

$$\omega = kc \tag{5.1}$$

We will now show that the usual theory of lasers is also applicable even when the medium is anisotropic and the dispersion relation differs from equation (5.1).

The effect of a medium on the propagation of electromagnetic radiation is taken into account by the dispersion relation of the photon: only those modes with a nonvanishing value of k can propagate. The dispersion relation specifies the relation between the photon frequency ω, the photon momentum $\mathbf{p} = \hbar\mathbf{k}$, and their relations with other variables (field strength, direction of propagation, and so on). The dispersion relation appears conspicuously in two quantities: $d\omega/dk$ and the normalization of the wave function of the photon (Adams, Ruderman, and Woo 1963). The first quantity affects the expression for the intensity of radiation in equilibrium (which is used to determine the induced emission coefficient B, as will be discussed later), and the latter affects the calculation of the emission coefficient so that photons of certain categories cannot be produced at all.

The radiative transfer equation is

$$\frac{dI}{ds} = j - \kappa I \tag{5.2}$$

where I is the intensity of radiation, j is the mass emission coefficient, and κ the mass opacity coefficient. Equation (5.2) is a general relation even when a magnetic field is present since a magnetic field does not *directly* affect the wave function nor other dynamical properties of the photon. Very generally we can write the mass emission coefficient j as

$$j = N_2 W_2 \hbar\omega + BI \tag{5.3}$$

where N_2 is the number density of particles in state 2 which are capable of making a transition to the state 1 by emitting a photon of energy $\Delta E = \hbar\omega$, and W_2 is the transition probability of this emission process. BI represents the stimulated emission and B is the stimulated emission coefficient. The opacity coefficient κ is likewise written as

$$\kappa = N_1 \sigma_1 \tag{5.4}$$

where N_1 is the particle density in state 1 which can absorb the photon $\hbar\omega$ to make a transition to the state 2, and σ_1 is the absorption cross section.

It remains to determine B. In equilibrium $dI/ds = 0$ and we have

$$\frac{dI}{ds} = N_2 W_2 \hbar\omega + BI - N_1 \sigma_1 I = 0 \quad . \tag{5.5}$$

Solving for I, we find

$$I = \frac{N_2 W_2 \hbar\omega / B}{\frac{N_1 \sigma_1}{B} - 1} \quad . \tag{5.6}$$

Comparing this expression with that in equilibrium, that is,

$$I = g \hbar\, \omega\, k^2 (2\pi)^{-3} / (\exp \frac{\hbar\omega}{kT} - 1) \tag{5.7}$$

we obtain

$$B = \frac{N_2 W_2}{g k^2} (2\pi)^3 \quad . \tag{5.8}$$

Here g is statistical weight of the photon. Equation (5.8) thus fixes B. The theorem of detailed balance relates the emission transition probability to the absorption cross section as follows:

$$\frac{W_2}{g k^2} (2\pi)^3 = \sigma_1 \quad . \tag{5.9}$$

The radiative transfer equation can now be written as

$$\frac{dI}{ds} = N_2 W_2 - N_1 \sigma_1 (1 - \frac{N_2}{N_1}) I \quad , \tag{5.10}$$

the excitation temperature T_x can be defined such that

$$\frac{N_2}{N_1} = e^{-\hbar\omega / kT_x} = e^{-\Delta E / kT_x} \quad , \tag{5.11}$$

and we then have the usual equation of radiative transfer

$$\frac{dI}{ds} = N_2 W_2 - \kappa' I \tag{5.12}$$

where the opacity is replaced by the one with stimulated coefficient

$$\kappa' = \kappa [1 - \exp(-\frac{\hbar\omega}{kT})] \quad . \tag{5.13}$$

Let us go back to equation (5.10) and consider the case $N_2 > N_1$. The level excitation temperature T_x is now negative. The stimulated emission is stronger than the absorption and the overall opacity (taking into account stimulated emission) becomes negative. All terms in equation (5.10) are positive and the medium emits at a natural rate $N_2 W_2 \hbar\omega$, plus the stimulated emission term $(N_2/N_1 - 1)\, N_1\sigma_1$. It is therefore proven generally that when a population inversion takes place between two states N_2 and N_1 (whether they belong to a continuum or to a set of discrete levels does not matter) the absorption coefficient becomes negative and there is only pure emission.

Thermodynamically speaking, a negative temperature has an infinite store of free energy and it can emit at a rate governed only by the emission coefficient. This has been demonstrated in the case of masers and lasers.

In a continuum emission process such as the bremsstrahlung process the absorption coefficient is summed over pairs of electron energy states, and the result for the bremsstrahlung process in a magnetic field is (Chiu and Canuto 1970)

$$\kappa = \frac{4h\nu^3}{c^2}\, mh\nu \int_0^\infty gh^{-1}\, dp \qquad (5.14)$$
$$\left[f(p)\frac{\partial}{\partial p}\frac{p}{p'}\frac{W(p,p')}{(p+p')} + f(-p)\frac{\partial}{\partial p}\frac{W(p,p')}{(p'-p)} \right]$$

where g is the statistical weight factor, p the electron momentum after absorption of a photon h, $f(p)$ the distribution function, p' the momentum before absorption of a photon, $W(p, p')$ is the emission transition probability per photon state and is given by equation (4.25)

Under conditions such that $f(p) >> f(-p)$ (a coherent streaming motion of the one type of charge with respect to the other) one finds that

$$\kappa < 0 \text{ when } \frac{\partial}{\partial p}\frac{W(p,p')}{p+p'}\frac{p}{p'} < 0 \quad ; \qquad (5.15)$$

since $p \sim p'$ is in the radio regime, one can conclude that the absorption coefficient will become negative if $W(p, p')$ does not rise faster than the first power of p. This condition is not fulfilled for an electron gas in the absence of a field but is fulfilled for an electron gas in an intense field.

A treatment by Simon and Strange (1969) gives the criterion

$$\frac{\partial}{\partial p} W(p) < 0 \tag{5.16}$$

for the condition for a negative absorption coefficient. This disagrees with our result. The origin of this disagreement has been traced to their improper way of extracting differentials.

This electric field can cause charge separation and can accelerate particles (electrons and ions) away from the surface to form a magnetosphere. According to their theory, acceleration of particles to an energy of 10^{18} eV is not impossible.

6. Gold-Goldreich-Julian Theory (Magnetosphere of Neutron Stars)

Gold (1969) proposed a mechanism by which acceleration of particles can take place. The essence of this theory is as follows: In a magnetic field the particle motions are locked with the magnetic field lines. If a magnetosphere exists, this magnetosphere will corotate with the star. The velocity due to rotation increases with distance, and eventually when the particles reach a radius R_c such that $R_c = c/\Omega$ the particle velocity reaches the velocity of light. The particles then become ejected into the interstellar medium. This distance (called the velocity-of-light circle, or cylinder) is about 1500 km away from the star in the case of the Crab nebula pulsar.

Goldreich and Julian (1969) advanced Gold's idea further. They developed a theory based on the principle of unipolar induction. A rotating magnet will, due to unipolar induction, generate in its vicinity an electric field E of a strength approximately given by

$$E \sim \frac{vH}{c} = 10^{11}\left(\frac{H}{10^{13}}\right)\left(\frac{v}{10\ \text{km/sec}}\right)\ \text{volt/cm}\ . \tag{6.1}$$

This electric field can cause charge separation and can accelerate particles (electrons and ions) away from the surface to form a magnetosphere. According to their theory, acceleration of particles to an energy of 10^{18} eV is not impossible.

This is an interesting and promising theory in the problem of cosmic ray acceleration and in the origin of magnetic fields in general. In particular, the Crab nebula contains too much magnetic field to be of stellar origin. According to Goldreich and Julian's theory, the field of the Crab nebula can be generated by the charged particles passing through the velocity-of-light circle. In their theory the field in the Crab nebula is of the torroidal shape, and this is consistent with the fact that the mass in the nebula is about 0.8 $\mathfrak{M}_\odot$. If the field is not of the torroidal shape, the minimum mass needed to maintain this field gravitationally will be over 30 $\mathfrak{M}_\odot$.* These charged particles which are accelerated through the velocity-of-light circle can also explain other activities of the Crab nebula.

7. Ostriker-Gunn-Pacini Theory (Magnetic-Dipole Radiation from Neutron Stars)

Based on an earlier idea of Pacini (1968), Gunn and Ostriker (1969) proposed a theory based on the classical radiation from rotating magnetic dipoles (Ostriker and Gunn 1969).

A rotating magnetic dipole will generate in its vicinity a time-varying magnetic field. However, at distances larger than the velocity-of-light circle this time-varying field is transformed into an electromagnetic wave, as the magnetic field lines can no longer co-rotate with the magnet. The frequency of this electromagnetic wave is the same as that of the rotating dipole.

Classically a rotating magnetic dipole in a vacuum will emit radiation at the rate (Gunn and Ostriker 1969; Landau and Lifshitz 1951, p. 295)

$$\frac{dE}{dt} = -\frac{2\overline{M}\sin^2\theta\Omega^4}{3c^2} \qquad (7.1)$$

where $\overline{M}$ is the magnetic dipole moment and is given by

$$\overline{M} = \frac{\Phi^2 a^2}{4\pi^2} = \frac{B_P{}^2\, a^6}{4} \quad . \qquad (7.2)$$

Here B_P is the field at the pole, Φ the magnetic flux through the half sphere about the pole, and θ the angle between the dipole moment and the axis of rotation. Taking the following parameters,

$$\Omega = 10\ \text{sec}^{-1},\ \theta = \frac{\pi}{4},\ a = 10\ \text{km},\ B_P = 10^{12}\ \text{gauss}, \qquad (7.3)$$

Ostriker and Gunn found that the radiation rate is 3×10^{38} ergs/sec, which is comparable to the energy loss rate of the Crab nebula.

---

*P. Sturrock, discussion during this symposium. See page 557.

From this Ostriker and Gunn obtained a slowing down rate of pulsars as follows:

$$\frac{dj}{dt} = -\frac{1}{\Omega}\frac{dE}{dt}; \qquad \frac{d\Omega}{dt} = \frac{\dot{j}}{I} = -\frac{|\Phi|^2 \sin^2\theta\, a^2\Omega^3}{6\pi^2 I c^2} \tag{7.4}$$

where I is the moment of inertia of the system. This equation can be solved to give

$$\Omega = \Omega_0\left(1 + 2\frac{t}{\tau_0}\right)^{-1/2} \quad ;$$

$$-\frac{1}{2\pi}\frac{d\Omega}{dt} = \frac{dP}{dt} = \frac{\Delta P}{P} = \frac{|\Phi|^2 \sin^2\theta\, a^2\Omega^3}{12\pi^3 I c^2} \tag{7.5}$$

so that $\Delta P/P$ is proportional to the cube of the angular frequency of rotation.

Based on this theory and the observed value of $\Delta P/P$ they found that for the pulsar NP 0532 the second derivative of P is given by

$$6 \times 10^{-24}\ \mathrm{sec}^{-1} = 0.016 \times 10^{-9}\ \mathrm{day}^{-1}\ \mathrm{yr}^{-1} \leqslant \frac{d^2P}{dt^2}$$

$$\leqslant 0.039 \times 10^{-9}\ \mathrm{sec\ day}^{-1}\ \mathrm{yr}^{-1} = 1.3 \times 10^{-24}\ \mathrm{sec}^{-1} \quad , \tag{7.6}$$

which agrees well with the observed value, which is 2×10^{-24} sec^{-1}.* However, this observed value is quite preliminary.

Ostriker and Gunn concentrated their efforts on the problem of cosmic ray production. Because of the long wavelength, the wave should have a nearly 100 percent efficiency in accelerating particles near the radiation zone. Their suggested mechanism makes use of the fact that, in sufficiently strong wave fields, a charged particle is accelerated in the propagataion direction to nearly the velocity of light in a small fraction of a wavelength, and thereafter rides the wave at essentially constant phase, slowly taking energy from the wave. The characteristic maximum energy available for the electron can be written as

$$E_{\max} \approx mc^2\left[2\omega_L r_0 c^{-1} \ln \frac{r_c}{r_0}\right]^{2/3} \quad , \tag{7.7}$$

*Reported by T. Gold at this symposium.

where r_c is the radius at which the wave energy is exhausted, r_0 is the inner wave radius ($\sim c/\Omega$), and ω_L is the cyclotron frequency eB/mc in the wave magnetic field at maximum amplitude at r_0. For the Crab nebula the magnetic field is 10^6 gauss at $r = 10^8$ cm and $E_{max} \approx 5 \times 10^{13}$ eV if $r_c \approx 10^{12}$ cm. Such a low-frequency wave can pass through an interstellar medium (whose plasma frequency is of the order of 1000 Hz if $N_e = 0.01$ cm^{-3}) if there is a magnetic field, and this process will operate when conditions are such that

$$\omega_g\Omega >> \omega_P{}^2 = 3 \times 10^8 N_e \tag{7.8}$$

where ω_g is the gyrofrequency of the wave field and ω_P is the plasma frequency. At a field strength of 10^6 gauss, this process will accelerate injected plasmas with densities at r_0 of up to about 10^7 particles per cubic centimeter.

Beyond the velocity-of-light circle the magnetic field is transformed into an electromagnetic wave. Synchrotron radiation loss will be absent for those particles riding on the electric component of the wave. This mechanism should be a very efficient process in the acceleration of particles which may come from interstellar space, or which may come from within the velocity-of-light circle through the magnetosphere proposed by Goldreich and Julian (1969).

8. Ginzburg-Zaitsev Theory (Pulsational and Rotational Models)

Ginzburg and Zaitsev (1969) favor the pulsation theory. Drake and Craft (1968) discovered that there are second periods in the pulses with values in the neighborhood of 10^{-2} sec. Ginzburg and Zaitsev then associate the second period with the pulsation periods of neutron stars and the principal period with the period of rotation. In their model, pulses are produced by pulses and the dimension of the region of emission chosen is close to 3×10^8 cm.

This theory is reminiscent of the pulsational theory itself, and, therefore, criticism applicable to the pulsation theory applies here. In particular, the second period is too long to be accounted for by the pulsation periods of neutron stars.

9. Chiu-Canuto Theory (1969*a*, 1969*b*, 1970)

The particular model they have in mind is that of an oblique rotator; that is, the direction of the magnetic field of the star is inclined at an angle with respect to rotation. The field strength at the surface is of the order of 10^{12} gauss. This model is the same as that of Pacini (1968) and Ostriker and Gunn (1969).

In the corotating frame an electric field parallel to the magnetic field is present. This field originates from the rotation of the star, as suggested by Goldreich and Julian (1969). The field strength is of the order of 10^{10} volt/cm for a pulsar rotating with a period of 1 sec.

This electric field causes particles of one type of charge to be accelerated away from the surface. A surface electric current along the magnetic field exists, and the effect of electrons at the surface therefore possesses a coherent streaming motion with a macroscopic velocity v_m (the macroscopic momentum is p_m). The distribution function of the electron, $f(p)$, is easily seen to be that of a displaced Maxwellian distribution:

$$f(p) = C \exp \left[- \frac{(p - p_m)^2}{2mKT} \right] \tag{9.1}$$

C = normalization constant, and p is the electron momentum.

This coherent motion gives rise to a negative temperature in parts of the electron spectrum where $p < p_m$. (A negative temperature is understood to be present when $df(p)/dp > 0$.) As has been discussed earlier, if the transition probability $W(p)$ does not increase faster than the first power of p, i.e.,

$$\frac{d}{dp}[W(p)/p] < 0 \quad , \tag{9.2}$$

the absorption coefficient of radiation (including stimulated emission) can become negative so that amplification of radiation via stimulated emission (maser) is possible. This maser differs from the ordinary one in that it can amplify radiation over a continuum. Hereafter this continuum maser will be referred to as a C maser.

Electrons in fields of the order of 10^{12} G are essentially one-dimensional particles, and the transition probability for bremsstrahlung in the ground-state Landau level has a different dependence on p than for the three-dimensional case. We have shown that in the strong magnetic field case (Canuto and Chiu 1970),

$$W(p) \propto p^{-1} \quad . \tag{9.3}$$

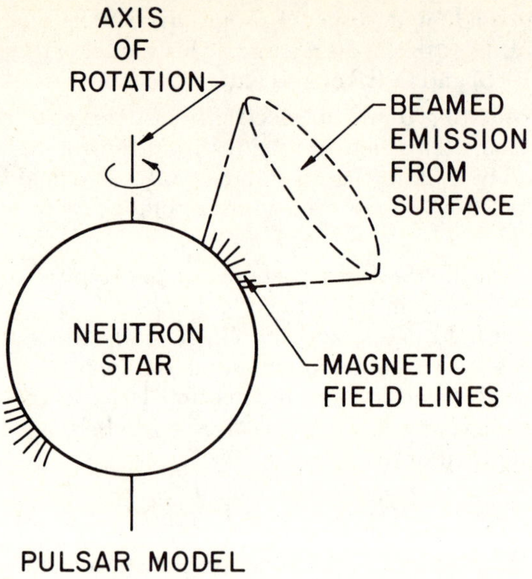

FIG. 6 — Configuration of emission. Note that the beaming is highly exaggerated here. The actual beaming is less pronounced and is discussed in Section 9.

Hence equation (9.2) is satisfied, and an electron gas in a strong field becomes a C maser if an electric current is introduced. The spectrum of this C maser is of the $\nu^{-\alpha}$ type where α is a positive number depending on the field strength.

The finite lifetime of the electron states poses a maximum power limit at which a medium can amplify radiation (this is known as saturation phenomenon in the theory of lasers). Corresponding to this maximum power is a maximum brightness temperature $T_B{}^M$. For a field strength of 10^{12} G and a surface neutron star particle density of $10^{24}/\text{cm}^3$, the value of $T_B{}^M$ is roughly $10^{24\circ}$ K. As it turns out the condition for saturation is always fulfilled for nearly all rotating neutron star models, thus in this theory all pulsars have approximately the same range of brightness temperature, which agrees with observation.

The propagation of the emitted radiation is strongly affected by the imbedding plasma and the magnetic field. At $H \sim 10^{12}$ G, a neutral plasma with particle density less than $10^{28}/cm^3$ is transparent to all frequencies up to the ion Larmor frequency, which is of the order of 10^{16} Hz As the presence of a magnetic field and particle motion is important in causing the C-maser effect, only the pole regions can emit radiation. A conical beam results from this anisotropic emission, and the rotation of the star sweeps the radiation through space, causing an observer to detect pulses as the beam sweeps by in beacon-like fashion (Figure 6).

This model, crude as it is, can explain semiquantitatively the following features of pulsar observation:

(1) That the spectrum of radiation is of the $\nu^{-\alpha}$ type where $\alpha \simeq 2$.

(2) That the radiation has approximately the same brightness temperature at $10^{24°}$ K, relatively independent of the pulsar periods (saturation phenomenon).

(3) That the radiation is beamed preferentially along the magnetic field lines to form a sharp beam, and qualitatively this theory can explain

(4) That the radiation has a component showing linear polarization and the degree and angular inclination of polarization change with the phase of the pulse. (This radiation is emitted more or less isotropically with a forward peak.)

In arriving at the above conclusions, they have not introduced any free paramenters into this theory. The particle densities and electron energies chosen were those taken from existing neutron star models constructed prior to the discovery of pulsars; the magnetic field strength used was the same as that derived by Ostriker and Gunn (1969) based on their theory of classical rotating magnetic dipoles. The uniform brightness temperature is the result of laser saturation, the condition for saturation being determined from fundamental interactions only.

Details of this theory are being published.

10. Summary and the Future of the Pulsar Problem

These four theories more or less summarize the current status of the pulsar problem. It is clear that the mechanism that is responsible for the acceleration of cosmic rays (e.g., in the Crab nebula) is different from that responsible for emitting radio and X-ray pulses. With this in mind, all theories appear to be in accord with each other.

Among these four theories the pulsating-rotation theory of Ginzburg and Zaitsev appears to have the weakest footing. In order to make their theory attractive it must be shown that neutron stars may plausibly have pulsational periods compatible with the second period. Further, a mechanism must be found to maintain the pulsation, and to couple an energy source (e.g., rotational) to the emission of radio and optical pulses.

Among the remaining theories, each appears to be satisfactory within its own account and claim. The Gold-Goldreich-Julian theory deals with the existence of a magnetosphere and the acceleration of particles through the velocity-of-light circle. These particles, even suffering synchrotron radiation loss in transit, will regain energy after being accelerated by the mechanism suggested in the Ostriker-Gunn-Pacini theory. In fact, the magnetosphere should be an important source for the supply of particles in the Ostriker-Gunn-Pacini theory. On the other hand, if both these theories attempt to explain the radio and optical pulses, several difficulties—notably the constancy of the pulsar periods, the stability of the time of arrival of pulses, and the synchrotron radiation loss for particles in transit to the velocity-of-light circle—must be overcome. The Chiu-Canuto theory attempts to explain the observed pulses from pulsars as a laser radiation from bremsstrahlung transition between two magnetic levels. This theory, on the other hand, is not designed for discussion of the acceleration of high-energy particles that evidently must exist, at least in the case of the Crab nebula.

So far, none of these theories have been developed to such a stage that a comparison with details of currently available observational data is possible. Several accounts have been given to show that a given theory can explain some observational data. But, such agreement with the observational data is, as Morrison pointed out*, an indication that the concept of a rotating magnetic neutron star model, first proposed by Gold, is essentially correct. It remains for the authors of each theory to develop their theories into more quantitative versions so that each pulsar may be studied and analyzed on the basis of its individual data. Until then, the nature of pulsars remains open.

*P. Morrison, see page 544 of this symposium.

I am grateful to Drs. V. Canuto, C. Chiuderi, S. Maran, and F. Occhionero for general discussions.

REFERENCES

Adams, J. B., Ruderman, M. A., and Woo, C. H. 1963, *Phys. Rev.* **129**, 1383.
Arnett, D. 1966, Dissertation, Yale University.
Baade, W., and Zwicky, F. 1934, *Phys. Rev.* **45**, 138.
Bondi, H. 1965, in *Quasi-Stellar Sources and Gravitational Collapse*, I. Robinson, A. Schild, and E. L. Schucking, eds. (Chicago: University of Chicago Press), p. 393.
Bradt, H., Rappaport, S., Mayer, W., Nather, R. E., Warner, B., MacFarlane, M., and Kristian, J. 1969, *Nature* **222**, 728.
Burbidge, E. M., Burbidge, G. R., Fowler, W. A., and Hoyle, F. 1957, *Rev. Mod. Phys.* **29**, 547.
Cameron, A. G. W. 1970, in *Stellar Evolution—Theory and Observation, Proc. III Summer Institute for Astronomy*, H.-Y. Chiu and A. Muriel, eds. (to be published).
Canuto, V., and Chiu, H.-Y. 1968*a*, *Phys. Rev.* **173**, 1210.
—— 1968*b*, *Phys. Rev.* **173**, 1220.
—— 1968*c*, *Phys. Rev.* **173**, 1229.
—— 1970, *Phys. Rev.* (to be published).
Canuto, V., Chiu, H.-Y., Chiuderi, C., Lee, J. J. 1970, *Nature* **225**, 47.
Canuto, V., Chiu, H.-Y., and Fassio-Canuto, L. 1969, *Phys. Rev.* **185**, 1607.
Chiu, H.-Y. 1964, *Annals of Phys.* **26**, 364.
—— 1966, *Annual Rev. of Nuclear Science* **16**, 591.
Chiu, H.-Y., and Canuto, V. 1968*a*, *Phys. Rev. Letters* **21**, 110.
—— 1968*b*, *Ap. J. (Letters)* **153**, L157.
—— 1969, *Phys. Rev. Letters* **22**, 415.
—— 1970 (to be published).
Chiu, H.-Y., Canuto, V., and Fassio-Canuto, L. 1969, *Nature* **221**, 529.
Chiu, H.-Y., and Fassio-Canuto, L. 1969, *Phys. Rev.* **185**, 1614.
—— 1970, *Phys. Rev.* (to be published).
Chiu, H.-Y., and Salpeter, E. E. 1964, *Phys. Rev. Letters* **12**, 413.
Cocke, W. J., and Cohen, J. M. 1968, *Nature* **218**, 1009.
Colgate, S. A. 1970, in *Stellar Evolution—Theory and Observation, Proc. III Summer Institute for Astronomy*, H.-Y. Chiu and A. Muriel, eds. (to be published).
Colgate, S. A., and White, R. H. 1966, *Ap. J.* **143**, 626.
Drake, F., Lecture delivered at Fourth Texas Symposium on Relativistic Astrophysics, December 15-20, 1968 (to be published).
Drake, F. D., and Craft, H. D. 1968, *Nature* **220**, 231.
Erber, T. 1966, *Rev. Mod. Phys.* **38**, 626.
Fowler, W. A., and Hoyle, F. 1964, *Ap. J. Suppl.* **9**, 201 (No. 91).
Friedman, H. 1967, *Annual Rev. of Nuclear Science* **17**, 317.
Giacconi, R., Gursky, H., and Van Speybroeck, L. P. 1968, *Annual Rev. of Astr. and Astrophysics* **6**, 373.
Ginsburg, V. L., and Zaitsev, V. V. 1969, *Nature* **222**, 230.
Gold, T. 1969, *Nature* **221**, 25.

Goldreich, P., and Julian, W. H. 1969, *Ap. J.* **157**, 869.
Gunn, J., and Ostriker, J. P. 1969, *Nature* **221**, 454.
Hartle, J. B. 1967, *Ap. J.* **150**, 1005.
Hartle, J. B., and Thorne, K. S. 1968, *Ap. J.* **153**, 807.
Hewish, A., Bell, S. J., Pilkington, J. D. H., Scott, P. F., and Collins, R. A. 1968, *Nature* **217**, 709.
Jackson, J. D. 1962, *Classical Electrodynamics* (New York, John Wiley & Sons, Inc.), chap. 14.
Jancovici, B. 1969, *Phys. Rev.* **187**, 2275.
Kelly, D. C. 1964, *Phys. Rev.* **134**, A641.
Klepikov, N. P. 1952, Dissertation, University of Moscow.
Landau, L. 1932, *Physikalische Zeitschrift der Sowjetunian* **1**, 285.
Landau, L., and Lifshitz, E. M. 1951, *Classical Theory of Fields* (Reading, Mass.: Addison-Wesley), p. 295.
—— 1958, *Statistical Physics* (Reading, Mass.: Addison-Wesley).
Lee, H. J., Canuto, V., Chiu, H.-Y., Chiuderi, C. 1969, *Phys. Rev. Letters* **23**, 390.
Levinger, J. S., and Simmons, L. M. 1961, *Phys. Rev.* **124**, 916.
Maran, S. P., and Cameron, A. G. W. 1969, *Earth and Extraterrestrial Sciences* **1**, 1.
Misner, C. W., and Zapolsky, H. S. 1964, *Phys. Rev. Letters* **12**, 635.
Morton, D. C. 1964, *Nature* **201**, 1308.
Oppenheimer, J. R., and Volkoff, G. M. 1939, *Phys. Rev.* **55**, 374.
Ostriker, J. P., and Gunn, J. 1969, *Ap. J.* **157**, 1395.
Pacini, F. 1967, *Nature* **216**, 567.
—— 1968, *Nature* **219**, 145.
Rabi, I. I. 1928, *Zs. f. Physik* **49**, 507.
Reichley, P. E., and Downs, G. S. 1969, *Nature* **222**, 229.
Roberts, J. A., and Fahlman, G. G. 1969, *Nature* **222**, 862.
Ruderman, M. 1968, *Nature* **218**, 1128.
—— 1969, *Nature* **223**, 597.
Salpeter, E. E. 1960, *Annals of Phys.* **11**, 393.
—— 1961, *Ap. J.* **134**, 669.
Schwartz, R. A. 1967, *Annals of Phys.* **134**, 669.
Shen, C. S. 1970 (to be published).
Simon, M., and Strange, D. L. D. 1969, *Nature* **224**, 49.
Stix, T. H. 1962, *The Theory of Plasma Waves* (New York: McGraw-Hill Book Co.).
Thorne, K. S. 1967, in *High Energy Astrophysics*, Vol. III, C. DeWitt, E. Schatzman, and P. Veron, eds. (New York: Gordon and Breach, Science Publishers), p. 259.
—— 1970, in *Stellar Evolution—Theory and Observation, Proc. III Summer Institute for Astronomy*, H.-Y. Chiu and A. Muriel, eds. (to be published).
Tolman, R. C. 1934*a*, *Phys. Rev.* **55**, 364.
—— 1934*b*, *Relativity, Thermodynamics, and Cosmology* (Oxford, Clarendon Press), chap. 7.
Tsuruta, S. 1965, Dissertation, Columbia University.
Tsuruta, S., and Cameron, A. G. W. 1966, *Canadian J. of Phys.* **44**, 1863.
Wheeler, J. A. 1966, in *Annual Rev. of Astr. and Astrophysics* **4**, 393.
Woltjer, L. 1964, *Ap. J.* **140**, 1309.
Zwicky, F. 1939, *Phys. Rev.* **55**, 726.

13

PHOTOMETRY OF FIELD HORIZONTAL-BRANCH STARS

A. G. Davis Philip*

Dudley Observatory
State University of New York at Albany

1. Introduction

Recently, there have been a number of studies (Graham and Doremus 1968; Newell, Rodgers and Searle 1969a,b) made of globular cluster blue horizontal-branch stars (hereafter referred to as HB stars). The consensus of these studies seems to be that the HB stars are less massive than the sun and occupy a wedge-shaped area in the theoretical HR diagram. Many questions concerning the evolution, and the helium and metal abundance of these stars remain to be answered. A major difficulty is that such stars are faint and thus difficult to study spectroscopically.

Photometry of A stars in several areas presently under investigation, to determine the stellar density distribution perpendicular to the galactic plane, has turned up a number of stars which are like the globular cluster HB stars. Since these stars are brighter than the cluster stars they are accessible to detailed spectroscopic investigation which may lead to important findings concerning the chemical composition of such highly evolved stars. Matching of four-color and H_β indices to atmospheric models allow the determination of θ_{eff} and log g. If one assumes that the field HB stars have the same absolute magnitude as the globular cluster HB stars, then their masses can be calculated.

If further investigation indeed shows that globular cluster HB stars are identical to the field HB stars, then the cluster HB stars can be used as probes to determine reddening. Thus distance moduli could be obtained in an independent way as a check on the conventional methods employing RR Lyrae absolute magnitudes and main sequence fitting.

For these and other reasons, it was decided that the characteristics of the field HB stars should be determined

Visiting Astronomer, Kitt Peak National Observatory and Cerro Tololo Inter-American Observatory, which are operated by the Association of Universities for Research in Astronomy, Inc., under contract with the National Science Foundation.

and compared to the characteristics of globular cluster HB stars. This article describes a program of identification, photometry, and spectroscopy of a group of 39 stars which have been classified as field HB stars. Strömgren four-color photometry has been especially useful since Graham (1966) and Philip (1966b, 1967, 1968a) have shown that Population II, early-type stars can be detected easily photoelectrically.

2. Detection of the Population II Component in the 1 HLF 2 Area (ℓ^{II} = 76°, b^{II} = -30°)

During the photometry of sequence stars in the 1 HLF 2 area at galactic latitude -30°, it was noted that there were some faint A stars which had B-V colors which were too blue for their spectral type (Philip 1965). In Figure 1, the color excesses of stars of spectral classes B9-A3 from Tables II and III (Philip 1966a) which have been measured in the UBV system, are plotted versus their distance moduli, m-M. Stars which have been identified as Population II stars, (Philip 1968a, 1969a, 1969b) by means of Strömgren four-color photometry (Strömgren 1963) are plotted as triangles. The rest, presumably Population I stars, are plotted as points. The point for one star, a possible λ Boo star, is circled. The dashed line indicates the estimate of the absorption in the 1 HLF 2 region. In Figure 2, the intrinsic two-color plot is presented for the stars plotted in Figure 1. The Population I stars scatter about the zero age main sequence; the HB stars generally fall below the Population I stars.

Spectral types used as a basis for the color excesses and distance moduli in Figure 1 were obtained from Schmidt spectral plates at a dispersion of 280 Å/mm at H_γ and classified according to the criteria set up by Nassau and Seyfert (1946). The color excesses were computed from intrinsic colors listed by Schmidt-Kaler (1965). Due to the errors in spectral classification at this low dispersion, the position of individual stars in a color excess-distance modulus plot cannot be determined with great precision; however, the absorption obtained from many such determinations should be quite good as long as the errors in classification are random. Some preliminary spectroscopic evidence indicates that low dispersion spectral types are approximately one MK subclass too early for A stars in the magnitude range V = 6 to 10. If confirmed, this would have the effect of decreasing the color excess in this range by approximately 0.04 mag. Such an effect may explain why some of the points fall too high in Figure 1. The more significant diagram therefore is Figure 2, where the spectral types do not enter into the calculations in a major way. An error of one subclass in the spectral type leads to a difference in the calculated color of ≃ 0.04. The corresponding difference in the distance modulus is ≃ 0.30 magnitudes. The UBV data for each star are corrected for the assumed reddening and then plotted. Except for one field HB star which is peculiar, the trend is for the points representing field HB stars as a group to fall to the left of the zero age main sequence. Later on, it will be seen that this is a characteristic of the majority of the field HB stars found so far.

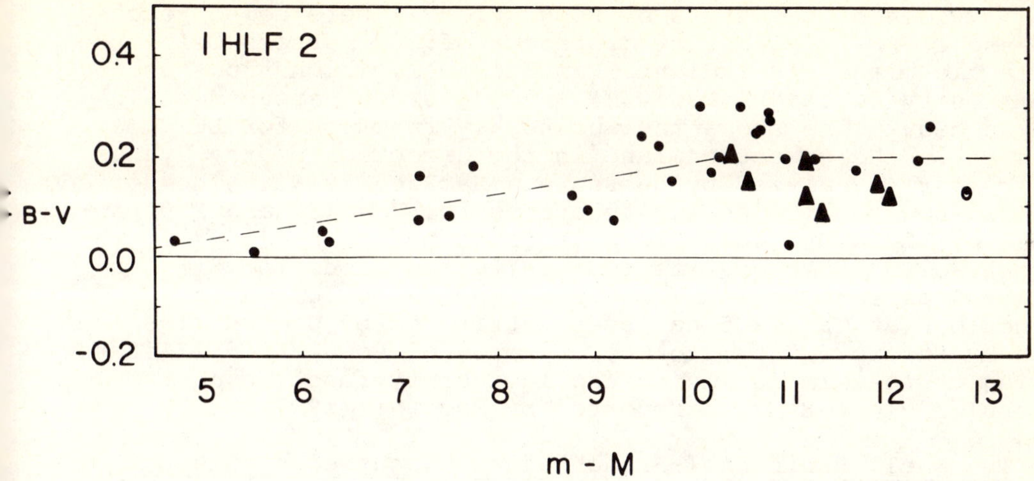

Figure 1. E_{B-V} versus m-M. B9A3 stars from the 1 HLF 2 area for which UBV photometry has been done are plotted. The points represent normal Population I stars, triangles represent field horizontal-branch stars. The circled point represents a possible λ Boo star.

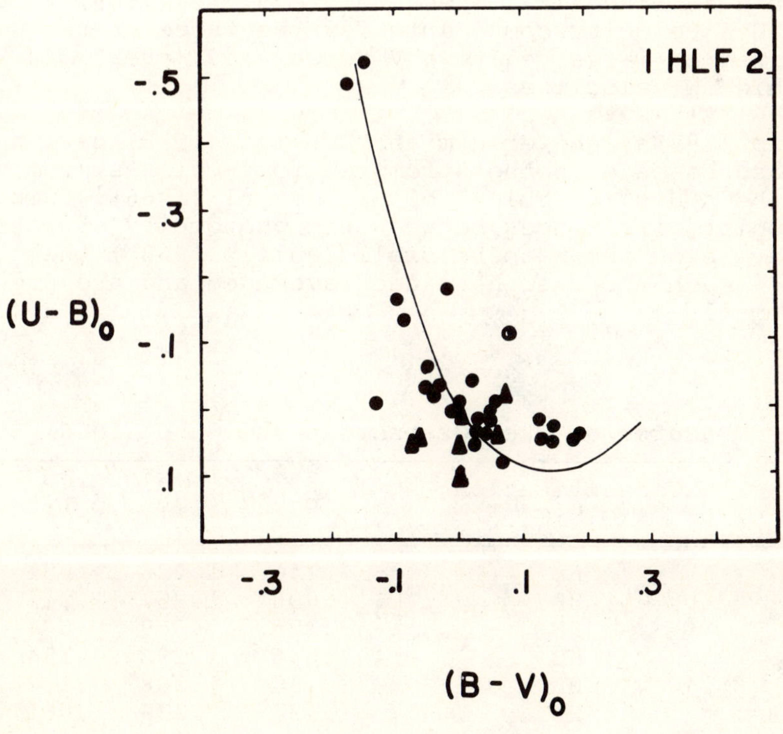

Figure 2. $(U-B)_o$ versus $(B-V)_o$ for the stars plotted in figure 1.

Evidence that stars of Population II were being reached in the l HLF 2 region, at distances of 200 parsecs below the galactic plane, is indicated in Figure 3, which shows the luminosity function, Log $\Phi(M) + 10$, plotted versus M_B. The solid heavy line shows the luminosity function for $LF3_b$, a region in the galactic plane in the direction of the Cygnus spiral arm, analyzed by Nassau and MacRae (1949). The dotted lines show the luminosity functions for Z = 100 and 200 parsecs at the North Galactic pole found by Upgren (1963). The light solid lines are the luminosity functions for Z = 100 and 200 parsecs in l HLF 2. The dip in the luminosity functions at $M_B = 2.5$ has been attributed by Upgren (1963) and by Philip (1966a) to the presence of Population II stars owing to the similarity to the luminosity function of the globular cluster M3 as reported by Sandage (1957).

3. The Field Horizontal-Branch Stars Discussed by Oke, Greenstein, and Gunn

Oke, Greenstein and Gunn (1966) discussed ten highly evolved stars for which H_γ profiles and photoelectric spectrum scans had been obtained. For evolved A stars, the spectra can be compared with model atmospheres to obtain θ_{eff} and log g. In Table I of Oke et al., which lists θ_{eff} and log g, six stars were listed as field HB stars on the basis of low values of log g for their effective temperatures. HD 86986, HD 109995, and HD 161817 have had chemical abundance analyses made (Kodaira 1964; Wallerstein and Hunziker 1964; Kodaira, Greenstein and Oke 1969) which show that the metal to hydrogen ratio is down from that of the sun by a factor greater than 10. Kodaira, Greenstein and Oke (1969) tabulated the UVW velocity components for the three stars, characterized by large negative V velocities (over -240 km/sec) and high W velocities.

Philip (1968a) and Graham and Doremus (1968) have published measures made in the Strömgren four-color system of seven of the stars in Table I of Oke, et al (1966). Since the zero point differences between the photometry of these authors have been shown to be small (Philip 1968b) the measures of each observer have been averaged and are presented in Table I.

TABLE I

Four-Color Measures of Oke, Greenstein and Gunn (1966) Stars

Star	Classification O.G.G.	b-y	c_1	m_1	n
BD-6° 86	HB	.146	1.226	.091	1
HD 86986	HB	.096	1.267	.121	12
HD 106223	HB	.229	.669	.084	6
HD 109995	HB	.046	1.292	.134	10
HD 161817	HB	.135	1.208	.112	13
BD17° 4708	HB	.337	.336	.061	5
BD39° 4926	Variable temperature	.179	1.607	.028	7

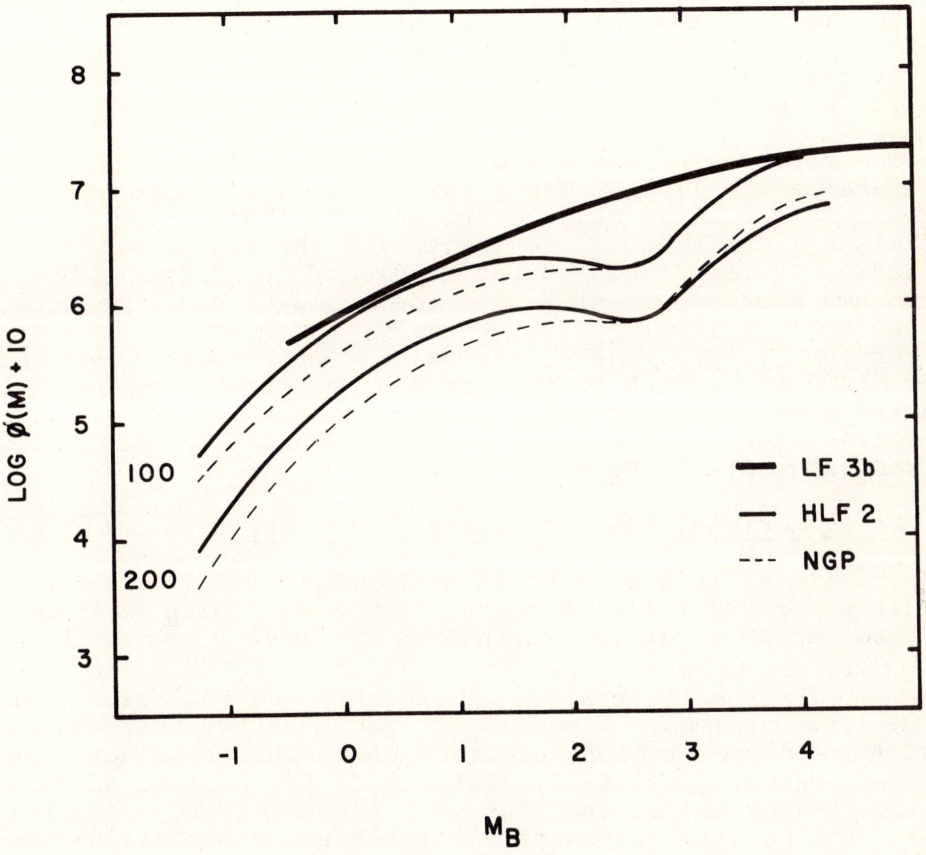

Figure 3. Log Φ(M) +10 versus M_B. The heavy lines show the luminosity functions for stars in the galactic plane. The dotted lines show the luminosity function for the NGP at distances of 100 and 200 parsecs above the galactic plane (Upgren 1963). The light solid lines are the luminosity functions for 1 HLF 2 and the same distances from the galactic plane. The dip at M_B = 2.5 is attributed to the presence of Population II stars in the sample.

Four of these stars have very similar characteristics; namely c_1 indices that are 0.2 higher than normal for A stars of Population I and m_1 indices about 0.08 less than normal. BD39° 4926 was noted as having the largest spectroscopically known measured Balmer jump. Photoelectrically, it has a very large c_1 index. HD 106223 and BD17° 4708 have low m_1 indices, but their c_1 indices are nearly normal for their b-y colors. Slettebak, Wright, and Graham (1968) have classified HD 106223 as a λ Bootis star of rather late type. BD17° 4708 would seem to be a star of the same class. The four stars, BD-6 86, HD 86986, 109995, and 161817 have been taken as prototypes of the field HB population.

Spectral differences between normal Population I, A stars and the field HB stars are illustrated in figure 4, which shows six spectra taken with the Cassegrain spectrograph (dispersion = 128 Å/mm) on the No. 1, 36" telescope at Kitt Peak.

The top two spectra are of HD 109995 and 86986, two of the stars in the list of Oke et al. They show somewhat narrower Balmer lines and the presence of lines closer to the Balmer limit than in the spectra of the two normal Population I stars shown in the middle of the figure. The bottom two spectra are of two of the field HB stars reported earlier (Philip 1967). Unfortunately, for most of the interesting objects, these spectral differences are too fine to be discernible with confidence at the dispersion used with the Schmidt telescopes employed in this project. One must first identify the stars as A stars and then photometry can separate out the HB stars.

4. Field Horizontal-Branch Stars at the North Galactic Pole

The aim of this project is to identify and then study the nature of the field HB stars. The best region to conduct the search is at the North Galactic Pole since it is there that the greatest numbers of HB stars relative to the normal Population I stars can be expected and the absorption will be at a minimum. A summary of photometric work done in a a 30 square degree region, centered on the North Galactic Pole was given by Philip (1968b). Fifteen of the A stars measured had normal uvby colors and thus were presumably Population I stars. The spectral types of Slettebak and Stock (1959) combined with UBV photometry (Philip 1968b) gave color excesses and distance moduli. The estimate of the absorption through the galactic disk, perpendicular to the plane, obtained using these fifteen stars is $E_{B-V} = 0.05$. In four-color photometry the corrections to the b-y, c_1 and m_1 indices are thus very small: -0.035, 0.007, and +0.006 respectively.

Ten field HB stars were identified in this 30 square degree region on the basis that their four-color indices matched those of the four prototype field HB stars. The intrinsic colors (the observed colors have been corrected for the reddening mentioned above) are presented for this group of stars in Table II, along with those of three brighter NGP stars, close to, but not in the central region. Two

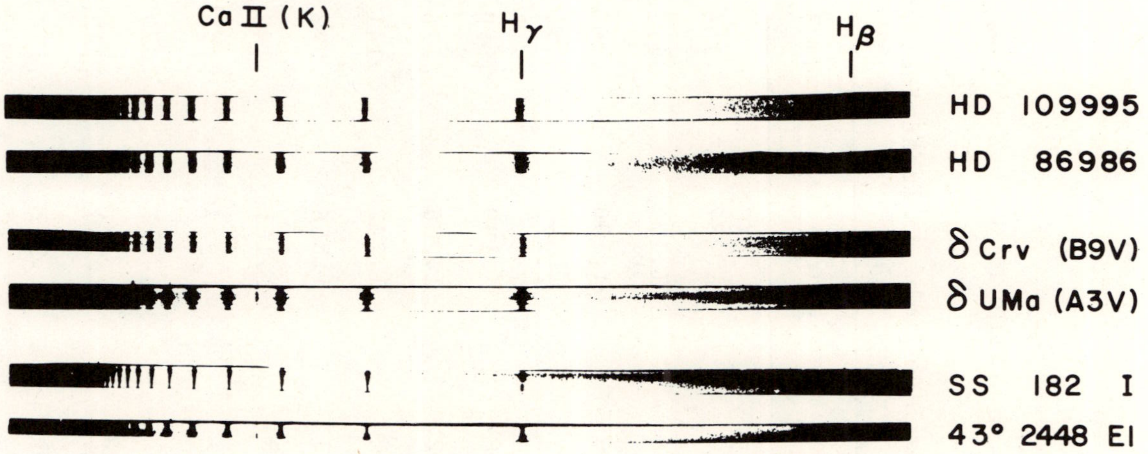

Figure 4. Spectra of two known field horizontal-branch stars, two normal Population I,A stars and two possible field horizontal-branch stars. The field horizontal-branch stars can be identified by their slightly narrower Balmer lines and by the presence of more hydrogen lines towards the Balmer limit.

of the stars in Table II differ from the rest: SS 227 II is unusual in that it has a very high c_1 index and a very low m_1 index and seems to be an extreme member of the group; SS 182 I has a low m_1 but a normal c_1. Graham and Doremus (1968) found HB stars in NGC 6397 that match the characteristics of SS 182 I; however, among the field stars so far identified it is the only one of this type found in the group of 39.

TABLE II

Intrinsic Colors of NGP Field Horizontal-Branch Stars

Name*	$(b-y)_o$	$(c_1)_o$	(m_1)
SS 191 II	.088	1.237	.156
SS 193 II	.008	1.179	.123
SS 194 II	.086	1.190	.107
SS 199 II	.024	1.273	.164
SS 202 II	.108	1.199	.090
SS 208 II	.052	1.236	.116
SS 209 II	.051	1.247	.141
SS 227 II	.072	1.356	.043
SS 229 II	-.028	1.161	.143
SS 234 II	.052	1.277	.123
SS 182 I	-.022	.992	.070
SS 287 I	.006	1.253	.132
SS 320 I	.046	1.292	.134†

*SS star numbers are from Slettebak and Stock (1959)

†SS 320 I = HD 109995

Figures 5, 6 and 7 indicate where the field HB stars fall in the $(c_1)_0$, $(m_1)_0$; $(b-y)_0$, $(c_1)_0$; and $(b-y)_0$, $(m_1)_0$ diagrams respectively. The line in each diagram marks the location of the zero age main sequence. The four prototype field HB stars are plotted as triangles, the NGP stars are indicated by points. In each diagram it can be seen that the HB stars occupy areas well away from those occupied by normal Population I A stars; and that the prototype field HB stars and the NGP stars fall together.

Preliminary H_β measures are plotted in Figure 8 for field HB stars (triangles) and normal Population I stars (points). The field HB stars have H_β indices that average 0.08 less than those of Population I. The smaller H_β values confirm the observed spectral appearance of the hydrogen lines in the field stars (cf. Figure 4). If all HB stars are physically homologous, then the (b-y), H_β diagram will provide an independent way of obtaining the reddening at large distances from the galactic plane. Once the relationship between H_β and absolute magnitude has been determined for the HB stars, a new method for calculating the distance moduli of globular clusters far away from the galactic plane will be available.

5. The List of Stars

The 39 field HB stars so far identified in the selected areas at high galactic latitudes are listed in Table III. For each star, the name is listed in column 1, the right ascension and declination in columns 2 and 3, and the B magnitude in column 4. The PS stars are taken from Philip and Sanduleak (1968); the SS stars are from Slettebak and Stock (1959). The remaining stars which are not in the Henry Draper Catalogue or the Bonner or Cordoba Durchmusterungen are named by giving them the number of the nearest BD or CD star followed by a letter and number. The letter indicates the direction of the star relative to the Durchmusterung star and the number differentiates among several stars which might be found close together in the same chart area.

The UBV and four-color indices for the stars in Table III are plotted in Figures 9 through 14. In each area, normal A stars were selected by means of four-color photometry (their indices fell on or close to the zero age main sequence. From the Schmidt spectral type and UBV photometry, the color excess and distance modulus for each star were calculated, which then gave the absorption as a function of distance. The intrinsic colors of the field HB stars could then be found in each region. The $(c_1)_0$, $(m_1)_0$ diagram is shown in Figure 9. The points cluster about a line at $(c_1)_0 = 1.22$. The one low point represents SS 182 I. Much the same sort of picture is presented in Figure 10 which shows $(c_1)_0$ versus $(b-y)_0$. The three blue stars [$(b-y)_0$ from -0.15 to -0.08] are located at the South Galactic Pole. The reddening in this region has not been well-determined as yet and therefore the $(b-y)_0$ colors of the SGP stars are not as precise as those determined for stars at the North Galactic Pole. The $(m_1)_0$, $(b-y)_0$ diagram is shown in Figure 11. The scatter exhibited in this diagram is related to the determination of the reddening and

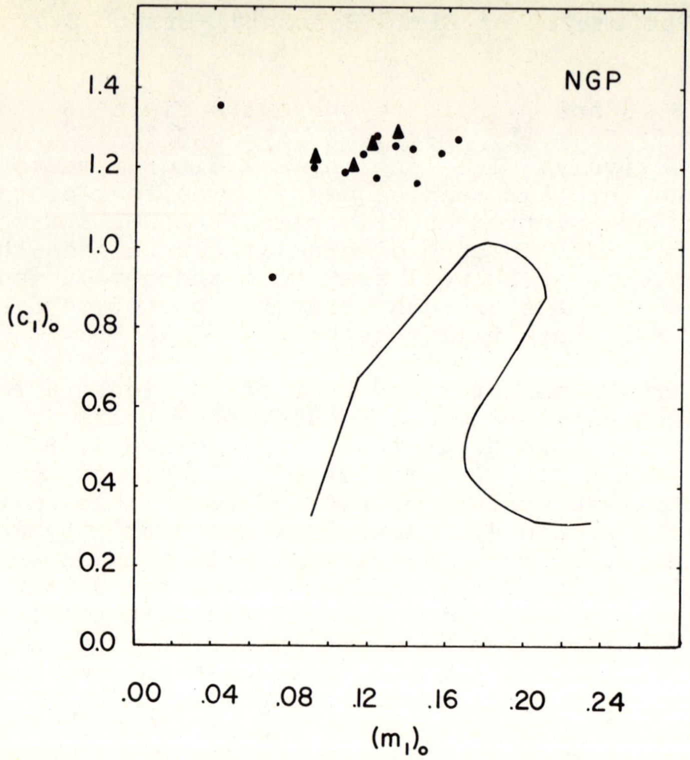

Figure 5. $(c_1)_o$ versus $(m_1)_o$. Field horizontal-branch stars from the NGP region are plotted as points. The four prototype field horisontal-branch stars are plotted as triangles. The line marks the zero age main sequence.

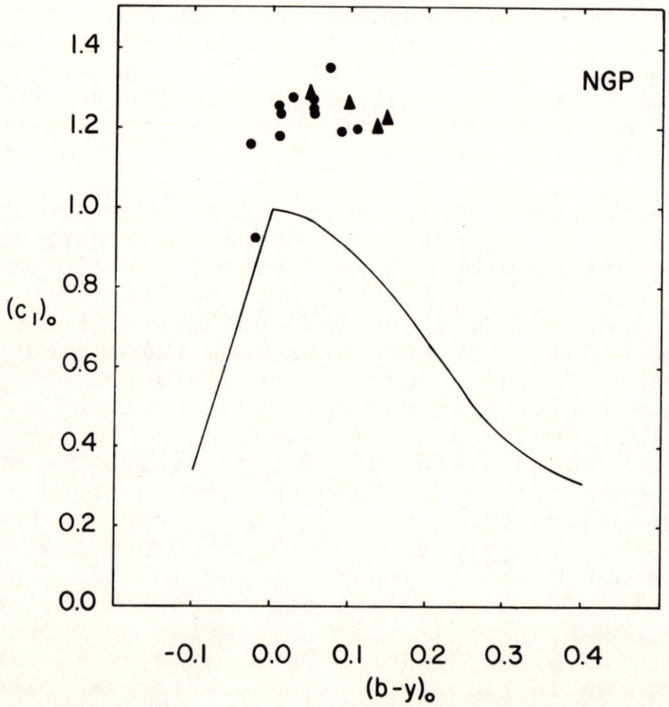

Figure 6. $(c_1)_o$ versus $(b-y)_o$. Symbols as in figure 5.

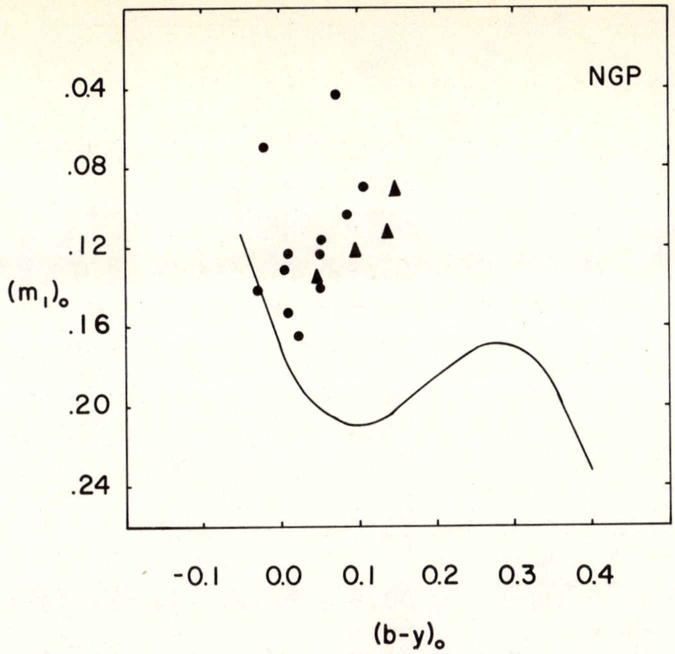

Figure 7. $(m_1)_o$ versus $(bpy)_o$. Symbols as in figure 5.

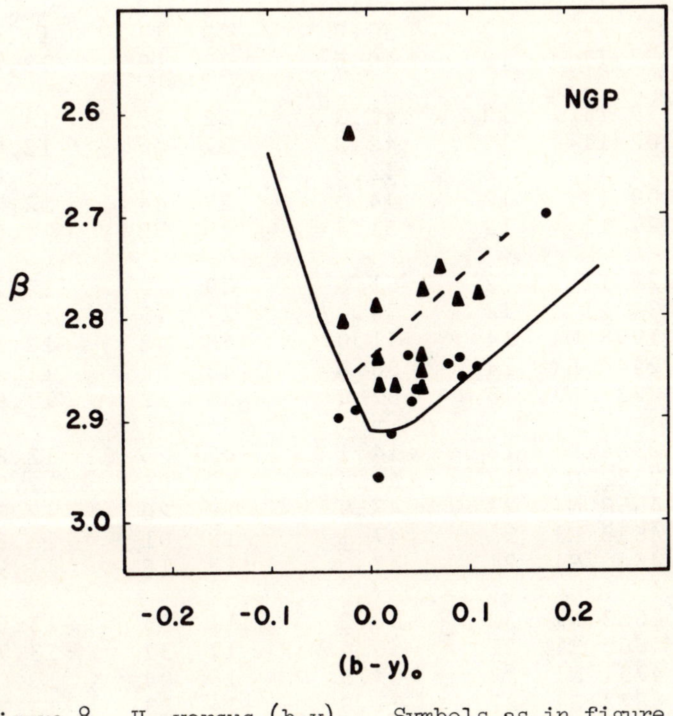

Figure 8. H_β versus $(b\text{-}y)_o$. Symbols as in figure 5.

TABLE III

Possible Field Horizontal-Branch Stars

Star	α	(1950)	δ		B Mag
PS 23 II	0	$38\overset{m}{.}7$	-26°	13'	14.0
PS 27 II	0	42.3	-27	27	13.1
PS 35 II	0	51.9	-27	12	13.1
PS 36 II	0	52.2	-28	30	13.7
PS 52 II	1	08.1	-26	21	13.1
PS 53 II	1	08.9	-21	54	13.1
-9° 392 Nl	2	00.7	- 8	24	12.7
-12° 411 El	2	11.0	-11	51	13.0
HD 14829	2	20.8	-10	54	10.3
-11° 487 Wl	2	33.4	-11	15	13.4
8° 578 Nl	3	47.9	+ 9	16	11.7
HD 57336	7	18.6	-79	20	8.0
40° 2179 El	9	13.1	+40	28	11.7
40° 2325 Nl	10	22.5	+39	47	12.8
SS 182 I	11	44.5	+32	07	10.7
SS 287 I	12	25.8	41	46	10.8
SS 191 II	12	38.9	29	27	13.9
SS 193 II	12	39.7	31	13	12.8
SS 194 II	12	39.8	32	13	11.9
SS 197 II	12	40.6	32	19	12.8
SS 199 II	12	41.3	32	38	13.1
SS 202 II	12	42.3	32	46	12.8
SS 208 II	12	43.8	27	27	13.8
SS 209 II	12	44.3	27	44	12.7
SS 227 II	12	51.4	29	52	14.0
SS 229 II	12	51.6	29	10	12.2
SS 234 II	12	54.0	27	45	13.3
46° 1998 Wl	14	51.0	45	42	12.1
43° 2448 El	14	58.8	42	54	11.2
50° 2236 El	15	57.0	50	12	13.0
47° 2324 Nl	16	14.1	47	25	12.8
48° 2522 Nl	17	27.8	48	05	12.5
18° 4858 Nl	21	43.9	19	20	12.3
17° 4638 El	21	47.1	18	01	13.2
17° 4640 Sl	21	47.5	18	16	11.9
16° 4652 Nl	22	00.2	17	05	13.5
17° 4665 El	21	56.4	17	37	13.7
14° 4751 Nl	22	09.5	15	04	13.0
13° 4873 Sl	22	11.8	+14	13	13.7

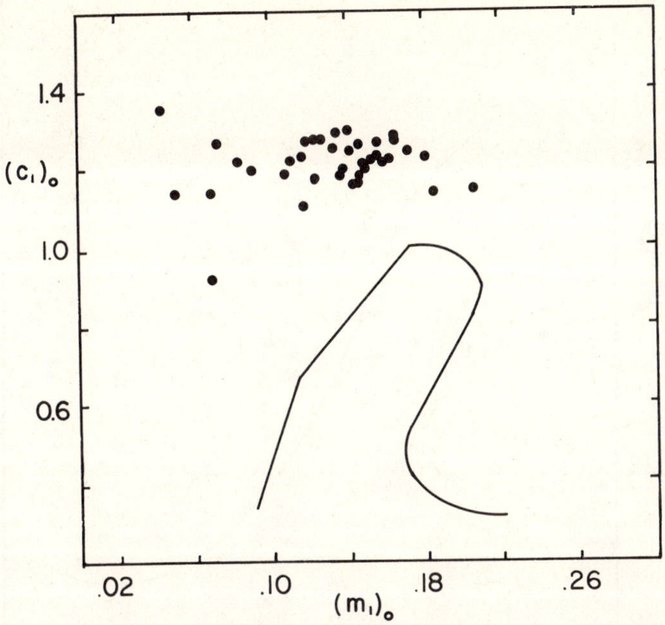

Figure 9. $(c_1)_o$ versus $(m_1)_o$. Thirty-eight field horizontal-branch stars are plotted as points. The line indicates the zero age main sequence.

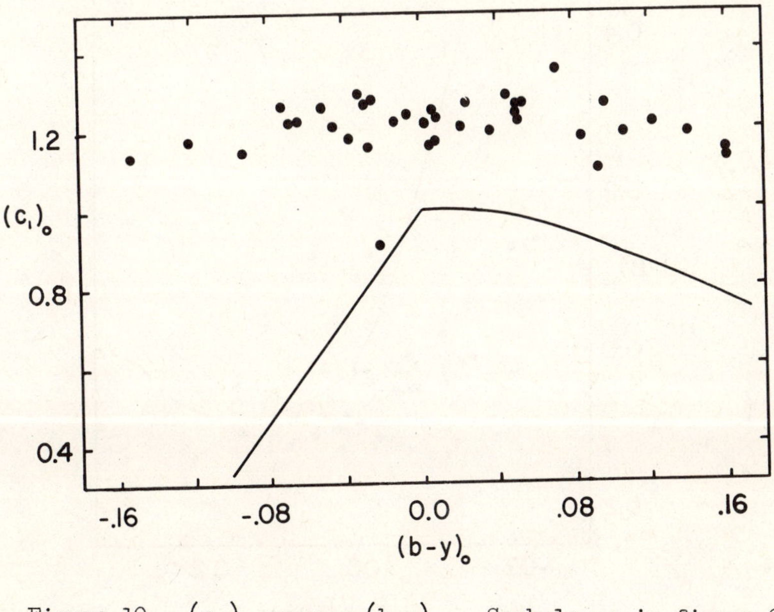

Figure 10. $(c_1)_o$ versus $(b-y)_o$. Symbols as in figure 9.

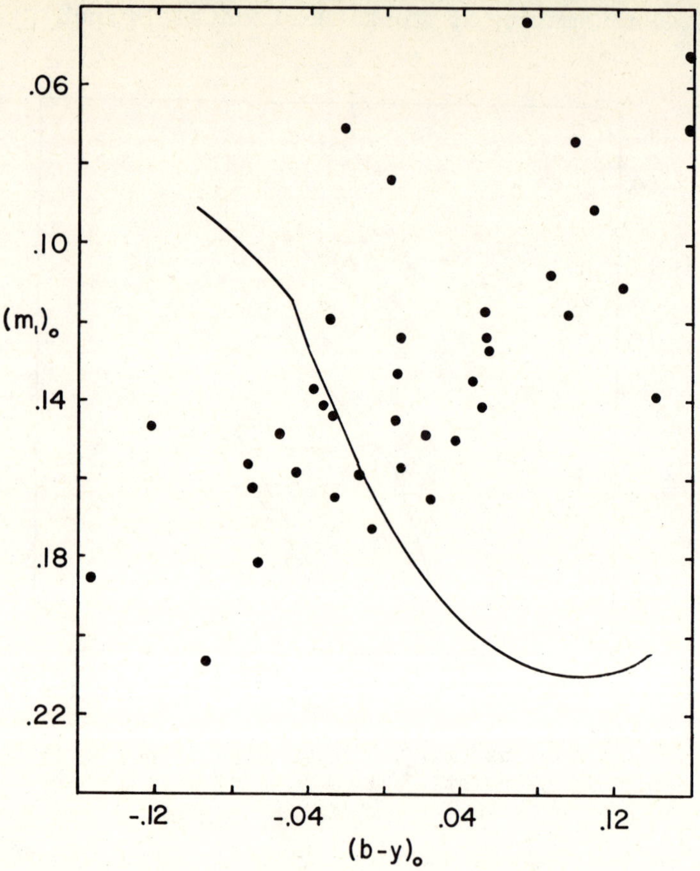

Figure 11. $(m_1)_o$ versus $(b-y)_o$. Symbols as in figure 9.

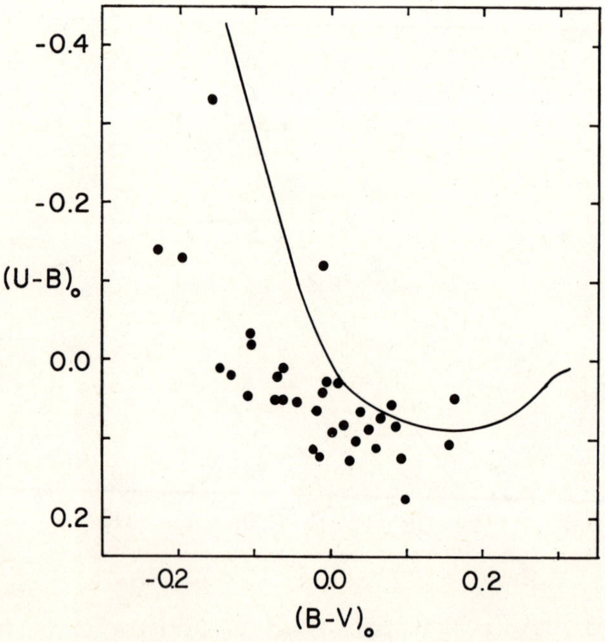

Figure 12. $(U-B)_o$ versus $(B-V)_o$. Symbols as in figure 9.

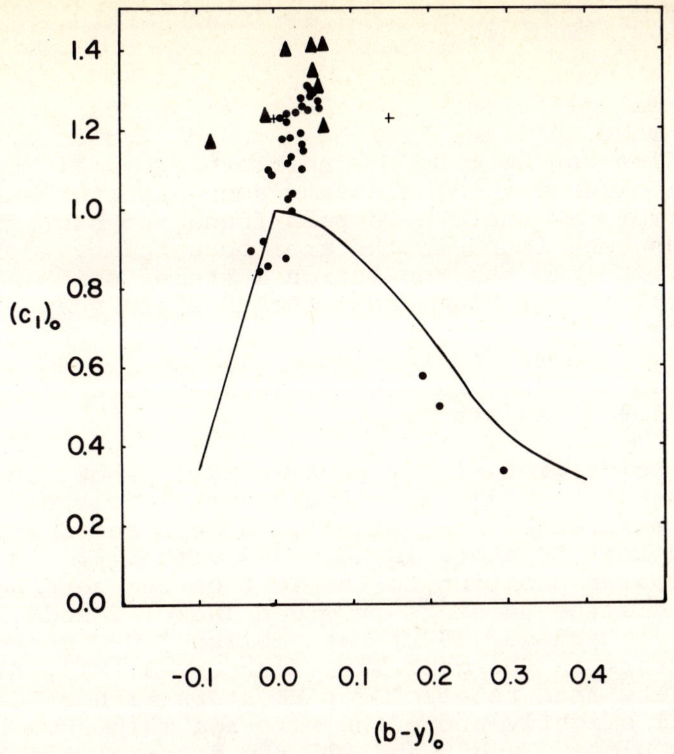

Figure 13. $(c_1)_o$ versus $(b-y)_o$ for NGC 6397 stars (points) and NGC 6809 stars (triangles).

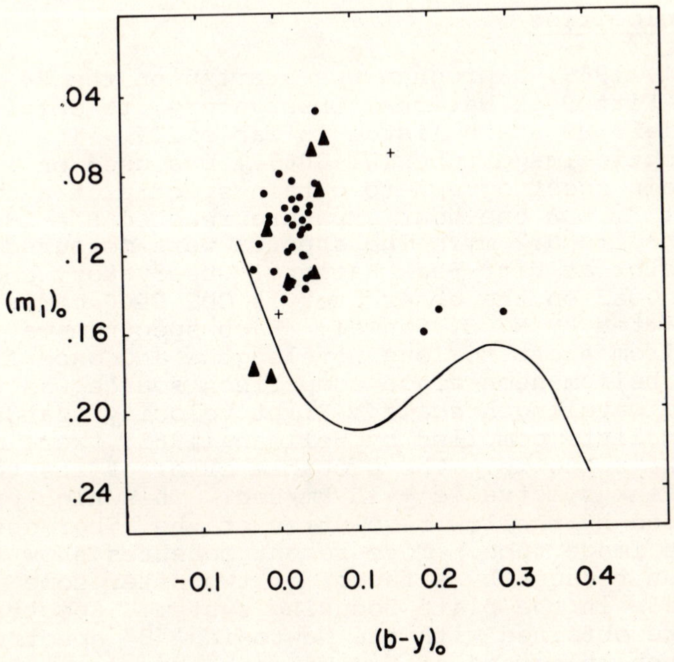

Figure 14. $(m_1)_o$ versus $(b-y)_o$ for NGC 6397 stars (points) and NGC 6809 stars (triangles).

possibly to real differences in the metal abundance of these stars. When more HB stars have been found in each of the regions, studies can be made of their average position in the $(m_1)_0$, $(b-y)_0$ diagrams. At present, among the three areas in which six or more HB stars have been found, no significant difference has been found in the area occupied by each group in Figure 11. Unlike the Population I stars, which scatter about the zero age main sequence, the HB stars scatter about a line running diagonally with a slope $\simeq -1/2$. Near $(b-y)_0 = 0$, the $(m_1)_0$ indices for Population I, and II, A stars are the same and so this diagram cannot be used as a population discriminant for B9-A1 stars.

The two-color plot of UBV photometry is shown in Figure 12. Here, the great majority of points fall to the left of the zero age main sequence. Newell, Rodgers and Searle (1969a) found that HB stars in NGC 6397 with $\theta_{eff} < 0.5$ occupied a similar position to the left of the zero age main sequence, as did the HB stars in ω Cen (Newell, Rodgers and Searle 1969b). Sandage (1969) has published UBV photometry for blue HB stars in the globular clusters M3, M13, M15 and M92. In M92 the mean relation for HB stars with B-V = 0.0 to 0.2 would fall slightly above the zero age main sequence in the two-color plot. In M13 and M15 the mean relation would fall slightly below; in M3 the mean relation would fall on the zero age main sequence. A program is planned to measure a selection of these stars in the four-color and H_β systems. In a $(u-b)_0$, $(b-y)_0$ plot, the effect noted in UBV photometry for the field HB stars is much more pronounced. The large Balmer jump depresses the u reading, making the stars fall below the zero age main sequence. In the UBV system, the $(B-V)_0$ and $(U-B)_0$ colors are affected by the Balmer jump and by blanketing effects due to differences in the metal abundance.

6. Radial Velocities

In April, 1969, a program was started on the 84 inch telescope at Kitt Peak National Observatory, to obtain spectra of all the field HB stars listed in Table III. The Westinghouse fiber optic image tube (W1-30677) was used on the 84 inch Cassegrain spectrograph to obtain spectra at a dispersion of 69 Å/mm at H_γ. A one-hour exposure reached B = 13.5 for spectra widened to 0.2 mm. The spectra were measured on the Grant Comparator at Kitt Peak National Observatory and the data were reduced on the observatory's CDC 6400 computer using a program written by N. B. Sanwal. Each spectrum was measured twice, once from short to long wavelengths and once from long to short. A helium-neon-argon comparison source was used to establish the wavelength scale. Eight velocity standards, selected from lists compiled by Wellman (1965), were measured along with the unknowns. The probable error of the velocities of the Kitt Peak spectra is ± 15 km/sec. This high probable error is caused by the low resolution of the fiber optic system in the image tube. More recent measures show that this error has been reduced by a factor of two after some improvements were made in the plate focusing system. Spectra of a few stars were obtained with the Newtonian "B" spectrograph on the 100 inch telescope at Mt. Wilson; these spectra were measured in the same way as those obtained at Kitt Peak.

The radial velocity data are presented in Table IV in which are listed the name, spectral type (Schmidt telescope), B mag., radial velocity, and galactic coordinates for each of the program stars (Philip 1969c, 1970). The velocity dispersion for this group of stars is ± 126 km/sec, thus confirming their classification as Population II stars. The Z velocity dispersion for the seven NGP stars is ± 98 km/sec, which is a marked increase over the dispersions of ± 8 and ± 18 km/sec found for A stars in the magnitude range of V < 8 and 8.0 < V < 10.00 respectively (Woolley and Stewart 1967).

TABLE IV

Radial Velocities of Field Horizontal-Branch Stars

1HLF 2 Stars†	Spectral Type	B mag	Radial Velocity	ℓ^{II}	b^{II}
14° 4751 N1	A2	13.0	-160 km/sec	75°.7	-32°.6
13° 4873 S1	A0	13.7	-123	75.5	-33.6
16° 4652 N1	A0	13.5	-253	75.4	-29.6
17° 4640 S1	A0	11.9	+ 76	73.9	-26.6
17° 4665 E1	A3	13.7	-212	75.1	-28.6
18° 4858 N1	B9	12.3	- 53	74.1	-25.3
NGP Stars					
SS 182 I	B8	10.7	+ 98 km/sec	190°.5	+75°.3
SS 287 I	A0	10.8	-123 "	139.6	+74.7
SS 194 II	A2	11.9	+ 36 "	144.9	+84.8
SS 199 II	A0:	13.1	+150 "	140.2	+84.5
SS 202 II	A0	12.8	+123 "	137.7	+84.4
SS 227 II	A0:	14.0	+ 83 "	111.1	+87.5
SS 229 II	A0	12.2	- 90* "	105.2	+88.1
Other Areas					
47 2324 N1	A0	12.8	- 56 km/sec	74°.0	+45°.6
43 2448 E1	A2	11.2	-228 (-233)*	73.3	+60.0
50 2236 E1	A3	13.0	- 89 km/sec	78.8	+47.8
40 2179 E1	A0	11.7	+ 70 "	182.0	+44.1
40 2325 N1	A0	12.8	+ 83 "	181°.2	+57°.4

†14° 4751 N1 = S70, 13° 4873 S1 = 13° 110, 16° 4652 N1 = 16° 111, 17° 4640 S1 = 17° 24, 17° 4665 E1 = 17° 136, 18° 4858 N1 = 18° 21 (Philip 1966).

*Spectra obtained with the 100 inch telescope at Mt. Wilson.

It would be interesting if proper motions could be found for these stars so that the space velocities could be determined. A few of the NGP stars have been placed on the observing list at the U.S. Naval Observatory, but it will be some time before proper motions become available.

7. Horizontal-Branch Stars in Globular Clusters

Four-color measures of blue HB stars have been published in only one globular cluster, NGC 6397 (Graham and Doremus 1968). It is important that blue HB stars be measured in many globular clusters of differing helium and metal abundances. In this way the effects of changes in Y and Z can be calibrated against the four-color indices which will then allow the field HB stars to be classified relative to the globular cluster stars. Preliminary measures have been made of blue HB stars in NGC 6809, a little-studied globular cluster in the Southern Hemisphere. The HB stars were identified by photometry of a set of UBV direct places obtained with the Michigan Curtis Schmidt telescope at Cerro Tololo Inter-American Observatory. Four-color photometry of the selected stars was done with the 36 inch and 60 inch telescopes.

In order to find the absorption through the galactic disc at the position of the cluster, foreground A stars are currently being measured photoelectrically. At a galactic latitude of $b^{II} = -24°$, one would expect some absorption but not a large amount. Van den Bergh (1967) has determined a color excess $E_{B-V} = 0.09$ for NGC 6809. The $(c_1)_0$ and $(m_1)_0$ indices are plotted versus the $(b-y)_0$ indices in Figures 13 and 14 for NGC 6397 and NGC 6809. (Points are used for the 6397 stars and triangles for the 6809 stars.) A reddening of $E_{b-y} = 0.06$ was assumed for NGC 6809 in Figures 13 and 14. Two crosses in each graph locate the mean values at $(b-y)_0 = 0.00$ and 0.15 found for $(c_1)_0$ and $(m_1)_0$ for the field HB stars. The NGC 6809 stars have higher c_1 indices than the field stars. Note that in the $(m_1)_0$, $(b-y)_0$ graph the mean line through the points for each cluster has a higher slope than the mean line for the field HB stars. The reddening correction of $E_{(b-y)} = 0.06$ brings the NGC 6809 stars in coincidence with the NGC 6397 stars in this diagram. Over twenty blue HB stars have been identified in NGC 6809 and these are currently being measured. The eight confirmed HB stars which are plotted in Figures 13 and 14 are identified in the chart given in Figure 15.

8. Summary

One of the best ways to systematically identify field HB stars in a selected area is to make four-color measures of A stars in the magnitude range V = 10 - 14. The HB stars stand out from the normal Population I, A stars by virtue of their high c_1 indices and low m_1 indices. Three groups of stars have been shown to have similar properties in the four-color system: 1. four well-studied field HB stars, three of which have spectroscopically-determined metal abundances less than 1/10 the solar abundance; 2. blue HB stars in two globular clusters; and 3. HB stars found in the general field. The classification of the third group of stars as members of Population II is confirmed by their high radial velocities.

The field HB stars should be compared to the blue HB stars in globular clusters of varying metal and helium abundances

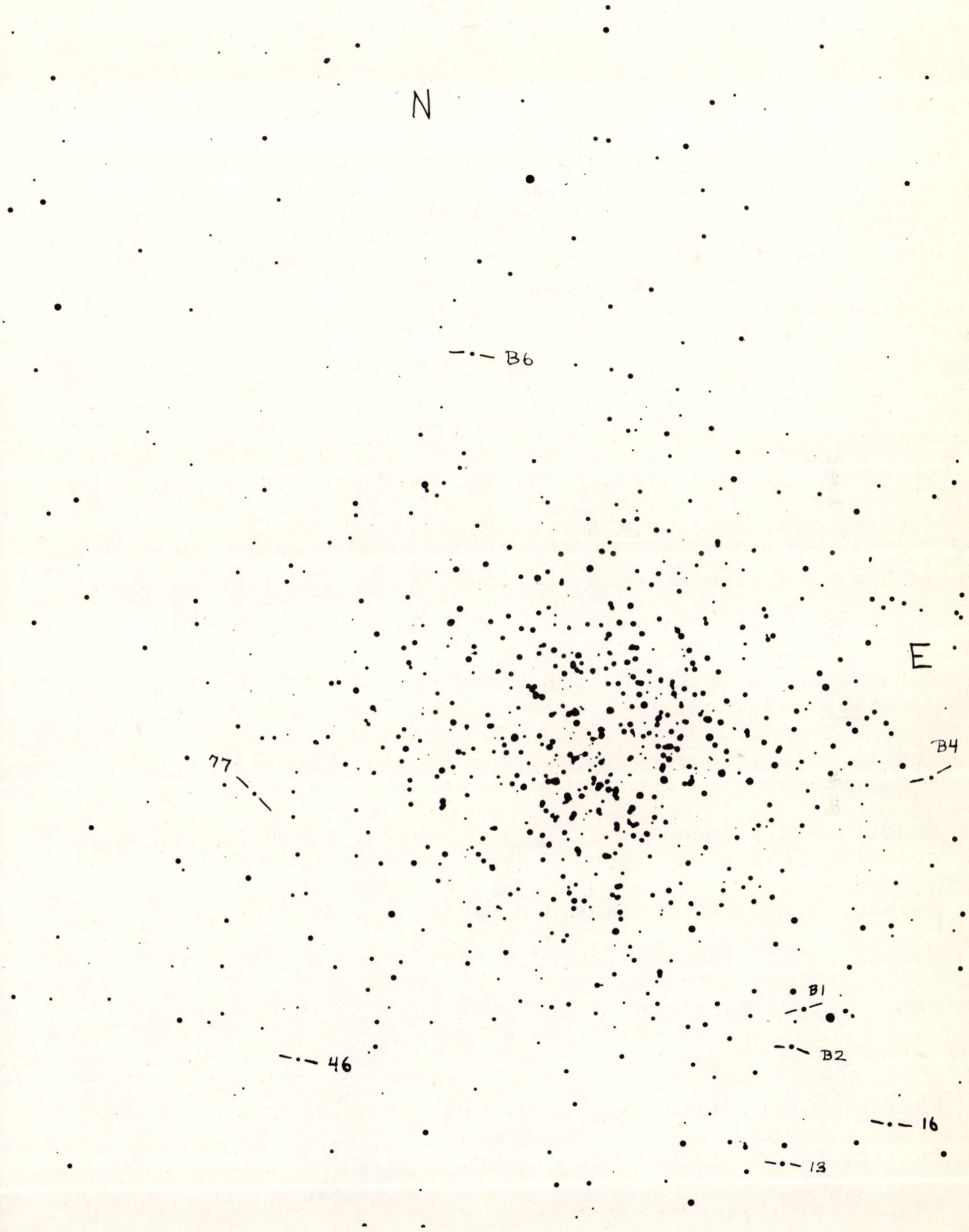

Figure 15. Chart for eight blue horizontal-branch stars in NGC 6807.

to determine if the field stars are members of a heterogeneous group like the clusters or whether they are all alike. Such studies will lead to a better understanding of the old population component near the Carina-Cygnus spiral arm.

Acknowledgements

This research was conducted under grants form the National Science Foundation and the Research Corporation. Observations were made at the Tonantzintla Observatory, Mt. Wilson and Palomar Observatories, Cerro Tololo Inter-American Observatory, and Kitt Peak National Observatory. I am indebted to the foundations for the financial support to carry on the research and to the directors of the observatories for making telescope time available to this project.

References

Bergh, S. van den 1967, A.J., 72, 70.

Graham, J.A. 1966, P.A.S.P., 78, 433.

Graham, J.A. and Doremus, C. 1968, A.J., 73, 226.

Kodaira, K. 1964, Zeitschrift for Astrophys., 59, 139.

Kodaira, K. Greenstein, J.L., and Oke, J.B. 1969, Ap. J., 155, 525.

Newell, E.B., Rodgers, A.W., and Searle, L. 1969a, Ap.J., 156, 597.

Newell, E.B., Rodgers, A.W., and Searle, L. 1969b, Ap. J., 158, 699.

Nassau, J.J., and MacRae, D.M. 1949, Ap. J., 110, 40.

Nassau, J.J., and Seyfert, C.K. 1946, Ap. J., 103, 117.

Oke, J.B., Greenstein, J.L., and Gunn, J. 1966, Stellar Evolution, ed. R.F. Stein and A.G.W. Cameron (New York: Plenum Press).

Philip, A.G.D. 1965, A.J., 70, 687.

Philip, A.G.D. 1966a, Ap. J. Supp., 12, 391.

Philip, A.G.D, 1966b, A.J., 71, 395.

Philip, A.G.D. 1967, Ap. J. Letters, 148, 143.

Philip, A.G.D. 1968a, Ap. J., 152, 1107.

Philip, A.G.D. 1968b, A.J., 73, 1000.

Philip, A.G.D. 1969a, A.J., 74, 209.

Philip, A.G.D. 1969b, A.J., 74, 812.

Philip, A.G.D. 1969c, Ap. J. Letters, 158, 113.

Philip, A.G.D. 1970, A.J., 75, 246.

Philip, A.G.D. and Sanduleak, N. 1968, Bol. Obs. Ton. y Tac., 4, 253.

Sandage, A.R. 1957, Ap. J., 125, 427.

Sandage, A.R. 1969, Ap. J. 157, 515.

Slettebak, A., and Stock, J. 1959, Astron. Abh. Hamburg, No.5.

Slettebak, A. and Wright, R.R., and Graham, J.A. 1968, A.J., 73, 152.

Schmidt-Kaler 1965, Landolt-Börnstein, Numerical Data and Functional Relationships, New Series, Group VI, Astronomy, Astrophysics, and Space Research, Vol. 1, ed. H.H. Voight, (Berlin: Springer Verlag), p. 298.

Strömgren,B. 1963, Stars and Stellar Systems, K.Aa. Strand, Ed. (University of Chicago Press, Chicago, Ill.), Vol. 3, Chap. 7.

Upgren, A.R. 1963, A.J., 68, 475.

Wallerstein, G. and Hunziker, W. 1964, Ap. J. 140, 214.

Wellman, P. 1965, Landolt-Börnstein, Numerical Data and Functional Relationships, New Series, Group VI, Astronomy, Astrophysics, and Space Research, Vol. 1, ed. H.H. Voight, (Berlin: Springer Verlag), p. 273.

Woolley, R. v.d. R., and Stewart, J.Q. 1967, MNRAS, 136, 329.

14

POPULATION I HELIUM ABUNDANCES

S.E. Strom

State University of New York
Stony Brook, N.Y.

The following discussion results from work performed jointly by the author and H.L. Shipman, presently a graduate student at the California Institute of Technology. The results presented are preliminary in nature and await confirmation from observations currently underway.

Recent investigations (Johnson and Poland 1968; Shipman and Strom 1970; Norris 1969; Poland 1970) have shown that the strong, well-observed He I diffuse lines are formed in LTE. In the last three of these papers it is also shown that the so-called singlet-triplet anomaly in the He I spectrum of B stars can be explained within the context of LTE line-formation theory. Furthermore, improved broadening theories for the strong He I λ4471 (Griem 1968; Barnard, Cooper and Shamey 1969) have enabled us to compute theoretical profiles for diffuse series lines for B stars. The work of Shipman and Strom (1970) demonstrates that helium-abundance values based on the line strengths computed from these broadening theories and the assumption of LTE are quite reliable. They conclude that accurate values of helium abundance can be obtained from observations of strong helium lines even at moderate dispersions.

Consequently, it now seems possible to examine published observations of B-star spectra and to determine the helium abundance for large numbers of stars. The survey of Walker and Hodge (1966) is ideal for this purpose since it contains measurements of λ4471 line strength for a significant sample of O and B stars. Moreover, because their survey encompasses a selection of faint B stars, it is possible to study the large-scale behavior of helium content with galactic position.

In order to derive abundances for these B stars, we must first determine their atmospheric parameters, θ_{eff} = (5040/Teff) and log. We chose to determine these parameters from, respectively, the observables $Q[Q = (U - B) - 0.71 (B - V)]$ and $W(H\gamma)$, the equivalent width of $H\gamma$. We computed the reddening-independent parameter Q from the colors extracted from the U.S.

Naval Observatory compilation of UBV observations (Blanco, Demers, Douglass, and Fitzgerald 1968). Most of these colors were in fact measured in Hiltner's (1956) survey. We found no systematic differences between Hiltner's measurements and the measurements of other observers whose colors we used. The observations of W(Hγ) were taken from the survey of Petrie and Lee (1966).

Hydrogen-line-blanketed models were used to compute UBV colors, from which a relation between Q, θ_{eff}, and log g was obtained. The temperature scale for stars if near main-sequence surface gravities derived in this way is in excellent agreement with the temperature scales of Morton and Adams (1968), Wolff, Kuhi, and Hayes (1968), and Hanbury Brown, Davis, Allen, and Rome (1967). We also computed equivalent widths for Hγ, using the Edmonds, Schluter, and Wells (1967) broadening theory, for models covering the appropriate θ_{eff} and log g range. This allowed choice of log g from the observed values of W(Hγ) and Q. Values of He I λ 4471 equivalent widths, W(4471), were computed from our models; the Barnard et al. (1969) broadening theory (cf. Shipman and Strom 1970) was adopted for these calculations. The observed value of W(4471) was then used to choose a value for the helium abundance, N(He), from the appropriate model calculation.

Because the Walker and Hodge (1966) survey includes many stars of luminosities higher than those appropriate to class V, we had to determine the range over which LTE line-formation theory is applicable. We note that the derived helium content in our survey should be independent of θ_{eff} and log g if LTE holds. Plots of our derived helium-abundance values against θ_{eff} and log g showed that departures from LTE are unimportant for log g > 3.5 since we found no systematic variation of N(He) with either of these parameters in this range. This confirms and strengthens the conclusions already reached by Shipman and Strom (1970). However, non-LTE effects become quite important for log g < 3.5, especially in the temperature regions (T_{eff} > 25,000°K and T_{eff} < 18,000°K) where a significant fraction of the equivalent width of the line is contributed by the line core; we excluded such stars from our survey. For log g < 3.5 and 18,000°K < T_{eff} < 25,000°K, there is a small dependence of abundance on log g; we chose to include stars in this range in our survey since the effects were small (<2% in N(He)). It is interesting to note that we found remarkably close agreement between the plots of extrapolation of the Johnson and Poland (1968) estimates of the effects of departures from LTE.

Results

The mean abundance we obtain for the 94 B stars remaining in the sample is N(He) = 0.105 ± 0.003 (p.e.) by number fraction. As a guide to the influence of errors in the observables on the derived mean helium abundance, we note that a systematic error of 0.02 in N(He) will result from either a systematic error of $0\overset{m}{.}04$ in Q or a systematic error of 10% in either Hγ of the He I λ 4471 equivalent widths. We do not expect the errors to exceed these values. The possibility of any significant error in N(He) introduced by changes in the mean slope of the reddening line, which would enter as an error in the Q value, appears to be ruled out by the absence of any correlation of abundance with color excess.

In order to examine the possibility of local variations in helium content, we assigned to each star a value for ΔN, the difference between the abundance of the star and the mean abundance of the sample at the appropriate value of surface gravity; at least-squares line was used to define a mean relation between N(He) and log g. We used this method in order to avoid introducing the spurious scatter in our abundance survey, which might arise from the (small) dependence of abundance on surface gravity. Also, because of the homogeneity of the color and line-strength data, we expect high internal accuracies in N(He). While systematic errors in the absolute values of N(He) may exist, ΔN should be significant.

The plots of these abundance deviations, ΔN, as a function of galactic position are shown in Figs. 1 and 2. In Fig. 1, we plot only the stars with the highest ΔN, while in Fig. 2 we plot the entire sample. The distances were obtained from the absolute magnitudes of Petrie and Lee (1966) and the assumption that $A_V = 3.0_{B-V}$. The apparent abundance excesses and deficiencies in these plots seem to congregate in groups, some of which correspond to well-known OB associations. Not only do these groups differ in helium abundance from the sample mean, but the internal consistency of the abundance determinations within a group is greater than that found for the complete sample. None of the groups is large enough to provide totally convincing evidence of real He abundance enhancements or deficiencies. However, we include the following discussion as a guide to future observation.

The group of apparently helium-deficient stars at $l^{II} =$ 200° includes the OB association I Mon. The mean abundance deficiency ΔN is -0.012 ± 0.005 (p.e.) and the standard error is 0.024 for the seven stars in the association. This difference between group and complete sample mean values is significant at the 90% confidence level, and the difference in standard deviations is significant at the 95% confidence level.

Another group of stars with negative ΔN is found in the direction of the constellation Perseus. This group falls between the Cygnus and Perseus arms with $130° < l^{II} < 150°$ and with heliocentric distances between 1.0 and 1.8 kps. For this group, $\Delta N = -0.029 \pm 0.004$ (p.e.). The difference between ΔN and its standard deviation from the values for the complete sample is significant at the 99.99% confidence level.

It is noteworthy that the average ΔN is $+0.013 \pm 0.06$ for all stars in the Perseus arm. This excess is significant at the 92% confidence level. In this regard, it is important to note with caution that the stars in the Perseus arm are, in the mean, slightly more luminous than the sample used for the Cygnus arm.

Finally, we note that the three stars in our sample from the association I Lac are highly overabundant. ($\Delta N = +0.027$, $+0.072$, and $+0.048$.)

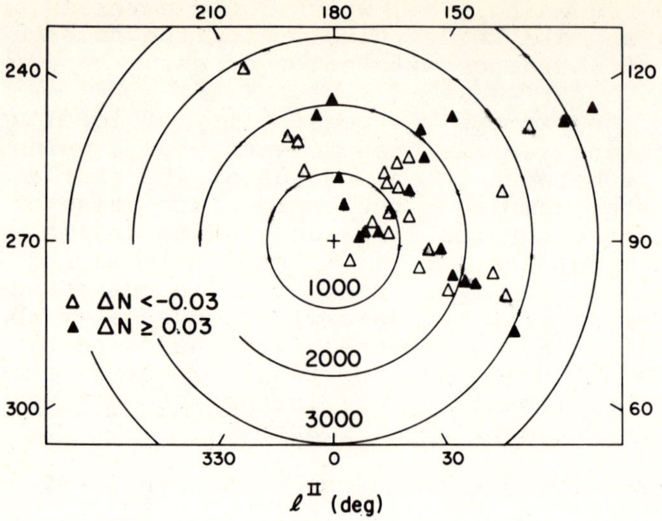

Figure 1. The galactic locations of the extreme overabundant and underabundant stars. Heliocentric distances are in parsecs; the shaded outlines represent the outlines of the galactic arms as determined by the 21-cm data (Oort et al. 1958).

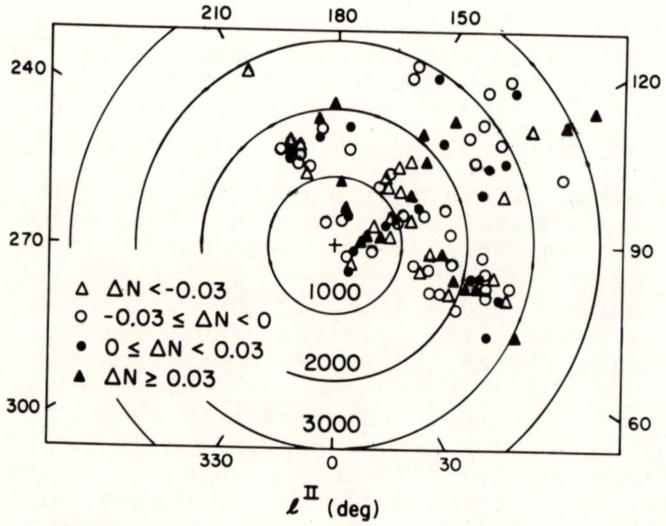

Figure 2. The galactic locations and abundance deviations of all the stars in the sample. Notation as in Fig. 1.

The above results seem to suggest that the helium abundance of B stars may vary from one part of the galaxy to another. Because of the small number of observations involved, however, we cannot regard these results as conclusive. Consequently, we searched for an independent check on these apparent variations.

In the context of the theory of Cepheid pulsation, Christy (1969) has remarked on the behavior of stellar evolutionary tracks in the vicinity of the Cepheid instability strip. He notes that the evolutionary tracks of Iben (1967) and Hoffmeister, Kippenhahn, and Weigert (1964) indicate that stars after reaching the red-giant tip will loop back farther to the left in the H-R diagram the higher their initial helium abundance. Consequently, with higher initial N(He), more stars of low mass will reach the instability strip, thus increasing the number of low-luminosity, short-period variables. Furthermore, more short-period Cepheids will reach the center of the instability strip with resulting increases in their amplitudes. Thus the behavior of the period-amplitude and period-frequency relations can act as a diagnostic for the initial stellar helium content. Because this diagnostic depends on a statistical treatment of Cepheid data, the consequent averaging over large sections of the galaxy tends to obscure the cluster-to-cluster variations mentioned above, although galactic arm-to-arm variations can be detected. Another difficulty is that an increase in initial helium content is mimicked by a decrease in metal content.

An increase in the frequency of short-period Cepheids in the distant northern Milky Way, roughly corresponding to the Perseus arm, was first noticed by Bahner, Hiltner, and Kraft (1962). This behavior in the period-frequency relation could indicate an increase in the helium abundance of that region, in agreement with the spectroscopic results for B stars. In order to rule out the possibility of selection effects in the period-frequency relation due primarily to patchy interstellar obscuration, we have examined the period-amplitude relations for Cepheids in various parts of the galaxy.

The number of high-amplitude, short-period Cepheids definitely increases in the Perseus arm and possibly in the Carina arm and decreases in the Cygnus and Sagittarius arms. According to Christy's (1969) interpretation, where high-amplitude, short-period Cepheids are correlated with helium content, these results indicare a higher helium abundance in the Perseus and Carina arms. Unfortunately, no reliable data are available for B stars in the Carina arm; the Cepheid data for the Perseus arm support the B-star results. From the tracks of Hoffmeister et al. (1964) we guess that the scale of the helium-abundance variations must be 0.02 - 0.03 in N(He), or roughly the same scale as implied by the B-star data; a more accurate determination of the scale of these variations must await more detailed calculations of evolutionary tracks. The likelihood that changes in the metal abundance might influence this conclusion is diminished by the determination of normal metal content for the Cepheid variable TV Cam by Abt, Osmer, and Kraft (1966). TV Cam is an extreme representative of the large-amplitude, short-period Cepheids in the Perseus arm; no data are yet available for analogous Cepheids in the Carina arm.

Radio observation of recombination lines in H II regions give a helium abundance of N(He) = 0.077 (by number fraction) while optical determinations have varied, with a mean of N(He) $\approx$0.11 (c.f. Palmer, Zuckefman, Penfield, Lilley, and Mezger 1969). We do not regard the difference between our value of N(He) and values obtained by other methods as significant, for systematic errors may exist in all methods of deriving the helium abundance. The possible sources of systematic error in our method are described above. In the optical analysis of nebular spectra, uncertainties in the corrections for interstellar reddening as well as uncertainties in the line-transfer problem and ionization equilibrium can affect the results. In the radio determinations, the assumption that the H II and He II regions coincide is the most crucial to the accurate determination of N(He). The calculations that test this assumption (Rubin 1968) indicate a strong dependence of the ionization structure of the H II region on the ultraviolet radiation field from the central star. This radiation field cannot be regarded as well-known. The models of Mihalas (1965), which Rubin used in his calculations, are unblanketed and do not include the important far-ultraviolet opacity arising from C, N, O, or Ne. Consequently, we must consider the validity of the assumption of coincident H II and He II regions as provisional. It should be emphasized, finally, that our results on abundance variations will be unaffected by any possible systematic error in our absolute value of N(He).

While either the Cepheid results or the B-star spectroscopic results might alone be regarded as merely tentative, we feel that the combination of these two independent methods of determining helium abundances provides strong evidence in favor of a variation of helium abundance with galactic position. Any firm conclusions, however, must await a more detailed investigation of the abundance variations from one OB association to another. In such cases, much more accurate estimates can be made for errors arising from reddening-law variations and stellar rotation. In particular, reddening-law variations in the vicinity of nearby associations can be important, thereby introducing errors into our determinations of θ_{eff}. We are examining data for the OB associations, I Lac, I Ori, I Per, and the Pleiades, using plate material generously supplied by Dr. Helmut A. Abt, which should allow rigorous study of all possible sources of error. However, if our suggestions of variations in helium content with galactic position are confirmed, they will present interesting problems in describing the chemical history of the galaxy.

References

Abt, H.A., Osmer, P.S. and Kraft, R.P. 1966, Ap.J. 145, 479.

Bahner, K., Hiltner, W.A. and Kraft, R.P. 1962, Ap. J. Suppl. 6, 319.

Barnard, A.J., Cooper, J., and Shamey, L.J. 1969, Astron. and Astrophys. 1, 28.

Blanco, V.M., Demers, S., Douglass, G.G., and Fitzgerald, M.P. 1968, Magnitudes and Colors of Stars in the U, B, V and Uc, B, V Systems. (U.S. Naval Observatory Publications, vol. 21).

Christy, R.F. 1969, Lecture notes from Summer Institute for Astronomy and Astrophysics, State University of New York at Stony Brook.

Edmonds, F.N., Jr., Schluter, H. and Wells, D.C., III. 1967, Mem. R.A.S. 71, 271.

Griem, H.R. 1968, Ap. J. 154, 1111.

Hanbury Brown, R., Davis, J., Allen, L.R., and Rome, J.M. 1967, M.N.R.A.S. , 137, 393.

Hoffmeister, E., Kippenhahn, R. and Weigert, A. 1964, Z. Ap., 60, 57.

Hiltner, W.A. 1956, Ap. J. Suppl., 2, 389.

Iben, I., Jr. 1967, Ann. Rev. Astron. Astrophys., 5, 571.

Johnson, H.R., and Poland, A.I. 1968, in Resonance Lines in Astrophysics. (Boulder, Colorado: National Center for Atmospheric Research), pp. 413-430.

Mihalas, D. 1965, Ap. J. Suppl., 9, 321.

Mitchell, R.I., Iriarte, B., Steinmetz, D. and Johnson, H.L. 1964, Photoelectric photometry of Cepheid variable stars. Boletin de los Observatorios Tonanzintla y Tacubaya 3, pp. 153-304.

Morton, D.C. and Adams, T.F. 1968, Ap. J., 151, 611.

Norris, J.E. 1969, Ap. J., in press.

Oort, J.H., Kerr, F.J. and Westerhout, G. 1958, M.N.R.A.S., 118, 379.

Palmer, P., Zuckerman, B., Penfield, H., Lilley, A.E., and Mezger, P.G. 1969, Ap. J., 156, 887.

Petrie, R.M. and Lee, E.K. 1966, Pub. Dom. Ap. Obs., 12, 435.

Poland, A.I 1970, Ap. J., in press.

Rubin, R.H. 1968, Ap. J., 153, 761.

Shipman, H.L. and Strom, S.E. 1970, Ap. J., in press.

Walker, G.A.H., and Hodge, S.M. 1966, Pub. Dom. Ap. Obs., 12 401.

Wolff, S.C., Kuhi, L.V., and Hayes, D.S. 1968, Ap. J., 152, 871.

15

STELLAR OPACITY

T. Richard Carson

University Observatory
St. Andrews

1. INTRODUCTION

It is now just a century since the pioneering work on stellar structure by Lane (1869), in which he considered the equilibrium of a stellar mass. In his original work, Lane assumed that the star (the Sun) was in a state of "convective equilibrium," as it was understood on the basis of the then current ideas of Kelvin. It was not until a quarter of a century later that Sampson (1894) introduced the concept of radiative equilibrium, but its full development had to await progress in the theory of the thermodynamics of radiation. The concept was adopted, in relation to stellar atmospheres, by Schuster (1903) and Schwarzschild (1906), the latter of whom also introduced the idea of local thermodynamic equilibrium. Later still, and nearly half a century after Lane's work, Eddington (1916) applied the same principle of radiative equilibrium to stellar interiors.

An early and important result of Eddington's work was the derivation of the mass-luminosity relation, which contained the radiative absorption coefficient, or opacity, as a parameter. In the absence of any theory of radiative absorption Eddington, in his earliest work, used the stellar model to derive the absorption coefficient. The value thus obtained was found to be about equal to the experimental value for hard x-rays absorbed by solid material. Because of the dissimilarity of the physical conditions this agreement was considered to be probably accidental. On the other hand, the only work which seemed to be relevant, namely, Thomson's theory of the scattering of x-rays by electrons, gave much too small a value. The subsequent development of the theory of radiative absorption, according to the Kramers theory and later on the basis of quantum mechanics, had therefore a profound influence on the theory of stellar structure. One of the most important problems to be resolved was that of the composition of stellar material. This manifested itself in the discrepancy between the opacity obtained from observations of mass and luminosity, when related via the mass-luminosity relation, and that obtained from theory (Eddington,

1926). The removal of this discrepancy was only made possible by adopting for stellar material a composition with a much higher abundance of hydrogen than had previously been assumed. This in turn led to the interpretation by Strömgren (1932, 1933) of the Hertzsprung-Russell diagram, and thus to the first serious steps in the development of a theory of stellar evolution. The importance of radiative energy transport, and therefore of the radiative opacity, in the theory of stellar structure and evolution may be seen from a consideration of the various transport mechanisms operative in stars.

2. ENERGY TRANSPORT

It is a principle of physics that whenever the energy density, or what is the same thing, the temperature, varies from one place to another, energy will flow from the hotter region to the cooler region, that is, in the opposite direction to that in which the energy density or temperature is increasing. There are three principal mechanisms by which the energy is transported, namely, conduction, convection and radiation, in which the energy is carried by the motion of individual particles, aggregates of particles and photons, respectively. Thus the total energy flux may be written as the sum of three components

$$F = F_{cond} + F_{conv} + F_{rad}$$

In the case of conduction the flux is found to be proportional to the negative of the temperature gradient, i.e.

$$F_{cond} = -K_{cond}\left[\frac{dt}{dr}\right]$$

where T is the temperature, r is the distance measured in the direction of transport and K_{cond} is the (thermal) conductivity. Generally speaking, under most stellar conditions, K_{cond} is very small, except in the case of degenerate matter when conduction by electrons becomes very efficient (Schatzman, 1958; Chiu, 1968; Clayton, 1968). In the case of convection, simple considerations show that it may occur only under certain conditions, namely

$$\Delta\nabla T = (1 - \frac{1}{\gamma})\frac{T}{P}\frac{dP}{dr} - \frac{dT}{dr} = -\frac{g}{C_p} - \frac{dT}{dr} > 0$$

with

$$\gamma = \frac{C_p}{C_v}$$

where T, P and g are the temperature, pressure and acceleration due to gravity and C_p and C_v are the specific heats at constant pressure and constant volume. The simple "mixing-length" theory (Schwarzschild, 1958; Chiu, 1968; Clayton, 1968) then gives for the energy flux carried by convection

$$F_{conv} = \frac{1}{4}C_p\rho\left(\frac{g}{T}\right)^{1/2}(\Delta\nabla T)^{3/2}\ell^2$$

where ρ is the density and ℓ the mixing length.

Even when conduction and convection are negligible, the transport of energy by radiation is always possible. Furthermore, since stars radiate, the energy flux must at some time, if only ultimately, be carried by radiation. The rate of transport of energy will be limited only by the interactions of the photons with matter. The general form of the relation between the radiative energy flux and the temperature gradient may be seen from a simple physical argument due to Clayton (1968). We consider two points at which the local temperatures are T and $T + \delta T$ and which are separated by a distance $\bar{\lambda}$ which is the average distance traversed by a photon before it is absorbed (mean free path). We suppose also that the matter at each point emits and absorbs radiation like a black body at the local temperature, i.e., the integrated radiative emissivity for a body at temperature T is σT^4 per unit surface where σ is the Stefan-Boltzmann constant, while all incident radiation is absorbed. The net radiative energy flux between the two points is therefore

$$F_{rad} = -\sigma\delta T^4 = -4\sigma T^3\delta T$$

or, putting

$$\delta T = \bar{\lambda}\frac{dT}{dr}$$

where dT/dr is the temperature gradient, we have

$$F_{rad} = -4\bar{\lambda}\sigma T^3\frac{dT}{dr}$$

Since

$$\sigma = \frac{ac}{4}$$

where a is the radiation constant, the expression for the flux becomes

$$F_{rad} = -ac\bar{\lambda}T^3\frac{dT}{dr}$$

Denoting by $\bar{\kappa}$ the average mass absorption coefficient, or absorption cross-section per unit mass, we may write

$$\bar{\lambda} = \frac{1}{\rho\bar{\kappa}}$$

so that we then obtain

$$F_{rad} = -\frac{ac}{\rho\bar{\kappa}}T^3\frac{dT}{dr} = -K_{rad}\left(\frac{dt}{dr}\right)$$

Thus the radiative energy flux is proportional to the negative of the temperature gradient, with the proportionality constant, which we may regard as the "radiative conductivity," given by

$$K_{rad} = \frac{ac}{\rho\bar{\kappa}}T^3$$

i.e., proportional to the reciprocal of the average mass absorption coefficient. However, the above argument does not show how the average mass absorption coefficient is to be formed. To do this we must consider the process of radiative transfer in more detail.

3. RADIATIVE TRANSFER

The study of radiative transfer, in common with other transport phenomena, may be approached via the Boltzmann transport equation. Thus, following Chiu (1968), we may introduce the photon distribution function $f(\underline{q},\underline{p})$, where $\underline{q}$ and $\underline{p}$ are the co-ordinates in configuration and momentum space, respectively, such that f is the photon occupation number per quantum state, i.e., the density of photons in phase-space is $2f/h^3$. The Boltzmann transport equation then takes the form

$$\frac{\partial f}{\partial t} + (\dot{\underline{q}}\cdot\underline{\nabla}_q)f + (\dot{\underline{p}}\cdot\underline{\nabla}_p)f = \left[\frac{df}{dt}\right]_i$$

in which the interaction term $[df/dt]_i$ (often referred to as the collision term) represents the processes of the emission or creation, the absorption or annihilation, and the scattering of photons. Because of the constancy of the velocity of photons, we may write

$$\dot{\underline{q}} = c\underline{\omega}$$

where ω is a unit vector in the direction of propagation $\omega = (\theta,\phi)$. Also since, if we ignore gravitational effects, the momentum of a photon can only be changed by interactions we have

$$\dot{\underline{p}} = 0$$

so that the transport equation becomes

$$\frac{\partial f}{\partial t} + c(\underline{\omega}.\underline{\nabla})f = \left[\frac{df}{dt}\right]_i$$

For a steady state we put

$$\frac{\partial f}{\partial t} = 0$$

and thus obtain

$$(\underline{\omega}.\underline{\nabla})f = \frac{1}{c}\left[\frac{df}{dt}\right]_i$$

or, denoting by s the distance measured in the direction of $\underline{\omega}$

$$\frac{df}{ds} = \frac{1}{c}\left[\frac{df}{dt}\right]_i$$

It has been more usual in the past to work, not with the photon distribution function f as defined above, but instead with the specific intensity of radiation $I(\nu)$, which is the energy flux per unit frequency per unit solid angle, and given by

$$I(\nu) = h\nu c n(\nu)$$

where $n(\nu)$ is the number density of photons per unit frequency interval per unit solid angle. From the relation

$$n(\nu) = 2\frac{\nu^2}{c^3}f$$

we then have

$$I(\nu) = 2\frac{h\nu^3}{c^2}f$$

In terms of the intensity the transport equation then becomes

$$\frac{dI(\nu)}{ds} = \frac{1}{c}\left[\frac{dI(\nu)}{dt}\right]_i = S$$

where the interaction term S is the net effect of sources and sinks (considered as negative sources) for photons. We may write S as the sum of three parts

$$S = S_e + S_a + S_s$$

the first of which represents the contribution from emission (including stimulated emission), the second represents that from absorption, and the third represents that due to scattering (including stimulated scattering) both into and out of the beam.

Introducing the Einstein transition probability coefficients A_{ba}, B_{ba}, B_{ab} defined by the relation

$$N_b A_{ba} + N_b B_{ba}\rho(\nu) = N_a B_{ab}\rho(\nu)$$

where N_a and N_b are the number densities of absorbers and emitters in the states a and b, ν is the frequency of radiation associated with a transition between these states, and $\rho(\nu)$ is the radiation energy density, we have

$$S_e = \frac{h\nu}{4\pi}\left[N_b A_{ba} + N_b B_{ba}\rho(\nu)\right] = \frac{h\nu}{c}\left[\frac{c}{4\pi} N_b A_{ba} + N_b B_{ba} I(\nu)\right]$$

$$S_a = -\frac{h\nu}{4\pi} N_a B_{ab}\rho(\nu) = -\frac{h\nu}{c} N_a B_{ab} I(\nu)$$

where

$$I(\nu) = \frac{c}{4\pi}\rho(\nu)$$

Adding S_e and S_a and rearranging the terms

$$S_e + S_a = -N_a\frac{h\nu}{c} B_{ab}\left(1 - \frac{N_b}{N_a}\frac{B_{ba}}{B_{ab}}\right)\left[I(\nu) - \frac{c}{4\pi}\frac{N_b}{N_a}\frac{A_{ba}}{B_{ab}}\left(1 - \frac{N_b}{N_a}\frac{B_{ba}}{B_{ab}}\right)^{-1}\right]$$

In equilibrium at temperature T the intensity is given by the Planck function

$$I(\nu) = B(\nu,T) = \frac{2h\nu^3/c^2}{e^{h\nu/kT}-1}$$

and the Einstein coefficients satisfy the relations

$$\frac{N_b}{N_a}\frac{B_{ba}}{B_{ab}} = e^{-h\nu/kT}$$

and

$$\frac{N_b}{N_a}\frac{A_{ba}}{B_{ab}} = \frac{8\pi h\nu^3}{c^3}$$

Substituting in the expression for $S_e + S_a$, that is assuming local thermodynamic equilibrium, we obtain

$$S_e + S_a = -N_a\frac{h\nu}{c}B_{ab}\left(1 - e^{-h\nu/kT}\right)[I(\nu) - B(\nu,T)]$$

or, introducing the absorption cross-section $\sigma_a(\nu)$ given by

$$\sigma_a(\nu) = \frac{h\nu}{c}B_{ab}$$

we have

$$S_e + S_a = -N_a\sigma_a(\nu)\left(1 - e^{-h\nu/kT}\right)\left[I(\nu) - B(\nu,T)\right]$$

The scattering term S_s may be written

$$S_s = N_s\left[\int\sigma_s(\omega',\omega)I(\nu,\omega')d\omega' - I(\nu,\omega)\int\sigma_s(\omega,\omega')d\omega'\right]$$

where N_s is the number density of scatterers, and $\sigma_s(\omega,\omega')$ is the differential cross-section for scattering from ω to ω'. The first term on the right-hand side represents direct scattering into the beam, while the second term represents direct scattering out of the beam. Strictly speaking, we should also include terms which represent induced scattering both into and out of the beam. In general, these induced terms do not compensate each other. However, in the case of coherent scattering (i.e., with no change of frequency) and in an isotropic medium (where the refractive index is independent of the direction of propagation), the induced scattering terms cancel each other exactly (Cox, 1968). Writing

$$\sigma_s(\omega,\omega') = \sigma_s(\nu)p(\omega,\omega')$$

where $\sigma_s(\nu)$ is the total cross-section and $p(\omega',\omega)$ is a phase function such that

$$\int p(\omega,\omega')d\omega' = 1$$

and

$$p(\omega,\omega') = p(\omega',\omega)$$

so that the scattering cross-section is a function only of the angle between ω and ω', we then have, for coherent scattering

$$S_s = -N_s\sigma_s(\nu)\left[I(\nu) - \int p(\omega',\omega)I(\nu,\omega')d\omega'\right]$$

Putting

$$N_a\sigma_a(\nu) = \mu_a(\nu) = \rho\kappa_a(\nu)$$

$$N_s\sigma_s(\nu) = \mu_s(\nu) = \rho\kappa_s(\nu)$$

where ρ is the density of matter and $\mu_{a,s}(\nu)$, $\kappa_{a,s}(\nu)$ are the volume and mass absorption and scattering coefficients, the transport equation takes the form of the equation of radiative transfer

$$\frac{dI(\nu)}{ds} = -\rho\kappa_a(\nu)\left(1 - e^{-h\nu/kT}\right)\left[I(\nu) - B(\nu,T)\right]$$
$$- \rho\kappa_s(\nu)\left[I(\nu) - \int p(\omega',\omega)I(\nu,\omega')d\omega'\right]$$

Alternatively we may write it in the more familiar form

$$\frac{dI(\nu)}{ds} = -\rho\kappa_e(\nu)I(\nu) + \rho j_e(\nu)$$

where

$$\kappa_e(\nu) = \kappa_a'(\nu) + \kappa_s(\nu), \quad \kappa_a'(\nu) = \kappa_a(\nu)\left(1 - e^{-h\nu/kT}\right)$$

and

$$j_e(\nu) = \kappa_a'(\nu)B(\nu,T) + \kappa_s(\nu)\int p(\omega',\omega)I(\nu,\omega')d\omega'$$

Thus the effective absorption coefficient $\kappa_e(\nu)$ is the absorption coefficient $\kappa_a(\nu)$ corrected for stimulated emission by the factor $(1 - e^{-h\nu/kT})$ plus the scattering coefficient $\kappa_s(\nu)$. The correction factor for stimulated emission was first introduced by Rosseland (1930). That it is not also applied to the scattering coefficient has been pointed out by Mayer (1947) (who attributed it to Maria G. Mayer) and by Rudkjobing (1947). Similarly, the effective source term $j_e(\nu)$ in the equation of transfer is the Kirchhoff emission $\kappa_a(\nu)B(\nu,T)$ also corrected for stimulated emission (which we have seen contributes negatively to the absorption) plus a contribution from scattering.

3.1 The Radiative Flux

For stellar applications we introduce spherical co-ordinates (r, θ) where r is the radial co-ordinate and θ is the angle between the direction of s and the radius vector. From the relations

$$dr = ds\cos\theta, \quad d\theta = -\frac{1}{r}\sin\theta ds$$

and

$$\frac{d}{ds} = \frac{\partial r}{\partial s}\frac{\partial}{\partial r} + \frac{\partial \theta}{\partial s}\frac{\partial}{\partial \theta} = \cos\theta \frac{\partial}{\partial r} - \frac{1}{r}\sin\theta \frac{\partial}{\partial \theta}$$

the transfer equation becomes

$$\cos\theta\frac{\partial I(\nu)}{\partial r} - \frac{1}{r}\sin\theta\frac{\partial I(\nu)}{\partial \theta} = - \rho\kappa_e(\nu)I(\nu) + \rho j_e(\nu)$$

which in the limit $r \to \infty$ goes over into the plane parallel geometry case. We now define

$$J(\nu) = \frac{1}{4\pi}\int I(\nu)d\omega = \frac{c}{4\pi} E(\nu)$$

$$H(\nu) = \frac{1}{4\pi}\int I(\nu)\cos\theta d\omega = \frac{1}{4\pi} F(\nu)$$

$$K(\nu) = \frac{1}{4\pi}\int I(\nu)\cos^2\theta d\omega = \frac{c}{4\pi} P(\nu)$$

where $d\omega = 2\pi\sin\theta d\theta$ and $E(\nu)$, $F(\nu)$ and $P(\nu)$ are the radiation energy density, net flux and pressure, respectively. By successively multiplying the equation of transfer by $(\cos\theta)^n$ where $n = 0,1$ and integrating over all angles using the results

$$\frac{1}{4\pi}\int \sin\theta\frac{dI(\nu)}{d\theta}d\omega = - 2H(\nu)$$

and

$$\frac{1}{4\pi}\int \sin\theta\cos\theta\frac{dI(\nu)}{d\theta}d\omega = - [3K(\nu) - J(\nu)]$$

we obtain

$$\begin{aligned}\frac{dH(\nu)}{dr} + \frac{2}{r}H(\nu) &= -\rho [\kappa_e(\nu) - \kappa_s(\nu)] J(\nu) + \rho\kappa_a'(\nu)B(\nu,T)\\ &= -\rho\kappa_a'(\nu) [J(\nu) - B(\nu,T)]\end{aligned}$$

and

$$\begin{aligned}\frac{dK(\nu)}{dr} + \frac{1}{r} [3K(\nu) - J(\nu)] &= -\rho \left[\kappa_e(\nu) - \kappa_s(\nu)\bar{\mu}\right] H(\nu)\\ &= -\rho \left[\kappa_a'(\nu) + \kappa_s'(\nu)\right] H(\nu)\end{aligned}$$

where

$$\kappa_s'(\nu) = \kappa_s(\nu)(1 - \bar{\mu})$$

and

$$\bar{\mu} = \int p(\omega,\omega')\cos\theta'' d\omega'$$

θ'' being the angle between ω and ω'. In deriving the above, we have made use of the additional relations

$$\frac{1}{4\pi}\iint p(\omega,\omega')I(\nu,\omega')d\omega' d\omega = \frac{1}{4\pi}\int I(\nu,\omega')d\omega' = J(\nu)$$

and

$$\frac{1}{4\pi}\iint p(\omega',\omega)I(\nu,\omega')d\omega'\cos\theta d\omega = \bar{\mu}H(\nu)$$

To obtain the last result, we refer $\omega = (\theta,\phi)$ to a new polar axis about ω' so that the new co-ordinates are $\omega" = (\theta",\phi")$ and using the relation

$$\cos\theta = \cos\theta'\cos\theta" + \sin\theta'\sin\theta"\cos(\phi'-\phi")$$

we have

$$\frac{1}{4\pi}\iint p(\omega',\omega)I(\nu,\omega')d\omega'\cos\theta d\omega =$$

$$= \left(\int p(\omega',\omega")\cos\theta" d\omega"\right)\left(\frac{1}{4\pi}\int I(\nu,\omega')\cos\theta' d\omega'\right)$$

$$= \bar{\mu}H(\nu)$$

the term in ϕ' and $\phi"$ giving zero on integrating since $I(\nu,\omega')$ is a function only of (ν and) θ', and $p(\omega',\omega")$ a function only of $\theta"$. If the phase function $p(\omega',\omega")$ is an even function of $\cos\theta"$, then $\bar{\mu} = 0$ and

$$\kappa_s'(\nu) = \kappa_s(\nu)$$

The quantity of primary interest in the stellar problem is the net flux of radiative energy

$$F(\nu) = 4\pi H(\nu)$$

and, more particularly, the total flux integrated over all frequencies

$$F = \int F(\nu)d\nu = 4\pi\int H(\nu)d\nu = 4\pi H$$

From the equation of transfer we have, on rearrangement

$$H(\nu) = -\frac{1}{\rho\kappa'(\nu)}\left[\frac{dK(\nu)}{dr} + \frac{1}{r}\{3K(\nu) - J(\nu)\}\right]$$

where

$$\kappa'(\nu) = \kappa_a'(\nu) + \kappa_s'(\nu)$$

We may eliminate the $\frac{1}{r}$ term by setting the intensity equal to the Planck function, i.e.

$$I(\nu) = B(\nu,T)$$

from which it follows that

$$J(\nu) = 3K(\nu) = B(\nu,T)$$

so that the equation of transfer reduces to

$$H(\nu) = -\frac{1}{\rho\kappa'(\nu)}\frac{dK(\nu)}{dr}$$

or, on integrating

$$H = -\frac{1}{\rho}\int\frac{1}{\kappa'(\nu)}\frac{dK(\nu)}{dr}\,d\nu$$

3.2 The Rosseland Mean Opacity

Introducing the temperature T and using the relation

$$\frac{dK(\nu)}{dr} = \frac{dK(\nu)}{dT}\frac{dT}{dr}$$

we then have

$$H = -\frac{1}{\rho}\left(\int\frac{1}{\kappa'(\nu)}\frac{dK(\nu)}{dT}d\nu\right)\frac{dT}{dr}$$

If $\kappa'(\nu) = \kappa$, i.e., the absorption coefficient is frequency independent, we can write

$$H = -\frac{1}{\rho\kappa}\left(\int\frac{dK(\nu)}{dT}d\nu\right)\frac{dT}{dr} = -\frac{1}{\rho\kappa}\frac{dK}{dT}\frac{dT}{dr}, \quad K = \int K(\nu)d\nu$$

When $\kappa'(\nu)$ is frequency dependent, we may retain the above equation by defining a mean coefficient of absorption, or mean opacity, κ by the relation

$$\frac{1}{\kappa} = \int\frac{1}{\kappa'(\nu)}\frac{dK(\nu)}{dT}d\nu \Big/ \frac{dK}{dT}$$

If we now make the substitution

$$K(\nu) = \tfrac{1}{3}B(\nu,T)$$

we obtain

$$\frac{1}{\kappa} = \int\frac{1}{\kappa'(\nu)}\frac{dB(\nu,T)}{dT}d\nu \Big/ \frac{dB(T)}{dT}, \quad B(T) = \int B(\nu,T)d\nu$$

The above form of harmonic averaging with respect to the temperature derivative of the Planck function was first introduced by Rosseland (1924), and the value of κ thus defined is usually known as the Rosseland mean coefficient of absorption or Rosseland mean opacity. Substituting for the Planck function giving

$$\frac{dB(\nu,T)}{dT} = \frac{2h^2\nu^4}{c^2kT^2}\frac{e^{h\nu/kT}}{(e^{h\nu/kT}-1)^2}$$

and using

$$B(T) = \frac{acT^4}{4\pi}, \quad \frac{dB(T)}{dT} = \frac{acT^3}{\pi}, \quad a = \frac{8\pi^5 k^4}{15c^3h^3}$$

we have for the weighting function

$$\frac{dB(\nu,T)}{dT} \Bigg/ \frac{dB(T)}{dT} = \frac{15}{4\pi^4} \frac{h^5\nu^4}{k^5T^5} \frac{e^{h\nu/kT}}{(e^{h\nu/kT}-1)^2}$$

Putting

$$u = \frac{h\nu}{kT}$$

the Rosseland mean opacity is then given by

$$\frac{1}{\kappa} = \frac{15}{4\pi^4}\int \frac{u^4e^u}{\kappa'(u)(e^u-1)^2}\,du$$

We also have

$$H = -\frac{1}{\rho\kappa}\frac{dK}{dT}\frac{dT}{dr} = -\frac{acT^3}{3\pi\rho\kappa}\frac{dT}{dr}$$

and

$$F = -\frac{4acT^3}{3\rho\kappa}\frac{dT}{dr}$$

Thus the radiative conductivity is given by

$$K_{rad} = \frac{4acT^3}{3\rho\kappa}$$

i.e., proportional to the reciprocal of the Rosseland mean opacity. The Rosseland weighting function

$$\frac{15}{4\pi^4}\frac{u^4e^u}{(e^u-1)^2}$$

which is normalized to unity, attains a maximum at $u =$ 3.830016 ...... and therefore gives the greatest weight to $\kappa'(u)$ for that value of u. The Planck function has its maximum at $u = 2.821439$.... Now

$$\kappa'(u) = \kappa_a(u)\left(1 - e^{-u}\right) + \kappa_s(u)(1 - \bar{\mu})$$

which, on factoring out $(1 - e^{-u})$, may be written in the form

$$\kappa'(u) = \kappa(u)\left(1 - e^{-u}\right)$$

where

$$\kappa(u) = \kappa_a(u) + \kappa_s(u)(1 - \bar{\mu})\left(1 - e^{-u}\right)^{-1}$$

The Rosseland mean opacity is then given by

$$\frac{1}{\kappa} = \frac{15}{4\pi^4}\int \frac{u^4 e^{2u}}{\kappa(u)(e^u-1)^3}\,du$$

The new weighting function

$$\frac{15}{4\pi^4}\,\frac{u^4 e^{2u}}{(e^u-1)^3}$$

no longer normalizes to unity but to the value

$$\frac{1}{2}\left[1 + \frac{\xi(3)}{\xi(4)}\right] = 1.055313.....$$

where

$$\xi(n) = \frac{1}{\Gamma(n)}\int_0^\infty \frac{\kappa^{n-1}}{e^\kappa-1}\,d\kappa = \sum_{r=1}^{\infty} r^{-n}, \quad \xi(4) = \frac{\pi^4}{90}$$

is the Riemann Zeta function. Thus, if the opacity is due to frequency independent scattering alone, an incorrect application of the stimulated emission correction would result in a mean opacity which is too low by a factor of the above value.

4. ABSORPTION AND SCATTERING CROSS-SECTIONS

The interaction of radiation with matter may take place in a variety of ways, depending upon the radiation energy and the state in which the matter occurs, e.g., elementary particles, nuclei, atoms, molecules. The actual processes of interest in opacity will therefore depend upon the physical conditions of the system under consideration. For stellar matter the temperature may vary from about 10^9 °K at the center of the hottest stars to about 10^3 °K at the surface of the coolest, with corresponding variations in density as determined by the equation of state and the requirements of hydrostatic equilibrium. Under such conditions the chief interaction is essentially that between photons and electrons. But even then there may be a number of different processes depending upon the states of the electron before and after the interaction, in particular, whether the electron is free or bound in another system such as an atom or molecule. Over a wide range of stellar conditions, the main radiative processes are (Thomson or Compton) scattering of photons by free electrons, the absorption of photons by free electrons (free-free absorption or inverse bremsstrahlung) and the absorption of photons by bound atomic electrons in bound-free transitions (photo-electric effect) or bound-bound transitions (line absorption). At the highest temperatures other processes such as pair production, photo-nuclear absorption and photon-photon scattering may occur, while at the lowest temperatures negative ion absorption, molecular absorption and Rayleigh and Raman scattering may become important. In general, not all of the above processes operate simultaneously, so that under given conditions only the several most important need to be considered. The relevant theory for the calculation of the absorption and scattering cross-sections is provided by quantum mechanics, although in certain cases the results of classical electrodynamics are approximately valid.

4.1 Classical and Semiclassical Formulae

4.1.1 Scattering

The cross-section for the scattering of radiation by free electrons may be obtained classically by considering the motion of an electron in the electric field of an incident electromagnetic wave. The electron undergoes forced vibrations, giving rise to the emission of a secondary (scattered) wave, with the frequency of the incident wave. The differential cross-section is given by (Heitler, 1954)

$$\sigma(\theta,\phi) = \frac{1}{2} r_o^2 (1 + \cos^2\theta)$$

where θ is the angle of scattering and r_o is the classical radius of the electron given by

$$r_o = \frac{e^2}{mc^2} = 2.81785 \;.\; 10^{-13}\text{cm}$$

The total cross-section then becomes

$$\sigma = \iint \sigma(\theta,\phi)\sin\theta d\theta d\phi = \frac{8\pi}{3} r_o^2 = 0.665203 \;.\; 10^{-24}\text{cm}^2 \;.$$

We may therefore write

$$\sigma(\theta,\phi) = \sigma p(\theta,\phi)$$

where now $p(\theta,\phi)$ is the phase function

$$p(\theta,\phi) = \frac{3}{16\pi}(1 + \cos^2\theta)$$

or, on integrating over ϕ

$$p(\theta) = \int p(\theta,\phi)d\phi = \frac{3}{8}(1 + \cos^2\theta)$$

It will be noticed that the cross-section as given above is independent of the frequency of the radiation. This is no longer the case if we take account of the reaction of the field produced by a moving electron on the electron itself. According to classical electrodynamics, this self-force which has the effect of damping the motion is given by

$$\underline{F}_s = \varepsilon m \dddot{\underline{r}}, \quad \varepsilon = \frac{2e^2}{3mc^3} = 6.2818 \;.\; 10^{-24}\text{sec}$$

When this term is introduced into the equation of motion of the electron, the cross-section for scattering of radiation of frequency ν becomes

$$\sigma(\theta,\phi) = \frac{1}{2}r_o^2\frac{1}{1+\varepsilon^2\omega^2}(1 + \cos^2\theta)$$

where $\omega = 2\pi\nu$. We also have

$$\sigma = \frac{8\pi}{3}r_o^2\frac{1}{1+\varepsilon^2\omega^2}$$

Now $\varepsilon\omega = (4\pi/3)(r_o/\lambda)$ and is therefore exceedingly small for all wavelengths down to those of γ-rays, and consequently may often be neglected. The above cross-section (with and without the damping correction) was first obtained by Thomson and is known as the Thomson cross-section.

Classical theory may also be used to derive the cross-section for scattering of radiation by an oscillator consisting of an electron which is elastically bound to a center of force with a fundamental frequency ω_o. We thus obtain

$$\sigma(\theta,\phi) = \sigma p(\theta,\phi), \quad \sigma = \frac{8\pi}{3}r_o^2\frac{\omega^4}{(\omega^2-\omega_o^2)^2+\varepsilon^2\omega^6}$$

where $p(\theta,\phi)$ is as given above. In the limit of very high frequencies ($\omega >> \omega_o$), the Thomson cross-section is recovered, while for very low frequencies ($\omega << \omega_o$), we have

$$\sigma = \frac{8\pi}{3}r_o^2\frac{\omega^4}{\omega_o^4}$$

which is known as the Rayleigh cross-section.

4.1.2 Absorption

For free-free and bound-free absorption, the cross-sections may be obtained from the Kramers (1923) semiclassical theory of emission. According to this theory the radiation from electrons in atoms is divided into two parts: one due to electrons which change from one hyperbolic orbit to another (free-free emission), and another in which the electrons change from a hyperbolic orbit to an elliptical one (free-bound emission). On the basis of the principle of detailed balancing, the Kramers theory gives the absorption cross-sections for electrons moving in the Coulomb field of a nucleus of charge number Z. For free-free absorption the cross-section per nucleus per unit volume is given by

$$\sigma_{ff}^{K}(\nu,v) = \frac{4\pi e^6 Z^2 N_e}{3\sqrt{3}\ hcm^2\nu^3 v}$$

where ν is the frequency of the radiation, N_e is the number density of free electrons and v their velocity. Averaging over a Maxwellian velocity distribution

$$f(v) = \sqrt{\frac{2}{\pi}} \left(\frac{m}{kT}\right)^{3/2} v^2 e^{-mv^2/2kT}$$

for temperature T we obtain

$$\overline{\frac{1}{v}} = \int \frac{1}{v} f(v)\,dv = \sqrt{\frac{2m}{\pi kT}}$$

so that

$$\sigma_{ff}{}^K(\nu) = \sigma_{ff}{}^K(\nu,\bar{v}) = \frac{4\sqrt{2\pi}\; e^6 Z^2 N_e}{3\sqrt{3}\; hcm^{3/2}(kT)^{1/2}\nu^3}$$

For bound-free absorption, the cross-section per electron in an energy level of principal quantum number n is

$$\sigma_{bf}{}^K(\nu,n) = \frac{64\pi^4 me^{10} Z^4}{3\sqrt{3}\; ch^6 n^5 \nu^3}$$

A classical formula for the cross-section for bound-bound absorption of radiation may be derived by considering the energy transfer from an incident wave to a bound oscillating electron. In the absence of any damping the result is (Heitler, 1954)

$$\sigma_{bb}(\omega) = \frac{2\pi^2 e^2}{mc}\delta(\omega-\omega_o) = \frac{\pi e^2}{mc}\delta(\nu-\nu_o)$$

where $\omega_o = 2\pi\nu_o$, ν_o being the fundamental frequency of the oscillator. If account is taken of the electrodynamic self-force, the resulting damping leads to a broadened absorption line profile such that

$$\sigma_{bb}(\omega) = \frac{2\pi^2 e^2}{mc} F(\omega)$$

with

$$F(\omega) = \frac{\gamma/\pi}{(\omega-\omega_o-\delta)^2 + \gamma^2}$$

where $\gamma = \frac{1}{2}\varepsilon\omega_o{}^2$ and $\delta = -\frac{5}{8}\varepsilon^2\omega_o{}^3$.

4.2 Quantum Mechanical Formulae

According to quantum mechanics, the interaction of radiation with matter is described by introducing into the Hamiltonian for the complete system an additional term--the interaction Hamiltonian--which represents the coupling of the electronic charge with the electromagnetic field. The interaction Hamiltonian is given by

$$H' = -\frac{e}{mc}\,\underline{p}.\underline{A} + \frac{e^2}{2mc^2}A^2$$

where $\underline{p}$ is the electron momentum and $\underline{A}$ is the electromagnetic potential which has the form

$$\underline{A} = \underline{A}_o \exp[i(\underline{k}.\underline{r} - 2\pi\nu t)]$$

where ν is the frequency, $\underline{k}$ the wave number vector ($k = 1/\lambda = c/\nu$) and $\underline{A}_o$ is a constant amplitude vector related to the radiation intensity by

$$I(\nu) = \frac{2\pi\nu^2}{c}\,|\underline{A}_o|^2$$

The rate at which transitions occur is then obtained by the application of quantum mechanical time-dependent perturbation theory. The transition probability per unit time for a transition between two states a and b of the complete (radiation field + electron) system with energies ε_a and ε_b is then given by the standard formula

$$P_{ab} = \frac{2\pi}{\hbar}\,|H'_{ab}|^2\,\delta(\varepsilon_a - \varepsilon_b)$$

where H'_{ab} is the transition matrix element. For simple absorption where only a single quantum is involved, the term in A^2 may be neglected. When the wavelength concerned is sufficiently large that, for those regions of space where the interaction is appreciable, $\underline{k}.\underline{r} \ll 1$, we may make the expansion

$$\exp(i\underline{k}.\underline{r}) = 1 + i\underline{k}.\underline{r} + \ldots$$

The dipole approximation is obtained by retaining only the first term in this expansion in the matrix element. It may then be shown that the transition probability for the absorption of radiation of frequency ν per unit solid angle, after averaging over all orientations of the electron co-ordinate system, is given by

$$P_{ab} = \frac{2\pi e^2}{3m^2ch^2\nu^2}\,I(\nu)\,|\underline{p}_{ab}|^2\,\delta(\nu - \nu_{ab})$$

where $\underline{p}_{ab}$ is the momentum operator matrix element, taken between the electron states, and ν_{ab} is the frequency associated with the transition, i.e.

$$h\nu_{ab} = E_b - E_a$$

where E_a and E_b are the initial and final electron energies. We may express the transition rate as a probability per unit incident flux of photons, or cross-section, obtaining

$$\sigma_{ab}(\nu) = \frac{P_{ab}}{I(\nu)/h\nu} = \frac{2\pi e^2}{3m^2 ch\nu} |\underline{p}_{ab}|^2 \delta(\nu - \nu_{ab})$$

Before we can evaluate this expression, a knowledge of the electron wave functions is required.

4.2.1 Wave functions

The wave function ψ and energy E of an electron are related by the Schrödinger equation

$$\nabla^2\psi + \frac{2m}{\hbar^2}(E + eV)\psi = 0$$

where V is the electrostatic potential in which the electron moves. For spherically symmetric potentials, i.e.

$$V = V(r)$$

ψ is separable in spherical polar co-ordinates (r,θ,ϕ), so that we can write

$$\psi_{n,\ell,m}(r,\theta,\phi) = R_{n\ell}(r)Y_{\ell m}(\theta,\phi) = R_{n\ell}(r)\Theta_{\ell m}(\theta)\Phi_m(\phi)$$

where n, ℓ and m are the principal, azimuthal and magnetic quantum numbers. The angular part of the wave function is just a spherical harmonic given by

$$Y_{\ell m}(\theta,\phi) = \left[\frac{2\ell+1}{2}\frac{(\ell-m)!}{(\ell+m)!}\right]^{1/2} P_\ell^{\,m}(\cos\theta)\left[\frac{1}{2\pi}\right]^{1/2} e^{im\phi}$$

where

$$P_\ell^{\,m}(x) = (1 - x)^{m/2}\frac{d^m P_\ell(x)}{dx^m}$$

$P_\ell(x)$ being the Legendre polynomial of degree ℓ. The spherical harmonics are orthonormal in that

$$\int_0^{2\pi}\int_0^{\pi} Y^*_{\ell' m'}(\theta,\phi)Y_{\ell m}(\theta,\phi)\,d\omega = \delta(\ell,\ell')\delta(m,m')$$

where δ is the Kronecker symbol. Only the radial part of the wave function depends on the actual form of the potential, being given by the solution of the differential equation

$$\frac{d^2R}{dr^2} + \frac{2}{r}\frac{dR}{dr} + \left[\frac{2m}{\hbar^2}(E+eV) - \frac{\ell(\ell+1)}{r^2}\right]R = 0$$

with appropriate boundary conditions imposed by requirements of regularity. There are two distinct types of solution: (i) bound state wave functions for which E < 0 and (ii) free state wave functions for which E < 0. For bound states the normalization is such that

$$\int R^*_{n'\ell'}(r)R_{n\ell}(r)r^2dr = \delta(n,n')$$

or

$$\int |R_{n\ell}(r)|^2 r^2 dr = 1$$

For free states a different normalization is required, since the functions do not vanish sufficiently rapidly at large distances to make the above integrals convergent. For our purposes it is convenient to normalize "on the energy scale," i.e., per unit energy, according to the specification

$$\int R^*_{E'\ell}(r)R_{E\ell}(r)r^2dr = \delta(E - E')$$

or

$$\lim_{\Delta E\to 0} \frac{1}{\Delta E}\int \left| \int_{E-\Delta E/2}^{E+\Delta E/2} R_{E\ell}(r)dE\right|^2 r^2dr = 1$$

It can be shown that with this normalization the radial wave function for a free state has the asymptotic form

$$R_{E\ell} \approx \sqrt{\frac{2mk}{\pi\hbar^2}}\,\frac{\sin(kr+\delta_\ell)}{kr}$$

where k is defined by

$$E = \hbar^2k^2/2m$$

and δ_ℓ is a uniquely determined phase shift. Alternative normalizations with respect to some other parameter may be imposed in accordance with the above prescription (Bethe and Salpeter, 1956). Thus, if T is any function of the energy E, we have

$$R_{T\ell}(r) = \left(\frac{dT}{dE}\right)^{-1/2} R_{E\ell}(r)$$

4.2.2 Bound-bound absorption

For transitions between discrete states, i.e., bound-bound or line absorption, we may write the cross-section in the form

$$\sigma_{bb}(\nu) = \frac{\pi e^2}{mc} f_{ab}\delta(\nu - \nu_{ab})$$

which differs from the classical expression only in the factor f_{ab}--the oscillator strength--defined by

$$f_{ab} = \frac{2}{3mh\nu} |\underline{p}_{ab}|^2$$

The matrix element $\underline{p}_{ab}$ may be transformed by making use of the quantum mechanical relations

$$\underline{p}_{ab} = 2\pi i m \nu_{ab} \underline{r}_{ab} = -i \underline{\nabla}_{ab} = \frac{e}{2\pi i \nu_{ab}} (\underline{\nabla} V)_{ab}$$

which give the dipole moment, momentum and potential gradient forms--sometimes referred to as the dipole length, dipole velocity and dipole acceleration formulae. Thus the oscillator strength becomes in the dipole length form

$$f_{ab} = \frac{4\pi m \nu_{ab}}{3\hbar} |\underline{r}_{ab}|^2$$

When the initial and final levels are degenerate, we may sum over the final substates and average over the initial substates to obtain

$$f_{ab} = \frac{4\pi m \nu_{ab}}{3\hbar} \frac{1}{g_a} S_{ab}$$

where g_a is the degeneracy (including spin) of the level a and

$$S_{ab} = \sum_{\alpha} \sum_{\beta} |\underline{r}_{a\alpha b\beta}|^2$$

in which α and β denote the degenerate substances (including spin) of the levels a and b.

For spherically symmetric potentials, where the angular wave functions are already given by the spherical harmonics, we may immediately carry out the appropriate angular integration in the matrix element

$$\underline{r}_{ab} = \int \psi_b^*(\underline{r}) \underline{r} \psi_a(\underline{r}) d\underline{r}$$

to obtain as the only non-vanishing components

$$\left.\begin{aligned} |\underline{r}_{n\ell m, n'\ell' m\pm 1}|^2 &= \frac{(\ell \pm m+1)(\ell \pm m+2)}{2(2\ell+1)(2\ell+3)} (r_{n\ell, n'\ell'})^2 \\ |\underline{r}_{n\ell m, n'\ell' m}|^2 &= \frac{(\ell+m+1)(\ell-m+1)}{(2\ell+1)(2\ell+3)} (r_{n\ell, n'\ell'})^2 \end{aligned}\right\} \quad \ell' = \ell+1$$

$$\left.\begin{aligned} |\underline{r}_{n\ell m, n'\ell' m\pm 1}|^2 &= \frac{(\ell \mp m)(\ell \mp m-1)}{2(2\ell+1)(2\ell-1)} (r_{n\ell, n'\ell'})^2 \\ |\underline{r}_{n\ell m, n'\ell' m}|^2 &= \frac{(\ell+m)(\ell-m)}{(2\ell+1)(2\ell-1)} (r_{n\ell, n'\ell'})^2 \end{aligned}\right\} \quad \ell' = \ell-1$$

where

$$r_{ab} = \int R_b^*(r) r R_a(r) r^2 dr$$

Thus the electric dipole selection rules are $\ell - \ell' = \pm 1$ and $m-m' = 0, \pm 1$. Performing the summation over the substates (m), we obtain

$$S_{ab} = 2\ell_{max}\delta(\ell_a - \ell_b, \pm 1)(r_{ab})^2$$

where

$$\ell_{max} = \text{maximum}(\ell_a, \ell_b)$$

ℓ_a and ℓ_b being the angular momentum quantum numbers of the states a and b, and where account is taken of the fact that in this (dipole) approximation spin changes are forbidden. Thus the oscillator strength is determined if the radial matrix element r_{ab} is known or calculable. It may also be expressed in the other two alternative forms

$$r_{ab} = -\frac{\hbar}{2\pi m\nu_{ab}} \int R_b^*(r) \left[\frac{d}{dr} + (\ell_a - \ell_b)(\ell_{max} + \ell_a - \ell_b)\frac{1}{r}\right] R_a(r) r^2 dr$$

and

$$r_{ab} = -\frac{e}{4\pi^2 m\nu_{ab}^2} \int R_b^*(r) \left(\frac{dV}{dr}\right) R_a(r) r^2 dr$$

It is convenient to introduce a (fictitious) oscillator strength density $df/d\nu$, or oscillator strength per unit frequency, defined by the relation

$$\sigma(\nu) = \frac{\pi e^2}{mc} \frac{df}{d\nu}$$

so that for bound-bound transitions

$$\frac{df}{d\nu} = f_{ab}\delta(\nu - \nu_{ab})$$

When line broadening is taken into account, the δ-function profile is replaced by broadened line profile $F(\nu)$ such that

$$\frac{df}{d\nu} = f_{ab}F(\nu)$$

where

$$\int F(\nu) d\nu = 1$$

and thus the integrated cross-section remains unchanged. The above form of the cross-section is particularly useful when either or both of the initial and final energy states belong to the continuous spectrum.

4.2.3 Bound-free absorption

In the case of bound-free absorption when the final energy E_b lies in the continuum, we take for the cross-section

$$\sigma_{bf}(\nu) = \frac{\pi e^2}{mc}\frac{df}{d\nu} = \frac{\pi e^2 h}{mc}\frac{df}{d(h\nu)}$$

where now, on multiplying by the density of final states (which is unity by virtue of our normalization) and integrating, we have

$$\frac{df}{d(h\nu)} = \frac{4\pi m\nu}{3\hbar}\frac{1}{g_a} S_{ab}$$

Thus the cross-section becomes

$$\sigma_{bf}(\nu) = \frac{8\pi^3 e^2 \nu}{3c} \ell_{max}\delta(\ell_a - \ell_b, \pm 1)(r_{ab})^2$$

$$= \frac{e^4}{6\pi m^2 c\nu^3} \ell_{max}\delta(\ell_a - \ell_b, \pm 1)\left(\frac{dV}{dr}\right)^2_{ab}$$

This gives the cross-section for bound-free absorption in the transition from a bound energy level E_a (with quantum numbers n_a and ℓ_a) to a free energy level E_b (with quantum number ℓ_b). For electrons moving in a Coulomb potential, following Gaunt (1930) the cross-section may be expressed in terms of the Kramers result by writing

$$\sigma_{bf}(\nu,n) = \sigma_{bf}^{K}(\nu,n)g_{bf}(\nu,n)$$

where g_{bf} is the so-called Gaunt factor given by

$$g_{bf}(n_a\ell_a;E_b\ell_b) = \frac{3\sqrt{3}\, h^7\nu^3 n^5}{64\pi^3 m^2 e^8 Z^4}\frac{df}{d(h\nu)}$$

or in terms of the matrix elements

$$g_{bf}(n_a\ell_a;E_b\ell_b) = \frac{\sqrt{3}\, h^6\nu^4 n^5}{8\pi m e^8 Z^4}\frac{\ell_{max}}{2\ell_a+1}\delta(\ell_a - \ell_b, \pm 1)(r_{ab})^2$$

$$= \frac{\sqrt{3}\, h^6 n^5}{128\pi^5 m^3 e^6 Z^4}\frac{\ell_{max}}{2\ell_a+1}\delta(\ell_a - \ell_b, \pm 1)\left(\frac{dV}{dr}\right)^2_{ab}$$

Summing over final ℓ states we define

$$g_{bf}(n_a\ell_a;E_b) = \sum_{\ell_b} g_{bf}(n_a\ell_a;E_b\ell_b)$$

and, if the initial ℓ states are degenerate, we may form the average

$$g_{bf}(n_a;E_b) = \frac{1}{n_a^2} \sum_{\ell_a=0}^{n_a-1} (2\ell_a+1) g_{bf}(n_a\ell_a;E_b)$$

$$= \frac{1}{n_a^2} \sum_{\ell_a=0}^{n_a-1} (2\ell_a+1) \sum_{\ell_b} g_{bf}(n_a\ell_a;E_b\ell_b)$$

and thus finally obtain

$$g_{bf}(\nu,n) = g_{bf}(n,E)$$

4.2.4 Free-free absorption

For free-free absorption, when both the initial and final energy levels are in the continuum, we may again write

$$\sigma_{ff}(\nu) = \frac{\pi e^2 h}{mc} \frac{df}{d(h\nu)}$$

Once more integrating over final states, we obtain for the oscillator strength per unit (final) energy E_b per electron per unit initial energy E_a, with initial and final angular momentum quantum numbers ℓ_a and ℓ_b

$$\frac{df}{d(h\nu)} = \frac{4\pi mv}{3\hbar} \frac{1}{g_a} S_{ab}$$

Multiplying by the number of electrons $N_e(E_a,\ell_a)$ in the state with energy E_a and angular momentum quantum number ℓ_a given by

$$N_e(E_a,\ell_a) = g_a p(E_a)$$

where $p(E_a)$ is the probability of finding an electron with energy E_a (i.e., the average occupancy per state) and bearing in mind that the normalization of the initial free wave function provides for only one state of given E_a and ℓ_a per unit energy, we obtain

$$\frac{df}{d(h\nu)} = \frac{4\pi mv}{3\hbar} S_{ab} p(E_a)$$

Now for the electron distribution in momentum, velocity and energy

$$N_e(p)\,dp = N_e(v)\,dv = N_e(E)\,dE = \frac{8\pi p^2}{h^3}\,p(E)\,dp$$

$$= \frac{8\pi m^3 v^2}{h^3}\,p(E)\,dv = \frac{8\pi m^2 v}{h^3}\,p(E)\,dE$$

or

$$p(E) = \frac{h^3 N_e(p)}{8\pi p^2} = \frac{h^3 N_e(v)}{8\pi m^3 v^2} = \frac{h^3 N_e(E)}{8\pi m^2 v}$$

Substituting for $p(E)$ and summing over ℓ_a and ℓ_b we then have

$$\frac{df}{d(h\nu)} = \frac{\pi h^2 \nu N_e(E)}{3mv} \sum_{\ell_a} \sum_{\ell_b} S_{ab}$$

$$= \frac{2\pi h^2 \nu N_e(E)}{3mv} \sum_{\ell_a} \sum_{\ell_b} \ell_{max}\,\delta(\ell_a - \ell_b, \pm 1)\,(r_{ab})^2$$

so that the cross-section becomes

$$\sigma_{ff}(\nu,E) = \frac{2\pi^2 h^3 \nu e^2 N_e(E)}{3m^2 vc} \sum_{\ell_a} \sum_{\ell_b} \ell_{max}\,\delta(\ell_a - \ell_b, \pm 1)\,(r_{ab})^2$$

or in the dipole acceleration form

$$\sigma_{ff}(\nu,E) = \frac{h^3 e^4 N_e(E)}{24\pi^2 m^4 c v \nu^3} \sum_{\ell_a} \sum_{\ell_b} \ell_{max}\,\delta(\ell_a - \ell_b, \pm 1) \left(\frac{dV}{dr}\right)^2_{ab}$$

As in the case of bound-free absorption we may express the cross-section in terms of the Kramers cross-section and write

$$\sigma_{ff}(\nu,E) = \sigma_{ff}^{K}(\nu,E)\,g_{ff}(\nu,E)$$

where the Gaunt factor g_{ff} is given by

$$g_{ff}(\nu,E) = \frac{\sqrt{3}\,\pi h^4 \nu^4}{2e^4 Z^2} \sum_{\ell_a} \sum_{\ell_b} \ell_{max}\,\delta(\ell_a - \ell_b, \pm 1)\,(r_{ab})^2$$

$$= \frac{\sqrt{3}\,h^4}{32\pi^3 m^2 e^2 Z^2} \sum_{\ell_a} \sum_{\ell_b} \ell_{max}\,\delta(\ell_a - \ell_b, \pm 1) \left(\frac{dV}{dr}\right)^2_{ab}$$

Averaging over the electron distribution function

$$N_e(E) = N_e f(E)$$

where N_e is the number density of electrons of all energies, we obtain

$$\sigma_{ff}(\nu) = \frac{4\pi e^6 Z^2}{3\sqrt{3}hcm^2\nu^3}\int \frac{N_e(E)}{v} g_{ff}(\nu,E)\,dE$$

$$= \frac{4\pi e^6 Z^2 N_e}{3\sqrt{3}hcm^2\nu^3}\int \frac{1}{v} g_{ff}(\nu,E) f(E)\,dE$$

which may be written in the form

$$\sigma_{ff}(\nu) = \frac{4\pi e^6 Z^2 N_e}{3\sqrt{3}hcm^2\nu^3\bar{v}} g_{ff}(\nu) = \sigma_{ff}{}^K(\nu) g_{ff}(\nu)$$

where

$$\sigma_{ff}{}^K(\nu) = \sigma_{ff}{}^K(\nu,\bar{v}), \quad g_{ff}(\nu) = \bar{v}\int \frac{1}{v} g_{ff}(\nu,E) f(E)\,dE$$

and

$$\frac{1}{\bar{v}} = \int \frac{1}{v} f(E)\,dE$$

For a Maxwell-Boltzmann velocity distribution

$$\bar{v} = \left(\frac{\pi kT}{2m}\right)^{1/2}, \quad \sigma_{ff}{}^K(\nu,\bar{v}) = \frac{4\sqrt{2}\pi e^6 Z^2 N_e}{3\sqrt{3}hcm^{3/2}(kT)^{1/2}\nu^3}$$

as defined earlier and

$$g_{ff}(\nu) = \frac{\sqrt{2}}{m^{1/2}kT}\int \frac{1}{v} g_{ff}(\nu,E) E^{1/2} e^{-E/kT}\,dE$$

$$= \frac{1}{kT}\int g_{ff}(\nu,E) e^{-E/kT}\,dE$$

or on putting $x = E/kT$

$$g_{ff}(\nu) = \quad g_{ff}(\nu,x) e^{-x}\,dx$$

If instead of Boltzmann statistics we use Fermi-Dirac statistics, taking for the distribution function

$$N_e(E) = \frac{8\pi m^2 v}{h^3} p(E)$$

with

$$p(E) = \frac{1}{e^{\eta+E/kT}+1}$$

where η is the degeneracy parameter, we then have

$$N_e = \int N_e(E)\,dE = \frac{4\pi(2mkT)^{3/2}}{h^3} I_{1/2}(-\eta)$$

and

$$\frac{1}{\bar{v}} = \left(\frac{m}{2kT}\right)^{1/2} \frac{\ell n(1+e^{-\eta})}{I_{1/2}(-\eta)}$$

where $I_{1/2}(x)$ is the Fermi-Dirac integral defined by

$$I_n(x) = \int_0^\infty \frac{y^n dy}{\exp(y-x)+1}$$

We then obtain

$$\sigma_{ff}^K(\nu) = \sigma_{ff}^K(\nu,\bar{v}) = \frac{2\sqrt{2}\pi e^6 Z^2 N_e \ell n(1+e^{-\eta})}{3\sqrt{3}hcm^{3/2}\nu^3(kT)^{1/2} I_{1/2}(-\eta)}$$

$$= \frac{32\pi^2 e^6 Z^2 kT}{3\sqrt{3}h^4 c\nu^3} \ell n(1+e^{-\eta})$$

and

$$g_{ff}(\nu) = \frac{1}{kT\ell n(1+e^{-\eta})} \int g_{ff}(\nu,E) \frac{dE}{e^{\eta+E/kT}+1}$$

$$= \frac{1}{\ell n(1+e^{-\eta})} \int g_{ff}(\nu,x) \frac{dx}{e^{\eta+x}+1}, \quad x = \frac{E}{kT}$$

In the classical limit where $\eta \to \infty$, $\ell n(1 + e^{-\eta}) \to e^{-\eta}$ and $I_{1/2}(-\eta) \to (\sqrt{\pi}/2)e^{-\eta}$, we recover the Boltzmann formulae.

In the case of high degeneracy, i.e., at very high density, it may be necessary to take account of the (un)availability of the final state by putting

$$q(E+h\nu) = 1-p(E+h\nu) = 1 - \frac{1}{e^{\eta+(E+h\nu)/kT}+1}$$

We then obtain

$$g_{ff}(\nu) = \frac{1}{kT\ell n(1+e^{-\eta})}\int g_{ff}(\nu,E)p(E)q(E+h\nu)dE$$

$$= \frac{1}{\ell n(1+e^{-\eta})}\int g_{ff}(u,x)\left(\frac{1}{e^{\eta+x}+1}\right)\left(1-\frac{1}{e^{\eta+x+u}+1}\right)dx$$

where $u = h\nu/kT$.

4.2.5 Evaluation of matrix elements, oscillator strengths and Gaunt factors

In the case of a Coulomb field $V = Ze/r$ the radial wave equation can be solved in terms of known functions. For the bound states with energy $E = -Z^2e^2/2n^2a_o$ we have, with the appropriate normalization

$$R_{n\ell}(r) = -\left[\frac{(n-\ell-1)!}{\{(n+1)!\}^3 2n}\right]^{1/2}\left(\frac{2Z}{na_o}\right)^{3/2}\left(\frac{2Zr}{na_o}\right)^{\ell} e^{-\frac{Zr}{na_o}} L_{n+\ell}^{2\ell+1}\left(\frac{2Zr}{na_o}\right)$$

where $a_o = \hbar^2/me^2$ and $L_a^b(x)$ is the associated Laguerre polynomial defined by the relations

$$L_a(x) = e^x\frac{d^a}{dx^a}(e^{-x}x^a) = \sum_{s=o}^{a}(-1)^s\frac{(a!)^2x^s}{(a-s)!(s!)^2}$$

$$L_a^b(x) = \frac{d^b}{dx^b}L_a(x) = (-1)^b\sum_{s=o}^{a-b}(-1)^s\frac{(a!)^2x^s}{s!(b+s)!(a-b-s)!}$$

$L_a^b(x)$ may also be expressed in terms of a hypergeometric function

$$L_a^b(x) = (-1)^b\frac{(a!)^2}{b!(a-b)!}F[-(a-b),b+1,x]$$

where

$$F(\alpha,\beta,x) = 1 + \frac{\alpha}{\beta.1!}x + \frac{\alpha(\alpha+1)}{\beta(\beta+1).2!}x^2\ldots$$

$$= \sum_{s=o}^{\infty}\frac{(\alpha+s-1)!(\beta-1)!x^s}{(\alpha-1)!(\beta+s-1)!s!}$$

so that we have

$$R_{n\ell}(r) = \frac{1}{(2\ell+1)!}\left[\frac{(n+\ell)!}{(n-\ell-1)!2n!}\right]^{1/2} \cdot$$

$$\cdot \left(\frac{2Z}{na_o}\right)^{3/2}\left(\frac{2Zr}{na_o}\right)^{\ell} e^{-\frac{Zr}{na_o}} F\left[-(n-\ell-1),(2\ell+2),\frac{2Zr}{na_o}\right]$$

For the free states where the energy $E = \hbar^2k^2/2m$ is positive, the wave function may be obtained by making the substitution

$$n = \frac{iZ}{ka_o}$$

Since this parameter is imaginary, the Laguerre polynomials have no longer any meaning. However, the hypergeometric function is still defined, and we obtain with the appropriate normalization

$$R_{E\ell}(r) = \sqrt{\frac{2mk}{\pi\hbar^2}}\,\frac{\left|\Gamma\left(\ell+1-i\,\frac{Z}{ka_o}\right)\right|}{(2\ell+1)!}\, e^{\frac{\pi Z}{2ka_o}}(2kr)^{\ell}e^{ikr} \cdot$$

$$\cdot\, F\left[\ell+1-\frac{iZ}{ka_o},2\ell+2,-2ikr\right]$$

which has the asymptotic form

$$R_{E\ell}(r) \approx \sqrt{\frac{2mk}{\pi\hbar^2}}\,\frac{\sin(kr+\delta_\ell)}{kr}$$

where

$$\delta_\ell = -\ell\frac{\pi}{2} + \frac{Z}{ka_o}\,\ell n(2kr) + \arg\Gamma\left(\ell+1-i\frac{Z}{ka_o}\right)$$

Using the exact hydrogenic (Coulomb) wave functions, the transition matrix elements have been evaluated by a number of workers. The first results were obtained by Sugiura (1927,1929), Nishina and Rabi (1928), Reiche (1929) and Gordon (1929). Gaunt (1930) computed the Gaunt factors--as they have subsequently been called--for some free-free and bound-free transitions. Further results have been given by Stobbe (1930), and retardation and relativistic effects have been treated by Sommerfeld and Schur (1930), Schur (1930), Fischer (1931), Sauter (1931) and Hall and Oppenheimer (1931). Menzel and Pekeris (1935) have derived general formulae for oscillator strengths and Gaunt factors. Analytical formulae have also been given by Sommerfeld (1953), Bethe and Salpeter (1956) and Biedenharn (1956). Numerical tabulations have been made by Berger (1956,1957), Green, Rush and Chandler (1957) and Herdan and Hughes (1961). The most extensive computations are those of Karzas and Latter (1961) who give, in tabular form, bound-bound oscillator strengths for all transitions between states with principal quantum numbers up to n = 16, and bound-free Gaunt factors for transitions from all bound states up to n = 6, and averages over ℓ-degenerate sub-states up to n = 15, to a selected series of free states extending from $E/(Z^2e^2/2a_o) = 10^{-16}$ to 10^{+12}. Also, in graphical form, are given free-free Gaunt factors for a series of initial electron energies $E/(Z^2e^2/2a_o) = 10^{-6}$ to 10^{+7} and photon energies $h\nu/(Z^2e^2/2a_o) = 10^{-3}$ to 10^{+8}, together with temperature averaged Gaunt factors for a series of temperatures given by $(Z^2e^2/2a_o)/kT = 10^{-3}$ to 10^{+3} and photon energies $h\nu/kT$ = 1 to 30. The bound-bound oscillator strengths are independent of (the nuclear charge number) Z, while for bound-free and free-free transitions the scaling of the electron free energy and the photon energy by Z^2 renders the computed results independent of Z. More recently, interest in transitions involving very highly excited bound states has led to the extension of the tabulation of some bound-bound oscillator strengths for principal quantum numbers up to n = 500 (Goldwire, 1968), and up to n = 900 (Menzel, 1969).

The use of the Born approximation leads to particularly simple formulae for free-free Gaunt factors. The wave functions are those appropriate to a zero field V = 0, so that the radial wave function becomes, with the same normalization as before

$$R_{E\ell}(r) = \sqrt{\frac{2mk}{\pi\hbar^2}}\; j_\ell(kr)$$

where $j_\ell(x)$ is the spherical Bessel function of the first kind defined by

$$j_\ell(x) = \left(\frac{\pi}{2x}\right)^{\frac{1}{2}} J_{\ell+\frac{1}{2}}(x)$$

The asymptotic form of $R_{E\ell}(r)$ is now

$$R_{E\ell}(r) \approx \sqrt{\frac{2mk}{\pi\hbar^2}}\; \frac{\sin(kr-\ell\frac{\pi}{2})}{kr}$$

which may be obtained from the Coulomb wave functions by putting Z = 0. The Gaunt factor is then given by the series

$$g(\nu,E) = \frac{\sqrt{3}h^4}{32\pi^3m^2Z^2e^2} \sum_{\ell_a}\sum_{\ell_b} \ell_{max}\,\delta(\ell_a-\ell_b,\pm 1)\left(\frac{dV}{dr}\right)^2_{ab}$$

where now $\left(\frac{dV}{dr}\right)_{ab}$ is the matrix element formed with the above radial wave functions.

Alternatively we may represent the wave function for the electron with momentum $\hbar \underline{k}$ in a zero field by a plane wave, obtaining with our normalization

$$\psi_{E(\underline{k})}(\underline{r}) = \sqrt{\frac{mk}{2\pi^2\hbar^2}}\, e^{i\underline{k}\cdot\underline{r}}$$

For the transition from the state with wave vector $\underline{k}_a$ to that with wave vector $\underline{k}_b$ the cross-section is then given by

$$\sigma^B(\nu,\underline{k}_a,\underline{k}_b) = \frac{\pi e^2 h}{mc}\,\frac{8\pi^2 m\nu}{3h}\, 2P(E)\frac{e^2}{16\pi^4 m^2\nu^4}(\underline{\nabla}V)^2_{ab}$$

$$= \frac{e^4 P(E)}{3\pi c m^2\nu^3}(\underline{\nabla}V)^2_{ab} = \frac{e^4 h^3 N_e(E)}{24\pi^2 c m^4 v\nu^3}(\underline{\nabla}V)^2_{ab}$$

where

$$P(E) = \frac{h^3 N_e(E)}{8\pi m^2 v}$$

is the probability of a particular (initial) state (of given spin) with energy E being occupied, and $N_e(E)$ is the number density of electrons per unit energy and

$$(\underline{\nabla}V)_{ab} = \int \psi^*_{E_b(\underline{k}_b)}(\underline{r})(\underline{\nabla}V)\psi_{E_a(\underline{k}_a)}(\underline{r})\,d\underline{r} = \frac{m\sqrt{k_a k_b}}{2\pi^2\hbar^2}\int e^{i\underline{K}\cdot\underline{r}}(\underline{\nabla}V)\,d\underline{r}$$

with

$$\underline{K} = \underline{k}_b - \underline{k}_a$$

For a central potential $V = V(r)$ we have

$$\underline{\nabla}V = \frac{\underline{r}}{r}\frac{dV}{dr}$$

and using the expansion

$$e^{i\underline{K}\cdot\underline{r}} = \sum_{\ell=0}^{\infty} i^\ell (2\ell+1) P_\ell(\cos\Theta)\, j_\ell(Kr)$$

where Θ is the angle between $\underline{K}$ and $\underline{r}$, we obtain

$$\int e^{i\underline{K}\cdot\underline{r}}(\underline{\nabla}V)\,d\underline{r} = 4\pi i M \frac{\underline{K}}{K}$$

with

$$M = \int_0^\infty j_1(Kr)\frac{dV}{dr}\, r^2 dr$$

We therefore have

$$(\underline{\nabla}V)^2_{ab} = \frac{64\pi^2 m^2 k_a k_b}{h^4} M^2$$

and

$$\sigma^B(\nu,\underline{k}_a,\underline{k}_b) = \frac{8e^4 N_e(E) k_a k_b}{3m^2 hc\nu\nu^3} M^2$$

To obtain the cross-section for the transition from energy state $E_a(k_a)$ to energy state $E_b(k_b)$ we average over the orientations of $\underline{k}_a$ and $\underline{k}_b$. Now

$$K^2 = k_a{}^2 + k_b{}^2 - 2k_a k_b \cos\theta$$

where θ is the angle between $\underline{k}_a$ and $\underline{k}_b$, and hence

$$d\omega = -2\pi d\cos\theta = \frac{2\pi}{k_a k_b} KdK$$

Thus

$$\overline{(\underline{\nabla}V)^2_{ab}} = \frac{1}{4\pi}\int(\underline{\nabla}V)^2_{ab} d\omega = \frac{32\pi^2 m^2}{h^4}\int_{k_b-k_a}^{k_b+k_a} M^2 KdK$$

and

$$\sigma^B(\nu,k_a,k_b) = \frac{4e^4 N_e(E)}{3m^2 hc\nu\nu^3}\int_{k_b-k_a}^{k_b+k_a} M^2 KdK$$

The Gaunt factor therefore becomes

$$g^B(\nu,k_a,k_b) = \frac{\sqrt{3}}{\pi Z^2 e^2}\int_{k_b-k_a}^{k_b+k_a} M^2 KdK$$

To evaluate M we integrate by parts and use the relations

$$\frac{d}{dx}[x^2 j_1(x)] = x^2 j_o(x) = x \sin x$$

so that

$$M = -K\int_0^\infty V j_o(Kr) r^2 dr = -\int_0^\infty V\sin(Kr) r dr$$

Substituting $V = Ze/r$ and using the result

$$\int_0^\infty e^{-ax}\sin(bx+c)\,dx = \frac{a\sin c + b\cos c}{a^2 + b^2}$$

we have

$$M = -\frac{Ze}{K}$$

so that we finally obtain

$$g^B(\nu,k_a,k_b) = \frac{\sqrt{3}}{\pi}[\ell nK]_{k_b-k_a}^{k_b+k_a} = \frac{\sqrt{3}}{\pi}\,\ell n\left(\frac{k_b+k_a}{k_b-k_a}\right)$$

which diverges logarithmically for very low frequencies where $k_b \to k_a$. Another interesting case is that of the shielded Coulomb field

$$V = \frac{Ze}{r}\,e^{-\alpha r}$$

for which we obtain

$$M = -Ze\frac{K}{K^2+\alpha^2}$$

Thus we have

$$\int M^2 K dK = Z^2e^2\int\frac{K^3 dK}{(K^2+\alpha^2)^2} = \frac{Z^2e^2}{2}\,\ell n(K^2+\alpha^2) + \frac{\alpha^2}{K^2+\alpha^2}$$

and

$$g^B(\nu,k_a,k_b) = \frac{\sqrt{3}}{2\pi}\left[\ell n\left\{\frac{(k_b+k_a)^2+\alpha^2}{(k_b-k_a)^2+\alpha^2}\right\} - \frac{4\alpha^2k_ak_b}{\{(k_b+k_a)^2+\alpha^2\}\{(k_b-k_a)^2+\alpha^2\}}\right]$$

It should be noted that in both cases the Born Gaunt factor is independent of Z. It should also be remarked that using the dipole length matrix element instead of the dipole acceleration matrix element we would have

$$M = -\frac{e}{4\pi^2 m\nu^2}M'$$

where

$$M' = \int_0^\infty j_1(Kr)r^3dr = \left(\frac{\pi}{2K}\right)^{1/2}\int_0^\infty r^{5/2}J_{3/2}(Kr)\,dr$$

Using the result

$$\int_0^\infty e^{-at^2} t^{n+1} J_n(bt)\,dt = \frac{b^n}{(2a)^{n+1}} e^{-\frac{b^2}{4a}}$$

we see that in the limit $a \to 0$ the integral vanishes. Thus M' and therefore M and the Gaunt factor are zero. However, there is no inconsistency because the relation between the various forms of the matrix element apply only to the exact matrix elements, and not to the approximate ones as constructed here with the plane wave approximate wave functions. These wave functions would be exact for a zero field, which if also used in the dipole acceleration formula would give a zero result. This merely demonstrates that a truly free electron cannot absorb (or emit) a photon, due to the fact that it is impossible to conserve both momentum and energy simultaneously in such an interaction. However, in the presence of another particle (such as a nucleus) the free-free transition can occur.

4.2.6 Scattering

According to quantum mechanics scattering is a two quantum process and therefore involves the interaction terms which are of second order. For scattering by free electrons at non-relativistic photon energies, i.e., $\gamma = h\nu/mc^2 \ll 1$, the frequencies of the incident and scattered photons are the same and the cross-section is given by the classical Thomson formula

$$\sigma(\theta,\phi) = \sigma^T p^T(\theta,\phi), \quad \sigma^T = \frac{8\pi}{3} r_o^2, \quad r_o = \frac{e^2}{mc^2}$$

where the (Thomson) phase function is

$$p^T(\theta,\phi) = \frac{3}{16\pi}(1+\cos^2\theta)$$

The total cross-section is, as before

$$\sigma = \sigma^T$$

In general, for the relativistic case (Compton scattering) the incident and scattered photon frequencies are not the same. For an electron initially at rest they are related by the formula

$$\frac{\nu'}{\nu} = \frac{1}{1 + \gamma(1 - \cos\theta)}$$

and the cross-section is given by the Klein-Nishina formula

$$\sigma(\theta,\phi) = \sigma^T p^C(\theta,\phi),$$

$$p^C(\theta,\phi) = \frac{1}{\Phi}\, p^T(\theta,\phi)\left[\frac{1}{1+\gamma(1-\cos\theta)}\right]\left[1+ \frac{\gamma^2(1-\cos\theta)^2}{(1+\cos^2\theta)\{1+\gamma(1-\cos\theta)\}}\right]$$

The total cross-section then becomes

$$\sigma = \sigma^T \Phi ,$$

$$\Phi = \frac{3}{4}\left[\frac{1+\gamma}{\gamma^3}\left\{\frac{2\gamma(1+\gamma)}{1+2\gamma} - \ln(1+2\gamma)\right\} + \frac{1}{2\gamma}\ln(1+2\gamma) - \frac{1+3\gamma}{(1+2\gamma)^2}\right]$$

For small values of γ, Φ may be developed as a direct power series

$$\Phi = 1 - 2\gamma + \frac{26}{5}\gamma^2 - \frac{133}{10}\gamma^3 + \frac{1144}{35}\gamma^4 - \frac{544}{7}\gamma^5 + \ldots$$

while for large values of γ we obtain the inverse power series

$$\Phi = \frac{3}{4}\left[\frac{1}{\gamma}\left\{\frac{1}{4} + \frac{1}{2}\ln(2\gamma)\right\} + \frac{1}{\gamma^2}\left\{\frac{9}{4} - \ln(2\gamma)\right\} + \ldots\right]$$

The cross-section for moving electrons may be obtained by a Lorentz transformation, and has been given by Milford (1957) and Sampson (1959). Stimulated scattering effects have been considered by Sampson (1959) and Chin (1965).

For bound electrons a general expression for the scattering cross-section is given by Heitler (1954). In the case of coherent scattering Rosseland (1936) gives, for an electron in the state n, in the absence of damping

$$\sigma = \frac{8\pi}{3} r_o^2 \left(\sum_m \frac{f_{nm}\omega^2}{\omega^2 - \omega_{nm}^2} \right)^2$$

where f_{nm} and ω_{nm} are the oscillator strength and angular frequency associated with a transition between the states n and m. In the high frequency limit we obtain again the Thomson result

$$\sigma = \frac{8\pi}{3} r_o^2 \left(\sum_m f_{nm} \right)^2 = \frac{8\pi}{3} r_o^2$$

on making use of the sum rule whereby $\sum_m f_{nm} = 1$. In the low frequency limit we obtain

$$\sigma = \frac{8\pi}{3} r_o^2 \left(\sum_m f_{nm} \frac{\omega^2}{\omega_{nm}^2} \right)^2$$

as the quantum mechanical analogue of the Rayleigh formula.

5. OPACITY CALCULATIONS

5.1 Composition of Stellar Material

In order to calculate the opacity of stellar material, it is necessary to specify the composition of the material and the physical conditions. The composition is specified when the atomic number Z, the atomic weight M_Z, and the concentration of each of the different chemical elements, and/or isotopes of each element, which are present are given. While isotope abundances are vital in determining the rate of energy generation by nuclear reactions in a star, they are of much lesser importance for the opacity. Generally speaking, the isotope effects in atomic spectra and in the electronic spectra of molecules, which appear as fine structure of the energy levels due to the dependence of the reduced mass of the electron on the nuclear mass, are negligible. On the other hand, the isotope effects in molecular vibration and rotation spectra may be more significant because of their consequences for the reduced nuclear mass which enters the vibrational and rotational constants.

The concentration of each atomic species (element or isotope) may be given either by number or by mass. If the matter density is otherwise given, it is only necessary to specify the concentrations as fractional concentrations. If α_Z and β_Z are the number and mass fractions for the atomic species (Z, M_Z), it may be shown that

$$\alpha_Z = \frac{\beta_Z \bar{M}}{M_Z}, \qquad \beta_Z = \frac{\alpha_Z M_Z}{\bar{M}}$$

where

$$\bar{M} = \sum_Z \alpha_Z M_Z = 1 \Big/ \sum_Z (\beta_Z / M_Z)$$

If N_Z is the number density of atoms of type (Z, M_Z) then the number density N_A of atoms of all types is

$$N_A = \sum_Z N_Z, \qquad N_Z = \alpha_Z N_A$$

and the mass density is

$$\rho = \sum_Z N_Z M_Z m_H = N_A m_H \sum_Z \alpha_Z M_Z = N_A m_H \bar{M}$$

where m_H is the mass of the hydrogen atom (the unit of atomic weight). Thus we have

$$N_A = \frac{\rho}{m_H \bar{M}}$$

so that, given the relative or fractional concentrations and the matter density, the absolute concentrations can be found. The physical conditions may be given by specifying the temperature T and either the matter density ρ or the pressure P, in the latter case the density being found from the equation of state

$$P = P(T,\rho)$$

5.2 Statistical Mechanics

Before the opacity can actually be calculated, it is still necessary to know the energy levels of each atom and the number of electrons occupying each energy level--the occupation numbers. The energy levels determine the absorption spectrum, i.e., at what photon energies absorption may occur, while the occupation numbers determine (together with the transition probabilities or cross-sections) the absorption strength, by giving the total number of electrons per unit volume which can participate in the various absorption and scattering processes. The determination of the energy levels and the occupation numbers is in itself quite a difficult exercise in quantum and statistical mechanics, and is an essential part of the calculation of the equation of state, which now becomes a preliminary in any opacity calculation.

The simplest case arises at the highest temperatures when there is complete ionization so that all the electrons are free. The free electron number density is then

$$N_e = \sum_Z N_Z Z = N_A \sum_Z \alpha_Z Z = N_A \bar{M} \sum_Z \beta_Z \frac{Z}{M_Z} = \frac{\rho \bar{Z}}{m_H \bar{M}},$$

$$\bar{Z} = \sum_Z \alpha_Z Z = \bar{M} \sum_Z \beta_Z \frac{Z}{M}$$

It is customary in stellar contexts to denote the mass fractions of hydrogen, helium and the aggregate of the heavier elements by X, Y and Z. Introducing these values for the β_Z and putting

$$z_1 = \sum_{Z>2} \beta_Z \frac{Z}{M_Z} \Big/ \sum_{Z>2} \beta_Z$$

we obtain since $Z = \sum_{Z>2} \beta_Z$

$$N_e = \frac{\rho}{m_H} \left[X + \tfrac{1}{2}Y + z_1 Z \right]$$

Since $z_1 \simeq \frac{1}{2}$ we may write, using $X + Y + Z = 1$

$$N_e \simeq \frac{\rho}{m_H}(X + \tfrac{1}{2}Y + \tfrac{1}{2}Z) = \frac{\rho}{2m_H}(1 + X)$$

At lower temperatures where there is incomplete ionization, the problem of determining the energy levels and occupation numbers is much more difficult. There are basically two different methods which have been applied. The first method--the average atom or atomic method (Mayer, 1947)--which is applicable at high temperatures (but still low enough for ionization to be incomplete), disregards the distinction between the various stages of ionization of an atom, and further assumes that the energy levels and occupation numbers may be represented by average values $\bar{E}_n$ and $\bar{n}_n$. Using the subscript n to denote all the quantum numbers (e.g., both the principal quantum number n and the azimuthal quantum number ℓ), we may therefore write in accordance with Fermi-Dirac statistics

$$\bar{n}_n = \frac{g_n}{e^{\eta+\bar{E}_n/kT} + 1}$$

in which g_n is the statistical weight of the level, and $\bar{E}_n$ is given by perturbation theory

$$\bar{E}_n = \bar{E}_n(\bar{n}_n) = E_n^o + \Delta E_{nb} + \Delta E_{nf}$$

where E_n^o represents the unperturbed energy and ΔE_{nb} and ΔE_{nf} the perturbations due to the bound and free electrons, respectively. Thus we may write

$$E_n^o = - \frac{Z^2e^2}{2n^2a_o}$$

and

$$\Delta E_{nb} = \sum_{m\neq n} \bar{n}_m V_{nm} + \bar{n}_n\left(1 - \frac{1}{g_n}\right)V_{nn}$$

where the V_{nm} are the two-electron interaction energies and may be expressed in terms of Slater integrals. To obtain the free electron perturbation ΔE_{nf} we enclose the atom in a (Wigner-Seitz) sphere of radius a_Z determined in such a way that the residual charge on the atom is just neutralized by the free electrons contained within the sphere, assuming they are uniformly distributed. Thus

$$Z' = Z - \sum_n \bar{n}_n = \frac{4}{3}\pi a_Z^3 N_e$$

and ΔE_{nf} becomes

$$\Delta E_{nf} = \frac{Z'e^2}{2a_Z}\left[3 - \left(\frac{r^2}{a_Z^2}\right)_{nn}\right]$$

where for bound hydrogenic wave functions for nuclear charge number Z

$$(r^2)_{nn} = \frac{n^2 a_o^{\,2}}{Z^2}\left[5n^2 + 1 - 3\ell(\ell+1)\right]$$

Finally, η is a degeneracy parameter which is related to the density of free electrons by

$$\frac{4\pi(2mkT)^{3/2}}{h^3} I_{1/2}(-\eta') = N_e = \sum_Z N_Z Z' = \frac{\rho}{m_H \bar{M}} \sum_Z \alpha_Z Z'$$

with

$$\eta' = \eta - \Delta E_{ff}/kT$$

where ΔE_{ff} is a correction term representing the depression of the continuum states and given by

$$\Delta E_{ff} = -\frac{3e^2}{10N_e}\sum_Z \frac{N_Z Z'^2}{a_Z} = -\frac{3e^2\rho}{10N_e m_H \bar{M}}\sum_Z \left(\alpha_Z Z'^2/a_Z\right)$$

Clearly on account of the coupled nature of the equations for $\bar{E}_n$ and $\bar{n}_n$ they must be solved by iteration until self-consistency is obtained. It still remains to take account of the fact that in reality the energy levels and occupation numbers will not actually be the average values thus calculated. Instead each atom will exist in a wide range of various electronic configurations corresponding to different sets of integral occupation numbers n_n, each configuration having a corresponding energy E_n. If the temperature is sufficiently high that the Boltzmann term $e^{-E_n/kT}$ does not vary appreciably with the configuration, then the probability p_n of finding an electron in the level n is also virtually independent of the actual number in the level. We may then write

$$p_n = \frac{\bar{n}_n}{g_n}, \quad q_n = 1 - p_n$$

and the probability of finding n_n electrons in any level n is therefore

$$p(n_n) = \frac{g_n!}{n_n!(g_n-n_n)!} p_n^{\,n_n} q_n^{\,(g_n-n_n)}$$

Hence the probability of the configuration in which there are n_n electrons in each level n is

$$P(n_n) = \prod_n p(n_n)$$

and the energy associated with an electron in the level n is then

$$E_n = E_n(n_n) = E_n^o + \Delta E_{nb} + \Delta E_{nf}$$

where now

$$\Delta E_{bn} = \sum_{m \neq n} n_m V_{nm} + (n_n - 1) V_{nn}$$

In the second method--the ionic method (Cox, 1965)--which is applicable at lower temperatures, account is taken of the fact that each atom may exist in a number of different ionization stages. The ratio of the number of ions in two adjacent stages (r, r+1) of ionization is then given by the Saha formula

$$r_{Z,r+1} = \frac{N_{Z,r+1}}{N_{Z,r}} = \frac{2}{N_e}\left[\frac{U_{Z,r+1}}{U_{Z,r}}\right]\left[\frac{2\pi mkT}{h^2}\right]^{3/2} e^{-\chi_{Z,r}/kT}$$

where $U_{Z,r}$ is the partition function

$$U_{Z,r} = \sum_n g_{Z,r,n} e^{-E_{Z,r,n}/kT}$$

for the ion (Z,r), $\chi_{Z,r}$ is the ionization energy and $g_{Z,r,n}$ and $E_{Z,r,n}$ are the statistical weight and energy of the level n. The fraction of all atoms of type Z in the ionization stage r is then

$$f_{Z,r} = \frac{\prod^r r_{Z,r}}{\sum_r \prod^r r_{Z,r}}, \quad \prod^o = 1 \quad \text{and} \quad \prod^r r_{Z,r} = r_{Z,1} r_{Z,2} \cdots r_{Z,r}$$

If Z'_r is the residual charge on an ion of type (Z,r), the average residual charge per ion is

$$Z' = \sum_{r=o}^{Z} f_{Z,r} Z'_r$$

and the free electron number density becomes

$$N_e = N_A \sum_Z \alpha_Z Z' = N_A \sum_Z \alpha_Z \sum_r f_{Z,r} Z'_r$$

The existence of the various electron configurations is allowed for simply by letting n now denote a configuration with specified occupation numbers n_n, energy $E_{Z,r,n}$ and statistical weight

$$g_{Z,r,n} = \prod_n \frac{g_n!}{n_n!(g_n-n_n)!}$$

The number of electrons per ion (Z,r) in the level n belonging to the configuration is then given by the Boltzmann formula

$$n_{Z,r,n} = n_n \frac{g_{Z,r,n}}{U_{Z,r}} e^{-(E_{Z,r,n}+\chi_{Z,r})/kT}$$

By summing over configurations the previous one-electron level result may be recovered. As in the case of the average atom method the energy $E_{Z,r,n}$ of a configuration may be calculated by perturbation theory or, if it is an observable quantity, taken as given at least in the zero approximation. Also as before the equations must be solved by iteration until self-consistent results are obtained.

Once the energy levels and occupations numbers have been calculated the absorption coefficient as a function of the photon energy is obtained with the aid of the relevant cross-sections. The Rosseland mean opacity may then be found. Because the Rosseland opacity is a harmonic mean of the absorption coefficient, it is not equal to the sum of the Rosseland opacity for each separate process or atom contributing to the absorption coefficient. Thus in general the absorption coefficients for all the processes contributing at each photon energy must be summed before the harmonic mean is formed. However, under different physical conditions, different processes may be dominant. Thus at the highest temperatures the most important process may be scattering, while at progressively lower temperatures it is free-free absorption or bound-free and bound-bound absorption. It is therefore instructive to derive expressions for the opacity by considering various processes alone.

5.3 Scattering Opacity

For electron scattering the mass scattering coefficient is

$$\kappa_s(u) = \frac{1}{\rho} N_e \sigma_s(u)$$

If we take for $\sigma_s(u)$ the Thomson cross-section, then we have

$$\bar{\mu} = \int p^T(\theta,\phi)\cos\theta d\omega = -\frac{3}{8}\int (1+\cos^2\theta)\cos\theta d\theta$$

$$= -\frac{3}{8}\left[\frac{\cos^2\theta}{2} + \frac{\cos^4\theta}{4}\right]_0^\pi = 0$$

so that

$$\kappa_s'(u) = \kappa_s(u)(1-\bar{\mu}) = \frac{1}{\rho} N_e \sigma^T = \kappa_s'$$

Since this is frequency independent the Rosseland mean given by

$$\frac{1}{\kappa_s} = \frac{15}{4\pi^4}\int_0^\infty \frac{u^4 e^u du}{\kappa_s'(u)(e^u-1)^2} = \frac{1}{\kappa_s'}$$

is the same, so that we obtain for the opacity due to scattering alone

$$\kappa_s = \frac{1}{\rho} N_e \sigma^T = 6.65203 \, . \, 10^{-24} \frac{N_e}{\rho}$$

For a fully ionized gas the electron density is

$$N_e = \frac{\rho}{m_H}\frac{\bar{Z}}{\bar{M}} \simeq \frac{1}{2}\frac{\rho}{m_H}(1 + X)$$

We have therefore

$$\kappa_s = \frac{\sigma^T}{m_H}\frac{\bar{Z}}{\bar{M}} = 0.400776\,\frac{\bar{Z}}{\bar{M}} \simeq 0.200388(1 + X)$$

Thus the scattering opacity is independent of both temperature and density and depends only on the composition. If the frequency dependent Klein-Nishina formula is used for the scattering cross-section the Rosseland mean can no longer be so simply evaluated.

5.4 Free-Free Opacity

For free-free absorption alone we have

$$\kappa(u) = \kappa_{ff}(u) = \frac{1}{\rho}\sum_Z N_Z \sigma_{ff,Z}(u)$$

$$= \frac{1}{m_H \bar{M}}\sum_Z \alpha_Z \sigma_{ff,Z}(u) = \frac{1}{m_H}\sum_Z \frac{\beta_Z}{M_Z}\,\sigma_{ff,Z}(u)$$

Taking into consideration the frequency dependence of the free-free absorption coefficient, we put

$$\kappa_{ff}(u) = \frac{D_{ff}(u)}{u^3}, \qquad D_{ff}(u) = \sum_Z D_{ff,Z}(u)$$

so that the Rosseland mean becomes

$$\frac{1}{\kappa_{ff}} = \frac{15}{4\pi^4}\int_0^\infty \frac{u^7 e^{2u} du}{D_{ff}(u)(e^u-1)^3} = \int_0^\infty \frac{R(u)}{D_{ff}(u)}\,du$$

Adopting the Kramers formula for the free-free cross-section we then have

$$D_{ff,Z}(u) = D^K_{ff,Z} = \frac{4\sqrt{2\pi}e^6h^2Z^2N_eN_Z}{3\sqrt{3}cm^{3/2}(kT)^{7/2}\rho} = C_{ff}\,\frac{N_eN_ZZ^2}{\rho T^{7/2}}$$

which is independent of u. Hence taking D^K_{ff} outside the integral we obtain

$$\kappa_{ff} = \kappa^K_{ff} = D^K_{ff}/S(\infty)$$

where

$$S(u) = \int_0^u R(u)\,du$$

and

$$S(\infty) = \frac{15}{4\pi^4}\,7!\{\zeta(7) + \zeta(6)\} = 196.51957$$

Thus we have

$$\kappa^K_{ff} = \frac{C_{ff}N_e\sum_Z N_ZZ^2}{S(\infty)\rho T^{7/2}} = \frac{C_{ff}N_e}{S(\infty)m_HT^{7/2}}\sum_Z \frac{\beta_Z}{M_Z}Z^2$$

Introducing the mass concentrations X, Y, Z and putting

$$z_2 = \sum_{Z>2}\frac{\beta_Z}{M_Z}Z^2\Big/\sum_{Z>2}\beta Z$$

we have

$$\sum_Z \frac{\beta_Z}{M_Z}Z^2 = X + Y + z_2Z$$

Using our previous expression for N_e we then obtain

$$\kappa^K_{ff} = \frac{C_{ff}}{S(\infty)}(X+\tfrac{1}{2}Y+z_1Z)(X+Y+z_2Z)\frac{\rho}{m_H{}^2T^{7/2}}$$

Thus the free-free opacity, in the Kramers approximation, is characterized by a temperature and density dependence of the form

$$\kappa = \kappa_1\rho/T^{7/2}$$

If we now adopt the quantum mechanical cross-sections by introducing the appropriate Gaunt factors we have

$$D_{ff,Z}(u) = D^K_{ff,Z} g_{ff,Z}(u)$$

where now g_{ff} is a function of both u and Z. Defining an average Gaunt factor by the relation

$$\frac{1}{g_{ff}} = \frac{\sum_Z \frac{\beta_Z}{M_Z} Z^2}{S(\infty)} \int_0^\infty \frac{R(u)\,du}{\sum_Z \frac{\beta_Z}{M_Z} Z^2 g_{ff,Z}(u)}$$

we may then write

$$\kappa_{ff} = \frac{C_{ff} N_e \sum_Z \frac{\beta_Z}{M_Z} Z^2 g_{ff}}{S(\infty) m_H T^{7/2}} = \kappa^K_{ff} g_{ff}$$

which, if we ignore the temperature dependence of g_{ff}, is again of the Kramers form. In the Born approximation the free-free Gaunt factor is independent of Z so that we then have for the average Gaunt factor as defined above

$$\frac{1}{g_{ff}} = \frac{1}{g^B_{ff}} = \frac{1}{S(\infty)} \int_0^\infty \frac{R(u)\,du}{g^B_{ff}(u)}$$

and we obtain

$$\kappa_{ff} = \kappa^B_{ff} = \kappa^K_{ff} g^B_{ff}$$

5.5 Bound-Free Opacity

Bound-free absorption differs from scattering and free-free absorption, which contribute to the opacity at all photon frequencies, in that it only contributes at frequencies above that corresponding to the (photo-)ionization threshold for the initial bound level, i.e., for some level n with ionization energy χ_n

$$\frac{h\nu}{kT} = u > u_n = \frac{\chi_n}{kT}$$

Thus the absorption coefficient is now a discontinuous function, with discontinuities at each ionization edge. We may therefore write

$$\kappa_{bf}(u) = \sum_Z \sum_n \kappa_{bf,Z,n}(u)\,\theta(u - u_{Z,n})$$

where $\kappa_{bf,Z,n}(u)$ is the bound-free absorption coefficient at reduced frequency u for a level n of atom (or ion) Z and θ is a unit step function such that

$$\theta(u - u_{Z,n}) = \begin{cases} 0 & u < u_{Z,n} \\ 1 & u > u_{Z,n} \end{cases}$$

Given the absorption cross-section, the absorption coefficient is only determined when the number of electrons $N_{Z,n}$ in each initial level of each atom is known, i.e., we have

$$\kappa_{bf,Z,n}(u) = \frac{1}{\rho} N_{Z,n}\sigma_{bf,Z,n}(u)$$

If $n_{Z,n}$ is the average number of electrons per atom Z in level n, then

$$N_{Z,n} = N_Z n_{Z,n}$$

As an example, since it enables us to derive a simple expression for the opacity when bound-free transitions contribute, we follow Strömgren (1932) and put, for any atom

$$n_{Z,n} = \frac{g_{Z,n}}{e^{-\chi_{Z,n}/kT+\eta} + 1}$$

where

$$e^{\eta} = \frac{2(2\pi mkT)^{3/2}}{h^3 N_e}$$

Assuming all the levels are hydrogenic we have

$$g_{Z,n} = 2n^2$$

and

$$\chi_{Z,n} = \frac{2\pi^2 me^4 Z^2}{h^2 n^2}$$

where n denotes the principal quantum number. Writing

$$\kappa_{bf}(u) = \frac{D_{bf}(u)}{u^3} = \frac{1}{u^3}\sum_Z \sum_n D_{bf,Z,n}(u)\,\theta(u - u_{Z,n})$$

and adopting the Kramers bound-free cross-section, we obtain, on extracting D^K_{ff} as a factor

$$D_{bf,Z,n}(u) = D^K_{bf,Z,n}(u) = D^K_{ff}\tau_{Z,n} = C_{ff}\frac{N_e N_Z Z^2}{\rho T^{7/2}}\tau_{Z,n}$$

with

$$\tau_{Z,n} = \frac{2}{n}\frac{u_{Z,n}}{e^{-u_{Z,n}} + e^{-\eta}}$$

Strictly speaking, where there are bound electrons as is necessary for bound-free absorption to occur, the nuclear charge as seen by free and bound electrons will not be the same, nor equal to the full nuclear charge. Introducing Z_f and Z_n for the effective nuclear charge numbers for free electrons and bound electrons in the level n, then

$$\tau_{Z,n} = \frac{Z_n^2}{Z_f^2} \frac{2}{n} \frac{u_{Z,n}}{e^{-u_{Z,n}} + e^{-\eta}}$$

Summing over all levels n and atomic species Z with the appropriate weights we obtain

$$D_{bf}(u) = D_{bf}^K(u) = C_{ff} \frac{N_e}{m_H T^{7/2}} \sum_Z \sum_n \frac{\beta_Z}{M_Z} Z_f^2 \tau_{Z,n} \theta(u - u_{Z,n})$$

Including the free-free absorption and introducing the free-free and bound-free Gaunt factors we then have

$$D(u) = D_{ff}(u) + D_{bf}(u)$$

$$= C_{ff} \frac{N_e}{m_H T^{7/2}} \sum_Z \frac{\beta_Z}{M_Z} Z_f^2 \left[g_{ff,Z_f}(u) + \sum_n \tau_{Z,n} \theta(u-u_{Z,n}) g_{ff,Z_n,n}(u) \right]$$

Writing

$$D(u) = \sum_i D_i(u) \theta(u-u_i) = \sum_i D_{i,i+1}(u) \theta(u-u_i) \theta(u_{i+1}-u)$$

where i is an index which sequentially orders according to increasing u all the ionization edges of all the atomic species, and

$$D_{i,i+1}(u) = D(u), \quad u_i < u < u_{i+1}$$

The Rosseland mean opacity then may be written

$$\frac{1}{\kappa} = \sum_i \int_{u_i}^{u_{i+1}} \frac{R(u)\,du}{D_{i,i+1}(u)}$$

If the Gaunt factors are taken to be unity or are replaced by some constant average values between ionization edges, then the $D_{i,i+1}(u)$ are independent of u, i.e.

$$D_{i,i+1}(u) = D_{i,i+1}$$

and may therefore be taken outside the integration sign to give

$$\frac{1}{\kappa} = \sum_i \frac{1}{D_{i,i+1}} \left[S(u_{i+1}) - S(u_i) \right]$$

If we now define

$$\frac{1}{\tau} = \frac{g_{ff} \sum_Z \frac{\beta_Z}{M_Z} Z_f^2}{S(\infty)} \int_0^\infty \frac{R(u)\,du}{\sum_Z \frac{\beta_Z}{M_Z} Z_f^2 \left[g_{ff,Z_f}(u) + \sum_n \tau_{Z,n} \theta(u-u_{Z,n}) g_{bf,Z,n}(u) \right]}$$

we obtain

$$\kappa = \kappa_{ff} \tau$$

The multiplying factor τ which we have introduced bears some analogy with the reducing or "guillotine" factor t introduced by Eddington. On the basis of classical electromagnetic theory, applied to electrons which could occupy a continuous spectrum of bound (i.e., negative energy) states, Eddington (1926) derived an expression for the opacity in the form

$$\kappa = \kappa_1 \rho / T^{7/2}$$

where κ_1 is a constant. According to the Kramers theory in which, in accordance with the Bohr theory of the atom, only a discrete spectrum of bound states was permitted, the absorption and emission processes were cut off at frequencies corresponding to the ionization edges. The subsequent modification of the opacity formula was to reduce the opacity by a factor t, i.e.

$$\kappa = \frac{\kappa_1}{t} \frac{\rho}{T^{7/2}}$$

where the guillotine factor t allowed for the existence of the ionization edges which were not present in the classical theory. For small values of $u_{Z,n}$ the occupancy of the bound states is low, t is large and τ and κ are small, while for large values of $u_{Z,n}$ the occupancy of the bound states is high, t is small and τ and κ are large.

Our formulae show that if the Gaunt factors and the $\tau_{Z,n}$ are not very dependent on temperature and density, then so also is τ and the opacity including bound-free absorption has the same temperature and density dependence as the opacity due to free-free absorption alone. This will be true for high temperatures when the ionization is almost complete so that N_e is also approximately independent of T and ρ. At lower temperatures N_e decreases rapidly with decreasing T especially where the most abundant elements (usually H and H_e) are partially ionized, so that the product $N_e T^{-7/2}$ eventually begins to decrease as T decreases. The temperature at which the maximum--and the maximum opacity--is reached depends upon the density, being lower at lower density and higher at higher density. The opacity including bound-free absorption is also sensitive to the composition, and particularly the heavy element component, especially when H and H_e are completely ionized and therefore not themselves contributing to the bound-free absorption. In more detailed opacity calculations it is necessary to abandon the simple formulae of Strömgren for the level energies and occupation numbers based on the assumption that the electrons and nuclei are completely independent, and use instead the more elaborate statistical mechanical methods.

5.6 Bound-Bound Opacity

The bound-bound or line absorption coefficient is given by

$$\kappa_{bb}(u) = \frac{1}{\rho}\sum_Z \sum_n N_{Z,n}\sigma_{bb,Z,n}(u) = \frac{1}{\rho}\sum_Z N_Z \sum_n n_{Z,n}\sigma_{bb,Z,n}(u)$$

where $\sigma_{bb}(u)$ is the bound-bound absorption cross-section at reduced frequency u. Now for a bound-bound transition from an initial level n to a final level m the cross-section is given by

$$\sigma_{nm}(u) = \frac{\pi e^2}{mc} f_{nm}F(u)$$

where f_{nm} is the oscillator strength for the transition and F(u) is the line profile in reduced frequency units, which is related to the profile $F(\nu)$ in frequency units and $F(\omega)$ in angular frequency units by

$$F(u) = \frac{kT}{h} F(\nu) = \frac{kT}{\hbar} F(\omega)$$

with normalizations

$$\int F(u)du = \int F(\nu)d\nu = \int F(\omega)d\omega = 1$$

In all the early calculations of opacity, line absorption was ignored largely on the basis of the supposition that, since the absorption was appreciable only in the relatively narrow regions within the line profiles, the overall effect on the Rosseland mean opacity would be small. If the opacity were formed as a straight or arithmetic mean of the absorption coefficient, then since the integrated strength of a line is independent of the actual profile, the contribution of the line to the opacity would also be almost independent of the profile, if the weighting function did not vary appreciably across it. On the other hand, for a harmonic mean, such as that of Rosseland, the absorption due to a line, if concentrated in a narrow interval, and since it appears in the denominator of the (Rosseland) harmonic integral, was expected to make only a small contribution, which would indeed vanish in the limit of a δ-function profile. However it was pointed out by Menzel and Pekeris (1935) that under stellar interior conditions, atomic levels would be so broadened that line absorption would probably be spread into a broad band. A consideration of the distribution of oscillator strength between the discrete (line) and continuous spectrum indicated that the effect of line absorption would be to approximately double or treble the opacity. Similar considerations have been attributed to Teller (Mayer, 1947).

Any calculation of opacity which includes line absorption requires a knowledge of the line profiles. The profile of a line is the result of a number of "broadening" mechanisms whereby, in a given transition between two bound levels, absorption takes place no longer at a single photon energy but for a range of energies. The broadening processes may be considered in three categories: Doppler effect, damping phenomena (which includes the effects of spontaneous decay--both radiative and radiationless, e.g., autoionization--and induced radiative decay) and pressure broadening. Doppler broadening results from the thermal motions of the atoms and depends only on the temperature and the atomic mass. The damping arising from spontaneous decay is a property only of the atomic system and independent of external influences, while that arising from induced radiative decay depends on the radiation flux and therefore also on the temperature. Pressure broadening results from inter-particle interactions and, as its name suggests, depends mainly upon the pressure or density, as well as on the temperature.

5.6.1 Doppler broadening

The shift of a spectral line due to the Doppler effect, as the result of the relative motion of source and observer, gives rise to the effective broadening of the line when there is a relative motion velocity distribution, as in the case of the thermal motions of atoms. A photon, whose angular frequency is ω_o in the relative rest frame, will have an apparent angular frequency $\omega' = \omega'(v)$ when the line-of-sight velocity is v. If f(v) is the distribution of v, the line profile is then given by

$$F(\omega) = \int f(v)\,\delta\{\omega-\omega'(v)\}dv$$

Using the formula

$$\omega'(v) = \omega_o\left[\frac{1-\frac{v}{c}}{1+\frac{v}{c}}\right]^{\frac{1}{2}} \simeq \omega_o\left(1-\frac{v}{c}\right)$$

we then obtain, in the non-relativistic approximation

$$F(\omega) = \int f(v)\,\delta\left\{\omega-\omega_o\left(1-\frac{v}{c}\right)\right\}dv = \int f(v)\,\delta\left\{(\omega-\omega_o)+\frac{\omega_o}{c}v\right\}dv = \frac{c}{\omega_o}f\left\{\frac{c}{\omega_o}(\omega-\omega_o)\right\}$$

Adopting the Maxwell-Boltzmann distribution

$$f(v) = \frac{1}{\sqrt{2\pi}}\left(\frac{M}{kT}\right)^{\frac{1}{2}} e^{-Mv^2/2kT}$$

where M is the mass of the atom, the resulting line profile takes the form of a Gauss distribution

$$F(\omega) = \frac{1}{\sqrt{2\pi}\omega_o}\left(\frac{Mc^2}{kT}\right)^{\frac{1}{2}} e^{-\frac{Mc^2}{2kT}\left(\frac{\omega-\omega_o}{\omega_o}\right)^2} = \frac{1}{\sqrt{2\pi}\sigma} e^{-\frac{(\omega-\omega_o)^2}{2\sigma^2}}$$

with dispersion

$$\sigma = \left[\frac{kT}{Mc^2}\right]^{\frac{1}{2}} \omega_o$$

5.6.2 Damping broadening

According to classical electrodynamics, due to the retarding self-force of a radiating electron the line profile for a classical oscillator, consisting of an electron elastically bound with a fundamental angular frequency ω_o, takes the form of a Cauchy or Lorentz distribution

$$F(\omega) = \frac{\gamma/\pi}{(\omega-\omega_o-\delta)^2 + \gamma^2}$$

where

$$\gamma = \gamma_c = \frac{1}{2}\,\varepsilon\omega_o^2 = \frac{1}{3}\,\frac{e^2\omega_o^2}{mc^3}, \qquad \varepsilon = \frac{2e^2}{3mc^3} = 6.2818 \,.\, 10^{-24}\text{sec}$$

and

$$\delta = \delta_c = -\,\frac{5}{8}\,\varepsilon^2\omega_o^3 = -\,\frac{5e^4\omega_o^3}{18m^2c^6} = -\,\frac{5\gamma^2}{2\omega_o}$$

On the basis of quantum mechanics (Heitler, 1954) the same kind of profile is obtained with $\gamma = A/2$ where, for a two-state atom, A is the spontaneous radiative transition probability for the line and therefore the spontaneous radiative decay probability for the upper state of the transition, given by

$$A = 6|f_{em}|\gamma_c$$

where f_{em} is the emission oscillator strength. In general for a many state atom, for a transition between the state i and the state j

$$A = A_i + A_j$$

where A_i is the total probability for decay of the state i.

In the case of spontaneous radiative decay

$$A_i = A_{i,r} = \sum_{k<i} A_{ik}$$

where A_{ik} is the Einstein A-coefficient for the transition indicated by the subscripts. In the presence of a radiation field, stimulated transitions (both emission and absorption) will increase the total radiative decay probability, so that now

$$A_{i,r} = \sum_{k<i} A_{ik} + \sum_{k<i} B_{ik}\rho(\nu_{ik}) + \sum_{k>i} B_{ik}\rho(\nu_{ik})$$

where we have introduced the Einstein B-coefficients for stimulated emission and absorption and $\rho(\nu)$ is the radiation energy density. In the case of upward transitions (absorption) the summation includes an integration over the continuum. Making use of the relations between the A- and B-coefficients we then have

$$A_{i,r} = \sum_{k<i} \left[1 + \frac{c^3}{8\pi h\nu_{ik}^3}\,\rho(\nu_{ik})\right] A_{ik} + \sum_{k>i} \frac{g_k}{g_i}\,\frac{c^3}{8\pi h\nu_{ik}^3}\,\rho(\nu_{ik})A_{ki}$$

Assuming local thermodynamic equilibrium, so that we may replace $\rho(\nu)$ by the Planck expression, this becomes

$$A_{i,r} = \sum_{k<i} \frac{e^{h\nu_{ik}/kT}}{e^{h\nu_{ik}/kT} - 1}\,A_{ik} + \sum_{k>i} \frac{g_k}{g_i}\,\frac{1}{e^{h\nu_{ik}/kT} - 1}\,A_{ki}$$

and introducing the (absorption) oscillator strengths we obtain

$$A_{i,r} = \frac{2e^2}{mc^3}\left(\frac{kT}{\hbar}\right)^2 \left[\sum_{k<i} \frac{g_k}{g_i}\,f_{ki}\,\frac{u_{ki}^2}{e^{u_{ki}}-1} + \sum_{k>i} \frac{u_{ik}^2}{e^{u_{ik}}-1}\,f_{ik}\right]$$

The integral over the continuum may be evaluated approximately by writing

$$\int_{u_i}^{\infty} \frac{u^2}{e^u-1}\,\frac{df}{du}\,du = C(u_i)\int_{u_i}^{\infty} \frac{du}{u(e^u-1)} = C(u_i)\sum_{r=1}^{\infty} E_1(ru_i)$$

where we have put $C(u) = u^3 df/du$ and, since it is slowly varying, we have replaced it by its value at the ionization threshold u_i and taken it outside the integral and $E_1(x)$ is the exponential integral

$$E_1(x) = \int_x^{\infty} \frac{e^{-t}}{t}\,dt$$

Generally speaking the radiative damping effects are small and may often be neglected. However, if a level i has an energy lying above the lowest ionization threshold, it is subject to autoionization whereby it makes a radiationless decay with a transition probability $A_{i,a}$ to the adjoining (overlapping) continuum. The total transition probability for the level i is then

$$A_i = A_{i,r} + A_{i,a}$$

The autoionization transition probability $A_{i,a}$ is typically many orders of magnitude greater than the radiative transition probability $A_{i,r}$, so that, when autoionization occurs the latter may be neglected in comparison with the former. A more detailed theory of the autoionization effect (Fano, 1961; Fano and Cooper, 1965; Shore, 1967) shows that the line profile deviates from the Lorentz profile, and has instead a profile which shows the asymmetry that is characteristic of resonance phenomena. This may be represented in the parameterized form

$$F(\omega) = \frac{a(\omega-\omega_o) + b\gamma/\pi}{(\omega-\omega_o)^2 + \gamma^2}$$

where a and b are energy dependent parameters, determined by the interacting discrete and continuous states, and approximately constant in the neighborhood of $\omega = \omega_o$, such that the profile and the integrated absorption coefficient have the same normalization as before. According to the magnitude of a the profile may show an absorption dip or "window" adjoining the usual peak. For small values of the ratio a/b the profile may again be represented by the Lorentz profile.

5.6.3 Pressure broadening

The pressure broadening of spectral lines results from the perturbations of the emitting or absorbing atom by the surrounding particles. Since the particles are in motion, their configuration and hence their interaction with the atom, which determines the distribution in frequency of the photons emitted or absorbed, is time dependent. In general, therefore, the problem of determining the line profile involved multiple time-dependent interactions, which may be further complicated by the possibility of perturber correlations as well as the presence of different kinds of perturbers.

The earliest classical theories were confined to the two limiting extremes of very fast and very slow perturbing collisions--the impact and static approximations. In the Lorentz (1906) theory, the collisions were assumed to be so fast and strong that the radiative process is instantaneously terminated. Fourier analysis of a monochromatic wave with angular frequency ω_o and duration T, which represents the time between two collisions, and averaging over the Poisson distribution

$$f(T) = \frac{1}{\tau} e^{-T/\tau}$$

where τ is the mean time between collisions gave the Lorentz profile in which the shift $\delta = 0$ and the width $\gamma = 1/\tau$, i.e., the mean frequency of collisions. Defining an "optical" cross-section σ, we may write $\gamma = Nv\sigma$, where N is the number density of perturbers and v their velocity. To calculate the effective cross-section it is necessary to take account of the disturbing influence of the colliding perturbers. If the perturbation is a function of the perturber-atom distance R of the form $\Delta\omega = C_n/R^n$ then assuming the perturber follows a classical (rectilinear) path with velocity v and impact parameter ρ, so that $R^2 = \rho^2 + v^2t^2$, the time-integrated perturbation or phase shift along the entire trajectory is

$$\phi(\rho) = \int \Delta\omega dt = \int_{-\infty}^{\infty} \frac{C_n dt}{(\rho^2+v^2t^2)^{n/2}} = \frac{a_n C_n}{v\rho^{n-1}}, \quad a_n = \sqrt{\pi}\,\frac{\Gamma\left(\frac{n-1}{2}\right)}{\Gamma\left(\frac{n}{2}\right)}$$

Weisskopf (1933) then defined

$$\sigma = 2\pi \int_0^{\rho_c} \rho d\rho = \pi\rho_c^2, \quad \phi(\rho_c) = 1$$

Lindholm (1942) showed that, taking all impact parameters or phase shifts into account, the Lorentz profile is recovered with

$$\gamma + i\delta = Nv\sigma = 2\pi Nv\int\{1-e^{-i\phi(\rho)}\}\rho d\rho = A_n N\Gamma\left(\frac{n-3}{n-1}\right)e^{i\frac{\pi}{n-1}}, \quad n > 2$$

where

$$A_n = \pi v^{\frac{n-3}{n-1}}(a_n C_n)^{\frac{2}{n-1}}$$

For n = 2 the integral is not convergent for large ρ unless a cut-off to take account of screening is introduced. Assuming a Maxwellian distribution for v we obtain

$$\gamma + i\delta = 2\sqrt{\pi}N\left(\frac{2kT}{M}\right)^{\frac{n-3}{2(n-1)}}\Gamma\left(\frac{n-3}{n-1}\right)\Gamma\left(\frac{2n-3}{n-1}\right)(a_n C_n)^{\frac{2}{n-1}}e^{i\frac{\pi}{n-1}}$$

In the Holtsmark (1919) theory the perturbers are assumed to move so slowly that they do not change their positions appreciably during the radiative process. The line profile is then given by the statistical distribution of the perturbations arising from the distribution of the perturber configurations. Assuming that the effect of the perturbers on the atom is via the static Stark effect of the instantaneous resultant $\underline{E}$ of their electric fields, so that the frequency of the line is given by $\omega'(\underline{E})$, the line profile is then

$$F(\omega) = \int P(\underline{E})\,\delta\{\omega-\omega'(\underline{E})\}d\underline{E}$$

where $P(\underline{E})$ is the probability distribution of $\underline{E}$. Holstmark treated the case of independent perturbers for which the field $\underline{E}_i$ of an individual perturber with position vector $\underline{r}_i$ is given by a power law of the form $\underline{E}_i = q_n\underline{r}_i/r_i^{n+1}$, with $n = 2$, 3 and 4 corresponding to the fields produced by ions (monopoles), dipoles and quadrupoles, respectively. Chandrasekhar (1949) has given the result for the general power law, obtaining

$$P(\underline{E}) = \frac{1}{4\pi E^2}P(E) = \frac{1}{4\pi E_n^2\beta^2}P(\beta)$$

where $E = \beta E_n$ and $E_n = \gamma_n q_n N^{n/3}$ with

$$\gamma_n = \left[\frac{2\pi^2 n}{3(n+3)\Gamma(3/n)\sin(3\pi/2n)}\right]^{n/3}$$

and

$$P(\beta) = \frac{2}{\pi}\beta\int_0^\infty \exp(-y^{3/n})y\sin\beta y\,dy$$

Field distributions, taking into account shielding and perturber correlations have been calculated by a number of workers including Mozer and Baranger (1960) who, using a cluster expansion to represent increasing orders of correlation, have carried out calculations to the first order (pair correlations).

The interaction parameters entering all the theories must be determined from quantum mechanics. Under the influence of perturbations the change in frequency of a line is given by

$$\omega' = \omega_o + \Delta\omega, \qquad \Delta\omega = (\Delta E_j - \Delta E_i)/\hbar$$

where ΔE_i, ΔE_j are the changes in the energies of the levels. For a given perturbing Hamiltonian we obtain

$$\Delta E_i = H'_{ii} + \sum_k \frac{|H'_{ik}|^2}{E_i-E_k} + \ldots$$

in the absence of degeneracy, and

$$|H'_{\ell m} - \delta_{\ell m}\Delta E_i| = 0, \qquad \ell,m = 1,2\ldots n$$

for an n-fold degenerate level. For the case of a charged perturber with charge Z'e we have

$$H' = -Z'e^2 \frac{1}{|\underline{R}-\underline{r}|}$$

where $\underline{R}$ and $\underline{r}$ are the position vectors of the perturber and the perturbed electron. Putting

$$\frac{1}{|\underline{R}-\underline{r}|} = \begin{cases} \frac{1}{R} \sum_{\ell=o}^{\infty} \left(\frac{r}{R}\right)^{\ell} P_{\ell}(\cos\theta) & r < R \\ \frac{1}{r} \sum_{\ell=o}^{\infty} \left(\frac{R}{r}\right)^{\ell} P_{\ell}(\cos\theta) & r > R \end{cases}$$

where θ is the angle between $\underline{R}$ and $\underline{r}$, we may for large R expand ΔE_i as a series

$$\Delta E_i = \sum_{n=1} \frac{a_n}{R^n}$$

The first term is simply the Coulomb energy, while the second and fourth terms give the linear and quadratic Stark effects in the electric field $E = Z'e/R^2$ of the perturber. The third term corresponds to the effect of the inhomogeneity of the electric field. For a homogeneous field we may alternatively make the expansion

$$\Delta E_i = \sum_{n=1} b_n E^n$$

in which the first two terms again give the linear and quadratic Stark effects. For hydrogenic levels the coefficients are

$$b_1 = \frac{3n}{2Z}(n_1 - n_2) e a_o$$

$$b_2 = -\frac{1}{16}\left(\frac{n}{Z}\right)^4 \left[17n^2 - 3(n_1 - n_2) - 9m^2 + 19\right] a_o^3$$

where Z is the atomic nuclear charge number, n_1 and n_2 the parabolic quantum numbers, m the magnetic quantum number, and n is the principal quantum number such that

$$n = n_1 + n_2 + |m| + 1$$

For non-hydrogenic levels the first non-vanishing term is the quadratic one with coefficient $b_2 = a_4/Z'^2e^2$ where

$$a_4 = -Z'^2 e^4 \sum_k \frac{|(r\cos\theta)_{ik}|^2}{E_i - E_k}$$

$$= -Z'^2 \frac{e^4\hbar^2}{2m} \sum_k \frac{f_{ik}}{(E_i - E_k)^2} = -Z'^2 \frac{e^4\hbar^2}{2m(kT)^2} \sum_k \frac{f_{ik}}{u_{ik}^2}$$

in which the summation over k includes an integration over the continuum

$$\int_{u_i}^{\infty} \frac{1}{u^2}\frac{df}{du}\,du \simeq C(u_i)\int_{u_i}^{\infty}\frac{du}{u^5} = \frac{1}{4}\,C(u_i)\,u_i^{-4}$$

When any of the classical assumptions made regarding the interaction of the perturbers with the atom is no longer valid the corresponding effect must be treated quantum mechanically (Baranger, 1962; Griem, 1964). Using the time-evolution operator technique to describe the development of the perturbed quantum mechanical system, the Fourier transform of the line profile is obtained in terms of the autocorrelation function of the transition matrix element or amplitude. The calculation is again greatly simplified in the two extreme limits corresponding to the impact and static approximations. In the first case the line profile is again of the Lorentz type

$$F(\omega) = \frac{\gamma/\pi}{(\omega-\omega_o-\delta)^2 + \gamma^2}$$

while in the second case we have, also as before

$$F(\omega) = \int \delta(\omega-\omega')P(\omega')d\omega' = P(\omega)$$

where $P(\omega')$ is the probability distribution of the line frequency resulting from the distribution of perturber configurations. Generally speaking, the impact approximation is applicable to electron perturbers while the static approximation is valid for heavier particles. It is also found that the classical path approximation is independently valid in both limits.

Baranger (1958) has shown that quantum mechanically the width and shift of a line in the impact approximation may be expressed in terms of the amplitudes and phase shifts for scattering of the perturbers

$$\gamma+i\delta = N\left[-i\,\frac{2\pi\hbar}{m}\left\{f_i(o)-f_j^*(o)\right\} - v\int f_j^*(\Omega)f_i(\Omega)d\Omega\right] = N[v\sigma]$$

where $f_i(\Omega)$ is the amplitude for elastic scattering of the perturber in the direction $\Omega = (\theta,\phi)$ by the perturbed atom in the level i, N is the density of perturbers and the square brackets indicate that the statistical average is to be taken. The relation of the classical result to the above may be seen by writing the scattering amplitude in terms of the phase shifts η_ℓ

$$f(\Omega) = \frac{1}{2ik}\sum_{\ell=o}^{\infty}(2\ell+1)(e^{2i\eta_\ell}-1)P_\ell(\cos\theta)$$

where $\hbar k = mv$ and ℓ is the angular momentum quantum number.

Substituting for $f(\Omega)$ we then obtain

$$\gamma+i\delta = N\left[-\frac{\pi\hbar}{mk}\sum_{\ell=o}^{\infty}(2\ell+1)\left\{e^{2i(\eta_{i,\ell}-\eta_{j,\ell})}-1\right\}\right]$$

Putting $\phi_\ell = -2(\eta_{i,\ell}-\eta_{j,\ell})$ this becomes

$$\gamma+i\delta = N\left[\frac{\pi\hbar^2}{mv}\sum_{\ell=o}^{\infty}(2\ell+1)\left(1-e^{-i\phi_\ell}\right)\right]$$

from which, using the relation $mv\rho = \hbar k\sqrt{\ell(\ell+1)}$ and replacing the summation by an integration, we recover the Lindholm formula. Returning to the quantum mechanical formula we have therefore

$$\gamma = N[v\mathscr{R}(\sigma)], \qquad \delta = N[v\,\mathscr{I}(\sigma)]$$

Making use now of the optical theorem in the form

$$\mathscr{I}f(o) = \frac{k}{4\pi}\sigma_t = \frac{k}{4\pi}\left(\sigma_{e\ell}+\sigma_{in}\right), \qquad \sigma_{e\ell} = \int|f(\Omega)|^2 d\Omega$$

where σ_t is the total scattering cross-section and $\sigma_{e\ell}$ and σ_{in} are the elastic and inelastic components, we then find

$$\mathscr{R}(\sigma) = \frac{1}{2}\left\{\sigma_{ij,e\ell} + \sigma_{i,in} + \sigma_{j,in}\right\}$$

where

$$\sigma_{ij,e\ell} = \int|f_i(\Omega) - f_j(\Omega)|^2 d\Omega$$

Thus we may also write γ as the sum of three terms

$$\gamma = \gamma_{ij,e\ell} + \gamma_{i,in} + \gamma_{j,in}$$

Thus the width of the line is determined as if by damping collisions for which the cross-section is the sum of the inelastic cross-sections for scattering on the two levels involved, plus an "elastic" cross-section for which the scattering amplitude is the difference of the amplitudes for elastic scattering on the two levels. The calculation of the electron scattering amplitudes, phase shifts and cross-sections is a difficult problem. Carson, Mayers and Stibbs (1968) have used the Born approximation cross-sections of Carson (1966) to obtain

$$\sigma_{i,in} = \sum_k \sigma_{ik,in}, \qquad \sigma_{ik}(E) = \frac{\pi e^4 f_{ik}}{E(E_k-E_i)}\ln\left[\frac{2E}{|E_k-E_i|} \mp 1\right]$$

where E is the electron energy and the upper and lower signs refer to excitation and de-excitation. Thus we have

$$\gamma_{i,in} = \sqrt{2\pi}N\left(\frac{kT}{m}\right)^{\frac{1}{2}} \frac{e^4}{(kT)^2}\sum_k \frac{|f_{ik}|}{u_{ik}} I(|u_{ik}|)$$

with

$$I(x) = e^{\mp(x/2)}E_1(x/2)$$

The extension of the summation on k to an integration over the continuum gives

$$\int_{u_i}^{\infty} \frac{1}{u}\frac{df}{du} I(u)du \simeq C(u_i)\int_{u_i}^{\infty} u^{-4}e^{-u/2}E_1(u/2)du = \frac{1}{8}C(u_i)I_4(u_i/2)$$

where

$$I_4(x) = \int_x^{\infty} t^{-4}e^{-t}E_1(t)dt$$

$$= \frac{1}{3}\left[x^{-3}\frac{e^{-x}}{2}\left\{E_1(x)(x^2-x+2) - \frac{e^{-x}}{6}(20x^2-7x+4)\right\} + \frac{10}{3}E_1(2x) - \frac{1}{4}\left\{E_1(x)\right\}^2\right]$$

Similarly for the "elastic" contributions to the line width, using

$$\sigma_{ij,e\ell} = 4\pi\left(\frac{me^2}{\hbar^2}\right)^2 |(r^2\cos^2\theta)_{ii} - (r^2\cos^2\theta)_{jj}|^2$$

we obtain

$$\gamma_{ij,e\ell} = 4\sqrt{2\pi}N\left(\frac{kT}{m}\right)^{\frac{1}{2}}\left(\frac{me^2}{\hbar^2}\right)^2 |(r^2\cos^2\theta)_{ii} - (r^2\cos^2\theta)_{jj}|^2$$

with

$$(r^2\cos^2\theta)_{ii} = \frac{2\ell_i^2 + 2\ell_i - 2m_i + 1}{(2\ell_i - 1)(2\ell_i + 3)}(r^2)_{ii}$$

where ℓ_i and m_i are the angular momentum and magnetic quantum numbers associated with the level i. The line width may alternatively be expressed in terms of S scattering matrices instead of the scattering amplitudes. Expressions for the width have also been given by Baranger (1962) and Griem (1964) in the classical path approximation.

In the static approximation the line profile is obtained when the field distribution P(E) is known according to the relation

$$F(\omega) = P(\omega) = P(E)\frac{dE}{d\omega}$$

In the case of the linear Stark effect

$$\omega - \omega_o = c_1 E$$

where c_1 is a constant, and we have

$$F(\omega) = \frac{1}{c_1} P(E)$$

For the quadratic Stark effect

$$\omega - \omega_o = c_2 E^2$$

with c_2 constant, and thus

$$F(\omega) = \frac{1}{2c_2 E} P(E)$$

5.6.4 Superposition of profiles

When a number of different broadening processes operate simultaneously, the resultant line profile is formed by combining the profiles for each of the separate processes. Assuming that the various processes are uncorrelated, the resultant profile is then given by the superposition or folding of the independent profiles by considering each point on a profile as the origin of another profile, thus

$$F(\omega_o,\omega) = \int F_o(\omega_o,\omega_1) F_1(\omega_1,\omega_2) \ldots F_n(\omega_n,\omega) d\omega_1 d\omega_2 \ldots d\omega_n$$

where $F_i(\omega_i,\omega)$, i = 1, 2 ..... n represent the n independent profiles with origins at ω_i and $F_o(\omega_o,\omega) = \delta(\omega-\omega_o)$. In certain cases the folding may be performed analytically. Thus for two Lorentz profiles with widths and shifts (γ_1,δ_1) and (γ_2,δ_2) the resultant is also a Lorentz profile with width $\gamma = \gamma_1+\gamma_2$ and shift $\delta = \delta_1+\delta_2$. For two Gauss profiles with dispersions and shifts (σ_1,δ_1) and (σ_2,δ_2) the resultant is also a Gauss profile with dispersion and shift given by $\sigma^2 = \sigma_1^2+\sigma_2^2$, $\delta = \delta_1+\delta_2$. The superposition of a Lorentz profile with width γ and a Gauss profile of dispersion σ leads to the so-called Voigt profile

$$F(\omega) = \frac{1}{\sqrt{\pi}b} H(a,x)$$

where $b = \sqrt{2}\sigma$, $a = \gamma/b$, $x = (\omega-\omega_o)/b$ and $H(a,x)$ is defined by

$$H(a,x) = \frac{a}{\pi}\int_{-\infty}^{\infty} \frac{e^{-y^2}dy}{(x-y)^2+a^2}, \quad H(o,o) = 1$$

5.6.5 Opacity including line absorption

The importance of line absorption in opacity has been firmly established by many workers. Most calculations have followed the general methods of Mayer (1947), in which the energy levels and occupation numbers are derived from perturbation theory and statistical mechanics, and the cross-sections are obtained by assuming that the oscillator strengths and Gaunt factors (for some appropriate effective nuclear charge) are given by their hydrogenic values. Such calculations have been performed by Moszkowski and Meyerott (1951), Meyerott and Moszkowski (1951), Cox and Stewart (1962), Arking and Herring (1962, 1963), Cox (1965), Cox, Stewart and Eilers (1956) and by Cox and Stewart (1965). A critical study of the hydrogenic approximation has been made by Carson and Hollingsworth (1968). In addition, Carson, Mayers and Stibbs (1968) have carried out calculations in which the cross-sections are obtained directly from wave functions, derived by solving the Schrödinger equation in a Thomas-Fermi field, which is taken to represent the potential field in the neighborhood of each atom. They find that in this model the atomic energy levels are generally higher than those obtained from perturbation theory, with the result that the ionization edges occur at lower energies and thus lead to higher values of the calculated opacity. All the above calculations agree in that the effect of line absorption on opacity is to increase it, as originally predicted, by a factor which depends upon the physical conditions but which may be as high as three or more. Watson (1968) has studied the effect of autoionization and finds that the opacity is also increased by the inclusion of autoionization lines.

5.7 Other Sources of Opacity

5.7.1 Negative ion absorption

At low temperatures most atoms will be neutral, and in certain cases some of the neutral atoms may each acquire one or more additional electrons to form negative ions. The system consisting of a neutral atom and an electron, in a free or a bound state, will give rise to the usual free-free or bound-free absorption. However, on account of the nature of such systems, the field in which the electron moves is very far from being Coulomb, and hence they cannot be described in hydrogenic terms. In such cases special methods have to be applied to derive wave functions and absorption cross-sections.

By far the most important negative ion in astrophysics is probably the negative hydrogen ion, H^-. Because of the competing requirements of a supply both of neutral atoms and electrons, and also competition from other absorption processes, negative hydrogen ion absorption is most important under conditions appropriate to the atmospheres of solar-type stars. Free-free absorption by H^- (i.e., a free electron in the field of a neutral hydrogen atom) can take place at all frequencies while bound-free absorption may take place from the only one discrete energy level of H^-, which has an ionization potential of $I_{H^-} = 0.7551$ eV, so that the threshold for bound-free absorption is at a wavelength $\lambda = 16{,}420Å$. Free-free and bound-free cross-sections for H^- have been computed by, among others, John (1960, 1961), Geltman (1962, 1965) and Doughty et al. (1966). The Saha ionization equation may be used to determine the numbers of negative hydrogen ions N_{H^-} relative to the number of neutral (ground-state) hydrogen atoms N_{H^o}, thus

$$\frac{N_{H^o}}{N_{H^-}} N_e = \frac{U_{H^o}}{U_{H^-}} 2 \left[\frac{2\pi mkT}{h^2}\right]^{3/2} e^{-I_{H^-}/kT}$$

where U_{H^o}, U_{H^-} are the partition functions, which under conditions of sufficiently low temperature may be replaced by the ground state statistical weights $g_{H^o}(1s) = 2$, $g_{H^-}(1s^2) = 1$.

5.7.2 Molecular absorption

At sufficiently low temperatures atoms will associate to form molecules which may then contribute to the absorption coefficient according to their characteristic spectra. The formation of molecules may be described by the equation of dissociative equilibrium, which is simply an expression of the Guldbert and Waage Law of Mass Action and may be regarded as a generalization of the Saha relation for ionization equilibrium. For the case of two atoms A, B combining to form a diatomic molecule AB, the equation of dissociative equilibrium is

$$\frac{N_A N_B}{N_{AB}} = \frac{U_A U_B}{U_{AB}} \left[\frac{2\pi MkT}{h^2}\right]^{3/2} e^{-D/kT}$$

where N_A, N_B, N_{AB} and U_A, U_B, U_{AB} are the number densities and partition functions of the various species, D is the dissociation energy and M is the reduced mass

$$M = \frac{M_A M_B}{M_A + M_B}$$

in which M_A, M_B are the masses of the two atoms. The molecular partition function may be written in the usual form of a summation over the rotational, vibrational and electronic energy levels

$$U_{AB} = U_s \sum_{r,v,e} g_{r,v,e}\, e^{-E_{r,v,e}/kT}$$

If all the modes of motion can be regarded as independent we may express the energy as the sum of three parts

$$E_{r,v,e} = E_r + E_v + E_e$$

and the statistical weight as the product of the statistical weights

$$g_{r,v,e} = g_r g_v g_e$$

in which the three terms in each case correspond to the three modes of motion. We may therefore write

$$U_{AB} = U_S U_r U_v U_e$$

where U_S is a factor arising from symmetry considerations such that

$$U_S = \begin{cases} \frac{1}{2} & A = B \\ 1 & A \neq B \end{cases}$$

and U_r, U_v and U_e are given by

$$U_r = \sum_r g_r e^{-E_r/kT} \simeq \sum_{J=o}^{\infty} (2J+1) e^{-BJ(J+1)hc/kT}$$

$$\simeq \int_0^{\infty} (2J+1) e^{-BJ(J+1)hc/kT} \, dJ = \frac{kT}{hcB}$$

in which B is the rotational constant, $B = h/8\pi^2 cI$, I being the moment of inertia of the molecule,

$$U_v = \sum_v g_v e^{-E_v/kT} \simeq \sum_o^{\infty} e^{-nh\nu_o/kT} = \left(1 - e^{-h\nu_o/kT}\right)^{-1}$$

where ν_o is the fundamental frequency of vibration,

$$U_e = \sum_e g_e e^{-E_e/kT} = \sum_{\Lambda S} g_\Lambda (2S+1) e^{-E_e(\Lambda,S)/kT} \simeq g_{\Lambda_o}(2S_o+1)$$

where Λ and S denote the electronic (axial) angular momentum and spin quantum numbers classifying the electronic states (zero subscripts referring to the ground state), and

$$g_\Lambda = \begin{cases} 1 & \Lambda = 0 \\ 2 & \Lambda \neq 0 \end{cases}$$

The equations of dissociative equilibrium must be solved simultaneously with the equations of ionization equilibrium. Because of the possibility of changes in the rotational and vibrational motions as well as in the electronic motions, the spectra of molecules are much more complex than those of atoms. Also on account of the non-central nature of the force field, in only a few simple cases is it possible to calculate the oscillator strengths for the electronic transitions, so that in general it is necessary to rely on experimentally determined results. Because of these difficulties only a few attempts have been made to investigate the contribution of molecular absorption to stellar opacity (Vardya, 1964, 1966; Auman, 1967).

5.7.3 Electron thermal conduction

When electron conduction takes place we have already seen that the energy flux carried by this mechanism is

$$F_{cond} = -K_{cond} \frac{dT}{dr}$$

where K_{cond} is the thermal conductivity. The total flux carried by both radiation and conduction is therefore

$$F_{tot} = -\left(K_{rad} + K_{cond}\right)\frac{dT}{dr} = -K_{tot} \frac{dT}{dr}$$

where the radiative conductivity K_{rad} is given by

$$K_{rad} = \frac{4acT^3}{3\rho\kappa_r}$$

in which κ_r is the Rosseland mean radiative opacity. It is convenient to define a conductive opacity κ_c by a similar relation

$$K_{cond} = \frac{4acT^3}{3\rho\kappa_c}$$

It therefore follows that the total conductivity is

$$K_{tot} = \frac{4acT^3}{3\rho\kappa_t}$$

where the "total" effective opacity κ_t is given by

$$\frac{1}{\kappa_t} = \frac{1}{\kappa_r} + \frac{1}{\kappa_c}$$

i.e., the harmonic mean of the radiative and conductive opacities. Expressions for the thermal conductivity have been given by Marshak (1940, 1941), Lee (1950), Mestel (1950), Schatzman (1958), Chiu (1968), Clayton (1968) and Iben (1968).

REFERENCES

Arking, A., and Herring, J. 1962, A.J., 67, 110; 1963, PASP, 75, 226.

Auman, J. 1967, Ap.J. Supp., 14, 171.

Baranger, M. 1958, Phys.Rev., 111, 481; 1958, Phys.Rev., 111, 494.

———. 1962, Atomic and Molecular Processes, ed. D. R. Bates, Academic Press.

Berger, J. M. 1956, Ap.J., 124, 550; 1957, Phys.Rev., 105, 35.

Bethe, H. A., and Salpeter, E. E. 1956, Handbuch der Physik, Springer, Atome 1.

Biedenharn, L. C. 1956, Phys.Rev., 102, 262.

Carson, T. R. 1966, J.Q.S.R.T., 6, 563.

Carson, T. R., and Hollingsworth, H. M. 1968, M.N.R.A.S., 141, 77.

Carson, T. R., Mayers, D. F., and Stibbs, D. W. N. 1968, M.N.R.A.S., 140, 483.

Chandrasekhar, S. 1949, Proc.Camb.Phil.Soc., 45, 219.

Chin, C-W. 1965, Ap.J., 142, 1481.

Chiu, H-Y. 1968, Stellar Physics, Blaisdell, Vol. 1.

Clayton, D. D. 1968, Principles of Stellar Evolution and Nucleosynthesis, McGraw-Hill.

Cox, A. N. 1965, Stars and Stellar Systems: Vol. 8, Stellar Structure, ed. L. H. Aller and D. B. McLaughlin, Chicago.

Cox, A. N., and Stewart, J. N. 1962, A.J., 67, 113.

———. 1965, Ap.J. Supp., 11, 22.

Cox, A. N., Stewart, J. N., and Eilers, D. D. 1965, Ap.J. Supp., 11, 1.

Cox, J. P. 1968, Principles of Stellar Structure, Gordon and Breach, Vol. 1.

Doughty, N. A., et al. 1966, M.N.R.A.S., 132, 255; 1966, M.N.R.A.S., 132, 267.

Eddington, A. S. 1916, M.N.R.A.S., 77, 16.

———. 1926, The Internal Constitution of the Stars, Cambridge U.P. (and Dover, 1959).

Fano, U. 1961, Phys.Rev., 124, 1866.

Fano, U., and Cooper, J. W. 1965, Phys.Rev., 137, A1364.

Fischer, J. 1931, Ann.der Phys., 8, 821.

Gaunt, J. A. 1930, Proc.Roy.Soc., A, 126, 654; 1930, Phil. Trans.Roy.Soc., A, 229, 163.

Geltman, S. 1962, Ap.J., 136, 935; 1965, Ap.J., 141, 376.

Goldwire, H. C., Jr. 1968, Ap.J. Supp., 17, 445.

Gordon, W. 1929, Ann.der Phys., (5) 2, 1031.

Griem, H. R. 1964, Plasma Spectroscopy, McGraw-Hill.

Green, L. C., Rush, P. P., and Chandler, C. D. 1957, Ap.J. Supp., 3, 37.

Hall, H., and Oppenheimer, J. R. 1931, Phys.Rev., 38, 57.

Heitler, W. 1954, The Quantum Theory of Radiation, Oxford.

Herdan, R., and Hughes, T. P. 1961, Ap.J., 133, 294.

Holtsmark, J. 1919, Ann.der Phys., 58, 577.

Iben, I., Jr. 1968, Ap.J., 154, 557.

John, T. L. 1960, M.N.R.A.S., 121, 41; 1961, M.N.R.A.S., 128, 93.

Karzas, W. J., and Latter, R. 1961, Ap.J. Supp., 6, 167.

Kramers, H. A. 1923, Phil.Mag., 46, 836.

Lane, J. H. 1869, Amer.J.Sci.2nd Ser., 50, 57.

Lee, T. D. 1950, Ap.J., 111, 625.

Lindholm, E. 1942, Über die Verbreiterung und Verschiebung von Spektralinien, Almquist und Wiksells.

Lorentz, H. A. 1906, Proc.Amst.Acad., 8, 591.

Marshak, R. E. 1940, Ap.J., 92, 321.

———. 1941, Ann.New York Acad.Sci., 41, 49.

Mayer, H. 1947, Los Alamos Scientific Laboratory, LA-647.

Menzel, D. H. 1969, Ap.J. Supp., 19, 1.

Menzel, D. H., and Pekeris, C. L. 1935, M.N.R.A.S., 96, 77.

Mestel, L. 1950, Proc.Camb.Phil.Soc., 46, 331.

Meyerott, R. E., and Moszkowski, S. A. 1951, Argonne National Laboratory, ANL-4594.

Milford, S. N. 1957, Ap.J., 125, 213.

Moszkowski, S. A., and Meyerott, R. E. 1951, Argonne National Laboratory, ANL-4743.

Mozer, B., and Baranger, M. 1960, Phys.Rev., 118, 626.

Nishina, Y., and Rabi, J. 1928, Verh.d.Deut.Phys.Ges., 9, 8.

Reiche, F. 1929, Z.für Phys., 53, 168.

Rosseland, S. 1924, M.N.R.A.S., 84, 525.

———. 1930, Handbuch der Astrophysik, 3, Part I, 443.

———. 1936, Theoretical Astrophysics, Oxford.

Rudkjobing, M. 1947, Pub.Copenhagen Obs., No. 145.

Sampson, D. H. 1959, Ap.J., 129, 734.

Sampson, R. A. 1894, Mem.R.A.S., 51, 123.

Sauter, F. 1931, Ann.der Phys., 9, 217.

Schatzman, E. 1958, White Dwarfs, North-Holland.

Schur, G. 1930, Ann.der Phys., 4, 433.

Schuster, A. 1903, Ap.J., 17, 165.

Schwarzschild, K. 1906, Gött. Nach., p. 41.

Schwarzschild, M. 1958, Structure and Evolution of the Stars, Princeton.

Shore, B. W. 1967, Rev.Mod.Phys., 37, 439.

Sommerfeld, A. 1953, Atombau und Spektralinien, Ungar.

Sommerfeld, A., and Schur, G. 1930, Ann.der Phys., 4, 409.

Stobbe, M. 1930, Ann.der Phys., 7, 661.

Strömgren, B. 1932, Z.Ast., 4, 118; 1933, Z.Ast., 7, 222.

Sugiura, Y. 1927, J.de Phys., 8, 113; 1929, Sci.Pap.Inst.Phys. Chem.Res.No. 193.

Vardya, M. S. 1964, Ap.J. Supp., 8, 277.

———. 1966, M.N.R.A.S., 134, 347.

Watson, W. D. 1968, Doctoral thesis, M.I.T.

Weisskopf, V. 1933, Phys.Z., 34, 1.

16

TRANSPORT MECHANISMS IN STARS

E. A. Spiegel*

New York University
New York, N. Y.

I. Elementary Theory of Radiative Transfer

A. Description of the Radiation Field

To describe the flow of radiant energy from a star we have the option of describing the field of radiation either by e.m. wave theory or by photons. In the transfer problem the latter view is preferred since this makes the description of absorption and scattering much simpler. We consider then a photon gas diffusing through the star. In most stellar problems the photon density is large enough that we can neglect quantum fluctuations. Typically, the gravitational fields are weak enough that we can neglect general relativity, and the velocities are usually small enough that even special relativity is not important, except for first order terms such as Doppler effect. Thus we consider a purely classical gas of photons.

We describe the photon gas by its density in the one-particle phase space (μ-space), whose coordinates are the position and momentum of the photon. The momentum of a photon is $p_i = (h\nu/c)\mu_i$, $i = 1,2,3$, where ν is the frequency of the photon and it moves in the direction given by the unit vector $\mu_i (\mu_i\mu_i = 1)$. In the phase space we assign a representative point to each photon and the state of the photon gas is described by the density of representative points in the phase space, $D(p_i, x_j, t)$. The volume element in phase space is

$$d_3\vec{p}d_3\vec{x} = \left(\frac{h^3\nu^2}{c^3}\right) d\nu d\Omega d_3\vec{x}, \qquad (1)$$

and it is convenient to introduce the function

$$f(\nu, \mu_i, x_j, t) = \left(\frac{h^3\nu^2}{c^3}\right) D(p_i, x_j, t). \qquad (2)$$

*Now at Columbia University, Astronomy Dept.

Then the density of photons in physical space is written

$$n(x_j, t) = \int_0^\infty d\nu \int f(\nu, \mu_i, x_j, t) d\Omega. \tag{3}$$

Often it is convenient to use the monochromatic photon density

$$n_\nu(x_i, t) = \int f d\Omega. \tag{4}$$

Astronomers do not normally use the one-particle distribution function, f; rather they deal with the specific intensity,

$$I_\nu(x_j, \mu_i, t) = h\nu c\, f(\nu, x_j, \mu_i, t). \tag{5}$$

(Note that I am not consistent about whether I put ν as a subscript or an argument; this is just that I prefer it as an argument but in some cases, as I_ν, suffix form is forced by usage.)

The specific intensity is normally defined as follows. Consider an arbitrary surface containing x_j with unit normal, $\vec{n}$, at x_j. Let dA be an infinitesimal area of this surface containing x_j and let $\Theta = \cos^{-1}(\vec{\mu}\cdot\vec{n})$.

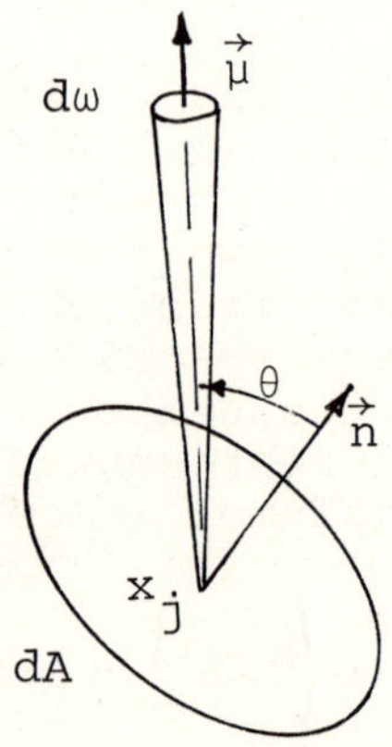

Finally let dE_ν be the radiant energy flowing through dA in direction $\vec{\mu}$, in frequency interval $d\nu$ about ν, and in solid angle $d\omega$ about $\vec{\mu}$. Then

$$I_\nu(x_j, \mu_i; t) = \lim_{\substack{dA,dt \\ d\omega,d\nu \\ \to 0}} \frac{dE_\nu}{dA \cos\theta \, d\omega d\nu dt} . \qquad (6)$$

Having defined the basic quantities that characterize the radiation field we may introduce some of the standard derived quantities. These are

(1) the integrated intensity,

$$I(x_j, \mu_i; t) = \int_0^\infty I_\nu(x_j, \mu_i; t) d\nu, \qquad (7)$$

(and likewise for an integrated anything),

(2) the mean intensity,

$$J_\nu(x_j, t) = \frac{1}{4\pi} \int I_\nu d\Omega = \frac{hc\nu}{4\pi} n_\nu = \frac{c}{4\pi} u_\nu, \qquad (8)$$

where u_ν is the monochromatic energy density, and

(3) the energy flux,

$$\mathscr{F}_i(\nu, x_j, t) \equiv \pi F_i = \int I_\nu \mu_i d\Omega, \qquad (9)$$

which is clearly a vector quantity, and

(4) the radiation pressure tensor,

$$P_{ij} = \frac{1}{c} \int I \mu_i \mu_j d\Omega. \qquad (10)$$

All these quantities are to be calculated from f (or I) for which we must now find the governing equation.

B. The Transfer Equation

We now write down an equation for I_ν (or f) from which follow the properties of the radiation field. To do this consider a beam of photons moving in direction μ_i. In traversing a distance ds the beam loses an amount of energy per unit area of beam,

$$dE_\nu = I_\nu d\nu d\Omega dt$$

= energy emitted by medium in ds in direction μ_i

+ energy scattered from other beams into μ_i in ds

- energy scattered from μ_i into other beams in ds

- energy absorbed in ds

The energy emitted into the beam is

$$\rho j_\nu ds \frac{d\Omega}{4\pi} dt d\nu$$

where j_ν is the emission coefficient, i.e. the energy emitted per unit mass, per unit time, per unit frequency in all directions. (We assume isotropic emission.)

The energy scattered into the beam from other directions is

$$\int \rho\sigma_\nu(\mu_j', \mu_i) I_\nu(\mu_j') \frac{d\Omega'}{4\pi} dt d\nu ds d\Omega$$

where $\sigma_\nu(\mu_j', \mu_i)$ is the cross section per unit mass for scattering from μ_j' into μ_i. I have neglected here what astronomers call noncoherent scattering, i.e. scattering with frequency shifts. If the scattering is isotropic σ is independent of angle (henceforth I shall assume this) and the scattering source is just,

$$\rho\sigma_\nu J_\nu dt d\nu ds d\Omega.$$

Likewise the loss from the beam by scattering is

$$\rho\sigma_\nu I_\nu dt d\nu ds d\Omega.$$

Finally loss by absorption is

$$\rho k_\nu I_\nu dt d\nu ds d\Omega$$

where k_ν is the cross section per unit mass for absorption. Thus we have

$$dI_\nu = \left[\rho\sigma_\nu(J_\nu - I_\nu) + \frac{j_\nu}{4\pi}\rho - \rho k_\nu I_\nu\right] ds. \tag{11}$$

The beam traverses the distance ds in dt = ds/c, so that

$$\frac{1}{c}\frac{dI_\nu}{dt} = -\rho\kappa_\nu I_\nu + \rho\left(\sigma_\nu I_\nu + \frac{j_\nu}{4\pi}\right) \tag{12}$$

where $\kappa_\nu = \sigma_\nu + k_\nu$ is the total cross section for the removal of a photon from the beam. Now

$$\begin{aligned} \frac{dI_\nu}{dt} &= \frac{\partial I_\nu}{\partial t} + \frac{\partial x_i}{\partial t}\frac{\partial I_\nu}{\partial x_i} \\ &= \frac{\partial I_\nu}{\partial t} + c\mu_i \frac{\partial I_\nu}{\partial x_i}, \end{aligned} \tag{13}$$

so the transfer equation is

$$\frac{1}{c}\frac{\partial I_\nu}{\partial t} + \mu_i \frac{\partial I_\nu}{\partial \chi_i} = \rho\kappa_\nu(\mathscr{J}_\nu - I_\nu) \tag{14}$$

where

$$\mathscr{J}_\nu = \frac{\sigma_\nu}{\kappa_\nu} J_\nu + \frac{j_\nu}{\kappa_\nu 4\pi} \tag{15}$$

is called the source function.

Of course, you recognize that this is the conservation equation for photons (actually representative points) in phase space. In this space, the representative point has a 6-velocity $(\dot{p}_i, \dot{x}_i)$, and making use of the result that for a Hamiltonian system the phase fluid is incompressible, we can write at once that

$$\frac{\partial f}{\partial t} + \dot{x}_i \frac{\partial f}{\partial x_i} + \dot{p}_j \frac{\partial f}{\partial p_j} = \text{sources - sinks.} \tag{16}$$

In most stellar problems the gravitational field is low enough, the density is low enough, and the index of refraction is close enough to unity that we can set $\dot{p}_i = 0$. Thus, the transport equation we have derived is in the familiar form normally encountered in kinetic theory.

C. Consequences of the Transfer Equation

(1) Thermodynamic Equilibrium

In equilibrium we wish to check that the transfer equation satisfies the usual concepts. Thus in a closed cavity, for example, we expect I_ν to be independent of space, time, and direction, so that $I_\nu = J_\nu$. The transfer equation reduces to

$$j_\nu = 4\pi k_\nu I_\nu \tag{17}$$

Now in thermodynamic equilibrium we expect that the intensity should be the Planckian intensity

$$B_\nu = \frac{2h\nu^3}{c}\left(e^{h\nu/kT} - 1\right)^{-1} \tag{18}$$

thus we find

$$j_\nu = 4\pi k_\nu B_\nu \tag{19}$$

which is the Kirchhoff-Planck law.

In the first approximation it is usual to assume that to each point in a star one can assign a temperature, and thus we can use the Kirchhoff-Planck law to estimate j_ν. Clearly this estimate must be bad near the edge of a star since radiation is lost directly into space from there and the conditions are not at all like those in a hohlraum. For our purposes though, this crude estimate, called the local thermodynamic equilibrium approximation (LTE), will have to suffice.

(2) Energy Conservation

Let us integrate the transfer equation over all angles; then we obtain

$$\frac{4\pi}{c}\frac{\partial J_\nu}{\partial t} + \frac{\partial}{\partial x_i}\mathscr{F}_i(\nu) = 4\pi\rho\kappa_\nu(\mathscr{J}_\nu - J_\nu)$$

$$= 4\pi\rho k_\nu(B_\nu - J_\nu). \tag{20}$$

The energy density, we saw, is $u_\nu = \frac{4\pi}{c}J_\nu$, so

$$\frac{\partial u_\nu}{\partial t} + \nabla\cdot\vec{\mathscr{F}}(\nu) = 4\pi\rho k_\nu(B_\nu - J_\nu), \tag{21}$$

which is the conservation equation for radiant energy. The right hand side is a net source or sink for a given frequency depending on whether the main intensity at the point is greater or less than the Planckian intensity for the local temperature. If we integrate over frequency we have

$$\frac{\partial u}{\partial t} + \nabla\cdot\vec{\mathscr{F}} = 4\pi\rho\int_0^\infty k_\nu(B_\nu - J_\nu)d\nu \tag{22}$$

If the right hand side is zero the total radiant energy is conserved; this condition is called radiative equilibrium.

(3) Momentum Conservation

This time let us multiply the transfer equation by μ_j and integrate over $d\Omega$. We find

$$\frac{1}{c}\frac{\partial\mathscr{F}_j(\nu)}{\partial t} + c\frac{\partial}{\partial x_i}P_{ij} = -\rho\kappa_\nu\mathscr{F}_j(\nu). \tag{23}$$

$\mathscr{F}_j$ is the flux of radiant energy and we can regard $\mathscr{F}_j/c$ as a momentum density. The pressure tensor describes the momentum flux. Thus this equation describes momentum conservation and is analogous to the equation of motion of a fluid. The right hand side is the negative of the force exerted by the radiation on the ambient matter and therefore is a source for the momentum of the radiation field.

In the theory of stellar atmospheres it is necessary to solve the full equations with few simplifications, though it is entirely safe to neglect time derivatives since, as you can readily convince yourselves, these only describe retardation effects and are usually not important. Also in stellar atmospheres one usually assumes in first approximation that the medium is horizontally stratified. The early work in that field assumed LTE but modern treatments try to do better. It is also not always valid to assume radiative equilibrium. My task in my later lectures is to say why not and to try to explain why that makes things difficult.

With the neglect of time derivatives we find that the flux is

$$\mathscr{F}_j = - \frac{c}{\rho\kappa_\nu} \frac{\partial P_{ij}}{\partial x_i} \tag{24}$$

Well inside the star where the mean free path of photons $(1/(\rho\kappa_\nu))$ is small, photons arriving at a point have interacted with matter near the point. If the matter properties do not vary much in a photon free path, then the radiation field is nearly isotropic and we have

$$P_{ij} = \frac{1}{c}\int I_\nu \mu_i \mu_j d\Omega \approx \frac{J_\nu}{c}\int \mu_i \mu_j d\Omega = \frac{4\pi}{3}\frac{J_\nu}{c}\delta_{ij} = \frac{1}{3} u_\nu \delta_{ij}, \tag{25}$$

where δ_{ij} is the Kronecker delta. Then

$$\vec{\mathscr{F}}(\nu) \approx - \frac{c}{3\kappa_\nu \rho} \nabla u_\nu ,$$

which is the diffusion approximation. In most stellar structure work it is usual to replace κ_ν by a weighted average over all frequencies and one uses the LTE approximation to write

$$u = \int_0^\infty u_\nu d\nu = aT^4 \tag{26}$$

where a is the radiation constant = $4\sigma/c$ where σ is the Stefan-Boltzmann constant. We then have

$$\vec{\mathscr{F}} = - K\nabla T \tag{27}$$

where the radiative conductivity is

$$K = \frac{16\sigma T^3}{3\kappa\rho} \tag{28}$$

In stellar atmospheres you have to do better than this and other lectures in the course will explain how. What I have given here has been known for years and as a preamble to the modern things I urge you to read Milne's article in Handbuch d. Astrophys. and Chap. II of Chandrasekhar's Introduction to Stellar Structure. To learn how transfer problems are solved see Chandrasekhar's Radiative Transfer.

D. The Radiative Heat Equation

When one talks about heat flow one usually has in mind the heat equation governing the temperature of the medium. It is natural to wonder how the formalism we have described relates to this. In the theory of radiative transfer the energy equation has sources and sinks from the matter--

$$\frac{\partial u}{\partial t} + \nabla\cdot\mathscr{F} = 4\pi\rho\int_0^\infty k_\nu(B_\nu - J_\nu)\,d\nu \tag{29}$$

The heat equation, on the other hand, is for the energy of the medium, and comes from the first law of thermodynamics--

$$\frac{de}{dt} + p\frac{dV}{dt} = \frac{đQ}{dt} \tag{30}$$

where V is volume, e is internal energy.
If the rate of heat input is due only to the radiative effects, then conservation of energy demands

$$\frac{đQ}{dt} = -4\pi\rho\int_0^\infty k_\nu(B_\nu - J_\nu)\,d\nu. \tag{31}$$

An example may clarify the physics a bit. Consider an infinite homogeneous medium at temperature T. Suppose that we specify the temperature everywhere at t = 0, as $T(\vec{x},o) = T_o + \theta(\vec{x},o)$ where T_o is constant and $|\theta| \ll T_o$. We can ask what is $T(\vec{x},t)$ or equivalently what is $\theta(\vec{x},t)$? This is a standard heat conduction type problem.

Suppose the medium cannot move, so that dV/dt = 0 and the density ρ is constant. We can also write $de = \rho c_\nu dT$, and we need to express đQ/dt in terms of T. To do this we need the formal solution of the transfer equation (see Chandrasekhar's Radiative Transfer). But in terms of LTE it is not difficult to see what đQ/dt is if we assume the medium is grey, i.e. that κ is independent of ν. Then per unit volume the medium emits $\rho j = 4\pi kB = 4\rho k\sigma T^4$. Now at $\vec{x}$, emitted light arrives from $\vec{x}'$ in the amount $\rho j(\vec{x}',t)e^{-k\rho|\vec{x}-\vec{x}'|}/4\pi|\vec{x}-\vec{x}'|^2$ where the exponential factor expresses the attenuation between $\vec{x}$ and $\vec{x}'$ and the denominator expresses the inverse square law. In the present approximation the undisturbed ρk is constant.

So we have with all our approximations--and let me also leave out scattering--

$$c_v\frac{\partial T}{\partial t} = -4k\sigma T^4 + k\rho\int d\vec{x}'\left(4k\sigma T^4(\vec{x}',t)\right)\frac{e^{-k\rho|\vec{x}-\vec{x}'|}}{4\pi|\vec{x}-\vec{x}'|^2} \tag{32}$$

where I have integrated over all $\vec{x}'$ to get the energy arriving at $\vec{x}$ and multiplied by ρk to get the fraction of the incident energy which is absorbed there. With $T(\vec{x},t) = T_o + \theta(\vec{x},t)$ and $|\theta| \ll T_o$ we can linearize and find,

$$\frac{\partial\Theta}{\partial t} = \frac{16\sigma T_0^3 k}{c_V}\left\{\int d\vec{x}'\frac{\Theta(\vec{x}',t)}{4\pi|\vec{x}-\vec{x}'|}\, e^{-k\rho|\vec{x}-\vec{x}'|} - \Theta(\vec{x},t)\right\} \qquad (33)$$

The quantity

$$q = \frac{16\sigma T_0^3 k}{c_V} = \frac{16\sigma T_0^4\, k\rho}{\rho c_V T_0} \qquad (34)$$

is a flux, σT_0^4, divided by an energy density ($\rho c_V T_0$) divided by a length $(k\rho)^{-1}$, hence $1/q$ is a characteristic time for radiation to smooth out bumps in Θ. Thus we have,

$$\frac{\partial}{\partial t}\,\theta(\vec{x},t) = q\int K(\vec{x},\vec{x}')\Theta(\vec{x}',t)\,d\vec{x}' \qquad (35)$$

as the analogue to the heat equation where

$$K(\vec{x},\vec{x}') = \frac{e^{-k\rho|\vec{x}-\vec{x}'|}}{4\pi|\vec{x}-\vec{x}'|^2}\, k\rho - \delta(\vec{x}-\vec{x}') \qquad (36)$$

As $k\rho \to \infty$ the equation for Θ becomes

$$\frac{\partial\Theta}{\partial t} = \frac{q}{3(k\rho)^2}\,\nabla^2\Theta \qquad (37)$$

(usual diffusion equation)

while as $k\rho \to 0$

$$\frac{\partial\Theta}{\partial t} = -q\Theta \qquad \text{(Newton's law of cooling)} \qquad (38)$$

So the analogy with heat equation holds for this highly simple case; indeed there is a Green function for this problem such that

$$\Theta(\vec{x},t) = \int G(\vec{x},\vec{x}';t)\,\Theta(\vec{x}',0)\,d\vec{x}', \qquad (39)$$

where G is not as simple as in the diffusion case, but plays the same role as in normal heat conduction.

II. Convection

A. The Schwarzschild Criterion

If a star model is constructed on the assumption that heat is transported by radiation it is not considered a realistic model unless it is stable. Unfortunately, it is quite difficult to demonstrate stability, but at least we can check to see whether any of the known instabilities can occur. The most prevalent instability is the convective instability which has been known at least as early as the last century. Indeed, prior to the development of the theory of radiative transfer it had been believed that convective transfer was the dominant transport mechanism. When radiative transfer was developed and applied to stars, K. Schwarzschild then applied the test for convective instability and showed how to use it in the stellar case. The criterion is quite simple and can be obtained from the following physical argument.

We consider a stratified medium with given variation of material properties in the radial direction, r. Fix attention on a parcel of matter in equilibrium with its surroundings. We ask what happens if we displace the parcel vertically by a small amount--say upward. The parcel will expand slightly since the ambient pressure decreases upwards. If we assume that the motion is highly subsonic, the pressure of the parcel has time to equilibrate to the ambient value, so that if we know the displacement, we know how much the pressure has decreased in the parcel. Thus, for fixed molecular weight, we know the change in ρT.

Clearly, if the parcel finds itself lighter than its surroundings, it will float upward, while if it is heavier it will sink back toward its equilibrium position. The former case represents an instability.

Since ρT has the same value in the parcel as in the surrounding medium, a density defect is equivalent to a temperature excess, and a relatively hot parcel rises. The criterion for convective instability is then that the temperature in the blob should change per unit length of vertical displacement less than does the ambient temperature, i.e.,

$$\left|\frac{dT}{dr}\right|_{\text{parcel}} \le \left|\frac{dT}{dr}\right|_{\text{ambient}} \qquad \text{for instability}$$

Since we are testing the stability of a model in which transport is by radiation, one writes $|dT/dr|_{RAD}$ for the ambient gradient. The gradient in the parcel must be computed, and the result depends on how you make the displacement. If it is on a time short compared to the thermal time of the parcel, the displacement is adiabatic and we can say that instability occurs if

$$\left|\frac{dT}{dr}\right|_{RAD} \ge \left|\frac{dT}{dr}\right|_{AD}$$

which is the Schwarzschild criterion. Incidentally, in assuming that an adiabatic displacement is possible we have assumed that the thermal time of the parcel is much less than the time taken for a sound wave to traverse it, which means roughly that $\ell^2/\kappa << \ell/c$ where ℓ is the size of the parcel, κ is the thermometric conductivity, and c is the sound speed. Thus, so long as we displace parcels small enough that $\ell c/\kappa << 1$, we have not done anything inconsistent. Of course, if we make ℓ too small we might have to consider viscosity, but that is virtually never a consideration in stellar cases.

Now all this can be made mathematically precise and you can see how it goes in Lamb's book and for the spherical case in Lebovitz' article. What becomes successively more difficult is the study of competing, often stabilizing mechanisms, such as rotation, magnetic fields, composition gradients, etc. Fortunately, Chandrasekhar has also written a book on such matters.

We can put the Schwarzschild criterion in more familiar terms. For an adiabatic gas,

$$T = \text{const.} \times p^{\frac{\gamma-1}{\gamma}} \tag{40}$$

hence

$$\left|\frac{dT}{dr}\right|_{AD} = \left(\frac{\gamma-1}{\gamma}\right) \frac{T}{p} \left|\frac{dp}{dr}\right| , \tag{41}$$

which, because of the hydrostatic equation becomes,

$$\left|\frac{dT}{dr}\right|_{AD} = \left(\frac{\gamma-1}{\gamma}\right) \frac{\rho T}{p} g, \tag{42}$$

where g is the local value of the gravitational acceleration. Recall that for a perfect gas

$$\frac{p}{\rho T} = c_p - c_v = c_v\left(\frac{c_p}{c_v} - 1\right) = (\gamma - 1)\ c_v , \tag{43}$$

so that

$$\left|\frac{dT}{dr}\right|_{AD} = \frac{g}{c_p} . \tag{44}$$

We can also here settle for the approximation

$$\left|\frac{dT}{dr}\right|_{RAD} = \frac{\pi F}{K} \tag{45}$$

except in the energy producing regions (πF is the radiative flux).

Hence, in a stellar envelope, we have convective instability if

$$\frac{gK}{\pi c_p F} < 1.$$

Thus convection will occur when the conductivity is low or the specific heat is high. This is what happens when we have an ionization zone of (say) hydrogen. Ionization raises the specific heat and favors convection. In addition, in weakly ionized hydrogen, H^- forms and the opacity goes up. This means that convection is favored. For this reason, stars with surface temperatures $\lesssim 10{,}000$ °K, have H-ionization zones near their surfaces and have extensive convective envelopes.

In a similar way, in the energy generating cores, convection can arise if the energy generation is a sufficiently rapidly varying function of position. For in that case $|dT/dr|_{RAD}$ gets large. This happens when energy generation is very temperature-sensitive, as it is on the C-N-O cycle.

B. Laboratory Convection

Just to get some idea of what convection is like we should look at the laboratory situation. This differs from the stellar case in many respects, but it would probably be optimistic to suppose that we could master stellar convection without an understanding of laboratory convection.

The standard laboratory situation consists of a fluid confined between two horizontal plates maintained at fixed temperatures T_1 and T_2.

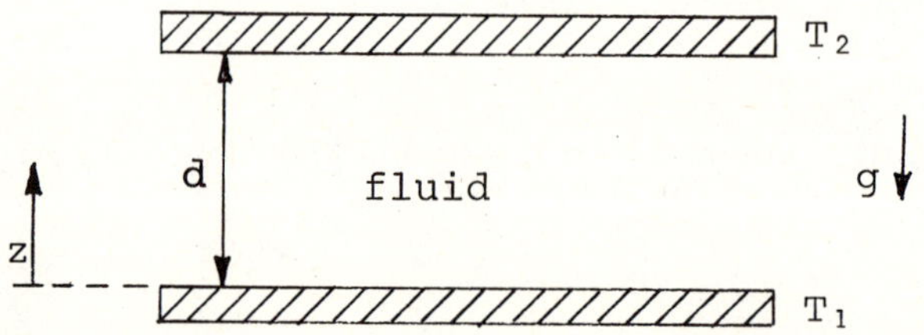

It is usual to try to keep the ratio of the plate separation, d, to the plate size small, since most theories take this ratio to be small. The case of most interest is, of course, $T_1 > T_2$.

Suppose we start the experiment and wait a long time so that it has time to reach a steady state (if it wants to). It will certainly do so if we do not permit the fluid to move. In that case a steady, constant flux will flow upward by conduction. (If the flux were not constant heat would accumulate somewhere and the state would not be steady.) For most laboratory cases, the conductivity is constant so that the temperature gradient,

$$\frac{dT}{dz} = - \frac{1}{K}\mathscr{F}, \qquad (46)$$

is constant. Thus, in the language of the stellar problem

$$\left|\frac{dT}{dz}\right|_{RAD} = \frac{T_1 - T_2}{d} = \frac{\Delta T}{d} \, . \qquad (47)$$

The Schwarzschild criterion then leads us to expect convection when

$$\frac{\Delta T}{d} > \frac{g}{c_p} \, .$$

In many laboratory situations one deals with liquids, for which c_p is very large. So the right-hand side of this criterion is essentially zero and the criterion would become $\Delta T > 0$ for instability. This is quite familiar and is just the usual result that warm material rises. The correction, g/c_p, that is used in stars simply adds that if warm compressible material rises it also expands and cools on expanding. In that case it must be warm enough to provide energy for expansion and still be warm enough to rise. For the moment, though, let us neglect this correction and speak of liquids.

However, there are other corrections which matter in the laboratory case. First recall that we treated the displaced parcel as adiabatic, the idea being that we can always find a parcel of the right size to move adiabatically. But this still supposes a relatively quick motion, which in fact cannot always be realized if the fluid is too viscous. Thus if the fluid is very viscous, the parcel moves so slowly that its excess heat leaks out before it can go anywhere, and this produces stability. In short, the combined action of viscosity and conductivity can stabilize convection, but neither can do it by itself. In stars the viscosity is very small so that it is practically negligible. But in the lab, this is not always so and the Schwarzschild criterion must be modified. As Rayleigh showed (see also Chandrasekhar), the criterion then becomes

$$\frac{\Delta T}{d} > \frac{\kappa\nu}{g\alpha d^4} \times 1708$$

for instability where $\kappa = k/\rho c_p$; ν is the viscosity divided by ρ (kinematic viscosity) and α is the coefficient of thermal expansion. (If you want to correct for compressibility, just add g/c_p to the right-hand side.)

More than this, Rayleigh showed that even if the layer is unstable, not all scales of motion are unstable. His discussion is in terms of cells, i.e., motions which in cross section are like,

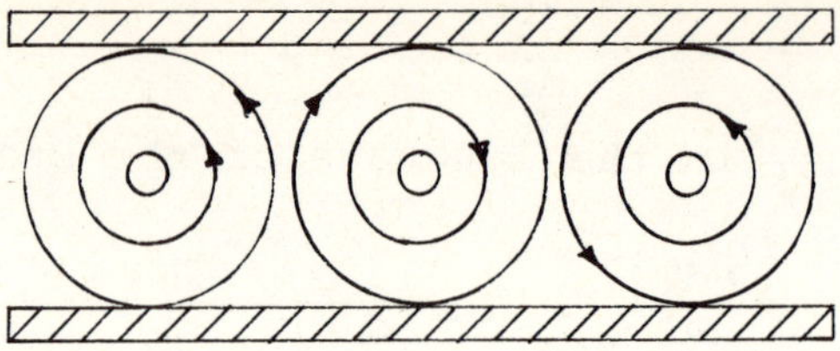

while in plan they may have any of a variety of shapes. People usually discuss only close-packed shapes, i.e., polygons, and these include rolls (which are pure two-dimensional motions), rectangles and hexagons. All of these shapes are equivalent in their instability properties, but different size motions have different instabilities. The usual way of expressing these results is to let $2\pi d \div$ the horizontal scale of the cells = a, and to express the temperature gradient in units $\kappa\nu/g\alpha d^4$; the radiative gradient in these units is

$$R = \frac{g\alpha d^4}{\kappa\nu} \cdot \frac{\Delta T}{d} \tag{48}$$

which is known as the Rayleigh number.

Rayleigh's main results then can be simply expressed graphically--

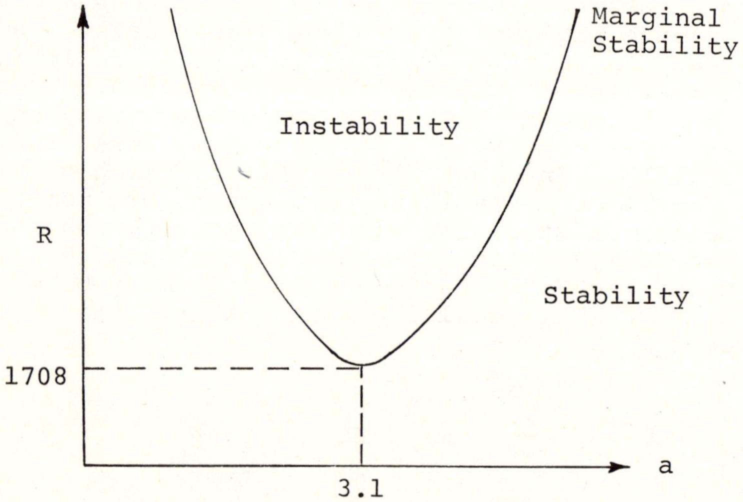

Thus, if $R \geq 1708$ convection can occur, but what happens in detail? Rayleigh's study does not answer this question.

In fact what Rayleigh did was to consider motions of infinitesimal amplitude, so that he was able to linearize the equations of motion. But homogeneous linear equations do not give amplitudes, and Rayleigh's results simply imply that for linear solutions in the instability zone, the amplitudes grow exponentially. Moreover in linear theory we have a superposition principle so that polygons of different shapes and sizes can all begin to grow at once. However, when one does experiments, for R just in excess of 1708 one sees only one polygon (usually a roll) and it is in a steady state (after time is allowed for the steady state to be achieved). This has to be understood.

The first step in understanding the experimental results at high Rayleigh number is to try to find out what the limiting amplitude of a given cell is. This was done by Malkus and Veronis (JFM ∿ 1958) and others. They treated the amplitude of the motion as small and used this as a perturbation parameter. In zeroth order they obtain just the state of conduction without motion. In first order they recover Rayleigh's stability theory. They take as their first order solution a single polygon which is marginally stable and then calculate higher order corrections to it. In this calculation, R is not given, but the value of R corresponding to a given amplitude of motion is to be found. This can be later inverted and for a given R (not too much bigger than 1708), one finds the amplitude. Thus one has solutions for a variety of cell shapes and sizes.

Now one can again perturb these nonlinear solutions and see if they are stable. This has been done by Busse and others for the case of infinite Prandtl number, σ, where

$$\sigma = \frac{\nu}{\kappa} \quad .$$

One finds, in this case, that all cells except some rolls are unstable, and even rolls are unstable for $R \gtrsim 20{,}000$.

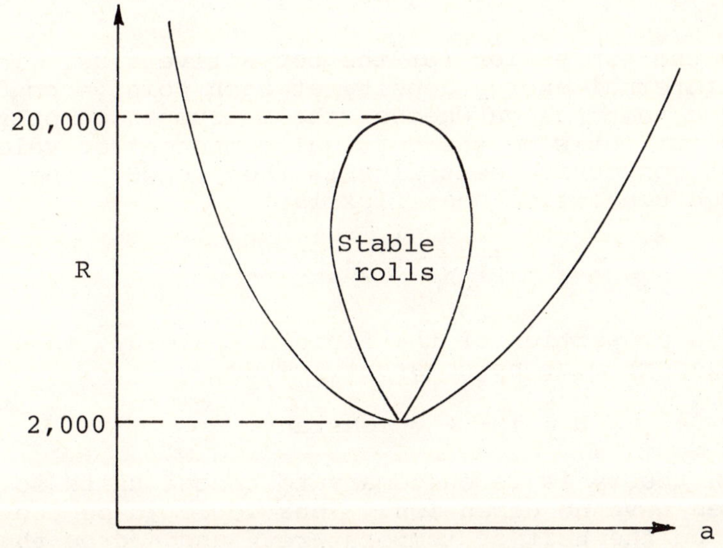

It is also true that for $R \gtrsim 20{,}000$, the rolls are seen experimentally to break down into three dimensional motions. It is not clear how to do theory beyond this point. One method might be to try to take a superposition of two cells as the first order solution, and perturb it further, but the matter rests there for now.

As the Rayleigh number is further increased, the observed motions become time dependent, first in a periodic way, then aperiodically. The Rayleigh number at which this takes place depends on the Prandtl number and this is just being pinned down now, and time scales of variation being obtained experimentally and theoretically. These time-dependent motions may have a bearing on what one sees in the sun, and their understanding may lead to some improvements in our notions about details of solar convection.

But of more immediate relevance to our considerations here are the questions of heat transport and average temperature gradients. To go into this we must first agree on an averaging procedure, and the one that is most widely used is the average over all horizontal coordinates. Thus, we write

$$T(\vec{x},t) = \bar{T}(z,t) + \Theta(\vec{x},t) \tag{49}$$

where $\bar{T}$ is the horizontally averaged temperature. In general, $\bar{T}$ depends on z and t, but in the case discussed here, with fixed boundary conditions, the average quantities show no time dependence as far as one can tell. Fluctuating quantities like Θ, can show marked time-dependence as we have seen.

Now the mean conductive (or radiative) flux is given at each depth by

$$\mathscr{F} = -K\frac{d\bar{T}}{dz} \tag{50}$$

We need also the expression for the convective flux. The motion produces an internal energy density at each point = $\rho c_p\Theta$ (The use of c_p or c_V depends on whether the medium is compressible). At a given point let w be the vertical component of velocity. The local net convective heat flux is then $\rho c_p w\Theta$. The horizontally averaged convective heat flux is

$$\Phi = \overline{\rho c_p w\Theta} \cong \rho c_p \overline{w\Theta}. \tag{51}$$

If the average properties of the fluid are steady, then the total heat transport must be constant. Thus,

$$H = \Phi + \mathscr{F} = \text{constant}. \tag{52}$$

In convection theory it is customary to choose units so that all quantities have no dimension. Thus we let d be the unit of length, ΔT be the unit of temperature, and κ/d be the unit of velocity. Further the Nusselt number,

$$N = \frac{H}{\kappa\frac{\Delta T}{d}}, \tag{53}$$

which measures the ratio of the total heat transport with convection, H, to the conductive heat transport without convection is a useful parameter characterizing the problem. The heat transport relation then gives

$$\overline{w\Theta} + \beta = N \tag{54}$$

where $\beta = -d\bar{T}/dz$, and the quantities on the left hand side are in the units mentioned. In astrophysics we are mainly interested in N and β since if we know β we can integrate into the star. But, unfortunately to get β we need $\overline{w\Theta}$ which means that we must solve the equations of motion, a feat which is not possible in general. From experiments, however, we can get some idea of what happens.

First we have to note that for given boundary temperatures, i.e., for given ΔT, we have specified the Rayleigh number for a given fluid. Also the fluid has a particular Prandtl number. These two quantities are all the input needed for the idealized experiment I described and the question to be decided is, how does N depend on R and σ. For fixed σ we get something like

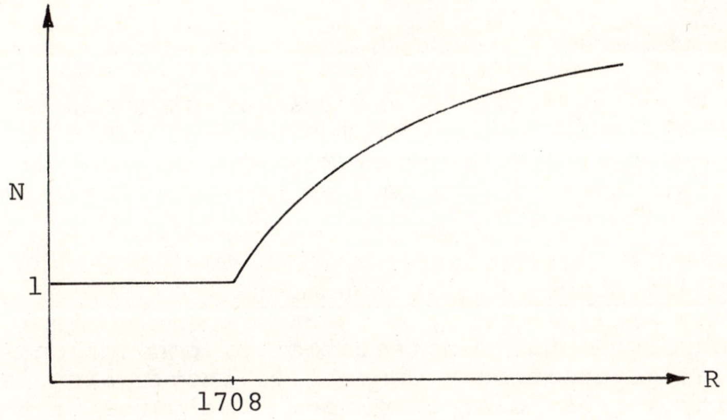

That is for $R < 1708$ there is pure conduction, $N = 1$, while for $R > 1708$, $N > 1$ showing the convective enhancement of the flux. Unfortunately, few good data exist for $R > 10^7$ and, I believe, no reliable data exist for $R > 10^{10}$. Whereas in the sun, for example, the equivalent value of $R \sim 10^{20}$, so that a real need for experiments exists. Such data as we have indicate that as $R \to \infty$, $N \to \text{const.} \times R^{1/3}$. The exponent is not certain and a more reliable value is probably .32; the 1/3 is an approximation to the measured .32 as influenced by theoretical arguments.

There is even less that is known experimentally about the Prandtl number dependence of N. One reason is that the most interesting σ-dependence occurs for $\sigma \ll 1$ while most fluids have $\sigma > 1$. For example $\sigma_{H_2O} \sim 7$, $\sigma_{air} \sim .8$. Some work has been done with mercury for which $\sigma \sim .025$, but this is not extensive enough. On the other hand in the sun $\sigma \sim 10^{-9}$, since the conduction is by radiation while the viscosity is atomic, so that again, we do not have data where we need them most. About the only thing the experiments seem to indicate is that N is practically independent of σ for $\sigma \gtrsim 1$ and tends to fall off with decreasing σ for $\sigma < 1$.

The experiments also give some information for $\bar{T}$. At large R we have something like

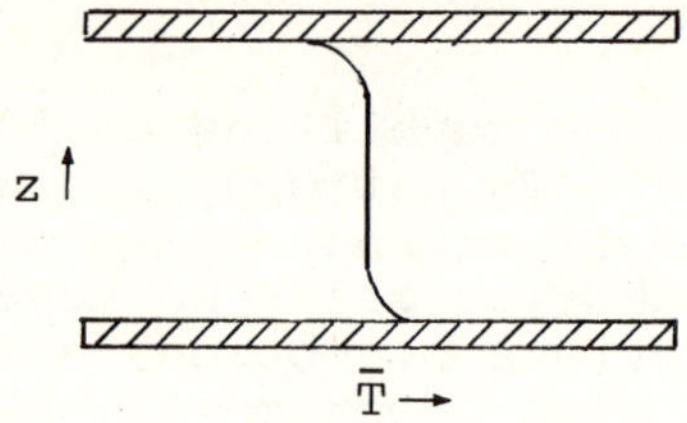

That is, away from the boundaries the fluid is nearly isothermal and near the boundaries we have rapid variation of $\bar{T}$. These regions of rapid variation are called boundary layers and correspond to the transition layers (see below) of stellar convection zones. The isothermal region becomes an adiabatic region in the compressible case.

The form of $\bar{T}(z)$ can be understood quantitatively. We know that

$$N = \beta + \overline{w\Theta}.$$

For large R we have large N and this is produced by a large convective flux $\overline{w\Theta}$. Thus over most of the fluid $N \sim \overline{w\Theta}$ and $\beta = 0$. However at the boundaries $w = 0$, so that in the boundary region $\beta \sim N$. Since $\beta = -d\bar{T}/dz$ we see that the behavior of β gives the nature of $\bar{T}(z)$.

C. Qualitative Solar Convection

An observational manifestation of convection in the sun is the solar granulation. Convective instability is caused by H^-opacity in the solar envelope. Estimates of the depth of the convective envelope range from 1/7 to 1/3 $R_\odot$.

Photographs of the solar surface show a granular appearance. Typical sizes of the granules are about 1000 km although some are seen at all sizes down to the resolution limit of the stratosphere project telescope ($\sim$ 300 km). The density scale height is a few hundred kilometers in this region. Perhaps that is the factor which determines the scale size of the granules.

Leighton and collaborators observed the solar surface with a two-slit spectroheliograph. The slits were arranged to fall on either side of the core of a medium strong line. They could then detect mass motion by noting intensity differences caused by Doppler shifts of the line and by clever photographic techniques, were able to make kinematic pictures of the sun. In such pictures large "granules" called supergranules appear prominently. These have diameters $\sim$30,000 km. but remarkably enough, on ordinary white photographs they do not appear. The supergranules are roughly polygonal and have upward flows in their centers and downward flows at their edges.

The convective activity near the solar surface is important in estimating the heat flux. It also has far-reaching effects on other aspects of the sun, one example of which is the formation of the solar corona which is well above the surface and has a temperature of 10^6 °K. Bierman suggested that the corona is heated by mechanical energy transport of noise generated by turbulence in the photosphere. Noise from turbulence can be of several types. There may be quadrupole noise such as arises from turbulence from jet planes. Vertical oscillation of the convective blobs is a dipole mechanism which gives rise to the type of noise generated by the wind blowing past a telephone wire. Squeezing of the blobs produces monopole noise--the same type of noise that is characteristic of a babbling brook. One then tries to calculate the total noise from all of these effects, as outlined first by Lighthill. Also of importance in the process are magnetic fields on the sun which are mainly confined to the regions between the supergranules, as shown in the figure below. The mean field estimated for the sun has been increasing as the resolution becomes better. Some current estimates of the mean solar magnetic field are 100-500 gauss. This field probably does not extend to the center of the sun because the field flips every 11 years. The diffusion time for a field from the center is 10^9 years so that this would seem to imply that the field is confined to outer layers.

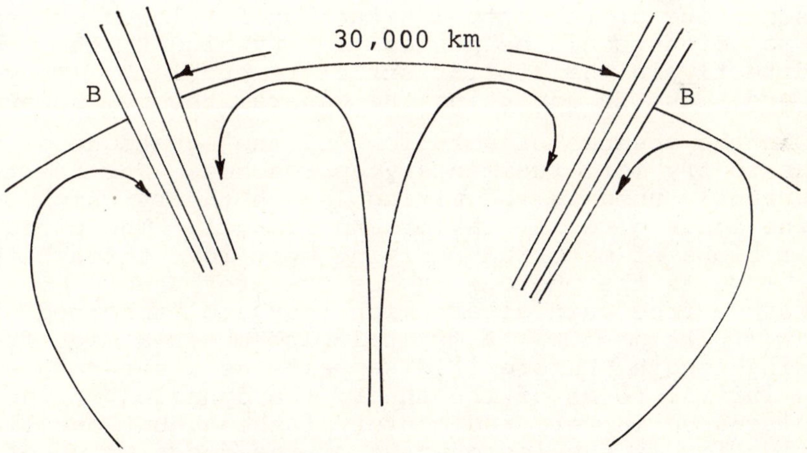

Schematic View of Solar Magnetic Fields

The field seems to be contained by dynamical pressure from the motion of the supergranules. We have that

$$\tfrac{1}{2}\rho v^2 \sim \frac{B^2}{8\pi}$$

Taking $\rho \sim 10^{-7}$ gms/cm^3 and $v \sim 10^5$ cm/sec gives a mean field of about 100 gauss, an order of magnitude agreement with the observed fields. The importance of fields in the generation and propagation of solar waves was discussed by Osterbrock (Ap. J.).

In the qualitative story of coronal heating the noise starts as a small amplitude sound wave propagating outward in the sun. During the motion, the wave conserves ρv^2. The density falls off markedly as the wave moves outward, hence, its velocity increases and it develops into a shock. The shock rapidly dissipates its energy by heating and ionizing the surrounding medium. Low density of the medium results in a large increase in temperature by dissipation of small amounts of energy. Since the material is ionized, it cannot cool itself by radiation so it must expand. In fact, it expands right off the sun causing the solar wind as Parker explained a decade ago. The expanding material pulls the magnetic field with it. Mass loss by the solar wind is small (10^{-13} $M_\odot$/yr) but the associated angular momentum losses are definitely non-neglibible.

D. Mixing-Length Theory

The problem of computing heat transport by convection involves intimately the solution of the equations of motion of a fluid. Naturally, one is tempted to try to achieve this by numerical methods, but even for only moderately large Rayleigh numbers, this has not been achieved, except in the case of two dimensional motion. Unfortunately there are arguments that suggest that two dimensional solutions are not representative of the actual situation for low σ, so even if these solutions could be pushed to very high R, they would not help directly in the stellar structure problem. (They would be of great use in advancing the general theory, however.)

In the absence of solutions of the equations of motion, one has to try more rudimentary approaches. The one that is now currently used is the mixing length theory which comes from the ideas of G. I. Taylor and Prandtl. The picture there is that blobs of material carrying heat move through the ambient medium much as the photons in the transfer theory do. Thus they have a free path after which they are "absorbed" and are mixed with the medium. A typical blob moves a mean free path, or mixing-length, before it disappears as a separate entity. In the initial forms of the theory the dynamics of the blobs were discussed in very rudimentary fashion, but one line of investigation is the improvement of the description of their dynamics. Nevertheless, the theory remains rudimentary and alternate approaches are being sought. Here, however, I should like to discuss the mixing length theory since that is what is normally used in stellar structure calculations.

Suppose that we are interested in convection in an envelope. Then, roughly speaking, curvature is small and we can make the plane-parallel approximation of stellar atmospheres. If z is the vertical coordinate, a blob of mass M has the equation of motion.

$$M\frac{d^2z}{dt^2} = -g(\rho-\bar{\rho})V, \tag{55}$$

where V is the volume of the blob, ρ is its density and $\bar{\rho} = \bar{\rho}(z)$ is the local value of the horizontally averaged density. This equation is already simplified in that it omits viscosity and treats the denisty of the blob as uniform. It has some more subtle omissions such as the neglect of hydrodynamic mass, neglect of exchange of mass with the surroundings, and the assumption that the blob size is much less than $(d\ell n\bar{\rho}/dz)^{-1}$.

We assume as before that the pressure fluctuation is not significant so that

$$\delta p = 0 = \delta\left(\frac{k\rho T}{\mu m_H}\right). \tag{56}$$

For constant μ, we have

$$\delta\rho = -\frac{\rho}{T}\,\delta T \tag{57}$$

If, as is usually assumed, the density fluctuations are not large,

$$\rho - \bar{\rho} = \delta\rho = -\rho\alpha(T - \bar{T}) = -\rho\alpha\Theta \tag{58}$$

where $\alpha = 1/\bar{T}$, $\Theta = T - \bar{T}$. Since $M/V \sim \rho \sim \bar{\rho}$, we find

$$\frac{d^2z}{dt^2} = +g\alpha\Theta \tag{59}$$

to leading order in $\rho - \bar{\rho}$. (The neglect of higher order in $\rho - \bar{\rho}$ is part of what is sometimes called the anelastic approximation.)

We now need an equation for Θ. To get one we recall that for an ideal gas the first law of thermodynamics can be written,

$$\rho c_p \frac{dT}{dt} - \frac{dp}{dt} = \frac{đQ}{dt}. \tag{60}$$

We also recall our decomposition of temperature,

$$T = \bar{T}(z) + \Theta \tag{61}$$

whence,

$$\frac{dT}{dt} = \frac{d\Theta}{dt} + \frac{d\bar{T}}{dz}\frac{dz}{dt}. \tag{62}$$

Likewise, with the decomposition

$$p = \bar{p} + \delta p, \tag{63}$$

we find

$$\frac{dp}{dt} = \frac{d\bar{p}}{dz}\frac{dz}{dt} + \frac{d}{dt}(\delta p). \tag{64}$$

Our procedure has been to neglect δp (Boussinesq approximation) and if we also use the hydrostatic equation for the mean state,

$$\frac{d\bar{p}}{dz} = -g\bar{\rho} \tag{65}$$

we obtain

$$\frac{dp}{dt} = -g\bar{\rho}\frac{dz}{dt} \tag{66}$$

Finally, as is usual in mixing length theory, we take

$$\frac{dQ}{dt} = -\rho c_p q\Theta, \tag{67}$$

where $1/q$ is the radiative lifetime of the blob. This differs from Newton's law of cooling only in that q may depend on the size of the blob. We then find that

$$\frac{d\Theta}{dt} - \beta\frac{dz}{dt} = -q\Theta \tag{68}$$

where

$$\beta = -\left(\frac{d\bar{T}}{dz} + \frac{g}{c_p}\right). \tag{69}$$

At the moment β is not known and we cannot find it until we know the convective heat transport. We proceed as if it were a given function of z, but clearly we have a self-consistency problem; the motions of blobs depend on β and β depends on the motions of blobs. With this proviso we have as our equations of motion

$$\frac{d^2z}{dt^2} = +g\alpha\Theta \tag{70}$$

$$\frac{d\Theta}{dt} - \beta\frac{dz}{dt} = -q\Theta. \tag{71}$$

These equations can be combined into

$$\frac{d^3z}{dt^3} + q\frac{d^2z}{dt^2} - g\alpha\beta\frac{dz}{dt} = 0. \tag{72}$$

This equation has the integral

$$\frac{d^2z}{dt^2} + q\frac{dz}{dt} + g\alpha\left[\overline{T} + \frac{g}{c_p}z\right] = \text{const.} \equiv g\alpha T_o \tag{73}$$

which looks something like the equation of motion of a damped nonlinear oscillator. To see this suppose the temperature, $\overline{T}$, is written as the derivative of some other function; i.e., introduce a "potential" such that

$$\frac{d\psi}{dz} = g\alpha\left(\overline{T} - T_o + \frac{g}{c_p}z\right) \tag{74}$$

which gives our equation the form

$$\frac{d^2z}{dt^2} + q\frac{dz}{dt} + \frac{d\psi}{dz} = 0. \tag{75}$$

Of course, we cannot solve this in general, but if the motion were adiabatic ($q = 0$) there would be another integral of the motion,

$$\tfrac{1}{2}\left(\frac{dz}{dt}\right)^2 + \psi = E. \tag{76}$$

What is the meaning of this oscillator? In a typical convection zone, $\beta > 0$, while in the adjoining layers $\beta < 0$. So we have something like:

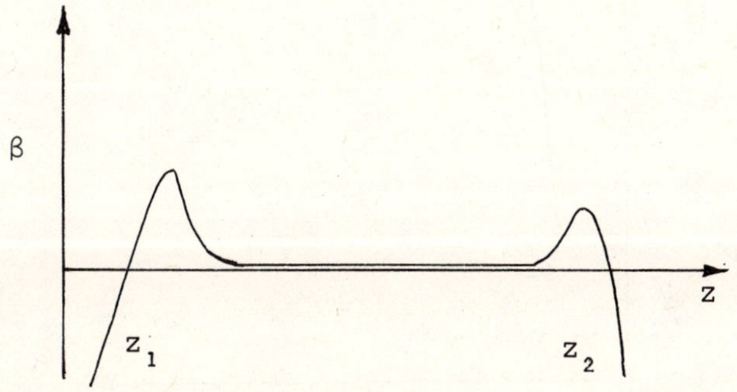

That is in most of the convection zone, the gradient is nearly adiabatic, $\beta \sim 0$ but slightly positive. Near the edges, β is fairly large, just as the lab case. If the blob moves adiabatically it has a variety of possible motions depending on its initial temperature, etc., but the most typical motion is something like what is shown in the following figure,

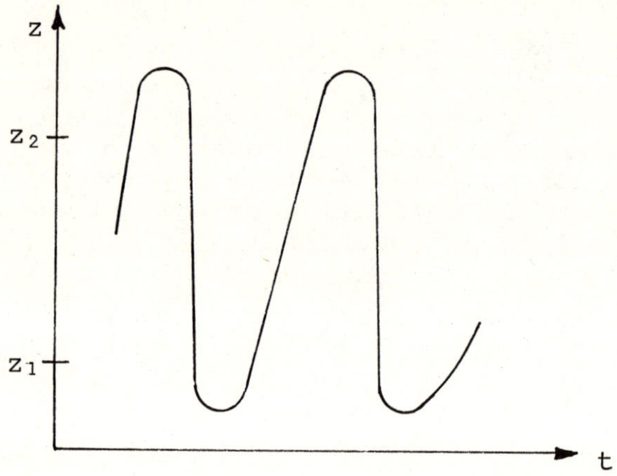

This simply indicates how if the blob starts out in the convection zone and is rising it will shoot upward and into the stable region. There it is turned back by the stable gradient and dips back into the unstable regions. Once again it zips through, only to be turned back again by the lower stable region. To get a better feel for this, it is useful to play around with analytic models; I recommend $\beta = \beta_o\left(1 - \frac{z^2}{L^2}\right)$ for a start (or see Moore and Spiegel, Ap.J.).

If we now put the heat loss term back in, $q \neq 0$, the problem is harder to solve, but tractable. What happens then, typically, is shown in the following figure:

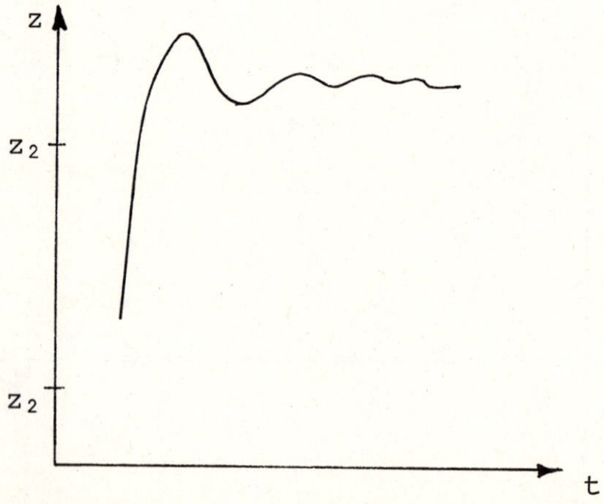

In other words, in the stable region the radiative damping may succeed in trapping the blob in equilibrium. Sometimes this happens after a pass or two more through the unstable zone. Thus the blob in reality does not keep going for long.

If there are many blobs, they will collide and that is not easy to calculate. In the simplest approximation, such an inelastic collision brings them both to rest and they mix with the surroundings and are no longer recognizable. The mean free path for such collisions is the mixing length in this model. It is generally assumed that the mixing length in stellar convection zones is sufficiently small that the blobs do not traverse great variations in temperature. This clearly is not true near the edges of the convection zone, and this approximation is bad in the sense that it is bad to use a diffusion approximation for radiation near the edge of a star. It is possible to remove this approximation by going to a full transfer theory based on the blobs, but that is usually not done.

Let us now see how these notions are actually applied in stellar models. First we estimate Θ from equation (71). We suppose that the blob starts out at z at t = 0 with $\Theta \sim 0$. It moves a distance, ℓ, which is its mixing length, and then disappears. This takes place at time t. From (71) we obtain

$$\Theta(t) = \int_z^{z+\ell} \beta dz - q \int_0^t \Theta dt. \tag{77}$$

Over a distance ℓ we let β be effectively constant, so that

$$\Theta = \beta\ell - q \int_0^t \Theta dt. \tag{78}$$

The last term represents the accumulated change in the blob temperature as it moves a distance ℓ. At this moment we cannot evaluate it, but let us assume that the change in Θ with time can be made equivalent to a change in the blob temperature with z as it moves along. Then, formally, we can write

$$q \int_0^t \Theta dt = -\ell\beta_{blob} = -\ell\left(\frac{dT}{dz} + \frac{g}{c_p}\right) . \tag{79}$$

Thus

$$\Theta = \ell\left(\frac{dT}{dz} - \frac{d\bar{T}}{dz}\right) . \tag{80}$$

Now we consider the vertical velocity w = dz/dt. Multiply (71) by dz/dt and integrate,

$$\frac{1}{2}\left(\frac{dz}{dt}\right)^2 = \frac{1}{2}w^2 = g\alpha \int_0^t \Theta\frac{dz}{dt}dt$$

$$= g\alpha \int_z^{z+\ell} \Theta dz \approx g\alpha\ell^2\left(\frac{dT}{dz} - \frac{d\bar{T}}{dz}\right) ,$$

hence,

$$w^2 = g\alpha\ell^2\left(\frac{dT}{dz} - \frac{d\bar{T}}{dz}\right) \tag{81}$$

(Note that I do not bother to keep factors of 2; they, in fact, accumulate, but I am really too unsure about them to write them in.)

Now we can write the convective heat flux as

$$\Phi = \overline{\rho c_p w\Theta} \simeq +\rho c_p w\ell\left(\frac{dT}{dz} - \frac{d\bar{T}}{dz}\right) . \tag{82}$$

so that the total heat flux is

$$H = -K\frac{d\bar{T}}{dz} + \rho c_p w\ell\left(\frac{dT}{dz} - \frac{d\bar{T}}{dz}\right) . \tag{83}$$

All that we have to do now is to specify the blob temperature gradient. A rough physical argument gives us this. If the blob moved adiabatically we would get from (82)

$$\Phi_{AD} = \rho c_p w\ell\left[\left(\frac{dT}{dz}\right)_{AD} - \frac{d\bar{T}}{dz}\right], \tag{84}$$

thus the loss in convective heat flux due to nonadiabatic effects is

$$\Delta\Phi = \Phi - \Phi_{AD} = \rho c_p w\ell\left[\frac{dT}{dz} - \left(\frac{dT}{dz}\right)_{AD}\right] . \tag{85}$$

This loss in flux is a result of the energy leak from the blob, which is proportional to the area of the blob. If the size of the blob $\sim$L the leak rate is

$$\left(K\frac{\Theta}{L}\right)L^2 ,$$

and to convert this to a flux we divide by an area. Thus,

$$\Delta\Phi = K\frac{\Theta}{L} \sim \frac{K\ell}{L}\left(\frac{dt}{dz} - \frac{d\bar{T}}{dz}\right) . \tag{86}$$

If we assume $L \sim \ell$ and equate the two expressions for $\Delta\Phi$, we find,

$$\frac{\frac{dT}{dz} - \frac{d\bar{T}}{dz}}{\left(\frac{dT}{dz}\right)_{AD} - \frac{dT}{dz}} = \frac{\rho c_p w \ell}{K} \quad , \tag{87}$$

where the quantity on the right-hand side is called the Péclet number of the blob. These equations constitute the Vitense theory currently in use in stellar structure theory (apart from some numerical factors and some improvements introduced by various authors). The procedure is clearly approximate, and there are some inconsistencies in it. On the other hand, no one has yet produced a scheme which is simple and which is markedly better.

One of the main difficulties which still exist is in the choice of ℓ, which is both the mean free path and the size of the blobs. The natural procedure might seem to be to apply the theory to laboratory case and to see what choice of ℓ gives agreement. But this does not appear to work. The reason is that in the laboratory convection the size of the system is much less than a scale height. That means that ℓ is limited by the geometry of the apparatus; for example, might be taken as the distance to the nearest boundary. In stars, however, it may well be that ℓ is limited by other considerations.

For example, Bierman has argued that a blob cannot maintain integrity if it extends over more than a scale height of pressure, so that $\ell \sim (d \ell n\ p/dz)^{-1}$. Schwarzschild, on the other hand, prefers the density scale height. Other choices have been offered. In any case, once you have chosen the intrinsic scale of the system which fixes the mixing length, you can set ℓ proportional to this and choose the proportionality constant to make a solar model agree with the sun. You then have a determinate system with which to calculate other models. The results of this will no doubt be presented by some other lecturers.

17

THERMONUCLEAR REACTIONS AND NUCLEOSYNTHESIS*

J. W. Truran†

California Institute of Technology
Pasadena, California

I. INTRODUCTION

The subject of nucleosynthesis has received considerable attention over the past decade. Detailed studies of the abundance distributions realized in various thermonuclear burning processes have been carried out, built upon improved determinations of the appropriate nuclear cross sections. Several possible sites of element synthesis have been considered.

a) stellar interiors

Burbidge, Burbidge, Fowler and Hoyle (1957) first discussed in detail the various nuclear processes which participate in the synthesis of elements in stellar interiors. According to their model, charged-particle reactions are mainly responsible for element production through the vicinity of iron, beyond which neutron capture becomes the predominant mechanism. Subsequent studies by these authors and by Cameron (1957; 1963) and others have modified various aspects of this picture.

It is clear that the production of all of the heavier elements ($A \lesssim 40$) cannot take place during an extended stable phase of stellar evolution. In the interior of a star, nuclear-energy generation will cease when nuclei of minimum energy are produced. In this respect the iron-group nuclei are the most stable, ^{56}Fe having the maximum binding energy per nucleon of any nucleus. An increase in the temperature of a stellar core composed primarily of iron will result in the endoenergetic breakdown of iron to helium. This absorption of energy by the system can lead to gravitational collapse. Furthermore, energy losses by neutrino and antineutrino emission may be sufficient to initiate the collapse prior to the realization of an iron equilibrium peak. These considerations suggest

*Supported in part by the National Science Foundation [GP-9433, GP-9114] and the Office of Naval Research (Nonr-220(47)].

†Current address: Belfer Graduate School of Science, Yeshiva University, New York, N.Y.

that some other mechanism must be responsible for the production of the heavier elements ($A > 60$), and perhaps for the intermediate mass nuclei ($20 \lesssim A \lesssim 44$) and the iron group elements as well.

b) the "big bang"

Interest in the subject of the universal synthesis of the elements has recently been revived by the discovery of Penzias and Wilson (1965) of evidence for a universal background temperature of approximately 3°K. Subsequent calculations have demonstrated that the production of a helium mass fraction of 0.20-0.30 is possible (Peebles 1966; Greenstein 1968; Wagoner, Fowler and Hoyle 1967). Substantial abundances of ^{2}D, ^{3}He and ^{7}Li may also be produced in a universal fireball. However, for a present day temperature of 3°K and a present day universal density of 10^{-28}gm/cm^{-3} (this upper limit being inferred from cosmological red-shift observations), the universal synthesis of the heavy elements in the big bang is not possible (Wagoner *et al.* 1967). The production of heavier elements would require either a significantly higher present day density or a lower present day temperature.

c) super-massive objects

Fowler and Hoyle (1964) and Wagoner *et al.* (1967) have considered the possible synthesis of heavy elements in extremely massive stars, $M \gtrsim 10^3 M_\odot$. The existence of such objects in substantial numbers during the early history of the galaxy was proposed in an attempt to explain the small abundances of heavy elements (of the order of one percent of the solar value) observed in the oldest Population II stars. While the heavy element abundances produced in this manner were found to agree quantitatively with the observed mass fraction ($\approx$ 1 percent of solar values), the relative elemental and isotopic abundances were found to be quite different from the solar values. This suggests again that some other means of heavy element production may be necessary. A helium mass fraction of 0.2-0.4 may also be realized in supermassive objects which have bounced through peak temperature conditions $T_9 \lesssim 20$.

d) supernova explosions

The difficulties encountered in the attempts to account for the observed heavy element abundances in the galaxy by either of the previous mechanisms tend to support the view that the main source of heavy element synthesis is stellar evolution through the entire history of the galaxy, with gradual enrichment of the interstellar gas by the material ejected in supernova explosions. Further, the generally endoenergetic character of the various processes of heavy element synthesis suggests that the supernova event might itself be the site of substantial heavy element production.

Recent studies of supernova hydrodynamics have been concerned with two specific mechanisms: thermonuclear explosions (Fowler and Hoyle 1964; Arnett 1969a; Hansen and Wheeler 1969) and the neutrino energy-transport mechanism (Colgate and White 1966; Arnett 1966, 1967; Schwartz 1967). Considerations of nucleosynthesis for supernova explosions triggered by thermonuclear processes are extremely model dependent. Specifically, the resulting abundances will depend in a sensitive manner both upon the run of temperature and density through the star and upon the particular nuclear fuel involved (presumably ^{12}C, ^{16}O, or ^{28}Si will provide the required energy).

For supernova models based upon the neutrino-transport mechanism, the temperature and density conditions predicted for much of the ejected material are somewhat more restricted. The density of the regions immediately surrounding the core (at the"mass cut") must be large enough to enable neutrinos leaving the core to interact before escape, and thus to heat this material sufficiently to provide expansion velocities in excess of the escape velocity. Recent studies of the dynamics of the cores of highly evolved massive stars (Colgate and White 1966; Arnett 1966, 1967) predict conditions for the ejected material of $\rho > 10^{10}$g/cc and $T > 10^{10}$°K. In both investigations, it was found that a considerable fraction of the core mass should be ejected.

These lectures will deal primarily with problems associated with nucleosynthesis in supernovae. Specifically, Section III presents the results of a recent examination of the nucleosynthesis conditions predicted by current hydrodynamic models of the supernova mechanism (Arnett and Truran 1970; Arnett, Truran and Woosley 1970). In Section Four (IV), the possible synthesis of intermediate mass nuclei ($28 \lesssim A \lesssim 48$) by an explosive oxygen burning process is considered (Truran and Arnett, 1970a). The methods employed in obtaining estimates of the thermonuclear reaction rates appropriate to these nucleosynthesis studies are outlined in the second section. It should be emphasized that no attempt has been made to survey recent studies of other mechanisms of nucleosynthesis. For this purpose, the reader is referred to the review articles on nuclear astrophysics by Cameron (1958) and Burbidge (1962) and to the extensive bibliography of this subject compiled by Kuchowicz (1968), as well as to the recent literature.

II. THERMONUCLEAR REACTION RATES

In this section we shall be concerned with the determination of reaction rates for reactions proceeding under the conditions prevailing in stellar interiors. The number of reactions per unit volume per second, r, of two nuclear species with number densities n_1 and n_2 can be written in the form:

$$r = n_1 n_2 \quad \langle\sigma v\rangle \qquad (1)$$

Here $\langle\sigma v\rangle$ is an appropriate average of the product of the reaction cross section, $\sigma(v)$, and the relative velocities of the nuclei, v,

$$\langle\sigma v\rangle = \frac{\int_{v_t}^{\infty} \sigma(v)\ v\ N(v)\ dv}{\int_0^{\infty} N(v)dv} \qquad (2)$$

where v_t is the threshold velocity for the reaction and N(v) is the number density of nuclei having relative velocities v. Assuming that the velocity distributions of the two species are Maxwellian, the above equation for $\langle\sigma v\rangle$ takes the form:

$$\langle\sigma v\rangle = \left(\frac{8}{\pi\mu}\right)^{\frac{1}{2}} (kT)^{-3/2} \int_{E_t}^{\infty} E\sigma(E)\exp\left(-\frac{E}{kT}\right) dE \qquad (3)$$

In this expression k is the Boltzman constant, T is the temperature, E is the kinetic energy of relative motion, E_t is the threshold energy, and μ is the reduced mass:

$$\mu = \frac{m_1 m_2}{m_1 + m_2} \qquad (4)$$

The evaluation of this expression for $\langle\sigma v\rangle$ will be the subject of this section. As an introduction to this study we will first consider several limiting cases of particular interest in astrophysics. Numerous examples of the application of these formulas may be found in Fowler, Caughlan and Zimmerman (1967).

1. Non-resonant reactions, charged particles

At low bombarding energies and far away from resonances the cross sections for reactions involving charged particles can be written in the form

$$\sigma(E) = \frac{S(E)}{E}\exp(-2\pi\eta) \qquad (5)$$

where S(E) is typically a slowly varying function of energy and the Coulomb parameter, η, is expressed in terms of the energy, the reduced mass and the charges of the two species:

$$\eta = \frac{Z_1 Z_2 e^2}{\hbar}\left(\frac{\mu}{2E}\right)^{\frac{1}{2}} \qquad (6)$$

The term $[\exp(-2\pi\eta)]$ is the usual Gamow penetration factor. In this approximation $\langle\sigma v\rangle$ can be written as:

$$\langle\sigma v\rangle = \left(\frac{8}{\pi\mu}\right)^{\frac{1}{2}} \frac{1}{(kT)^{3/2}} \int_{E_t}^{\infty} \exp\left(-\frac{E}{kT} - \frac{(2\mu)^{\frac{1}{2}}\ \pi Z_1 Z_2 e^2}{\hbar\ E^{\frac{1}{2}}}\right) S(E)dE \qquad (7)$$

The factor, S(E), can be determined by a fit to the experimental cross section. For non-resonant interactions, S(E) is usually expressed in the form

$$S(E) = S(O) + \left\langle \frac{ds}{dE} \right\rangle E + \ldots\ldots \qquad (7a)$$

In some instances, both the energy independent term S(O) and the slope <dS/dE> can be inferred from experimental measurements. For many cases of interest in astrophysics, however, the expression S(E) = S(O) must be used.

Equation (7) is not integrable in its present form. We note that the integrand is a sharply peaked function: at low energies the Coulomb penetration probability increases rapidly while at high energies the Maxwell distribution falls off exponentially. The usual procedure (Salpeter 1952) is to approximate the integrand by a Gaussian form, which is integrable. The maximum of the integrand defines the 'optimum bombarding energy'

$$E_o = 0.122 \left[Z_1^2 \; Z_2^2 \; \mu T_9^2 \right]^{1/3} \; \text{MeV.} \qquad (8)$$

where μ is in amu and T_9 is the temperature in units of 10^9 °K. The full width at half maximum of the Gaussian peak is given by:

$$\Delta E_o = .198 \left[Z_1^2 \; Z_2^2 \; \mu T_9^5 \right]^{1/6} \; \text{MeV.} \qquad (9)$$

This defines the energy region from which we can expect the maximum contribution to the integrand. Substituting equation (7a) into equation (7), we obtain an expression for $<\sigma v>$ in terms of these parameters

$$<\sigma v> = \left(\frac{2}{\mu}\right)^{\frac{1}{2}} \frac{1}{(kT)^{3/2}} \Delta E_o e^{-3E_o/kT} \; S_{eff} \qquad (10)$$

where

$$S_{eff} = S(O) \left[1 + \frac{5}{36} \frac{kT}{E_o} + \frac{<dS/dE>}{S(O)} \left\{ E_o + \frac{35}{36} kT \right\} + \ldots\ldots \right] \qquad (10a)$$

Reeves and Salpeter (1959) have improved the approximation to the cross section, equation (5), with the inclusion of one further term in the expansion of the Coulomb penetrability factor:

$$\sigma(E) = \frac{S(E)}{E} \exp(-2\pi\eta - y^{3/2}/6\eta^2) \qquad (11)$$

where

$$y^{3/2}/6\eta^2 = 0.122 [\mu R^3 E^2 / Z_1 Z_2]^{\frac{1}{2}} \qquad (11a)$$

and R is the nuclear interaction radius. This correction term improves the approximation at somewhat higher bombarding energies. Proceeding as before we can define a corrected optimum bombarding energy E_0' and width $\Delta E_0'$ in terms of the parameter

$$\varepsilon = 1.05 \times 10^{-2} (\mu T_9^2 R^3/Z_1 Z_2)^{1/2} \quad (12)$$

where R is expressed in fermis:

$$E_0' = E_0(1 - 2\varepsilon/3) \quad (13)$$

$$\Delta E_0' = \Delta E_0(1 - 5\varepsilon/6) \quad (14)$$

These approximations are found to be extremely good for interactions involving light nuclei at low energies. They find wide application in astrophysics due to the fact that hydrogen and helium thermonuclear reactions predominate for most of the lifetime of a star. It should be emphasized that these approximations have meaning only when the behavior of the nuclear cross sections is determined by the Gamow factor. At higher energies the partial widths for charged particle emission may dominate the neutron and photodisintegration partial widths. Under these conditions, the cross sections for (p,γ), (α,γ), (p,n) and (α,n) reactions are proportional to Γ_n and Γ_γ and the concept of an optimum bombarding energy as defined above loses its meaning.

These equations are restricted to regions far away from a resonance. The presence of a resonance can increase the cross section significantly over that value predicted by the non-resonant formula. The increased nuclear level densities for heavier nuclei suggest that the non-resonant approximation will no longer be applicable.

2. Resonant reactions, charged particles and neutrons

In the vicinity of a resonance the reaction rate can be determined from a knowledge of the resonance parameters. The cross section at resonance is assumed to be well represented by the Breit-Wigner single level formula. Specifically, consider a resonance of energy E_r and spin J. Designating the widths for the entrance and exit channels by Γ_α and Γ_β respectively, the cross section is given by:

$$\sigma_{\alpha\beta}(E) = \pi \lambdabar_\alpha^2 \, g \, \frac{\Gamma_\alpha \Gamma_\beta}{(E - E_r)^2 + (\Gamma/2)^2} \quad (15)$$

In this expression $\lambdabar_\alpha$ is the reduced de Broglie wavelength for the incident channel, Γ is the total width for the decay of this compound nuclear state and g is the statistical factor

$$g = \frac{(2J + 1)}{(2s_\alpha + 1)(2I_\alpha + 1)} \quad (16)$$

where s_α is the spin of the bombarding particle and I_α is the spin of the target nucleus. In general Γ_α and Γ_β represent summations over all values of the orbital angular momentum ℓ consistent with spin and parity restrictions.

Substituting into equation (3) we obtain the following expression for $\langle\sigma v\rangle$:

$$\langle\sigma v\rangle = \left(\frac{8}{\pi\mu}\right)^{\frac{1}{2}} \frac{\pi g}{(kT)^{3/2}} \int_{E_t}^{\infty} \frac{\lambda_\alpha^2\ \Gamma_\alpha \Gamma_\beta}{(E-E_r)^2 + (\Gamma/2)^2}\, e^{-E/kT}\, EdE \qquad (17)$$

If we assume that the resonance is extremely sharp, we can approximate those terms in the integrand which vary slowly with energy by their value at resonance, hence:

$$\langle\sigma v\rangle = \left(\frac{8}{\pi\mu}\right)^{\frac{1}{2}} \frac{\pi g}{(kT)^{3/2}}\, ƛ_\alpha^2(E_r) E_r e^{-E_r/kT}\ \Gamma_\alpha \Gamma_\beta \int_{E_t}^{\infty} \frac{dE}{(E-E_r)^2 + (\Gamma/2)^2} \qquad (18)$$

As the contributions from the resonance tails are small, we can extend the limits of integration from $-\infty$ to ∞ and evaluate the integral by the method of residues:

$$\int_{E_t}^{\infty} \frac{dE}{(E-E_r)^2 + (\Gamma/2)^2} \sim \int_{-\infty}^{\infty} \frac{dE}{(E-E_r)^2 + (\Gamma/2)^2} = \frac{2\pi}{\Gamma} \qquad (19)$$

Thus the contribution to the rate at resonance is given by:

$$\langle\sigma v\rangle = \left(\frac{32}{\mu}\right)^{\frac{1}{2}} \left(\frac{\pi}{kT}\right)^{3/2} ƛ_\alpha^2(E_r) E_r\ e^{-E_r/kT} \left(\frac{g\Gamma_\alpha \Gamma_\beta}{\Gamma}\right) E_r \qquad (20)$$

This indicates that resonance contributions to the rates can be determined from experimental values of $(g\Gamma_\alpha\Gamma_\beta/\Gamma)$ at resonance. If there are many resonances occurring in the energy region of interest for which experimental values of the resonance parameters are available, the rates can be computed as a sum of these contributions.

3. Non-resonant reactions, neutrons

The cross sections for neutron reactions at low energies are well represented in the vicinity of a resonance by the Breit-Wigner single level formula

$$\sigma_{n,\beta} = \pi\lambda_n^2 g\, \frac{\Gamma_n\ \Gamma_\beta}{(E-E_r) + (\Gamma/2)} \qquad (21)$$

In this expression Γ_β ($=\Gamma_\gamma$, Γ_p or Γ_α) is a representative partial width and Γ is the total width. For a number of reactions involving neutrons on light nuclei the resonance energy,

E_r, will be that of the lowest excited state and will be such that

$$|E_r| >> E \sim kT \tag{22}$$

In this limit, the neutron cross section (assuming s-waves dominate) will obey the (1/v) law, viz:

$$\left.\begin{aligned} &\lambda\!\!\!^{-}{}_n^2 \propto \frac{1}{E} \propto \frac{1}{v^2} \\ &\Gamma_n(\text{s-wave neutrons}) \propto v \\ &\Gamma_\beta = \Gamma_\beta(E + Q_{n\beta}) \approx \Gamma_\beta(Q_{n\beta}) \approx \text{constant} \end{aligned}\right\} \tag{23}$$

$$\sigma \approx \frac{\pi\lambda_n^2\,\Gamma_n\Gamma_\beta}{E_r^{\,2} + (\Gamma/2)^2} \propto \left(\frac{1}{v^2}\right)v \propto \frac{1}{v} \tag{24}$$

In this limit, our expression for $\langle\sigma v\rangle$ gives

$$\langle\sigma v\rangle_{n,\beta} = \text{constant}$$

One can demonstrate as well that for the cases σ = constant and $\sigma \propto 1/E$, the reaction rate $\langle\sigma v\rangle$ = constant.

It should be emphasized that the (1/v) approximation discussed above is strictly valid only when the first excited state is far above the thermal region. At somewhat higher neutron energies, particularly for intermediate mass nuclei, this restriction is not met and the cross section behavior will be dominated by the resonances which fall in the thermal energy region. The reaction rates for these cases must be determined as sums over the various resonance contributions (equation 20). For still more massive nuclei, many resonances may appear in the thermal energy region. When the averaged cross section dependence obeys $\sigma \propto 1/E$, the constancy of the calculated rates, $\langle\sigma v\rangle$, with energy evidenced in the thermal region may again be realized.

4. Averaged cross sections

For thermonuclear reactions involving medium and heavy nuclei and proceeding at temperatures in excess of $T_9 \approx 1$, many closely spaced levels may appear in the appropriate energy range in the compound nucleus. In some instances, sufficient experimental information is available concerning the strengths of the individual resonances to enable the reaction rates to be computed as sums of resonance contributions (Fowler, Caughlan and Zimmerman 1967). Unfortunately, both studies of oxygen burning and silicon burning nucleosynthesis and studies of neutron capture processes involve many nuclei for which no cross section data is available. In this section we will consider the problems involved in attempts to estimate these reaction rates.

We shall assume that the cross section due to an individual resonance is well represented by the Breit-Wigner single level formula. The cross section for a reaction proceeding from channel α to channel β via a compound nuclear state of energy U and spin J is therefore (equation 15)

$$\sigma_{\alpha\beta}(E_\alpha, J) = \pi \lambda\!\!\!^{-2}_{\alpha} g_\alpha \frac{\Gamma_\alpha \Gamma_\beta}{(E_\alpha - E_r)^2 + (\Gamma/2)^2} \tag{25}$$

This expression is valid for either a particle-particle reaction or a particle-capture reaction ($\Gamma_\beta \equiv \Gamma_\gamma$). The particle widths represent summations over all values of the orbital angular momentum ℓ_α and channel spin S_α consistent with spin and parity restrictions (for convenience in notation only, the parity dependence is not explicitly contained in these formulas):

$$\Gamma_\alpha = \sum_{S_\alpha = |I_\alpha - s_\alpha|}^{I_\alpha + s_\alpha} \; \sum_{\ell_\alpha = |J - S_\alpha|}^{J + S_\alpha} \Gamma_\alpha(\ell_\alpha, E_\alpha) \tag{26}$$

We are interested in the case for which many resonances contribute to the cross section in a narrow energy range. Taking the average level spacing to be D(U, J), where $U = Q_\alpha + E_\alpha$ is the energy of excitation of the compound nucleus, the average cross section is

$$\begin{aligned} \overline{\sigma_{\alpha\beta}}(E_\alpha, J) &= \frac{1}{D(U, J)} \pi g_\alpha \int_{\Delta E} \frac{\Gamma_\alpha \Gamma_\beta}{(E - E_r)^2 + (\Gamma/2)^2} dE_\alpha \\ &\approx 2\pi^2 \lambda\!\!\!^{-2}_{\alpha} g_\alpha \left(\frac{\Gamma_\alpha \Gamma_\beta}{D(U, J)\Gamma} \right)_{E_r} \end{aligned} \tag{27}$$

Here we have assumed that an individual resonance is extremely sharp, hence that only the resonance denominator varies rapidly over the resonance peak (equation 19). The total 'average' cross section as a function of energy is obtained by summing over all values of the compound nuclear spin (and parity). The reaction rates can then be determined as

$$\begin{aligned} \langle\sigma v\rangle_{\alpha\beta} &= \frac{(2.51 \times 10^{-13})}{(\mu_\alpha T_9)^{3/2}} \frac{1}{(2s_\alpha + 1)(2I_\alpha + 1)} \times \\ &\sum_{J=0}^{\infty} (2J + 1) \int_{\varepsilon_\alpha}^{\infty} dE_\alpha e^{-E_\alpha/kT} \left(\frac{\Gamma_\alpha \Gamma_\alpha}{D(U, J)\Gamma} \right) \text{cm}^3\,\text{sec}^{-1} \end{aligned} \tag{28}$$

This equation provides a general expression for the rates of particle-particle and particle-capture reactions as a function of temperature. In the rest of this section we will be concerned with the extent to which this relation can be used to provide realistic estimates of thermonuclear reaction rates. The nuclear parameters which must be estimated are the partial widths for emission of neutrons, protons and α-particles and the radiation widths.

a) Partial widths for particle emission

The average partial widths for emission of neutrons, protons and α-particles of orbital angular momentum ℓ can be expressed in the form (Vogt 1968)

$$\frac{\langle\Gamma_\ell\rangle}{D} = 2(fP_\ell)\,\frac{\langle\gamma_\ell^2\rangle}{D} \tag{29}$$

where P_ℓ is the coulomb penetration factor

$$P_\ell = \rho/\,[F_\ell^{\,2}(\rho,\,\eta) + G_\ell^{\,2}(\rho,\,\eta)] \tag{30}$$

D is the average level distance and $\langle\gamma_\ell^{\,2}\rangle/D$ is the strength function. The regular and irregular solutions of the Coulomb wave equation, F_ℓ and G_ℓ, are expressed in terms of the parameters η (Equation 6) and $\rho = kR$. The strength function may be expressed in terms of the penetration factor and the transmission coefficient (T_ℓ) as

$$\frac{\langle\gamma_\ell^{\,2}\rangle}{D} = \frac{T_\ell}{4\pi fP_\ell} \tag{31}$$

where, from continuum theory assumptions,

$$T_\ell = \frac{4KR\ fP_\ell}{S_\ell^{\,2} + (kR + fP_\ell)^2} \tag{32}$$

and f is a reflection factor which takes into account the diffuse surface of the nucleus (Vogt, Michaud and Reeves 1965). In the above equation for the transmission coefficient S_ℓ is the shift factor (the primes denote differentiation with respect to ρ)

$$S_\ell = P_\ell[F_\ell(\rho,\,\eta)F_\ell'(\rho,\,\eta) + G_\ell(\rho,\,\eta)G_\ell'(\rho,\,\eta)] \tag{33}$$

R is the nuclear radius and K is the total wave number

$$K^2 = K_o^{\,2} + k^2 \tag{34}$$

where K_o is the wave number associated with the mutual interaction potential V_o, $K_o = (2\mu V_o)^{\frac{1}{2}}/\hbar$, and k is the wave number associated with the relative kinetic energy.

b) Nuclear radiation widths

On the assumption that electric-dipole transitions dominate, the nuclear radiation width for a state of energy E and spin J can be written in the form (Blatt and Weisskopf 1952)

$$\Gamma_\gamma(E,\,J) = 0.296\,\frac{A^{2/3}}{D_o}\sum_{J'}\frac{1}{\rho(E,\,J)}\int_0^E E'^{\,3}\rho(E - E',\,J')dE' \tag{35}$$

where A is the mass number and D_0 is effectively a normalization factor (although it may be interpreted in some sense as the single particle level spacing). If we assume that the angular momentum dependence of the nuclear level density is well represented by $\rho(E, J) = (2J + 1)\rho(E)$, the summation over allowed values of J' consistent with electric-dipole selection rules ($\Delta J = \pm 1, 0$; not $0 \rightarrow 0$; parity change) introduces a factor 3 into the expression for the width, hence

$$\Gamma_\gamma(E, J) = 0.89 \frac{A^{2/3}}{D_0} \frac{1}{\rho(E)} \int_0^E E'^3 \rho(E - E') dE' \quad \text{MeV} \tag{36}$$

Employing the level density prescription of Gilbert and Cameron (1965), the widths have been calculated from this expression and compared to experimental values taken from the Nuclear Data Sheets and from Hughes, Magurno and Brussel (1960). The values of D_0 inferred from these studies (Truran, Hansen, Cameron and Gilbert 1966) are: $D_0 \approx 200$ for nuclei in the mass range $A > 40$ and $D_0 \approx 400$ for $A \lesssim 40$. The ratio Γ_γ (calculated)/ Γ_γ (experimental) is plotted as a function of mass number in Fig. 1. Recent studies indicate that a value $D_0 \approx 800$ is more appropriate for the low mass region, $A \leq 40$.

These average estimates of the radiation widths clearly fail to reproduce the experimental values at the neutron closed shell positions N = 50 and 82 ($A \approx 90$ and 140). This situation is not improved when one calculates radiation widths in the context of the giant dipole resonance model (Axel 1962). The values of Γ_γ (calculated)/ Γ_γ (experimental) determined by Oliva and Prosperi (1967) show even more pronounced discontinuities at the closed shell positions due, primarily, to uncertainties in their estimates of the giant resonance parameters.

c) Choice of the nuclear radius

The partial widths for charged particle emission are extremely sensitive to the choice of the nuclear radius. The appropriate radii for the proton and alpha-particle interactions may be estimated for the cases of interest in astrophysics in the following manner. The cross sections for a number of 'threshold' (p, n) and (α, n) reactions have been determined experimentally in the energy range immediately above the reaction threshold (Johnson 1968; Stelson and McGowan 1964). For these cases, the partial widths for neutron emission generally dominate the total widths, and the (p, n) and (α, n) cross sections will closely approximate the total proton and alpha-particle cross sections. This provides a rather direct measure of the nuclear radius in the appropriate energy range.

In Fig. 2, the total proton cross section

$$\sigma_p^{\text{total}} = \pi \lambda\!\!\!^{-}{}_p^2 \sum_{\ell=0}^{\infty} (2\ell + 1) T_\ell \tag{37}$$

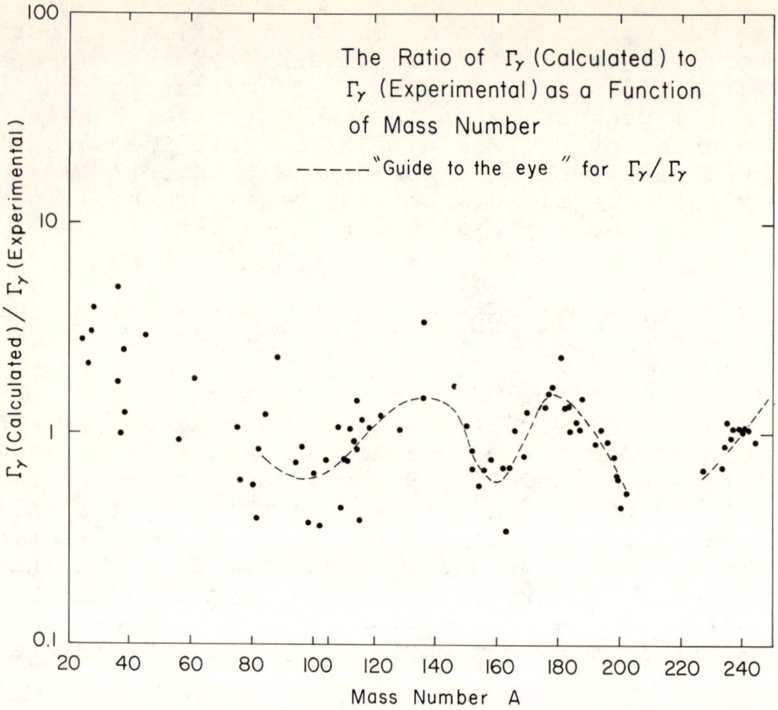

Figure 1. The ratio Γ_γ(calculated)/Γ_γ(experimental) is plotted as a function of mass number. From Truran, Hansen, Cameron, and Gilbert, Can. J. Phys., 44, 151 (1966).

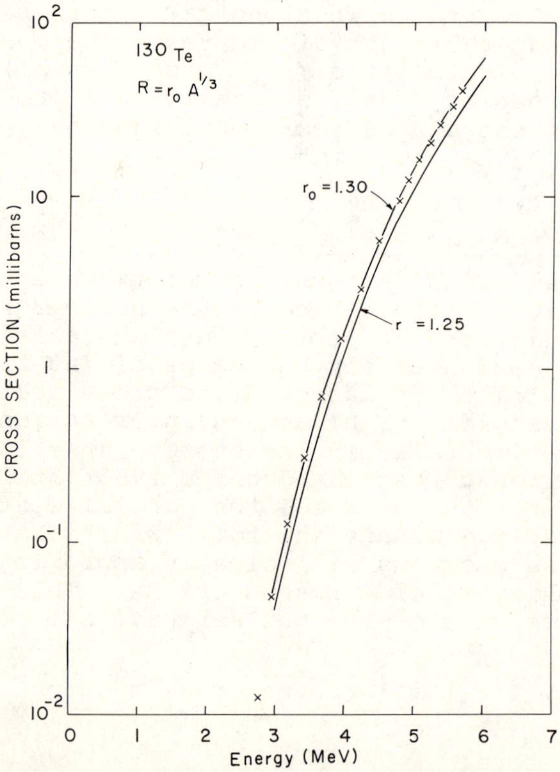

Figure 2. The total proton cross section for ^{130}Te, calculated for two choices of the nuclear radius parameter, is compared to the experimental ^{130}Te(p,n)^{130}I cross section (Johnson 1968). A reflection factor f = 3.8 was used in both cases.

is compared to the experimental $^{130}Te(p,n)^{130}I$ cross section (Johnson 1968) for two choices of the nuclear radius. For both cases a reflection-factor of 3.8 was employed. The reflection-factor is given approximately by (Peaslee 1957; Vogt et al. 1965)

$$f = \pi K_0 a \coth(\pi K_0 a) \qquad (38)$$

where a is the surface thickness and K_0 the wave number corresponding to the well depth V_0. The value of f = 3.8 used here is that predicted for a potential V_0 = 50 MeV and a surface thickness a = 0.65 fermi.

It is clear from Fig. 2 that the choice for a proton channel radius $R = 1.30A^{1/3}$ provides an extremely good fit to the experimental (p, n) cross section. This is true as well for a variety of other cases as is illustrated in Figs. 3 and 4. The good agreement over a range of mass number $65 \leq A \leq 130$ gives us confidence that the use of these factors for masses $A < 60$ should not introduce large errors. (Similar comparisons carried out for masses $A < 60$ are complicated by the presence of resonance structure in the (p, n) cross sections (Truran, Hansen, Cameron and Gilbert 1966).)

A similar analysis has been carried out to determine the alpha-particle channel radius, using the experimental (α, n) cross sections of Stelson and McGowen (1964). The situation here is somewhat more complex as, for a number of cases, the proton partial widths are comparable to or larger than the neutron widths. The best choice for the alpha-channel radius was found to be

$$R = 1.25A^{1/3} + 2.2$$

This is somewhat larger than the value given by Michaud, Scherk and Vogt (1970). The reflection factor for this case was f = 4.7. (A more extensive discussion of this work is in preparation (Truran 1970).)

d) Calculations of $\langle\sigma v\rangle$

For a number of charged particle reactions on light nuclei, experimental determinations of the resonance parameters are available for a large number of levels in the energy region of interest. These provide a means of testing the methods outlined in this section for the determination of $\langle\sigma v\rangle$. The ratios of the reaction rates calculated from equation (28) to the experimental values (Fowler *et al*. 1967; Fowler 1969) at temperatures T_9 = 1, 3 and 5 are presented in Tables 1 and 2.

In general, the calculated reaction rates agree to within factors of two or three with the experimental values. There are several notable exceptions, specifically: $^{20}Na(p, \alpha)^{20}Ne$, $^{23}Na(\alpha, p)^{26}Mg$ and $^{31}P(\alpha, p)^{34}S$. For these three cases, the ratios $\langle\sigma v\rangle_{calc.}/\langle\sigma v\rangle_{exp.}$ are seen to increase with increasing temperature. It is not certain that this behavior is due to a systematic error in the calculations; it might,

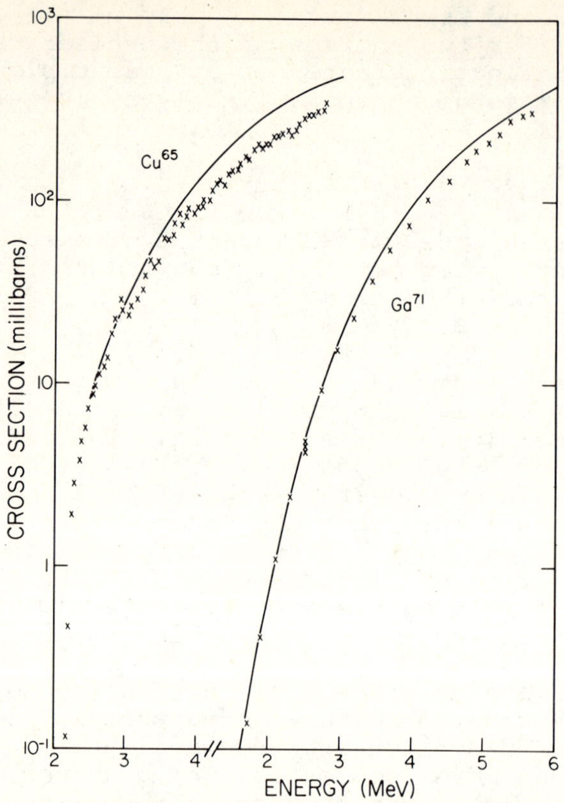

Figure 3. The total proton cross sections for ^{65}Cu and ^{71}Ga are compared to the experimental (p,n) cross sections (Johnson 1968). The nuclear radius was taken to be $R = 1.30A^{1/3}$ fermi and the reflection factor f = 3.8.

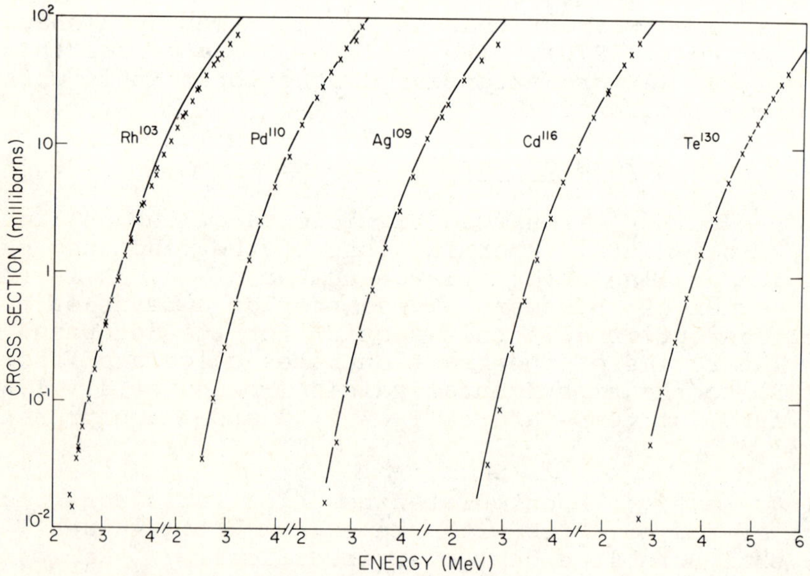

Figure 4. The total proton cross sections for ^{103}Rh, ^{110}Pd, ^{109}Ag, ^{116}Cd and ^{130}Te are compared to the experimental (p,n) cross sections (Johnson 1968). The nuclear radius was taken to be $R = 1.30A^{1/3}$ fermi and the reflection factor f = 3.8.

TABLE 1

(p,γ) and (α,γ) REACTIONS

$[\langle\sigma v\rangle_{calculated} / \langle\sigma v\rangle_{experimental}]$

Reaction	$T_9=1$	$T_9=3$	$T_9=5$
$^{23}Na(p,\gamma)^{24}Mg$	0.88	1.07	1.05
$^{26}Mg(p,\gamma)^{27}Al$	1.42	2.46	2.18
$^{27}Al(p,\gamma)^{28}Si$	4.32	1.52	1.01
$^{29}Si(p,\gamma)^{30}P$	0.62	2.47	3.40
$^{30}Si(p,\gamma)^{31}P$	1.55	1.95	2.13
$^{31}P\ (p,\gamma)^{32}S$	2.90	3.46	2.90
$^{34}S\ (p,\gamma)^{35}Cl$	6.55	3.23	3.74
$^{35}Cl(p,\gamma)^{36}Ar$	1.45	1.53	1.66
$^{37}Cl(p,\gamma)^{38}Ar$	1.25	0.80	0.75
$^{39}K\ (p,\gamma)^{40}Ca$	6.86	1.92	1.56
$^{52}Cr(p,\gamma)^{53}Mn$	7.86	2.50	2.62
$^{20}Ne(\alpha,\gamma)^{24}Mg$	1.70	1.68	2.34
$^{24}Mg(\alpha,\gamma)^{28}Si$	0.82	1.35	1.30
$^{30}Si(\alpha,\gamma)^{34}S$	1.46	1.68	2.19
$^{32}S\ (\alpha,\gamma)^{36}Ar$	4.02	2.09	2.26
$^{34}S\ (\alpha,\gamma)^{38}Ar$	0.87	0.52	1.34

TABLE 2

(α,p) REACTIONS

$[\langle\sigma v\rangle_{calculated} / \langle\sigma v\rangle_{experimental}]$

Reaction	$T_9=1$	$T_9=3$	$T_9=5$
$^{23}Na(p,\alpha)^{20}Ne$	1.32	7.99	22.2
$^{27}Al(p,\alpha)^{24}Mg$	1.87	0.68	1.28
$^{31}P\ (p,\alpha)^{28}Si$	1.25	0.64	1.34
$^{35}Cl(p,\alpha)^{32}S$	0.48	0.16	0.34
$^{37}Cl(p,\alpha)^{34}S$	0.88	1.08	1.58
$^{23}Na(\alpha,p)^{26}Mg$	1.84	2.83	10.6
$^{27}Al(\alpha,p)^{30}Si$	0.59	0.86	3.44
$^{31}P\ (\alpha,p)^{34}S$	1.30	1.58	12.8

rather, be attributable to missing resonances at higher energies. The calculated rates for (p, γ) reactions on ^{34}S, ^{39}K and ^{52}Cr at $T_9 = 1$ are also somewhat high; again the source of these discrepancies is difficult to determine.

Macklin and Gibbons (1965) have published a rather extensive compilation of experimental determinations of neutron-capture cross sections in the energy range of importance in astrophysics. Their results afford a good test of the statistical methods employed in our calculations over a wide range in mass number. The ratios $\langle\sigma v\rangle$ (calculated)/ $\langle\sigma v\rangle$ (experimental) are plotted as a function of mass number in Fig. 5. Again, agreement to within a factor of three is obtained over the entire mass range $54 \leq A \leq 238$. There are, however, rather pronounced discrepancies at the neutron closed shell positions(neutron numbers N = 50, 82 and 126). The average widths calculated by our methods clearly do not provide reasonable estimates of the partial widths for nuclei in these regions.

III. NUCLEOSYNTHESIS CONDITIONS IN SUPERNOVA MODELS

1) The Neutrino-Transport Model

A critical analysis of the nucleosynthesis conditions predicted by supernova models built upon the neutrino energy-transport mechanism has recently been presented (Arnett and Truran 1970). The abundance configurations formed in the expanding core material are sensitive to two critical "freezing" processes. The first of these involves the weak interactions; for temperature-density conditions in the range $T > 10^{10}$ °K, $\rho > 10^{10}$g/cc (demanded in these models to assure that neutrinos leaving the core will interact with the surrounding medium), electron and positron capture reactions proceed rapidly on the free neutrons and protons which comprise most of the mass. The reaction lifetimes are lengthened substantially by the subsequent cooling and expansion of the medium, eventually exceeding the hydrodynamic expansion time scale. The final neutron to proton ratio may be roughly approximated by the equilibrium value at the temperature and density where the hydrodynamic expansion time equals the reaction lifetime.

Contours of constant neutron-to-proton ratio, n/p, are presented in Fig. 6 as a function of temperature and mass density ρ. As will be discussed below, the range of n/p ratios (and hence $\overline{N}/\overline{Z}$ ratios) which produce abundances consistent with observation is given by $(n/p - 1) \sim 10^{-2}$ to 10^{-3}. The region in the ρ-T plane having such a (n/p) ratio is exceedingly narrow, as may be seen in Fig. 6. In fact, this extreme narrowness makes it exceedingly unlikely that the supernova models discussed here could always be expected to expand along such a contour. Note that even the region between $0.95 < (n/p) < 1.05$, ten times wider than the desired region, is quite narrow. The dashed line in Fig. 6 represents that temperature and density combination for which the equilibrium of strong interactions produces equal parts by mass of ^{4}He and of nucleons. Below this line nuclear burning to more complex nuclei occurs, and this effect must be included in an estimate

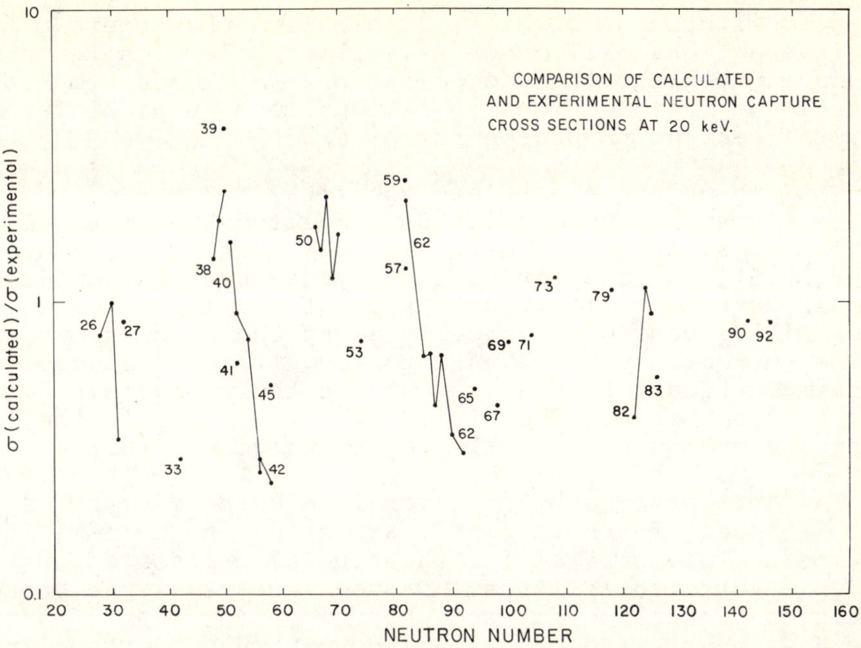

Figure 5. The ratio σ(calculated)/σ(experimental) is plotted as a function of mass number. The 20 keV. experimental neutron capture cross sections are those of Macklin and Gibbons (1965).

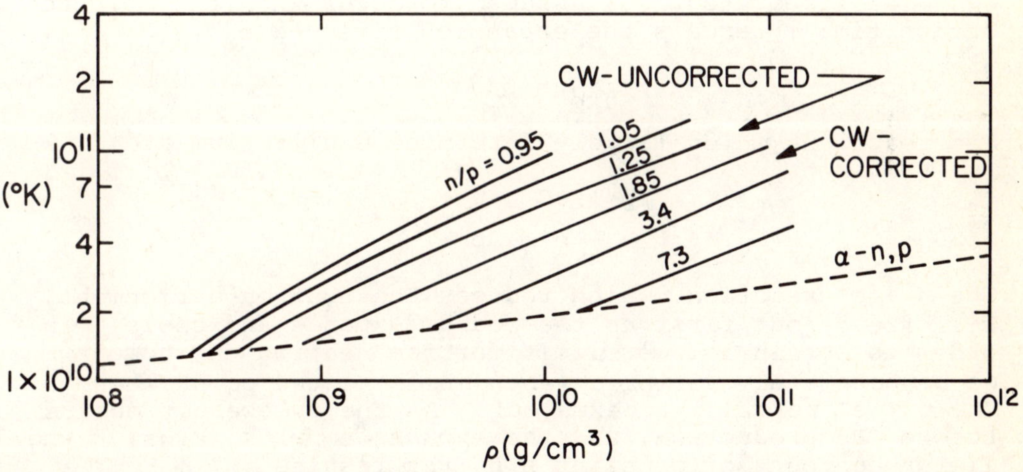

Figure 6. Contours of constant neutron/proton ratio determined in weak-interaction equilibrium are plotted in the temperature-density plane. The adiabat followed by a typical zone at the mass-cut of the model of Colgate and White (1966) is indicated. The strong-interaction equilibrium contour for $^4He \rightleftarrows 2p + 2n$ is shown for that case for which half the protons are in ^4He. From Arnett and Truran, Ap. J., 160, 959 (1970).

of the equilibrium n/p ratio. Typically, the freeze-out of weak interactions will occur at sufficiently high temperatures that we may neglect this complication. At lower temperatures the ratio of free neutrons to free protons, n/p, differs from the total neutron-to-proton ratio, $\bar{N}/\bar{Z}$.

Now, in order to estimate the final $\bar{N}/\bar{Z}$ ratio, we must examine the explosive conditions predicted by numerical calculation of supernova models. For the models of Colgate and White (1966), the characteristic temperature and density at the mass cut are $T \simeq 2 \times 10^{11}$ °K and $\rho \simeq 2 \times 10^{11}$ g/cc. Such matter will expand adiabatically along the arrow shown in Fig. 6. The corrected path takes into account an error due to the approximate equation of state used in their models. For this case, the peak temperature is $T \sim 1.3 \times 10^{11}$°K and the "frozen" neutron-to proton ratio falls in the range $n/p \sim 2 - 3$.

In Fig. 7 the paths of several relevant mass zones of the 4 $M_\odot$ model of Arnett (1967) are shown in the ρ-T plane. These paths, illustrated for the crucial explosive phase, are depicted by solid lines. Dashed lines show the phase-transition boundaries for the photodisintegration of iron ($^{56}Fe \rightarrow \alpha$, n, p), and of helium($^{4}He \rightarrow$ p, n). Dotted lines are contours of constant n/p. The expansion of this material is essentially adiabatic, and parallels the contours of constant n/p. Note that all paths lie between n/p = 1.85 and n/p = 3.4, consistent with the results obtained above for the models of Colgate and White.

The "freeze-out" is assumed to occur when the mean reaction time τ_r equals the expansion time scale

$$\tau_{exp} = \rho/\dot{\rho}$$

The reactions which largely determine equilibrium are

$$p + e^- \rightarrow n + \gamma_e$$
$$n + e^+ \rightarrow p + \bar{\gamma}_e$$

The $n + e^+$ reaction, which reduces the neutron enrichment, will freeze-out first as the medium expands and cools. In order to obtain a lower limit for the neutron enrichment, we compute the equilibrium abundances at the point where the $p + e^-$ reaction freezes out. As the freeze-out occurs before ^{4}He production, and the expansion paths (Figs. 6 and 7) lie on contours of constant n/p, the results are not sensitive to this choice. Using typical conditions obtained from the hydrodynamic computations, we find that the condition

$$\tau_r = \tau_{exp}$$

implies a density $\rho \sim 5 \times 10^9$g/cc. The neutron/proton ratios at freezing for the various models (including the 2 $M_\odot$ model of Arnett) are collected in Table 3. For all cases, a neutron enrichment $n/p \gtrsim 1.8$ was realized.

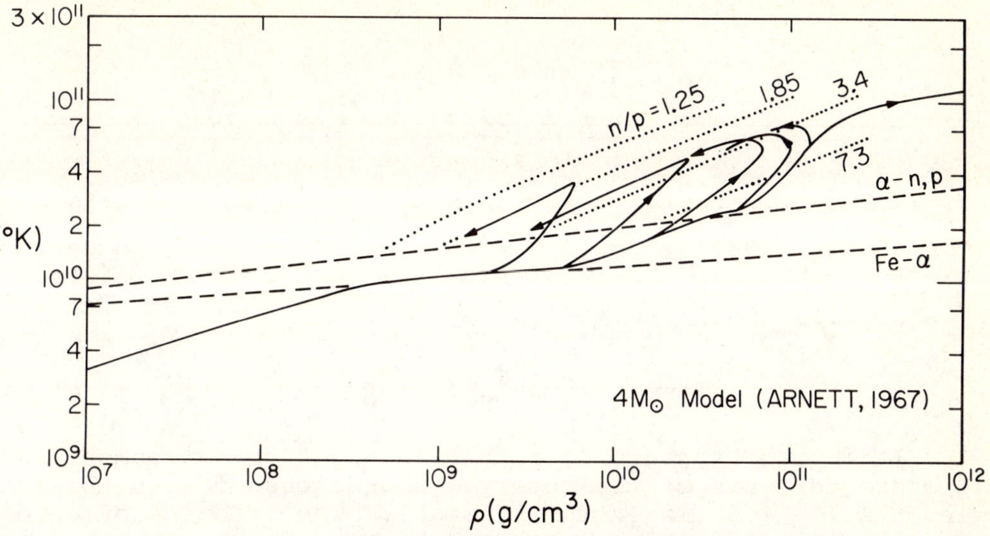

Figure 7. The evolutionary paths followed by typical mass zones during the collapse and explosion of the 4 $M_{\odot}$ core model of Arnett (1967) are illustrated. The n/p and $\alpha \rightarrow$ n,p contours are as shown in Figure 6. The contour defined by $X(^{56}Fe) = X(^{4}He)$ in equilibrium is also shown. From Arnett and Truran, Ap. J., 160, 959 (1970).

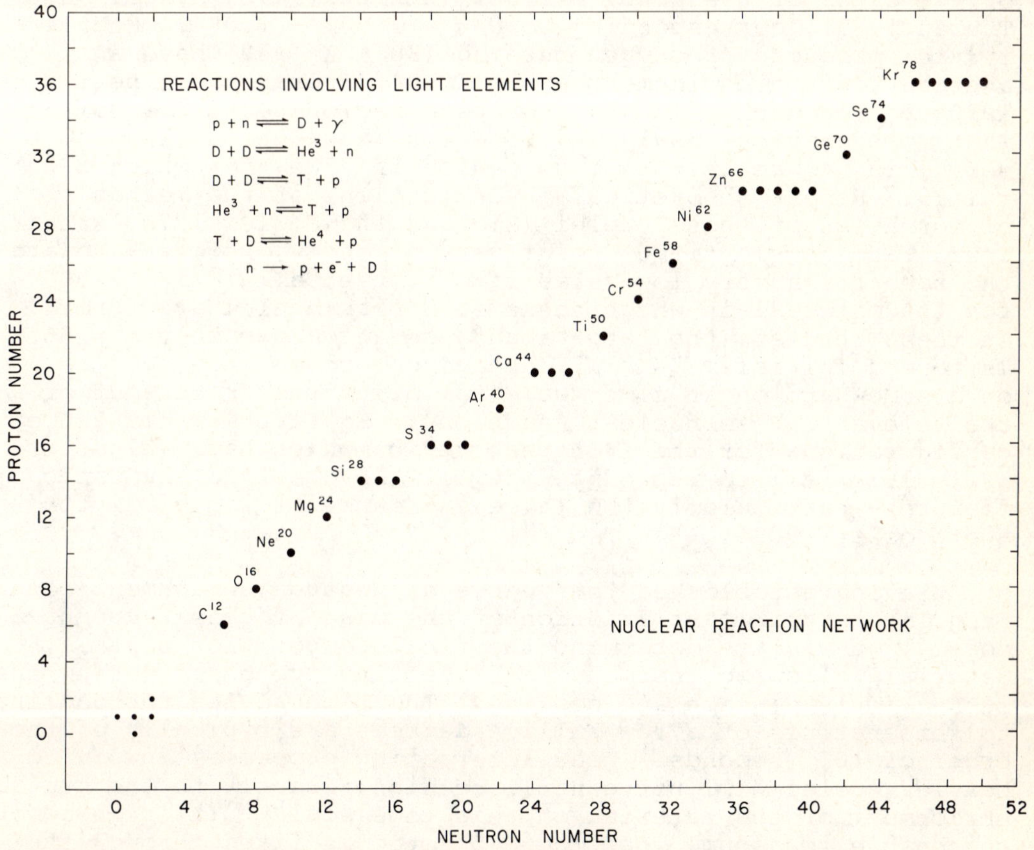

Figure 8. Schematic of the nuclear reaction network employed in studies of the buildup of heavy nuclei in the mass zones ejected in the 2 $M_{\odot}$ supernova core model of Arnett (1967). From Truran, Arnett, Tsuruta, and Cameron, Astrophys. Space Sci., 1, 129 (1968).

TABLE 3

Source	Core Mass	n/p
Arnett (1967)	2.0	5 to 8
-	4.0	1.8 to 3.4
Colgate and White (1966)	1.5	
	2.0	$\gtrsim$ 2.0
	10.0	

It is important to note that the neutron enrichment which is characteristic of the observed abundances of less massive elements ($A \lesssim 60$) is extremely small. The products of carbon burning ($20 \leq A \lesssim 30$) possess a neutron enrichment

$$\eta = \frac{\bar{N} - \bar{Z}}{\bar{N} + \bar{Z}}$$

of the order of $\eta = 0.001 - 0.002$ (Arnett 1969b, Arnett and Truran 1969), corresponding to $\bar{N}/\bar{Z} = 1.002 - 1.004$. Similarly, the products of oxygen burning ($30 \leq A \lesssim 42$) have an average neutron enrichment $\bar{N}/\bar{Z} \sim 1.004$ (Truran and Arnett 1970a). The nuclei in the iron peak region require a somewhat more cautious analysis. The most abundant nucleus observed in nature in this mass region is ^{56}Fe, for which $\bar{N}/\bar{Z} = 1.15$. Recent theoretical investigations of the mechanism of formation of these nuclei (Michaud and Fowler 1970; Arnett, Truran and Woosley 1970) point very strongly, however, toward the need for a very low value of $\bar{N}/\bar{Z}$, typically $\bar{N}/\bar{Z} \lesssim 1.006$, for the material in which these iron peak nuclei are formed. In these studies, the large abundance at mass number A = 56 is formed initially as ^{56}Ni, decaying through ^{56}Co to ^{56}Fe on a somewhat longer time scale following the freeze-out of the relevant thermonuclear reactions. Equilibrium abundance configurations for the iron peak region which have values of $\bar{N}/\bar{Z} \sim 1.10$ fail to reproduce many of the observed abundance features--particularly the large abundance of ^{58}Ni (Clayton and Woosley 1969).

Having established the degree of neutron enrichment occurring in a significant fraction of the mass of these supernova models, we wish to determine the final composition of this material following thermonuclear reaction processes during the subsequent expansion and cooling. The time scales for cooling to temperatures of a few billion degrees are typically of the order of 10^{-1} seconds. Weak interaction processes involving heavier nuclei will not contribute significantly to the neutron enrichment of the material on this time scale. It is therefore reasonable to assume that the $\bar{N}/\bar{Z}$ ratio is fixed through the thermonuclear burning stage by the value "frozen in" at higher densities.

It has been demonstrated (Truran, Arnett, Tsuruta and Cameron 1968) that the time scale of the expansion is sufficient to allow a true nuclear statistical equilibrium to be approached and maintained. In these studies the point of view adopted was that if nuclear statistical eqilibrium can be approached by any set of reaction pathways during the expansion along the adiabat, then it should certainly be attained by the fastest pathway. For the initial neutron-rich conditions considered here, the actual reaction paths presumably lie in the region of very neutron-rich heavy nuclei. The chosen pathway, illustrated by the network in Fig. 8, lies close to the valley of β-stability so that the calculated reaction rates should be fairly reliable. This network is chosen conservatively in that the charged-particle addition reactions involve α-particle capture at least up to nuclear charge numbers equivalent to those encountered in nuclear statistical equilibrium. Experimental determinations of the reaction rates were employed for the triple-α reaction and for the α-capture reactions on ^{12}C, ^{16}O, ^{20}Ne and ^{24}Mg.

The nuclear transmutations which will take place in the expanding core material have been investigated for representative mass zones ejected for the 2 $M_\odot$ model of Arnett (1967). The initial temperature considered was 10^{10}°K, at which point the weak interactions have frozen out and the formation of helium from neutrons and protons will proceed rapidly. The decrease in the temperature as a function of time is determined by the detailed hydrodynamic calculations. The change in density accompanying the cooling is given by $\rho \propto T^3$. Two expansion adiabats were considered:

$$\rho = 7 \times 10^5 \, T_9{}^3$$

$$\rho = 7 \times 10^4 \, T_9{}^3$$

The mass fractions of the more important nuclear constituents are plotted as a function of time in Fig. 9 for the high density adiabat. The initial neutron to proton ratio was 8. It is evident from the behavior in the early stages that an equilibrium has been rapidly established among neutrons, protons, α-particles and ^{12}C. As the temperature decreases, and the buildup by α-particle capture past ^{12}C becomes more rapid, the abundances of the light elements and ^{12}C decrease, while the abundances of intermediate mass nuclei and ultimately ^{86}Kr are rapidly increasing. At a temperature of approximately 4 billion degrees (approximately 2×10^{-2} seconds) the reactions involving the heavy elements freeze out in the sense that the nuclear-reaction time scale becomes long compared to the expansion time scale. The final abundance of ^{86}Kr is 20% by mass.

The results of a similar calculation performed for the lower-density profile are shown in Fig. 10. Here the buildup of ^{12}C is somewhat impeded by the lower density. However, the final ^{86}Kr abundance still approaches 10% by mass.

The effects of varying the total neutron to proton ratio realized in the weak interaction freezing are shown in Fig. 11.

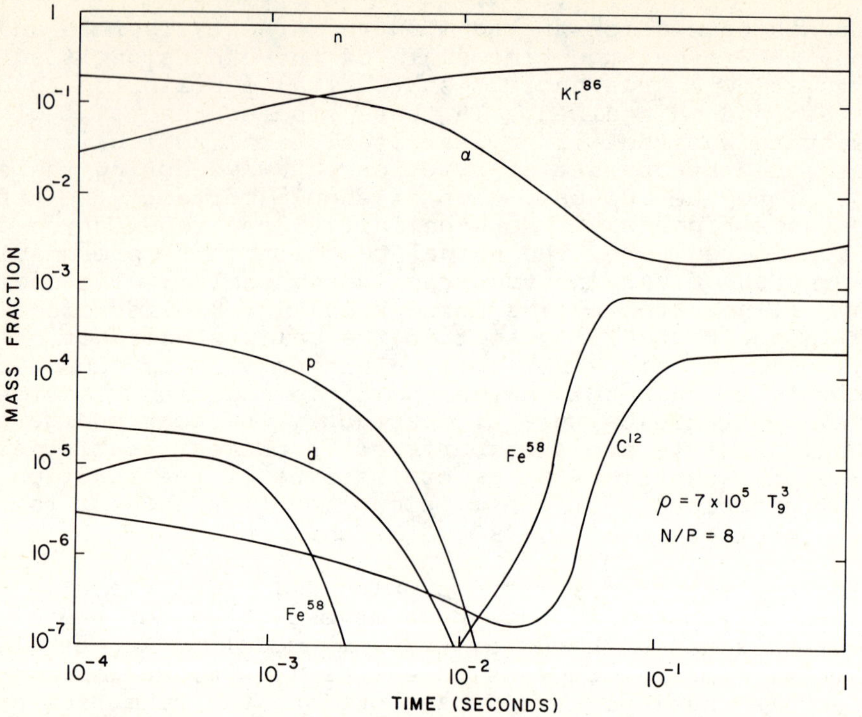

Figure 9. The evolution of a region composed initially of free neutrons and protons in the ratio n/p = 8 is shown for the high density adiabatic cooling curve. The mass fractions of the more abundant constituents are plotted as a function of time. From Truran, Arnett, Tsuruta, and Cameron, Astrophys. Space Sci., 1, 129 (1968).

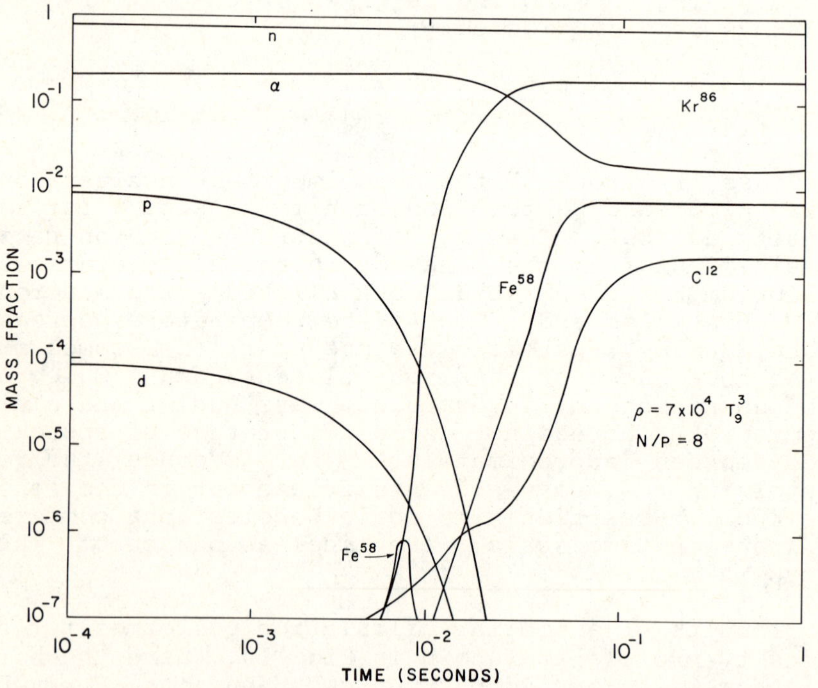

Figure 10. The evolution of a region composed initially of free neutrons and protons in the ratio n/p = 8 is shown for the low density adiabatic cooling curve. The mass fractions of the more abundant constituents are plotted as a function of time.From Truran, Arnett, Tsuruta, and Cameron, Astrophys. Space Sci., 1, 129 (1968).

Here the ratio $\bar{N}/\bar{Z} = 3$, and the low-density adiabat is employed. Generally, the protons will be converted completely to helium; a larger fraction of the mass is therefore converted to heavy elements. The ^{86}Kr abundance realized in this calculation approaches 50% by mass.

These results suggest that the breakthrough past ^{12}C to heavy elements can take place readily under the conditions predicted for the innermost ejected mass zones for these neutrino energy-transport supernova models. Having established that nuclear statistical equilibrium can be approached and maintained, the detailed final abundance configuration may be estimated by determining the equilibrium abundances at the charged-particle freezing temperature ($T_9 \sim 3 - 4$). For the very neutron rich conditions we are considering, the abundant nuclei typically lie too far from the valley of beta stability for experimental determinations of their masses and spins to have been made. The binding energies used in these studies are those predicted by the "exponential" mass formula of Cameron and Elkin (1965). Other reasonable choices would have given qualitatively similar results. For the partition functions, W_{ZA}, the following "average" values are used:

$$W_{ZA} = 1 \quad \text{for even-even nuclei}$$

$$= 6 \quad \text{for odd-A nuclei}$$

$$= 7 \quad \text{for odd-odd nuclei}$$

In Table 4 are presented the mass fractions of the five most abundant nuclei calculated in this manner for $T = 3 \times 10^9$ °K, densities of 10^6, 10^7 and 10^8 g/cc and $\bar{N}/\bar{Z}$ ratios of 1.25, 1.5, 2.0, 4.0 and 8.0. At $\rho = 10^6$ and 10^8 and $\bar{N}/\bar{Z} = 1.25$, these results agree reasonably well with those of Clifford and Tayler (1965). In Table 5 a similar set of abundances, but for $T = 4 \times 10^9$ °K, is presented for comparison. Note that the qualitative features of the distributions at the two temperatures are similar. In the range $2 \lesssim \bar{N}/\bar{Z} \lesssim 8$, the most abundant nuclei at any plausible density ($10^6 \lesssim \rho \lesssim 10^8$ g/cm^3) have an atomic number $A \gtrsim 80$. The mass fraction of free neutrons is large. Upon further cooling the charged particle reactions will cease to be effective, but the neutrons, encountering no coulomb barrier, may continue to be captured. As the cooling will inhibit the inverse (γ, n) reactions, some significant further neutron addition may be realized.

We have established both that the interior mass zones for the neutrino-transport supernova models will emerge with a substantial neutron enrichment ($\bar{N}/\bar{Z} \gtrsim 2$) and that this enrichment implies the buildup of nuclei well past the iron-peak region ($A \gtrsim 80$). The implications of this heavy element production with regard to nucleosynthesis will now be examined.

It is clear from the equilibrium abundance distributions shown in Tables 4 and 5 that an extremely distorted version

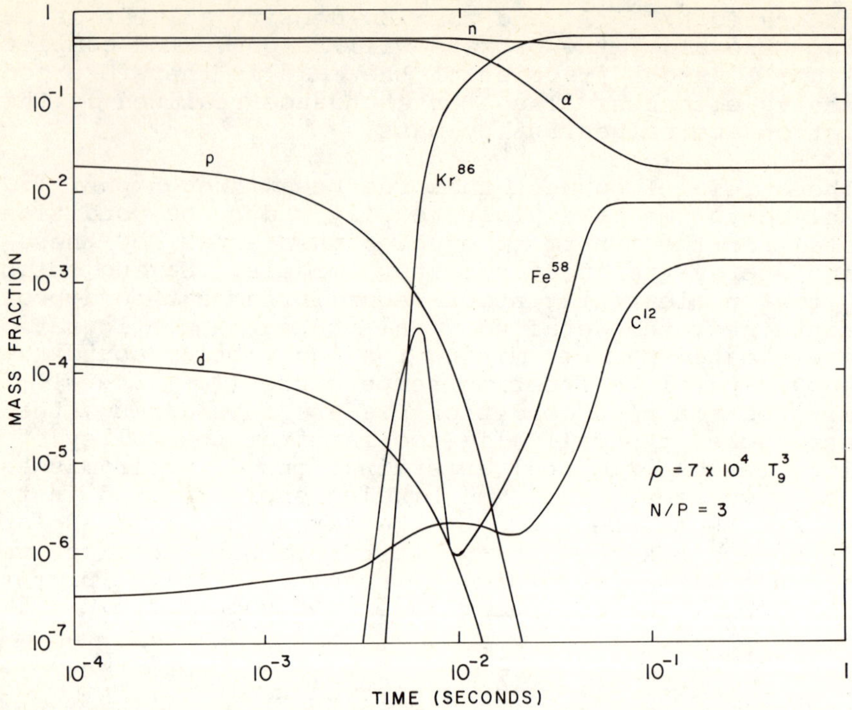

Figure 11. The evolution of a region composed initially of free neutrons and protons in the ratio n/p = 3 is shown for the low density adiabatic cooling curve. The mass fractions of the more abundant constituents are plotted as a function of time. From Truran, Arnett, Tsuruta, and Cameron, Astrophys. Space Sci., 1, 129 (1968).

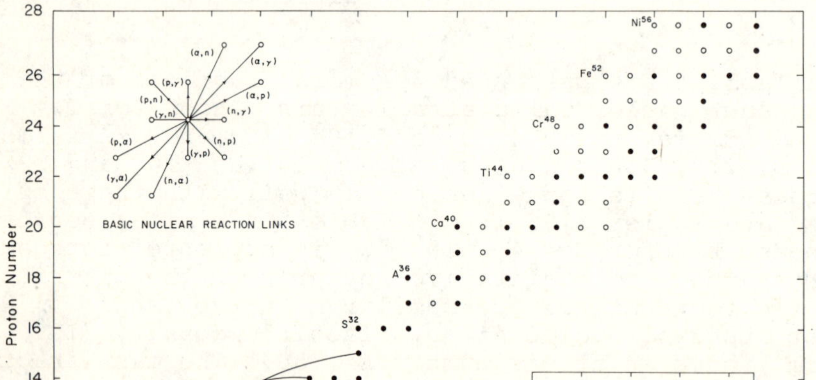

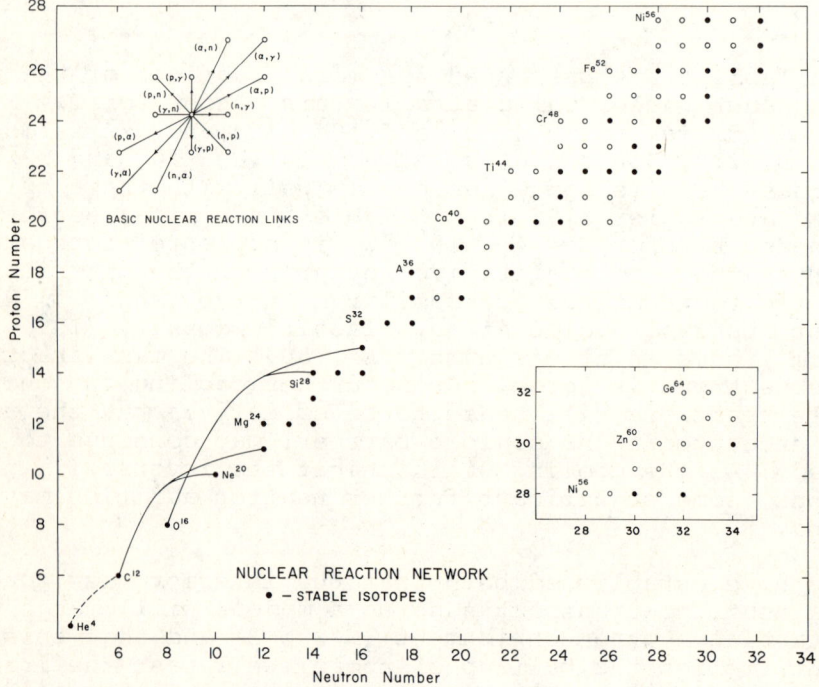

Figure 12. Schematic of the nuclear reaction network employed in studies both of the nucleosynthesis conditions predicted for the mass zones ejected by the ^{12}C detonation model and of explosive oxygen burning nucleosynthesis.

TABLE 4

EQUILIBRIUM ABUNDANCES AT T = 3.0 x 10^9 °K

$\overline{N}/\overline{Z}$	$\rho = 10^6$		$\rho = 10^7$		$\rho = 10^8$	
1.25	^{54}Cr	0.296	^{58}Fe	0.304	^{58}Fe	0.309
	^{58}Fe	.294	^{54}Cr	.253	^{64}Ni	.266
	^{64}Ni	.193	^{64}Ni	.231	^{54}Cr	.211
	^{50}Ti	.081	^{62}Ni	.093	^{62}Ni	.114
	^{62}Ni	.074	^{50}Ti	.057	^{50}Ti	.039
1.5	^{84}Se	.638	^{84}Se	.639	^{84}Se	.640
	^{82}Ge	.302	^{82}Ge	.303	^{82}Ge	.303
	^{83}As	.026	^{83}As	.026	^{83}As	.027
	^{85}Se	.012	^{85}Se	.013	^{85}Se	.014
	^{80}Ge	.012	^{80}Ge	.011	^{80}Ge	.010
2.0	^{78}Ni	.857	^{78}Ni	.792	^{80}Ni	.481
	n	.073	^{79}Ni	.085	^{78}Ni	.155
	^{79}Cu	.044	n	.069	^{79}Ni	.117
	^{79}Ni	.010	^{80}Ni	.050	^{82}Ni	.092
	^{80}Zn	.009	^{77}Cu	.001	^{81}Ni	.056
4.0	^{78}Ni	.507	n	.431	^{118}Kr	.417
	n	.442	^{80}Ni	.281	n	.352
	^{79}Ni	.035	^{78}Ni	.116	^{120}Sr	.190
	^{80}Ni	.013	^{79}Ni	.077	^{119}Rb	.017
	^{79}Cu	.010	^{82}Ni	.042	^{117}Br	.011
8.0	n	.845	n	.836	n	.817
	^{78}Ni	.126	^{80}Ni	.052	^{118}Kr	.151
	^{79}Ni	.017	^{120}Sr	.034	^{117}Br	.028
	^{80}Ni	.012	^{82}Ni	.030	^{120}Sr	.001
	^{77}Co	<0.001	^{76}Fe	0.011	^{117}Kr	0.001

TABLE 5

EQUILIBRIUM ABUNDANCES AT T = 4.0 x 10^9 °K

$\overline{N}/\overline{Z}$	$\rho = 10^6$		$\rho = 10^7$		$\rho = 10^8$	
1.25	^{54}Cr	0.305	^{54}Cr	0.279	^{54}Cr	0.248
	^{58}Fe	.192	^{58}Fe	.210	^{58}Fe	.223
	^{50}Ti	.180	^{50}Ti	.139	^{64}Ni	.162
	^{64}Ni	.094	^{64}Ni	.127	^{50}Ti	.103
	^{62}Ni	.043	^{62}Ni	.056	^{62}Ni	.072
1.5	^{84}Se	.598	^{84}Se	.556	^{84}Se	.558
	^{82}Ge	.193	^{82}Ge	.245	^{82}Ge	.251
	^{83}As	.065	^{83}As	.070	^{83}As	.071
	^{85}Se	.036	^{85}Se	.039	^{85}Se	.041
	^{80}Ge	.031	^{80}Ge	.029	^{80}Ge	.027
2.0	^{80}Zn	.626	^{80}Zn	.439	^{78}Ni	.775
	^{81}Ga	.132	^{79}Cu	.230	^{79}Ni	.078
	n	.117	^{78}Ni	.145	n	.071
	^{82}Ge	.061	n	.099	^{79}Cu	.042
	^{79}Zn	.014	^{81}Zn	.058	^{80}Ni	.012
4.0	n	.465	^{78}Ni	.451	n	.436
	^{80}Zn	.411	n	.444	^{78}Ni	.235
	^{79}Cu	.067	^{79}Cu	.051	^{79}Ni	.145
	^{81}Zn	.026	^{79}Ni	.028	^{80}Ni	.131
	^{78}Ni	.013	^{80}Zn	.007	^{81}Ni	.020
8.0	n	.850	n	.845	n	.841
	^{80}Zn	.079	^{78}Ni	.129	^{80}Ni	.054
	^{79}Cu	.037	^{79}Ni	.015	^{79}Ni	.031
	^{78}Ni	.021	^{79}Cu	.004	^{78}Ni	.026
	^{81}Zn	0.009	^{80}Ni	0.003	^{81}Ni	0.016

of the usual abundance distribution might result from these events. Unfortunately, plausible and detailed nucleosynthesis calculations (primarily concerned with the r-process) have not as yet been carried out within the context of these models. For our present purposes, however, it is sufficient to assume that the abundance distribution formed in zones processed to masses $A > 60$ is in agreement with the solar-system abundances. The arguments in the ensuing discussion will be testing the predictions of these supernova models with regard to the total mass processed to heavy elements ($A > 60$); the detailed abundance features in this region will not be considered.

Solar-system abundance studies show elements of mass number $A > 60$ to constitute 5.6×10^{-6} by mass of primordial solar-system material (Cameron 1968). This is considerably smaller than the mass fraction of 1.5×10^{-3} in the form of iron-peak nuclei ($21 \leq Z \leq 28$). It is important to note that observations of the nearest supernova remnant, the Crab Nebula (Woltjer 1958), do not suggest any great discrepancies in abundance relative to the solar system. The same conclusion follows from recent studies of abundances of heavy cosmic rays (Price, Rajan and Tamhane 1968; Fowler, Adams, Cowen and Kidd 1967). This point is particularly interesting because the neutrino-transport supernova model has been suggested as a direct source of cosmic rays (Colgate and Johnson 1960; Colgate and White 1965). Pagel (1968) has found that the ratio of heavy elements to Fe remains surprisingly constant even in the oldest stars. It appears that any argument based on an overabundance of heavy elements ($A > 60$) relative to iron will be valid for many other objects besides the solar system.

The total production of heavy elements predicted by the neutrino energy-transport supernova models discussed above can be estimated in the following manner. The preceding arguments indicate that a mass $M > 0.5\ M_\odot$ will be processed in each such event to nuclei of mass number $A > 60$. The current rate of occurrence of supernova has been estimated to be approximately one per 300 years by Zwicky (1965) and more recently, one per 30 years by Katgert and Oort (1967). We will conservatively choose 1/(100 years). Salpeter (1959) has estimated, on the basis of his stellar birth-rate function, that the average rate of supernova activity has been substantially higher over the history of our Galaxy than is presently observed. He estimates the total number of supernova events to have been

$$(\text{number of supernova events}) = (\text{average current rate})(10) \times (\text{galactic age}).$$

The total mass processed to heavy elements is then given by

$$\begin{aligned} M &= (\text{number of supernova events})(0.5\ M_\odot \text{ per event}) \\ &= \left(\frac{1}{100}\right)(10)(10^{10})(0.5) = 5 \times 10^{9}\ M_\odot \end{aligned}$$

where the galactic age is taken to be 10^{10} years. The total mass of our Galaxy is about 2×10^{11} $M_\odot$. Thus, if we assume that all supernova events are of the type discussed in this section, the total mass fraction in the Galaxy in the form of elements of mass A > 60 is

$$X_{A>60} = \frac{5 \times 10^8}{2 \times 10^{11}} = 2.5 \times 10^{-3}$$

These models therefore predict a total mass fraction of heavy elements which is at least comparable to the mass fraction of solar material in the form of iron-peak nuclei, exceeding by two orders of magnitude the observed concentration of these heavy elements in the solar system and in stars.

This discrepancy involves the explosive nature of the model--its ejection of mass. There is no obvious inconsistency between these arguments and the idea that gravitational collapse occurs to form neutron stars (or singularities). Also, it is clear from the foregoing arguments that if the matter ejected was composed of less massive nuclei ($A \leq 60$) in solar-system abundance ratios, the discrepancy would disappear. Such a situation might occur, for instance, in the model of Fowler and Hoyle (1964), but not the neutrino-transport model.

Arnett and Truran (1970) argue that the general character of these results points strongly to the conclusion that supernova explosions triggered by the neutrino energy-transport mechanism cannot comprise a substantial fraction of the observed supernova events. They argue further, that these difficulties with nucleosyntheis cannot be easily surmounted within the framework of these models. The critical consideration in their arguments is that the densities in the innermost ejected mass zones must be in excess of 10^{10} g/cc if sufficient energy is to be deposited in these regions by the interaction of neutrinos leaving the supernova core. The general character of the abundance distribution formed in the further expansion and cooling of this material is not sensitive to uncertainties in the nuclear parameters, as nuclear statistical equilibrium is readily achieved at these high temperatures. From these arguments they conclude that, from the point of view of nucleosynthesis, the neutrino-transport model cannot be a general explanation of supernovae.

2) The ^{12}C Detonation Model

A critical analysis of the nucleosynthesis conditions predicted for the mass zones ejected in the ^{12}C detonation model (Arnett 1969a) will soon be published (Arnett, Truran and Woosley 1970). In this section, preliminary results pertaining to only the interior mass zones are summarized. This supernova event is triggered by the explosive ignition of ^{12}C at high densities (central density $\rho_c \sim 2 \times 10^9$ g/cc). The star is totally disrupted, leaving no condensed remnant. The ejected core material is subjected to temperatures well in excess of $T \sim 6 \times 10^9$ °K and densities $\rho > 10^8$ g/cc. The various physical characteristics of representative interior mass zones ejected on this model (following the passage of the detonation front) are summarized in Table 6.

TABLE 6

PHYSICAL CHARACTERISTICS OF MASS ZONES STUDIED

Internal Mass ($M_r/M_\odot$)	Peak Temperature (°K)	Peak Density (g/cc)	Expansion Time Scale (sec.)	Neutron Enrichment (7×10^3) Low η	High η
0.071	8.3×10^9	1.70×10^9	0.260	7.1	30
0.36	8.25×10^9	1.03×10^9	0.277	3.9	14
0.65	8.1×10^9	6.20×10^8	0.296	2.7	8.0
0.93	7.98×10^9	4.17×10^8	0.266	2.2	5.6
1.22	7.6×10^9	2.20×10^8	0.210	1.8	3.5

The reaction network employed in these nucleosynthesis studies is shown in Fig. 12. This network is identical to that used in previous studies of silicon burning (Truran, Cameron and Gilbert 1966; Truran, Arnett and Cameron 1967). The approximate methods of solution used in those studies have been modified extensively as described by Arnett and Truran 1969).

The various nuclei included in the reaction network are related to their neighbors by the appropriate (n, γ), (p, n), (p, γ), (α, p), (α, n) and (α, γ) reactions and their corresponding inverse reactions. The triple-alpha reaction and the heavy ion reactions, $^{12}C + ^{12}C$ and $^{16}O + ^{16}O$, are also included. Where possible, experimental determinations of the relevant reaction rate parameters have been used in these calculations. Specifically, the rates for a number of charged particle reactions have been determined as sums over experimental resonance strengths (Fowler, Caughlan and Zimmerman 1967; Fowler 1969). The rate for the $^{12}C + ^{12}C$ reaction is taken from the recent study by Patterson, Winkler and Zaidins (1969) and that for the $^{16}O + ^{16}O$ reaction from Patterson, Winkler and Spinka (1969). For these heavy ion reactions, only the proton and α-particle channels have been considered; the neutron and deuteron branches have no significant influence on the results of silicon burning under explosive conditions.

The network shown in Fig. 12 includes nuclei through ^{60}Ni. In order to assure that the network provides sufficient reaction links in the mass region $56 \leq A \leq 60$, these calculations were checked with a somewhat enlarged network including several isotopes of copper and zinc (^{59}Cu, ^{60}Cu, ^{61}Cu, ^{60}Zn, ^{61}Zn, ^{62}Zn). The inclusion of these additional nuclei made no substantial difference in the abundances of ^{59}Co and ^{60}Ni formed in the interior mass zones.

The results of the thermonuclear reaction studies of the interior mass zones ejected on the ^{12}C detonation model are summarized in Table 7.

TABLE 7

X_{CALC}/X_{OBS} FOR VARIOUS CHOICES OF THE NEUTRON ENRICHMENT

Nucleus	η = .002	η = 0	Low η	High η
^{56}Fe	1.00	1.00	1.00	1.00
^{54}Fe	0.54	$< 10^{-8}$	1.24	6.10
^{58}Ni	0.46	0.077	0.58	3.30
^{57}Fe	0.65	0.27	0.74	1.38
^{60}Ni	$< 10^{-3}$	$< 10^{-3}$	$< 10^{-3}$	0.0066
^{52}Cr	1.48	1.03	1.36	1.09
^{55}Mn	0.53	$< 10^{-3}$	1.00	2.15
^{58}Fe	$< 10^{-8}$	$< 10^{-8}$	$< 10^{-8}$	$< 10^{-3}$
^{59}Co	$< 10^{-3}$	$< 10^{-3}$	0.0010	0.030
^{48}Ti	0.50	0.44	0.57	0.41
^{53}Cr	0.42	0.052	0.67	1.05
^{51}V	0.23	$< 10^{-8}$	0.42	0.72
^{50}Cr	0.22	$< 10^{-8}$	0.53	2.10
^{54}Cr	$< 10^{-8}$	$< 10^{-8}$	$< 10^{-8}$	$< 10^{-3}$
^{46}Ti	$< 10^{-3}$	$< 10^{-8}$	0.0021	0.0072
^{47}Ti	$< 10^{-3}$	$< 10^{-8}$	$< 10^{-3}$	$< 10^{-3}$
^{49}Ti	0.19	0.022	0.31	0.42
^{50}Ti	$< 10^{-8}$	$< 10^{-8}$	$< 10^{-8}$	$< 10^{-8}$
^{50}V	$< 10^{-8}$	$< 10^{-8}$	$< 10^{-8}$	$< 10^{-3}$

The ratios of the calculated mass fractions to the solar system values (Cameron 1968), normalized at ^{56}Fe, are presented for several choices for the neutron enrichment parameter, η. The calculated mass fractions represent appropriate averages over the mass zones defined in Table 6. The most abundant nucleus formed in all cases was ^{56}Fe. The mass fractions of nuclei not included in this table never exceeded $\sim$ 1 percent; it is clear that this model produces only the iron peak nuclei.

The importance of obtaining realistic estimates of the degree of neutron enrichment of the ejected mass is clearly demonstrated by these results. In the limit of no neutron enrichment (η = 0) in any of the mass zones, only three nuclei in the iron peak are formed in abundances relative to ^{56}Fe (formed as ^{56}Ni) consistent with their solar values to within a factor of four; these are ^{52}Cr (formed as ^{52}Fe), ^{48}Ti (formed as ^{48}Cr) and ^{57}Fe (formed as ^{57}Ni). This is in fact a rather unrealistic case within the context of current stellar evolutionary models. Even for a first generation star, some degree of neutron enrichment should result from weak interactions taking place at the high densities realized in the central zones.

The abundances resulting from these calculations on the assumption that the neutron enrichment throughout the core is $\eta = .002$ are in substantially better agreement with the solar system abundances. For this case, seven nuclei are produced in abundances relative to ^{56}Fe with a factor of two of their observed values: ^{54}Fe, ^{58}Ni, ^{57}Fe, ^{52}Cr, ^{55}Mn (formed as ^{55}Co), ^{48}Ti and ^{53}Cr. Furthermore, the abundances of ^{50}Cr, ^{49}Ti (formed as ^{49}Cr) and ^{51}V (formed as ^{51}Mn) are low by only a factor of five.

Estimates have been made of the contributions of nuclear position decays, proceeding at the high temperatures and densities realized in this model, to the neutron enrichment of this material. A rather firm lower limit can be established by considering only the contributions from transitions for which experimental ln $(f\tau)$ values are available. The abundances realized for this case are those given in the column entitled 'low η' in Table 7. (The neutron enrichment realized in each mass zone is as indicated in Table 6.) These abundances are illustrated as well in Fig. 13.

For this case, a total of ten nuclei are produced in abundances relative to ^{56}Fe within a factor of three of the observed values. These include three isotopes of chromium, two of which (^{52}Cr and ^{53}Cr) were formed as proton rich isotopes of iron, decaying via position emission following the freezing of this material. Similarly, two of the three isotopes of iron (^{56}Fe and ^{57}Fe) were formed initially as proton rich isotopes of nickel.

The abundance of ^{51}V is low by a factor of approximately two. The abundance of this odd-Z nucleus, as well as that of ^{55}Mn, is extremely sensitive to the details of the freezing process. This sensitivity is illustrated in Fig. 14, where the mass fractions of a number of nuclei are plotted as a function of temperature in the expansion and cooling of the outermost mass zone studied ($T_O = 7.6 \times 10^9$ °K; $\rho_O = 2.2 \times 10^8$ g/cc). At temperatures above $\sim 5 \times 10^9$ °K, a substantial proton abundance is maintained in equilibrium and the concentrations of the odd-Z nuclei ^{51}Mn, ^{55}Co and even ^{59}Cu are relatively large. As the material cools further, the rates for the (γ, p) reactions decrease rapidly and proton capture reactions as well as (p, n) and (p, α) reactions act to reduce the ^{55}Co and ^{51}Mn abundances. It is important to note as well that the abundance of ^{59}Cu falls off rapidly with decreasing temperature; the final abundance is too low (following position decay) to account for the observed abundance of ^{59}Co. A somewhat shorter expansion time scale might serve to increase

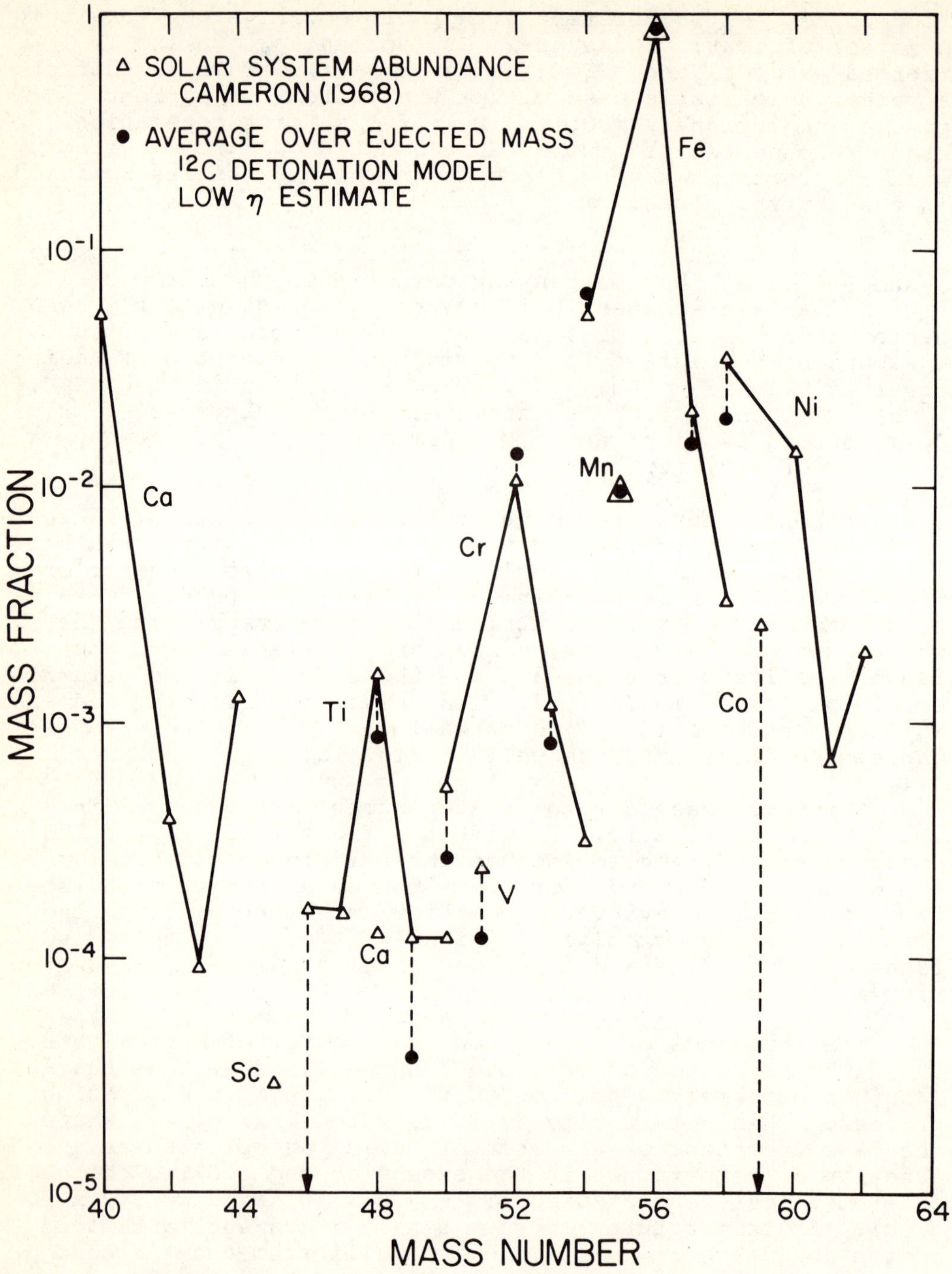

Figure 13. The nuclear abundances by mass produced in the mass zones ejected in the ^{12}C detonation model are compared to the solar system abundances. The two distributions are normalized at ^{56}Fe. The "low η" estimate of the neutron excess was employed.

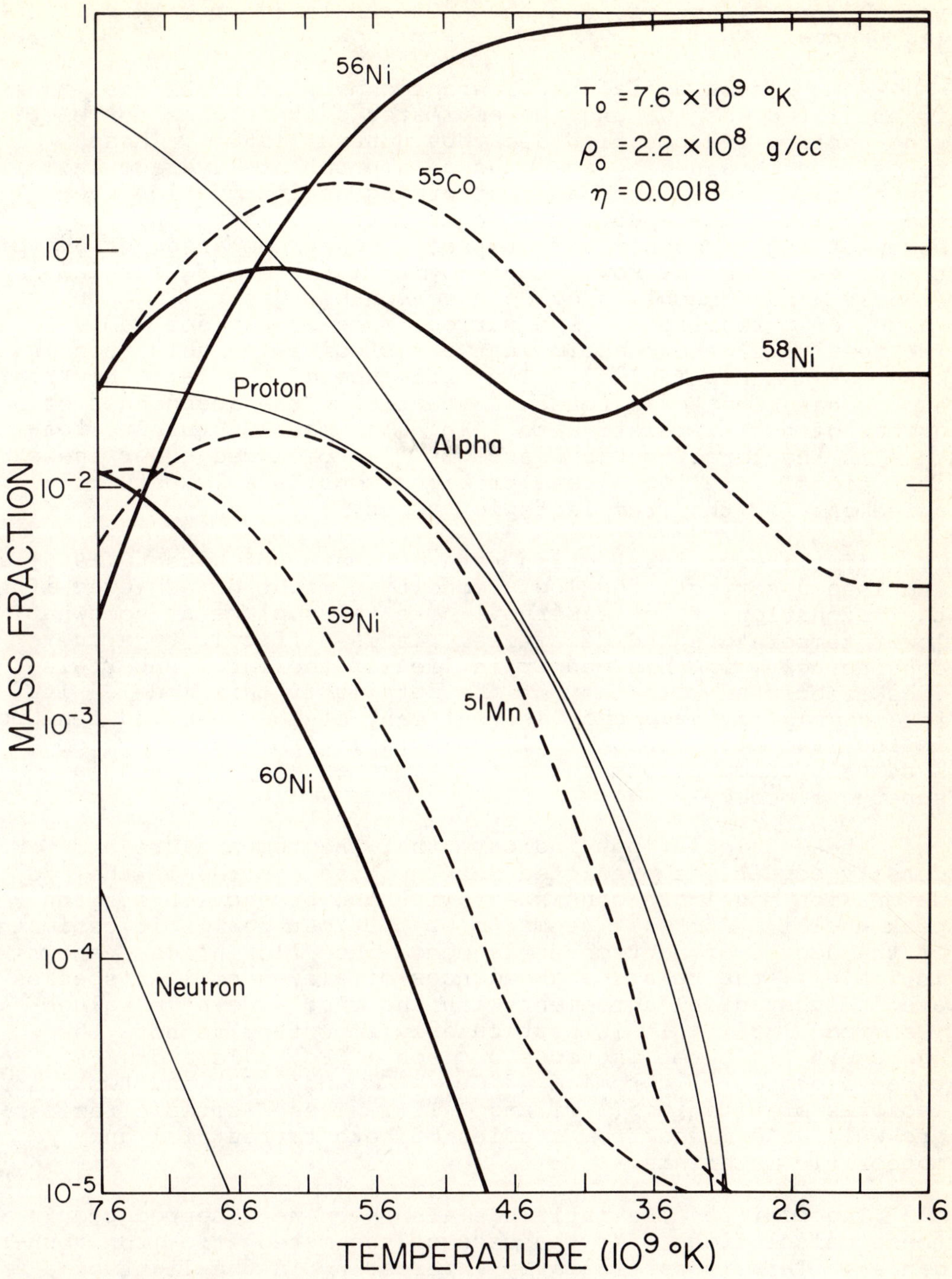

Figure 14. The detailed evolution for the exterior mass zone is traced through the expansion and cooling. The mass fractions of representative nuclei are plotted as a function of decreasing temperature.

the concentrations of ^{51}Mn and ^{59}Cu resulting from this freezing process.

A more substantial neutron enrichment of these mass zones is predicted when we use the estimates of the rates for electron capture and position decay by Hansen (1968). These rates were obtained by assuming an appropriate average ln (fτ) for allowed transitions and performing an integration over the excited state spectrum. These estimates are typically somewhat high (roughly a factor of 5) when compared for example to the estimates of Fowler and Hoyle (1964) or of Tsuruta (1969). These should provide a reasonable upper limit on the values of η realized in the various mass zones (see Table 6). The results of the reaction network studies for this 'high η' case are given in Table 7. The influence of the large neutron excess has been to increase substantially the abundances of a number of nuclei relative to ^{56}Fe. The most disturbing feature is the large overabundance of ^{54}Fe produced under these conditions. The 'low η' case provides considerably better agreement for the iron isotopic ratios.

If the weak interaction rate estimates of Hansen (1966) are indeed correct, the low η condition might be achieved if the detonation of the supernova core takes place at somewhat lower temperature and densities. It is difficult to argue convincingly in this manner in lieu of the large uncertainties associated with many of the details of this model. It is encouraging, nevertheless, that many of the observed features in the iron peak region follow naturally from these nucleosynthesis calculations for the mass zones ejected on this supernova model.

These calculations indicate that the temperature and density conditions predicted for the mass ejected on the ^{12}C detonation model are consistent with the production of iron peak nuclei. For what seems to be the most realistic estimate of the degree of neutron enrichment, the 'low η' case shown in Table 7, the relative abundances of eleven nuclear species are in substantial agreement with their solar system values (Cameron 1968). It is particularly important to note the agreement for three isotopes of iron (^{54}Fe, ^{56}Fe and ^{57}Fe), three isotopes of chromium (^{50}Cr, ^{52}Cr and ^{53}Cr) and two isotopes of titanium (^{48}Ti and ^{49}Ti). These isotopic ratios are well determined from studies of both terrestrial and meteoritic materials.

The ^{58}Ni to ^{56}Fe ratio has also been well reproduced in these calculations (for the case of the meteoritic iron abundance). This is particularly interesting in the light of a recent study of this problem by Clayton and Woosley (1969). Their results lead to the following conclusions with regard to the equilibrium process, if the relative abundances of ^{54}Fe, ^{56}Fe and ^{58}Ni are to be consistent with their solar system values: (1) the large abundance at mass number A = 56 must be formed initially as ^{56}Ni in quasi-equilibrium and (2) this equilibrium must be dominated by ^{56}Ni to the extent that the abundances of the lighter alpha-nuclei (^{28}Si, ^{32}S, ^{36}Ar and ^{40}Ca) are not formed in their observed solar system ratios.

Both of these restrictions are well satisfied for the nucleosynthesis conditions realized in this model. This suggests that some other process must be responsible for the production of nuclei in the silicon-to-calcium region. Studies of the abundance features resulting from explosive oxygen burning nucleosynthesis indicate that this may provide a promising alternative mechanism for the production of elements in the mass range $28 \lesssim A \lesssim 46$. (Truran and Arnett 1970a). This subject will be considered in the next section.

The preliminary results summarized in this section indicate that nuclei in the mass range $48 \lesssim A \lesssim 58$ can be formed in a ^{12}C detonation supernova event. Nuclei on the high mass side of the iron peak ($A > 58$) are not formed in amounts consistent with solar system abundances. These results follow directly from the predictions of the hydrodynamic study of Arnett (1969a) of the temperature-density character of the ejected matter.

The calculations of Arnett (1969a) did not, however, treat the outer core zones exactly. Arnett, Truran and Woosley (1970) have recalculated the nucleosynthesis in the ejected mass zones for an initial model based on the integrations of Paczynski (1969). These studies treat the outer parts of the core more carefully. The temperature-density structure of the outer core zones were found to be consistent with the production of substantial abundances of ^{59}Co, ^{60}Ni, ^{61}Ni, ^{62}Ni, ^{63}Cu, ^{64}Zn and ^{65}Cu (formed as ^{59}Cu, ^{60}Zn, ^{61}Zn, ^{62}Zn, ^{63}Ga, ^{64}Ge and ^{65}Ge, respectively). The integrated abundances for the revised core structure are in substantial agreement with the solar system abundances for nuclei in the mass range $48 \leq A \lesssim 65$. For a detailed discussion of these results, the reader is referred to the paper by Arnett, Truran and Woosley (1970; see also Truran and Arnett 1970b).

IV. EXPLOSIVE OXYGEN BURNING NUCLEOSYNTHESIS

The synthesis of nuclei in the mass range 20 _ A _ 60 is generally attributed to charged particle reaction mechanisms (Burbidge, Burbidge, Fowler and Hoyle 1957). Extensive studies of carbon, oxygen and silicon-burning have been carried for constant temperature conditions consistent with the view that these processes take place as stable energy generation stages of stellar evolution. The typical abundance distributions resulting from such calculations do not, in general, provide detailed agreement with the observed solar system abundance features.

Recent studies of carbon and silicon burning nucleosynthesis under explosive conditions have proved considerably more successful in their predictions of detailed abundance features. Explosive carbon burning taking place at temperatures $T \sim 2 \times 10^9$ °K has been shown to reproduce extremely well many of the observed abundance features in the mass range $20 \lesssim A \lesssim 30$ (Arnett 1969b). Studies of explosive silicon burning (described in the previous section) for peak temperatures exceeding $T \sim 6 \times 10^9$ °K have been found to account very well for most of the abundance features in the mass range $48 \leq A \leq 58$ (Arnett, Truran and Woosley 1970). It is significant that it was not possible, in either of these cases, to account for the observed abundance features in the silicon-to-calcium region.

In this regard, the choice of appropriate nucleosynthesis conditions for explosive silicon burning requires further elaboration. Calculations of incomplete silicon burning have been found to produce relative abundances of elements from silicon through the iron peak ($14 \leq Z \lesssim 26$) in excellent agreement with the solar system abundances (Bodansky, Clayton and Fowler 1968; Truran, Arnett and Cameron 1967). A major difficulty with these calculations is the abundance of ^{58}Ni. Clayton and Woosley (1969) have shown that the abundance at masses ^{54}Fe, ^{56}Fe, ^{58}Ni can be produced in the correct relative amounts only for the case of complete equilibrium, if other features are not to be distorted. The high peak temperature quoted above for explosive burning, $T > 6 \times 10^9$ °K, assures that silicon burning will go to completion. Under these conditions, only a small fraction of the mass remains in the form of nuclei in the mass region $28 \leq A \lesssim 48$. This suggests that some other mechanism must be responsible for the synthesis of elements in this mass range. Clayton and Woosley have suggested that incomplete silicon burning taking place in zones located farther out in the star, and thus subjected to somewhat less extreme temperature-density conditions, may be responsible for the production of these intermediate mass nuclei. Michaud and Fowler (1970) have demonstrated that such an approach, making use of a range of quasi-equilibrium burning conditions, can provide excellent fits to the observed abundances.

Truran and Arnett (1970a) have been concerned with a somewhat different approach to the problem of the synthesis of intermediate mass nuclei. They argue that these nuclei may have been formed in the explosive burning of mass zones composed initially either of carbon and oxygen (the major products of helium burning) or of oxygen, magnesium and silicon (the major products of constant temperature carbon burning (Arnett and Truran 1969). Such initial conditions are more consistent with current stellar evolutionary models than is an extensive core region composed primarily of ^{28}Si.

The reaction network employed in these oxyten burning studies is illustrated in Fig. 12. This network is identical to that used in previous investigations of silicon burning. The inclusion of nuclei through the iron-peak region in these oxygen burning studies is necessary both to obtain a precise measure of the loss of neutrons required to build the neutron rich isotopes in the silicon-to-calcium region (a neutron excess $\eta = 0.0015$ is completely absorbed by a ^{54}Fe mass fraction $X(^{54}\mathrm{Fe}) = 0.04$) and to assure that discrepantly large abundances of iron-peak nuclei are not produced.

We wish now to estimate the conditions of temperature, density and time scale which are appropriate to considerations of explosive oxygen burning. These nucleosynthesis calculations must ultimately be carried out in the context of realistic stellar evolutionary models. In the absence of such detailed models, we can proceed in the manner described by Arnett (1969b).

We approximate the explosive hydrodynamic time scale by (Fowler and Hoyle 1964)

$$\tau_{HD} = (d \ln\rho/dt)^{-1}$$

$$\simeq (1/\sqrt{24\pi G\rho})$$

$$\simeq 446/\rho^{1/2} \text{ seconds}$$

The restriction that an appreciable fraction of the oxygen is burned may be imposed by demanding that the oxygen burning lifetime be of the order of the hydrodynamic time scale. The subsequent evolution may then be calculated if we relate $d\rho/dt$ to dT/dt, the time derivative of the temperature. Truran and Arnett (1970a) choose:

$$\frac{d \ln\rho}{dt} = 3 \frac{d \ln T}{dt}$$

We are concerned with the possible role of explosive oxygen burning in the synthesis of intermediate mass nuclei. The conditions under which this process might take place are uncertain due, primarily, to uncertainties in models of highly evolved stars. The estimates of the temperature and density conditions appropriate to explosive oxygen burning presented above define a reasonable range of burning temperatures. The approach taken by Truran and Arnett (1970a) was both to survey the dependence on temperature and density within this range and to consider the influence of the initial composition and the degree of neutron enrichment on the resulting abundance distribution.

The abundances resulting from explosive oxygen burning for a peak temperature $T_9 = 3.6$, a density $\rho = 5 \times 10^5$ g/cc and a neutron enrichment

$$\eta = \sum_i (N_i - Z_i) \frac{X_i}{A_i} = 0.0015$$

are compared to the solar system abundances (Cameron 1968) in Fig. 15. The initial abundances were chosen to be consistent with the view that the star had previously undergone both a carbon burning phase (Arnett and Truran 1969) and a brief neon burning phase, specifically: 54 percent ^{16}O, 30 percent ^{24}Mg, 2 percent ^{26}Mg (providing a neutron excess) and 14 percent ^{28}Si by mass. In this figure, the two abundance distributions have been normalized at ^{28}Si.

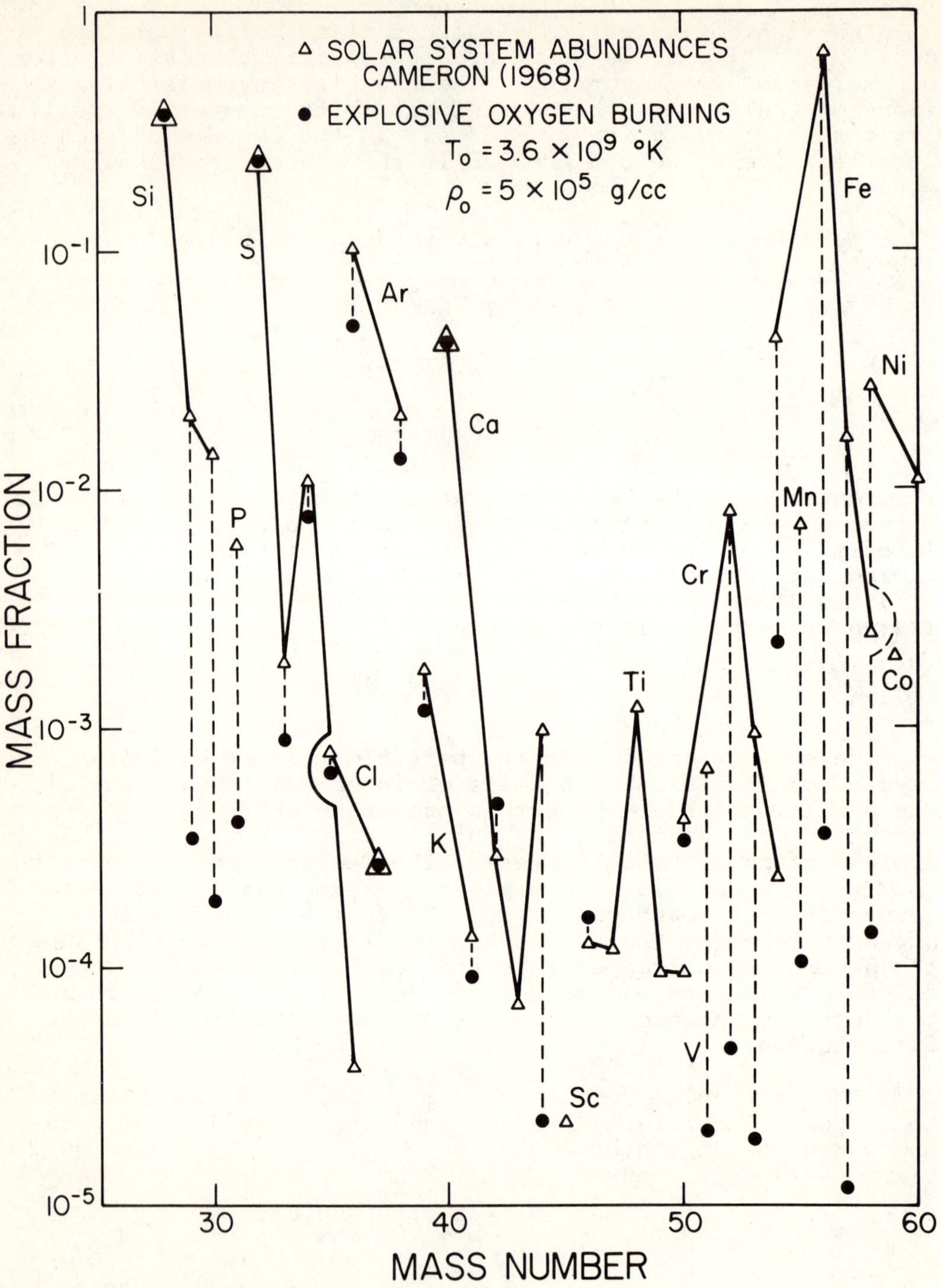

Figure 15. The nuclear abundances by mass resulting from the explosive burning of oxygen for an initial temperature and density of $T = 3.6 \times 10^9$°K and $\rho = 5 \times 10^5$g/cc are compared to the solar system abundances. The two distributions are normalized as ^{28}Si. From Truran and Arnett, Ap. J., 160, 181 (1970a).

These results indicate that an explosive oxygen burning process can account very well for many of the abundance features in the silicon-to-calcium region. The relative abundances of the α-particle nuclei ^{28}Si, ^{32}S and ^{40}Ca are well reproduced for these conditions. Agreement for these abundance ratios can be achieved, as we shall see, rather independent of the choice of initial composition or of the neutron enrichment. It is sensitive to the burning temperature and density insofar as these factors determine both the extent to which the initial ^{16}O abundance is depleted and the concentrations of free protons and α-particles in the medium.

It is more significant that these calculatons have reproduced the relative concentrations both of the neutron rich isotopes ^{33}S, ^{34}S and ^{42}Ca and of the isotopes of the odd-Z nuclei (^{35}Cl, ^{37}Cl, ^{39}K and ^{41}K). The isotopic ratios ^{41}K/^{39}K, ^{37}Cl/^{35}Cl, ^{33}S/^{32}S, ^{34}S/^{32}S and ^{42}Ca/^{40}Ca are all in agreement with accurately determined values obtained for terrestrial material to well within a factor of two. This is true, as well, for the ratio of the abundant isotopes of argon (^{38}Ar/^{36}Ar).

The calculated abundances for the neutron rich isotopes of silicon (^{29}Si and ^{30}Si) and for ^{31}P fail by more than an order of magnitude to reproduce their solar system values relative to ^{28}Si. This behavior is characteristic of oxygen burning conditions which provide agreement for the isotopes of sulfur, chlorine, argon, potassium and calcium. When a solar system ^{40}Ca/^{28}Si ratio is achieved, the excess neutrons in the system will generally distribute themselves among the neutron rich isotopes of nuclei from sulfur to calcium (assuming no substantial buildup of iron-peak nuclei occurs). The production of ^{29}Si, ^{30}Si and ^{31}P presumably will take place in material composed of carbon and oxygen and subjected to less extreme temperature and density conditions (see for example, Arnett 1969b).

The mass region $A > 42$ presents several interesting problems relevant to oxygen burning and silicon burning nucleosynthesis. The abundances of ^{46}Ti and ^{50}Cr have been produced in these calculations in their correct (solar system) ratios relative to ^{28}Si. In contrast, the calculated abundances for ^{43}Ca, ^{44}Ca, ^{45}Sc and ^{47}Ti are more than an order of magnitude lower than their solar system values relative to ^{28}Si. As both oxygen burning and silicon burning processes produce various of these isotopes, this intermediate mass region ($43 \leq A \leq 50$) may provide significant clues as to the appropriate nucleosynthesis conditions fot these two processes. With regard to the oxygen burning calculations illustrated in Fig. 15, we note finally that the abundances of nuclei in the mass range $A > 50$ are not explained for these conditions.

We wish now to explore the dependence of the abundances produced in explosive oxygen burning both on the initial composition of the medium and on the temperature and density conditions under which this burning process takes place.

a) composition dependence

Studies of the influence of the initial composition on the abundances resulting from these explosive oxygen burning calculations are summarized in Table 8 (Truran and Arnett 1970a)

TABLE 8

$X_{calculated}/X_{observed}$ FOR VARIOUS INITIAL COMPOSITIONS

Nucleus	Comp A $\eta = .0015$	Comp B $\eta = .0015$	Comp C $\eta = .0015$	Comp A $\eta = .0005$	Comp A $\eta = .0045$
^{16}O	.41	.48	.0014	.42	.38
^{28}Si	1.00	1.00	1.00	1.00	1.00
^{29}Si	.017	.0097	.10	.0093	.035
^{30}Si	.014	.0044	.52	.0035	.070
^{31}P	.068	.043	.046	.035	.15
^{32}S	1.09	1.48	.0064	1.15	.86
^{33}S	.45	.37	.014	.26	.74
^{34}S	.72	.36	.078	.21	2.74
^{36}S	.010	.0050	$< 10^{-3}$	.0030	.035
^{35}Cl	.86	.76	.0011	.47	1.32
^{37}Cl	1.00	1.21	$< 10^{-3}$	.64	1.22
^{36}Ar	.47	.86	$< 10^{-3}$	.52	.29
^{38}Ar	.67	.46	$< 10^{-3}$	.21	1.81
^{40}Ar	$< 10^{-3}$	$< 10^{-5}$	$< 10^{-5}$	$< 10^{-5}$	$< 10^{-3}$
^{39}K	.67	.84	$< 10^{-5}$	.40	.73
^{40}K	.018	.012	$< 10^{-5}$	.0058	.043
^{41}K	.63	.92	$< 10^{-5}$	.42	.61
^{40}Ca	.95	2.47	$< 10^{-5}$	1.16	.46
^{42}Ca	1.65	1.52	$< 10^{-5}$	.55	3.80
^{43}Ca	.0021	.0018	$< 10^{-5}$	$< 10^{-3}$	.010
^{44}Ca	.022	.075	$< 10^{-5}$	.028	.0085
^{46}Ca	.0058	$< 10^{-5}$	$< 10^{-5}$	$< 10^{-5}$	$< 10^{-3}$
^{48}Ca	$< 10^{-5}$	$< 10^{-5}$	$< 10^{-5}$	$< 10^{-5}$	$< 10^{-5}$
^{45}Sc	.032	.10	$< 10^{-5}$	.036	.014
^{46}Ti	1.28	1.70	$< 10^{-5}$	.54	2.03
^{47}Ti	.0050	.0033	$< 10^{-5}$	.0011	.034
^{50}Cr	.84	1.68	$< 10^{-5}$	.40	.72
^{54}Fe	.054	.47	$< 10^{-5}$	.028	.037

The ratios of the calculated mass fractions of the stable isotopes in the silicon-to-calcium region to their solar system mass fractions (Cameron 1968), normalized at ^{28}Si, are presented for a variety of initial compositions. For all cases, the initial temperature and density conditions were the same: $T = 3.6 \times 10^9$ °K and $\rho = 5 \times 10^5$ g/cc. The initial mass fractions for the various cases are given by:

Composition A $X(^{16}O) = 0.56 - 13\eta$
$X(^{24}Mg) = 0.30$
$X(^{26}Mg) = 13\eta$
$X(^{28}Si) = 0.14$

Composition B $X(^{12}C) = X(^{16}O) = 0.49$
$X(^{26}Mg) = 0.02$

Composition C $X(^{27}Al) = 0.04$
$X(^{28}Si) = 0.96$

For Composition A, three choices of the neutron enrichment parameter (η) have been studied: a value consistent with the solar system abundances ($\eta_{ss} = 0.0015$) and limiting values chosen to be $3\eta_{ss}$ and $\eta_{ss}/3$.

The abundance ratios shown for Composition A, $\eta = 0.0015$, correspond to the case illustrated in Fig. 15. The abundance features resulting for these initial conditions have been discussed in detail. We note, in addition, that the very neutron rich isotopes ^{36}S, ^{40}Ar, ^{46}Ca and ^{48}Ca as well as ^{40}K are not formed under these conditions. The production of these isotopes is generally attributed to neutron capture processes (Burbidge et al. 1957; Cameron 1963; Bodansky et al. 1968).

The abundance distribution (Composition B) resulting from the explosive burning of a region composed initially of equal concentrations of ^{12}C and ^{16}O (plus 2 percent ^{26}Mg to provide a neutron excess $\eta = .0015$) is weighted more heavily toward the heavier elements. In particular, the final ^{54}Fe mass fraction is only a factor of two below that value necessary to account for the $^{54}Fe/^{28}Si$ ratio found in the solar system. This increased production relative to composition A results in part from the release of protons and alpha-particles by the more rapid $^{12}C + ^{12}C$ reaction. Although the $^{40}Ca/^{28}Si$ ratio is somewhat high, the agreement for the concentrations of the isotopes ^{32}S, ^{33}S, ^{34}S, ^{35}Cl, ^{37}Cl, ^{36}Ar, ^{38}Ar, ^{39}K, ^{41}K, ^{40}Ca, ^{42}Ca, and ^{46}Ti relative to ^{28}Si is generally good. The abundances resulting from the explosive burning of zones composed initially of ^{12}C and ^{16}O have been found to be rather insensitive to the initial $^{12}C/^{16}O$ ratio. A slightly lower temperature might be preferred in order to reduce the ^{54}Fe mass fraction.

In contrast, the buildup of intermediate mass nuclei will not proceed readily under these burning conditions for mass zones composed primarily of ^{28}Si (Composition C). For this case, the initial ^{28}Si abundance is only slightly depleted and no significant buildup of even the isotopes of sulfur is achieved. Substantially higher temperatures ($T \gtrsim 4.2 \times 10^9$ °K) are required if zones composed of ^{28}Si are to burn sufficiently to account for the silicon-to-calcium abundances (Bodansky, Clayton, and Fowler 1968; Michaud and Fowler 1970).

The influence of the degree of neutron enrichment on the resulting abundances is illustrated in the last two columns of Table 8. The abundance ratios are given here for η = 0.0005 0.0005 and 0.0045, bracketing the value η = 0.0015 consistent with that of solar material. The main difference noted for the low value of η (0.0005) is the low abundance of the neutron rich isotopes ^{33}S, ^{34}S, ^{38}Ar and ^{42}Ca (relative to Composition A, η = 0.0015). The relative abundances of the chlorine and potassium isotopes are also somewhat low. The general agreement remains quite good, however, due to the more limited buildup of the heavier nuclei (^{46}Ti, ^{50}Cr, etc.) which would absorb much of the neutron excess.

For the case η = 0.0045, the agreement of the calculated abundances with some of the observed features is poor. In particular, it is clear that the isotopes ratios $^{34}S/^{32}S$, $^{38}Ar/^{36}Ar$ and $^{42}Ca/^{40}Ca$ are substantially increased relative to their solar system values. It is also apparent that this tendency to over-produce the more neutron rich isotopes has not significantly improved the situation with regard to ^{36}S, ^{40}Ar, ^{40}K, ^{43}Ca, ^{44}Ca, ^{46}Ca and ^{48}Ca. It seems unlikely that any significant amount of the solar system material could have been processed under such neutron rich conditions. Unfortunately, studies of the spectra of other stars would probably not be able to discriminate between even the η = 0.0005 and the η = 0.0045 cases, as the relative elemental abundances of silicon, sulfur, chlorine, argon, potassium and calcium are the same to better than a factor of two.

b) temperature-density dependence

The sensitivity of the resulting abundance distribution to the initial conditions of temperature and density is illustrated in Table 9. Again we give the ratios of the calculated mass fractions to the solar system mass fractions, normalized at ^{28}Si. For all cases, the neutron enrichment was taken to be η = 0.0015 and Composition A was used.

For a peak temperature of 3.6 x 10^9°K, the abundances have been calculated for densities of 10^5g/cc, 5 x 10^5g/cc (the standard case shown in Fig. 15) and 2 x 10^6g/cc. A comparison of these three abundance distributions reveals a strong density dependence of this oxygen burning process. While good agreement is achieved for many of the silicon-to-calcium nuclei at the intermediate density (5 x 10^5g/cc), rather severe distortions appear at slightly lower and higher densities. For the low density case, a significant mass fraction of ^{54}Fe is formed which tends to absorb the excess neutrons. The abundances of ^{33}S, ^{34}S and ^{38}Ar are therefore seen to be reduced. Moreover, the ratios $^{32}S/^{28}Si$ and $^{40}Ca/^{28}Si$ tend to be somewhat high, indicating that the burning may have gone too far. The enhanced buildup of heavier nuclei results in part from the larger proton and α-particle concentrations achieved in an oxygen burning "quasi-equilibrium" at this lower density. The longer expansion time scale also plays a role.

TABLE 9

$X_{calculated}/X_{observed}$ FOR VARIOUS TEMPERATURE DENSITY CONDITIONS

Nucleus	$3.6\times10^9\,°K$ 10^5g/cc	$3.6\times10^9\,°K$ 5×10^5g/cc	$3.6\times10^9\,°K$ 2×10^6g/cc	$3.9\times10^9\,°K$ 2×10^6g/cc	$3.3\times10^9\,°K$ 2×10^6g/cc
^{16}O	.34	.41	.22	.010	1.14
^{28}Si	1.00	1.00	1.00	1.00	1.00
^{29}Si	.0070	.017	.032	.0019	.69
^{30}Si	.0026	.014	.047	$< 10^{-3}$	1.21
^{31}P	.031	.068	.10	.0012	.19
^{32}S	1.53	1.09	.47	1.08	.010
^{33}S	.28	.45	.38	.053	.061
^{34}S	.23	.72	.84	.0011	.17
^{36}S	.0024	.010	.013	$< 10^{-3}$	.018
^{35}Cl	.57	.86	.42	.026	.0058
^{37}Cl	.93	1.00	.32	.13	.0014
^{36}Ar	.90	.47	.087	.47	$< 10^{-3}$
^{38}Ar	.32	.67	.28	.0014	$< 10^{-3}$
^{40}Ar	$< 10^{-5}$	$< 10^{-3}$	$< 10^{-3}$	$< 10^{-5}$	$< 10^{-3}$
^{39}K	.67	.67	.12	.031	$< 10^{-3}$
^{40}K	.0064	.018	.0078	$< 10^{-3}$	$< 10^{-3}$
^{41}K	.74	.63	.091	.074	$< 10^{-3}$
^{40}Ca	2.70	.95	.072	1.01	$< 10^{-5}$
^{42}Ca	1.05	1.65	.33	.0078	$< 10^{-5}$
^{43}Ca	.0010	.0021	$< 10^{-3}$	$< 10^{-3}$	$< 10^{-5}$
^{44}Ca	.073	.022	$< 10^{-3}$	.025	$< 10^{-5}$
^{46}Ca	$< 10^{-5}$	.0058	$< 10^{-5}$	$< 10^{-5}$	$< 10^{-5}$
^{48}Ca	$< 10^{-5}$	$< 10^{-5}$	$< 10^{-5}$	$< 10^{-5}$	$< 10^{-5}$
^{45}Sc	.090	.032	.0013	.017	$< 10^{-5}$
^{46}Ti	1.34	1.28	.13	.031	$< 10^{-5}$
^{47}Ti	.0019	.0050	.0019	$< 10^{-3}$	$< 10^{-5}$
^{50}Cr	1.51	.84	.037	.90	$< 10^{-5}$
^{54}Fe	.31	.054	$< 10^{-3}$	.65	$< 10^{-5}$

At the higher density, $\rho = 2 \times 10^6$g/cc, the buildup of heavy nuclei is seen to be substantially reduced. The abundance ratio $^{32}S/^{28}Si$ is low compared to the solar system value by approximately a factor of two, and the abundances of ^{36}Ar and ^{40}Ca fail by more than an order of magnitude to reproduce their solar system values relative to ^{28}Si. This behavior follows both from the low proton and α-particle concentrations maintained at this higher density and from the shorter expansion time scale.

The temperature dependence of the explosive oxygen burning abundance distribution is evident from the results shown in the last three columns of Table 9. For a density of 2×10^6 g/cc and a neutron enrichment $\eta = 0.0015$, the final abundances resulting from explosive oxygen burning for peak temperatures of $T_9 = 3.3$, 3.6 and 3.9 are compared to the solar system values. At the lower temperature, the production of nuclei heavier than ^{28}Si is restricted both by the slower rate of the $^{16}O + {}^{16}O$ reaction and by the low proton and α-particle abundances maintained in the system. The abundance of ^{32}S is low by approximately a factor of ten, and the situation is much worse for ^{36}Ar and ^{40}Ca. The heavy silicon isotopes, ^{29}Si and ^{30}Si, are produced in the correct ratios relative to ^{28}Si.

The abundances resulting from high temperature burning ($T_9 = 3.9$) indicate a substantial buildup of iron peak nuclei [$X_{calc.}(^{54}Fe)/X_{solar}(^{54}Fe) = 0.65$]. While the ratios $^{32}S/^{28}Si$ and $^{40}Ca/^{28}Si$ are in good agreement with the solar system values, the abundances of the neutron rich isotopes of sulfur, argon and calcium as well as those of the chlorine and potassium isotopes are low by more than a factor of ten.

It is apparent that, if agreement is to be achieved for the abundances of odd-Z nuclei and of the neutron-rich isotopes of sulfur, argon, and calcium, no significant production of iron-peak nuclei can take place in these mass zones. The most abundant nucleus produced in the mass range $A > 40$ is seen to be ^{54}Fe. The mass fraction of ^{54}Fe sufficient to exhaust the entire neutron excess is given by

$$\eta = \sum_i (N_i - Z_i)\frac{X_i}{A_i} = \frac{X(^{54}Fe)}{27}$$

For a neutron enrichment $\eta = 0.0015$, a ^{54}Fe mass fraction of 0.0405 will therefore absorb all of the excess neutrons. When the ^{54}Fe mass fraction approaches this value, the abundances of the neutron rich isotopes in the silicon-to-calcium region will be low relative to their solar system values.

The survey of explosive oxygen burning nucleosynthesis conditions described in this section indicates that the abundances of nuclei in the silicon-to-calcium region as well as that of ^{46}Ti can be accounted for by this mechanism. Many of the abundance features have been found to be sensitive to the choice of temperature and density conditions as well as to the initial composition. For this reason, detailed abundance predictions may be possible only in the context of realistic stellar models for stars in the advanced stages of evolution. The explosive ignition of zones composed largely of ^{12}C and ^{16}O in amounts consistent with post helium burning or post carbon burning conditions should provide a range of temperature and density conditions for the various mass zones consistent with those considered in this section and with the explosive carbon burning conditions considered by Arnett (1969b). It is possible as well that zones in which temperatures of $T_9 \gtrsim 5$ are achieved may account for many of the abundance features in the iron-peak region.) The final abundance distribution is then determined by summing over the various elements of ejected mass.

The "quasi-equilibrium" character of the oxygen burning process referred to in the previous discussion is particularly interesting. Bodansky et al (1968) have considered in detail the quasi-equilibrium features of the silicon burning process. They have demonstrated that, as silicon burning proceeds, the abundances of nuclei heavier than ^{28}Si are maintained in quasi-equilibrium with the remaining ^{28}Si and the number densities of free neutrons, protons, and α-particles. The α-particle and nucleon number densities are determined by the effective rate of silicon photodisintegration.

Preliminary analyses of the evolution of the abundances resulting from explosive oxygen burning indicate that the abundances of nuclei heavier than ^{28}Si again approach their quasi-equilibrium values relative to the ^{28}Si abundance and the α-particle and nucleon number densities. (This behavior is evident particularly at higher temperatures.) The feature which distinguished this oxygen burning "quasi-equilibrium" behavior from that of silicon burning is the source of the free nucleons and α-particles. During oxygen burning, these concentrations are determined by the rates both of the ^{16}O + ^{16}O reaction and the ^{12}C + ^{12}C reaction and of various photodisintegration and capture reactions on light nuclei ($A < 28$). This general behavior is partially responsible for the very different abundances resulting from oxygen burning at densities of 10^5 g/cc and 2×10^6 g/cc (Table 9). For the same temperature, the "quasi-equilibrium" concentrations of nucleons and α-particles tend to be smaller at the higher density; the abundance ratios $^{36}Ar/^{28}Si$ and $^{40}Ca/^{28}Si$ are therefore low compared to the solar system values.

While these calculations have demonstrated that the observed distribution of the more abundant isotopes in the silicon-to-calcium region can be accounted for by an explosive oxygen burning process, it is not clear that this mechanism of formation is unique. The silicon quasi-equilibrium studies of Bodansky et al (1968) and, more recently, those of Michaud and Fowler (1970) indicate that the relative abundances of the intermediate mass nuclei may be reproduced in zones composed primarily of ^{28}Si. The answer to the question as to which of these mechanisms is most appropriate must await further studies of stellar models.

The abundance features in the mass region $42 < A < 48$ may, however, provide some clues regarding the appropriate nucleosynthesis conditions. The abundances of ^{43}Ca, ^{44}Ca, ^{45}Sc and ^{47}Ti relative to ^{28}Si produced in these calculations are low in every case when compared to the solar system ratios. This seems to imply that the burning must proceed further, effectively exhausting the initial ^{16}O (this might take place for example in a mass zone subjected to a temperature in excess of $T_9 \sim 4$.) This condition would lead to an abundance configuration entirely equivalent to a silicon burning quasi-equilibrium distribution. It is encouraging that the abundances at masses A = 43, 44, 45, and 47 have been shown to be formed in quasi-equilibrium with ^{28}Si (Bodansky et al 1968; Michaud and Fowler 1970), in the form of ^{43}Sc, ^{44}Ti, ^{45}Ti, and ^{47}V, at somewhat higher temperatures ($T_9 > 4$). The abundances of the more massive nuclei ($48 \leq A \leq 58$) can then be accounted for on the basis of a complete statistical equilibrium distribution dominated (prior to beta decay) by ^{56}Ni (Section III).

REFERENCES

Arnett, W.D. 1966, Can. J. Phys., 44, 2553.

_______. 1967, Can. J. Phys., 45, 1621.

_______. 1969a, Astrophys. Space Sci., 5, 180.

_______. 1969b, Ap. J., 157, 1369.

Arnett, W.D. and Truran, J.W. 1970, Ap. J., 160, 959.

_______. 1969, Ap. J., 157, 339.

Arnett, W.D., Truran, J.W. and Woosley, S.E. 1970, "Nucleosynthesis in Supernova Models: II. The ^{12}C Detonation Model", in preparation.

Axel, P. 1962, Phys. Rev., 126, 671.

Blatt, J.M. and Weisskopf, V.F. 1952, Theoretical Nuclear Physics (New York: Wiley and Sons).

Bodansky, D., Clayton, D.D. and Fowler, W.A. 1968, Ap. J. Suppl., No. 148, 16, 299.

Burbidge, E.M., Burbidge, G.R., Fowler, W.A. and Hoyle, F. 1957, Rev. Mod. Phys., 29, 547.

Burbidge, G.R. 1962, Ann. Rev. Nucl. Sci., 12, 507.

Cameron, A.G.W. 1957, Chalk River Report, CRP-690.

_______. 1958, Ann. Rev. Nucl. Sci., 8, 299.

_______. 1963, Nuclear Astrophysics, Yale University lecture notes.

_______. 1968, The Abundances of the Elements, in The Origin and Distribution of the Elements, ed. L.H. Ahrens (New York: Pergamon Press).

Cameron, A.G.W. and Welkin, R.M. 1965, Can. J. Phys. 43, 1288.

Clayton, D.D. and Woosley, S.E. 1969, Ap. J., 157, 1381.

Clifford, F.E. and Tayler, R.J. 1965, Mem. R.A.S., 69, 21.

Colgate, S.A. and Johnson, H.J. 1960, Phys. Rev. Letters, 5, 235.

Colgate, S.A. and White, R.H. 1965, International Conference on Cosmic Rays, Juipur, India, 1963.

_______. 1966, Ap. J., 143, 626.

Fowler, P.H., Adams, R.A., Cowen, V.G. and Kidd, J.M. 1967, Proc. Roy. Soc., A301, 30.

Fowler, W.A. and Hoyle, F. 1964, Nucleosynthesis in Massive Stars and Supernovae (Chicago: University of Chicago Press).

Fowler, W.A. 1969, private communication.

Fowler, W.A., Caughlan, G.R. and Zimmerman, B.A. 1967, Ann. Rev. Astr. Ap., 5, 525.

Gilbert, A. and Cameron, A.G.W. 1965, Can. J. Phys., 43, 1446.

Greenstein, G.S. 1968, Astrophys. Space Sci., 2, 155.

Hansen, C.J. 1968, Astrophys. Space Sci., 1, 499.

Hansen, C.J. and Wheeler, J.C. 1969, Astrophys. Space Sci., 3, 464.

Hughes, D.H., Magurno, B.A. and Brussel, M.K. 1960, BNL-325.

Johnson, C.H. 1968, private communication.

Katgert, P. and Oort, J.H. 1967, Bull. Astron. Inst. Neth., 19, 239.

Kuchowicz, B. 1968, Nuclear Astrophysics, A Bibliographical Survey (New York: Gordon and Breach).

Macklin, R.L. and Gibbons, J.H. 1965, Rev. Mod. Phys., 37, 166.

Michaud, G., Scherk, L. and Vogt, E. 1970, Phys. Rev., Cl, 753.

Michaud, G. and Fowler, W.A. 1970, in preparation.

Oliva, P. and Prosperi, D. 1967, Nuo. Cim., 49B, 161.

Pagel, B.E.J. 1968, Russell Memorial Volume of Vistas in Astronomy, ed. A. Beer (New York: Pergamon Press).

Patterson, J.R., Winkler, H. and Spinka, H. 1969, in preparation.

Patterson, J.R., Winkler, H. and Zaidins, C.S. 1969, Ap. J., 157, 367.

Peebles, P.J.E. 1966, Phys. Rev. Lett., 16, 410.

Penzias, A.A. and Wilson, R.W. 1965, Ap. J., 142, 419.

Price, P.B., Rajan, R.S. and Tamhane, A.S. 1967, J. Geophys. Res., 72, 1377.

Reeves, H. and Salpeter, E.E. 1959, Phys. Rev., 116, 1505.

Salpeter, E.E. 1952, Phys. Rev., 88, 547.

_______. 1959, Ap. J., 129, 608.

Schwartz, R.A. 1967, Ann. Phys., 43, 42.

Stelson, P.H. and McGowan, F.K. 1964, Phys. Rev., 133, B911.

Truran, J.W. 1970, "Charged Particle Thermonuclear Reaction Rates", in preparation.

Truran, J.W. and Arnett, W.D. 1970a, Ap. J., 160, 181.

_______. 1970b, "Explosive Nucleosynthesis and the Composition of Metal-Poor Stars", in preparation.

Truran, J.W., Arnett, W.D. and Cameron, A.G.W. 1967, Can. J. Phys., 45, 2315.

Truran, J.W., Cameron, A.G.W. and Gilbert, A. 1966, Can. J. Phys., 44, 563.

Truran, J.W., Hansen, C.J., Cameron, A.G.W. and Gilbert, A. 1966, Can. J. Phys., 44, 151.

Truran, J.W., Arnett, W.D., Tsuruta, S. and Cameron, A.G.W. 1968, Astrophys. Space Sci., 1, 129.

Tsuruta, S. 1969, private communication.

Vogt, E. 1968, Adv. Nucl. Phys., 1, 261.

Vogt, E., Michaud, G. and Reeves, H. 1965, Phys. Letters, 19, 570.

Wagoner, R.V., Fowler, W.A. and Hoyle, F. 1967, Ap. J., 148, 3.

Woltjer, L. 1958, Bull. Astron. Inst. Neth., 14, 39.

Zwicky, F. 1965, Stellar Structure, Vol. VIII of Stars and Stellar Systems, eds. L.H. Aller and C.B. McLaughlin (Chicago: University of Chicago Press).

INSTABILITY PROBLEM IN NUCLEAR BURNING SHELLS

W. K. Rose

Massachusetts Institute of Technology
Cambridge, Mass.

I. Nature of Thermal Instability in Stars

The evolution of a star is characterized by three fundamental time scales. The dynamical time scale,

$$\tau_D \sim \left(\frac{R^3}{GM}\right)^{\frac{1}{2}} , \tag{1}$$

is the free fall time for the stellar configuration. The virial theorem implies that for a star in quasi-hydrostatic equilibrium the dynamical time scale is approximately twice the time for a sound wave to traverse the star. The Kelvin time scale,

$$\tau_K \sim \frac{GM^2}{RL} , \tag{2}$$

represents the approximate lifetime for a star of luminosity L if gravitational contraction is the only energy source. In discussions of unstable stars it will be convenient to define a local Kelvin time, $\tau \sim (\frac{3p}{2\rho})/\varepsilon$ where ε is the rate of energy generation per gram. The nuclear lifetime of a star is approximately

$$\tau_N \sim \frac{E_N}{L_N} \tag{3}$$

where E_N is the energy content of the dominant nuclear source and L_N is the total rate of energy generation due to this source. Because the energy content of hydrogen ($\sim 10^{52}$ erg/$M_\odot$) and helium ($\sim 10^{51}$ erg/$M_\odot$) are both much greater than the gravitational energy source ($\sim 10^{50}$ erg/$M_\odot$) for normal evolution, it is generally anticipated that evolutionary changes will proceed slowly (i.e., on a time scale $\sim \tau_N$). However, theoretical studies of stellar models have shown that with a suitable physical environment nuclear energy generation can take place in a thermally unstable manner and thus lead to significant evolutionary changes in a time scale that is much shorter than τ_N. The onset of nuclear burning under degenerate conditions (Mestel 1952) was the first suggested example of such an unstable environment in stars. The helium core flash and carbon core flash are examples of such instabilities. The burning of nuclear fuel under degenerate conditions is expected to lead to very high rates of nuclear energy generation because the pressure and density of degenerate matter are relatively insensitive to changes in temperature but the nuclear burning rates of most nuclear reactions are very temperature sensitive (e.g., $\varepsilon_{3\alpha} \propto T^n$, $\varepsilon_{C^{12}+C^{12}} \propto T^n$, $\varepsilon_{CN} \propto T^m$ where $n \sim 30$ and $m \sim 15$). Under degenerate conditions expansive cooling does not accompany increased rates of nuclear energy generation until degeneracy is lifted or an explosion has taken place. Recently Arnett (1968, 1969) and Rose (1968b, 1969) have pointed out that if a direct $e^- - \nu$ interaction exists, carbon core burning should take place under explosive conditions for stars of intermediate mass.

That degeneracy is not a necessary condition for unstable burning of nuclear fuel in stars was first pointed out by Schwarzschild and Härm (1965). They studied models for a 1 $M_\odot$ red giant star and found that the burning of helium in a shell takes place under thermally unstable conditions (i.e., the rate of nuclear energy generation increases on a local Kelvin time scale). Subsequently it was shown (Weigert 1966, Rose 1966) that thermal instability causes a star to undergo a series of relaxation oscillations of increasing peak amplitude. Figure 1 shows such relaxation oscillations for a .75 $M_\odot$ hydrogen-deficient star.

Although most of the results that we will discuss are fairly recent, the question of the stability of nuclear burning sources within a star is a very old one. The observed secular stability of the Sun on geological time scales indicates that the Sun behaves as a self-regulated system. Any increase (decrease) in the rate of nuclear energy generation will be compensated for by an expansion (contraction) that will return the overall rate of nuclear energy generation to its unperturbed value. Jeans argued that main sequence stars such as the Sun should be secularly stable (i.e., the rate of nuclear energy generation should be essentially constant over a Kelvin time scale). In carrying out this calculation, Jeans assumed that the perturbation about the equilibrium state is homologous (i.e., $\frac{\delta P}{P} = \frac{-4\delta r}{r}$, $\frac{\delta T}{T} = \frac{-\delta r}{r}$, $\frac{\delta \rho}{\rho} = \frac{-3\delta r}{r}$) and $\frac{\delta r}{r} =$ constant throughout the star. During the early development of a thermal instability, the star is able to maintain quasi-hydrostatic equilibrium because $\tau_K \gg \tau_D$. For this reason, the above approximation of homologous perturbations is reasonably correct for stellar models that approximate a polytrope of index 3. For these configurations a homology transformation exists such that the hydrostatic equation and the equation of mass conservation remain satisfied under the transformation. To show that this is true one first combines the hydrostatic equation and equation of mass conservation.

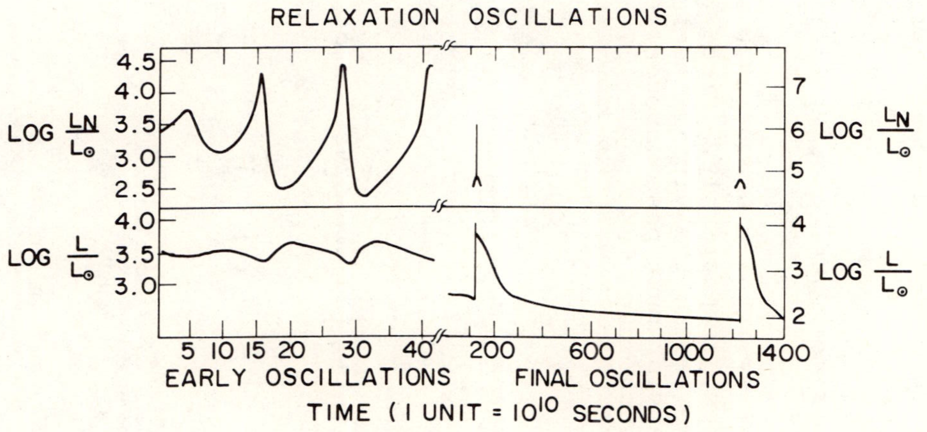

Figure 1. The nuclear energy generation, L_N, and the luminosity, L, are shown as a function of time for a thermally unstable, hydrogen-deficient, helium shell-burning star. The left-hand side of the figure shows a few of the early relaxation oscillations of the .75 $M_\odot$ star. The right-hand side shows the final two oscillations. The star contracts toward the white dwarf state after the final oscillation.

$$\frac{1}{r^2}\frac{d}{dr}\left(\frac{r^2}{\rho}\frac{dP}{dr}\right) = 4\pi G\rho \quad (4)$$

For a polytrope of index 3,

$$P = K\rho^{4/3} \quad (5)$$

Equations 4 and 5 can be transformed into the usual Lane-Emden equation and it can readily be shown that the solutions are invariant under a homology transformation.

Following Ledoux (1965) let us investigate the question of the secular stability of a simple stellar model for which the assumption of a homologous perturbation is valid. The virial theorem for a spherical star can be written

$$\frac{1/2\, d^2I}{dt^2} = 2\,K + V + 3\int_0^M P/\rho\, dM_r \quad (\text{if } \gamma = 5/3) \quad (6)$$

where I is the moment of inertia and K and V are the kinetic and potential energies, respectively. Since the velocities and accelerations are small, the virial theorem implies that a thermally unstable star must pass through a series of quasi-equilibrium states such that

$$\delta V + 3\int_0^M \delta(P/\rho)\, dM_r = 0 \quad (7)$$

where $\delta V = \int \frac{GM_r}{r}(\delta r/r)\, dM_r$.

For small changes in time the conservation of energy implies

$$\delta t\left[\int_0^M \varepsilon dM_r - L\right] = \delta V + \int_0^M \delta(3\frac{P}{\rho})\, dM_r - \frac{3}{2}\int_0^M \delta(\frac{\beta P}{\rho})\, dM_r \quad (8)$$

If

$$\varepsilon = \varepsilon_o\rho^{\mu}T^{\nu}$$

$$\kappa = \kappa_o\rho^{m}T^{-n} \quad (9)$$

$$\frac{L}{(4\pi r^2)^2} = \frac{4ac}{3\kappa}\, T^3\frac{dT}{dM_r}$$

then the variations of these quantities can be written

$$\frac{\delta\varepsilon}{\varepsilon} = \mu\frac{\delta\rho}{\rho} + \nu\frac{\delta T}{T}$$

$$\frac{\delta\kappa}{\kappa} = m\,\frac{\delta\rho}{\rho} - n\,\frac{\delta T}{T}$$

$$\frac{\delta L}{L} = 4\,\frac{\delta r}{r} + (4+n)\,\frac{\delta T}{T} - m\,\frac{\delta\rho}{\rho} + \frac{\frac{d}{dM_r}\left(\frac{\delta T}{T}\right)}{\frac{1}{T}\,\frac{dT}{dM_r}} \tag{10}$$

Assuming β and the molecular weight remain constant, and recalling that $\int \varepsilon_o dM - L_o = 0$ in the unperturbed state

$$\frac{\beta}{2}\,\frac{d}{dt}\,\delta V = \frac{\beta}{2}\int \frac{GM_r}{r}\,\frac{d}{dt}\left(\frac{\delta r}{r}\right) dM_r$$

$$= \int\left(\mu\frac{\delta\rho}{\rho} + \nu\frac{\delta T}{T}\right)\varepsilon dM_r \tag{11}$$

$$- L\left[4\,\frac{\delta r}{r} + (4+n)\,\frac{\delta T}{T} - m\,\frac{\delta\rho}{\rho} + \frac{\frac{d}{dM_r}\left(\frac{\delta T}{T}\right)}{\frac{1}{T}\,\frac{dT}{dM_r}}\right] \text{surface.}$$

The last term in equation 11 approaches zero at the surface.

If we assume that a homology transformation exists (i.e., if r, ρ, p, and T are solutions, then so are $r' = \alpha r$, $\rho' = \alpha^{-3}\rho$, $p' = \alpha^{-4}p$ and $T' = \alpha^{-1}T$) equation 11 becomes (for $\frac{\delta r}{r}$ = constant)

$$\frac{\frac{d}{dt}\left(\frac{\delta r}{r}\right)}{\delta r/r} = \frac{3}{\beta}\quad -(3\mu+\nu) + (n-3m)_{\text{surface}}\,\frac{L}{(-V)} \tag{12}$$

where

$$V = -\int \frac{GM_r}{r}\,dM_r$$

The above result implies that the perturbation will grow if $(n-3m) < 3\mu + \nu$ or decay if $3\mu+\nu > n-3m$. It suggests that a nuclear fuel with a high temperature sensitivity will tend to stabilize a star. This latter conclusion, which depends on the assumption of homologous modes, is precisely the reverse of what we shall conclude below for shell burning stars.

Although nondegenerate stars are stable against homologous perturbations, they may become unstable if the perturbation that preserves quasi-hydrostatic equilibrium is non-homologous and certain additional physical requirements are satisfied. When an increased release of nuclear energy occurs within the nuclear burning shell of a star, the shell expands. This expansion causes the pressure to drop within the shell source since the condition that the star remain in hydrostatic equilibrium implies that the pressure at a particular layer is proportional to the weight of the layers above it. However, if the shell source is sufficiently thin it can expand by an appreciable amount without producing a significant fractional change in pressure inside the shell source. For a nondegenerate gas the fractional changes in pressure, temperature and density have the following relation

$$\frac{\delta p}{p} = \frac{\delta \rho}{\rho} + \frac{\delta T}{T} \tag{13}$$

If the fractional change in pressure is small then

$$\frac{\delta \rho}{\rho} \sim -\frac{\delta T}{T} \tag{14}$$

For this above type of perturbation a temperature rise can accompany an expansion. On the other hand, the fractional change in temperature and density have the same sign for a homologous perturbation. If the nuclear energy source is the 3α process or the CN cycle then the rate of nuclear energy generation is highly temperature sensitive and a thermal runaway can be initiated in an expanding shell source so long as the energy input into the burning region exceeds the radiative losses from the region. We conclude that it is possible to have a thermal instability occur under nondegenerate conditions if the following two conditions are satisfied:

(1) The shell source must be sufficiently <u>thin</u> that the fractional perturbation in its density will produce a much smaller fractional perturbation in its pressure.

(2) The shell source must be sufficiently <u>thick</u> for radiative energy losses to be less than energy gains. For a nuclear fuel with a given temperature sensitivity, we can say that the shell source must have a certain critical mass.

Let us assume that condition (1) is satisfied and that the perturbation takes place under isobaric conditions. The equation of radiative transfer can be written

$$\frac{L}{4\pi r^2} = -\frac{4acT^3}{3\kappa\rho}\frac{dT}{dr} = -\frac{\sigma dT}{dr} \tag{15}$$

where σ = the effective thermal conductivity. If we neglect radiatin pressure, the equation of energy conservation can be written

$$\frac{3}{2}\rho^{2/3}\frac{d(P/\rho^{5/3})}{dt} = \varepsilon - \frac{dL}{dM_r} \tag{16}$$

where $dM_r = 4\pi r^2 \rho dr$. Since the temperature sensitivity of the 3α process and the CN cycle are quite high we can write

$$\frac{\delta\varepsilon}{\varepsilon} \sim \frac{\nu\delta T}{T} \quad (\nu \text{ large in eq. 9}) \tag{17}$$

Assume that the initial temperature distribution in the shell is as shown in Figure 2 and that radiative losses occur equally from both sides of the shell sources. If we further assume that the variation in the temperature gradient is much greater than either the variation in σ or r then we obtain

$$\delta \frac{dL}{dM_r} \approx \frac{2L_N}{(\frac{1}{2}\Delta M)} \quad \frac{\delta T}{(T-T_o)} \tag{18}$$

where ΔM is the mass inside the shell and $\varepsilon = L_N/\Delta M$. Equations 16, 17, and 18 imply (Schwarzschild and Härm, 1965)

$$\left[\nu - \frac{4T}{(T-T_o)}\right] \frac{\delta T}{T} = \left[(\frac{3}{2} \frac{P}{\rho})/\varepsilon\right] \frac{d\delta \ln (P/\rho^{5/3})}{dt} \tag{19}$$

where it is assumed that the unperturbed value of the time derivative given in equation 16 is small. If the left hand side of equation 19 is positive then the nuclear energy input will exceed the radiative losses and a positive temperature perturbation will lead to an increase in the entropy. Equation 19 indicates that the e-folding time for the development of a thermal instability (i.e., local Kelvin time) is

$$\tau \sim (\frac{3}{2} \frac{P}{\rho})/\varepsilon \tag{20}$$

If we neglect changes in r and σ as well as the time derivative in equation 16, equation 15 and 16 can be combined to give

$$\sigma \frac{d^2T}{dr^2} = - \rho\varepsilon \tag{21}$$

For the 3α process $\varepsilon_{3\alpha} = 3.9 \times 10^{11}\ f\rho^2\ x_4^3 \left(\frac{10^8}{T}\right)^3 e^{-\alpha/T}$. f is usually of order unity and $\alpha = 42.9 \times 10^8$. If $(T-T_o) << T$, and $T_o << 1.5 \times 10^9$, then

$$\varepsilon \approx \varepsilon_o\ e^{\alpha(T-T_o)/T_o^2} . \tag{22}$$

The theory of thermal explosions asserts that the absence of stationary solutions to Equation 21 determines the necessary condition for a thermal explosion. It can be shown (Landau and Lifshitz, 1959) that this requirement leads to the condition

$$\lambda \approx \rho\varepsilon_o \ (\frac{\alpha}{T_0^2})\ (\frac{\ell}{2})^2/\sigma \geq \lambda_{cr} \approx 1 \tag{23}$$

for instability to occur. If the computed physical parameters for a helium shell-burning star are substituted into equation 23, the condition for instability is satisfied by a number of order unity. An equivalent criterion for instability can be obtained by comparing the diffusion time for energy to leave

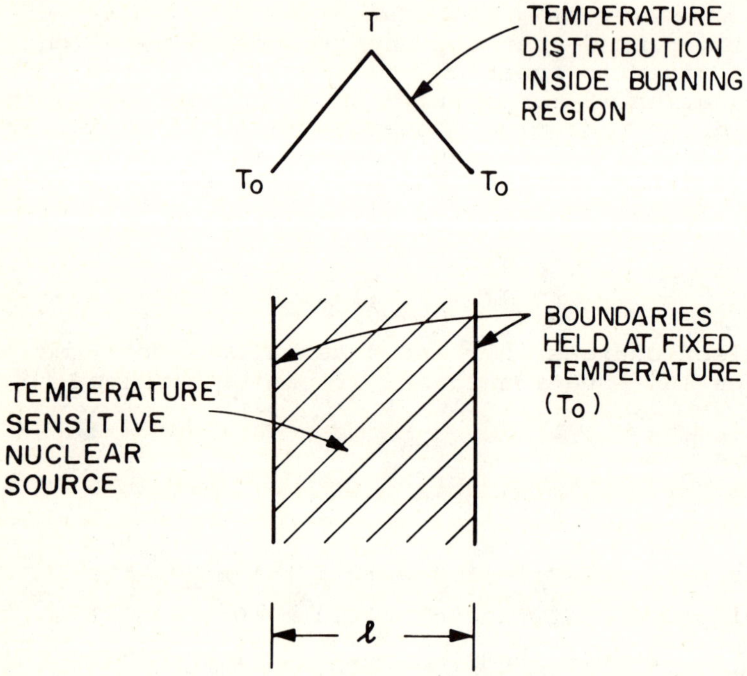

Figure 2. A simplified model for thermal instability is shown.

the shell with the thermal e-folding time τ. Instability is predicted if $\tau < \tau_{Diff.}$.

Condition 2 is not sufficient for instability to occur since it assumes that the readjustments that a star must undergo to preserve hydrostatic equilibrium produce very small fractional changes in the pressure. That a star must preserve quasi-hydrostatic equilibrium during the early phases of a thermal instability follows because $\tau_D << \tau_K$. Following Schwarzschild and Härm (1965) we estimate how a star maintains hydrostatic equilibrium during a thermal instability by linearizing the equation for hydrostatic equilibrium and mass conservation and then solving for the variations with the unperturbed model given by the calculations. The linearized equations are

$$M_r \frac{d\frac{\delta r}{r}}{dM_r} = \frac{1}{U}\left[-3\frac{\delta r}{r} - \frac{3}{5}\frac{\delta P}{P} + \frac{3}{5}\delta \ln(P/\rho^{5/3})\right]$$

$$M_r \frac{d\frac{\delta P}{P}}{dM_r} = \frac{V}{U}\left[4\frac{\delta r}{r} + \frac{\delta P}{P}\right] \tag{24}$$

where U and V are the usual homology invariants (defined below). Consider $\delta \ln(P/\rho^{5/3})$ as the inhomogeneous term. The homogeneous pair of equations has two independent solutions. One is regular at the center, the other at the surface. The solutions of the inhomogeneous equations can be expressed in terms of the homogeneous solutions and the unperturbed red giant models. Schwarzschild and Härm (1965) solved these equations and found that for their helium shell-burning red giant models, the pressure variation within the shell can be approximated as

$$\frac{\delta P}{P} \approx \frac{3}{5} Q \frac{\Delta r}{r} \delta \ln(P/\rho^{5/3}) \tag{25}$$

where Δr is the thickness of the shell source and Q is ~ -6 for their computed models. Since $\frac{\delta P}{P} = \frac{\delta T}{T} + \frac{\delta \rho}{\rho}$ equation 25 implies

$$\frac{\delta T}{T} = \frac{3}{5}\left(1 + \frac{2}{3} Q \frac{\Delta r}{r}\right) \delta \ln(P/\rho^{3/5}) \tag{26}$$

where Q is negative and depends on the computed models. The above equations show that for a sufficiently thin shell, the pressure variation will be small and the associated temperature variation positive if the entropy change $\delta \ln(P/\rho^{5/3})$ is positive. This result demonstrates that the occurrence of thermal instability in shell burning stars is possible during certain phases of stellar evolution.

The above discussion of thermal instability is somewhat approximate since we have not solved all four equations of stellar interiors simultaneously. To develop a more precise test for thermal instability we first replace the four differential equations with 4n-4 difference equations that are defined between n mass shells, add 4 boundary conditions and obtain solutions in the usual manner. We then see if we can add terms of the form $\delta P_i e^{t/\tau}$, $\delta T_i e^{t/\tau}$, $\delta L_{ri} e^{t/\tau}$, $\delta R_i e^{t/\tau}$ (i = 1, ...; n denotes mass shells, and τ is positive) to the existing solution and obtain a new solution. If such a positive τ exists then a determinant formed from the coefficients of 4n homogeneous equations vanished and we can solve for the eigensolutions δP_i, δT_i, δL_{ri} and δR_i. The eigenvalue, τ, corresponds to the e-folding time for the development of the thermal instability. This procedure is very similar to the Henyey method of obtaining solutions and therefore convenient to carry out in practice if it is recognized that the range of eigenvalues for which the determinant dips toward zero is very narrow. It is worth mentioning that an indefinitely large number of solutions for which τ is $\leq$ some negative number can always be found. We can say that these stable thermal modes describe the relaxation of the system if it is perturbed from its equilibrium state. Figure 3 shows a typical set of eigensolutions. These modes are not homologous and satisfy the physical requirement that we discussed previously.

Before discussing detailed solutions for stellar models, it is worth reviewing some elementary properties of shell sources. (Schwarzschild 1958; Hayashi, Hōshi and Sugimoto 1962). The hydrogen or helium burning shell source is sufficiently narrow that the variation of most physical variables is small across it. We make the simplifying assumption that there is no nuclear fuel for $r < r_1$ and a constant mass fraction of fuel for $r > r_1$. Define

$$V = -\frac{r}{P}\frac{dP}{dr} = \frac{\rho}{P}\frac{G M_r}{r}$$

$$n+1 = \frac{T}{P}\frac{dP}{dT} \tag{27}$$

Since these variables are nearly constant inside the shell we can write

$$\frac{P}{P_1} = \left(\frac{r}{r_1}\right)^{-V_1} \qquad \frac{T}{T_1} = \left(\frac{r}{r_1}\right)^{-V_1/(n_1+1)} \tag{28}$$

and $$\frac{\rho}{\rho_1} = \left(\frac{r}{r_1}\right)^{-n_1 V_1/(n_1+1)} \quad \text{(if shell is nondegenerate)}$$

The expression for the nuclear energy generation is

$$L_N = \int_{r_1}^{\infty} 4\pi r^2 \rho\varepsilon dr = \frac{4\pi r_1^3 \varepsilon_1 (n_1+1)}{[n_1(1+\mu)+\nu]V_1 - 3(n_1+1)} \tag{29}$$

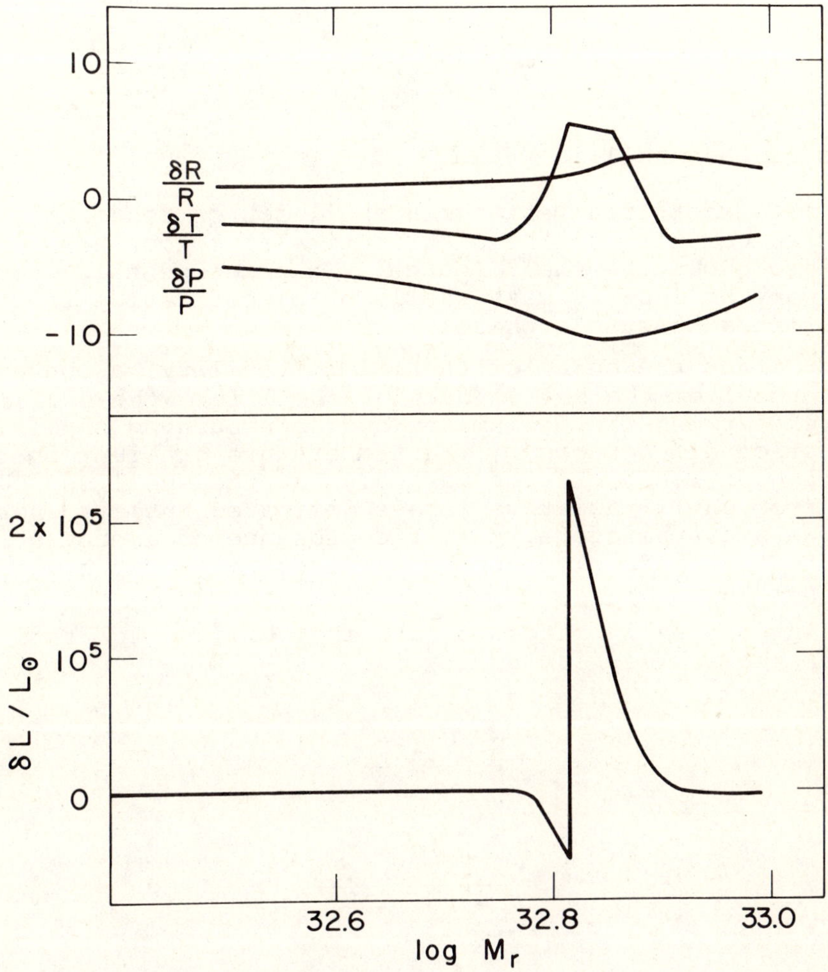

Figure 3. The relative amplitudes of the four dependent variables that describe an unstable thermal mode are shown as a function of mass point. The mass is measured in grams.

If the above expressions for the variation of the physical variables inside the shell are substituted into equation 29 L_N is determined as a function of the variable defined at the discontinuity r_1. The thickness of the shell, Δr, can be estimated from equation 29 to be $<0.1r_1$ and therefore since

$$\frac{\Delta M}{M_1} \approx U_1 \frac{\Delta r}{r_1} \tag{30}$$

where

$$U = \frac{4\pi r^3 \rho}{M_r} < 1$$

the mass of the shell is $<0.1M_{core}$.

A more exact treatment for a shell source requires a detailed calculation of the distribution of molecular weight throughout the shell source. In practice this requires that as many as 100-200 mass shells be placed inside the shell during an unstable phase.

Soon after the onset of thermal instability, a convection zone extends from the helium-burning shell into the helium-rich layer between the helium and hydrogen burning shells. If convection did not occur, the temperature gradient in the burning region would steepen markedly and the shell source would narrow until radiative losses dominated nuclear energy input. We see, therefore, that the presence of a convective layer affects the development of the instability in a very fundamental way.

During the early stages of the instability, the rate of change of the nuclear luminosity is

$$\frac{dL_N}{dt} = +\frac{1}{\tau} L_N \tag{31}$$

but

$$\tau = \frac{\frac{3}{2}\frac{P}{\rho}\,\Delta M}{\varepsilon\,\Delta M} \equiv \frac{A}{L_N}$$

so

$$\frac{dL_N}{dt} = +\frac{1}{A} L_N^2 \tag{32}$$

If L_O is the value of L_N at the onset of the instability (usually $\sim 100\ L_\odot$) when $t = 0$, we obtain

$$L_N = \frac{L_O}{1 - \frac{L_O\ t}{A}} \tag{33}$$

Equation (33) indicates that a small increase in the local time, t, for which a star remains thermally unstable (so that equation 31 applies) can make a very large increase in the value of L_N that is attained. The total nuclear input during the unstable period can be expressed as a function of L_N

$$E_N = \int_0^t L_N dt = A(\ln L_N - \ln L_0) \quad (34)$$

Thus the total energy input during a shell flash (about 10^{49} ergs) depends only logarithmically on L_N, the peak rate of nuclear energy generation.

II. Some possible consequences of thermal instabilities:

1. Complicates attempt to follow evolution of star in detail and makes unambiguous predictions.
2. Leads to pulsational instability in blue stars. (Novae?)
3. Mixing - S-process nucleosynthesis.
4. Shell flashes that take place under nearly degenerate conditions near peak of the red giant branch may lead to mass ejection. (Planetary nebulae?)
5. Possible connection with origin of Population II Cepheids.

Relaxation Oscillations and S-process Nucleosynthesis

In their classic paper Burbidge, et al., (1957) discussed several mechanisms by which heavy elements might be formed in stars. They introduced the S-process to explain the formation of certain elements which did not appear to be produced in the normal nuclear burning stages. In the S-process, the buildup of the heavy elements takes place by the capture of neutrons on a time scale long compared to the average beta-decay lifetimes of the isotopes involved. Free neutrons are captured and gamma rays emitted in the chain of captures. S-process nucleosynthesis implies (see Clayton, et al., 1961) that the abundances of isotopes of about the same A should be inversely proportional to the neutron capture cross sections. This mode of nucleosynthesis is believed responsible for the production of many nuclei with A between 23 and 46 (except those produced by silicon burning) and for a considerable proportion of the isotopes with A in the range 63 to 209.

The reaction

$$C^{13}(\alpha,n)O^{16} \quad (35)$$

is a likely source of neutrons in stellar interiors. The C^{13} to make these neutrons is produced when hydrogen is introduced into the helium-burning region by convection.

During a shell flash, a convective zone extends from the helium-burning shell into the helium rich layer between the helium and hydrogen-burning shells. The maximum extent of this convection is reached sometime after the peak shell flash. Schwarzschild and Härm (1967) showed that this convective zone can even extend into the hydrogen layer, and therefore mix a small amount of hydrogen into the helium-rich layer (which also contains considerable C^{12} and O^{16}). This mixing produces C^{13} in the chain of reactions

$$C^{12}(p,\gamma)N^{13}(\beta^+,\bar{\nu})C^{13}$$

and (36)

$$C^{13}(p,\gamma)N^{14} \text{ or } C^{13}(\alpha,n)O^{16}$$

The latter is more likely, since the helium abundance is greater. The rate equations for this chain are (Sanders, 1967)

$$\frac{dN(C^{12})}{dt} = -A\ N(C^{12})\ N(p)$$

$$\frac{dN(C^{13})}{dt} = A\ N(C^{12})\ N(p) - B\ N(C^{13})\ N(p)$$

$$\frac{dN(N^{14})}{dt} = B\ N(C^{13})\ N(p) \tag{37}$$

$$\frac{dN(p)}{dt} = -A\ N(C^{12})\ N(p) - B\ N(C^{13})\ N(p)$$

where N(x) is the number of nuclei of type x per cm^3, and A and B are the rate constants of the $C^{12} \rightarrow N^{13}$ and $N^{13} \rightarrow C^{13}$ reactions, respectively. Solving for $N(C^{13})$ as a function of $N(C^{12})$ and the initial $N(C^{12}) = N_0(C^{12})$,

$$\frac{N(C^{13})}{N_0(C^{12})} = \left(\frac{1}{\alpha-1}\right)\left[\frac{N(C^{12})}{N_0(C^{12})} - \left[\frac{N(C^{12})}{N_0(C^{12})}\right]^{\alpha}\right] \tag{38}$$

where $\alpha = B/A \cong 5$ for Carbon.

The maximum amount of C^{13} is produced when

$$\frac{N(C^{12})}{N_0(C^{12})} = \alpha^{1/(1-\alpha)} \cong 0.67 \tag{39}$$

Then, the maximum amount is

$$\frac{N(C^{13})}{N_0(C^{12})} = 0.13 \tag{40}$$

N^{14} is also produced. However, the principal result of convective mixing is to produce C^{13}. Most C^{13} then reacts with alpha particles to produce neutrons. These free neutrons can react with Fe^{56} to produce heavy elements. The abundance of N^{14} at the time of mixing is quite important since N^{14} has a large neutron capture cross section. However, much of the N^{14} can be destroyed by

$$N^{14}(\alpha,\gamma)F^{18} \tag{41}$$

since the helium-rich layer is convective, and much of the matter can be mixed with interior temperatures over 1.5 x 10^8 °K.

Although the detailed models are so complicated that one should not hope to predict very exactly how many protons are mixed with C^{12} and the precise physical conditions at the time of this mixing, it would appear that the succession of neutron fluxes required by Clayton, et al. (1961) could be readily attained during the relaxation oscillations. It is of course necessary to find an additional mechanism to eject this synthesized matter into the interstellar medium. The time interval for successive neutron fluxes is the period of the relaxation oscillation, i.e., about 3 x 10^5 years.

Final Evolution of a Low Mass Star

The present discussion (based on a paper by Rose and Smith 1970) describes a series of calculations that follows the evolution of a .85 $M_\odot$ star in an approximate but hopefully physically interesting manner from the initial stage of helium burning in a shell (a hydrogen-burning shell is also present) up to the point where the star has evolved into the white dwarf state. During most of its final evolution, the star is slowly ascending the red giant branch for the second time. The star's first evolution up the red giant branch is characterized by hydrogen burning in a shell and is ended by the helium flash (Härm and Schwarzschild 1964). Subsequently, the star evolves through a stage of helium burning in the core and hydrogen burning in a shell. This evolutionary stage, which takes place at a much lower luminosity ($\sim 100 L_\odot$), is associated with the origin of the horizontal branch (Faulkner and Iben 1966). After the star has burned the helium in its convective core, a helium-burning shell is formed, and the star ascends the red giant branch for the second time. Our calculations begin at this point.

During the early stages of helium burning in a shell, the helium-burning shell provides more than 50 percent of the total energy output from the star. However, the relative importance of hydrogen burning as the principal energy source increases as the star evolves up the red giant branch. The helium-burning shell does not burn in a stable manner, but undergoes a series of relaxation oscillations of increasing amplitude. Each relaxation oscillation consists of one or more helium shell flashes. In several previous papers (Rose 1966, 1967, 1968a), helium-shell burning has been studied for stellar models that are somewhat more idealized than those that we consider in this paper. In the first two papers only hydrogen deficient stars were studied. In the first paper, it was shown that thermal instability leads to relaxation oscillations such that the final oscillation is the most violent one. In the second paper, it was shown that thermal instability could under rather special circumstances lead to pulsational instability. In addition, it was shown that the high luminosities and short evolutionary time scales that appear to be associated with the central stars of planetary nebulae might be associated with the final relaxation oscillation of a thermally unstable star. In this picture for the formation of planetary nebulae, the progenitor is a red giant which has ejected a shell of mass as a result of very high rates of nuclear energy generation. The blue central stars of planetary nebulae are the remaining cores of these red giants. In the third paper, it was demonstrated that the presence of a hydrogen envelope would allow the helium-burning shell to evolve to higher densities than would otherwise be reached. It was pointed out that only shell flashes that were initiated under nearly degenerate densities could lead directly to the ejection of mass shells, such as might be necessary to explain the formation of planetary nebulae. In the present paper, it is shown that the density of the helium-burning shell increases as the star ascends the red giant branch. The shell flashes that occur during the early and less luminous stages of evolution are not sufficiently violent to lead to mass ejection. However, near the tip of the red giant branch ($L \sim 10^4 L_\odot$ for the stellar models in question) the shell flashes may be sufficiently violent to lead directly to mass ejection. This point will be studied more carefully in a future paper.

Hayashi, Hōshi, and Sugimoto (1965) found two helium shell flashes in their study of a low mass star. Their solutions differ from ours because they assumed that the C-O core is isothermal and did not find thermal instability under nondegenerate conditions. Schwarzschild and Härm (1967) have carried out very extensive calculations for the early shell flashes of a star very similar to the one studied in this paper. The good agreement between their results and the results we have obtained for the early shell flashes constitutes a test of the adequacy of our approximations.

The mass of the computed models ($M = .85\ M_\odot$) was chosen to be sufficiently close to $1.0\ M_\odot$ that the solutions may approximate the eventual evolution of the sun and at the same time be close to some estimates for the masses of population II stars. The estimated average mass for a white dwarf (Greenstein and Trimble 1967, Weidemann 1968) is also close to this mass. The assumed helium abundance is $Y = .29$, and the heavy metal abundance is $Z = .01$. Although these choices for the chemical composition are somewhat arbitrary, the final evolutionary stages do not appear to depend very sensitively on the assumed values.

The computed evolutionary sequence was not followed all the way from the main sequence, but was started near the beginning of helium burning in a shell. At this stage, the helium-burning shell supplies somewhat more than 50 percent of the energy output from the star. The lower energy content of helium ($6\text{-}8 \times 10^{17}$ erg/gm) as compared to hydrogen (6×10^{18} erg/gm) implies that at this stage the helium-burning shell approaches the surface much more rapidly than the hydrogen-burning shell. For this reason and also because the nuclear lifetime of the star is much longer than a Kelvin time, the initial choice of the mass point for the helium-burning shell is not very important. Previous calculations for earlier evolutionary stages (Faulkner and Iben 1966, Hartwick, Härm and Schwarzschild 1968) were used to fix the mass point of the hydrogen-burning shell at $.56 M_\odot$ for the initial model.

Figure 4 shows the total rate of helium burning, the total rate of hydrogen burning, and the surface luminosity as a func-

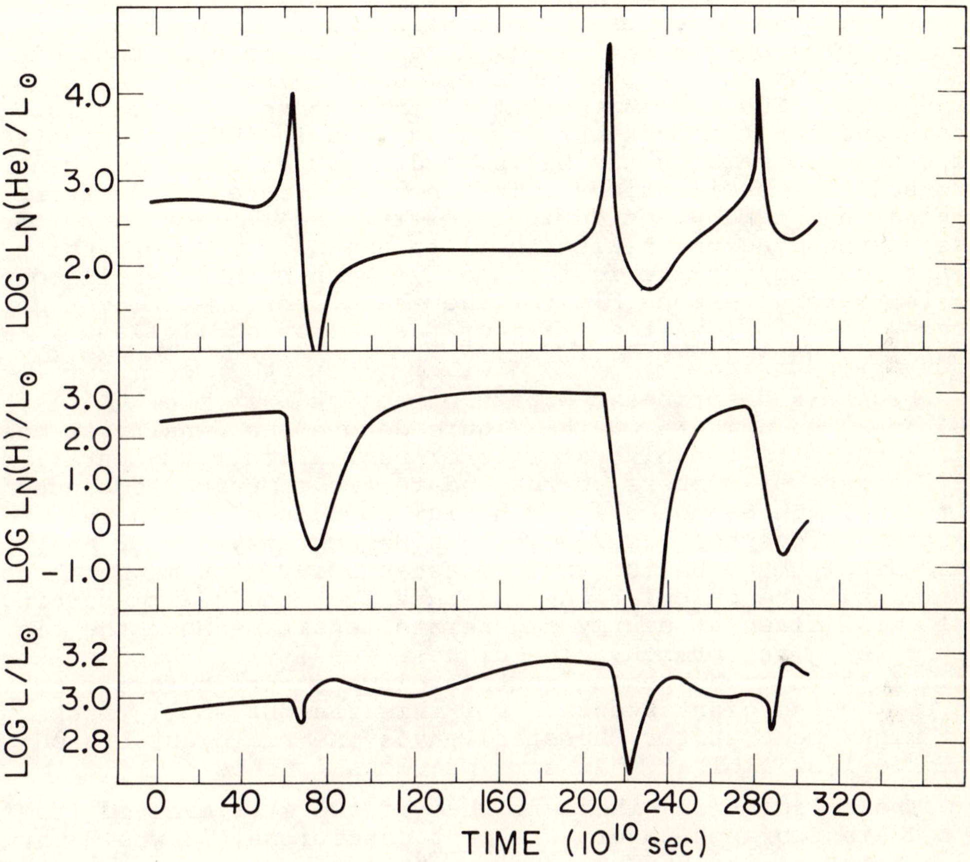

Figure 4. $L_N(He)/L_\odot$, $L_N(H)L_\odot$ and $L/L_\odot$ are shown as a function of time for the first two relaxation oscillations.

tion of time for the first two relaxation oscillations. Perhaps the most striking behavior of the oscillations is the appearance of two helium-burning peaks associated with the second relaxation oscillation. Presumably all subsequent oscillations have more than one peak. Relaxation oscillations of this type can take place only when the hydrogen-burning shell is a significant energy source. This type of relaxation oscillation has been obtained by Schwarzschild and Härm (1967). The shortest e-folding time for the rate of increase of nuclear burning is $\sim 10^9$ seconds for the early oscillations, and consequently, the star can easily maintain itself in hydrostatic equilibrium. During each shell flash, a convection zone extends from the helium-burning shell source into the helium-rich layer. The maximum extent of this convection zone is attained after the peak burning stage and includes about one half the mass between the helium and hydrogen shell sources for the early oscillations. The increased rates of nuclear burning that take place during a helium-shell flash cause the mass surrounding the helium shell to expand. This expansion lowers the temperature inside the hydrogen-burning shell source and is thereby responsible for the sharp decrease in hydrogen burning that follows the peaks of the helium shell flashes shown in Figure 4.

Figure 5 describes the path of the star in the H-R diagram. The numbers on the Figure denote the density at the mass point with the highest rate of nuclear energy generation. The two sets of numbers correspond to two computed sequences of models. In Sequence 1 the helium abundance within the helium-burning shell is Y = .5. In Sequence 2, Y = .2 and, consequently, the helium shell flashes occur at somewhat higher densities. The calculations indicate that the helium-burning shell will flash at nearly degenerate densities near the tip of the red giant branch. The highest density at which helium shell flashes are expected to take place occurs before reaching the tip of the giant branch. For this reason, shell flashes that might occur after the star leaves the red giant branch would be less violent than some previous flashes.

The calculations indicate that if the star evolved from the red giant branch under quiescent conditions, it would burn $\sim$ 2 percent of its nuclear energy content before reaching the blue end of its evolutionary path. The computed time scale for the star to evolve from the tip of the red giant branch to the region in the H-R diagram where the central stars of planetary nebulae are observed is approximately 70,000 years. It is evident that the calculated models, for which helium shell flashes have been suppressed, do not agree with the observationally indicated evolutionary path for the central stars of planetary nebulae (O'Dell 1963, Harmon and Seaton 1964, Seaton 1966).

Figure 6 shows the interior temperature distributions for a selected group of models. The relatively low temperatures in the core due to neutrino cooling, as shown in the Figure, can be seen to be far short of the 6-8 x 10^8°K necessary for carbon burning. The very effective cooling of the interiors of the computed models explains the very sudden drop in luminosity that the star experiences as soon as it has exhausted its nuclear energy sources.

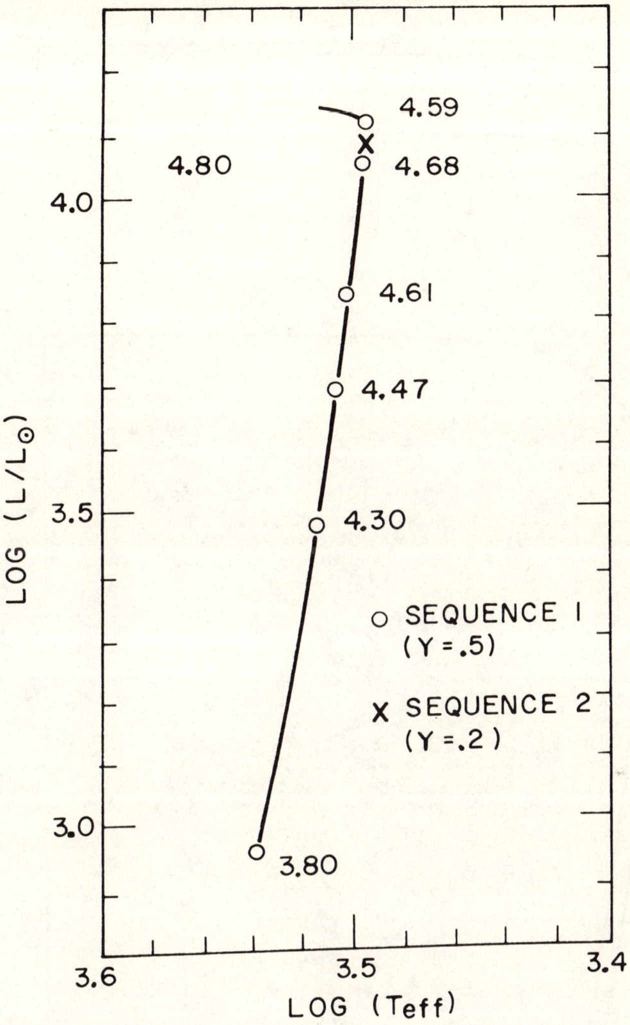

Figure 5. The path of the star on the H-R diagram is shown for evolution up red giant branch. The numbers given refer to log ρ at the helium burning shell prior to a shell flash.

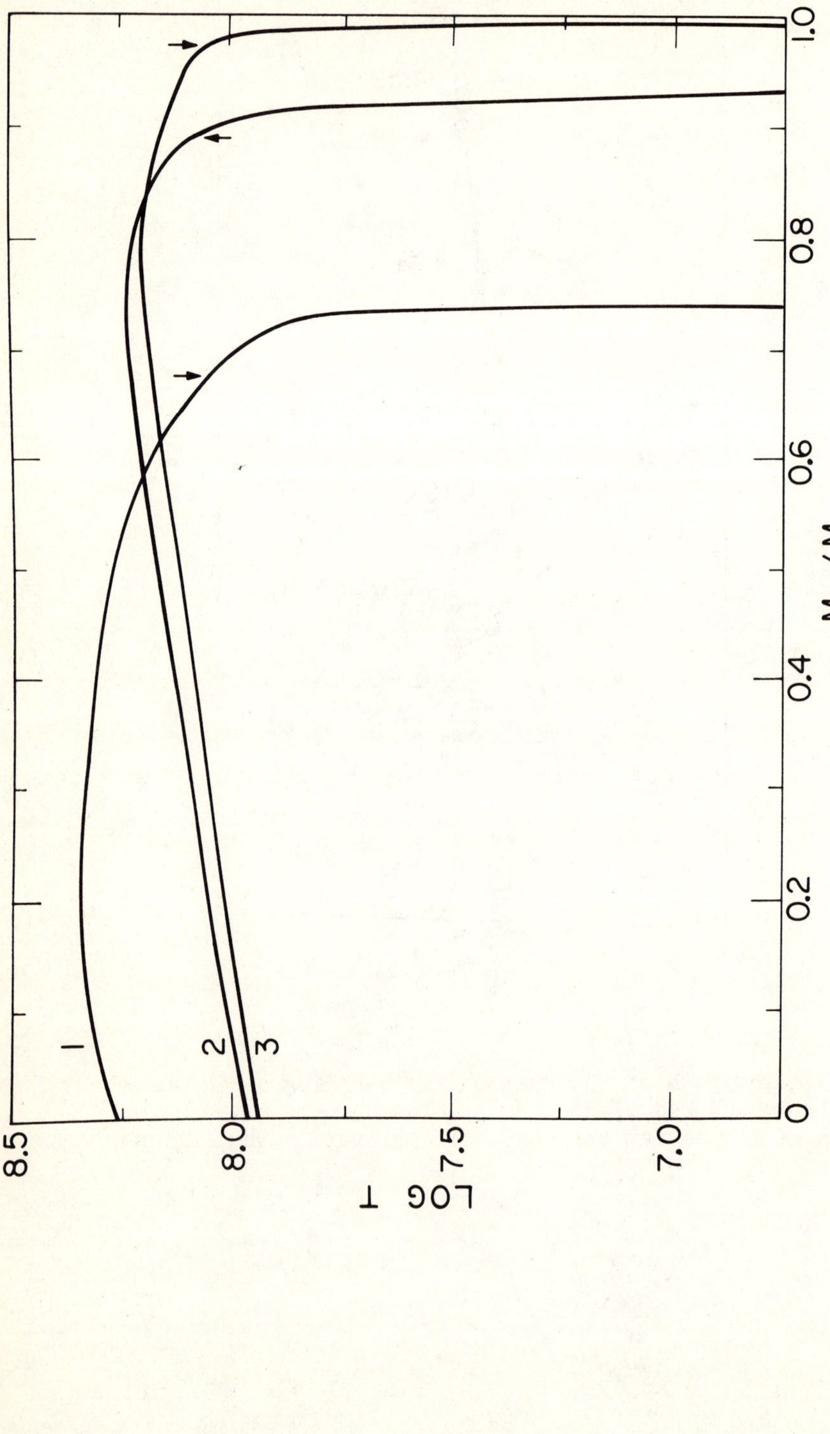

Figure 6. The interior temperature distribution is shown for selected models. Curve 1 shows the interior temperature distribution on the red giant branch when $L = 4000\ L_{\odot}$. Curve 2 shows the interior temperature distribution at the tip of the red giant branch. Curve 3 shows the temperature distribution as the star enters the region of observed central stars of planetary nebulae. The arrows show the position of the helium burning shell.

The calculations indicate that thermal instability is present during the early stages of helium-burning in a shell. At this early evolutionary stage, the shell flashes are much too weak to lead directly to mass loss or mixing. At a somewhat later evolutionary stage, when the luminosity has risen to a $\sim$4000 $L_\odot$ and the density in the helium-burning shell to $\sim 3 \times 10^4$ gm/cm^3, the peak nuclear burning becomes $\sim 2.6 \times 10^7$ $L_\odot$ as compared to $\sim 5 \times 10^4$ $L_\odot$ during the second shell flash. However, the star can still maintain quasi-hydrostatic equilibrium throughout the flash. In the present calculations, we have not investigated the question of mixing during this intermediate evolutionary stage. The calculations of Schwarzschild and Härm (1967) indicate that a small amount of mixing probably takes place during some of the oscillations of intermediate amplitude.

For the initial stages of helium burning in a shell, the total amount of nuclear energy released during a single relaxation oscillation is comparable for helium and hydrogen burning. However, as the evolution proceeds, the total amount of hydrogen burning must eventually exceed that due to helium burning, since the energy content of hydrogen is 8 to 10 times greater than that of helium. The calculations indicate that helium burning proceeds in relaxation oscillations of increasing amplitude. When the helium-burning shell approaches the hydrogen-burning shell too closely, the temperature is no longer sufficient for helium burning to continue. During those intervals when the helium shell is extinguished, the hydrogen-burning shell becomes the principal energy source. The helium-rich layer contracts as the hydrogen-burning shell approaches the surface. Contraction continues until the temperature within the helium shell becomes sufficiently high for helium burning to begin again. The most interesting question concerning shell flashes is whether or not they will lead directly to the ejection of a shell of mass and therefore perhaps to the formation of planetary nebulae. The calculations indicate that as the density within the helium-burning shell immediately prior to a shell flash increases, the peak rates of nuclear energy generation also increase. Although we intend to deal more completely with this question in a future paper, it would appear than an explosive event will result from a shell flash if the e-folding time for the development of the thermal instability ($\tau \sim (3P/2\rho)/\varepsilon_{3\alpha}$) becomes shorter than the convective mixing time,

$$\tau_{mix} = \frac{(\Delta R)^2}{Hv} \qquad (42)$$

ΔR is the thickness of the convective zone at the peak of a shell flash, v is the convective velocity, and H is the pressure scale height. For the anticipated conditions during a shell flash ($V \sim 10$ km/sec, $H \sim 2 \times 10^8$ cm and $\Delta R \sim 10$ H), $\tau_{mix} \sim 2 \times 10^4$ seconds. The calculations indicate that the condition $\tau < \tau_{mix}$ will probably be satisfied for the computed models if the rate of nuclear energy generation becomes as high as $\sim 10^9$-10^{10} $L_\odot$. Such high rates of nuclear energy generation are anticipated if the density in the helium-burning shell at the onset of a shell flash exceeds about 6-8 x 10^4 gm/cm^3. Figure 4 indicates that such high densities are attained near the tip of the red giant branch. The more detailed treatment of the helium-burning shell source given by Schwarzschild and Härm (1967) indicates that the highest rate of nuclear energy generation occurs as a mass point within the shell for which $Y \sim .1$. This result indicates that Sequence 2 in Figure 5 probably gives a more realistic estimate of the relevant density in the shell than Sequence 1. If the development of a thermal instability proceeds on a time scale that is shorter than the convective mixing time, the shell source will narrow, and consequently the thermal e-folding times will be greatly reduced. This latter point has been checked by means of a sequence of computed models. Under these conditions, an explosive event is likely to occur. It is important to emphasize that for the luminous red giants studied in this paper, the gravitational binding energies of the hydrogen-rich envelopes ($\sim 10^{45}$ erg) are much less than either the binding energy of the core ($\sim 10^{49}$ erg) or the expected nuclear energy release during a shell flash ($\sim 10^{48}$-10^{49} erg).

We have carried out the present calculations for the final evolutionary stages in the approximation that the helium-burning shell flashes can be neglected. Because the energy content of hydrogen is much greater than that of helium, this approximation is realistic so long as helium burning in a shell does not lead to an explosive event or cause very extensive mixing. The computed models evolve off the red giant branch when the hydrogen-rich envelope has been reduced to about 2 percent of the mass of the star. As discussed above, the subsequent evolution of the computed models do not give good agreement with the evolutionary path estimated for the central stars of planetary nebulae. This result indicates that theories for the origin of planetary nebulae that depend on extensive ionization zones on the red giant branch (Lucy 1967, Paczynski and Ziolkowski 1968, Roxburgh 1967) will likewise not agree with the observations. The calculations indicate that a low mass star with a hydrogen-rich envelope will enter its final evolutionary stages with low interior temperatures ($<2 \times 10^8$ °K) as a consequence of the neutrino cooling that takes place on the red giant branch. Because the interior temperatures are low, the star will evolve through the region of the H-R diagram occupied by the central stars of planetary nebulae in a time scale ($\sim$4,000 years) that is short compared to that given by the observations ($\sim$17,000 years). This discrepancy between observations and calculations can be removed if a helium shell flash takes place at about the time of the formation of the planetary nebula (Rose, 1967, 1968a).

If neutrino cooling were not included in the calculations, the behavior of the shell flashes would remain substantially unchanged. However, the time scale for the luminosity of the star to be reduced from $\gtrsim 10^4 L_\odot$ to $\lesssim 10^2 L_\odot$ would very likely exceed that estimated for the central stars of planetary nebulae. The present calculations suggest (see also Chin, Chiu, and Stothers 1966, Stothers 1966) that the existence of central stars of planetary nebulae with $L < 100\ L_\odot$ may constitute one of the most interesting astrophysical tests for the existence of a direct electron-neutrino interaction. This important point needs further study.

Thermal Instability and the Origin of Population II Cepheids

Recent calculations (Schwarzschild and Härm 1970) have shown that helium shell-burning stars of low mass (.65$M_{\odot}$) and high helium content (Y = .3) will after each shell flash pass through a loop in the H-R diagram that reaches into the strip occupied by Population II Cepheids. They suggest that the occurrence of such loops might explain the origin of Population II Cepheids. Models of stars with higher mass ($\gtrsim$ .75$M_{\odot}$) or lower helium content (Y $\lesssim$ .1) do not leave the red giant branch after each shell flash (Schwarzschild and Härm 1970, Rose and Smith 1970).

It has recently been pointed out (Wallerstein 1970) that the available data on Cepheids in globular clusters supports the results of these calculations because Cepheids are found only in globular clusters with blue horizontal branches. Blue horizontal branches require low mass and probably high helium abundances to be explained on the basis of stellar evolutionary calculations.

References

Arnett, W.D. 1968, Nature, 219, 1344.

_______. 1969, Ap. J., in press.

Chin, C.W., Chiu, H.Y., and Stothers, R. 1966, Annals of Physics, 39, 280.

Faulkner, J., and Iben, I. 1966, Ap. J., 144, 995.

Greenstein, J., and Trimble, V. 1967, Ap. J., 149, 283.

Härm, R., and Schwarzschild, M. 1964, Ap. J., 139, 594.

Harmon, R., and Seaton, M. 1964, Ap. J., 140, 824.

Hartwick, F., Härm, R, and Schwarzschild, M. 1968, Ap. J., 151, 389.

Hayashi, C., Hōshi, R., and Sugimoto, D. 1965, Progr. Theoret. Phys., 34, 885.

Landau and Lifshitz 1959, Fluid Mechanics (Reading: Addison-Wesley).

Ledoux, P. 1965, Stellar Structure (Chicago: Chicago University Press).

Lucy, L. 1967, A. J., 72, 813.

Mestel, L. 1952, M.N.R.A.S., 112, 598.

O'Dell, C.R. 1963, Ap. J., 138, 67.

Paczynski, R., and Ziolkowski, J. 1968, Proc. I.A.U. Symp. 34: Planetary Nebulae.

Rose, W. K. 1966, Ap. J., 146, 838.

_______. 1967, Ap. J., 150, 193.

_______. 1968a, Proc. I.A.U. Symp. 34: Planetary Nebulae.

_______. 1968b, A. J., 73, S116.

_______. 1969, Ap. J., 155, 491.

Rose, W. K. and Smith, R. L. 1970, Ap. J. 159, 903.

Roxburgh, I. W. 1967 Nature, 215, 838.

Schwarzschild, M. 1958, Structure and Evolution of the Stars, (Princeton, N.J.: Princeton University Press).

Schwarzschild, M. and Härm, R. 1965, Ap. J., 142, 855.

_______. 1967, Ap. J., 150, 1961.

_______. 1970, Ap. J., 160, 341.

Seaton, M. 1966, M.N.R.A.S., 132, 961.

Stothers, R. 1966, A. J., 71, 943.

Wallerstein, G. 1970, Ap. J., 160, 346.

Weidemann, V. 1968, Ann. Rev. of Astron. and Astrophys., 6 (Palo Alto, Calif: Annual Reviews, Inc.).

Weigert, A. 1966, Zs, f. Ap., 64, 395.

19

RELATIVISTIC STARS AND GRAVITATIONAL WAVES
--AN ACCOUNT FOR NON-RELATIVISTS--

K. S. Thorne

California Institute of Technology
Pasadena, Calif.

1. Introduction

Recent astronomical discoveries (quasars, pulsars, cosmic microwave radiation, gravitational waves, ...) are forcing theoretical astrophysicists to confront situations in which the Newtonian theory of gravity is inapplicable or is, at best, a poor approximation to the correct, relativistic theory of gravity. This series of lectures will describe some of the new phenomena which the astrophysicist encounters when he turns his attention from the Newtonian theory to relativistic theories. These lectures, like current research, will concentrate almost exclusively on Einstein's relativistic theory of gravity ("general relativity theory," or GRT), but they will mention occasionally other theories, such as the scalar-tensor theories of Jordan, Dicke, and Brans.

These lectures are not intended for experts in general relativity (except, perhaps, as bedtime reading). Rather, they are pitched at the level of a first-year graduate student in astronomy, who has never studied general relativity theory and who perhaps never will--at least not in depth. No attempt is made here to teach the formalism of general relativity. Instead, the physical ideas behind it are described in qualitative terms, and the quantitative and qualitative features of its application to astrophysics are presented, largely without derivation. The reader who wants derivations and greater mathematical detail will find them in the author's 1966 Les Houches lectures (Thorne 1967) and in a recent supplement to them (Thorne 1971a) in a monograph which the author is now writing (Thorne 1971b), in the book Gravitation Theory and Gravitational Collapse by Harrison, Thorne, Wakano and Wheeler (1965), and in the treatise Relativistic Astrophysics by Zel'dovich and Novikov (1967).

These lectures cover (much too briefly) three general-relativistic topics of interest to astrophysics: the relativistic theory of stellar structure; the stability of relativistic stellar models; and the theory of the emission and detection of gravitational waves. Important topics not included here are gravitational collapse, relativistic star clusters, and a variety of problems in relativistic cosmology. I thank J. Warman and V. L. Trimble for assistance with the preparation of this manuscript during the course of the Summer Institute.

2. Theories of Gravity

2.1 Special Relativity and Tidal Gravitational Forces

Einstein produced two theories of relativity--the special theory (SRT) in 1905, and the general theory (GRT) in 1915. SRT describes the laws of physics in the absence of gravity, while GRT describes the laws in the presence of gravity. Let us make this more explicit by discussing first SRT, then GRT.

The most fundamental concept of SRT is the inertial reference frame. An inertial reference frame is a rectilinear latticework of rods and clocks with the property that any uncharged test particle initially at rest in the latticework remains always at rest, and any particle initially in motion continues onward with ever constant velocity.[1]

Inertial reference frames enter into SRT by way of the fundamental principle of SRT, which states: "The laws of physics are the same in every inertial reference frame." The meaning of this principle is clarified by the following gedanken experiment: Write down a set of instructions for the construction of a particular (but arbitraty) apparatus, and for its use to perform a particular laboratory experiment. (The experiment must be entirely self-contained; it must involve only systems whose states are prepared by the experimenter.) Xerox many copies of these instructions and give them to many experimenters, who are in different inertial frames. If those experimenters are competent, they must all obtain identical results for the experiment--to within the accuracy permitted by the technology of the day.

From this fundamental principle of relativity, one can derive the various consequences of SRT such as time dilation, Lorentz contraction, and breakdown in simultaneity. [See Taylor and Wheeler (1967) for a careful and beautiful treatment.]

SRT is valid only in those regions of spacetime where inertial reference frames exist--for example, in interstellar space. However, in regions which contain inhomogeneous gravitational fields (e.g.,near the earth), inertial frames cannot exist. To see this, imagine two uncharged test particles initially at rest relative to each other in an external, inhomogeneous gravitational field. As time passes, the inhomogeneities ("tidal gravitational forces") will set the test particles into motion relative to each other--and will thus change the velocity of at least one of them relative to any reference frame which was claimed to be inertial. Consequently, no inertial frames can exist.

The tidal gravitational forces which prevent the existence of inertial frames become arbitrarily small when the test par-

---

[1]This operational definition of an inertial frame, together with the fundamental principle of SRT--see below--tells us that the clocks of an inertial frame are synchronized using light signals in the manner described in most textbooks on SRT.

ticles are put arbitratily close together. (On a small scale the gravitational inhomogeneities are undetectable.) This means that in sufficiently small regions of space, viewed for sufficiently short lengths of time, inertial reference frames exist despite the presence of inhomogeneous gravitational fields. However, these "local inertial frames" near various events in spacetime cannot be meshed together to form large-scale inertial frames.

This "non-meshability" of local inertial frames is analogous to a situation in the theory of curved two-dimensional surfaces. Near any point on any curved two-surface one can construct a tiny, local Cartesian coordinate system. However, the local Cartesian coordinates at various points cannot be meshed together to form large-scale Cartesian coordinates; the curvature of the surface prevents meshing here, just as tidal gravitational forces prevent meshing in spacetime.

2.2 Relativistic Theories of Gravity

This analogy can be used to construct a variety of geometric theories of gravity which incorporate SRT locally. The local validity of SRT in a gravitational field--as verified in part by the Dicke-Eotvös experiment (cf. Dicke 1964)--guarantees that the path of a particle through spacetime is independent of its rest mass, μ (so long as $\mu \neq 0$). Whatever the particle paths may be, one can use them to define the curved geometry of spacetime. The procedure is to define the spacetime geometry as that geometry whose geodesics (paths of extremal length) are identical to the paths of freely falling test particles. The local validity of SRT guarantees that the resultant geometry can be described by a metric:[2] Between any two neighboring events q and q + dq there is an invariant squared interval ds^2--which is precisely the squared interval that would be measured in a SRT local inertial frame with coordinates $\bar{x}^0, \bar{x}^1, \bar{x}^2, \bar{x}^3$;

$$ds^2 = (d\bar{x}^0)^2 - (d\bar{x}^1)^2 - (d\bar{x}^2)^2 - (d\bar{x}^3)^2 . \tag{1}$$

Because the inertial coordinates $\bar{x}^\mu$ at q cannot be extended throughout spacetime, one is forced to use a curvilinear coordinate system x^α everywhere except in an infinitesimal neighborhood of q. In terms of these curvilinear coordinates, the squared interval (or "line element") is written

$$ds^2 = \sum_{\alpha,\beta} g_{\alpha\beta} dx^\alpha dx^\beta . \tag{2}$$

---

[2]If SRT were not valid locally, but test-particle paths were independent of rest mass, one might still try to define the spacetime geometry by demanding that those paths be geodesics. However, one would not necessarily obtain a "Riemannian geometry" --i.e. a geometry described by a metric. Rather, one might obtain a more primitive type of geometry--a geometry with an "affine connection", with curvature, and with geodesics, but without a metric. For example, this is what results when one couches the Newtonian theory of gravity in geometric terms [see e.g. Trautman (1965)].

The symmetric matrix $g_{\alpha\beta}$, which is a function of position in spacetime, is called the metric tensor.[3]

Any theory of gravity compatible locally with SRT (any "relativistic" theory of gravity) can be put into this kind of a geometric form. The crucial difference between the various relativistic theories of gravity lies in the manner by which the matter content of the universe generates the geometry of spacetime--i.e. determines the metric tensor. Einstein's theory (GRT) chooses the simplest way for matter to generate geometry, subject to the constraint that the resultant theory reduce to Newtonian theory in the nonrelativistic limit: it says that a certain measure of curvature, called the Einstein curvature tensor $G_{\alpha\beta}$, must be proportional to the stress-energy tensor $T_{\alpha\beta}$ for all matter and (nongravitational) fields present in the Universe:

$$G_{\alpha\beta} = (8\pi\ G/c^4)T_{\alpha\beta}. \tag{3}$$

In the constant of proportionality, G is Newton's gravitational constant, and c is the speed of light.

In other relativistic theories the matter generates the curvature in more complicated ways. For example, in the Brans-Dicke theory the matter first generates a cosmological scalar field ϕ, and then the scalar field and the matter together generate the Einstein curvature tensor.

Because SRT has been tested to fantastic precision and has never been found wanting, physicists today believe that, whatever the correct classical (i.e. non-quantum) theory of gravity may be, it must be a relativistic theory. It must be expressible in a geometric form with a metric, and with particle paths being geodesics. Because GRT is the simplest such theory--and also for obvious historical reasons--physicists will continue to use GRT for most astrophysical calculations until (if ever) it is proved wrong by experiment. However, one should keep himself painfully aware of the dangers inherent in basing astrophysical judgments on the predictions of a particular theory, GRT, which has not been adequately tested by experiment. Perhaps the best way to maintain this awareness is to examine occasionally the predictions of other relativistic theories of gravity.

As our interpretations of quasars, pulsars, cosmology,

---

[3] By comparison of equations (1) and (2), one sees that the metric in the curvilinear coordinates x^α is related to the metric

$$\eta_{\mu\nu} = \begin{cases} 1 \text{ if } \mu=\nu=0 \\ -1 \text{ if } \mu=\nu=1,\ 2,\text{ or } 3 \\ 0 \text{ if } \mu\neq\nu \end{cases}$$

of the local inertial frame by

$$g_{\alpha\beta} = \sum_{\mu,\nu} (\partial\bar{x}^\mu/\partial x^\alpha)(\partial\bar{x}^\nu/\partial x^\beta)\eta_{\mu\nu}\ .$$

and gravitational waves come to rely more and more on GRT considerations, it becomes more and more crucial that GRT be tested to high precision. Fortunately, man's rapidly developing space technology should make this possible within the next decade or two. [See Thorne and Will (1970)].

2.3 Realm of Validity of Newtonian Theory

Calculations in the framework of GRT or any other relativistic theory of gravity are very difficult. For this reason, an astrophysicist would be a fool to use a relativistic theory in a context where Newtonian theory yields sufficient accuracy; and he would be equally foolish to trust Newtonian theory when relativistic effects are large. There is a simple prescription for avoiding foolishness: First analyze the system using Newtonian theory. Calculate, in particular, the maximum value, U_{max}, of the Newtonian potential inside the system. Then the dimensionless number

$$\varepsilon^2 \equiv U_{max}/c^2 \tag{4}$$

is a rough measure of the fractional error made by using Newtonian theory rather than the correct relativistic theory of gravity (whatever that may be).

Take the solar system as an example. There

$$\varepsilon^2 \simeq GM_\odot/R_\odot c^2 \simeq 2 \times 10^{-6}. \tag{5a}$$

Consequently, Newtonian theory cannot be trusted to give correctly such fine details as the deflection of light at the limb of the sun through an angle of

$$\alpha \simeq 4GM_\odot/R_\odot c^2 = 1.''75 \simeq 8 \times 10^{-6} \text{ radians}, \tag{5b}$$

or the anomalous perihelion shift of mercury at a rate of about

$$43''/\text{century} \sim 10^{-6} \text{ of an orbit } (360^\circ) \text{ per orbit}, \tag{5c}$$

or the gravitational redshift off the surface of the sun of

$$z \simeq GM_\odot/R_\odot c^2 \sim 2 \times 10^{-6}, \tag{5d}$$

or the anomalous fractional radar delay time of

$$\frac{\text{anomalous delay}}{\text{total delay}} \simeq \frac{2 \times 10^{-4} \text{ sec}}{20 \text{ min}} \sim 2 \times 10^{-7} \tag{5e}$$

for a signal between the Earth and Mercury which just grazes the limb of the sun.

In the interior of a massive ($\sim 1.5\ M_\odot$) neutron star, by contrast

$$\varepsilon^2 \sim U(\text{center of star})/c^2 \sim 2. \tag{6a}$$

Consequently, Newtonian models of massive neutron stars cannot be trusted to very much more than an order of magnitude. For the magnetosphere of a neutron star (pulsar model!), on the other hand,

$$\varepsilon^2 \sim GM/Rc^2 \sim 0.25, \tag{6b}$$

so Newtonian models of the magnetosphere are probably accurate to within 50 per cent.

When applying the U_{max}/c^2 test for relativistic effects one must be cautious about two points: First--some relativistic effects are sufficiently cumulative over time as to markedly alter the system in time intervals of interest. For example, in 10^6 years the relativistic perihelion shift of Mercury will alter the orientation of its orbit markedly from the orientation predicted by Newtonian theory. As another example, for two neutron stars in orbit around each other with a separation of $\sim$1000km, the relativity parameter is $\varepsilon^2 \simeq GM_\odot/(c^2 \times 10^3 \mathrm{km}) \simeq 10^{-3}$. This does not mean that relativistic effects can be treated as a small correction for long; in less than a year the two stars should spiral into each other and collide as a result of the emission of gravitational waves.

Second--when a system's stability is governed by a delicate balance between opposing Newtonian forces, even a very small relativistic correction can upset that balance and completely change the stability. For example, some models of supermassive stars and dense white dwarfs with $\varepsilon^2 \lesssim 10^{-3}$, which are stable according to Newtonian theory, are made unstable by general relativistic corrections.

For situations where relativistic effects are small ($\varepsilon^2 << 1$) but important, it is often easier to work with the post-Newtonian approximation to general relativity than with general relativity itself. Chandrasekhar (1965a) has developed the post-Newtonian approximation into a form well suited to astrophysical calculations.

2.4 Units

Relativity theorists usually work in "geometrized units", i.e., in units where the speed of light, c, and Newton's gravitation constant, G, are equal to one: G=c=1. In these units c is regarded as a conversion factor between seconds and centimeters

$$1 = c = 3\times10^{10} \text{ cm/sec}; \tag{7a}$$

G/c^2 is regarded as a conversion factor between grams and centimeters

$$1 = G/c^2 = 0.742\times10^{-28} \text{ cm/g}; \tag{7b}$$

and G/c^4 is regarded as a conversion factor between ergs and centimeters

$$1 = G/c^4 = 0.826\times10^{-49} \text{ cm/erg}. \tag{7c}$$

Thus, the mass of the sun is

$$M = 2x10^{33} \text{ g} = 1.5 \text{ km}. \tag{7d}$$

As a crutch we shall put asterisks on quantities when the equations in which they appear are written in geometrized units (i.e., with c and G omitted).

3. General-Relativistic Stellar Structure

In this section we present a brief overview of the general-relativistic equations of stellar structure and some of their implications. For a more detailed discussion of this subject see, e.g., chapter 3 of Thorne (1967).

3.1 The Spacetime Geometry for a Nonrotating Star

With an appropriate choice of coordinates, the line element describing the geometry of spacetime inside and outside a general-relativistic star takes the form

$$ds^2 = e^{2\Phi}dt^{*2} - \frac{dr^2}{1-2m^*/r} - r^2(d\theta^2+\sin^2\theta \, d\phi^2). \tag{8a}$$

Here, t^*, r, θ, ϕ are the coordinates, while Φ and m^* are functions of the radial coordinate, r

$$\Phi = \Phi(r), \; m^* = m^*(r). \tag{8b}$$

In effect, Φ and m^* are gravitational potentials, since a knowledge of them gives us a complete knowledge of the gravitational field--i.e. of the geometry of spacetime.

The uninitiated reader will now say: "You have told me the form of the line element; now tell me the nature of the coordinates in which you have chosen to write it." As Charles Misner (1969) emphasizes very strongly, the line element itself tells us the nature of the coordinates as well as the nature of the geometry of spacetime.

If we hold t^* and r fixed in the line element (8a), we obtain the line element, $-r^2(d\theta^2 + \sin^2\theta \, d\phi^2)$, for the surface of a sphere. This fact tells us three things: (i) the geometry of spacetime is spherically symmetric; (ii) the coordinates θ and ϕ are the familiar angular coordinates on each of the invariant spheres about the center of symmetry; and (iii) the radial coordinate r has been chosen such that $4\pi r^2$ is the surface area, and $2\pi r$ is the circumference of the invariant sphere at radius r. From the fact that the metric coefficients $g_{\alpha\beta}$ are independent of the time coordinate we learn two things: (iv) spacetime is stationary--i.e., the star is in a time-independent equilibrium state; and (v) the time coordinate t^* displays the stationary nature of the geometry--i.e., it has the property that the three-dimensional surfaces of constant t^* are indistinguishable from each other. From the fact that all off-diagonal g_{t^*j} terms in the metric vanish, we learn that (vi) the geometry of spacetime is invariant under time reversal (this would not be true if the star were rotating--cf. §3.9); and (vii) the time coordinate t^* has been constructed so as to display this time-reversal invariance.

It is important to notice that the radial coordinate r is not an accurate measure of the distance outward from the center of the star. According to the line element (8a), radial distance is given by

$$\ell = \int_0^r (1-2m^*/r)^{-\frac{1}{2}}\, dr. \qquad (9)$$

We shall see shortly that m^* is always positive (except at the center of the star), so that ℓ is always greater than r:

$$\text{radial distance from center} > \text{circumference}/2\pi \qquad (10)$$

This nature of the curved spacetime geometry is most easily visualized in terms of an "embedding diagram" for the two-dimensional equatorial slice (t^* = constant, $\theta = \pi/2$) through the star. The geometry of that two-dimensional surface,

$$d\sigma^2 = -\, ds^2 = \frac{dr^2}{1-2m^*/r} + r^2 d\phi^2, \qquad (11)$$

is identical to the geometry of a certain two-dimensional surface embedded in three-dimensional Euclidean space. For the case of a massive neutron star the embedded surface has the form shown in Figure 1.

3.2 Matter Variables

In the Newtonian theory of stellar structure one describes the matter of a star by its density of rest mass, ρ_0, its pressure, p, its temperature, T, its entropy per gram of rest mass, S, and its chemical abundances, X, Y, Z. In the relativistic theory, one obtains the simplest formalism by ignoring the rest mass and concentrating attention instead on: (i) a quantity which is conserved--the baryon number--, and (ii) a quantity which (unlike rest mass) generates the curvature of spacetime--the total mass-energy, including rest mass-energy, thermal energy, compressional energy, etc., but not including gravitational energy. Consequently, the relativistic variables describing the matter are its baryon number density (with antibaryons counted negatively), n; its density of total mass-energy, ρ; its pressure, p; its temperature T; its entropy per baryon, s, and its chemical abundances.

It is important to notice that, when a sample of matter is compressed, its total mass-energy changes, whereas its Newtonian rest mass and its number of baryons do not. If V is the volume of the sample, then the first law of thermodynamics says that

$$d(\rho^* V) = -p^* dV \Rightarrow d\rho^*/dV = -(\rho^*+p^*)/V; \qquad (12a)$$

while conservation of baryons says that

$$d(nV) = 0 \Rightarrow dn/dV = -\, n/V. \qquad (12b)$$

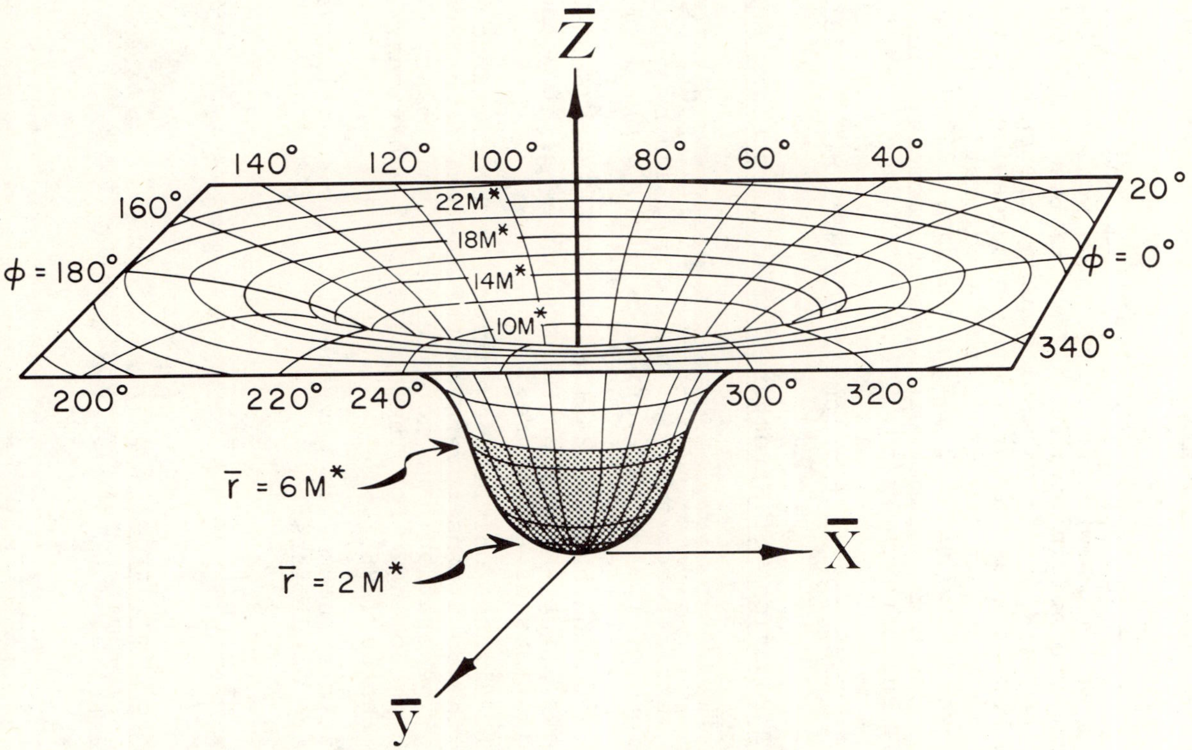

Figure 1. Embedding diagram for the geometry of the two-surface ($\theta = \pi/2$, t = constant) of a nonrotating star in hydrostatic equilibrium. The coordinates (r,ϕ) intrinsic to this two-surface are related to the coordinates $[\bar{r} = (\bar{x}^2 + \bar{y}^2)^{1/2}, \bar{\phi} = \tan^{-1}(\bar{y}/\bar{x}), \bar{z}]$ of the Euclidean embedding space by the equations $\bar{r} = r$, $\bar{\phi} = \phi$, $\bar{z} = \int_o^r [2m^*/(r-2m^*)]^{1/2} dr$. The interior of the star is distinguished from the exterior by stippling. Because the rotational degree of freedom associated with θ has been suppressed from the diagram, regions of constant radius, r, are here circles of circumference $2\pi r$ about the center, r = 0. The 3-geometry of space around the star at a particular moment of time, t, can be visualized by mentally replacing the circles of constant r in this diagram with spheres of area $4\pi r^2$. [This figure is taken from Thorne (1967).]

Combining these relations, we learn that for an adiabatic process (no energy transfer between adjacent fluid elements)

$$d\rho^*/dn = (\rho^*+p^*)/n, \tag{12c}$$

or, in conventional cgs units

$$\frac{d\rho}{dn} = \frac{\rho+p/c^2}{n} \tag{12c'}$$

The quantity ρ^*+p^* (or $\rho+p/c^2$) appears time and again in relativity theory where a Newtonian calculation would give rest-mass density ρ_0. The origin of the p^* term in the combination ρ^*+p^* is always the work (≡energy≡mass) done by pressure forces. We saw in detail how it arose above in the compression of matter A second context in which it arises is the acceleration of matter. It turns out [see e.g. §2.4.4 of Thorne (1967)] that the inertial mass per unit volume which resists acceleration is not ρ_0, nor ρ, but rather $\rho+p/c^2$, i.e.

$$(\text{inertial mass per unit volume})^* = \rho^*+p^*. \tag{13}$$

3.3 The GRT Equations of Structure for a Nonrotating Star

The structure of a general-relativistic stellar model is described by the metric functions $\Phi(r)$ and $m^*(r)$, and by the radial distributions of the matter variables $\rho(r)$, $p(r)$, $T(r)$, $s(r)$,... . These quantities are linked by the Einstein field equations (3), and by the usual constitutive equations (equation of state, etc.) for the matter--which, taken all together, are called the equations of structure.

Here we shall discuss only the equations of hydrostatic structure, since the equations governing the thermal structure and nuclear energy generation have not yielded significant, new, relativistic phenomena. For the full set of equations, see chapter 3 of Thorne (1967).

The hydrostatic structure is governed by four equations and associated boundary conditions:

$$m^*(r) = \int_0^r 4\pi r^2 \rho^* dr, \tag{14a}$$

$$\frac{dp^*}{dr} = -\frac{(\rho^*+p^*)(m^*+4\pi r^3 p^*)}{r(r-2m^*)}, \quad \begin{cases} p^*(r=0)=p_0{}^* \\ p^*(r=R)=0 \end{cases}, \tag{14b}$$

$$p^* = p^*(\rho^*), \tag{14c}$$

$$\frac{d\Phi}{dr} = \frac{m^*+4\pi r^3 p^*}{r(r-2m^*)}, \quad \Phi(r=\infty)=0. \tag{14d}$$

Let us discuss each of these equations in turn.

The form of equation (14a) suggests that we give the name "mass inside radius r" to $m^*(r)$. This mass can be divided into three parts:

$$m^*(r) = m_0^*(r) + E^*_{int}(r) + \Omega(r). \tag{15a}$$

Here $m_0(r)$ is the rest mass inside radius r

$$m_0(r) = \int \rho_0 d\text{ volume} = \int_0^r \rho_0\ 4\pi r^2 \left[(1-2m^*/r)^{-\frac{1}{2}} dr\right] \tag{15b}$$

(notice from the line element (8a) that the volume of the shell between the spheres of "radii" r and r + dr is $4\pi r^2 d\ell = 4\pi r^2 [(1-2m^*/r)^{-\frac{1}{2}}\ dr]$, rather than $4\pi r^2 dr$). $E_{int}(r)$ is the total internal energy inside radius r

$$E_{int}(r) = \int (\rho-\rho_0)c^2 d\text{ volume} = \int_0^r (\rho-\rho_0)c^2 4\pi r^2 \left[(1-2m^*/r)^{-\frac{1}{2}} dr\right] \tag{15c}$$

and $\Omega(r)$ is the gravitational potential energy inside radius r

$$\Omega(r) = \int \rho c^2 [4\pi r^2 dr - d\text{ volume}] = -\int_0^r \rho c^2\ 4\pi r^2 \left[(1-2m^*/r)^{-\frac{1}{2}} - 1\ dr\right]$$

$$\approx -\int_0^r (G\rho m/r)\ 4\pi r^2 dr \quad \text{In Newtonian limit.} \tag{15d}$$

The value of m^* at the surface of the star,

$$M^* \equiv m^*(r=R),$$

is called the star's total mass-energy. It is this quantity, not the total rest mass, which is conserved when the star undergoes a change of state without releasing energy to the outside; and it is this quantity that is decreased by an amount E/c^2 when an energy E is released to infinity in the form of radiation or particle rest mass and kinetic energy.

Turn attention now to the second equation of structure--equation (14b) for the pressure gradient:

$$\frac{dp^*}{dr} = -\frac{(\rho^*+p^*)(m+4\pi r^3 p^*)}{r(r-2m^*)} \tag{14b}$$

or, in cgs units

$$\frac{dp}{dr} = -\frac{G(\rho+p/c^2)(m+4\pi r^3 p/c^2)}{r(r-2Gm/c^2)}\ . \tag{16}$$

This equation is called the TOV equation of hydrostatic equilibrium because Tolman, Oppenheimer, and Volkoff made the pioneering studies of relativistic stellar structure in 1939. Notice that in the Newtonian limit ($p^* \ll \rho^*; m^* \ll r$), the TOV equation becomes the familiar equation of hydrostatic equilibrium

$$dp/dr = -G\rho m/r^2. \tag{17}$$

The TOV equation expresses the balance between the gravitational force and the pressure-buoyancy force that act on a volume element inside the star. The pressure-buoyancy force on a fluid element of volume V, that would be measured in the standard, everyday way by an engineer at radius r, is

$$F_{buoy} = -\,V dp/d\ell = -\,V(1-2m^*/r)^{-1/2} dp/dr. \tag{18a}$$

This outward force must be counterbalanced by the inward gravitational force

$$F_{grav} = [\text{inertial mass in volume V}] \times [\text{gravitational acceleration}]$$

$$= [(\rho+p/c^2)V][d\Phi/d\ell] = [(\rho+p/c^2)V]\,\frac{G(m+4\pi r^3 p/c^2)}{r^2(1-2m^*/r)^{1/2}} \tag{18b}$$

The form of the TOV equation (14b)--or, equivalently, the form of the buoyancy and gravitational forces (18)--has two very important consequences which are absent from Newtonian theory: (i) the maximum mass of a neutron star is remarkably insensitive to uncertainties in the equation of state; and (ii) no equilibrium stellar model can have $r \le 2m^*$ anywhere in its interior.

<u>Maximum mass of a neutron star</u>: One knows with fair confidence the equation of state, $p(\rho)$, for neutron stars at densities below the density of an atomic nucleus, $\rho_{nuc} \approx 3\times10^{14}\,g/cm^3$. Above that density one knows very little. By assuming a variety of widely different behaviors for the equation of state above nuclear densities, and calculating numerically the resultant stellar models, one finds that the stellar properties are remarkably insensitive to the uncertainty in the equation of state. For example, the maximum mass of a neutron star varies by at most a factor of ~ 4. The main reason for this is the presence of the two pressure terms in the gravitational force (18b): One might try to support a star of large mass against gravity by making its equation of state very stiff ($p^*/\rho^* \gtrsim 1$) at supernuclear densities. If one tries this, he will fail because the gravitational force (18b) increases <u>quadratically</u> with the pressure; when $p^* \gtrsim \rho^*$, an increase in the pressure causes an even greater increase in the gravitational force, which completely overwhelms the pressure and causes collapse.[4] One might instead try to support a massive star by making its equation of state very soft

[4]While this argument is essentially correct, the above version of it is much too simple-minded to be more than heuristic.

($p^*/\rho^* \ll 1$) at supernuclear densities. Again one will fail. This time the pressure contributions to the gravitational force are negligible, but the usual Newtonian-type force ($\propto G\rho m/r^2$) will overwhelm the relatively small pressure that the "soft" matter can generate.

The limit $r>2m^*$: The quantity $2m^*(r)$ is called the "gravitational radius" or "Schwarzschild radius" of that part of the star contained inside radius r. The TOV equation (14b) guarantees that for any spherical equilibrium configuration each portion of the star is outside its own gravitational radius,

$$r > 2m^*(r). \tag{19}$$

To understand this, imagine a numerical integration of the equations of structure (14) outward from the center of the star ($r=0$) to the surface of the star (point where p has decreased to zero). Suppose that in the course of this integration $2m^*/r$ approaches very close to unity. Then the gravitational force becomes extremely strong [cf. eq. (18b)] and causes the pressure gradient that balances it to become extremely negative [cf. TOV eq. (14b)]. As a result, the pressure plumets to zero, signalling that the surface of the star has been reached, before $2m^*/r$ can become unity. Outside the star $2m^*/r$ decreases, since $m^*=M^*=$constant there.

The limit $r>2m^*$ applies only to equilibrium stars. In a collapsing star, where the pressure gradient need not balance the gravitational force, any part of the star (including the whole star) can sink inside its gravitational radius--and once this has happened, the star can never again regain equilibrium. [See, e.g., the review of gravitational collapse by Misner (1969), or chapter 8 of Thorne (1967).]

In our discussion of the equations of structure (14), we now turn our attention to (14c), the equation of state. We here idealize the equation of state to be a unique relation between pressure and density. This is done only because we have chosen, in this brief review, not to discuss the equations of thermal structure.

In actuality, for white dwarfs and neutron stars one can easily introduce a one-parameter equation of state, $p=p(\rho)$, without loss of any crucial physics. This is because the thermal pressure in such stars is negligible by comparison with the nonthermal pressure. (See the lectures by Cameron in this volume.)

The remaining equation of structure (14d) determines the metric function $\Phi(r)$--which is often called the "gravitational redshift factor." In the Newtonian limit of general relativity, in any nearly inertial frame, the g_{tt} component of the metric is related to the Newtonian potential U by

$$g_{tt} \approx 1-2U/c^2 + O(U^2/c^4). \tag{20}$$

Consequently, in the Newtonian limit of our relativistic

stellar models, where $g_{tt} = e^{2\Phi} \approx 1+2\Phi$, we must obtain

$$\Phi = - U/c^2 + 0(U^2/c^4). \tag{21a}$$

That we do, indeed, find this, one sees from the Newtonian limit of the equation of structure (14d) for Φ:

$$d\Phi/dr = -m^*/r^2 = 0(U^2/c^4). \tag{21b}$$

The boundary condition $\Phi(r=\infty)=0$ (eq. 14d) is imposed so that proper time, $s = \int e^{\Phi} dt$, will agree with coordinate time, t, far from the star--i.e., so that the coordinates will become inertial far from the star.

3.4 The External Gravitational Field

The equations of structure (14) are easily solved outside the star to yield

$$m^* = m^*(r=R) \equiv M^* = \text{constant},$$
$$\Phi = \frac{1}{2} \ln(1-2M^*/r). \tag{22a}$$

These metric functions correspond to the line element

$$ds^2 = (1-2M^*/r)dt^{*2} - \frac{dr^2}{1-2M^*/r} - r^2(d\theta^2 + \sin^2\theta\, d\phi^2), \tag{22b}$$

which is called the Schwarzschild line element. Recall that, so long as the star is in hydrostatic equilibrium, no part of the star's exterior is inside the gravitational radius

$$r \geq R \geq 2M^*,$$

so one need not worry about the apparent singularity in the Schwarzschild line element at $r=2M^*$.

In general relativity theory the Schwarzschild line element (22b) describes the external spacetime geometry not only for static stars, but also for any star in spherically-symmetric motion (i.e. pulsation or collapse). This is not the case in Brans-Dicke theory or in other theories which make use of scalar fields. There a pulsating or collapsing star can emit monopole, scalar gravitational waves which cause spherical ripples in its exterior, spherical gravitational field. In general relativity theory, no monopole gravitational waves are possible.

Far from a relativistic star ($r>>2M^*$) Newtonian theory is very nearly valid and the Newtonian potential is $U=c^2M^*/r=GM/r$ [cf. Eqs. (20) and (22b)]. Consequently, <u>the total mass-energy of the star, M, is numerically identical to the mass which governs the Keplerian orbits of distant planets.</u>

3.5 The Construction of Relativistic Stellar Models

When a one-parameter equation of state $p=p(\rho)$, is available the construction of a relativistic stellar model is very straightforward: (i) Specify a central density ρ_c and calculate the corresponding central pressure $p_c=p(\rho_c)$. (ii) Specify the central value of Φ, $\Phi(r=0)$, for the purpose of numerical integrations; when the integrations are finished, Φ can be renormalized so that $\Phi(r=\infty)=0$. (iii) Numerically integrate the equations of structure (14) outward from $r=0$; the boundary conditions $p(r=0)=p_c$, $\rho(r=0)=\rho_c$, $\Phi(r=0)=0$ [and $m(r=0)=0$ as automatically incorporated into equation (14a)] are a complete set for the integration, so everything is precisely determined. (iv) Stop the integration when p reaches zero; that point is the surface of the star and determines its radius, $R\equiv r$, and mass, $M^* = m^*(R)$. (v) Renormalize Φ so that it satisfies the boundary condition $\Phi(r=\infty)=0$:

$$\Phi_{new}(r) = \Phi_{old}(r) + \frac{1}{2}\ \ell n(1-2M^*/R) - \Phi_{old}(R). \tag{23}$$

(vi) Adopt the Schwarzschild line element (22) to describe the star's external spacetime geometry.

When one does not have a one-parameter equation of state, he must couple the relativistic equations of thermal equilibrium to the equations of hydrostatic equilibrium in order to calculate the stellar structure and evolution. [See chapter 3 of Thorne (1967)].

3.6 Energy Redshift

When a particle of finite or zero rest mass moves outward through the spacetime geometry of a star--or through any other time-independent geometry, $g_{\alpha\beta}\neq f(t)$,--its energy suffers a redshift. This redshift is most simply understood in terms of the local observations made by a family of "static observers" --i.e. a family of observers who are at rest in the gravitational field, so that their world lines (trajectories through spacetime) have x^1, x^2, x^3 constant but $x^0 = ct$ varying. Let each such observer along the world line of a given particle measure the particle's energy E as it passes him. For a particle of finite rest mass μ, that energy will be

$$E = \mu c^2/[1-v^2/c^2]^{1/2}, \tag{24a}$$

where v is the velocity measured by the observer using his own physical meter sticks and clocks. For a photon or neutrino, the energy is

$$E = h\nu, \tag{24b}$$

where ν is the frequency measured with the static observer's clock.

The various static observers will not measure the same energy for the particle because the particle must "do work" in order to climb out of the gravitational field, and it gains energy when it falls downward. The equation which governs this "energy redshift" and "energy blueshift" is

$$E|g_{tt}|^{1/2} \equiv E_\infty = \text{constant} \tag{25a}$$

in a general time-independent metric and for any relativistic theory of gravity; and it reads

$$Ee^{\Phi} \equiv E_\infty = \text{constant} \tag{25b}$$

in our stellar metric (8a). E_∞ is the value of $E|g_{tt}|^{1/2}$ at $r = \infty$.

Notice that in the Newtonian limit, where $g_{tt} \approx 1-2U/c^2$, and $v/c<<1$, the energy-redshift equation (25a) becomes

$$E(1-U/c^2) = \text{constant}. \tag{26a}$$

For a particle this is the usual law of "conservation of energy" (including potential energy, $-\mu U$, which is left out of E since it is not measured by the local observers):

$$\mu c^2 + \frac{1}{2}\mu v^2 - \mu U = E_\infty = \text{constant}. \tag{26b}$$

For a photon equation (26a) is the familiar redshift equation

$$\frac{\Delta\nu}{\nu} = \frac{\nu(r_2)-\nu(r_1)}{\nu(r_1)} = \Delta U = U(r_2)-U(r_1). \tag{26c}$$

Often one wants to know the energy redshift between two observers, A_M and B_M, who are not static--i.e. who are moving relative to the space coordinates x^1, x^2, x^3. To calculate this, one first calculates the redshift between two static observers, A_S and B_S, who are at the same point in spacetime as A_M and B_M at the moments when A_M and B_M measure the particle in question:

$$E_{A_S}/E_{B_S} = [g_{tt}(B_S)/g_{tt}(A_S)]^{1/2} \tag{27a}$$

One then calculates the redshifts between A_M and A_S, and between B_M and B_S, using special relativity (Lorentz transformations if $\mu \neq 0$; Doppler shifts if $\mu=0$), which is valid locally at the two events where the measurements are made.

$$E_{A_S}/E_{A_M} = \text{SRT formula from one frame to another,}$$

$$E_{B_S}/E_{B_M} = \text{SRT formula from one frame to another.} \tag{27b}$$

One finally combines the gravitational effect (eq. 27a) with the SRT effect (eq. 27b) to get the desired redshift between moving observers.

Example: Calculate the redshift of a spectral line emitted radially by a collapsing star, whose radius $R(2\pi R \equiv$ circumference) is given in terms of the Schwarzschild time coordinate by

$$R = R(t). \tag{28}$$

The observer who sees the spectral line is far away on Earth; and before the collapse begins, he "measures" the star to move with speed v relative to him in a direction making an angle θ relative to the radial direction. Between the astronomer (A_M) and a "static observer" (A_S) (one at rest relative to the star but located near the astronomer) there is a Doppler shift of

$$\frac{E_{A_M}}{E_{A_S}} = \frac{\nu_{A_M}}{\nu_{A_S}} = \frac{(1-v^2/c^2)^{1/2}}{1+(v/c)\cos\theta}\,. \tag{29a}$$

Between the static observer B_S at the star's surface $r=R(t_{em})$, at the time t_{em} when the photon was emitted, and the distant static observer A_S, there is the gravitational shift

$$\frac{E_{A_S}}{E_{B_S}} = \frac{\nu_{A_S}}{\nu_{B_S}} = \left(\frac{g_{tt}(B_s)}{g_{tt}(A_s)}\right)^{1/2} = [1 - \frac{2M^*}{R}]^{1/2}. \tag{29b}$$

The observer B_M moving with the star's surface at the moment of emission has a locally measured velocity relative to the static observer B_S, which is given by

$$\beta = \frac{-d(\text{proper radial distance})}{d(\text{proper time})} = \frac{-|g_{rr}|^{1/2}\,dR}{|g_{tt}|^{1/2}\,dt} = \frac{-dR/dt}{1-2M^*/R}\,. \tag{29c}$$

Consequently, the Doppler shift between these two observers is

$$\frac{E_{B_S}}{E_{B_M}} = \frac{\nu_{B_S}}{\nu_{B_M}} = \frac{(1-\beta^2/c^2)^{1/2}}{1+\beta/c}\,. \tag{29d}$$

Combining all three shifts, we obtain

$$\frac{\nu_{\text{received}}}{\nu_{\text{emitted}}} \equiv \frac{\nu_{A_M}}{\nu_{B_M}} = \frac{(1-v^2/c^2)^{1/2}}{1+(v/c)\cos\theta} \frac{(1-\beta^2/c^2)^{1/2}}{1+\beta/c} \left(1 - \frac{2M^*}{R}\right)^{1/2}. \tag{29e}$$

For photons emitted from the surface of a static star ($\beta=0$) and received by a static, distant observed ($v=0$) only the gravitational redshift occurs; and it is independent of whether the photon is emitted radially or not:

$$z \equiv \frac{\Delta\lambda}{\lambda} = \frac{\lambda_{rec}-\lambda_{em}}{\lambda_{em}} = \frac{\nu_{em}}{\nu_{rec}} - 1 = \frac{1}{(1-2M^*/R)^{1/2}} - 1$$
$$\approx M^*/R \quad \text{in Newtonian limit.} \tag{30}$$

Notice that since R is always greater than the gravitational radius, $R \geq 2M^*$, the redshift from an equilibrium star cannot be infinite.

Bondi (1964, 1965), by a very careful and elegant analysis of the TOV equation of hydrostatic equilibrium, has put much more stringent limits on z: (See also Buchdahl (1959), a forerunner of Bondi's work.) For all configurations which are in principle physically realizable [$\rho^* \geq p^* \geq 0$ everywhere; cf. §2.4.1 of Thorne (1967)] Bondi finds

$$M^*/R \leq 0.432, \quad z \leq 1.71. \tag{31a}$$

If, in order to avoid Taylor instabilities, one imposes the additional restriction that ρ^* be monotonic decreasing from the center of the star to the surface, then one obtains (Bondi 1964, 1965)

$$M^*/R \leq 0.390, \quad z \leq 1.14. \tag{31b}$$

In most realistic situations one actually has $p^* \leq \frac{1}{3}\rho^*$ (cf. HTWW, Chap. 10) and an adiabatic or subadiabatic temperature gradient. When this is the case, Bondi's analysis yields the limits

$$M^*/R \leq 0.310, \quad z \leq 0.63. \tag{31c}$$

Limits very nearly equal to (31a,b,c) also exist in the Brans-Dicke theory of gravity according to an unpublished analysis by Richard Price and James Bardeen (1967).

Although the redshift of a photon originating at the <u>surface</u> of any equilibrium configuration is limited by $z \leq 1.71$, the redshift of a photon--or, more realistically, of a neutrino--which originates at the <u>center</u> can be arbitrarily large. An expression for the redshift of such a neutrino is [cf. equations (25b) and (14b,d)]

$$z = (1-2M^*/R)^{-1/2} \exp\left[\int_0^R (\rho^*+p^*)^{-1}(-dp^*/dr)\,dr\right] - 1. \tag{32}$$

For a configuration of arbitrarily high central pressure, this redshift is arbitrarily large.

3.7 Particle and Photon Orbits

In the Newtonian gravitational field outside a spherical star, photons move along straight lines, while particles of finite rest mass move along ellipses, parabolae, and hyperbolae. Small relativistic corrections to the gravitational field deform the orbits. For a photon they cause deflection through an angle

$$\alpha = 4GM/r_{min}c^2, \tag{33}$$

where r_{min} is the radius of closest approach to the star. For a particle of finite rest mass, they cause a slight excess in the normal bending of the orbit (relativistic gravitational forces are stronger than the corresponding Newtonian forces). As a result, for example, the periastrons of the elliptical orbits precess forward in time.

In highly relativistic regions ($r/2M^* \lesssim 10$) the orbits are so strongly deformed that they are unrecognizably different from their Newtonian counterparts. For example, a photon at $r=3M^*$ can be in a closed, circular orbit about the star (provided the star is smaller than that orbit, $R<3M^*$). For detailed discussions of the highly deformed orbits see §3.5.4 of Thorne (1967) and references cited therein; for even more detail see Hagihara (1931).

3.8 Rotating, Relativistic Stellar Models--Computational Techniques[5]

There are two dimensionless parameters which characterize a rotating, relativistic stellar model--the relativity parameter

$$\varepsilon_{Rel}{}^2 \approx GM/Rc^2 \sim \text{(Relativistic corrections to Newtonian theory)} \tag{34a}$$

and the rotational parameter

$$\varepsilon_{Rot}{}^2 \approx (\Omega/\Omega_{crit})^2 \equiv \Omega^2(R^3/GM) \sim \text{(Rotational corrections to non-rotating star)} \tag{34b}$$

Here Ω is a mean angular velocity of the star, and Ω_{crit} is the surface angular velocity required to deform it so badly that it sheds mass at its equator. For several stars of interest these parameters have the following values:

---

[5]For a very detailed review of the theory of rotating, relativistic stars as of July 1969, see Thorne (1971a).

Sun:

$$\varepsilon_{Rel}{}^2 \sim 10^{-6}, \ \varepsilon_{Rot}{}^2 \sim 10^{-8}; \tag{35a}$$

Rapidly rotating white dwarfs as studied by Ostriker and Bodenheimer (1968), Ostriker and Hartwick (1968), and Ostriker and Marks (1968):

$$\varepsilon_{Rel}{}^2 \sim 10^{-3}, \ \varepsilon_{Rot}{}^2 \sim 1; \tag{35b}$$

Neutron star that may be in the Crab pulsar:

$$\varepsilon_{Rel}{}^2 \sim 1, \ \varepsilon_{Rot}{}^2 \sim 10^{-3}; \tag{35c}$$

A neutron star during the first few hours after it is formed by stellar collapse:

$$\varepsilon_{Rel}{}^2 \sim 1, \ \varepsilon_{Rot}{}^2 \sim 1, \tag{35d}$$

The theoretical analysis of a rapidly rotating ($\varepsilon_{Rot}{}^2 \sim 1$) highly relativistic ($\varepsilon_{Rel}{}^2 \sim 1$) star is so difficult that no substantial progress has yet been made in it. The only results available are a few scattered items, such as (i) variational principles for the stellar structure due to Hartle and Sharp (1966) and Bardeen (1970); (ii) an exact solution for the external spacetime geometry in the special case where

$$2^{\ell} \text{ moment} \equiv \left[\text{coefficient of } r^{-(2\ell+1)} P_{2\ell}(\cos\theta) \text{ in distant Newtonian potential}\right]$$

$$= (-1)^{\ell+1} M(J/Mc)^{2\ell} \tag{36}$$

M = mass; J = angular momentum; c = speed of light

due to Kerr (1963); and (iii) several minor theorems about equipotentials, shapes of stellar surfaces, rotational energy, and convection due to Thorne (1967, 1971a) and Hartle (1970).

For the case of a slowly rotating ($\varepsilon_{Rot}{}^2 \ll 1$), highly relativistic ($\varepsilon_{Rel}{}^2 \sim 1$) star, the theory is in a much more complete shape. In a series of recent papers Hartle (1967, 1970), Hartle and Thorne (1968, 1969), Chitre, Hartle and Thorne (1970), and Hartle and Thorne (1970) have expanded the equations of structure and pulsation for such stars in powers of $\varepsilon_{Rot}{}^2$, to second order in the angular velocity (terms of order $\varepsilon_{Rot}{}^3$ and higher discarded.) The problems which the resulting formalism can handle include the effects of rotation on the stellar structure, the effects of rotation on the frequencies and stability of the radial modes of pulsation, and the rotational coupling between radial pulsations and quadrupole pulsations, which causes the radial pulsations to emit gravitational waves and damp out. Hartle and Thorne (1968, 1970) have used this formalism to calculate numerically the properties of slowly rotating, highly relativistic neutron stars, such as those which may be in pulsars.

Rapidly rotating ($\varepsilon_{Rot}{}^2 \sim 1$), nearly Newtonian ($\varepsilon_{Rel}{}^2 << 1$) stars are analyzed by expanding in powers of the relativity parameter but not in powers of the rotational parameter. Chandrasekhar's (1965a) post-Newtonian formalism is ideally suited to such an analysis. He [Chandrasekhar (1965b, 1967a, b,c, 1970a,b)] has used it to study relativistic effects on rapidly rotating McClaurin spheroids and Jacobi ellipsoids.

Slowly rotating ($\varepsilon_{Rot}{}^2 << 1$), nearly Newtonian ($\varepsilon_{Rel}{}^2 << 1$) stars are analyzed by expanding in powers of both parameters simultaneously. This has been done in a number of studies of the effects of rotation on slightly relativistic supermassive stars (stars with $M \gtrsim 10^5 M_\odot$) by Fowler (1966), Durney and Roxburgh (1967), and others. In this case of small rotation and small relativity effects, the spacetime geometry outside the star takes the form [Hartle and Thorne (1968), eq. (A4)]

$$ds^2 = \left[1 - \frac{2M^*}{r} + \frac{2Q^*}{r^3} P_2(\cos\theta)\right] dt^{*2} + \left[1 - \frac{2M^*}{r} + \frac{2Q^*}{r^3} P_2(\cos\theta)\right]^{-1} dr^2$$

$$+ \left[1 - \frac{2Q^*}{r^3} P_2(\cos\theta)\right] r^2 \left\{ d\theta^2 + \sin^2\theta \left[d\phi - \frac{2J^*}{r^3} dt\right]^2 \right\}. \quad (37)$$

Here M is the star's mass, Q is its quadrupole moment, J is its angular momentum, and the star rotates in the ϕ direction (from $\phi=0$ toward $\phi=\pi/2$). The terms omitted from this line element due to the assumption of slow rotation and small relativity effects are of the order of magnitude

$$(J^*/R^2)^2 \sim 10^{-24} \text{ for sun,}$$

$$(Q^*/R^3)^2 \sim 10^{-20} \text{ for sun,}$$

$$(J^*/R^3)(M^*/R) \sim 10^{-18} \text{ for sun,}$$

$$(Q^*/R^2)(M^*/R) \sim 10^{-16} \text{ for sun.} \quad (38)$$

3.9 Dragging of Inertial Frames by a Rotating, Relativistic Star

The most fascinating aspect of a rotating, relativistic star, which does not occur in Newtonian theory, is the dragging of inertial frames. The dragging of inertial frames is produced by the rotation of the star, and it is characterized by a quantity

$$\omega_D \equiv -g_{t\phi}/g_{\phi\phi}, \quad (39)$$

which is sometimes called the "angular velocity of cumulative dragging."[6] The physical significance of ω_D can be understood in terms of the following gedanken experiment (see Figure 2): Throw a particle directly at the axis of symmetry of the star, from a point very far away. If the star were not rotating (and if the particle, like a neutrino, did not interact directly with the stellar matter), then the particle would pass through the axis of symmetry. However, because the star is rotating and drags the inertial frames, Coriolis forces act on the particle at each point along its trajectory and deflect it away from the axis of symmetry. The cumulative effect of these Coriolis forces is embodied in the particle's nonzero angular velocity at each point along its trajectory. The value of the angular velocity at each point, $d\phi/dt$, is equal to the value of the function ω_D evaluated at that point.[7] This is actually the angular velocity measured with the clock of a distant observer (time t), not with the clock of a local, static observer (time $|g_{tt}|^{1/2}t$). The two angular velocities are related by the usual redshift factor

$$d\phi/d(\text{time of local static observer}) = |g_{tt}|^{-1/2} d\phi/dt$$
$$= |g_{tt}|^{-1/2}\omega_D. \tag{40}$$

The angular velocity of cumulative dragging outside a slowly rotating, fully relativistic star; and also outside a rapidly rotating, nearly Newtonian star; and also in the distant, nearly Newtonian region outside any rotating star is given by

$$\omega_D{}^* = 2J^*/r^3. \tag{41}$$

Here J is the star's total angular momentum. If the direction of rotation is reversed, (from $+\phi$ to $-\phi$), both J and ω_D become negative; thus, the dragging is in the same direction as the star's rotation.

The dragging which is measured by ω_D is a cumulative effect of the Coriolis forces at each point along the trajectory. These local Coriolis forces are due to the fact that local reference frames with their axes fixed relative to the "distant stars"--i.e. fixed relative to the static coordinate

---

[6]This quantity is also sometimes called the "angular velocity of the local inertial frames", but that is a misnomer; see below. For a beautiful discussion of ω_D from a different viewpoint see Bardeen (1970).

[7]This fact is easily proved: Because the metric is independent of t and ϕ, the covariant components u_t and u_ϕ of the four-velocity are conserved along the trajectory. Because the particle is thrown directly toward the rotation axis, $u_\phi(r=\infty)=0=u_\phi(\text{all } r)$. Consequently, at any point on the orbit $u^t=g^{tt}u_t$, and $u^\phi=g^{\phi t}u_t$ (we use coordinates where $g_{1t}=g_{1\phi}=g_{2t}=g_{2\phi}=0$). The angular velocity is then

$$d\phi/dt = u^\phi/u^t = g^{\phi t}/g^{tt} = -g_{\phi t}/g_{\phi\phi}.$$

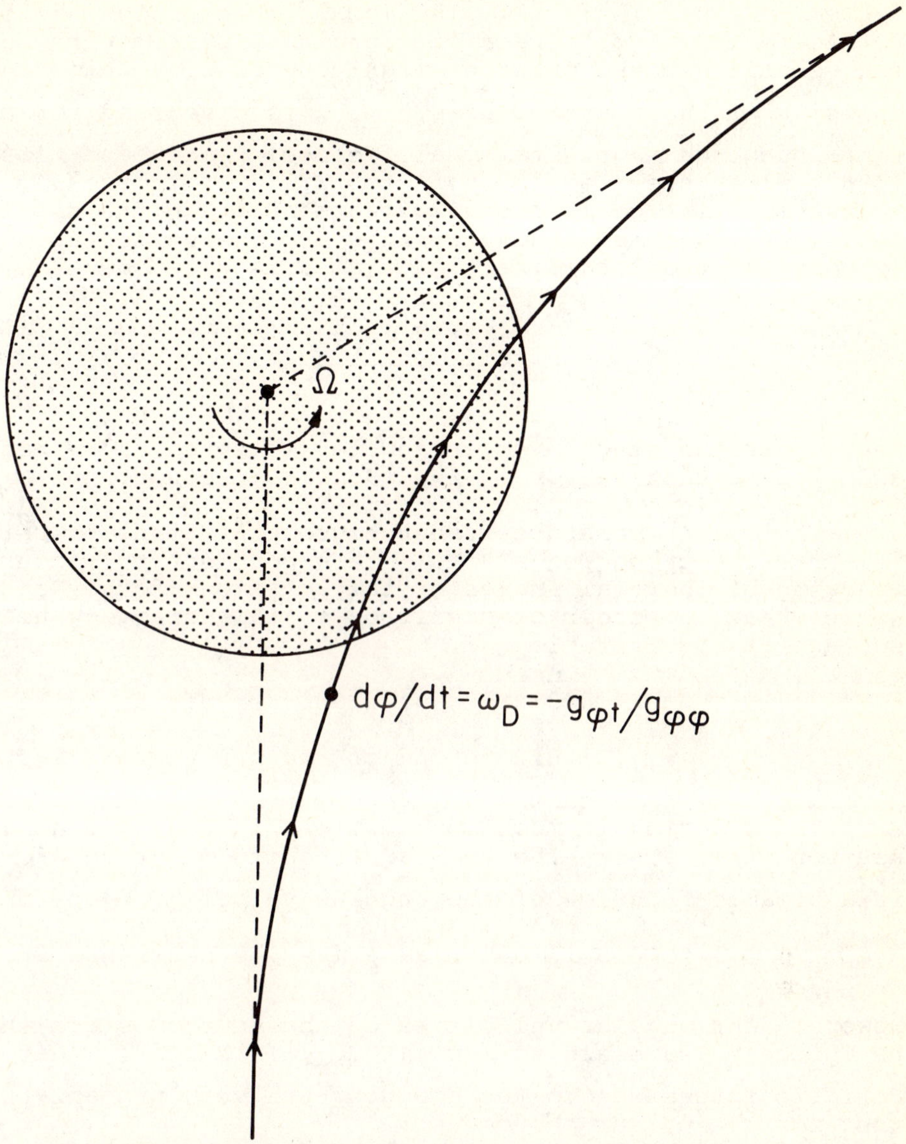

Figure 2. Orbit of a particle thrown from $r = \infty$ toward the rotation axis of a rotating, relativistic stellar model. Because of the dragging of inertial frames, the particle experiences Coriolis forces at each point along its trajectory. The integrated influence of these Coriolis forces is to give the particle, at any point on its orbit, the angular velocity $d\phi/dt = \omega_D \equiv -g_{\phi t}/g_{\phi\phi}$, where ω_D is to be evaluated at the given point. [This figure is taken from Thorne (1971a, p.252).]

system--are not inertial. The local frames which are inertial rotate relative to the static coordinate system with an angular velocity of the order of magnitude of ω_D. This "angular velocity of the inertial axes," $\underline{\omega}_{IA}$, is discussed for a general time-independent gravitational field by Landau and Lifshitz (1962, §89) [see also appendix to §3.1.7 of Thorne (1971a) for a more mathematically elegant treatment]. In the nearly Newtonian limit, outside a rotating star, the vector angular velocity of the local inertial frames is given by

$$\underline{\omega}^*{}_{IA} = \frac{3(\underline{J}^*\cdot\underline{e}_r)\underline{e}_r - \underline{J}^*}{r^3}, \tag{42}$$

where $\underline{J}$ is the star's vector angular momentum, r is radius, and $\underline{e}_r$ is a unit vector in the radial direction.

The low-temperature research group at Stanford University is currently preparing a satellite experiment to measure the dragging of inertial frames by the rotation of the Earth. Notice that the order of magnitude of this dragging near the surface of the Earth is

$$\omega^* \sim J^*/R^3 \sim 3 \times 10^{-9} \text{ radians per day on Earth} \tag{43a}$$

$$\sim 0.2 \text{ second of arc per year on Earth}$$

The dragging is much more extreme in and near a massive neutron star. There the angular velocity of dragging, ω_D, is a sizable fraction of the angular velocity, Ω, of the star

$$\omega_D/\Omega \sim 0.1 \text{ to } 0.5 \text{ in and near a massive neutron star.} \tag{43b}$$

However, the angular velocities of the neutron stars that may be in the pulsars are so low $[(\Omega/\Omega_{crit})^2 \lesssim 10^{-3}]$, that the Coriolis forces due to the dragging are very probably unimportant for the pulsar phenomenon:

$$\begin{pmatrix}\text{Coriolis forces}\\ \text{due to dragging}\end{pmatrix} \lesssim 10^{-3} \times \begin{pmatrix}\text{Dominant gravitational}\\ \text{forces}\end{pmatrix} \text{ in pulsars} \tag{43c}$$

4. Relativistic Stellar Stability and Pulsations

4.1 Dynamical Analysis of Stability Against Radial Perturbations

In 1963, S. Chandrasekhar (1964) and R.P. Feynman (unpublished work which greatly influenced the Hoyle-Fowler ideas on supermassive stars) discovered a relativistic effect of considerable importance in astrophysics:[8] They developed independently the mathematic theory of adiabatic radial pulsations of relativistic stellar models, and in the process they discovered that general relativity can catalyze a dynamical instability in stars, which is absent in Newtonian theory: In Newtonian theory a star is stable against radial perturbations if and only if its adiabatic index, suitably averaged over the stellar interior, is greater than 4/3. However, relativistic effects raise the critical value of the adiabatic index, Γ_1, above 4/3:

$$\Gamma_{crit} = 4/3 + \text{relativistic term of } 0(M^*/R). \tag{44}$$

(See Figure 3.) As a consequence, a star, in which relativistic influence on the structure is negligible, and which is just barely stable according to Newtonian theory ($< \Gamma_1 >$ only slightly greater than 4/3), may actually be unstable because of relativistic effects. Such is the case, for example, in a non-rotating supermassive star of 10^8 solar masses with $M^*/R \approx 10^{-4}$, and also in a nonrotating white dwarf near the Chandrasekhar limiting mass with $M^*/R \approx 10^{-3}$.

The reason that Γ_{crit} rises above 4/3 in the relativistic regime is quite simple: The nonlinearities of general relativity cause gravitational forces to be stronger than in Newtonian theory, and to vary more with the compression and expansion of a star. Consequently, the pressure forces needed to counterbalance gravity and stabilize the star must also vary more rapidly in general relativity--i.e., the value of

$$\Gamma_1 \equiv - \left(\frac{\partial \ell n p}{\partial \ell n V}\right)_s = \frac{\rho^* + p^*}{p^*} \left(\frac{\partial p^*}{\partial \rho^*}\right)_s \tag{45}$$

needed for stability is greater.

In Brans-Dicke theory the gravitational forces are also stronger and more sensitive than in Newtonian theory, so Γ_{crit} increases above 4/3 according to a law similar to equation (44). However, the increase of Γ_{crit} is not quite so great as in general relativity theory, so the relativistic instability is not quite so strong. (See Nutku 1969).

The calculation of a star's normal modes of adiabatic oscillation is not much more difficult in general relativity theory than in Newtonian theory. The problem is a Sturm-Liouville eigenvalue problem of precisely the same type as occurs in the Newtonian theory; the only difference is that the coefficients in the differential equation are more complicated. A detailed discussion of the relativistic eigenvalue problem and techniques for solving it is contained in Bardeen, Thorne, and Meltzer (1966).

---

[8]This effect had previously been noticed by Kaplan (1949); but his work went virtually unnoticed and had no influence on astrophysical thinking.

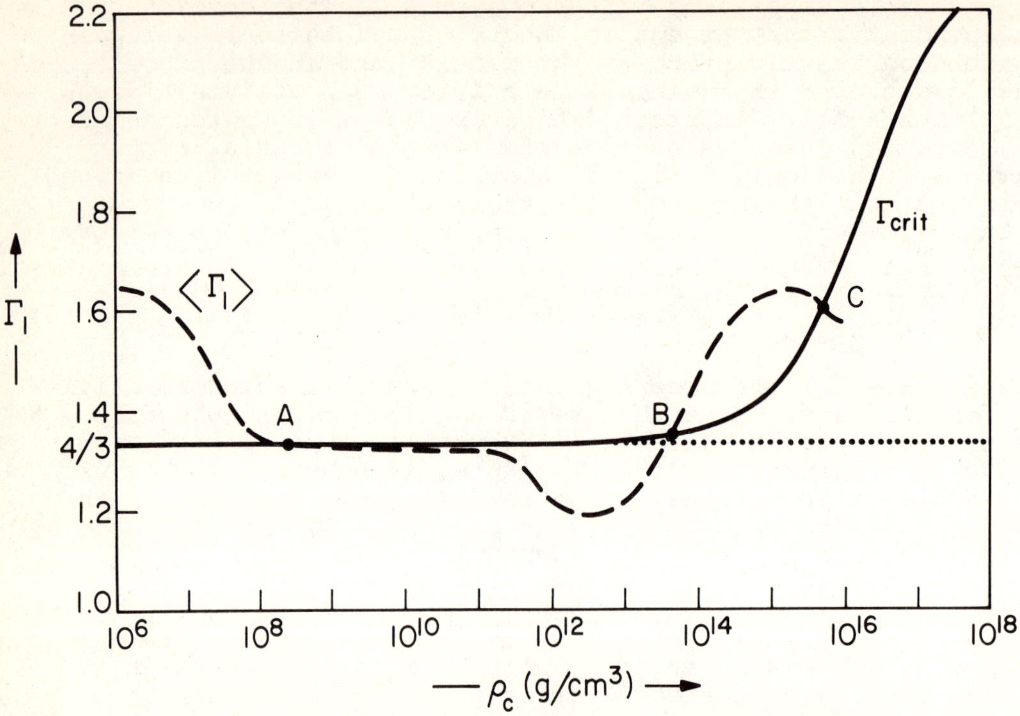

Figure 3. The critical adiabatic index Γ_{crit} and volume-averaged actual adiabatic index $\langle\Gamma_1\rangle$ for Harrison-Wakano-Wheeler (1958) models of stars at the endpoint of thermonuclear evolution. Models to the left of point A are stable white dwarfs, and models between points B and C are stable neutron stars. Models between A and B and to the right of C are all unstable. The onset of instability at point A is due to a combination of (i) the relativistic rise of Γ_{crit} very slightly above 4/3, and (ii) the approach of $\langle\Gamma_1\rangle$ toward 4/3 and below as the degenerate electrons become more and more relativistic, and as electron capture begins to occur. The recovery of stability at point B occurs when nucleon-nucleon forces push $\langle\Gamma_1\rangle$ above Γ_{crit}. The onset of instability at point C is due entirely to the steeply rising, relativistic value of Γ_{crit}. Analytic formulas for Γ_{crit} have been derived in the nearly Newtonian regime by Fowler (1964), and in the ultrarelativistic regime by Ipser (1970). [This figure is based on the work of Ipser (1970).]

4.2 Static Analysis of Stability Against Radial Perturbations

In certain cases of physical interest, one can determine the precise number of unstable normal radial modes of adiabatic oscillation for an equilibrium configuration without solving the eigenvalue problem described above. The "static" analysis, which makes possible this counting of unstable modes, is applicable in one form to zero-temperature stellar models (result due to J.A. Wheeler), and in a slightly different form to hot, isentropic models (result due to J.M. Bardeen). [For a review see chapter 4 of Thorne (1967).] More particularly, it concentrates its attention on stellar models which belong to one-parameter sequences of two types: (i) zero-temperature sequences whose models are all characterized by the same one-parameter equation of state, $p=p(\rho)$, and by the same adiabatic index, Γ_1, which must be related to the equation of state by $\Gamma_1=(\rho^*+p^*)p^{*-1}(dp^*/d\rho^*)$; and (ii) hot sequences, whose models all obey the same two-parameter equation of state, $p=p(\rho,s)$, all have constant entropy per baryon (s independent of r), and all have the same number of baryons (i.e. the same rest mass). White-dwarf and neutron-star models--even hot ones, since their thermal pressure is negligible--belong to sequences which are very nearly of the first type, while supermassive-star models belong to sequences of the second type. Each sequence can be parameterized by the central density, ρ_c.

For a cold sequence one diagnoses stability by making a plot of mass versus radius, $M(R)$, parameterized by central density, ρ_c. (See Figure 4.) At each peak and valley in the $M(R)$ curve one mode of radial pulsation changes stability. If the curve bends counterclockwise at a critical point, the mode becomes unstable with increasing ρ_c; if the curve bends clockwise, it becomes stable with increasing ρ_c. These facts, together with a knowledge of the stability of the Newtonian models of very low ρ_c, are sufficient to determine the precise number of unstable modes in each model along the sequence.

For a hot, isentropic sequence one usually plots minus the binding energy,

$$-E_B \equiv M - \mu A = (\text{total mass-energy}) - (\text{rest mass of baryons}), \tag{46}$$

against radius. In this plot of $-E_B(R)$, again one mode of radial oscillation changes stability at each peak and valley. However, in this case by contrast with the zero-temperature case, a clockwise spiral corresponds to loss of stability with increasing ρ_c, while a counterclockwise spiral corresponds to a regain of stability with increasing ρ_c.

These static analyses of stability have recently been

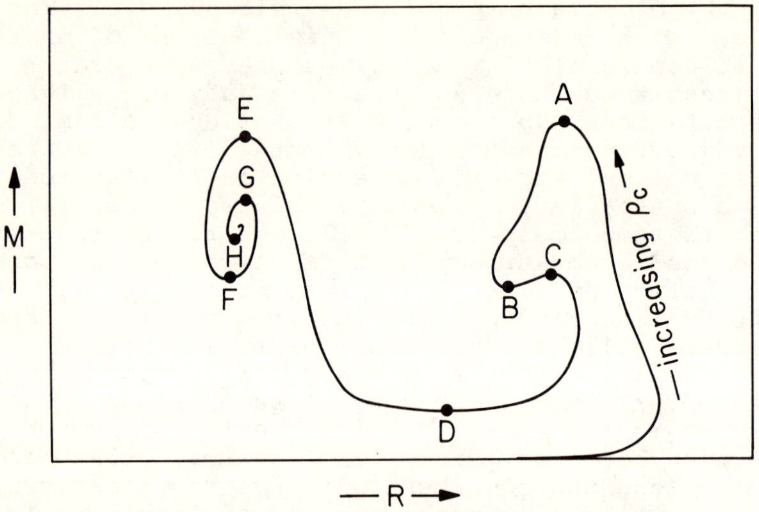

Figure 4. Idealized mass versus radius plot for a sequence of cold stellar models obeying a single equation of state $p = p(\rho)$ and having the same adiabatic index $\Gamma_1 = (\rho^* + p^*)p^{*-1}(dp^*/d\rho^*)$. The static analysis of stability predicts that, since all models (white dwarfs) are stable below point A, the fundamental mode of oscillation will go unstable at point A (counterclockwise bend); the first harmonic will go unstable at B (counterclockwise bend); the first harmonic will regain stability at C (clockwise bend); the fundamental will regain stability at D (clockwise bend); and the fundamental followed by successive harmonics will go unstable along the counterclockwise spiral through E, F, G, H, ... The actual M(R) plot for white dwarfs and neutron stars at the endpoint of thermonuclear evolution is qualitatively like this one; there the white dwarfs lie in the region below A, while the neutron stars lie between D and E.

generalized to rigidly rotating stellar models by Hartle and Thorne (1969).

4.3 Stability Against Convection

Although general-relativistic effects have a marked influence on stability against adiabatic, radial pulsations, they are totally unimportant for convection. This is because radial pulsation, being a global phenomenon, couples easily to the non-Newtonian curvature of spacetime, while convection, being a local phenomenon, cannot. A general-relativistic star, like a Newtonian star, begins to convect in any region where its temperature gradient is superadiabatic. More precisely, a star is unstable against convection in any region where the Schwarzschild discriminant

$$S(r) \equiv \frac{dp^*}{dr} - \frac{\Gamma_1 p^*}{\rho^*+p^*}\frac{d\rho^*}{dr} \tag{47}$$

is negative, and it is stable in any region where the discriminant is positive. Where $S(r)$ is zero, there is an infinity of nonradial zero-frequency modes corresponding to each choice of the spherical-harmonic indices (ℓ,m). [See Thorne (1965, 1969a).] [The reader puzzled by the form of $S(r)$ should recall the expression (45) for the adiabatic index, which follows from the first law of thermodynamics, equation (12a).]

4.4 Nonradial Pulsations of Relativistic Stellar Models

The theory of adiabatic, nonradial pulsations of general-relativistic stellar models has been developed recently in a series of five papers by Thorne and Campolattaro (1967), Price and Thorne (1969), and Thorne (1969a,b), and Campolattaro and Thorne (1970). The dipole pulsations are qualitatively the same as in Newtonian theory, but the quadrupole and higher-order pulsations are very different--they radiate gravitational waves. We shall delay discussing these pulsations until §5.6 after we have reviewed the basic features of gravitational radiation.

5. Gravitational Waves

5.1 Properties of Gravitational Waves in Various Relativistic Theories

A gravitational wave is a ripple in the geometry of spacetime--i.e. an inhomogeneity in the gravitational field--which propagates with the speed of light.

Gravitational waves are predicted by all relativistic theories of gravity--or at least by all that I know of. However, in the various theories the waves have different properties, and the rate at which a source emits them is different. The most important difference is in the multipolarity that the waves can have: In general relativity and other theories where gravity is due entirely to the curvature of spacetime (there are no other long-range fields present), gravitational waves are of quadrupole order and higher. In theories like Brans-Dicke, where there is a scalar field present along with the curvature of spacetime, gravitational waves can have monopole and dipole forms, as well as quadrupole and higher. This difference in multipolarity can be used, in principle, to distinguish experimentally between theories: Only waves with monopole features can set an elastic sphere (e.g., the Earth) into spherical pulsations; so a detector of waves which uses spherical modes of pulsation should not be excited if general relativity is correct, but can be excited if Brans-Dicke theory is correct.

Henceforth, we shall concentrate entirely on general-relativistic waves.

5.2 The Linearized, Weak-Field Theory of Gravitational Waves

The simplest context in which to calculate the properties of gravitational waves is when they are very weak and when their sources are nearly Newtonian. In this context, one usually linearizes the perturbations about flat spacetime--i.e. one introduces a nearly inertial reference frame in which the metric tensor has the form

$$g_{\alpha\beta} = \eta_{\alpha\beta} + h_{\alpha\beta} , \tag{48}$$

and one calculates only to first order in $h_{\alpha\beta}$. To simplify the calculations further, one makes an infinitesimal (i.e., of order $h_{\alpha\beta}$) coordinate transformation to guarantee that

$$\sum_\beta (\partial/\partial x^\beta)(h_\alpha^\beta - \tfrac{1}{2} h\delta_\alpha^\beta) = 0, \tag{49}$$

where the index β is "raised" by means of the special relativity metric $\eta^{\mu\nu} = \eta_{\mu\nu}$

$$h_\alpha^\beta = \sum_\mu \eta^{\beta\mu} h_{\alpha\mu}; \quad h = \sum_\alpha h_\alpha^\alpha , \tag{50}$$

and where δ_α^β is the Kronecker delta.

When this is done, the Einstein field equations take on a beautifully simple and familiar form (see, e.g., Landau and Lifshitz (1962) for details of calculations):

$$\Box h_{\alpha\beta} = (16\pi\, G/c^4)\,(T_{\alpha\beta} - \tfrac{1}{2}\eta_{\alpha\beta}\, T). \tag{51}$$

Here $\Box$ is the special-relativistic wave operator, $T_{\alpha\beta}$ is the stress-energy tensor, and T is the trace of the stress-energy tensor

$$\Box = \partial^2/\partial t^{*2} - \nabla^2 = \partial^2/\partial t^{*2} - \sum_j \partial^2/\partial x_j^{\,2}. \tag{52}$$

The analysis of the emission of gravitational waves, from equation (51) onward, is identical in form to the analysis of the emission of electromagnetic waves by a charged system. The key difference of interpretation is that an electromagnetic wave is produced by changing charge and/or current distributions, while a gravitational wave is produced by changing mass, energy, momentum, and/or stress distributions. Because the source of gravitational waves is the stress-energy tensor of all matter and (nongravitational) fields present, they can be generated not only by moving matter, but also by changing electromagnetic fields, neutrino fields, etc.

From equation (51) and a "stress-energy pseudotensor" for the gravitational field, one can derive a formula for the time-averaged rate at which gravitational waves carry energy away from a nearly-Newtonian mass distribution:

$$\left\langle -\frac{dE}{dt}\right\rangle = \frac{G}{45c^5} \sum_{i,j} \left\langle \left(\frac{\partial^3 D^{ij}}{\partial t^3}\right)^2 \right\rangle . \tag{53a}$$

Here D^{ij} is the quadrupole moment of the mass distribution, as calculated in a Cartesian coordinate system

$$D^{ij} \equiv \int_{\text{source}} \rho(3x^i x^j - r^2)\, d\,\text{volume}. \tag{53b}$$

Notice the order of magnitude of the power radiated: The quadrupole moment has $D^{ij} \lesssim MR^2$, where M is the mass of the system and R is its radius; and the characteristic rate at which the system changes, if it is a gravitationally-bound, nearly Newtonian system, is $\omega \lesssim (GM/R^3)^{1/2}$. [In some cases, e.g. rotation of the sun $\omega << (GM/R^3)^{1/2}$, and $D^{ij} << MR^2$]. Consequently,

$$\left\langle -\frac{dE}{dt}\right\rangle \lesssim \frac{1}{45}\frac{G}{c^5}\left[\left(\frac{GM}{R^3}\right)^{3/2} MR^2\right]^2 = \frac{1}{45}\frac{c^5}{G}\left[\frac{GM/c^2}{R}\right]^5 \tag{54}$$

$$\sim 10^{-3}(c^5/G)\;(R_g/R)^5 \sim (10^{57}\,\text{erg/sec})(R_g/R)^5 .$$

Here R_g is the gravitational radius of the system. Notice that the system can radiate significantly only if it is near its gravitational radius. For example, in the Earth-moon system R_g is about 1 cm, and R is about 3×10^{10} cm, so the power radiated is $\lesssim 10^5$ erg/sec. However, in a newly formed neutron star which is pulsating with large nonradial amplitude, $R_g \sim 2$ km and $R \sim 10$ km, so the power radiated is $\lesssim 10^{53}$ erg/sec.

Formula (54) suggests that there might be an absolute upper limit

$$dE/dt \text{ always less than } c^5/G \sim 10^{60} \text{ erg/sec} \tag{55}$$

on the power output of any source of gravitational waves. That there is, indeed, such a limit one can see most clearly as follows: Consider a source of mass M. Conservation of mass-energy guarantees that it cannot radiate more energy than $E = Mc^2$. Moreover, it cannot radiate anything if it is inside its gravitational radius; so when radiating, it must have a size $R > R_g \sim GM/c^2$. The time interval in which it radiates its energy must be greater than or of the order of the light travel time, R/c, across it. Consequently, its power output is limited:

$$\frac{dE}{dt} \lesssim \frac{Mc^2}{R/c} \lesssim \frac{Mc^2}{GM/c^3} = \frac{c^5}{G} \sim 10^{60}\text{erg/sec.}$$

Turn attention now to the forces which gravitational waves exert on matter as they pass through it. To analyze these forces in the linearized, weak-field approximation it is convenient to focus attention on a region of spacetime small enough that the waves appear to be plane-fronted, and to orient the z-axis of a nearly inertial frame along the direction of propagation of the waves. Then, after an appropriate infinitesimal coordinate transformation [see Landau and Lifshitz (1962), §101], the waves take the form

$$h_{xx} = - h_{yy} = h_{xx}(t-z/c); \quad h_{xy} = h_{yx} = h_{xy}(t-z/c);$$

$$\text{all other } h_{\alpha\beta} \text{ vanish.} \tag{56}$$

Here, $h_{xx}(t - z/c)$ and $h_{xy}(t - z/c)$ are solutions to the vacuum wave equation (51), whose particular form is determined by the nature of the source.

Notice that the waves (56) are "transverse" (i.e. $h_{z\alpha} = 0$ for all α) and "traceless" (i.e. $h_{xx} + h_{yy} + h_{zz} - h_{tt} = 0$). This transverse-traceless nature is also exhibited by the interaction between the waves and matter: The waves produce no tidal gravitational forces along the longitudinal (z) direction; and the tidal forces in the transverse (x and y) directions deform a body without changing its cross-sectional area.

Let us spell this out more explicitly: Consider a body at rest in the nearly-inertial reference frame of equation (56). Focus attention on two mass elements, each of mass μ, separated by a vector displacement

$$\underline{\eta} = \eta^x \underline{e}_x + \eta^y \underline{e}_y + \eta^z \underline{e}_z, \tag{57}$$

such that $|\underline{\eta}|$ is very small compared to the mean wavelength of the waves. As the gravitational waves of equation (56) pass the body, they produce a tidal gravitational force between the two mass-elements which is given by

$$\underline{F} = \begin{pmatrix}\text{force acting on the mass element at } \underline{x} + \underline{\eta}/2 \text{ relative} \\ \text{to the center of mass at } \underline{x}\end{pmatrix}$$

$$= - \begin{pmatrix}\text{force acting on the mass element at } \underline{x} - \underline{\eta}/2 \text{ relative} \\ \text{to the center of mass at } \underline{x}\end{pmatrix}$$

$$= F^x \underline{e}_x + F^y \underline{e}_y ; \tag{58a}$$

$$F^x = -\frac{1}{4}\mu[\partial^2 h_{xx}/\partial(t-z/c)^2]\eta^x - \frac{1}{4}\mu[\partial^2 h_{xy}/\partial(t-z/c)^2]\eta^y,$$

$$F^y = +\frac{1}{4}\mu[\partial^2 h_{xx}/\partial(t-z/c)^2]\eta^y - \frac{1}{4}\mu[\partial^2 h_{xy}/\partial(t-z/c)^2]\eta^x. \tag{58b}$$

A straightforward calculation reveals that these forces deform the body in the transverse plane without affecting its cross-sectional area--unless its shear modulus is anisotropic.

The gravitational waves of equation (56), like any gravitational wave in general relativity theory, have two independent modes of polarization corresponding to the two independent functions $h_{xx} = -h_{yy}$ and $h_{xy} = h_{yx}$. The various types of polarization (linear, circular, elliptical), which a monochromatic wave can have, are understood most clearly in terms of a "polarization ellipse": Position a small, circular ring of noninteracting test particles in the x-y plane. As the waves pass the ring, they will deform it into an ellipse of periodically changing shape and orientation. If the polarization ellipse does not rotate, then the waves are said to be plane polarized. If it rotates but changes its eccentricity with time, then they are elliptically polarized. If it rotates maintaining constant eccentricity, then the waves are circularly polarized (see Figure 5).

5.3 Critique of the Linearized, Weak-Field Theory

The linearized, weak-field approximation to general relativity, which was used in the preceding section, has several flaws that have worried theoretical physicists for nearly half a century. The three main flaws are these:

(i) The energy transported by a gravitational wave is analyzed using a "stress-energy pseudotensor" whose physical significance is highly questionable. A much more satisfactory way is needed to analyze the energy of a wave.

(ii) In the linearized, weak-field approximation one cannot calculate the back reaction of the gravitational field on its source. Neither the gravitational forces which bind the source together, nor the radiation reaction due to the emission of waves is contained in this approximation. The reason is that the two fundamental equations of the approximation, (49) and

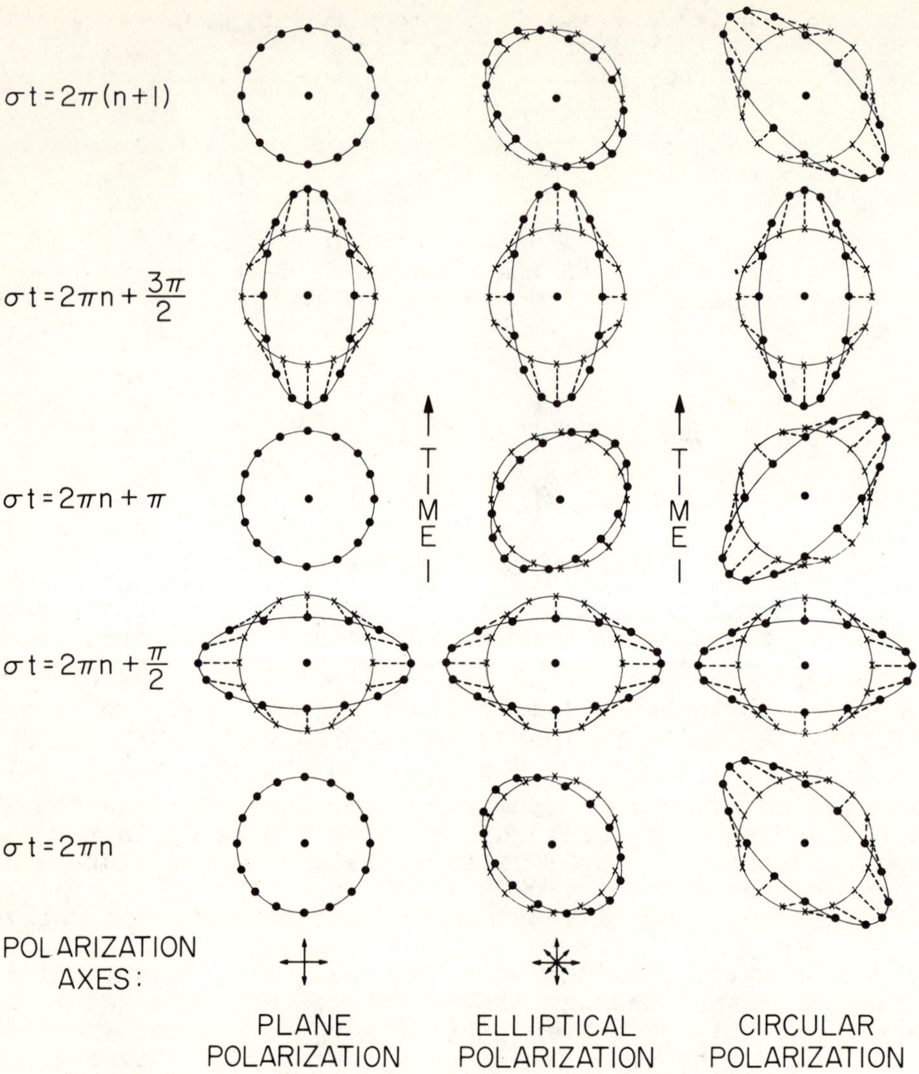

Figure 5. The polarization ellipse for various types of nearly plane, weak gravitational waves. This ellipse consists of a circular ring of test particles surrounding a central test particle and lying in a plane perpendicular to the direction of propagation of the waves. As the waves pass, they deform the ring into an ellipse of varying eccentricity and orientation. For plane-polarized waves the ellipse pulsates in and out but does not rotate, and each particle on the ring moves back and forth in a straight line relative to the central particle. For circularly polarized waves the ellipse rotates without ever changing its eccentricity; and each particle on the ring moves around a small circle relative to the central particle. Note, however, that just as the drops in an ocean wave do not move along with the wave, so the particles on the ring do not move around the central particle with the ellipse. For elliptically polarized waves the ellipse rotates but changes its shape in the process; and each particle on the ring performs small-amplitude elliptical motion relative to the central particle. [This figure is taken from Price and Thorne (1969).]

(51) guarantee that

$$\sum_{\beta} (\partial/\partial x^{\beta}) T^{\alpha\beta} = 0, \tag{59}$$

which is the special-relativistic version of the equation "F=ma" without any gravitational forces contained therein. Actually, the self-gravitational forces are absent because they are of an order which has been neglected in the linear-approximation procedure. In order to see them, including radiation-reaction forces, one must very carefully build up the approximation procedure into the nonlinear regime, order after order. Until that has been done, one cannot have full confidence in the conclusions of the elementary analysis, since it makes use of a non-self-consistent mixture of Newtonian theory (to calculate the motion of the source) and linearized weak-field theory (to calculate the radiation that the source emits).

(iii) Significant astrophysical sources of gravitational waves must be near their gravitational radii, where gravity is strong and the linearized weak-field approximation is far from valid. In order to understand these sources, one needs an alternative, strong-field method of calculating their motion, radiation, and radiation reaction.

Within the last two years, these three flaws in the elementary weak-field theory have been healed, at least in part. In the next three sections we shall review briefly the work which has accomplished this.

5.4 The Energy of a Gravitational Wave as Analyzed by Isaacson

Richard A. Isaacson (1968a,b) using techniques due to Brill and Hartle (1964), has recently given a satisfactory treatment of the energy in gravitational waves whose wavelength is short compared to the radii of curvature of the background spacetime in which they propagate (a condition always satisfied in practice). The basic idea of his work is to split the Einstein curvature tensor into two parts--a "ripply" part, $\underline{\underline{G}}_W$, associated with the waves, and a "smooth" part, $\underline{\underline{G}}_B$, associated with the background space; and to expand $\underline{\underline{G}}_W$ in powers of the ratio, a/R, of the mean wavelength of the waves a, to the characteristic radius of curvature R of the background space:

$$\underline{\underline{G}} = \underline{\underline{G}}_B = \underline{\underline{G}}_{W\text{ linear}} + \underline{\underline{G}}_{W\text{ quadratic}} + \dots \tag{60}$$

The total Einstein tensor must vanish. This is accomplished for the dominant, linear part of the waves by demanding

$$\underline{\underline{G}}_{W\text{ linear}} = 0 . \tag{61a}$$

This equation turns out to be a curved-space analogue of the flat-space wave equation $\Box h_{\alpha\beta} = 0$ for the ripply perturbation in the metric, $h_{\alpha\beta}$. The spacetime-averaged quadratic part of $\underline{\underline{G}}_W$ turns out to be of the same order of magnitude as the background curvature; so to get $\underline{\underline{G}} = 0$ one must impose

$$\underline{\underline{G}}_B + \langle \underline{\underline{G}}_{W\ \text{quadratic}} \rangle = 0, \tag{61b}$$

where < > denotes a particular average over several wavelengths of the waves. From the point of view of the background spacetime, $\langle \underline{\underline{G}}_{W\ \text{quadratic}} \rangle$ is not part of the Einstein tensor; rather, it is a foreign entity which produces the curvature $\underline{\underline{G}}_B$ by means of equation (61b). Put more precisely, as viewed by the background space, the quantity

$$\underline{\underline{T}}_W \equiv -(c^4/8\pi G)\ \underline{\underline{G}}_{W\ \text{quadratic}} \tag{62}$$

functions as a stress - energy which produces curvature by way of Einstein's field equation

$$\underline{\underline{G}}_B = (8\pi G/c^4)\ \underline{\underline{T}}_W \tag{63}$$

As a consequence, $\underline{\underline{T}}_W$ is the stress-energy tensor for the gravitational waves propagating in the background space.

It turns out that this stress-energy tensor for the waves is identical to the spacetime-average of the Landau-Lifshitz pesudotensor, which is usually employed in the linearized weak-field theory. Consequently, the energy calculations of the linearized weak-field theory are vindicated.

A useful formula for the stress-energy tensor of the plane gravitational waves of equation (56) is

$$T_W^{*zz} = T_W^{*tt} = \frac{1}{16\pi}\left\langle \left[\frac{dh_{xy}}{d(t^*-z)}\right]^2 + \left[\frac{dh_{xx}}{d(t^*-z)}\right]^2 \right\rangle \tag{64}$$

5.5 Slow-Motion Expansions in General Relativity[9]

The second difficulty with the linearized weak-field theory was its ignorance of the self-gravitational forces that act on a radiating system, including radiation-reaction forces. Since the 1930's relativity theorists have tried, with only partial success, to develop the higher-order corrections to the linearized theory in a self-consistent manner which yields self-gravitation and radiation reaction. The dominant, Newtonian part of the self-gravitation was a rather easy problem, but the radiation reaction seems to have defied satisfactory treatment until the 1969 work of William L. Burke and of S. Chandrasekhar. [See Burke (1969, 1970), Thorne (1969b), and Chandrasekhar and Esposito (1970) for full details; see Burke and Thorne (1969) for a detailed overview].

---

[9]Portions of this section are nearly identical to portions of Burke and Thorne (1969). I thank Burke and the two publishers for permitting this overlap.

A key idea that was missing from all earlier attempts to study radiation reaction was the idea of using "matched asymptotic expansions." In the region inside and near the almost Newtonian system one performs a "slow-motion" expansion of the Einstein field equations, such as the post-Newtonian expansion of Chandrasekhar (1965a, 1969) and Nutku and Chandrasekhar (1969). In the radiation zone one performs a "fast motion," weak-field expansion of general relativity, whose first order is identical to the linearized theory without sources, which was discussed in §5.2. In the intermediate region one matches the two expansions together in a manner that has been developed to great sophistication by applied mathematicians [see, e.g., Cole (1968)]. The result is a self-consistent approximation scheme that properly couples the radiation to its source and properly describes the radiation-reaction force on the source.

One result of Burke's analysis which should be very useful in astrophysical calculations is this: It is possible to incorporate the general-relativistic radiation-reaction forces into the Newtonian theory of gravity to an accuracy

$$\begin{aligned} \varepsilon^2 &\sim (U_{max}/c\ ) \sim (\text{velocities of matter}/c)^2 \\ &\sim (\text{size of source}/\text{wavelength of radiation})^2 . \end{aligned} \tag{65}$$

In order to do this one simply changes the boundary condition on the Newtonian potential at $r = \infty$. The resultant, modified Newtonian theory of gravity is as follows[10]:

Modified Newtonian Theory of Gravity, which gives correctly to accuracy ε^2 all general-relativistic radiation-reaction effects: Gravity is described by the usual Newtonian potential, $U(\underline{x},t)$, which produces forces on bodies in the usual way

$$\underline{F} = m\,\underline{\nabla}\,U. \tag{66}$$

The gravitational potential satisfies the usual source equation

$$\nabla^2 U = 4\pi G\rho, \tag{67}$$

and as usual it must be nonsingular throughout the system. However, by contrast with the usual Newtonian theory, the boundary condition at $r = \infty$ is not $U(\infty) = 0$. Rather, at a particular moment of time t, when the multipole moments of U are

---

[10]As of now (July 1969) there remain a few loose ends to be tied up in the rigorous mathematical proof that this modified Newtonian theory gives the correct general-relativistic radiation reaction. However, I am very confident that the result will not be unraveled by those loose ends--so confident, in fact, that a formal bet to that effect hangs, framed, on the wall of Norman Bridge Laboratory of Caltech: I wager twenty-five dollars against Burke's one dollar that this modified Newtonian theory will remain valid after he has tied up the loose ends. Note added in May 1970: Burke has paid up. The loose ends are tied together in his forthcoming paper, Burke (1970).

$$D^{\ell}_{m}(t) \equiv [4\pi/(2\ell+1)] \int_0^R \rho r^{\ell} Y^{\ell\dagger}_{m} \, d\,\text{volume} \tag{68}$$

where

$$\dagger \equiv \text{complex conjugate},$$

the form of U at large r must be

$$U = \frac{M}{r} + \sum_{\ell=2}^{\infty} \sum_{m=-\ell}^{\ell} \left[\frac{D^{\ell}_{m}(t)}{r^{\ell+1}} + \frac{(-1)^{\ell+1}(\ell+1)(\ell+2)}{\ell(\ell-1)(2\ell+1)[(2\ell-1)!!]^2} \frac{d^{2\ell+1}D^{\ell}_{m}}{d(ct)^{2\ell+1}} r^{\ell} \right] Y^{\ell}_{m} . \tag{69}$$

(Here c is the speed of light). As a consequence, anywhere in space the potential of the modified theory is related to that of the usual theory by

$$U = U_{\text{usual}} + U_{\text{reaction}}, \tag{70a}$$

where

$$U_{\text{reaction}} = \sum_{\ell=2}^{\infty} \sum_{m=-\ell}^{\ell} \frac{(-1)^{\ell+1}(\ell+1)(\ell+2)}{\ell(\ell-1)(2\ell+1)[(2\ell-1)!!]^2} \frac{d^{2\ell+1}D^{\ell}_{m}}{d(ct)^{2\ell+1}} r^{\ell} Y^{\ell}_{m} . \tag{70b}$$

The term $m\underline{\nabla}U_{\text{reaction}}$ in the force law represents radiation reaction. But notice that the radiation reaction force on a given particle is caused by the behavior of the system as a whole, rather than by the particle's behavior alone.

Two warnings must be made about this modified Newtonian theory of gravity: (i) Only over the long run, as its effects become cumulative, is the resultant radiation reaction larger than other, post-Newtonian effects (e.g. the perihelion shift of Mercury), which are neglected in this theory. (ii) The multipole terms with $\ell \geq 3$ in the radiation reaction potential (70b) are smaller than the $\ell = 2$ terms by a factor of order $\varepsilon^{2(\ell-2)}$. Since the reaction given by this theory is guaranteed to be accurate only to order ε^2, these $\ell \geq 3$ terms are "frosting on the cake" which have no real significance.

For the case $\ell=2$ Chandrasekhar and Esposito (1970) put U_{reaction} into a form that is simpler than, but equivalent to (70b).

As a particular application of Burke's modified Newtonian theory, one can calculate the rate at which energy is sapped from a system by the radiation-reaction forces:

$$\langle -dE/dt \rangle = \int (\rho \nabla U_{\text{reaction}}) \cdot v \, d\text{volume} \qquad (71)$$

Here ρ is the mass density and v is the velocity of the matter. After complicated manipulations involving equations (67),(68), and (70b), one finally arrives at the result

$$\langle -dE/dt \rangle = \frac{c^5}{G} \sum_{\ell=2}^{\infty} \sum_{m=-\ell}^{\ell} \frac{(\ell+1)(\ell+2)}{4\pi\ell(\ell-1)[(2\ell-1)!!]^2} \left\langle \left| \frac{G}{c^2} \frac{d^{\ell+1} D^{\ell}_{m}}{d(ct)^{\ell+1}} \right|^2 \right\rangle \qquad (72)$$

The dominant, $\ell=2$ part of this equation, rewritten in different notation, is identical to equation (53) for the energy carried off by the waves. Thus, there is a time-average equality between the rate at which the waves carry off energy, and the rate at which radiation-reaction forces sap energy from the source.

5.6 Nonradial Pulsations of Fully Relativistic Stellar Models

The third objection to the linearized weak-field approximation of §5.2, that it cannot treat the astrophysically important sources which are near their gravitational radii, is being answered at present by the development of alternative techniques for studying such sources. For one type of source --a nonradially pulsating, fully relativistic star; e.g., a newly formed neutron star--an alternative technique has been developed in full and has been applied numerically. The primary literature on this work is Thorne and Campolattaro (1967), Price and Thorne (1969), Thorne (1969a,b), and Campolattaro and Thorne (1970). Detailed overviews of this work will be found in Thorne (1968), and in Burke and Thorne (1969).

In this work the stellar material is treated as a perfect fluid, which is displaced by an amount $\xi(t,r,\theta,\phi)$ relative to the coordinate system. The gravitational field is characterized by the unperturbed Schwarzschild metric plus a perturbation, $h_{\mu\nu}(t,r,\theta,\phi)$. A particular choice of gauge (i.e., coordinate system) is made; and a computer is used (see Fletcher et al. 1967) to derive the explicit, analytic form for Einstein's field equations in that gauge. The field equations are linearized in ξ and $h_{\mu\nu}$, but no other approximation is made. (Note that the equilibrium configuration about which one perturbs is fully relativistic!)

Attention is concentrated initially on "standing-wave normal modes", in which the star pulsates sinusoidally and standing gravitational waves couple it to a perfect reflector at $r = \infty$. Each standing-wave normal mode is characterized by spherical-harmonic indices (ℓ,m) and by the pulsation angular frequency, ω. If the frequency, ω, is very near a natural pulsation frequency, σ, for the star, then the star's pulsation energy, E_M, is large compared to the energy, E_W, in one wavelength of the gravitational waves. (E_W is calculated using the Isaacson tensor.) As ω is allowed to vary near σ, the ratio of energy in matter to energy in waves goes through a resonance

$$\frac{E_M}{E_W} = \frac{\sigma/2\pi\tau}{(\omega-\sigma)^2 + (1/\tau)^2} \ . \tag{73}$$

From the physical viewpoint this resomance is associated with the natural mode of vibration of the star at $\omega=\sigma$; from the mathematical viewpoint it is associated with a pole in the S-matrix for scattering of gravitational waves off the star, which pole is located at

$$\omega_{\text{pole}} = \sigma + i/\tau \tag{74}$$

in the complex frequency plane. Thorne (1969a) has explored numerically the resonances for a variety of neutron-star models. (See Figure 6.)

The above discussion refers to the unrealistic problem of standing-wave modes. A more realistic situation is obtained by building a "wave packet" of standing waves peaked about a resonant frequency, σ. For example, the wave packet

$$\left\{h_{\mu\nu},\ \xi\right\} = \int \frac{\left\{ \underset{(\omega)}{h}{}_{\mu\nu},\ \underset{(\omega)}{\xi} \right\} e^{i\omega t}}{(\omega-\sigma)^2 + (1/\tau)^2}\, d\omega \tag{75}$$

represents a star in which gravitational waves flow in from $r = \infty$, exciting the star into pulsation; and then the star reradiates the waves. (See Figure 7.) The total energy carried in and back out by the waves is finite and can be arbitrarily small, so the solution (75) should be an excellent approximation to an exact solution of Einstein's field equations.

A close examination of the solution (75) reveals this, that as the star reradiates its gravitational waves, the amplitude of its pulsations (and hence also of the waves) is damped exponentially,

$$\text{Amplitude} \propto e^{-t/\tau} \tag{76}$$

with a damping rate, $1/\tau$, equal to the half-width of the standing-wave resonance. An analytic form for the reradiated waves in the radiation zone is

$$h_{\alpha\beta} = 0 \quad \text{if} \quad t < r + 2M^*\ell n(r-2M^*);$$

$$\frac{h_{\theta\theta}}{r^2} = -\frac{h_{\phi\phi}}{r^2\sin^2\theta} = A\left[\frac{\ell(\ell+1)}{2} P_\ell^m(\cos\theta) + \frac{d^2P_\ell^m(\cos\theta)}{d\theta^2}\right]\frac{1}{r}$$

$$\times \exp\left[-\frac{t-r-2M^*\ell n(r-2M^*)}{\tau^*}\right] \sin\{\sigma^*[t^*-r-2M^*\ell n(r-2M^*)]$$

$$+ m\phi + \delta\} + 0(1/r^2),$$

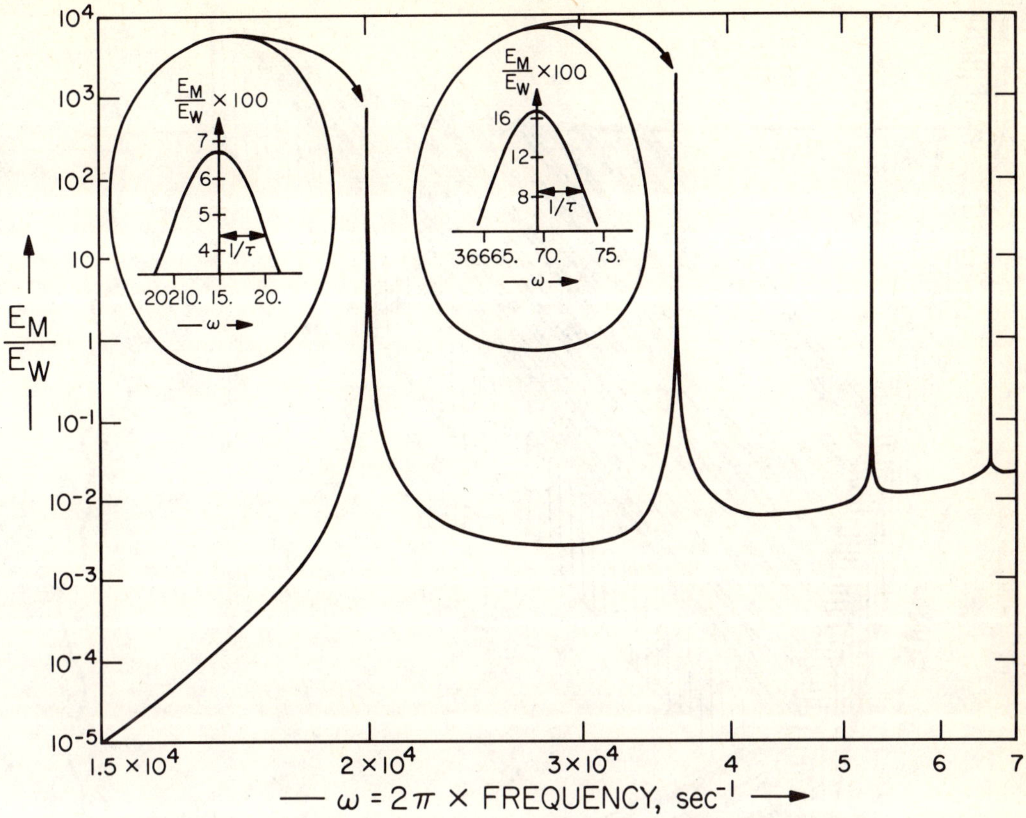

Figure 6. Resonances in the quadrupole standing-wave normal modes for the H-W-W neutron star of $M = 0.682\ M_\odot$ at the center of a huge ($r \gtrsim \infty$), perfectly reflecting cavity. Corresponding to each value of the angular frequency, ω, there are four independent standing-wave modes (m = -2, -1, 0, +1, +2). Plotted as a function of ω is the ratio of the total pulsation energy, E_M, of the matter in the star, to the energy, E_W, in one wavelength of the standing gravitational waves. This ratio is the same for all four modes (degeneracy in the quantum number m). [Figure based on numerical computations by Thorne (1969a).]

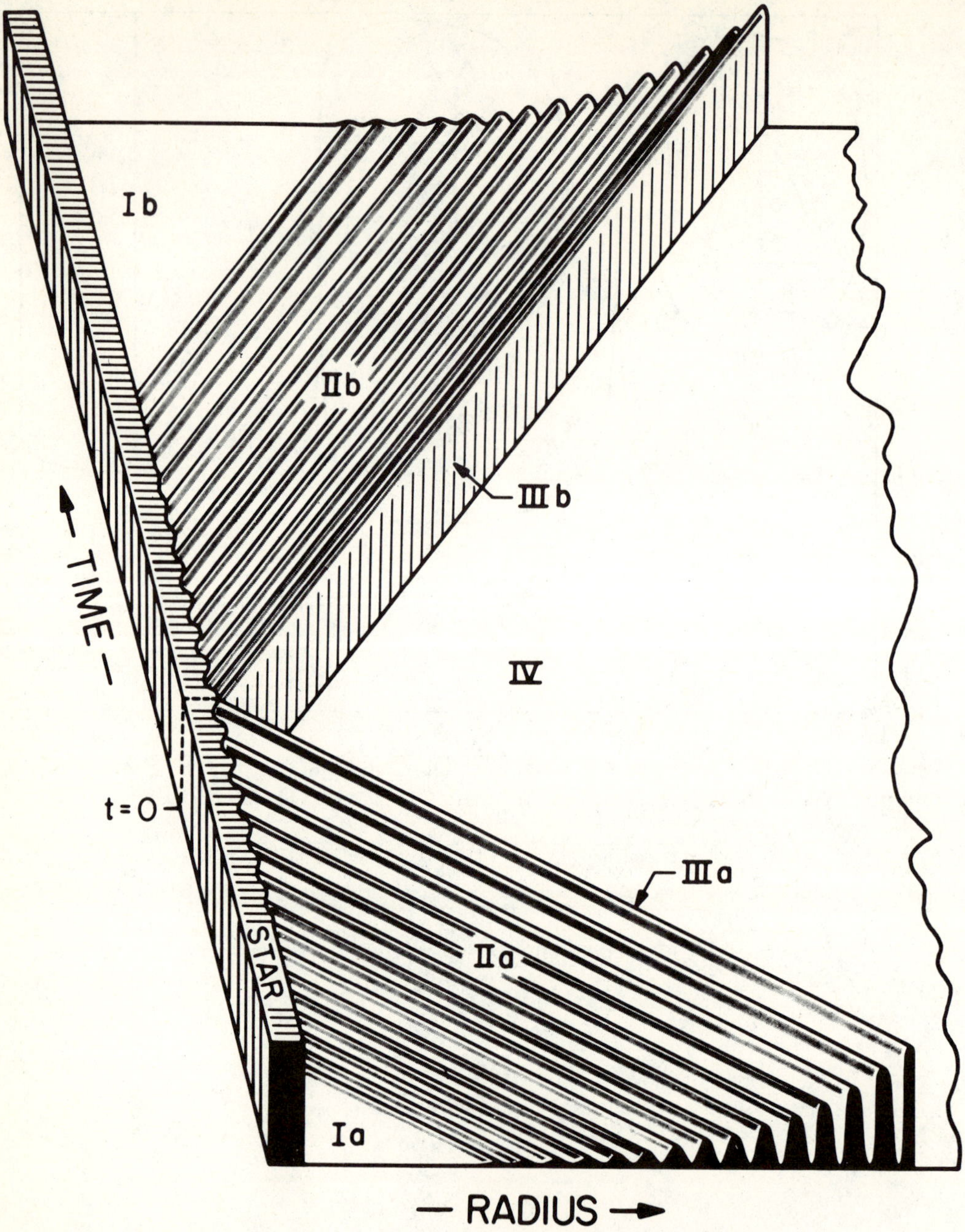

Figure 7. Schematic spacetime diagram for the absorption and reemission of gravitational waves as described by equations (75)-(77). [Based on the work of Price and Thorne (1968).]

$$\frac{h_{\theta\phi}}{r^2 \sin\theta} = Am\left[\frac{1}{\sin\theta}\frac{dP_\ell^m(\cos\theta)}{d\theta} - \frac{\cos\theta}{\sin^2\theta}P_\ell^m(\cos\theta)\right]\frac{1}{r}$$

$$\times \exp\left[-\frac{t^*-r-2M^*\ell n(r-2M^*)}{\tau^*}\right]\cos\{\sigma^*[t^*-r-2M^*\ell n(r-2M^*)]$$

$$+ m\phi + \delta\} + 0(1/r^2) \text{ if } t^* > r + 2M^*\ell n(r-2M^*). \tag{77}$$

Here $h_{\alpha\beta}$ is the perturbation of the Schwarzschild metric, and A and δ are constants. Notice that far from the star these waves are very nearly identical to the plane waves of the linearized weak-field theory (Landau and Lifshitz 1962, §101). They exhibit linear polarization if m = 0, and elliptical polarization if $m \neq 0$.

The pulsation angular frequencies, σ, and damping times, τ, have been calculated by Thorne (1969a) for the quadrupole pulsations of a variety of neutron-star models. The pulsation periods for neutron stars of mass $M \gtrsim 0.5\ M_\odot$ are typically

$$T = 2\pi/\sigma \sim 0.1 \text{ to } 0.3 \text{ milliseconds;} \tag{78}$$

and the damping times are typically

$$\tau \sim 0.1 \text{ to } 10 \text{ seconds.} \tag{79}$$

Consequently, immediately after a neutron star is formed in a supernova explosion, it should emit $\sim 10^{52}$ ergs in a burst of gravitational waves in the kilocycle frequency region; and the burst should last for only about 1 second.

5.7 Weber's Experiment

Very recently Joseph Weber (1969, 1970) has announced that he believes he is detecting bursts of gravitational radiation from cosmical sources. Each of his detectors consists of a massive ($\sim$ 1 ton) aluminum cylinder with a frequency of oscillation for the fundamental, end-to-end mode of

$$\nu_o = 1660 \text{ hz.} \tag{80a}$$

The Q of the fundamental mode is

$$Q \approx 10^5, \tag{80b}$$

corresponding to a bandwidth for exciting the oscillation of

$$\Delta\nu_o \approx \nu_o/Q \approx 0.01 \text{ hz.} \tag{80c}$$

When the oscillation is excited, it will decay due to internal damping with an e-folding time of

$$\tau_0 \approx 1/2\pi\Delta\nu_0 \sim 20 \text{ seconds.} \quad (80d)$$

The fundamental-mode oscillations of each cylinder are monitored by piezo-electric strain transducers attached to its middle. These strain transducers yield an output which fluctuates randomly due to the kT of thermal energy in the fundamental mode. Whenever an external force excites one of the cylinders into motion with an energy greater than

$$kT \approx 4 \times 10^{-14} \text{ ergs} \quad (80e)$$

(room temperature!), the output of the strain transducers rises markedly. (The threshold energy of 4×10^{-14} ergs corresponds to an end-to-end oscillation of the cylinder by $\sim 3 \times 10^{-14}$ cm.)

Weber has succeeded in isolating his cylinders so well from their surroundings that nothing seems capable of exciting them except gravitational waves. For example, neither large earthquakes nor nuclear tests excite the cylinders, though they do send most of Weber's anticoincidence, earth-measuring apparatus off scale.

Since December 1968, Weber has been seeing coincident excitations (coincident to within the resolution time of $\sim$ 0.4 seconds for the onset of an excitation) in cylinders located at Argonne National Laboratory near Chicago, and at the University of Maryland near Washington D.C. These "events" have all the characteristics that one would expect of excitations due to short bursts of gravitational waves, whose oscillating tidal forces set the cylinders into oscillation. The events come about once per day.

5.8 What is Weber Seeing?

For the purpose of discussion, let us accept Weber's conclusion that his events are due to cosmic burst of gravitational radiation. The characteristics of the bursts must be as follows:

(i) Frequency: In order to excite Weber's cylinders, the bursts must have sizable energy at his frequency of $\nu_0 = 1660$ hz, in his band width of $\Delta\nu_0 = .01$ hz. If they are narrow-band bursts, their probability of hitting Weber's frequency is very small. In order to avoid having too many bursts--and because it seems reasonable from the theory of astrophysical sources of gravitational waves--we shall assume that the bursts have a bandwidth at least as large as ν_0 itself:

$$BW \gtrsim \nu_0 . \quad (81)$$

(ii) Energy: The energy deposited in each detector by the type of burst which comes once per day exceeds the background noise (eq. 80e) by a factor of about 2. Consequently,

$$E_{deposited} \gtrsim 4 \times 10^{-14} \text{ ergs.} \tag{82}$$

The cross-section of any mechanical detector like Weber's, for extracting energy from a passing gravitational wave <u>in the detector's bandwith</u>, is

$$\sigma \simeq 10\ L^2 Q\ R_g/\lambda_o. \tag{83}$$

Here L^2 is the geometric size of the detector, Q is its quality factor, R_g is its gravitational radius, and λ_o is the wavelength of the radiation at its resonant frequency ν_o. [This formula for σ can be derived fairly easily from formulae (58) and (64) together with the elementary theory of damped harmonic oscillators.] The cross section (83) for each of Weber's detectors calculates out to $\sim 3 \times 10^{-19}$ cm^2.

Actually, the above cross section is relevant for a situation in which a detector at zero temperature is in a steady-state interaction with a long wave train (time of passage >> detector damping time). Weber's events are very different from this: the temperature and associated noise in the detector are high, and the wave train is short. In this case the theory of the interaction between waves and detector is much more complicated than in the steady-state case. For example, if the burst of waves hits the detector in phase with the thermal oscillations of its fundamental mode, a rather large amount of energy will be deposited:

$$\begin{aligned} \delta \text{ Energy} &\propto \delta(\text{Amplitude}^2) \\ &\approx 2 \times (\text{Thermal Amplitude}) \times \delta \text{ Amplitude.} \end{aligned} \tag{84}$$

clearly, the larger the thermal amplitude at the moment the burst hits, the greater the energy deposited! However, if the burst hits out of phase with the thermal oscillations, it deposits little energy--and it may even extract energy from the detector!

From studies of the detector response, Weber (private communication, March 1970) believes that he detects only about 10 per cent of all bursts of that type which produce about one detector coincidence per day. The other $\sim$ 90 per cent arrive sufficiently out of phase with one or both detectors, or arrive when the thermal amplitude is momentarily so low, that they cannot produce significant coincidences.

For those $\sim$ 10 per cent which do produce coincidence events, Weber estimates a cross section $\sim$ 20 times higher than equation (83)--i.e., $\sim 6 \times 10^{-18}$ cm^2.

The total energy which each burst carries across a unit area near the detector is

$$\frac{E_{emitted}}{4\pi \ell^2} = \frac{E_{deposited}}{\sigma} \quad \frac{BW}{\Delta\nu_o}, \tag{85}$$

where $E_{emitted}$ is the total energy emitted by the source, and ℓ is the distance of the source from Earth. Combining Weber's cross section with equations (80)-(82) and (84), we find

$$E_{emitted} = \left(\frac{4\pi\ell^2}{\sigma}\right)\left(\frac{BW}{\Delta\nu_o}\right)E_{deposited}$$

$$\sim (\ell_{pc})^2\left(\frac{10^{38}\ \text{cm}^2}{6\times10^{-18}\text{cm}^2}\right)\left(10^5\ \frac{BW}{\nu_o}\right)(4\times10^{-14}\ \text{ergs}) \qquad (86)$$

$$\sim (8\times10^{46}\ \text{ergs}) \times (BW/\nu_o) \times (\ell_{pc})^2 \ ,$$

where ℓ_{pc} is the distance to the source in parsecs. This is an enormous energy for reasonable distances!

(iii) Duration of the Bursts. Each burst is shorter than the decay of the detector, $\sim$20 sec, since no natural width to the events is seen.

From the above characteristics of the bursts, we can make the following comments: (1) The frequency, ν_o = 1660 hz, is reasonable if the bursts are due to the collapse that creates a neutron star, or to the collapse of a massive star through its gravitational radius. (2) In the formation of a neutron star the total energy released as gravitational waves is certainly less than one tenth the rest-mass energy of the sun, since this is the maximum binding energy of a neutron star. Hence, in this case the distance to the source must be less than 1500 pc (1/5 the distance to the nucleus of the galaxy). (3) In highly nonspherical collapse through the gravitational radius a star might emit a large fraction ($\sim$50 per cent perhaps) of its rest-mass energy as gravitational waves. Hence, in this case the source could be anywhere in our galaxy--but not outside it. However, it seems very implausible that stellar collapse could occur as frequently as Weber's data demand--$\sim$10 per day in our galaxy. A more reasonable figure would be one per year or less. (4) Whatever the source may be, the period of its gravitational waves ($\sim 10^{-3}$ seconds) must be greater than the light-travel time across its gravitational radius. Hence, its mass is limited by

$$2GM/c^3 < 10^{-3}\ \text{seconds;} \quad M < 100\ M_\odot . \qquad (87)$$

The total energy emitted must be less than the source's total mass-energy, $E_{emitted} < Mc^2 < 2\times10^{56}$ ergs. Combining this with equation (85) and assuming BW $\sim \nu_o$, we obtain an absolute upper limit on the distance to the source:

$$\ell < 10^5\ \text{pc}.$$

This means that the source, whatever it may be, must lie in or very near our galaxy!

Clearly, the explanation of Weber's events will be very difficult!

The author's research on this subject between 1964 and 1969 has been supported by the National Science Foundation [GP-9433, GP-9114, GP-7676, GP-5391], the U.S. Office of Naval Research [NOnr-220(47)], the U.S. Air Force Office of Scientific Research [AF-49-638-1545], and the Alfred P. Sloan Foundation.

REFERENCES

Bardeen, J.M. 1970, Ap. J., in press.

Bardeen, J.M., Thorne, K.S., and Meltzer, D.W. 1966, Ap. J., 145, 505.

Bisnovatyi-Kogan, G.C., Zel'dovich, Ya.B., and Novikov, I.D. 1967, Astr. Zhur., 44, 525. (English translation: Soviet Astronomy - A.J., 11, 419.

Bondi, H. 1964, Proc. Roy. Soc. London, A282, 303.

______. 1965, in Lectures on General Relativity, Vol. 1 of Proceedings of 1964 Brandeis Summer Institute in Theoretical Physics (Prentice-Hall, Englewood Cliffs, N.J.).

Brill, D.R. and Hartle, J.B. 1964, Phys. Rev. 135, B271.

Buchdahl, H.A. 1959, Phys. Rev., 116, 1027.

Burke, W.L. 1969, Ph.D. thesis, California Institute of Technology (available from University Microfilms, Inc., Ann Arbor, Michigan).

______. 1970, Phys. Rev., in press.

Burke, W.L. and Thorne, K.S.,1970, in Relativity, eds., M. Carmeli, S.I. Fickler, L. Witten (Plenum Press, N.Y.).

Campolattaro, A. and Thorne, K.S. 1970, Ap. J., 159, 847.

Chandrasekhar, S. 1964, Phys. Rev. Letters, 12, 114 and 437.

______. 1965a, Ap. J., 142, 1488.

______. 1965b, Ap. J., 142, 1513.

______. 1967a, Ap. J., 147, 334.

______. 1967b, Ap. J., 148, 621.

______. 1967c, Ap. J., 148, 645.

Chandrasekhar, S. and Esposito, F.P. 1970, Ap. J., 160, 153.

Chitre, S.M., Hartle, J.B. and Thorne, K.S. 1970, Ap. J., in preparation.

Cole, J. 1968, Perturbation Methods in Applied Mathematics (Ginn-Blaisdell, Waltham, Mass.).

Dicke, R.H. 1964, Relativity, Groups, and Topology, C. and B. Dewitt, eds. (Gordon and Breach, New York).

Durney, B. and Roxburgh, I. 1967, Proc. Roy Soc. London, A296, 189.

Fletcher, J.D., Clemens, R., Matzner, R., Thorne, K.S. and Zimmerman, B.A. 1967, Ap. J., 148, L91.

Fowler, W.A. 1964, Rev. Mod. Phys., 36, 545, and 1104.

______. 1966, Ap. J., 144, 180.

Hagihara, Y. 1931, Jap. J. Astr. Geoph., 8, 67.

Harrison, B.K., Thorne, K.S., Wakano, M. and Wheeler, J.A. 1965, Gravitation Theory and Gravitational Collapse, (University of Chicago Press, Chicago).

Hartle, J.B. 1967, Ap. J., 150, 1005.

______. 1970, Ap. J., in press.

Hartle, J.B. and Sharp, D. 1967, Ap. J., 147, 317.

Hartle, J.B. and Thorne, K.S. 1968, Ap. J., 153, 807.

______. 1969, Ap. J., 158, 719.

______. 1970, Ap. J., in preparation.

Hernandez, W.C. 1967, Phys. Rev., 159, 1070.

Ipser, J.R. 1970, Ap. and Space Sci., in press.

Isaacson, R.A. 1968a, Phys. Rev., 166, 1263.

______. 1968b, Phys. Rev., 166, 1272.

Kaplan, S.A. 1949, Uch. Zap. L'bobskovo Un-ta, 15, no. 4, 101.

Kerr, R.P. 1963, Phys. Rev. Letters, 11, 522.

Landau, L.D. and Lifshitz, E.M. 1962, The Classical Theory of Fields, (Addison-Wesley, Reading, Mass.) second edition.

Misner, C.W. 1969, in Astrophysics and General Relativity, eds. M. Chrétién, S. Deser, J. Goldstein, (Gordon and Breach, New York).

Nutku, Y. 1969, Ap. J., 155, 999.

Ostriker, J.P. and Bodenheimer, P. 1968, Ap. J., 151, 1089.

Ostriker, J.P. and Hartwick, F.D.A. 1968, Ap. J., 153, 797.

Ostriker, J.P. and Mark, J.W-K. 1968, Ap. J., 151, 1075.

Price, R. and Thorne, K.S. 1969, Ap. J., 155, 1075.

Price, R. and Bardeen, J.M. 1967, unpublished results.

Trautman, A. 1965, in Lectures on General Relativity, Vol. 1 of proceedings of 1964 Brandeis Summer Institute in Theoretical Physics (Prentice-Hall, Englewood Cliffs, N.J.)

Taylor, E.F. and Wheeler, J.A. 1967, Spacetime Physics, W.H. Freeman and Co., San Francisco).

Thorne, K.S. 1965, Ap. J., 144, 201.

______. 1967, in High Energy Astrophysics, Vol. 3, C. DeWitt, E. Schatzman and P. Veron, eds. (Gordon and Breach, New York).

______. 1968, Phys. Rev. Letters, 21, 320.

______. 1969a, Ap. J., 158, 1.

______. 1969b, Ap. J., 158, 997.

______. 1971a, in Relativity and Cosmology, Proceedings of Course 47 of the International School of Physics "Enrico Fermi", R. Sachs, ed. (Academic Press Inc., New York).

______. 1971b, Relativistic Stellar Theory, monograph in preparation.

Thorne, K.S. and Compolattaro, A. 1967, Ap. J., 149, 521, and 152, 673.

Thorne, K.S. and Will, C. 1970, Comments on Astrophysics and and Space Physics, 2, 35.

Weber, J. 1969, Phys. Rev. Letters, 22, 1320.

______. 1970, Phys. Rev. Letters, 24, 276.

Zel'dovich, Ya. B. and Novikov, I.D. 1967, Relyativistkaya Astrofisika (Nauka, Moscow) (English translation: Relativistic Astrophysics, University of Chicago Press, Chicago, in preparation.)

20

STELLAR MAGNETISM AND ROTATION

L. Mestel

University of Manchester
Manchester, England

Lecture I. ORIGIN OF STELLAR MAGNETISM

In this lecture, we summarize ideas on the origin of stellar magnetism. According to the "fossil" theory, a stellar magnetic field is a relic of the field present in the gas from which the star formed. If instead the magnetic field is built up in the star, its energy may come from the thermal energy of the electron gas (the "battery" theory) or from the kinetic energy of mass motions (the "dynamo" theory).

Summary of the Two-Fluid Plasma Model

As we are concerned primarily with low frequency problems in a nonrelativistic, highly conducting gas, we use the truncated Maxwell equations without displacement current:

$$\nabla \times \underset{\sim}{E} = -\frac{1}{c}\frac{\partial \underset{\sim}{B}}{\partial t}, \qquad \nabla \cdot \underset{\sim}{B} = 0,$$

$$\nabla \times \underset{\sim}{B} = \frac{4\pi}{c}\underset{\sim}{j}, \qquad \rho_e = \frac{\nabla \cdot \underset{\sim}{E}}{4\pi}. \tag{1}$$

These are invariant to the Galilean transformation

$$\underset{\sim}{r}' = \underset{\sim}{r} - \underset{\sim}{u}t, \qquad t' = t. \tag{2}$$

with the field variables transforming like

$$\underset{\sim}{B}' = \underset{\sim}{B}, \qquad \underset{\sim}{E}' = \underset{\sim}{E} + \frac{\underset{\sim}{u} \times \underset{\sim}{B}}{c},$$

$$\underset{\sim}{j}' = \underset{\sim}{j}, \qquad \rho_e' = \rho_e - \underset{\sim}{u}\cdot\underset{\sim}{j}/c^2 .$$

(As noted by Schlüter (1961), the retention of the relativistic term $\underset{\sim}{u}\cdot j/c^2$ in ρ_e' is required for consistency, since ρ_e and $\underset{\sim}{u}\cdot\underset{\sim}{j}/c^2$ are of the same order in a highly conducting gas).

The plasma is pictured as two gases, the ions and electrons, each with an isotropic partial pressure p_e, p_i, interacting by means of a mutual friction which tends to destroy the relative momentum of the two gases, and so yields electrical resistivity. The bulk velocity $\mathbf{v}$ of the gas satisfies the Eulerian equation of motion with the electromagnetic force density included. We can show at once that in the non-relativistic regime the electric forces are negligible compared with the magnetic. Anticipating that the perfect conductor condition $\mathbf{E} = -\mathbf{v} \times \mathbf{B}/c$ remains a very good approximation in most problems, we have from (1)

$$\rho_e \simeq \frac{vB}{4\pi cD}, \tag{4}$$

where D is a typical length-scale. The ratio of electric force density to inertial force density is of order

$$\frac{\rho_e E}{\rho v^2/D} \simeq \frac{\left(\frac{vB}{c}\right)^2 \frac{1}{4\pi D}}{\rho v^2/D} = \frac{B^2}{4\pi\rho c^2}, \tag{5}$$

which is small unless the magnetic energy density is comparable with the rest-mass energy density, implying that the Alfven speed has merged with the speed of light. By contrast, the corresponding ratio with the magnetic force density is of order

$$\frac{|\nabla \times \mathbf{B}|B}{4\pi\rho v^2/D} \simeq \frac{B^2}{4\pi\rho v^2} \tag{6}$$

if we assume that $|\nabla \times \mathbf{B}| \simeq B/D$. Thus the electric force density is usually smaller than the magnetic by the factor $(v/c)^2$, and so must be dropped in a consistent non-relativistic theory based on the transformations (2) and (3). Similarly, the net charge density will be small compared with the density of electrons or ions separately: for if $\mathbf{V} = -\mathbf{j}/n_e e$ is the drift velocity of electrons relative to ions,

$$\frac{\rho_e}{n_e e} \simeq \left(\frac{vB}{c}/4\pi D\right)(V/j) \simeq \left(\frac{v}{c}\right)\left(\frac{V}{c}\right)\frac{B}{D|\nabla \times \mathbf{B}|} \ll 1 \tag{7}$$

if again $|\nabla \times \mathbf{B}|D/B \simeq 1$. However, it should be noted that if $B^2/4\pi\rho c^2 \simeq 1$, then the regime is "relativistic" even if the material velocities are all well below c. Thus in a recent pulsar model (Goldreich and Julian 1969) the gas co-rotating with the neutron star but well within the light sphere has a velocity $v \ll c$, but its density is so low that $B^2/4\pi\rho c^2 > 1$, so that by (5) the electric force density is no longer negligible. But the inertial forces must be comparable with the dominant force acting, whereas the assumption $|\nabla \times \mathbf{B}| \simeq B/D$ yields a magnetic force density $(c/v)^2$ larger (cf. the ratio (6)). We conclude that in this regime the magnetic field must in fact approximate very closely to the curl-free form, with $|\nabla \times \mathbf{B}| \ll B/D$, as is indeed found in the authors' detailed integrations. It is also found that the net charge density is no longer small — the ratio (7) is in fact near unity in this region, with almost complete charge separation. But if $B^2 \ll 4\pi\rho c^2$, as in applications to normal magnetic stars, all terms in ρ_e -- including the convection current $\rho_e \underset{\sim}{v}$ -- are negligible.

The principle aim of the plasma model is to find an analogue of "Ohm's law" relating j to the field vectors. The equation of motion of the electrons is (e.g. Cowling 1957)

$$-\nabla p_e - n_e e\left(\underset{\sim}{E} + \frac{\underset{\sim}{v} \times \underset{\sim}{B}}{c} + \frac{\underset{\sim}{V} \times \underset{\sim}{B}}{c}\right) - \frac{n_e m_e \underset{\sim}{V}}{\tau_{ei}} = 0 \,. \tag{8}$$

The last term represents the ion-electron collisions by a collision rate τ_{ei} at which the drift momentum of the electrons is randomized. The gravitational force on the electrons is ignored because of the smallness of m_e, and the inertial force because we are concerned with problems with characteristic time-scales that are long compared with either the plasma oscillation period or the gyration period in the field B. However, we note that if the electrons are relativistic, the inertial terms may become important. Thus if the electron energy reaches $\gamma m_e c^2$, with $\gamma = [1 - (\underset{\sim}{v} + \underset{\sim}{V})^2/c^2]^{-\frac{1}{2}}$, then the inertial force $\gamma m_e \Omega|\underset{\sim}{v} + \underset{\sim}{V}|$ on an electron moving in a circle with angular velocity Ω will be comparable with $|e(\underset{\sim}{v} + \underset{\sim}{V}) \times \underset{\sim}{B}/c|$ if

$$\gamma\Omega/\omega_e \simeq 1, \tag{9}$$

where $\omega_e \equiv eB/m_e c$ is the (nonrelativistic) gyration frequency. For an electron accelerating parallel to its velocity the "longitudinal mass" must be used, and condition (9) is replaced by an analogue involving $\Omega \equiv |\nabla|\underset{\sim}{v} + \underset{\sim}{V}||$:

$$\gamma^3\Omega/\omega_e \simeq 1. \tag{10}$$

Since $\Omega/\omega_e << 1$ for quite modest fields B, we see that γ must be large compared with unity before the inertial term becomes large enough to interfere with the basic hydromagnetic approximation $\underset{\sim}{E} + \frac{\underset{\sim}{v} \times \underset{\sim}{B}}{c} = 0$. This again is relevant to the pulsar problem: electrons have to be highly relativistic before electron inertial forces are an important cause of deviation from e.g. the law of isorotation or the appropriate modifications due to gas streaming discussed later in the stellar wind problem.

We now write the equation (8) to the (nonrelativistic) electron gas in the standard form

$$\underset{\sim}{E} + \frac{\underset{\sim}{v} \times \underset{\sim}{B}}{c} = \frac{\underset{\sim}{j}}{\sigma} - \frac{\nabla p_e}{n_e e} + \frac{\underset{\sim}{j} \times \underset{\sim}{B}}{c n_e e} \,, \tag{11}$$

where the conductivity

$$\sigma = \frac{n_e e^2 \tau_{ei}}{m_e} \propto T^{3/2} \,, \tag{12}$$

T being the temperature. The three terms that cause departure from strict freezing of the field into the gas as a whole are respectively the Ohmic field, the partial pressure gradient (an analogue of the Peltier effect) and the Hall term.

The energy equation for the gas follows by taking the scalar product of (11) with $\underset{\sim}{j}$:

$$\underset{\sim}{j}\cdot\underset{\sim}{E} = \frac{\underset{\sim}{j}^2}{\sigma} - \frac{\nabla p_e}{n_e e}\cdot\underset{\sim}{j} + \frac{\underset{\sim}{j}\times\underset{\sim}{B}}{c}\cdot\underset{\sim}{v}\,. \tag{13}$$

Note that the Hall term (which can be treated as leading to anisotropic conductivity - cf. Cowling (1957)) disappears: the so-called "reduction of the conductivity" across a magnetic field does not correspond to any large increase in dissipation, which is in fact given by the familiar term $\underset{\sim}{j}^2/\sigma$, with σ the normal, "unreduced" conductivity. The second term may be positive or negative; if the electron partial pressure has the same sense as the current, it represents a possible source of energy for the magnetic field. The last term is the dynamo or electric motor term. If $j \times \underset{\sim}{B}/c$ is locally parallel to the velocity $\underset{\sim}{v}$ the magnetic forces do work on the gas at the expense of magnetic energy (the "electric motor"); if anti-parallel, the non-magnetic forces drive the gas against the magnetic forces and so pump energy into the field (the "dynamo").

We now couple equation (11) -- the analogue of Ohm's law-- with Faraday's law, to find the generalization of the flux-freezing constraint. Let C be a closed circuit moving with the local electron velocity $(\underset{\sim}{v} + \underset{\sim}{V})$, and consider the flux of $\underset{\sim}{B}$ over any surface S spanning C. Then

$$\frac{d}{dt}\int_S \underset{\sim}{B}\cdot\underset{\sim}{n}dS \equiv \int_S \left(\frac{\partial\underset{\sim}{B}}{\partial t} - \nabla\times\left[(\underset{\sim}{v}+\underset{\sim}{V})\times\underset{\sim}{B}\right]\right)\cdot\underset{\sim}{n}dS$$

(by a standard theorem in vector analysis)

$$= -c\int_S \nabla\times\left[\underset{\sim}{E} + (\underset{\sim}{v}+\underset{\sim}{V})\times\underset{\sim}{B}/c\right]\cdot\underset{\sim}{n}dS$$

$$= -c\oint_C \left[\underset{\sim}{E} + \frac{(\underset{\sim}{v}+\underset{\sim}{V})\times\underset{\sim}{B}}{c}\right]\cdot d\underset{\sim}{s} \quad \text{(by Stokes' theorem)}$$

$$= c\oint_C \left(\frac{\nabla p_e}{n_e e} - \frac{\underset{\sim}{j}}{\sigma}\right)\cdot d\underset{\sim}{s} \tag{14}$$

from (11) with the Hall term rewritten in its original form $\underset{\sim}{V}\times\underset{\sim}{B}/c$. We now see that the dynamo and Hall terms make no contribution to either the generation or the destruction of magnetic flux. The Ohmic term acts so as to destroy the flux through C, while the thermal "battery" term can create flux if its line-integral is non-zero -- i.e., if it has a non-zero "e.m.f.". This will be so if the electron pressure is not a pure function of the electron density.

The "Battery" Process in Rotating Stars

In a spherically symmetric star, the electron partial pressure is balanced by an electrostatic field, which merely transfers the electron pressure to act on the ions, and so to contribute towards equilibrium under the gravitational field. In a rotating star the battery integral becomes

$$\oint \frac{\nabla p_e}{n_e e} \cdot d\underset{\sim}{s} \propto \oint \frac{\nabla p}{\rho} \cdot d\underset{\sim}{s} \propto \oint \Omega^2 \underset{\sim}{\bar{\omega}} \cdot d\underset{\sim}{\bar{\omega}} \;, \qquad (15)$$

and this does not in general vanish if the centrifugal field is non-conservative, with Ω a function of z as well as $\bar{\omega}$. If the battery effect persists it will build up poloidal currents maintaining a toroidal field, limited asymptotically by the Ohmic resistance (Biermann 1950). The build-up is slow because of the large self-inductance of a star, but the asymptotic fields can be large——1000 gauss in a slowly rotating star like the sun, 10^7 - 10^8 gauss in a rapid rotator. However, the process is much less efficient if the star has a modest primeval poloidal field, able to keep the stellar rotation field close to the isorotational state (Mestel and Roxburgh 1962; Kato and Nakagawa 1969, 1970). A further uncertainty is introduced by the instabilities which affect non-conservative rotation fields (Goldreich and Schubert 1967; Fricke 1968). More recent studies (Kippenhahn 1969; James and Kahn 1970) suggest that the non-linear development of such instabilities will in fact be slow, so that an initial non-conservative centrifugal field could persist long enough for the "battery" to be able to build up a sizeable toroidal field, especially if the star has not retained a primeval poloidal field.

The relation of any such toroidal field to the fields observed in e.g. Ap stars remains highly problematic. However, one should at the very least bear in mind that spontaneous generation of even a weak magnetic field in initially nonmagnetic matter is important as supplying a "seed" field which can subsequently be amplified by dynamo action.

Fossil Fields

Consider now a primeval field in a star without mass motions, and so without any energy source to replace the energy dissipated by finite resistivity. A poloidal field $\underset{\sim}{B}_p$ is maintained by toroidal currents $\underset{\sim}{j}_t$, flowing under the self-induced electric field associated with the decay of the field:

$$\frac{\partial \underset{\sim}{E}_t}{\partial t} = \frac{1}{\sigma}\frac{\partial \underset{\sim}{j}_t}{\partial t} = \frac{c}{4\pi\sigma}\frac{\partial}{\partial t}(\nabla \times \underset{\sim}{B}_p) = -\frac{c^2}{4\pi\sigma}\nabla \times (\nabla \times \underset{\sim}{E}_t). \qquad (16)$$

A field of characteristic scale D will have a characteristic decay time

$$\tau = \frac{4\pi\sigma}{c^2}D^2 \;. \qquad (17)$$

For a dipolar field of scale comparable with the solar radius, τ is estimated to be longer than the solar age, and still longer for an A-star (Cowling 1945; Wrubel 1952). (The same enormous self-induction which cuts down the decay rate is also responsible for the long time-scale taken by a "battery" process such as Biermann's to build up a stellar field). In general, one can find a complete set of eigen-solutions of the decay equation for $\underset{\sim}{E}_t$:

$$\underset{\sim}{E}_t = f(r,\theta)\, e^{-t/\tau_{\underset{\sim}{t}}} , \qquad (18)$$

each corresponding to a toroidal current system j_t and its associated poloidal field $\underset{\sim}{B}_p$. Any initial poloidal field can be expanded in terms of these eigen-solutions and its decay followed. A similar theory exists for decaying toroidal fields.

The discovery that the general solar field reverses with the solar cycle forces us to look for an oscillatory dynamo rather than a "fossil" explanation of the solar field. However, the long decay-time of the largest-scale dipolar components does suggest that the fields of the magnetic Ap stars may be "fossils"——i.e., slowly-decaying remnants either of flux present in the gas clouds in which the stars formed, or of fields generated by dynamo action in an earlier epoch.

Dynamo Action

By (14) we see how large-scale laminar circulation will distort a magnetic field by dragging of the field-lines. In a fluid of finite conductivity the field is not strictly frozen in, and it is sometimes possible to ascribe a velocity to individual elements of a field-line, differing from the local velocity of the fluid. Thus consider an axisymmetric poloidal field $\underset{\sim}{B}_p$, maintained by toroidal currents $\underset{\sim}{j}_t = c\nabla \times \underset{\sim}{B}_p/4\pi$ which are necessarily perpendicular to $\underset{\sim}{B}_p$. The equation

$$\underset{\sim}{E} + \frac{\underset{\sim}{v} \times \underset{\sim}{B}}{c} = \frac{\underset{\sim}{j}}{\sigma} \qquad (19)$$

can then be replaced by

$$\underset{\sim}{E} + \frac{\underset{\sim}{v}' \times \underset{\sim}{B}}{c} = 0 , \qquad (20)$$

with

$$\underset{\sim}{v}' = \underset{\sim}{v} + \frac{c^2}{4\pi\sigma} \frac{(\nabla \times \underset{\sim}{B}) \times \underset{\sim}{B}}{\underset{\sim}{B}^2} . \qquad (21)$$

Thus from equation (20) coupled with the law of induction, the field can be regarded as moving with the velocity $\underset{\sim}{v}'$, the Ohmic term yielding a drift of the field through the gas in the direction of the magnetic force density. In particular, the decay of the field in a stationary medium can be described by a motion $(\nabla \times \underset{\sim}{B}) \times \underset{\sim}{B}/\frac{4\pi\sigma}{c^2}B^2$ of the field-lines. Near the O-type neutral point of a poloidal field, the magnetic force density is directed towards O, and there is a steady drift of the loops into O——a destruction of both the energy and the flux of the field. It is at once clear that axisymmetric motions such as meridian circulation cannot change this effect.

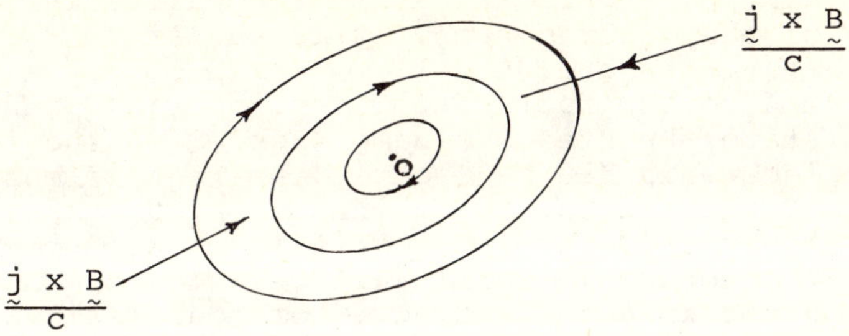

The term $\tilde{v} \times \tilde{B}/c$ implies a motion of the field-lines but not creation of flux to off-set the shrinkage into O: only a creation of matter at O with an outward-directed velocity field would be capable of off-setting the decay. From another point of view: the ratio of the induction term to the Ohmic term in (19) is the magnetic Reynolds number $4\pi\sigma vD/c^2$, where D is the characteristic scale of the magnetic field. This number is very large for stellar dimensions and conductivities, even if v is very small, and this might suggest that one can easily pump energy into the field to off-set Ohmic dissipation. But near the neutral point the relevant scale D becomes arbitrarily small, and the induction term is always swamped by the Ohmic term, leading to decay of the flux.

This result is Cowling's anti-dynamo theorem (1934). Similar arguments show that axisymmetric motions cannot prevent the decay of a toroidal axisymmetric field by shrinkage of the field-lines onto the axis. The most general theorem of this type (Bullard 1957) forbids dynamo maintenance of any field which contains a limiting closed curve C, with field-lines spiralling around C in the same sense all the way. The steady state dynamo implies

$$\frac{\tilde{j}}{\sigma} = -\nabla\phi + \frac{\tilde{v} \times \tilde{B}}{c} , \tag{22}$$

so that

$$\oint_C \frac{\tilde{j}}{\sigma} \cdot d\tilde{s} = \oint_C \frac{\tilde{v} \times \tilde{B}}{c} \cdot d\tilde{s} . \tag{23}$$

But the curve C is either a limiting field-line, or a neutral line; in either case the right-hand side of (23) vanishes, whereas the sense of the field-lines implies that $\tilde{j}$ is directed in the same sense all the way around C, and so the left-hand side is non-zero for finite σ. This contradiction implies that a field of such a structure would necessarily decay.

The upshot of these results is that any magnetic field that can be maintained against Ohmic decay must have a complicated structure. Rigorous existence theorems have been given by Backus (1958) and Herzenberg (1958) for very special velocity fields. Recent numerical work by Roberts (1969) suggests that some non-axisymmetric fields may be maintainable by axisymmetric motions.

In trying to understand at least qualitatively the complications of the solar dynamo, it is easier to think topologically rather than analytically. The loops of a poloidal field $\underset{\sim}{B}_p$ tend to contract into O-type neutral points; we may therefore state the steady dynamo problem in the form, how can we replace this flux? Clearly, we require mass motions that distort the initial field in such a way that the poloidal flux is regenerated. It is clear that no axisymmetric motions —poloidal or toroidal— will suffice. It is easy to picture non-uniform rotation generating a toroidal field $\underset{\sim}{B}_t$ from $\underset{\sim}{B}_p$. What we need now is to use the lines of $\underset{\sim}{B}_t$ to create a new poloidal field $\underset{\sim}{B}'_p$. But axisymmetric motions acting on $\underset{\sim}{B}_t$ merely move the circular loops of B_t around, without generating any new poloidal flux: analytically, $\nabla \times (\underset{\sim}{v}_p \times \underset{\sim}{B}_t)$ is a toroidal vector, while $\nabla \times (\underset{\sim}{v}_t \times \underset{\sim}{B}_t)$ is identically zero in axial symmetry. There is no way—using axisymmetric motions —to complete the cycle $\underset{\sim}{B}_t \rightarrow \underset{\sim}{B}_p$: Elsasser called this the "topological asymmetry" between poloidal and toroidal fields.

However, as soon as non-axisymmetric motions are allowed, this constraint is relaxed. This is the essential feature of the current semi-phenomenological models of the solar dynamo (Babcock 1961; Leighton 1969). The energy in the solar non-uniform rotation is used to generate a toroidal field, and magnetic buoyancy causes toroidal loops to rise and by buckling to give rise to sun-spot pairs. In Babcock's model, these pairs are now supposed twisted by Coriolis forces so that they have a largely north-south orientation, instead of the east-west orientation of the toroidal loops. The new contributions to the poloidal field in fact do not regenerate but <u>annihilate</u> and <u>reverse</u> the initial field, giving rise to a periodic rather than a steady dynamo. Much work remains to be done on both the kinematics and even more on the dynamics of the process, but one feels that the Babcock-Leighton model contains features of permanent importance for our understanding of the solar dynamo in particular, and the dynamo problem in general.

Turbulence and Magnetism

The Babcock-Leighton models do rely on systematic large-scale motions—non-uniform rotation, and the twisting of toroidal fields. Much discussion, largely inconclusive, has gone into the turbulent dynamo problem—whether motions that are largely chaotic can build up a large-scale field. Recent work, especially by Steenbeck, Krause and Rädler (1966), and by Moffatt (1970) suggests that turbulence with a sufficient degree of anisotropy can act as a dynamo, although strictly isotropic turbulence is incapable. It is not clear yet whether this work is relevant to large-scale stellar fields.

Earlier, various workers had suggested that turbulence acting on a primeval field would accelerate the decay, by systematic tangling of the field. Parker (1965) and Weiss (1966) have shown how inexorable motions of simple structure can expel the flux from the inner regions, concentrating it at the edges of the region—as indeed is observed at the junctions of the supergranular network. Earlier, Spitzer (1957) had argued that a weak large-scale magnetic field would certainly be expelled from the bulk of a convecting star, but that again the flux would not be destroyed, but merely concentrated in the axial and surface regions, the locally excessive rates of decay being off-set by the work done on the field by the motions. Essentially, the whole star acts as a Parker-Weiss cell. If this is correct, the Hayashi turbulence would not destroy a primeval field; with the decay of the turbulence, the field would leak back through the star into a more normal structure. Thus we have reason to think of the fields of Ap stars as primeval. Equally, the absence of observable primeval fields in later type stars may be due to the presence of the outer convection zone, which keeps such a field inside the radiative core.

References

Babcock, H.W. 1961, Astrophys. J., 133, 572.

Backus, G. 1958, Ann. Phys., 4, 372.

Biermann, L. 1950, Zs. f. Naturforsch., 5a, 65.

Bullard, E.C. 1957, Private communication.

Cowling, T.G. 1934, Mon. Not. R. Astr. Soc., 94, 39.

__________ 1945, Ibid. 105, 166.

__________ 1957, Magnetohydrodynamics, Interscience: New York.

Fricke, K. 1968, Z.Astrophys., 68, 317.

Goldreich, P. and Julian, W.H. 1969, Astrophys. J., 157, 869.

Goldreich, P. and Schubert, G., 1967, Astrophys. J., 150, 571.

Herzenberg, A., 1958, Phil. Trans. Roy. Soc. A. 250, 543.

James, R.E. and Kahn, F.D. 1970, Astron. Astrophys. (in press).

Kato, S. and Nakagawa, Y. 1969, Astrophys. Space Sci., 5, 171.

____________________ 1970, Astron. Astrophys. (in press).

Kippenhahn, R. 1969, Astron. Astrophys., 2, 309.

Leighton, R.B. 1969, Astrophys. J., 156, 1.

Mestel, L. and Roxburgh, I.W. 1962, Astrophys. J., 136, 615.

Moffat, H.K. 1970, J. Fluid Mech. (in press).

Parker, E.N. 1965, Stellar and Solar Magnetic Fields, ed. R. Lüst, 113. North Holland: Amsterdam.

Roberts, G.O. 1969, Ph.D. dissertation, Cambridge University.

Schlüter, A. 1961, Lectures delivered at Cambridge, England.

Spitzer, L. Jr. 1957, Astrophys. J., 125, 525.

Steenbeck, M., Krause, F. and Rädler, K.-H. 1966, Z. Naturforsch, 21a, 369.

Weiss, N.O. 1966, Proc. Roy. Soc. A., 293, 310.

Wrubel, M. 1952, Astrophys. J., 116, 291.

Lecture II

ROTATING MAGNETIC STARS

In this lecture we study some aspects of the theory of the structure of rotating magnetic stars, in the hope that understanding of large-scale properties may yield clues towards interpretation of the observations. As the observed strongly magnetic stars are mainly of type A, we shall be concerned primarily with early-type stellar models, with a central convective core surrounded by an envelope supposed stable against convection all the way to the photosphere.

We have noted in lecture I that there are plausible semi-phenomenological models of dynamo action in the surface layers of rotating solar-type stars. An essential feature of these models is the existence of non-axisymmetric motions, such as are certain to be present in a rotating sub-photospheric convective zone. It is not ruled out that laminar motions of sufficiently low symmetry may sometimes develop in the non-convective envelopes of early-type stars (cf. Parker 1970); however, we shall provisionally assume that the magnetic fields of these early-type stars are not currently being maintained by dynamo action, but are slowly decaying relics, e.g., of fields present in the gas from which the stars formed (cf. lecture VIII). [It can be plausibly argued (Mestel 1969) that the amount of magnetic flux surmised to be present within a strongly magnetic A star is sufficient to resist serious tangling by the pre-main sequence Hayashi turbulence.]

Early-type stars in uniform rotation

We shall be concerned with the mutual interaction of the star's thermal-gravitational field and the perturbing centrifugal and magnetic fields. It is convenient to begin with the problem that is the oldest and superficially the simplest--a star with a uniform angular velocity Ω, and with magnetic forces assumed everywhere negligible compared with the centrifugal. The interaction of the turbulence in the core with the rotation field remains a subject of controversy, but we shall assume that the turbulent viscosity keeps the core rotating uniformly with the radiative envelope. We concentrate attention on the envelope. The condition of hydrostatic equilibrium can be written

$$\nabla p = \rho \nabla (\phi + V) \equiv \rho \nabla \psi \ , \tag{1}$$

where ϕ is the gravitational potential and V the centrifugal potential:

$$V = \frac{1}{2} \Omega^2 \tilde{\omega}^2 \ . \tag{2}$$

[In general, such a potential exists if the angular velocity Ω is a function only of distance $\tilde{\omega}$ from the rotation axis:

$V = \int \Omega^2 \tilde{\omega} d\tilde{\omega}$.] The surfaces of constant pressure thus coincide with the surfaces of constant joint potential ψ (known as level surfaces), so that

$$p = p(\psi) \quad , \quad \rho = dp/d\psi \ . \tag{3}$$

These jointly imply $p = p(\rho)$--not an equation of state, but a consequence of the hydrostatics of rotating stars with a centrifugal potential. In a region of constant molecular weight μ, $T = T(\psi)$ also.

From the equation of radiative transfer

$$\underset{\sim}{F} = - \frac{4acT^3}{3\kappa\rho} \nabla T \ ,$$

we have

$$\underset{\sim}{F} = - \left(\frac{4acT^3}{3\kappa\rho} \frac{dT}{d\psi}\right) \nabla\psi \equiv - f(\psi)\nabla\psi \ , \tag{4}$$

since the opacity $\kappa = \kappa(\rho,T)$ and so is a function of ψ. This implies that in a region where heat transport is purely radiative, the flux at different points on a level surface is proportional to the net gravity. This is known as von Zeipel's theorem, or the law of gravity darkening (von Zeipel 1924).

Von Zeipel's celebrated "paradox" (see Eddington 1926) arises if we insist further that the star be everywhere in radiative equilibrium. Form the divergence of the flux:

$$\begin{aligned} - \nabla\cdot\underset{\sim}{F} = \nabla\cdot[f(\psi)\nabla\psi] &= f'(\psi)(\nabla\psi)^2 + f(\psi)\nabla^2\psi \\ &= f'(\psi)(\nabla\psi)^2 + f(\psi)(2\Omega^2 - 4\pi G\rho) \ . \end{aligned} \tag{5}$$

In a non-rotating stellar envelope in radiative equilibrium, ψ reduces to ϕ and $\nabla\cdot F = 0$ is equivalent to the mutual cancellation of $f'(\phi)(\nabla\phi)^2$ and $f(\phi)\nabla^2\phi$. But now $(\nabla\psi)^2$ is not constant on a level surface, because level surfaces are not equidistant at different points. According to von Zeipel, thermal balance has to be maintained by a local energy source ε per gram, such that

$$0 = \rho\varepsilon - \nabla\cdot\underset{\sim}{F} = \rho\varepsilon + f'(\psi)(\nabla\psi)^2 - f(\psi)(4\pi G\rho - 2\Omega^2) \ . \tag{6}$$

If ε is a function of local ρ,T values, it is also a function of ψ, and this equation can hold only if the parts respectively constant and variable over a level surface vanish separately. Thus

$$f'(\psi) = 0 \ : \tag{7}$$

the gravity-darkening proportionality constant is the same for different level surfaces. Further,

$$\varepsilon \propto \left(1 - \frac{\Omega^2}{2\pi G\rho}\right) \ , \tag{8}$$

implying a nuclear energy-generation rate that would go negative at low densities.

Eddington (1926, 1929) and Vogt (1925) recognized that the latter part of this argument is unphysical: a small rotation cannot require that the star has a spurious local energy source which balances one term in $\nabla \cdot \underset{\sim}{F}$, leaving the other to vanish separately. They saw that local radiative equilibrium is too strong a condition to impose; in general there is an excess or deficit of heat input into each element, and the consequent buoyancy drives a slow meridian circulation, which in a steady state will carry just the right amount of energy to maintain energy balance. We assume (to be verified a posteriori) that the velocities $\underset{\sim}{v}$ are small enough for the inertial forces to be ignorable, so that hydrostatic support remains an excellent approximation. The heat equation (with any genuine nuclear input given by ε) is

$$C_v \rho \frac{DT}{Dt} = \frac{p}{\rho} \frac{D\rho}{Dt} + (\rho\varepsilon - \nabla \cdot \underset{\sim}{F}) \ , \tag{9}$$

and this reduces in a steady state to

$$\rho A(\psi)(\underset{\sim}{v} \cdot \nabla)\psi = \rho\varepsilon - \nabla \cdot \underset{\sim}{F} \tag{10}$$

$$= \rho\varepsilon + f'(\psi)(\nabla\psi)^2 + f(\psi)(2\Omega^2 - 4\pi G\rho) \ , \tag{11}$$

where

$$A(\psi) = C_v \left[\frac{dT}{d\psi} - (\gamma - 1)\frac{T}{\rho}\frac{d\rho}{d\psi} \right] . \tag{12}$$

We now reduce $\nabla\psi$ to the unit normal $\hat{\underset{\sim}{n}}$ at the level surface by dividing through by $|\nabla\psi|$, and use the condition of zero net outflow of gas:

$$\int \rho(\underset{\sim}{v} \cdot \hat{\underset{\sim}{n}})dS = \rho \int (\underset{\sim}{v} \cdot \hat{\underset{\sim}{n}})dS = 0 \ . \tag{13}$$

Thus

$$\int \frac{dS}{|\nabla\psi|}[\rho\varepsilon + f(\psi)(2\Omega^2 - 4\pi G\rho)] + f'(\psi)\int |\nabla\psi| dS = 0 \ . \tag{14}$$

This defines the gravity-darkening function $f(\psi)$ throughout the radiative zone, replacing von Zeipel's spurious result (7).

We note from (10) that the mass-conservation condition (13) can be written

$$0 = \int_{\psi} \frac{(\rho\varepsilon - \nabla \cdot \underset{\sim}{F})}{|\nabla\psi|} dS = \int_{\psi} \frac{(\rho\varepsilon - \nabla \cdot \underset{\sim}{F})}{d\psi} dS d\ell \tag{15}$$

$$= \frac{1}{d\psi}\int (\rho\varepsilon - \nabla \cdot \underset{\sim}{F}) d\tau \ ,$$

where $d\tau \equiv dS d\ell$ is a volume element between the two level surfaces ψ, $\psi + d\psi$ (see Fig. 1). Thus, although radiative equilibrium breaks down locally, it holds in the mean.

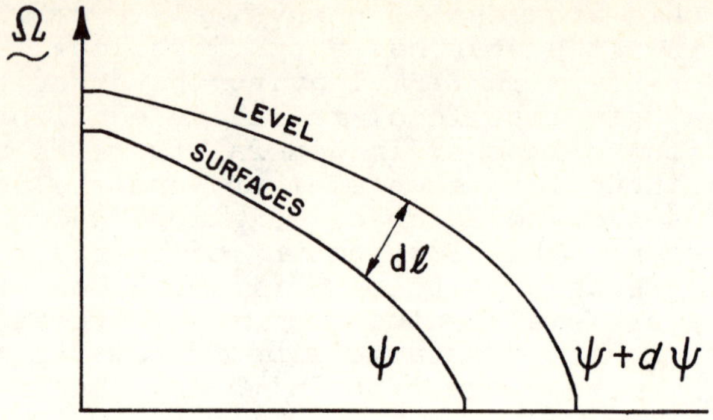

Fig. 1

We can write the circulation velocity normal to the level surface in the form

$$\rho A(\psi)\underset{\sim}{v}\cdot\nabla\psi = f'(\psi)\left[(\nabla\psi)^2 - \frac{\int|\nabla\psi|dS}{\int dS/|\nabla\psi|}\right]. \qquad (16)$$

The horizontal component is fixed by the equation of continuity. Since $|\nabla\psi|$ on any particular level surface has its maximum at the pole and decreases monotonically towards the equator, the square bracket on the right is positive at the pole and negative at the equator. In a convectively stable region, we must supply heat to make the gas rise: the circulation will therefore be up at the pole and down at the equator if $f'(\psi) > 0$, and reversed if $f'(\psi) < 0$. Now consider a radiative envelope, with $\varepsilon = 0$, so that by (14)

$$f'(\psi) = (4\pi G\rho)f(\psi)\left[1 - \frac{\Omega^2}{2\pi G\rho}\right]\frac{\int dS/|\nabla\psi|}{\int|\nabla\psi|dS} \propto f(\psi)\left[1 - \frac{2}{3}\frac{\bar{\rho}}{\rho}\eta\right], \qquad (17)$$

where $\bar{\rho}$ is very nearly the mean density of the star, and the parameter η, defined by $\Omega^2 = (4\pi/3)G\bar{\rho}\eta$, measures the ratio of centrifugal force to gravity at the surface. Since the effective gravity $\nabla\psi$ acts vertically downwards, from (4) we must have $f(\psi) > 0$, and the resulting circulation pattern breaks up into two zones (Öpik 1951; Mestel 1966); for provided the theory is valid all the way to the photosphere, there exists one level surface at which $f'(\psi) = 0$ for all realistic values of η.

The circulation pattern (given by the theory so far) is as in Fig. 2. Some of the most striking features--e.g., the flow into and out of the convective core--disappear on deeper analysis (cf. below).

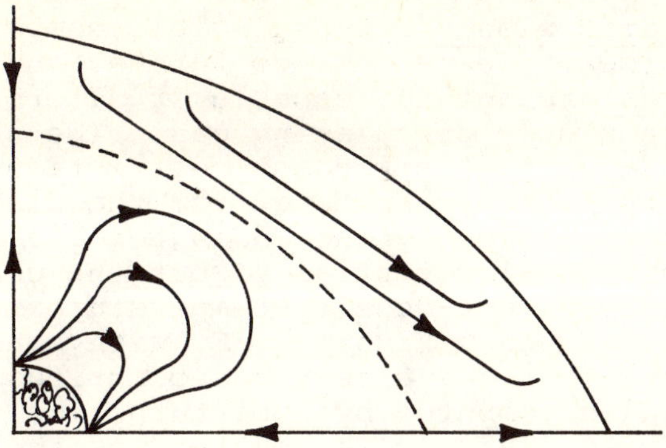

Fig. 2

The circulation speed over the bulk of the star is easily estimated to be of order

$$\frac{L}{Mg}\left(\frac{\Omega^2 r}{g}\right) \simeq 2 \times 10^{-5}\left(\frac{\Omega^2 r}{g}\right)\frac{\bar{L}\bar{R}^2}{\bar{M}^2} , \tag{18}$$

where g is the local value of gravity and barred quantities are measured in solar units. This is essentially the speed of overall thermal adjustment in the Kelvin-Helmholtz time-scale, reduced by the centrifugal distortion factor $\Omega^2 r/g$. Near the surface the speed is increased by the factor $(\bar{\rho}/\rho)\eta$. The theory cannot be applied at densities lower than the photospheric density, for in the stellar atmosphere the photon mean-free-path is too long for the local equation of radiative transfer (4) to be valid. The apparent singularity in the velocities at arbitrarily low density can be removed by returning to the integro-differential equation of radiative transfer (Smith 1970). However, the speeds remain very high, and this further complication emphasizes the need to scrutinize the implicit assumptions of the theory.

The effect of the circulation on the rotation field

If the inner radiative zone of the sun has essentially the surface rotation, then the time for the circulation currents to travel a fair fraction of the solar radius is of order 10^{12} years--much longer than the solar lifetime. But in a rapid rotator the current speeds are significant even deep in the star, while in the surface layers of an early-type star the extra factor $(\bar{\rho}/\rho)\eta$ can be as much as 10^7. The velocities are still small enough for the neglect of inertia to be fully justified. It is also well known that both radiative and ionic viscous forces are negligibly small (except for enormous velocity gradients). But the apparently successful resolution of the von Zeipel paradox immediately complicates the problem; for the thermally-driven circulation currents convect angular momentum, and so immediately destroy the assumed state of uniform rotation (ordinary viscous forces being equally impotent when acting on rotational shear).

The discussion naturally bifurcates at this point. The work of the last section assumed that any magnetic forces acting in meridian planes are negligible compared with centrifugal forces. We may now go further and demand that no magnetic torques are acting, capable of altering the angular momentum of the slowly circulating gas. (We shall see below that this is a much more stringent restriction.) The new angular velocity field--resulting from the distortion of the initial uniform rotation by the Eddington circulation--will not have a centrifugal potential V, but the centrifugal disturbance to the pressure-density-temperature field will again in general yield a non-vanishing $\nabla \cdot \underset{\sim}{F}$, with consequent thermally-driven currents. For any prescribed centrifugal field the circulation can be computed by a perturbation technique developed by Sweet (1950). The pattern of the circulation fields is found to be highly sensitive to the details of the angular momentum distribution (Kippenhahn 1958; Mahasweran 1968). The order of magnitude of the first-order circulation speeds is

$$\left(\frac{\Omega^2 r}{g}\right)\left(\frac{L}{Mg}\right)\left(\frac{\bar{\rho}}{\rho}\right) . \tag{19}$$

This differs from the estimate (18) by the factor $\bar{\rho}/\rho$, which becomes large at the surface (Baker and Kippenhahn 1959). [Uniform rotation is a singular case, in that the factor appears only in the second order--see equation (17).] If there develop sharp gradients in the angular velocity field, the resulting circulation speeds will be increased over the estimate (19) by the factor $r|\nabla\Omega^2|/\Omega^2$, a result relevant to the model proposed by Dicke (1970) for the internal solar rotation.

Over the bulk of the star, with $\bar{\rho}/\rho \sim 1$, the estimate (18) is generally valid; and since the Kelvin-Helmholtz time is about 1 percent of the nuclear evolution time, the circulation currents are negligibly slow over the bulk of the star, unless $\Omega^2 r/g$ is at least .01. But again the factor $\bar{\rho}/\rho$ enormously increases the speeds in the surface layers.

At one time it was thought that the evolution of the angular velocity field of a non-magnetic star could be discussed entirely in terms of the convection of angular momentum by the large-scale Eddington-Sweet circulation fields, themselves fixed by the instantaneous angular velocity fields. However, it is now clear that one must take account of instabilities to which non-conservative centrifugal fields are subject (Goldreich and Schubert 1967; Fricke 1968). The correct description should be in terms of the convection of angular momentum by large-scale circulation being partly offset by the macroscopic "friction" of the weak "turbulence" arising from the instabilities; but argument continues about the non-linear development of these instabilities (Kippenhahn 1969; James and Kahn 1970) and so about the effectiveness of the "turbulence" in redistributing angular momentum (see also Smith 1970).

The effect of a weak magnetic field

The problem is radically altered if the star retains some magnetic flux. We continue to assume provisionally that the centrifugal forces are everywhere stronger than the magnetic forces, so that the circulation field is given essentially by the theory of the last section. For simplicity, we suppose the magnetic field--called $\tilde{B}_p$--is symmetric about the rotation axis, and has initially only meridional ("poloidal") components. Then if the star is rotating uniformly--or, more generally, if the angular velocity is uniform along each field-line--the rotational velocities do not distort the frozen-in field. The only electrodynamic consequence is a slight charge separation: the $\tilde{v} \times \tilde{B}/c$ term in equation (11) of lecture I is canceled by a curl-free electric field $\tilde{E}$, maintained by the charge-density $\rho_e = \nabla \cdot \tilde{E}/4\pi = - \nabla \cdot (\tilde{v} \times \tilde{B}_p/4\pi c)$. This result--Ferraro's law of isorotation (1937)--is an immediate consequence of the flux-freezing constraint, although it was in fact discovered before the general formulation of the magnetohydrodynamic picture by Alfvén (1942, 1950). But if $(\tilde{B}_p \cdot \nabla)\Omega \neq 0$, the rotational shear acting on the frozen-in field will steadily generate an azimuthal ("toroidal") field component $\tilde{B}_t$. The associated magnetic force density has a toroidal component $(\nabla \times \tilde{B}_t) \times \tilde{B}_p/4\pi$ which reacts back on the rotation field. There results a magnetohydrodynamic wave, with the magnetic tensions transporting angular momentum along each field-line, and with energy being continually interchanged between the rotation field and the toroidal magnetic field. The wave travels at the local Alfvén speed $B_p/(4\pi\rho)^{1/2}$.

Now consider our uniformly rotating radiative envelope. It is subject to two competing processes--the slow Eddington-Sweet circulation, tending to upset the uniform rotation, and the torsional Alfvén waves, tending to restore it. It is clear that uniform rotation--at least along each field-line--will be closely preserved if the circulation speed v_p is much less than the Alfvén speed $B_p/(4\pi\rho)^{1/2}$. [This result may be derived rigorously by means of the formalism developed in the next two lectures--see Mestel (1961).] But we have assumed also that the centrifugal forces are the dominant non-spherical perturbation causing the circulation. This will be the case over the bulk of the zone if the rotational energy is larger than the magnetic. These two conditions combine into

$$\frac{1}{2}\,\rho v_p^2 << \frac{B_p^2}{8\pi} << \frac{1}{2}\,\rho\Omega^2\tilde{\omega}^2 \,, \qquad (20)$$

where $\tilde{\omega}$ represents the distance of a typical point from the rotation axis. Since the circulation speeds are far less than the rotational velocity over the bulk of the star, there is no difficulty in finding a magnetic field satisfying both conditions (20). And if the field is not strictly symmetric about the rotation axis, we may expect the magnetic pressures to resist relative shearing of individual field-lines, and so keep the zone as a whole rotating uniformly in the mean.

Having introduced the magnetic field as a means of preserving uniform rotation, we must now consider the effect of the meridian circulation on the poloidal field. As long as the centrifugal forces are the dominant perturbation and the rotation is kept close to uniformity, the circulation will retain the structure derived in the last section. In particular, deep in the star the flow will be up at the poles and down at the equator, and a field of dipolar structure, symmetric about the rotation axis, with field-lines crossing the equator, will be steadily distorted by the quadrupole-type circulation. A quasi-steady state may be reached with the field-lines sufficiently distorted near the equator for the slow circulation to be able to flow across the field by virtue of the finite resistivity [cf. equation (21) of lecture I].

More serious mathematical difficulties arise in the low-density regions near the surface, where the velocities become much faster because of the $\bar{\rho}/\rho$ term in (17) and (19). The two-zone structure of the Eddington-Sweet-Öpik circulation (Fig. 2) would yield such violent distortions to a dipolar field that neglect of the poloidal components of the magnetic force would break down. It is probable that in a fully self-consistent model of the surface regions, the hydrostatic equation (1) would have to be modified to include poloidal magnetic forces. The consequent modification of the thermal field may in fact remove the $\left(\frac{2}{3}\frac{\bar{\rho}}{\rho}\eta\right)$ term in (17), so destroying the uncomfortable two-zone structure of Fig. 2, and perhaps reducing the order of magnitude of the circulation speeds in the surface layers to values similar to those deeper down. We should note at this point the recent study by Michaud (1970) on the conditions for abundance anomalies to be produced at the surface: he requires the magnetic field to suppress all the mass motions that would otherwise offset the effects of gravitational settling and selective radiation pressure.

Rotational mixing

Given a large-scale circulation field driven by the centrifugal distortion of the thermal field, with the weak magnetic field maintaining the rotation uniform, it is *prima facie* possible that this could play a role in stellar evolution, by mixing helium-rich material from the convective core through some or all of the stellar envelope. Eddington's original overestimate (1929) of the circulation speed led workers in stellar evolution to assume that mixed models would be the rule, so that the inhomogeneous structures required to explain red giants had to result from subsequent accretion of interstellar hydrogen. Sweet's perturbation analysis (1950) corrected Eddington's estimate, leaving the (uniformly rotating) sun as an unmixed star, but predicted mixing for rapidly rotating O and B stars. However, gas streaming out of the core necessarily sets up a non-spherical distribution of mean molecular weight μ, requiring a corresponding non-spherical T-distribution, and yielding a disturbance to $\nabla\cdot\underset{\sim}{F}$ analogous to that due to rotation. Thus a horizontal μ-variation yields a "μ-current" velocity $\simeq (L/Mg)(\Delta\mu/\mu)$, and it is found that this tends to choke the circulation as it emerges from the core. If the circulation is in fact able to mix the star, the resulting horizontal μ-variations $\Delta\mu/\mu \sim .008/(\Omega^2R^3/GM)$ deep in the star. Thus if the μ-currents are not to choke the circulation, this must not exceed the value of Ω^2r/g at the convective core, which is $\sim (\Omega^2R^3/GM)/30$. We arrive at a critical value of $\Omega^2R^3/GM \sim 1/2$. Detailed numerical work (Mestel 1953, 1957) confirms that a uniformly rotating star capable of continuous steady mixing would need to be on the verge of rotational disruption. There has as yet been no attack on the formidable problem of how the star approaches a steady state; it is possible that some limited initial mixing could occur, sufficient perhaps to be relevant to the lithium abundance problem for a rapidly rotating primeval sun. In the steady state without mixing the circulation pattern of Fig. 2 is further modified, with the flow prevented from entering the core by the presence of a discontinuity in mean molecular weight, a "μ-barrier." The streamlines near the core are diverted horizontally in a boundary layer (Mestel 1953, 1965), but an adequate theory of the flow in the presence of the magnetic field has yet to be developed.

Strongly magnetic stars

Although we have seen that the structure of the surface layers of a rotating, weakly magnetic, early-type star awaits satisfactory treatment, it is clear that an "inexorable," rotationally-driven circulation will tend to distort an initial dipolar field, and may ultimately cause Ohmic disconnection of internal and external field, with trapping of the internal flux beneath the surface. In the next three lectures we shall be discussing the process of magnetic braking of stellar rotation, with the hope of explaining the abnormally show rotations of the strongly magnetic Ap stars. But this correlation of strong surface magnetic fields and low rotation may have a two-way significance, if rapid enough rotation generates circulation that ultimately prevents field-lines from emerging. The peculiar surface flux distributions that seem to be present in the Ap stars (Deutsch 1958; Peterson 1969) may also be explicable in terms of steady distortion of the field by circulation. Finally, there remains the question of the origin of the fields. If the "fossil" theory is correct (cf. lectures I and VIII), we are interested in the amount of flux possessed by the star as compared with upper limits given by the virial theorem, and so we need to know the relation between the surface field and the internal field.

It is instructive to approach these problems indirectly, by looking for the conditions in which there are no mass motions over the bulk of the star, tending to distort the flux emerging from the surface. So far, we have assumed that the magnetic forces over the bulk of the star have the modest strength necessary to maintain uniform rotation against the effect of circulation, but are much weaker than the centrifugal forces, so that the Eddington-Sweet circulation flows. We now go to the other extreme and require that the magnetic field over the star's bulk is so strong that the disturbance to radiative equilibrium due to the magnetic forces cancels that due to the rotation (again maintained uniform by the magnetic field). The field is again assumed symmetric about the rotation axis, and to have a dipolar angular dependence. The perturbations to the density, pressure and temperature fields are linearly superposed, and the radial variation of the magnetic field fixed by the conditions of hydrostatic support and the constraint of radiative equilibrium in the stellar envelope:

$$(\nabla \cdot \underset{\sim}{F})_{\Omega} + (\nabla \cdot \underset{\sim}{F})_{\underset{\sim}{B}} = 0 \tag{21}$$

in an obvious notation, with $(\nabla \cdot \underset{\sim}{F})_{\Omega}$ given by the linear Eddington-Sweet theory (Davies 1968; Wright 1969). (In the convective core the field is supposed strong enough to resist tangling by the convection.) Under these assumptions, there are now no mass motions tending to prevent field-lines from emerging from the surface.

The models depend on two parameters

$$\lambda_{\Omega} = \frac{\Omega^2 R^3}{GM} , \tag{22}$$

$$\lambda_B = \left(\frac{4\pi}{3} R^3\right)\left(\frac{\bar{B}^2}{8\pi}\right) \Big/ \frac{2}{3}\frac{GM^2}{R} = \frac{\bar{B}^2 R^4}{4GM^2} , \tag{23}$$

where $\bar{B}$ is the <u>surface</u> polar field. As the theory is linear, the ratio

$$\lambda = \frac{\lambda_{\Omega}}{\lambda_B} = \frac{4M\Omega^2}{\bar{B}^2 R} = 1.14 \times 10^{23}\left(\frac{\bar{M}}{\bar{R}}\right)\left(\frac{\Omega}{\bar{B}}\right)^2 \tag{24}$$

determines a given model.

The first model computed by Wright was with Ω and λ zero. The ratio of central field strength B_c to the surface value was about 40. The parameter λ was steadily increased, with the previous solution for ρ, p, T and $\underset{\sim}{B}$ being used as the zero-order approximation to the solution corresponding to the new value of λ. If Ω and $\bar{B}$ have values appropriate to typical observed rotating magnetic stars, λ must be increased to $\sim 2 \times 10^6$; for this value, $B_c/\bar{B}$ is found to be 1600. This large increase over the value for $\lambda = 0$ is due essentially to the enormous increase in density from surface to center: over the bulk of the star the centrifugal force density requires a correspondingly large magnetic energy density to stifle the Eddington-Sweet circulation and so to satisfy the constraint (21).

With $\bar{B} = 2 \times 10^3$ and $2\pi/\Omega \sim 9$ days--typical values for an Ap star--$\lambda = 2 \times 10^6$, and the central field strength is $\sim 1.6 \times 10^6$ gauss. It is significant that these internal fields, though high, are still well below the strengths required for them to make a significant contribution to hydrostatic support over the bulk of the star. As another example, consider the strongest surface field observed by Babcock, with $\bar{B} \sim 3.4 \times 10^4$ gauss. If this star were built on the model with $\lambda = 2 \times 10^6$, the central field would be 5.4×10^7 gauss, and the magnetic energy would be close to the virial theorem limit (cf. lecture VIII). But from (24), the corresponding value of Ω yields a period of only 12 hours: whereas in fact this star has a much longer rotation period, and the correct value of λ is much lower than 2×10^6, yielding an internal field that is again dynamically insignificant over the bulk. We seem driven to conclude that if any star were to form with a magnetic energy anywhere near the virial theorem limit, the surface flux would yield an integrated Zeeman effect far greater than is observed. This conclusion is clearly relevant to the theory of star formation in magnetic gas clouds (cf. lecture VIII).

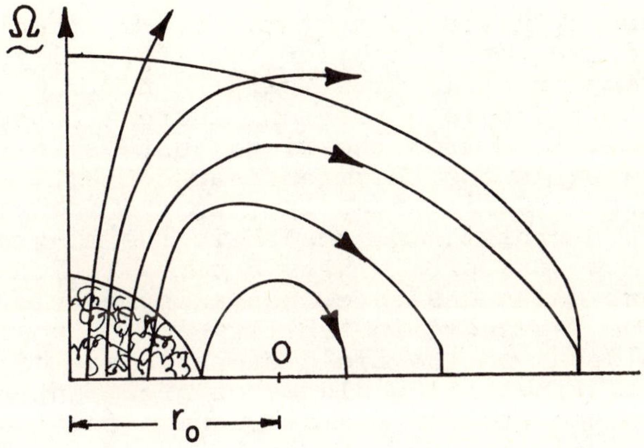

Fig. 3

The dipolar fields all have the general structure as in Fig. 3. The total flux F_t of the field is defined by the integral over the equator

$$F_t = \left| \int_0^{r_o} 2\pi r B_z dr \right| \tag{25}$$

where r_o is the radial distance of the neutral point. The <u>emergent</u> flux is

$$F_s = \left| \int_0^{R} 2\pi r B_z dr \right| = \pi \bar{B} R^2 . \tag{26}$$

For a given total flux F_t, an increase in Ω reduces F_s and $\bar{B}$ until at a critical rotation rate $\bar{B} = 0$: the field is completely submerged. With Ω increased still further there will be no states with zero circulation, satisfying (21)--there must flow a modified Eddington-Sweet circulation. Alternatively, if the star conserves its angular momentum but steadily loses flux

F_t through Ohmic decay, the emergent flux F_s will again decrease to zero at a critical value of F_t.

Neither these models, nor some oblique rotator generalization, are in fact likely to describe the interiors of the actual magnetic stars. However, the work does suggest that (1) the observed flux may be only a small fraction of the total magnetic flux; (2) the fields of even the strongest magnetic stars are dynamically insignificant over the bulk; (3) the correct description of the observed fields may be in terms of a slow but inexorable dragging of flux beneath the surface; and (4) rapidly rotating stars (viewed pole-on) may be observably non-magnetic because of the trapping of all the flux beneath the surface.

We conclude by noting that very little attention has been paid to the hydromagnetic stability or instability of large-scale stellar magnetic fields. Near the neutral ring of the field in Fig. 3 there is a toroidal current loop, and the magnetic forces act towards the ring, just as in the pinched discharges studied in the thermonuclear context. There is of course a crucial difference, in that the magnetic forces are at best a perturbation on the basic thermo-gravitational equilibrium. However, preliminary studies (Wright 1970) have shown that there exist unstable, non-axisymmetric motions--analogous to the "sausage" instability--which would tend to accelerate the decay of the field by shrinkage of flux-loops into the neutral ring. It would be of great interest if stable, large-scale stellar fields require toroidal flux to thread the poloidal loops, again as in the "stabilized pinch." These studies are likely to be significant in the star formation problem (cf. lecture VIII).

REFERENCES

Alfvén, H. 1942, Ark. f. Mat. Astr. o. Fysik, 29B, 1942.

______. 1950, Cosmical Electrodynamics (Oxford: Clarendon Press).

Baker, N. and Kippenhahn, R. 1959, Zs. f. Astrophys., 48, 140.

Davies, G.F. 1968, Austr. J. Phys., 21, 294.

Deutsch, A.J. 1958, Electromagnetic Processes in Cosmical Physics (Cambridge: Cambridge University Press), p. 209.

Dicke, R.H. 1970, Astrophys. J., 159, 1.

Eddington, A.S. 1926, Internal Constitution of the Stars (Cambridge: Cambridge University Press; Dover Edition, 1959).

______. 1929, Mon. Not. R. Astr. Soc., 90, 54.

Ferraro, V.C.A. 1937, Mon. Not. R. Astr. Soc., 97, 458.

Fricke, K. 1968, Zs. f. Astrophys., 68, 317.

Goldreich, P. and Schubert, G. 1967, Astrophys. J., 150, 571.

James, R.A. and Kahn, F.D. 1970, Astron. and Astrophys., 5, 232.

Kippenhahn, R. 1958, Zs. f. Astrophys., 46, 26.

______. 1969, Astron. and Astrophys., 2, 309.

Mahasweran, M. 1968, Mon. Not. R. Astr. Soc., 140, 93.

Mestel, L. 1953, Mon. Not. R. Astr. Soc., 113, 716.

______. 1957, Astrophys. J., 126, 550.

______. 1961, Mon. Not. R. Astr. Soc., 122, 473.

______. 1965, "Meridian Circulation in Stars," Stellar Structure, ed. L.H. Aller and D.B. McLaughlin (Chicago: University of Chicago Press).

______. 1966, Zs. f. Astrophys., 63, 196.

______. 1969, Liège Symposium.

Michaud, G. 1970, Astrophys. J., 160, 641.

Öpik, E.J. 1951, Mon. Not. R. Astr. Soc., 111, 278.

Parker, E.N. 1970, Astrophys. J., 160, 383.

Peterson, D. 1969 (private communication).

Smith, R.C. 1970, Mon. Not. R. Astr. Soc., 148, 275.

Sweet, P.A. 1950, Mon. Not. R. Astr. Soc., 110, 568.

Vogt, H. 1925, Astr. Nachr., 223, 229.

von Zeipel, H. 1924, Mon. Not. R. Astr. Soc., 84, 665.

Wright, G.A.E. 1969, Mon. Not. R. Astr. Soc., 146, 197.

______. 1970, Ph.D. Dissertation, Manchester University.

Lecture III

MAGNETIC BRAKING BY A STELLAR WIND I

In the last lecture we noted the fact that the strongly magnetic Ap stars have generally much lower rotations than the normal A stars. The suggested partial explanation is that most of the magnetic flux present in a rapidly rotating star may be prevented from emerging from the surface by circulation currents. We now study a complementary process -- the excess loss of angular momentum from stars which do have external magnetic fields. We shall see that the theory has also a direct application to the rotation of the sun and other solar-type stars in galactic clusters.

The basic idea goes back to Schatzman (1962) who pointed out that gas emitted from a star will carry away much more angular momentum per unit mass if it is constrained by magnetic field-lines to co-rotate with the star out to large distances. Schatzman thought in terms of corpuscular emission following flare-type disturbances, and of dynamo-built magnetic fields, both phenomena depending on the presence of sub-photospheric convection. We shall now develop the analogous theory for steady outflow of gas -- a stellar wind similar to the well-attested solar wind phenomenon. For much of the work we need not be too specific about the origin of the magnetic field -- it may be dynamo-built, as in the sun, or a relic, as may be the case for the fields of the magnetic Ap stars.

Outline of the theory of the spherically symmetric stellar wind

The high temperatures at the base of the solar corona are generally agreed to be due to the dissipation of the energy of waves emitted from the sub-photospheric convection zone. Chapman (1957) pointed out that the enormous scale-height forces us to treat the corona as extending into the interstellar medium. Suppose that the corona is at rest, and a supply of heat energy L is deposited at the coronal base r_s, to be carried away to infinity (radiation losses by free-free emission being small). With a thermal conductivity $\sigma = \sigma_0 T^{5/2}$, we require

$$-\sigma_0 T^{5/2} \frac{dT}{dr} = \frac{L}{4\pi r^2} , \qquad (1)$$

which immediately integrates to

$$T^{7/2} - T_u^{7/2} = \frac{7L}{8\pi\sigma_o}\left(\frac{1}{r} - \frac{1}{r_u}\right). \tag{2}$$

The radius r_u can be any reference level but we shall take it to be the radius beyond which the hypothetical non-expanding corona would become convectively unstable. The pressure and density for $r < r_u$ are given by combining (2) with the hydrostatic condition and the equation of state

$$\frac{dp}{dr} = - \frac{GM\rho}{r^2} = - \frac{GM\mu}{R}\left(\frac{p}{Tr^2}\right). \tag{3}$$

If the convection beyond $r = r_u$ is efficient enough for the adiabatic relation $p \propto \rho^{5/3}$ to hold, then

$$\frac{5}{3}\,\frac{1}{\rho}\,\frac{d\rho}{dr} = \frac{1}{p}\,\frac{dp}{dr} = - \frac{GM}{r^2}\,\frac{\rho}{p_u(\rho/\rho_u)^{5/3}}, \tag{4}$$

which integrates to

$$\left(\frac{GM}{r_u}\right)\left(\frac{r_u}{r}\right) = \frac{5}{2}\left(\frac{p_u}{\rho_u}\right)\left(\frac{\rho}{\rho_u}\right)^{2/3}, \tag{5}$$

where we have applied the boundary condition $p \to 0$ as $r \to \infty$. At $r = r_u$

$$\frac{GM}{r_u} = \frac{5}{2}\,\frac{p_u}{\rho_u} = \frac{5}{2}\,\frac{R}{\mu}\,T_u, \tag{6}$$

and for $r > r_u$

$$\left(\frac{T}{T_u}\right) = \left(\frac{\rho}{\rho_u}\right)^{2/3} = \left(\frac{r_u}{r}\right). \tag{7}$$

At $r = r_u$ the temperature gradient of the conduction zone $r \leq r_u$ must equal the adiabatic gradient, so that

$$L = -4\pi r_u^2 \sigma_o T_u^{5/2}\,(dT/dr)_{r_u}$$

$$= 4\pi r_u \sigma_o T_u^{7/2} \qquad \text{from (7)}, \tag{8}$$

$$= 4\pi\sigma_o \left(\frac{2\mu GM}{5R}\right)^{7/2} \Big/ r_u^{5/2} \qquad \text{from (6)}. \tag{9}$$

By substitution of (8), (2) simplifies to

$$\left(\frac{T}{T_u}\right)^{7/2} = \left(\frac{7}{2}\,\frac{r_u}{r} - \frac{5}{2}\right), \tag{10}$$

and from (10) and (6) the temperature at the coronal base becomes

$$T_s = \frac{2\mu GM}{5\mathcal{R}r_s}\left[\frac{r_s}{r_u}\left(\frac{7}{2}\frac{r_u}{r_s} - \frac{5}{2}\right)^{2/7}\right]. \tag{11}$$

We can obtain an upper limit for T_s by setting $r_u = r_s$:

$$(T_s)_{Max} = \frac{2\mu GM}{5\mathcal{R}r_s} . \tag{12}$$

For the sun $(T_s)_{Max} = 4 \times 10^6\,°K$, higher than the estimated value, so that such a model if applicable to the sun might yield a static corona. However, the observations of outflow imply that there is an invalid assumption in the above treatment. In fact, it is almost certain that heat from wave dissipation is fed directly into the corona for distances well above the coronal base, so that a better approximation would be to take $T \simeq T_s$ for all $r > r_s$. Then equation (3) yields

$$p = p(r_s)\exp\left[\frac{GM}{a^2}\left(\frac{1}{r} - \frac{1}{r_s}\right)\right] \tag{13}$$

and

$$p_\infty = p(r_s)\exp[-GM/a^2 r_s], \tag{14}$$

where

$$a^2 = \frac{\mathcal{R}T_s}{\mu} . \tag{15}$$

If we substitute the value for a appropriate to the observed coronal temperature, the pressure at infinity given by (14) is much larger than the pressure of the interstellar medium. The hot corona requires a wall to hold it in; without such a wall the corona must expand under its own thermal pressure (Parker 1963).

For simplicity we restrict the discussion to the isothermal case. The equation of hydrostatic support (3) is replaced by Bernoulli's equation for spherically symmetric radial flow

$$a^2\log\rho + \frac{1}{2}v^2 - \frac{GM}{r} = \text{constant}. \tag{16}$$

Eliminating the density ρ between (16) and the continuity equation

$$\rho v r^2 = \text{constant}, \tag{17}$$

we have

$$\left(\frac{1}{2}v^2 - a^2\log v\right) - \left(\frac{GM}{r} + 2a^2\log r\right) = \text{constant}. \tag{18}$$

The appropriate boundary conditions are:

(a) near the star v becomes small and Bernoulli's equation reduces to hydrostatic support;
(b) at infinity the pressure must be small.

These conditions yield a unique solution (Parker 1963) that passes smoothly through the critical point of the equation (18) at which the derivatives of the two brackets vanish:

$$\frac{v}{a} = 1, \qquad r = r_a \equiv \frac{GM}{2a^2}. \tag{19}$$

Thus the constant in (18) is fixed by the gas velocity becoming sonic at this critical point:

$$\left[\frac{1}{2}\left(\frac{v}{a}\right)^2 - \log\left(\frac{v}{a}\right)\right] - 2\left[\frac{r_a}{r} + \log\left(\frac{r}{r_a}\right)\right] = -\frac{3}{2}. \tag{20}$$

From (16) and (20), when $r/r_a \ll 1$, ρ increases and v decreases inwards exponentially. When $r/r_a \gg 1$, (20) yields

$$\frac{v}{a} \simeq 2\left[\log\,(r/r_a)\right]^{1/2} \tag{21}$$

--a very slow increase, so that by (17) $\rho \propto 1/r^2$ approximately.

It should be noted that the absolute value of the density in the stellar wind is not determined from the theory. It is convenient to introduce a reference level r_s, the coronal base, with a density ρ_s; then equations (17) and (20) determine ρ/ρ_s at higher levels. The value of ρ_s should be fixed (along with the details of the temperature distribution) by the ultimate theory of heat input into the corona. The numbers predicted by such a theory are likely to be sensitive to variations in surface temperature and gravity. We shall leave ρ_s as a free parameter in the theory, but noting that the deeper the sonic point, the greater is the mass loss, given by

$$-\frac{dM}{dt} = 4\pi\rho(r_a)ar_a^{\,2}. \tag{22}$$

For the sun at the present time the e-folding time for mass loss is estimated to be about 10^{13} years. In pre- and post-main sequence phases much greater loss is observed (Kuhi 1964, 1966; Deutsch 1960), though it may be questioned whether the process should then be described in terms of quasi-steady wind theory.

There are many ways in which the above simple theory can be made more realistic. The most obvious one is a replacement of the isothermal (or polytropic) equation of state by a genuine heat equation with distributed heat sources (corresponding to wave dissipation continuing above the coronal base), and with heat conduction included. Viscous effects appear to be small, at least for the solar wind. When the mean-free-paths become comparable with the radius it is no longer adequate to assume isotropic partial pressures or identical temperatures for electrons and ions (Hartle and Sturrock 1968). For comprehensive reviews of these and other problems, see Holzer and Axford (1969), and Parker (1969).

Interaction of the wind with rotation and magnetic fields

If gas flowing from a rotating star is unaffected by any torques it will remember the angular momentum it had when it left the star, so that at axial distance $\tilde{\omega}$ it will have an angular velocity Ω given by

$$\Omega\tilde{\omega}^2 = \Omega_s\tilde{\omega}_s{}^2. \tag{23}$$

The gas will therefore lag behind the rotating star, and the flow lines will form a spiral pattern. Now let there be a magnetic field emerging from the star. If the field is weak--i.e., if its energy is less than the kinetic energy of the wind--then the frozen-in field-lines will be pulled out by the wind, and will also be twisted by the non-uniform rotation (23) into a spiral pattern. The magnetic tensions try to straighten out the field-lines by sending out an Alfvén wave, with local speed $v_A = B/(4\pi\rho)^{1/2}$. If $v_A > v$ the Alfvén wave can catch up with the wind, and the field is strong enough to resist violent distortion; instead the field will constrain the gas so that it tends to co-rotate with the star (Cowling 1965). Thus we may anticipate that the magnetic field should enable the gas streaming to infinity to carry away roughly the angular momentum it would have if the star's surface extended as far as the Alfvénic surface S_A, defined by the wind speed equalling the local Alfvén speed.

It should be noted that the condition $v_A > v$ is equivalent to

$$\frac{B^2}{8\pi} > \frac{1}{2}\,\rho v^2: \tag{24}$$

within the surface S_A the magnetic energy must exceed the kinetic energy of the wind. The same criterion that enables the field to control the rotation of the gas also ensures that the field will resist pulling out by the wind, but will tend rather to control the wind flow.

In the following detailed analysis for the axisymmetric, steady-state problem (Chandrasekhar 1956; Mestel 1961, 1968), we give both the magnetic and velocity vectors poloidal (meridional) and toroidal (azimuthal) components:

$$\begin{aligned} \mathbf{B} &= \mathbf{B}_p + \mathbf{B}_t, \\ \mathbf{v} &= \mathbf{v}_p + \mathbf{v}_t \equiv \mathbf{v}_p + \tilde{\omega}\Omega\mathbf{t}. \end{aligned} \tag{25}$$

Here $\mathbf{v}_p$ is the generalized wind velocity, having both radial and transverse components; Ω is the local angular velocity, and $\mathbf{t}$ the unit toroidal vector. Because of axial symmetry, the magnetic flux condition restricts the poloidal component only:

$$\nabla \cdot \underset{\sim}{B}_p = 0. \tag{26}$$

From lecture I, we adopt the simplest (non-relativistic) analogue of Ohm's law for a fully ionized gas

$$0 = \underset{\sim}{E} + \frac{\underset{\sim}{v} \times \underset{\sim}{B}}{c} = -\nabla\phi + \frac{\underset{\sim}{v} \times \underset{\sim}{B}}{c} , \tag{27}$$

where ϕ is the electric scalar potential. (We may note here the ratio of the neglected Hall term to the induction term is at most of order

$$\left|\frac{\underset{\sim}{j} \times \underset{\sim}{B}}{cn_e e}\right| \Bigg/ \left|\frac{\underset{\sim}{v} \times \underset{\sim}{B}}{c}\right| \simeq \left(\frac{\rho v^2}{r}\right) \Bigg/ \left(\frac{vB}{c}\right)\left(\frac{\rho eZ}{Am_H}\right) \simeq \frac{\Omega}{\omega_i} \tag{28}$$

where ω_i is the Larmor gyration frequency of an ion of charge Ze and mass number A in the field B. This ratio of a macroscopic to a microscopic angular velocity is very small for quite moderate fields B. The pressure gradient term is of the same order; and the magnetic Reynolds number will certainly be large enough for Ohmic effects to be ignorable. We can therefore use (27) with confidence; if we had to take cognizance of the magnetic field's tendency to move with the electrons rather than the ions the theory would be considerably more difficult.)

In a steady state, (27) implies

$$\nabla \times (\underset{\sim}{v} \times \underset{\sim}{B}) = 0. \tag{29}$$

On substituting the definitions (25), equation (29) breaks up into poloidal and toroidal parts:

$$\nabla \times (\underset{\sim}{v}_p \times \underset{\sim}{B}_p) = 0, \tag{30}$$

$$\nabla \times (\underset{\sim}{v}_p \times \underset{\sim}{B}_t + \underset{\sim}{v}_t \times \underset{\sim}{B}_p) = 0. \tag{31}$$

Equation (30) implies

$$\underset{\sim}{v}_p \times \underset{\sim}{B}_p = c(\underset{\sim}{t} \cdot \nabla\phi)\underset{\sim}{t} = 0 , \tag{32}$$

since the (single-valued) scalar potential cannot have a gradient in the toroidal direction in an axisymmetric system. Thus

$$\underset{\sim}{v}_p = \kappa \underset{\sim}{B}_p , \tag{33}$$

where κ is a scalar: in an axisymmetric steady state, gas streaming in meridian planes--the wind in our problem--must be along the poloidal field-lines. This conclusion is of course a kinematic constraint, imposed by the steady-state condition on the field $\underset{\sim}{B}_p$ in the presence of a velocity $\underset{\sim}{v}_p$. It says nothing about the structure of the joint poloidal field-streamlines, or on whether the flow is constrained to follow the field or the field distorted by the flow; these questions are clearly dynamical.

The equation (31) can now be written

$$\nabla \times \left\{ \tilde{\omega}\left(\Omega - \frac{\kappa B_t}{\tilde{\omega}}\right) \underset{\sim}{t} \times \underset{\sim}{B}_p \right\} = 0. \tag{34}$$

When $\kappa = 0$ (pure rotation), we know that (34) implies Ferraro's law of isorotation -- $\Omega = \alpha$ = constant on field-lines; hence (34) implies

$$\Omega - \frac{\kappa B_t}{\tilde{\omega}} = \alpha \tag{35}$$

as a generalization. To prove this formally, we write

$$\underset{\sim}{B}_p = \nabla \times \left(\frac{P}{\tilde{\omega}}\underset{\sim}{t}\right) = \nabla P \times \left(\frac{\underset{\sim}{t}}{\tilde{\omega}}\right) + P\nabla \times \left(\frac{\underset{\sim}{t}}{\tilde{\omega}}\right) = \nabla P \times \left(\frac{\underset{\sim}{t}}{\tilde{\omega}}\right), \tag{36}$$

where the axisymmetric stream-function P satisfies

$$(\underset{\sim}{B}_p \cdot \nabla) P = 0 \tag{37}$$

and so is constant on field-lines. Then from equation (34)

$$\begin{aligned} 0 &= \nabla \times \left[\left(\Omega - \frac{\kappa B_t}{\tilde{\omega}}\right) \underset{\sim}{t} \times (\nabla P \times \underset{\sim}{t})\right] \\ &= \nabla\left(\Omega - \frac{\kappa B_t}{\tilde{\omega}}\right) \times \nabla P, \end{aligned} \tag{38}$$

and (35) follows from (37) and (38).

We may combine (33) and (35) into

$$\underset{\sim}{v} = \kappa \underset{\sim}{B} + \tilde{\omega}\alpha \underset{\sim}{t}. \tag{39}$$

Thus the general axisymmetric solution is a combination of an arbitrary motion along the field -- by (27) implying no charge separation--and a uniform rotation of each individual field-line. If all the field-lines emerge from a uniformly rotating star, then α will be constant over the whole field: seen in the frame rotating with angular velocity α , the flow $\underset{\sim}{v}$ is everywhere parallel to $\underset{\sim}{B}$.

In a steady state the continuity equation yields

$$0 = \nabla\cdot(\rho \underset{\sim}{v}) = \nabla\cdot(\rho \underset{\sim}{v}_p) = \nabla\cdot(\rho\kappa \underset{\sim}{B}_p) = (\underset{\sim}{B}_p \cdot \nabla)\rho\kappa , \tag{40}$$

where the result (33) has been substituted. Thus

$$\frac{\rho v_p A}{B_p A} = \frac{\rho v_p}{B_p} = \rho\kappa \tag{41}$$

$$= \eta = \text{constant along poloidal field-streamlines,}$$

where A is the varying area of cross-section of the infinitesimal poloidal flux-tube along the field-line considered. Written in this way, the condition $\rho\kappa = \eta$ is seen to be an immediate consequence of the conservation of both material and magnetic flux.

In the simplest geometry, with poloidal field-streamlines radial, and spherically symmetric, the continuity equation reduces to

$$\rho v_p r^2 \propto \rho v_p \tilde{\omega}^{-2} = \text{constant.} \tag{42}$$

Near to the star the field-lines will strongly resist distortion by the wind and a better approximation may be with the field's retaining a curl-free structure (see next lecture). Thus if the star has a dipolar surface flux distribution, the stream-function P of its curl-free extension is

$$P = -\frac{1}{2}\bar{B}_s r_s^{\,2}\left(\frac{\sin^2\theta}{r}\right) \tag{43}$$

in spherical polar coordinates, and the field-lines emerging at polar angle θ_s from the convenient reference level r_s are defined by

$$\frac{\tilde{\omega}^2}{r^3} \equiv \frac{\sin^2\theta}{r} = \frac{\sin^2\theta_s}{r_s} . \tag{44}$$

The total poloidal field-strength

$$B_p = \frac{\bar{B}_s}{(r/r_s)^3}\left(1 - \frac{3}{4}\frac{r}{r_s}\sin^2\theta_s\right)^{\frac{1}{2}} , \tag{45}$$

so that the two flux-conservation laws summarized by (41) reduce to

$$\rho v_p r^3 \propto \rho v_p \tilde{\omega}^2 \propto \left(1 - \frac{3}{4}\frac{r}{r_s}\sin^2\theta_s\right)^{\frac{1}{2}} . \tag{46}$$

The right-hand side of (46) varies fairly slowly, so that for both radial and dipolar field-lines (41) implies

$$\rho v_p \tilde{\omega}^2 \simeq \text{constant.} \tag{47}$$

REFERENCES

Chandrasekhar, S. 1956, Astrophys. J., 124, 232.

Chapman, S. 1957, Smithsonian Contrib. Astrophys., 2, 1.

Cowling, T.G. 1965, Stellar and Solar Magnetic Fields, ed. R. Lüst (Amsterdam: North-Holland Publishing Co.).

Deutsch, A.J. 1960, Stars and Stellar Systems, Vol. 6, ed. J. L. Greenstein (Chicago: University of Chicago Press).

Hartle, R.E., and Sturrock, P.A. 1968, Astrophys. J., 151, 1155.

Kuhi, L.V. 1964, Astrophys. J., 140, 1409.

______. 1966, Astrophys. J., 143, 991.

Holzer, T.E., and Axford, W.I. 1969, I.P.A.P.S., La Jolla preprint.

Mestel, L. 1961, Mon. Not. R. Astr. Soc., 122, 473.

______. 1968, Mon. Not. R. Astr. Soc., 138, 359.

Parker, E.N. 1963, Interplanetary Dynamical Processes (New York: Interscience).

______. 1969, Space Sci. Rev., 9, 325.

Schatzman, E. 1962, Ann. Astrophys., 25, 1.

Lecture IV

MAGNETIC BRAKING BY A STELLAR WIND II

Dynamics of the flow

In the preceding lecture we examined the kinematics of an infinitely conducting gas in an axially-symmetric steady state. We now examine the dynamics. The equation of motion in an inertial frame of a conducting inviscid gas is

$$-\frac{\nabla p}{\rho} + \nabla\phi + \frac{(\nabla\times\underset{\sim}{B})\times\underset{\sim}{B}}{4\pi\rho} = \nabla\left(\frac{1}{2}v^2\right) - \underset{\sim}{v}\times(\nabla\times\underset{\sim}{v}) \quad , \tag{1}$$

where ϕ is the gravitational potential. In a frame rotating with angular velocity $\alpha\underset{\sim}{k}$

$$\begin{aligned} -\frac{\nabla p}{\rho} + \nabla\left(\phi + \frac{1}{2}\alpha^2\omega^2\right) + (\nabla\times\underset{\sim}{B})\times\underset{\sim}{B}/4\pi\rho \\ = \nabla\left(\frac{1}{2}\underset{\sim}{V}^2\right) - \underset{\sim}{V}\times(\nabla\times\underset{\sim}{V}) + 2\alpha\underset{\sim}{k}\times\underset{\sim}{V}, \end{aligned} \tag{2}$$

where

$$\underset{\sim}{V} = \underset{\sim}{v} - \tilde{\omega}\alpha\underset{\sim}{t} = \kappa\underset{\sim}{B}. \tag{3}$$

[We recall that in a non-relativistic theory which neglects terms of order $(v/c)^2$, $\underset{\sim}{B}$ is invariant]. On taking the scalar product of (2) with $\underset{\sim}{V}$, and assuming an isothermal gas law with a (constant) sound speed a, we obtain the Jacobi integral

$$\begin{aligned} \frac{1}{2}\underset{\sim}{V}^2 - \phi - \frac{1}{2}\alpha^2\tilde{\omega}^2 + a^2\log\rho &= \delta \\ &= \text{constant on field-streamlines.} \end{aligned} \tag{4}$$

Transforming back to the inertial frame,

$$\frac{1}{2}v_p^{\ 2} + \frac{1}{2}\Omega^2\tilde{\omega}^2 - \alpha\Omega\tilde{\omega}^2 - \phi + a^2\log\rho = \delta. \tag{5}$$

One can readily verify that in the direct derivation of this modified Bernoulli integral in the inertial frame, the term $\alpha\Omega\tilde{\omega}^2$ results from the work $(\underset{\sim}{j}\times\underset{\sim}{B}/c)\cdot\underset{\sim}{v}$ done by the magnetic force on the gas flowing across the field. Near the star we shall find that $\Omega \simeq \alpha$ (approximate co-rotation), so that $\left(\alpha\Omega\tilde{\omega}^2 - \frac{1}{2}\Omega^2\tilde{\omega}^2\right) \simeq \frac{1}{2}\alpha^2\tilde{\omega}^2$. This excess energy is available to assist thermal pressure in driving the wind.

If we eliminate ρ between (5) and equation (41) of I, we obtain

$$\left(\frac{1}{2}v_p{}^2 - a^2\log v_p\right) - \left[\left(\phi - a^2\log B_p\right) - \left(\frac{1}{2}\Omega^2\tilde{\omega}^2 - \alpha\Omega\tilde{\omega}^2\right)\right] \tag{6}$$
$$= \text{constant along field-streamlines.}$$

For the moment let us suppose that the field B_p has a structure that is known a priori. Then equation (6) is the generalization of equation (18) of I. Along each field-line there will be a critical solution which accelerates from very small velocities deep in the corona to become sonic at the critical point, defined by the vanishing of the derivative along the field-line of the square bracket in (6). The terms in Ω are a complication; but provided the field is strong enough for the sonic point to be well within the Alfvénic surface we may safely put $\Omega = \alpha$. The differences from the simple radial flow case are the (as yet undetermined) geometrical factor in B_p and the net centrifugal term $\simeq \frac{1}{2}\alpha^2\tilde{\omega}^2$. This last term is important for rapidly rotating stars, leading to the idea of a "centrifugal wind" (Mestel 1968). In particular, it can move the sonic point inwards to a level at which the density is much higher, so that the mass and angular momentum loss can be correspondingly greater.

The toroidal component of the equation of motion can be written

$$\tilde{\omega}\left[\frac{(\nabla\times\underset{\sim}{B})\times\underset{\sim}{B}}{4\pi}\right]_t = \nabla\cdot(\rho\underset{\sim}{v}\Omega\tilde{\omega}^2): \tag{7}$$

the angular momentum given to an element of volume by the magnetic torques is off-set by the net efflux of angular momentum carried by the gas. Using $\underset{\sim}{v}_p = \kappa\underset{\sim}{B}_p$ this becomes

$$\nabla\cdot\left(\rho\kappa\Omega\tilde{\omega}^2\underset{\sim}{B}_p - \frac{\tilde{\omega}B_t}{4\pi}\underset{\sim}{B}_p\right) = 0, \tag{8}$$

or, since $\nabla\cdot\underset{\sim}{B}_p = 0$,

$$-\frac{\tilde{\omega}B_t}{4\pi} + \rho\kappa\Omega\tilde{\omega}^2 = -\frac{\beta}{4\pi} = \text{const. along field lines.} \tag{9}$$

The constant $-\beta/4\pi$ is interpreted as the rate of flow of z-angular momentum along a unit poloidal flux-tube. The flowing gas carries $\rho v_p\Omega\tilde{\omega}^2 A = (\rho\kappa\Omega\tilde{\omega}^2)B_pA = \rho\kappa\Omega\tilde{\omega}^2$ along a unit flux-tube; hence $-\tilde{\omega}B_t/4\pi$ is the amount carried by the magnetic stresses. This is a special case of a general result due to Lüst and Schlüter (1955). It is well-known that the magnetic force density can be written as the divergence of the Maxwell stress tensor:

$$f_k \equiv \left[\frac{(\nabla\times\underset{\sim}{B}) \times \underset{\sim}{B}}{4\pi}\right]_k = -\frac{\partial}{\partial x_l} T_{kl} \tag{10}$$

with

$$T_{kl} = \frac{B^2}{8\pi}\delta_{kl} - \frac{B_k B_l}{4\pi} ; \tag{11}$$

or

$$F_k \equiv \int f_k d\tau = -\int T_{kl} n_l dS \tag{12}$$

where n_l is the <u>outward</u>-drawn normal. Thus the total magnetic force F_k on a volume τ is equivalent to the <u>influx</u> of linear momentum $T_{kl}n_l dS$ over each element $n_l dS$ of the surrounding surface. Similarly we can write

$$d_k \equiv \left(\underset{\sim}{r} \times \left[\frac{(\nabla\times\underset{\sim}{B}) \times \underset{\sim}{B}}{4\pi}\right]\right)_k = -\frac{\partial}{\partial x_l} D_{kl} , \tag{13}$$

where

$$D_{kl} = \varepsilon_{kij}\, x_i\, T_{jl} ; \tag{14}$$

and

$$-\int d_k d\tau = \int D_{kl} n_l dS, \tag{15}$$

showing that D_{kl} represents the flow of the k-component of angular momentum across a unit area with normal pointing in the l-direction. In the present problem the k = 3 component reduces (because of the axial symmetry) to a flow $-\tilde{\omega}B_t B_p/4\pi$ per unit area of a flux-tube of the poloidal field, or $-\tilde{\omega}B_t/4\pi$ along a flux-tube of unit strength.

There remains one other component of the equation of motion (1), which may conveniently be taken as the poloidal component perpendicular to $\underset{\sim}{B}_p$:

$$\left[\nabla \times B_p - \frac{[(\nabla\times\underset{\sim}{B}_t)\times \underset{\sim}{B}_t] \times \underset{\sim}{B}_p}{\underset{\sim}{B}_p^{\,2}}\right] = \frac{4\pi\underset{\sim}{B}_p \times \underset{\sim}{F}_p}{\underset{\sim}{B}_p^{\,2}} , \tag{16}$$

where $\underset{\sim}{F}_p$ is the sum of all the non-magnetic poloidal forces (including inertial force). Written in this way, it defines the structure of $\underset{\sim}{B}_p$ required by the necessity for the magnetic and non-magnetic forces to balance. In general it is very difficult to deal with, except in the extreme case when the energy density of the magnetic field is much greater than that of the non-magnetic forces. In that case the right-hand side of (16) can be dropped, and the field approximates to a <u>force-free</u> structure, which will not differ much from the curl-free structure if $|\underset{\sim}{B}_t|/|\underset{\sim}{B}_p|$ << 1 (in the present problem, for slowly rotating stars). This approximation should be fairly good near to the star: the field-lines will not be distorted by the flow, but will have their structure determined essentially by the magnetic flux distribution over the stellar surface. In particular, some of the field-lines that emerge from regions of very high dipolar flux concentration may not extend to infinity but will return to the star, forming a dead zone within the general wind zone (Mestel 1966, 1968). Along the open field-lines in the wind zone the kinetic energy of the gas will certainly be comparable with the magnetic energy by the time the gas has reached the Alfvénic point, so that these field-lines will by then have become quasi-radial rather than quasi-dipolar.

The most conservative model (Mestel 1968) assumes that the field is strictly curl-free (in particular dipolar) all the way to the Alfvénic surface, with the wind zone and dead zone separated by a limiting field-line with its Alfvénic point on the equator. At the other extreme the wind is supposed to flow over the whole space, drawing out all the field-lines into a quasi-radial structure even in the regions near the star where the field energy is very high. As a consequence there must be a dense, pinched equatorial zone of gas separating the oppositely directed field-lines and carrying a high current. Outside this zone the non-magnetic force density is again small, and the quasi-radial field must adjust to a nearly curl-free structure, with the field-strength nearly independent of latitude. In the zone there is a strong thermal pressure gradient, built up by the external magnetic pressures to the strength required for equation (16) to be satisfied.

The numerical value of the braking torque depends quite strongly on the structure of the field and the associated extent of the wind zone, but the essence of the braking theory can fortunately be discussed for each field-line, independently of the precise field-structure. We shall note the observational evidence below.

We now combine the integrals (3), (9) and (41) of I to obtain

$$\Omega = (\alpha + \frac{\eta\beta}{\rho\tilde{\omega}^2})/(1 - \frac{4\pi\eta^2}{\rho}) \tag{17}$$

and

$$B_t = (\frac{\beta}{\tilde{\omega}} + 4\pi\eta\alpha\tilde{\omega})/(1 - \frac{4\pi\eta^2}{\rho}) \ . \tag{18}$$

The solutions are in general singular at any point where

$$1 = \frac{4\pi\eta^2}{\rho} \equiv \frac{v_p^2}{(B_p^2/4\pi\rho)} = \left(\frac{v_p}{v_A}\right)^2 \tag{19}$$

--i.e., where the wind speed equals the Alfvénic speed as determined by the poloidal field. We have seen that this ratio should enter in a sensitive way, since magnetic braking depends on Alfvén waves being able to keep up with the wind speed. Near the star where ρ is large, $4\pi\eta^2/\rho$ will be small if the magnetic field is at all sizeable, so that $v_p \ll v_A$. As ρ decreases outwards, there exists a critical point C on each field-streamline where $v_p = v_A$, i.e. where $\rho_c = 4\pi\eta^2$. In order that the solution should not blow up at the Alfvénic point C, we require that

$$\alpha + \frac{\eta\beta}{\rho_c \tilde{\omega}_c^2} = 0, \qquad \frac{4\pi\eta^2}{\rho_c} = 1 \qquad (20)$$

simultaneously, or

$$-\frac{\beta}{4\pi} = \frac{\alpha \rho_c \omega_c^2}{4\pi\eta} = \eta\alpha\tilde{\omega}_c^2. \qquad (21)$$

Thus the flow of angular momentum $-\beta/4\pi$ along each unit flux-tube emerges as an eigenvalue by the requirement that there be no singularities. Since η is the flow of gas along a unit flux-tube, we see from (21) that braking occurs at a rate equivalent to assuming that the magnetic field maintains strict co-rotation out to the Alfvénic surface. It should be noted that this result is not the same as asserting that the gas actually co-rotates with the star out to this surface: the gas will in fact always lag behind the star's rotation. It is the transport by the magnetic stresses and the gas flow jointly which yield a braking rate equivalent to co-rotation.

The angular velocity is now given by

$$\frac{\Omega}{\alpha} = \frac{\left(1 - \frac{\rho_c \tilde{\omega}_c^2}{\rho\tilde{\omega}^2}\right)}{\left(1 - \frac{\rho_c}{\rho}\right)} = \frac{\left(1 - \frac{v}{v_c}\right)}{\left(1 - \frac{v}{v_c}\frac{\tilde{\omega}^2}{\tilde{\omega}_c^2}\right)}, \qquad (22)$$

using equation (47) of I and dropping the suffix p. Near enough to the star--certainly within Parker's sonic point--the velocities are so low that both ρ_c/ρ and $\rho_c\tilde{\omega}_c^2/\rho\tilde{\omega}^2$ are small, so that $\Omega \simeq \alpha$--approximate (though not exact) co-rotation. If the centrifugal forces are weak all the way to S_A--by (6), essentially if $\alpha^2\tilde{\omega}_c^2/a^2 \ll 1$--then the flow has the characteristics of Parker's critical solution, with a very gentle acceleration once the gas has become supersonic (cf. equation (21) of I). The departure from co-rotation will therefore become noticeable well before S_A. If centrifugal forces are important they will cause a strong outward acceleration all the way to S_A, bringing S_A nearer to the star, but also yielding a larger region in which departures from co-rotation are small. Well beyond S_A (22) reduces to

$$\frac{\Omega}{\alpha} \simeq \left(1 - \frac{v_c}{v}\right)\left(\frac{r_c}{r}\right)^2 \qquad (23)$$

for nearly radial flow. The centrifugal terms in (6) are now seen to be negligible, and v increases slowly according to (21) of I. If we extrapolate the theory to infinity, then

$$\Omega r^2 \to \alpha r_c^{\,2} \qquad (24)$$

--the gas will have precisely the angular momentum it would have had if it were co-rotating with the star at S_A. In any realistic application, however, v_c/v will not be negligible, and the angular momentum of the gas at finite post-Alfvénic distances will be less than this. The value of Ω/α at the Alfvénic point can be found by use of l'Hôpital's rule--a simple procedure if the centrifugal terms in (6) can be ignored, but more difficult if the position of the Alfvénic point itself depends on Ω (see Weber and Davis (1967) for a discussion of the singularity of the Bernoulli integral (6) at the Alfvénic point).

The corresponding expression for B_t is

$$-\frac{\tilde{\omega}B_t}{4\pi} = \eta\alpha\tilde{\omega}_c^{\,2}\,\frac{\left(1 - \dfrac{\tilde{\omega}^2}{\tilde{\omega}_c^{\,2}}\right)}{\left(1 - \dfrac{\rho_c}{\rho}\right)} = \eta\alpha\tilde{\omega}_c^{\,2}\,\frac{\left(1 - \dfrac{\tilde{\omega}^2}{\tilde{\omega}_c^{\,2}}\right)}{\left(1 - \dfrac{v}{v_c}\dfrac{\tilde{\omega}^2}{\tilde{\omega}_c^{\,2}}\right)}. \qquad (25)$$

Again, near enough to the star

$$-\frac{\tilde{\omega}B_t}{4\pi} \simeq -\frac{\beta}{4\pi} = \eta\alpha\tilde{\omega}_c^{\,2} \qquad (26)$$

--effectively all the angular momentum is carried by the magnetic stresses. Far beyond the Alfvénic point (25) reduces to

$$-\frac{\tilde{\omega}B_t}{4\pi} \simeq \left(\eta\alpha\tilde{\omega}_c^{\,2}\right)\left(\frac{v_c}{v}\right). \qquad (27)$$

Again, extrapolation of the theory to infinity yields $\tilde{\omega}B_t \to 0$: ultimately none of the angular momentum would be carried by the field, as is required for consistency with the limit (24). But again at any realistic circumstellar distance--e.g. the orbit of the earth--v_c/v is not negligible, and in fact this simple theory predicts for the solar wind that the ratio v_c/v of angular momentum carried by the field to the total outflow is nearer unity than zero at the earth, although the Alfvénic points are well within the earth's orbit (cf. below).

The total rate of loss of angular momentum is given by integrating (21) over all the flux-tubes in the wind zone. The resulting formulae depend on the field structure and on whether the wind velocity near the Alfvénic surface is essentially thermally- or centrifugally-driven. In the simplest case, with all the quasi-radial field-lines partaking in the braking and with centrifugal forces small, the rate of loss is given by

$$-\frac{dJ}{dt} = -\Gamma(\alpha r_s^{\ 2})\frac{dM}{dt} = \frac{\left(B_s r_s^{\ 2}\right)^2}{a}\alpha I\ , \qquad (28)$$

where Γ and I are constants of order unity, α is again the stellar rotation, and -dM/dt is the rate of mass loss. It should be noted that there is a strong dependence on B_s but no dependence on the density factor ρ_s: an increase in ρ_s (with B_s given) reduces the radius of the Alfvénic sphere sufficiently to compensate for the increased mass loss. When the system has a dead zone, with only a fraction of the field-lines extended to infinity (Fig. 1), the analogue of (28) involves a weaker dependence on B_s, and some dependence on ρ_s.

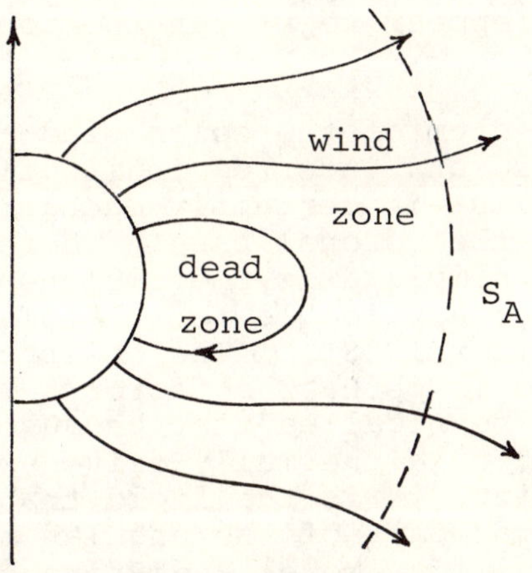

Fig. 1

In the most conservative case (Mestel 1968) with the field-lines dipolar all the way to S_A, then -dJ/dt is proportional to ρ_s but nearly independent of B_s (provided $B_s^{\ 2}/4\pi\rho_s a^2 >> 1$): an increase in B_s reduces the extent of the wind zone so much that it compensates for the increase in dimensions of S_A, so that a stronger field merely reduces the rate of mass loss required in order to achieve an angular momentum loss rate that is of order $8\pi\rho_s a\alpha r_s^{\ 4}$. In a realistic case we may expect a weaker but non-vanishing dependence on B_s than in (28), and some dependence on ρ_s. Centrifugal forces also introduce an explicit dependence on ρ_s. (For further discussion of the different cases, see Mestel 1968, 1969.)

The explicit dependence on α in (28) is linear, which would suggest a rather uncomfortable exponential braking law. However, it must be remembered that B_s will be independent of α only if the field is a decaying "fossil". If the field is dynamo-built, then its strength will almost certainly depend on α--possibly as strongly as α^2 (Cowling 1965). Further, we noted earlier that changes in the rotation rate of a rapidly rotating star can move the sonic point to regions of greater or lesser density, so altering the mass and angular momentum loss rates. A non-linear dependence on α gives the theory a good deal more flexibility.

Applications of the theory

1) The solar rotation

The most important application of the theory is to the braking of the solar rotation. Satellite observations (Wilcox 1968) have established a striking correlation between both the magnitude and sense of the interplanetary field near the earth and the field in the low-latitude solar regions (where the field is markedly stronger than in the polar regions). These observations and others of coronal structure at eclipse (Schatten 1969) seem also to confirm the general picture in which some emerging field-lines are strong enough to resist dragging out by the wind, returning instead to the photosphere; but the model with the field nearly curl-free all the way to the Alfvénic surface appear to be far too conservative (Schatten et al. 1968 a,b).

The angular momentum of the solar wind--found directly from satellite probes, and inferred indirectly from the orientation of comets' tails--is certainly much greater than that of gas emerging from the coronal base. There is, however, a discrepancy with the simple axisymmetric theory (which is certainly not strictly applicable to the solar field, with its sector structure (Wilcox 1968)). The observed azimuthal velocity of 10 km/sec is perhaps a factor 5 larger than that predicted. (It will be recalled that because of the very slow increase of the speed of a thermally-driven wind once it is through the sonic point, it is the field that should continue to transport the larger part of the angular momentum even in the post-Alfvénic regime.) Meyer and Pfirsch (1969) find a limited effect of the correct sign from pressure anisotropy along and across the field. Schubert and Coleman (1968) suggest that the emission of hydromagnetic waves from tilted sunspot fields more in the direction of the rotation of the sun than in the opposite direction is responsible.

The actual rate of braking of the sun is crucial for the controversy as to whether the interior of the sun rotates at essentially the same rate as the surface, or an order of magnitude more rapidly, as urged by Dicke (e.g. 1970), with its consequences for the theory of gravitation. Brandt's latest estimates (1970) give an e-folding time of 3-4 x 10^9 years for a rigidly rotating sun--comfortably close to the solar age; whereas if the torque decelerated only the convection zone the e-folding time would be a factor 10 shorter. However, it is certainly too early to quote this as a universally accepted figure.

Independent evidence for braking comes from Kraft's observations (1967) of the rapid surface rotation of G-type stars in the Pleiades and Hyades clusters, whose ages are estimated to be respectively 2 x 10^7 and 4 x 10^8 years. Further, as pointed out by Conti (1968), the observations do not appear to be consistent with a simple exponential decrease--the "e-folding time" is time-dependent. As already noted, such a non-linear dependence of the rate of braking on the instantaneous rotation is in at least qualitative accord with ideas on dynamo generation of the surface magnetic field.

2) The strongly magnetic stars

The strongly magnetic Ap stars are abnormally slow rotators, and it is immediately tempting to appeal to magnetic braking as the explanation. An A star does not have an extensive sub-photospheric convection zone, so that on current ideas it is unlikely to be surrounded by a hot corona. As long as the star rotates rapidly, the centrifugal forces will keep the sonic point deep enough down for a sizeable braking to occur; but as soon as the rotation drops, the sonic point in a cool corona will move out to a radius of low density, and the braking effectively disappears. Thus to explain the slow rotations of these stars we must appeal to the pre-main sequence Hayashi phase, when we have every reason to expect a hot corona and a strong thermally-driven wind, at least at the lower end of the Hayashi track. A star with a primeval field strong enough to withstand tangling by the Hayashi turbulence (cf. lecture II) may suffer so much excess braking that the increase of spin as the star subsequently contracts to the main sequence still leaves it rotating significantly more slowly than A stars born without primeval fields.

3) Close binary stars

We may reasonably expect that during the break-up of a gas cloud into a proto-star cluster some binary systems will be formed (cf. lectures 6-8); but as each star contracts towards the main sequence, the mutual distance will be unchanged as long as the system retains its angular momentum, and the result will be a wide binary system. The problem is how to form close or contact systems, such as the W Ursae Majoris and β-Lyrae systems. The alternatives seem to be either the break-up, presumably by rotational instability, of a star just before it reaches the main sequence; or loss of orbital angular momentum from a binary system of proto-stars; or possibly both processes.

The magnetic braking theory will be more complicated for a star in a binary system, because the flow now lacks an axis of symmetry, even if the stars are of equal mass. It is likely that the stresses exerted by the distorted magnetic field will now yield a net force as well as a couple on each star, so changing directly the orbital angular momentum. However, if we ignore this, and just apply the braking theory for single stars to each member of the binary system, it follows that the spin Ω_s of each star will begin to lag behind the orbital angular velocity $\bar{\Omega}$. Any process which tends to restore synchronization between $\bar{\Omega}$ and Ω_s will necessarily transfer orbital angular momentum into spin angular momentum, so that the two stars must come closer together. Magnetic coupling between the two stars may help (Huang 1966) but the classical process of synchrotization is "tidal friction" (Darwin 1908)

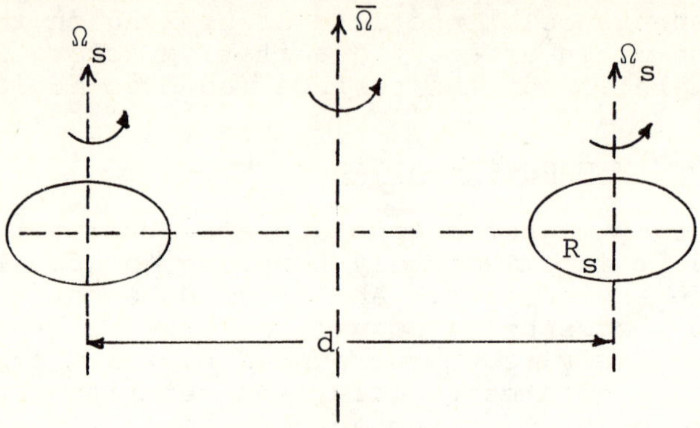

Fig. 2

In the absence of any friction, then as seen in a frame rotating with angular velocity $\bar{\Omega}$, there is a steady state in which each star is prolate with a common longest axis, and so with no mutual gravitational torque; but if $\bar{\Omega} \neq \Omega_s$, this structure is maintained by tidal motions of period $\pi/|\bar{\Omega} - \Omega_s|$ and with velocities of order (Cowling 1938)

$$R_s|\bar{\Omega} - \Omega_s|(R_s/d)^3 \ . \tag{29}$$

When friction is included, there develops a phase-lag between orbital and spin motion, with a consequent gravitational torque which tends to restore synchronization. A simple way of estimating the effect is to construct the friction-free mass-motions, and then compute the frictional energy dissipation. The result agrees with Darwin's estimate except for a factor of order unity. As the two stars approach each other, the gravitational energy released is destroyed by the friction. The effective coefficient of turbulent viscosity μ_t in the pre-main sequence convective phase can be estimated as $\mu_t = \rho v_t \lambda_t$, where v_t is the mean turbulent velocity, given by standard stellar structure theory, and λ_t is the mixing length. The time for approximate synchronization has been estimated as $(d/R_s)^6$ years, which should be compared with times of order 10^6 years spent in the Hayashi phase. Thus as long as the binaries stay close binaries, they should stay closely synchronized. The process therefore depends on the time of braking being at least as short as the time of contraction of individual stars. For further details, see Huang (1966), Mestel (1966, 1968) and Okamoto and Sato (1970). We may remark that it is significant that the old stars of low mass which rotate rapidly are mainly those in close binary systems, suggesting strongly that spin-orbit synchronization is maintained: according to our model, an essential part of the process of formation. Further, because centrifugal forces remain high in close binaries, the braking process probably works at its maximum efficiency.

REFERENCES

Brandt, J.C., and Heise, J. 1970, Astrophys. J., 159, 1057.

Conti, P. 1968, Astrophys. J., 152, 657.

Cowling, T.G. 1938, Mon. Not. R. Astr. Soc., 98, 734.

______. 1965, Stellar and Solar Magnetic Fields, ed. R. Lüst (Amsterdam: North-Holland Publishing Co.).

Darwin, G.H. 1908, Scientific Papers, 2 (Cambridge: Cambridge University Press).

Dicke, R.H. 1970, Internal Rotation of the Sun, Ann. Rev. Astron. Astrophys., (in press).

Huang, Su-Shu 1966, Ann. Astrophys., 29, 331.

Kraft, R.P. 1967, Astrophys. J., 150, 551.

Lüst, R., and Schlüter, A. 1955, Zs. f. Astrophys., 38, 190.

Mestel, L. 1966, Colloque d'Astrophysique, Liège, 351.

______. 1968, Mon. Not. R. Astr. Soc., 138, 359.

______. 1969, Colloque d'Astrophysique, Liège, Introductory Report (in press, 1970).

Meyer, F., and Pfirsch, D. 1969, Kleinheubacher Berichte, 13, 243.

Okamoto, I., and Sato, K. 1970, Pub. Astr. Soc. Japan (in press).

Schatten, K.H. 1969, Nature, 223, 652.

Schatten, K.H., Ness, N.F., and Wilcox, J.M. 1968a, Solar Phys., 5, 240.

Schatten, K.H., Wilcox, J.M., and Ness, N.F. 1968b, Series No. 9, Issue No. 26, Space Sciences Lab, University of California, Berkeley.

Schubert, G., and Coleman, P.J. 1968, Astrophys. J., 153, 943.

Weber, E.J., and Davis, L. Jr. 1967, Astrophys. J., 148, 271.

Wilcox, J.M. 1968, Space Sci. Rev., 8, 258.

Lecture V

MAGNETIC BRAKING AND THE OBLIQUE ROTATOR MODEL

We now want to examine the wind problem for rotating magnetic stars in which the axis of rotation and the axis of the magnetic field are not coincident. The results have application to the magnetic A stars, and an analogous theory in the relativistic regime should be relevant to current pulsar models.

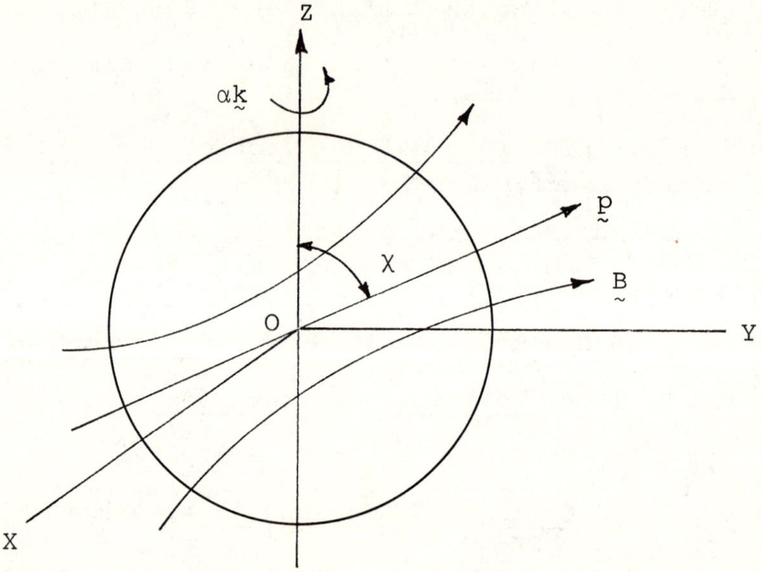

Fig. 1

In Fig. 1, OZ is the instantaneous axis of rotation, defined by the unit vector $\underset{\sim}{k}$; $\underset{\sim}{p}$ is the unit vector along the magnetic axis; OY is coplanar with OZ and $\underset{\sim}{p}$; and OX completes the right-handed triad. The magnetic field and so also the axis $\underset{\sim}{p}$ is frozen into the star, and so rotates with the star's angular velocity $\alpha\underset{\sim}{k}$ about OZ.

We seek solutions of the equations which are steady in the frame rotating with the star. The velocity in this frame is taken parallel to the field:

$$V_i = \kappa B_i, \tag{1}$$

consistent with the steady hydromagnetic equation. The equation of motion is

$$\frac{\partial}{\partial x_k}(\rho V_i V_k) + 2\alpha\varepsilon_{ijk}k_j\rho V_k = -\frac{\partial p}{\partial x_i} + \rho\frac{\partial \phi}{\partial x_i} - \frac{\partial T_{ij}}{\partial x_j} + \rho\varepsilon_{ijk}\alpha k_j\varepsilon_{klm}x_l\alpha k_m \tag{2}$$

where ε_{ijk} is the Levi-Civita alternating tensor. The first term incorporates the continuity equation

$$\frac{\partial}{\partial x_k}(\rho V_k) = 0 . \tag{3}$$

The second term is the Coriolis force $2\rho\alpha\underset{\sim}{k} \times \underset{\sim}{V}$; the third term is the pressure gradient; the fourth the gravitational force; the fifth is the magnetic force, written in terms of the Maxwell stress tensor $T_{ij} = \frac{B^2}{8\pi}\delta_{ij} - \frac{B_i B_j}{4\pi}$; and the last term is the centrifugal force

$$\rho\alpha^2\{\underset{\sim}{k} \times (\underset{\sim}{r} \times \underset{\sim}{k})\} \equiv \rho\alpha^2\{\underset{\sim}{r} - \underset{\sim}{k}(\underset{\sim}{r}\cdot\underset{\sim}{k})\} . \tag{4}$$

We want to find the net torque acting on the star, and so will have ultimately to make an attack on the equations. However, we first derive an integral result. Consider a volume τ, fixed in the rotating frame. Let S be the surface bounding τ, and let n_l be the outward normal

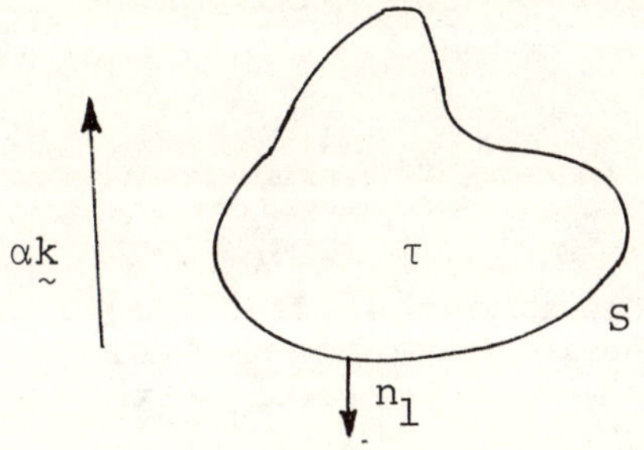

Fig. 2

Form the quantity,

$$\int_S F_{il} n_l ds \equiv \int_S \varepsilon_{ijk} x_j (T_{kl} + p\delta_{kl} + \rho V_l[V_k + \alpha(\underset{\sim}{k} \times \underset{\sim}{r})_k] n_l dS . \tag{5}$$

This is the outflow of the i-component of angular momentum across S due to magnetic stresses, thermal pressure and gas flow. Since the first three terms in the brackets are symmetric, those parts of the integral can be written

$$\int d\tau\varepsilon_{ijk}x_j\frac{\partial}{\partial x_l}\left(T_{kl} + p\delta_{kl} + \rho V_l V_k\right) , \tag{6}$$

which by the equation of motion (2) equals

$$\int d\tau\varepsilon_{ijk}x_j \left\{\rho\frac{\partial\phi}{\partial x_k} + \rho\alpha^2\left(\underset{\sim}{k} \times (\underset{\sim}{r} \times \underset{\sim}{k})\right)_k - 2\alpha\varepsilon_{kpl}k_p\rho V_l\right\} . \tag{7}$$

The centrifugal term in (7)

$$\int d\tau\varepsilon_{ijk}x_j\rho\alpha^2\left(\underset{\sim}{k} \times (\underset{\sim}{r} \times \underset{\sim}{k})\right)_k \tag{8}$$

reduces at once to

$$- \alpha\left[\underset{\sim}{k} \times \int \left(\underset{\sim}{r} \times \rho\alpha(\underset{\sim}{k} \times \underset{\sim}{r})\right)d\tau\right]_i . \tag{9}$$

The Coriolis term in (7)

$$-\int d\tau \varepsilon_{ijk} x_j 2\alpha \varepsilon_{kpl} k_p \rho V_l = -2\alpha \int d\tau \left(\rho V_l\right)\left(k_i x_l - k_j x_j \delta_{il}\right) \tag{10}$$

combines with the last term in the outflow integral (5)

$$\int_S \varepsilon_{ijk} x_j \rho V_l \alpha (\underset{\sim}{k} \times \underset{\sim}{r})_k n_l dS$$

$$= \alpha \int d\tau \rho V_l \frac{\partial}{\partial x_l}\left(\varepsilon_{ijk} x_j \varepsilon_{kpq} k_p x_q\right) \tag{11}$$

$$= \alpha \int d\tau \rho V_l \left[2k_i x_l - x_i k_l - k_p x_p \delta_{il}\right]$$

to yield

$$-\alpha\left[\underset{\sim}{k} \times \int (\underset{\sim}{r} \times \rho \underset{\sim}{V}) d\tau\right]_i . \tag{12}$$

The total outflow of the i-component of angular momentum is thus

$$\int_S F_{il} n_l dS \equiv \int_\tau d\tau (\underset{\sim}{r} \times \rho \nabla \phi)_i - \left(\alpha \underset{\sim}{k} \times \int \underset{\sim}{r} \times \rho (\underset{\sim}{V} + \alpha(\underset{\sim}{k} \times \underset{\sim}{r})) d\tau\right)_i . \tag{13}$$

The first term is the moment of the gravitational force density which vanishes if the gravitational field is radial. We will drop it in the present application, although it can be retained for higher order work.

To apply this transform, we take as our surface S a double surface consisting of S_1 at the base of the corona and S_A a general surface further from the star (later to be set at the Alfvénic surface).

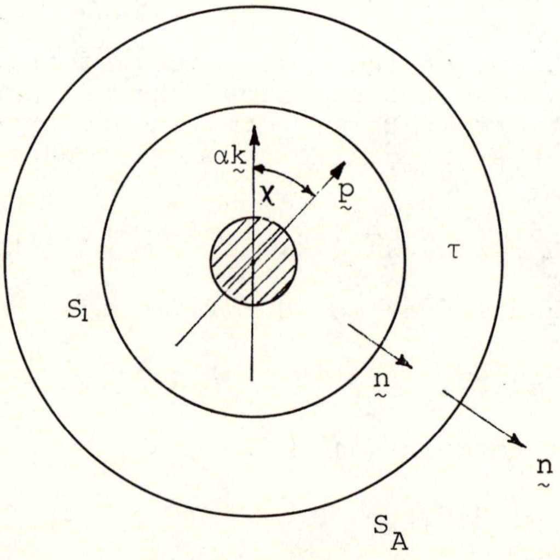

Fig. 3

Then

$$\int_{S_1} F_{il}n_l dS = \int_{S_A} F_{il}n_l dS + \left[\alpha k \times \int_\tau \underset{\sim}{r} \times \rho\left(\underset{\sim}{V} + \alpha(\underset{\sim}{k} \times \underset{\sim}{r})\right)d\tau\right]_i . \tag{14}$$

The term on the left hand side has had its sign changed because we are now taking the normal over S_1 away from the star. The second term on the right is a "flywheel term". The gas within τ has moment of momentum given by the volume integral, and an inertial observer sees this swung around at the rate $\alpha\underset{\sim}{k}$; this accounts for the difference between the angular momentum flow integrals over S_1 and S_A.

We now take S_A to be the Alfvénic surface, as defined by the velocity $\underset{\sim}{V}$ in the rotating frame. Thus on S_A,

$$4\pi\rho\kappa^2 = 1 \tag{15}$$

[κ being the scalar in equation (1)], and the torque due to the Reynolds stresses is balanced by that due to the magnetic tensions:

$$\int_{S_A} \varepsilon_{ijk}x_j n_l dS\left(\rho\kappa^2 - \frac{1}{4\pi}\right)B_k B_l = 0 . \tag{16}$$

Hence the total torque $\underset{\sim}{L}$ acting on the star is given by

$$\begin{aligned} -L_i = \int_{S_1} F_{il}n_l dS &= \left[\alpha\underset{\sim}{k} \times \int \underset{\sim}{r} \times \rho\left(\underset{\sim}{V} + \alpha(\underset{\sim}{k} \times \underset{\sim}{r})\right)d\tau\right]_i \\ &+ \int_{S_A} \left(p + \frac{B^2}{8\pi}\right)(\underset{\sim}{r} \times \underset{\sim}{n})_i dS \\ &+ \int_{S_A} \rho\underset{\sim}{V}\cdot\underset{\sim}{n}\left(\underset{\sim}{r} \times (\alpha\underset{\sim}{k} \times \underset{\sim}{r})\right)_i dS . \end{aligned} \tag{17}$$

The first term is the flywheel effect of the rotating gas between the coronal base and the Alfvénic surface. The second term is the moment of the thermal and magnetic pressures on the Alfvénic surface and disappears if the surface is spherical. The third term represents the angular momentum loss there would be if the gas had at the Alfvénic surface the angular velocity it had at the surface of the star. In the axisymmetric case only this last term survives, as seen earlier, and the total torque has only a Z-component. In a system of lower symmetry, all three terms survive, and the net torque has components in all three directions.

The angular momentum vector of the star stays fixed in space, apart from a negligible wobble with period $2\pi/\alpha$. The Z-component of torque again brakes the star's rotation. The X- and Y-components cause the instantaneous axis of rotation to precess through the star. In particular, the Y-component causes a precession about OX, so altering the angle χ between the rotation and magnetic axes, with a corresponding rotation in space of the frozen-in magnetic axis:

$$\dot{\chi} = - \left(\frac{L_Y}{L_Z}\right) \frac{d}{dt} [\log (C\alpha)] \qquad (18)$$

where C is the moment of inertia of the star about the rotation axis (Mestel and Selley 1970). The braking torque L_Z will depend only weakly on χ, but there will be special values of χ for which $L_Y = 0$. With the magnetic field at the stellar surface assumed symmetric about the magnetic axis $\underset{\sim}{p}$, then L_Y will vanish when the system is axisymmetric--$\chi = 0$--or equatorially symmetric--$\chi = \pi/2$ (Mestel 1968). One of these states will be unstable, in the sense that a small departure of χ from either 0 or $\pi/2$ will lead to torques that increase the departure. The principal motivation of this work is the chance that the same magnetic coupling may be responsible for both the abnormally slow rotations of the magnetic variable stars and for the large values of χ required if the oblique rotator model is to account for periodic field reversal. The hope is that the state with χ large will be the stable one, and that for general χ the torque ratio L_Y/L_Z will be large enough for an observationally reasonable loss in angular momentum substituted in (18) to cause a sizeable change in χ.

To compute L_Y/L_Z we need to solve the equations in some order of approximation. We still have a Bernoulli integral in the rotating frame:

$$\frac{1}{2}V^2 + a^2 \log \rho - \phi - \frac{1}{2}\alpha^2\tilde{\omega}^2 = \text{constant on field-streamlines}, \qquad (19)$$

where a is again the isothermal sound speed, and the other symbols have their usual meanings; also, by conservation of mass and magnetic flux

$$\rho\left({}^{V}\!/_{B}\right) = \text{constant}\ . \qquad (20)$$

Thus

$$\frac{1}{2} V^2 - a^2 \log V = \frac{GM}{r} - a^2 \log B + \frac{1}{2} \alpha^2\tilde{\omega}^2 + \text{constant} \qquad (21)$$

along each field-line, and we have Parker-type flow, with the constant for each field-line fixed by conditions at the sonic point, and with the gas speed small near the star and accelerating steadily outwards under thermal pressure and centrifugal force. But the rest of the problem is more difficult than the axisymmetric problem, because we no longer have a torque integral analogous to (9) of the previous lecture. This integral describes a steady transport of the Z-component of angular momentum along each field-streamline without any transfer from one field-line to another. In the oblique rotator problem, however, there is in general an interchange of <u>all three</u> components of angular momentum from one field-line to another by the action of the magnetic and thermal pressure gradients, so that the axisymmetric torque integral is replaced by a partial differential equation. (And of course there remain the difficulties associated with an accurate computation of the field structure even in the absence of rotation).

To make further progress we resort to a perturbation procedure. As the zero-order approximation we take the basic magnetic field of the star to be radial and independent of angle, except for changing sign at the magnetic equator. (This implies a pinched distribution of gas along the equator,

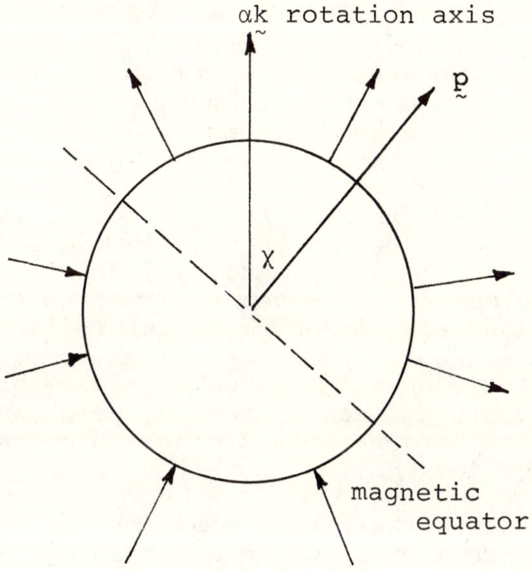

Fig. 4

exerting a pressure that keeps the oppositely-directed field-line from diffusing into each other and changing the field structure.) The rotation α is supposed slow enough for terms of order α^2 to be negligible, so that the wind is driven by thermal pressure only. Under these conditions the Alfvénic surface is a sphere of radius r_C. The perturbation $\underset{\sim}{B}'$ to the field of order $\alpha r_C/a$--the analogue of the toroidal field of the axisymmetric problem--can be found fairly easily. One can verify directly by integrating the Maxwell stresses over a surface S_1 near the star, or indirectly from the transform (17), that the precessional torques vanish to order $\alpha r_C/a$.

We then perturb the solution by allowing the field at the surface of the star to have an angle-dependent part of order ε. If we expand the flux distribution in surface harmonics, then a long and tedious calculation will yield the contribution to the field of order $(\alpha r_C/a)\varepsilon$, and from this we can compute a non-vanishing torque L_Y, proportional to $\sin 2\chi$. However, it is much simpler to use the transform (17), for the pressure and "flywheel" terms are of order ε^2, and both L_Y and L_Z are given to order $(\alpha r_C/a)\varepsilon$ by

$$- \int_{S_A} \rho(\underset{\sim}{V}\cdot\underset{\sim}{n})(\underset{\sim}{r} \times (\alpha\underset{\sim}{k} \times \underset{\sim}{r}))dS, \qquad (22)$$

just as in the axisymmetric case. The ρ and $\underset{\sim}{V}$ fields are required to the zero-order in $\alpha r_c/a$--i.e., they are given by the stellar wind theory for a non-rotating magnetic star, with the flux distribution over the stellar surface prescribed. Full details of both analysis and computations are given in Mestel and Selley (1970).

The principal qualitative result is that if the surface flux distribution is symmetric about the magnetic axis, then the precessional torque causes the instantaneous axis of rotation to seek out the region where the flux is strongest. Thus if the part of order ε is more concentrated to the magnetic poles, the two axes tend to align; if to the equator, the two axes tend to become orthogonal. Qualitative arguments suggest that this is a general result at least for cases where centrifugal forces are negligible.

For our doubly perturbed split monopole field, it is found that the ratio $|L_Y/L_Z|$ is usually too small for a sizeable change in χ to be associated by (18) with a plausible loss of angular momentum. However, approximate computations (Selley 1970) done assuming more plausible field structures show that this ratio can be large enough for the effect studied to be significant. Speaking roughly, the greater the departure of the field from spherical symmetry, the closer is the ratio to unity, and the larger the associated precession for a given angular momentum loss.

It is now clear that the present theory cannot by itself produce oblique rotators. Given a specific surface flux distribution, we can (in principle) determine whether coupling with a wind yields a precessional torque of the right sign and order of magnitude; but the explanation of a "favorable" flux distribution must be sought in the internal stellar magneto-hydrodynamics, as discussed in an earlier lecture. We may tentatively note that an inexorable rotationally-driven circulation would tend to concentrate any primeval flux towards the equator (defined by the rotation axis). If the star is emitting a powerful wind, then an aligned magnetic field would probably be unstable against the action of our precessional torque.

However, once a magnetic A-star has reached the main sequence, lost its convective envelope and the consequent powerful wind, then there is an argument against a permanently oblique rotator (Spitzer 1958). With the magnetic axis not coincident with the rotation axis, the body will not be symmetric in structure about its angular momentum axis, so that it undergoes a quasi-Eulerian nutation. Each element of gas moves subject to a fluctuating pressure, but the motions are not strictly adiabatic, because of dissipation by radiative conduction. This steady dissipation continues until the system (with fixed angular momentum) has reached a state of minimum energy--i.e., until the star rotates about the axis of maximum moment of inertia, with the magnetic axis rotated so as to coincide with the angular momentum vector. One would like an estimate of the time-scale of this effect.

The basic argument for a precessional torque component depends only on the system having a low enough symmetry for the tensions along the distorted magnetic field-lines to have a net non-vanishing moment about the axis OY in Fig. 1. We may therefore anticipate that a relativistic analogue of the present theory will again yield a precessional torque, and that the sign of the associated change in χ will depend critically on the distribution of flux over the pulsar surface. The question is particularly significant, if the coherent radiation mechanism requires a fluctuating macroscopic charge density and so also a finite obliquity.

REFERENCES

Mestel, L. 1968, Mon. Not. R. Astr. Soc., 140, 177.

Mestel, L. and Selley, C.S. 1970, Mon. Not. R. Astr. Soc., 149, 197.

Selley, C.S. 1970, in preparation.

Spitzer, L. Jr. 1958, in Electromagnetic Processes in Cosmical Physics, ed. B. Lehnert, (Cambridge: Cambridge University Press).

Lecture VI

STAR FORMATION I: GRAVITATIONAL COLLAPSE, FRAGMENTATION AND ACCRETION

In the last three lectures we discuss some of the problems that arise when we attempt to understand the formation of stars from the interstellar medium. We begin by considering the mutual interaction of self-gravitation and thermal pressure in diffuse gas clouds. It will become clear in lectures VII and VIII that neglect of the essentially anisotropic centrifugal and magnetic forces is certainly not justified; however, we have no hope of understanding the complex realistic problem without first studying gravitational collapse and fragmentation in the simplest context.

Interstellar gas clouds

We adopt provisionally the model of the interstellar medium and cloud formation developed by Field (1965, 1969) and Pikel'ner (1967). A crucial role in the thermal balance is played by cosmic rays (especially by the low-energy extrapolation of the spectrum) which ionize and heat up the gas (Hayakawa *et al*. 1961; Field 1962; Pikel'ner 1967).* The principal cooling process is by collisional excitation of hydrogen and helium at high temperatures, and of carbon, oxygen and silicon ions at low temperatures. Calculations of thermal equilibrium have been made by Pikel'ner (1967); Spitzer and Tomasko (1968); Spitzer and Scott (1969) and by Field, Goldsmith and Habing (1969 a,b). According to the last references, there exist three phases in pressure equilibrium, of which the hottest and coolest are thermally stable and so can persist (Field 1965). Assuming a cosmic ray ionization rate of 4×10^{-16}/sec, Field (1969) gives the following estimates: the intercloud medium has a temperature near 8000°, a density of $\simeq$ .2 heavy particles/cm^3, and an electron density of $\simeq$ .016/cm^3; the nearly neutral ("HI") clouds have temperatures below 300° and densities $\geq$ 10/cm^3.

In the most massive and densest clouds the conditions of thermal balance are altered somewhat, chiefly by the increased optical depth due to dust grains (Spitzer 1968; Field 1969), the general effect being to lower the temperature. Cooling by direct collision of particles with the grains is important at high densities. The grains act as loci for the formation of molecules, which are good coolants, and are saved from dissociation by extinction of the galactic ultraviolet field

*More recently it has been suggested (Pikel'ner--private communication) that cosmic X rays may be of comparable or greater importance.

by the grains. Further, as the mass per unit area through a cloud increases, the cosmic rays (especially those of low energy) are absorbed. As against this, coolant atoms and ions are depleted by attachment to the grains, while the process of molecule formation injects the binding energy into the gas. Field (1969) emphasizes that we still lack really reliable temperature estimates for massive clouds.

A lower limit of a fraction of a solar mass is set by thermal conduction, which clearly acts so as to wipe out temperature differences. Typical observed HI clouds have masses between $M_\odot$ and 100 $M_\odot$, densities of order $10 m_H/cm^3$ and temperatures near 100°. For these clouds self-gravitation is negligible, and if magnetic forces could be ignored (cf. lecture VIII), mechanical equilibrium would be just a balance of thermal pressures, as described above. However, inelastic collisions between clouds (Field et al. 1968) lead to the formation of a few massive clouds for which self-gravitation becomes steadily more important. Ebert (1955), Bonnor (1956) and McCrea (1957) have studied the equilibrium of self-gravitating isothermal spheres under external pressure (in our case, the pressure of the Field-Pikel'ner intercloud medium). Their results show that there is an upper limit to the mass of a cloud of given temperature and mean density, able to maintain itself in equilibrium against its self-gravitation and a fixed external pressure. Alternatively, we may say that a cloud of given mass may be brought to the point of gravitational collapse by an increase in external pressure, a reduction in temperature, or both. It is very plausible that gravitationally-bound clouds with masses of 10^3-10^4 $M_\odot$ ultimately form proto-star clusters; though it is apparently not ruled out that conditions may locally favor the formation of individual stars from clouds of low mass and very low temperature (Ebert 1955; Herbig 1969).

Gravitational collapse and fragmentation

For simplicity we shall assume that the cloud collapses spherically. When collapse starts, the mean density ρ_0, radius R_0 and temperature are related by a rough balance between thermal and gravitational energy per unit mass:

$$a^2 \equiv \frac{\mathcal{R}T}{\mu} \simeq \frac{GM}{R_0} = \frac{4\pi}{3} G\rho_0 R_0^2, \tag{1}$$

where a is the (isothermal) sound speed. The initial collapse will be nearly isothermal, since the compressional heat generated will be rapidly radiated away, and the ratio of thermal energy to gravitational energy steadily decreased. This has two distinct consequences. First of all, the bulk of the gravitational energy released is converted into macroscopic kinetic energy of collapse, very little being thermalized and radiated away. Secondly, it becomes possible for sub-masses within the cloud to be gravitationally-bound (Hoyle 1953). Thus when the cloud radius has shrunk to R and its density increased to ρ, with

$$\rho = \rho_0 \left(\frac{R_0}{R}\right)^3 , \tag{2}$$

the analogue of condition (1) applied to a sub-mass M' of radius R' is

$$\frac{4\pi}{3} G\rho R'^2 \simeq a^2 \simeq \frac{4\pi}{3} G\rho_0 R_0^2 , \tag{3}$$

so that

$$\left(\frac{R'}{R}\right)^2 = \left(\frac{R}{R_0}\right)^3 \left(\frac{R_0}{R}\right)^2 = \left(\frac{R}{R_0}\right) \tag{4}$$

and

$$M' = \frac{4\pi}{3} \rho R'^3 = \frac{4\pi}{3}\left(\frac{\rho_0 R_0^3}{R^3}\right)\left(\frac{R^3}{R_0}\right)^{3/2} = M\left(\frac{R}{R_0}\right)^{3/2} . \tag{5}$$

For example, if $R/R_0 \simeq 1/3$, then $R'/R \simeq 1/\sqrt{3}$ and $M'/M \simeq 1/5$: the cloud "can" break up into five gravitationally-bound "fragments." Hoyle (1953) pictures a hierarchical break-up process, with each successive generation of fragments becoming unstable as they contract isothermally.

Let us now suppose that simultaneous with this break-up the bulk of the kinetic energy of inward collapse, fed by the release of gravitational energy, is not only conserved but is also steadily randomized, so that we may think of the cloud of fragments as behaving ultimately like a monatomic gas and exerting the corresponding "pressure." The radius $\overline{R}$ of this proto-star cluster will be given by the virial theorem for a $\gamma = 5/3$ gas, relating the gravitational and kinetic energies $\overline{V}$ and $\overline{T}$, and by the condition of energy conservation; respectively

$$2\overline{T} + \overline{V} = 0 \tag{6}$$

and

$$\overline{T} + \overline{V} = V_0 , \tag{7}$$

so that

$$V_0 = \overline{V}/2 , \tag{8}$$

implying

$$\overline{R} \simeq R_0/2 . \tag{9}$$

This is a very important result, in spite of the obvious simplifications made, for a satisfactory model of the formation of star clusters must yield an average density that is not much greater than that of dense clouds, in accord with observation.

The result (9) depends critically on assuming small energy dissipation, which in turn depends on break-up of the collapsing cloud into high density fragments of small mutual collision cross-section. By contrast, if the cloud remains a more-or-less uniform gas, then we may expect continuous dissipation of energy, e.g., in shock formation. We therefore regard the dynamics of the fragmentation problem as the primary issue, and formulate the first question: is a cool collapsing cloud unstable against the formation of sub-condensations? We emphasize the contrast with the classical Jeans problem of gravitational instability in a uniform medium at rest. It is well known (e.g., Mestel 1965) that this formulation (sometimes irreverently referred to as the "Jeans swindle") is not strictly self-consistent, for the minimum unstable length-scale--the "Jeans length"--is necessarily comparable with the scale-height of a gas cloud in equilibrium. We are now not asking whether a small disturbance in a stationary medium will amplify, but whether a local density perturbation can grow more rapidly than the density of the collapsing background.

The following treatment illustrates how fragmentation can occur (Mestel 1965). The equation of motion of a cold, spherically symmetric gas sphere can be written

$$\ddot{r} = -\frac{Gm}{r^2}, \tag{10}$$

where the Lagrangian coordinate m is the mass within the sphere r, with

$$r = r(m,t). \tag{11}$$

For simplicity we suppose the whole sphere at rest at time t = 0, with

$$r(m,0) \equiv r_0(m), \quad \left(\frac{\partial}{\partial t} r(m,t)\right)_{t=0} = 0. \tag{12}$$

Then (10) integrates to

$$\dot{r}^2 = 2Gm\left(\frac{1}{r} - \frac{1}{r_0}\right), \tag{13}$$

which in turn has the parametric solution

$$r(m,t) = r_0(m)\cos^2\theta, \tag{14}$$

$$\theta + \frac{1}{2}\sin 2\theta = \left(\frac{2Gm}{r_0^3}\right)^{\frac{1}{2}} t = \left(\frac{8\pi}{3} G\bar{\rho}_0(m)\right)^{\frac{1}{2}} t, \tag{15}$$

with $\bar{\rho}_0(m)$ the mean density within $r_0(m)$ at t = 0. Thus the mass-shell m would achieve zero radius when $\theta = \pi/2$, i.e., at time

$$\tau(m) = \left[\frac{3\pi}{32G\bar{\rho}_0(m)}\right]^{\frac{1}{2}}. \tag{16}$$

For a cloud with an initially uniform density ρ_0

$$\overline{\rho}_0(m) = \rho_0, \tag{17}$$

and

$$\theta = \theta(t), \quad \frac{r(m,t)}{r_0(m)} = f(t), \tag{18}$$

$$\frac{\rho(m,t)}{\rho_0} = \sec^6\theta = F(t): \tag{19}$$

the whole cloud collapses homologously, the density remaining uniform, and with the free-fall time the same for all mass-spheres. Now suppose there is a spherical density fluctuation concentric with the cloud as a whole: i.e., let the initial density be again ρ_0 except for a small central region where it is $(\rho_0 + \delta\rho_0)$. Over the bulk of the cloud the initial mean density within r_0 is hardly changed, so that the solution given by (14) and (15) continues to apply. For the central blob, however, we may write

$$\rho + \delta\rho = (\rho_0 + \delta\rho_0)\sec^6\phi \tag{20}$$

with

$$\phi + \frac{1}{2}\sin 2\phi = \left[\frac{8\pi}{3} G\rho_0\left(1 + \frac{\delta\rho_0}{\rho_0}\right)\right]^{\frac{1}{2}} t. \tag{21}$$

If $\delta\rho_0/\rho_0 << 1$, then the collapse time of the central blob is hardly shorter than that of the cloud as a whole, but near the end of the collapse the blob density increases much more rapidly than the mean--a consequence of the inverse square law. To describe the asymptotic behavior, we write

$$\theta = \frac{\pi}{2} - \chi, \qquad \chi, \psi << 1, \qquad \phi = \frac{\pi}{2} - \psi, \tag{22}$$

so that

$$\left(\frac{\rho + \delta\rho}{\rho}\right)_t = \frac{(\rho_0 + \delta\rho_0)\sec^6\phi}{\rho_0\sec^6\theta} \simeq \left(1 + \frac{\delta\rho_0}{\rho_0}\right)\frac{\chi^6}{\psi^6}, \tag{23}$$

and (15) and (21) reduce to

$$\frac{\pi}{2} - \frac{2}{3}\chi^3 + O(\chi^5) = \left(\frac{8\pi}{3} G\rho_0\right)^{\frac{1}{2}} t, \tag{24}$$

$$\frac{\pi}{2} - \frac{2}{3}\psi^3 + O(\psi^5) = \left(\frac{8\pi}{3} G\rho_0\right)^{\frac{1}{2}}\left(1 + \frac{1}{2}\frac{\delta\rho_0}{\rho_0}\right) t. \tag{25}$$

Elimination of t yields

$$\psi^3 = \chi^3 - \frac{3\pi}{8}\left(\frac{\delta\rho_0}{\rho_0}\right), \tag{26}$$

so that (23) becomes

$$\left(\frac{\rho + \delta\rho}{\rho}\right)_t = \left(1 + \frac{\delta\rho_0}{\rho_0}\right)\Bigg/\left(1 - \frac{3\pi}{8\chi^3}\frac{\delta\rho_0}{\rho_0}\right)^2. \tag{27}$$

The upshot is that the relative amplitude becomes large when

$$\chi \simeq \left(\frac{3\pi}{8}\frac{\delta\rho_0}{\rho_0}\right)^{1/3} \ll 1, \tag{28}$$

i.e., very near the end of the collapse, when

$$\frac{r(m,t)}{r_0} = \cos^2\theta \simeq \chi^2 \simeq \left(\frac{3\pi}{8}\frac{\delta\rho_0}{\rho_0}\right)^{2/3}. \tag{29}$$

For example, if $\delta\rho_0/\rho_0 \simeq 10^{-3}$--a very modest fluctuation--(29) requires $r/r_0 \simeq 10^{-2}$: the cloud would need to shrink to 1 per cent of its initial radius for the blob density to have amplified.

A rigorous analysis along these lines has been given by Hunter (1962, 1964), without the perturbation being restricted to being concentric with the collapsing sphere. With the temperature finite, some account must be taken of the pressure gradients, which tend to restore stability. The temperature variation in a blob will in fact be determined by the difference between the compressional heat generated and the increased radiation loss. In a simplified analytical treatment Hunter assumes an adiabatic variation with a value of γ that is close to unity in the early, diffuse phases, but which will approach 5/3 at sufficiently high densities. As is to be expected from fundamentals of stellar structure theory, the pressure-free solution is asymptotically valid for $1 < \gamma < 4/3$, while for $\gamma > 4/3$ the perturbation's growth is ultimately reversed and the solution becomes oscillatory.*

A good deal of numerical work has been done recently on the collapse of non-uniform spheres, with various assumptions about the temperature variation (McNally 1964; Disney et al. 1968; Penston 1969 a,b; Hunter, J. H. 1969; Bodenheimer 1969; Larson 1969). All workers confirm that an initial central density condensation grows relative to the collapsing background. Penston points out that even though the sound speed and so also the total thermal energy are small, the pressure gradient $\nabla p = a^2\nabla\rho$ may still be important locally, since the

*For a critique of the fragmentation process and a reply, see Layzer (1963) and Hunter (1964).

small coefficient a^2 multiplies the term with the highest derivative; and in fact the integrations show that the pressure gradient is important in the central regions, reducing but not removing the tendency for central condensation. Of particular interest is one set of Penston's models which simulate the spherical collapse of a proto-galaxy out of hot ionized intergalactic gas, the principal cooling processes being bremsstrahlung and radiative recombination, and intergalactic cosmic rays the source of heat.

It should be noted that even though we have explicitly neglected the essentially anisotropic forces (magnetic and centrifugal), it does not therefore follow that isotropic collapse is the rule. On the contrary, it can be easily shown (Lin, Mestel and Shu 1965; Mestel 1965) that the collapse of a uniform sphere as described by equations (14), (15), (18) and (19) is essentially unstable to disturbances that leave the density uniform but change the shape into an ellipsoid. In particular, an oblate spheroid tends to become a "pancake" and a prolate spheroid a "cigar." However, qualitative arguments suggest strongly that local condensations of scale well above the instantaneous Jeans length will again spontaneously amplify during collapse.

As the density of a gas cloud (or of a sub-condensation within it) increases, the optical depth in the frequencies radiated by the cooling agents exceeds unity, and a temperature gradient is built up during collapse, sufficient to drive out the compressional heat generated. Free-fall accompanied by some fragmentation will continue as long as the outward pressure gradient associated with this temperature gradient is small compared with the gravitational force density. But once internal pressure increases more rapidly than a $\gamma = 4/3$ law, then gravitational energy released during collapse of a fragment is no longer converted into macroscopic kinetic energy, and fragmentation must stop; instead, the inward collapse is decelerated by the growing pressure field. This transition naturally occurs first in the central regions, which run ahead of the rest of the star during free-fall [cf. (15)], and so become opaque first: the core of the fragment "bounces" (Hayashi and Nakano 1965). The interaction between the bouncing core and the contracting outer regions gives rise to shock waves in which the macroscopic kinetic energy is thermally dissipated. Ultimately the proto-stellar "fragment" is converted into a pre-main sequence "star"--an opaque body nearly in hydrostatic equilibrium, with the energy transport (by radiation and convection) from center to surface yielding a time of contraction down the Hayashi track that is long compared with the free-fall time.

From a detailed study of opacities in the temperature range 50-3000°, Gaustad (1963) found that fragments of mass greater than $M_\odot/10$ could never achieve hydrostatic equilibrium in this phase--the energy loss is too high to maintain the necessary internal temperature. At slightly higher temperatures collisional ionization of hydrogen begins, and γ falls to near unity: hence although the supply of free electrons increases the opacity, the microscopic kinetic energy does not grow sufficiently for collapse to be halted (Cameron 1962). Once the bulk of the gas is fully ionized, γ returns to near 5/3 and the proto-star can achieve stable hydrostatic equilibrium (following dissipation of excess energy). The significant conclusion is that if fragmentation occurs with maximum efficiency, being halted only when hydrogen ionization is nearly complete, then the bulk of Type I stars should have masses near .2 $M_\odot$ It may be the case that most stars do form by some break-up process--certainly more complicated than simple fragmentation--that is indeed cut off only by the ultimate growth of opacity, while special conditions are required to build the stars of moderate and large mass that emit most of the observed light.

Accretion

Consider now a star or proto-star, kept in hydrostatic equilibrium by thermal pressure, and with any pre-main sequence contraction controlled by the comparatively slow rate of energy loss. If the diffuse gas remaining around the star is not heated to a high temperature like the solar corona, then instead of being driven away as a stellar wind (cf. lecture III), it will tend to flow into the star. This spherical accretion problem was in fact solved before the solar wind theory was developed (Bondi 1952). Equations (16) and (17) of lecture III are rewritten

$$4\pi\rho v r^2 = A = \text{accretion rate}, \tag{30}$$

and

$$a^2 \log\left(\frac{\rho}{\rho_\infty}\right) + \frac{1}{2} v^2 - \frac{GM}{r} = 0, \tag{31}$$

where again isothermality with sound speed a is assumed, and the appropriate boundary conditions at infinity

$$\rho = \rho_\infty, \qquad v = 0 \tag{32}$$

are used. Non-dimensional quantities are defined by

$$\frac{\rho}{\rho_\infty} = z, \quad v = ya, \quad r = r_B x, \tag{33}$$

$$r_B = \left(\frac{GM}{a^2}\right), \tag{34}$$

so that (30) and (31) reduce to

$$x^2 yz = \lambda, \tag{35}$$

$$A = 4\pi\rho_\infty a\left(\frac{GM}{a^2}\right)^2 \lambda, \tag{36}$$

and

$$\left(\frac{1}{2} y^2 - \log y\right) - \left(\frac{1}{x} + 2 \log x\right) = -\log \lambda \qquad (37)$$

--cf. with equation (18) of lecture III. In Parker's critical solution, gas that is effectively in hydrostatic equilibrium near the star is steadily accelerated outwards so that its speed becomes sonic at the critical point $r_a = GM/2a^2$. For the accretion problem the analogue of this is Bondi's critical solution, with the gas moving inwards from rest at infinity, reaching the sound speed at $x = \frac{1}{2}$--i.e., at $r = r_B/2 = GM/2a^2$--and flowing supersonically into the star. From (37) the critical value of λ is $\lambda_c = e^{3/2}/4$. There are no steady state solutions with $\lambda > \lambda_c$, so that the accretion rate is a maximum for $\lambda = \lambda_c$. If $\lambda < \lambda_c$, there are smoothly varying solutions which are subsonic everywhere, with the speed achieving its maximum at $x = \frac{1}{2}$, and which near the star again hardly differ from hydrostatic equilibrium. For $\lambda = \lambda_c$ there is also a solution subsonic everywhere except at $x = \frac{1}{2}$, but with a discontinuity in the acceleration at this point (cf. Fig. 1).

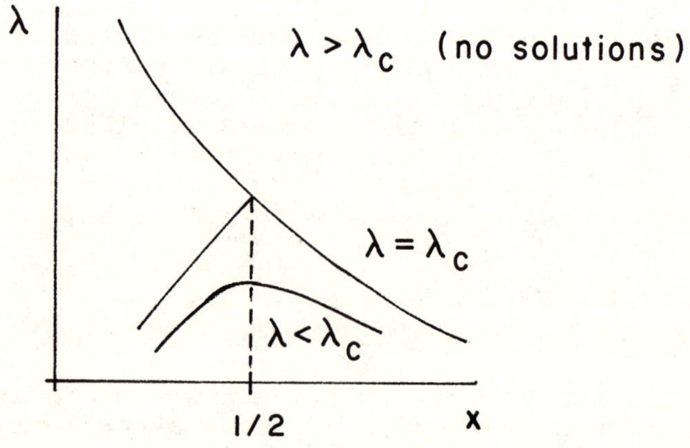

Fig. 1

However, when $r_B \gg$ the stellar radius R--implying from (34) that the temperature of the surrounding gas is well below the mean temperature through the star--then all these subsonic solutions have a density field varying like $\exp(r_B/r)$ and which therefore becomes unrealistically large near the star. The boundary condition requiring the flow of cool gas near the star to reduce to free-fall suffices to select the critical Bondi solution.

The flow has the following general characteristics. For $r \ll r_B$

$$v^2 \simeq \frac{2GM}{r}, \qquad (38)$$

$$\rho \propto r^{-3/2}, \qquad (39)$$

with pressure negligible until

$$1 \simeq \frac{-a^2\rho'/\rho}{GM/r^2} = \frac{\frac{3}{2}\,a^2}{GM/r}\,, \tag{40}$$

i.e., until r approaches r_B. Beyond r_B the flow is subsonic, with

$$a^2\rho' \simeq -\frac{GM\rho}{r^2}\,, \tag{41}$$

$$\frac{\rho}{\rho_\infty} \simeq \exp(r_B/r) \simeq 1, \text{ if } r >> r_B, \tag{42}$$

and

$$v \propto 1/r^2. \tag{43}$$

From (36) we see that the accretion rate is given roughly by assuming that gas with density ρ_∞ crosses the Bondi radius at the sound speed.

If the flow is not isothermal but adiabatic with the appropriate value of γ, the flow is qualitatively similar, except when $\gamma = 5/3$. For this limiting case the critical point goes to the origin, and the pressure gradient remains comparable with gravity all the way to the star.

The gravitational pull of the gas surrounding the star has been ignored. This is valid if

$$\frac{1}{M}\int_0^{r_B/2} 4\pi\rho r^2 dr << 1, \tag{44}$$

which from (39) and (36) is equivalent to

$$\frac{M}{\dot{A}} >> \left(\frac{GM}{a^2}\right)\Big/a: \tag{45}$$

the characteristic time of mass increase must be much greater than the time of flow from the sonic point to the star. This condition is easily satisfied in stellar cases, though not necessarily in all applications, e.g., in some quasar theories (Sturrock and Feldman 1968).

Granted the assumptions of the above model--the star at rest in a cool cloud with negligible magnetic energy and angular momentum--we expect the gas to settle into an accretion flow. We remember [cf. equations (13) and (14) of lecture III] that the condition for hydrostatic equilibrium of gas surrounding the star requires that the density and pressure fall off exponentially to a finite value at infinity. The argument for a stellar wind in this case depends on the coronal temperature being so high that the pressure far from the star greatly exceeds that in a normal HI cloud: the ram pressure of the expanding corona is able to drive away any surrounding cool gas. But, equally, if the temperature of the corona is substantially lowered, the scale-height becomes so small that the coronal pressure is quite unable to hold out the HI gas, which will therefore flow into the Bondi sphere and so into the star's potential well. [This, of course, assumes that there is no magnetically-controlled centrifugal field--cf. equation (6) of lecture IV.] Hunt (1970) has given a general treatment of the conditions under which a polytropic gas, subject to pressure boundary conditions at the star and at infinity, will adjust, respectively, to hydrostatic support, inflow, or outflow. He confirms that only high-temperature coronas can be expected to drive a wind. Stars with extensive sub-photospheric convection zones will almost certainly have hot coronas. However, in the opinion of some workers in the field, it is premature to draw the reverse conclusion that a star with a weakly convective envelope will not have a hot expanding corona: some believe that there will always be enough energy input to make the stellar wind a universal phenomenon. Thus the issue of accretion versus stellar wind must be considered open, even for early-type stars.

The problem is greatly complicated if the star is moving through the cloud, especially if its velocity is supersonic. The earliest treatments (Hoyle and Lyttleton 1939; Bondi and Hoyle 1944) assumed that gas pressure could be completely ignored, but they also recognized that the consequent flow necessarily had a singular region in which the pressure-free approximation must break down--the "accretion axis." There have been attempts recently to produce a proper gas-dynamical treatment of this problem (Ruderman and Spiegel 1966; Hunt 1970). Hunt's computer solution for the case $\gamma = 5/3$ has a bow shock ahead of the star, with the streamlines refracted at the shock either extending to infinity, or flowing into the star. The high value of γ prevents the density and velocity fields near the star from departing too much from spherical symmetry. It is not clear whether all the features of the solution--in particular, the bow shock ahead of the star and the absence of an "accretion shock" behind the star--will persist if γ falls much below 5/3. Hunt confirms Bondi's conjecture (1952) that the accretion rate is given by the formula (36) with the sound speed a replaced by $(a_\infty^2 + V^2)^{\frac{1}{2}}$, where a_∞ is the sound speed at infinity and V the speed of the star through the cloud. Thus if V is highly supersonic, the accretion rate is sharply reduced. The energy dissipation at the shock corresponds to a drag on the star, which steadily reduces V. As emphasized by McCrea (1953), if V is initially comparable with a_∞ in a cool cloud, the drag associated with the high accretion rate will be able to reduce the star to rest before it has passed through the cloud, so that the subsequent accretion should be described by the spherically symmetric theory.

The relevance of accretion on either a stellar or galactic scale clearly depends both on the plausibility of the assumptions made and on the application in mind. Recent attention has focused on galactic applications. The inwardly moving clouds observed at high galactic latitudes (Oort 1970) may be plausibly interpreted as tracers of galactic accretion (Hunt 1970). The density concentration at an accretion shock could be the source of soft X rays. Here, we are interested primarily in applications to star formation. The conditions of the Bondi problem are certainly the most favorable for accretion. If the cloud density ρ_∞ is fairly high--$10^3 m_H/cm^3$--and the temperature about 100°, typical of an HI region, then the rate of increase of the star's mass is sufficiently great for accretion to offer a possible explanation for the existence of short-lived O and B stars in the galaxy (McCrea 1953). This could be significant if it turns out that direct star formation is unable to produce enough O and B stars to explain the observations. Thus a star of low mass, formed as part of the disk population following the collapse of the galaxy, if reduced to rest in a dense cloud by the accretion drag, could evolve slowly into an O or B star. This clearly depends on the star's corona never being hot enough to drive a wind. It also requires that the cloud temperature should stay low enough for the formula (36) always to yield a significant rate of mass increase. The ultraviolet light emitted by an early-type star tends to ionize the surrounding gas and so heat it up to near 10^4° (Strömgren 1939). If this "HII" zone extends far into the cloud, the rate of accretion (36) is reduced by the factor $\simeq (10^4/10^2)^{3/2} = 10^3$ to a negligible value. However, the high density field (39) near the star, associated with the accretion process itself, greatly increases the destruction of ultraviolet photons and prevents the Strömgren sphere from developing until the star has grown into an O or B star (Mestel 1954; Schatzman 1955).

REFERENCES

Bodenheimer, P. 1969, Liège Symposium.

Bondi, H. 1952, Mon. Not. R. Astr. Soc., 112, 195.

Bondi, H., and Hoyle, F. 1944, Mon. Not. R. Astr. Soc., 104, 273.

Bonnor, W.B. 1957, Mon. Not. R. Astr. Soc., 117, 104.

Cameron, A.G.W. 1962, Icarus, 1, 13.

Disney, M.J., McNally, D., and Wright, A.E. 1968, Mon. Not. R. Astr. Soc., 140, 319.

Ebert, R. 1955, Zs. f. Astrophys., 37, 217.

Field, G.B. 1962, Interstellar Matter in Galaxies, ed. L. Woltjer (New York: Benjamin Inc.), p. 183.

______. 1965, Astrophys. J., 142, 531.

______. 1969, Introductory Report, Liège Symposium.

Field, G.B., Aanestad, P.A., Rather, J.D.G., and Orszag, S.A. 1968, Astrophys. J., 151, 953.

Field, G.B., Goldsmith, D.G., and Habing, H.J. 1969a, Astrophys. J., 155, L149.

______. 1969b, Astrophys. J., 158, 173.

Gaustad, J.E. 1963, Astrophys. J., 138, 1050.

Hayakawa, S., Nishimura, S., and Takayanagi, K. 1961, Pub. Astr. Soc. Japan, 13, 184.

Hayashi, C., and Nakano, T. 1965, Prog. Theoret. Phys., 34, 754.

Herbig, G. 1969, Introductory Report, Liège Symposium.

Hoyle, F. 1953, Astrophys. J., 118, 513.

Hoyle, F., and Lyttleton, R.A. 1939, Proc. Cam. Phil. Soc., 35, 405.

Hunt, R. 1970, Ph.D. Dissertation, Cambridge.

Hunter, C. 1962, Astrophys. J., 135, 594.

______. 1964, Astrophys. J., 139, 570.

Hunter, J.H. 1969, Mon. Not. R. Astr. Soc., 142, 473.

Larson, R.B. 1969, Liège Symposium.

Layzer, D. 1963, Astrophys. J., 137, 351.

Lin, C.C., Mestel, L., and Shu, F.H. 1965, Astrophys. J., 142, 1431.

McCrea, W.H. 1953, Mon. Not. R. Astr. Soc., 113, 162.

______. 1957, Mon. Not. R. Astr. Soc., 117, 562.

McNally, D. 1964, Astrophys. J., 140, 1088.

Mestel, L. 1954, Mon. Not. R. Astr. Soc., 114, 437.

______. 1965, Quart. J. R. Astr. Soc., 6, 161.

Oort, J.H. 1970, Astron. and Astrophys., 7, 381.

Penston, M.V. 1969a, Mon. Not. R. Astr. Soc., 144, 425.

______. 1969b, Mon. Not. R. Astr. Soc., 145, 457.

Pikel'ner, S.B. 1967, Astr. Zh., 44, 1915.

Ruderman, M.A., and Spiegel, E.A. 1966, Lectures by E.A. Spiegel, Interstellar Gas Dynamics Seminar, University of Wisconsin.

Schatzman, E. 1955, Gas Dynamics of Cosmic Clouds (Amsterdam: North-Holland Publishing Co.).

Spitzer, L., Jr. 1968, Diffuse Matter in Space (New York: Interscience).

Spitzer, L., Jr., and Tomasko, M.G. 1968, Astrophys. J., 152, 971.

Spitzer, L., Jr., and Scott, E.H. 1969, Astrophys. J., 158, 161.

Strömgren, B. 1939, Astrophys. J., 89, 526.

Sturrock, P.A., and Feldman, P.A. 1968, Astrophys. J., 152, L39.

Lecture VII

STAR FORMATION II: THE EFFECT OF ROTATION

The collapse and initial fragmentation of a rotating spherical cloud

We again consider a cloud with total thermal energy much less than the gravitational, but now let it have a finite angular momentum about its mass-center. For simplicity we suppose the cloud is spherical and of uniform density ρ, and that it is rotating uniformly with angular velocity $\tilde{\Omega}$. The cloud is now subject to the essentially anisotropic centrifugal

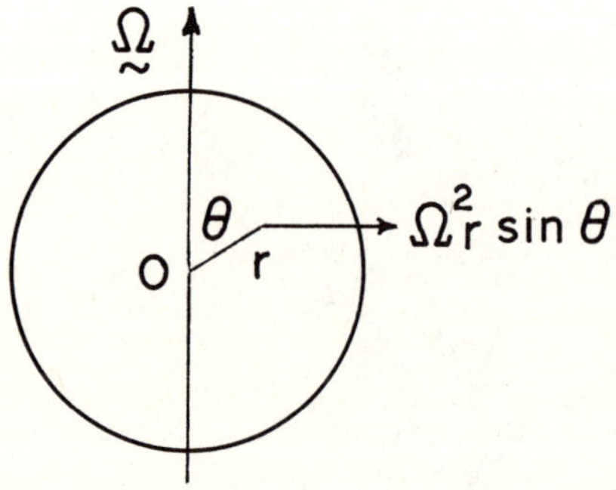

Fig. 1

force: the component of gravity perpendicular to the rotation axis is reduced by the factor (cf. Fig. 1)

$$\frac{\Omega^2 r \sin\theta}{G(\frac{4\pi}{3}\rho r^3)\sin\theta/r^2} = \frac{\Omega^2}{\frac{4\pi}{3}G\rho}. \qquad (1)$$

If the initial angular velocity Ω_0 is sufficiently small, the reduction factor (1) can be ignored, and the collapse will begin by being nearly isotropic and homologous. In the absence of strong torsional coupling with the external medium (cf. lecture VIII), angular momentum as well as mass will be conserved:

$$\Omega r^2 = \text{constant}, \qquad (2)$$

$$\rho r^3 = \text{constant}, \qquad (3)$$

so that under isotropic contraction

$$\Omega/\rho^{2/3} = \text{constant}, \qquad (4)$$

and the crucial ratio (1) increases like $\rho^{1/3}$. The centrifugal energy density $\frac{1}{2}\rho\Omega^2 r^2 \sin^2\theta \propto \rho^{5/3}$, so if we forget about the essential anisotropy, we may speak of the "centrifugal γ" being 5/3. Thus, even if the collapse is initially nearly isotropic, ultimately the cloud must begin to flatten; and once the eccentricity is markedly non-zero, the non-isotropic gravitational field accelerates the flattening (Lin, Mestel and Shu 1965).

Since the total thermal energy is much less than the gravitational, the cloud can flatten by a large factor before thermal pressure becomes important. As discussed in lecture VI, the asymptotic state depends critically on how much of the gravitational energy released is conserved as random kinetic energy of fragments. If the centrifugal forces interfere sufficiently with fragmentation, then we may expect the kinetic energy generated by the collapse to be dissipated in shocks, and the cloud to settle into a disk (cf. below); whereas if simultaneous with the flattening the cloud is able to break up into a number of discrete fragments with small mutual collision cross-section, then it will retain kinetic energy of z-motion comparable with its centrifugal energy, and the cloud of fragments will settle into a moderately flattened spheroidal structure.

As we are concerned first of all with the effect of <u>spin</u> rather than <u>shear</u>, we suppose that the sphere flattens through a series of states in uniform rotation. If the initial sphere is uniformly dense, uniformly rotating but otherwise at rest, then pressure-free gravitational collapse with detailed angular momentum conservation does in fact take place through a series of uniformly dense and uniformly rotating spheroids (Lynden-Bell 1962, 1964). This is due to the gravitational field of the ellipsoid's being linear in Cartesian coordinates; in particular, the field at a point in the equatorial plane distant r from the rotation axis is

$$-G\rho A(e) r \equiv -2\pi G\rho \frac{(1-e^2)^{\frac{1}{2}}}{e^3}\left\{\sin^{-1} e - e(1-e^2)^{\frac{1}{2}}\right\} r, \qquad (5)$$

where e is the eccentricity and ρ the instantaneous density. To bring out the effect of rotation, we maximize it by ignoring the inertia of the lateral motion, and suppose that at all stages of flattening centrifugal balance is maintained in the directions perpendicular to Oz, so that

$$\Omega^2 = A(e) G\rho. \qquad (6)$$

Then, as $e \to 1$, the ratio (1) becomes

$$\frac{\Omega^2}{\frac{4\pi}{3} G\rho} = \frac{3A(e)}{4\pi} \to \frac{3\pi}{4}(1-e^2)^{\frac{1}{2}} << 1. \qquad (7)$$

Now consider the fragmentation problem. If the initial rapidly rotating sphere has not yet flattened, then we can see at once that a spherical blob attempting to separate out runs up at once against centrifugal force of spin. By the elementary kinematics of rigid rotation (cf. the discussion below), such a blob has initially a spin Ω about its mass-center, and the ratio of centrifugal force of spin to the opposing component of self-gravitation is again $\Omega^2/\frac{4\pi}{3}G\rho$, which by (5) and (6) is near unity if $e \ll 1$. Thus, if the sphere rotates so rapidly that the centrifugal force of orbital motion is strong enough to prevent the overall contraction of the sphere, by the same criterion the centrifugal force of spin inhibits the formation of a sub-condensation. After flattening, however, even with the cloud as a whole in centrifugal balance, the corresponding ratio (1) for a spherical sub-condensation is small, and fragmentation can begin. Thus consider a spherical blob at the center of the spheroid, and let its density increase to $\rho + \delta\rho$, with an associated increase in spin [cf. (4)]

$$\Omega \rightarrow \Omega\left(1 + \frac{2}{3}\frac{\delta\rho}{\rho}\right). \tag{8}$$

The increase in the gravitational acceleration less the centrifugal force is

$$\begin{aligned} &\frac{4\pi}{3}G\delta\rho r + A(e)G\rho r - \Omega^2\left(1 + \frac{4}{3}\frac{\delta\rho}{\rho}\right)r \\ &= \frac{4\pi}{3}G\delta\rho\left(1 - \frac{A(e)}{\pi}\right)r \rightarrow \frac{4\pi}{3}G\delta\rho\left(1 - \pi(1-e^2)^{\frac{1}{2}}\right)r. \end{aligned} \tag{9}$$

Thus a spherical blob can now begin to condense. The reason for this instability is that the gravitational field (5) of the spheroid can be looked upon as the small difference between two large quantities, the inward- and outward-pulling parts, respectively, with the equilibrium rotation consequently much less than $(4\pi G\rho/3)^{\frac{1}{2}}$; under a small perturbation the net increase in the gravitational field therefore exceeds the increased centrifugal force (Mestel 1963).

As long as the cloud is in nearly uniform rotation, similar arguments apply to off-center condensations, which retain the bulk of their angular momentum in their orbital motion. The amount of z-kinetic energy retained (which fixes the ultimate shape of the cloud) will be determined by the value of the ratio $\Omega^2/\frac{4\pi}{3}G\rho$ in the state from which the cloud begins its collapse. If this ratio is negligible, then the collapse and fragmentation problem is initially hardly different from that of lecture VI: the first fragments condense with weak centrifugal forces of spin and form ultimately a quasi-spherical structure, with the centrifugal energy of orbital motion small compared with the random energy. With $\Omega^2/\frac{4\pi}{3}G\rho$ initially still well below unity but not negligible, we may expect the first fragments to be able to form and acquire small collision cross-sections before centrifugal forces of spin interfere, but the orbital centrifugal energy will now be comparable with the random kinetic energy (resulting from the randomization of the mainly z-kinetic energy of collapse). But, if $\Omega^2/\frac{4\pi}{3}G\rho \simeq 1$ initially, we expect the first fragmentation to occur only after the cloud has flattened and dissipated the bulk of its z-energy, so that the ultimate state will be disk-like (cf. below).

The "angular momentum problem"

We have seen how preferential flow down the rotation axis allows sub-condensations to contract out of the rotating cloud without running immediately into resistance from centrifugal force of spin. However, it must be remembered that again isotropic contraction cannot continue indefinitely, because with angular momentum conserved, the centrifugal forces will ultimately catch up with self-gravitation--the "$\gamma = 5/3$" effect. By following this line of argument, we can illustrate very strikingly the "angular momentum problem." For the moment we ignore thermal pressure completely, so that an arbitrary degree of flattening parallel to the rotation axis can occur. For simplicity we picture the condensation process as occurring in two stages (although this is not necessary for the argument): a cloud of initial radius R_0, density ρ_0 and angular velocity Ω_0 flattens into a spheroid of semi-thickness $\bar{z}$, with density $\bar{\rho} = \rho_0 R_0/\bar{z}$, and then a spherical sub-condensation of radius $\bar{z}$ contracts isotropically to a density ρ'. The mass of the condensation is

$$\bar{M} = \frac{4\pi}{3} \bar{\rho}\bar{z}^3 = \frac{4\pi}{3} \rho_0 R_0{}^3 \left(\frac{\bar{z}}{Ro}\right)^2 = M(\text{cloud}) \left(\frac{\bar{z}}{Ro}\right)^2 . \qquad (10)$$

By taking $\bar{z}$ sufficiently small, we can ensure that subsequent isotropic contraction to the high density ρ' is not prevented by centrifugal force, even though angular momentum is strictly conserved so that by (4)

$$\Omega' = \Omega_0 \left(\frac{\rho'}{\bar{\rho}}\right)^{2/3} . \qquad (11)$$

Thus we require

$$\Omega'^2 = \Omega_0{}^2 \left(\frac{\rho'}{\bar{\rho}}\right)^{4/3} \leq \frac{4\pi}{3} G\rho' . \qquad (12)$$

If in the initial spherical cloud $\Omega_0{}^2/\frac{4\pi}{3} G\rho_0 = \eta < 1$, then from (12)

$$\frac{\rho'}{\rho_0} \geq \left(\frac{\rho'}{\bar{\rho}}\right)^{4/3} \eta \qquad (13)$$

or

$$\eta \left(\frac{\rho'}{\rho_0}\right)^{1/3} \leq \left(\frac{\bar{\rho}}{\rho_0}\right)^{4/3} = \left(\frac{Ro}{\bar{z}}\right)^{4/3} . \qquad (14)$$

The larger the initial ratio η and the higher the ultimate density ρ', the greater the degree of flattening $R_0/\bar{z}$ necessary to ensure that the inequality (14) holds. The associated blob mass is from (10)

$$\bar{M} \leq M(\text{cloud})(\rho_0/\rho')^{1/2}/\eta^{3/2}. \qquad (15)$$

As an example, take $\rho' \simeq 10^{-7}$, by which density the blob would certainly have become opaque and entered the proto-stellar phase. Then with $\rho_0 \simeq 10^{-23}$, a typical HI-cloud density, and $\Omega_0 \simeq 10^{-15}$, close to the local angular velocity of galactic rotation, $\bar{z}/R_0 \leq 10^{-4}$, and the mass $\bar{M} \leq M(\text{cloud}) \times 10^{-8} \simeq 10^{-5}\ M_\odot$, if we take a reasonable cloud mass of $10^3\ M_\odot$.

This is one way of stating the angular momentum problem: with strict conservation of angular momentum, then even under anisotropic contraction and at zero temperature, it is nowhere near possible to form masses of stellar order and with proto-stellar densities from interstellar gas clouds rotating with an angular velocity comparable with that of the galactic rotation (Mestel 1965). And in fact the mass of $10^{-5}\ M_\odot$ is a reductio ad absurdum, for with a realistic cloud temperature the thermal energy would in fact prevent the enormous degree of flattening required for proto-stellar densities to be reached. For example, let us make conditions more favorable than those assumed above, and suppose the cloud is massive enough for a spherical blob of solar mass to form--by anisotropic motion--and reach a density of 10^{-16} before centrifugal force becomes comparable with gravity. At this density the radius $R \simeq 5/3 \times 10^{16}$ cm and the gravitational energy per unit mass $GM/R \simeq 8 \times 10^9$. At a temperature of $\simeq 100°$K the thermal energy per unit mass is $\simeq 8 \times 10^9$, so that a non-rotating blob would be in approximate hydrostatic equilibrium at this density. The crucial difference introduced by the rotation is that now the sphere cannot contract isotropically to a higher density (with the same temperature) and so into a state of gravitational collapse (cf. lecture VI). Instead it will flatten a little and achieve equilibrium as a spheroid of moderate eccentricity, with the isothermal pressure gradient balancing gravity parallel to the rotation axis Oz, and centrifugal force (assisted by a rather weaker pressure gradient) balancing gravity perpendicular to Oz. The point to be emphasized is that with reasonable initial conditions the difficulties arise long before densities appropriate to a proto-star or to a primeval solar nebula are reached.

Attempts to overcome the difficulty have been either geometrical or dynamical. Hoyle (1945) suggested that the mass that ultimately forms a star comes from a quasi-cylindrical region perpendicular to the rotation axis. Stars of small mass were supposed to form first, and subsequently to accrete more mass as they tunnel through the interstellar gas. Gas with too much angular momentum about the center of the star would simply not reach the surface. However, it is doubtful whether this type of accretion could be efficient enough for this model to be of general application, and in any case the formation of the original star remains unexplained. More recently McCrea (1961) has pictured the formation of a self-gravitating mass by the chance agglomeration of randomly moving density fluctuations--"floccules"--from all parts of the cloud. Again the angular momentum of spin is automatically kept down by virtue of the very mechanism by which the blob is pictured as growing: floccules with high angular momentum about the mass-center of the growing mass by definition will not hit it. The model is an interesting one, but it is highly unlikely that the basic picture would survive the introduction of the galactic magnetic field into the problem (lecture VIII), so we shall not consider it further.

Dynamical resolution of the problem looks for a process by which angular momentum is removed from a contracting body to the surrounding gas. The two obvious candidates are turbulent friction and magnetic torques. The first forms an essential feature of the cosmogonical scheme outlined by von Weizsäcker (1944, 1947). In lecture VIII we discuss the effects--both positive and negative--of the galactic magnetic field on star formation.

The angular momentum of a gas cloud and the galactic rotation

It may be suggested that we have made the angular momentum problem unnecessarily severe by taking the cloud to rotate initially with the local angular velocity of the galactic rotation. Since the galaxy rotates non-uniformly, the motions within a condensing cloud will certainly begin by being far more complicated than a uniform rotation about the mass-center of the cloud. However, if we for the moment assume that we can ignore any torques acting on the cloud during its formation (cf. lectures VI and VIII), then the angular momentum about the cloud mass-center of the internal motions will be conserved, and after dissipation of excess energy in shocks, it will take the form of spin (not necessarily uniform). We are now interested in whether the angular momentum of spin is substantially below that of the same body condensing from a hypothetical galaxy rotating locally with <u>uniform</u> angular velocity Ω_0.

Let us idealize the problem by ignoring completely random motions, so that the gas moves in the galactic disk with the local angular velocity $\underset{\sim}{\omega}(r) = \omega(r)\hat{\underset{\sim}{z}}$, where $r = |\underset{\sim}{r}|$ is the distance from the galactic center O, and $\hat{\underset{\sim}{z}}$ is the unit vector out of the plane in the right-handed sense. A sub-section Δ of the disk has angular momentum about its mass-center G

$$\underset{\sim}{H}(G) = \sum m[\underset{\sim}{R}\times(\underset{\sim}{\omega}\times\underset{\sim}{r})] = \sum m\{\underset{\sim}{\omega}(\underset{\sim}{r}\cdot\underset{\sim}{R}) - \underset{\sim}{r}(\underset{\sim}{\omega}\cdot\underset{\sim}{R})\}$$
$$= \sum m\underset{\sim}{\omega}(\underset{\sim}{r}\cdot\underset{\sim}{R}), \tag{16}$$

where an element of mass m has position vector $\underset{\sim}{r}$ relative to O and $\underset{\sim}{R}$ relative to G. Let GX,GY be the principal axes of inertia at G, with $\underset{\sim}{R} = (X,Y)$ referred to these axes, and let GX have direction cosines (λ,μ) relative to OG and the direction of motion GQ (Fig. 2).

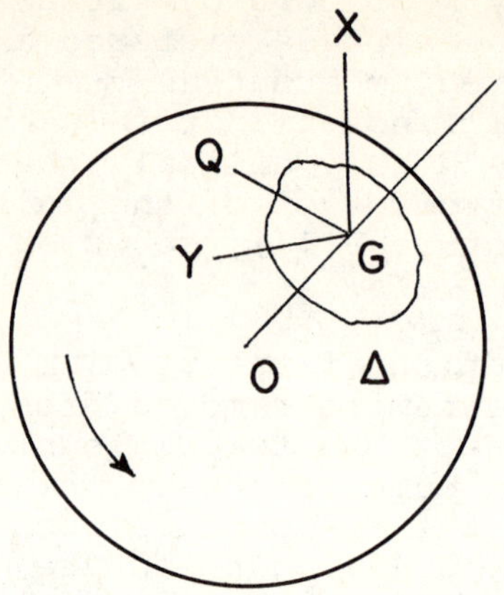

Fig. 2

Then locally

$$\omega(\underset{\sim}{r}) = \omega(\underset{\sim\sim}{OG} + \underset{\sim}{R}) = \omega_G + (\underset{\sim}{R} \cdot \nabla\omega)_G + \ldots$$
$$= \omega_G + (\lambda X - \mu Y)(d\omega/dr)_G + \ldots, \tag{17}$$

whence

$$\underset{\sim}{H}(G) = \hat{\underset{\sim}{z}}\sum m\left\{\left[\omega_G + (\lambda X - \mu Y)\omega'_G\right]\left[\{OG + (\lambda X - \mu Y)\}(\lambda X - \mu Y) + (\mu X + \lambda Y)^2\right]\right\}$$
$$= \hat{\underset{\sim}{z}}\sum mR^2\left[\omega_G + (r\omega')_G\left\{\frac{\sum m(\lambda^2X^2 + \mu^2Y^2)}{\sum m(X^2+Y^2)}\right\}\right]. \tag{18}$$

We have used here $\sum mX = \sum mY = 0$, because G is the mass-center and $\sum mXY = 0$ by definition of principal axes. Thus $\underset{\sim}{H}(G)$ depends not only on ω_G and ω'_G and the moment of inertia $I(G) = \sum mR^2$ but also on the shape of the condensation, through the term multiplying $(r\omega')_G$ (Mestel 1966).

We are explicitly assuming that there are no important external torques (turbulent, gravitational or magnetic) acting on the sub-condensation Δ, so that $\underset{\sim}{H}(G)$ is conserved. As Δ changes its shape and mass-distribution, its mass-center G moves through Δ--only in the case of a rigid body does it coincide with a particular material particle. Once Δ has relaxed into a rotating sub-disk or (taking account of finite thickness) into a rotating quasi-spheroid, then its angular momentum of spin will (under our assumptions) be identical with (18).

One case deserves special mention: if $\lambda = \mu = 1/\sqrt{2}$, (18) becomes

$$\underset{\sim}{H}(G) = \hat{\underset{\sim}{z}}I(G)\left(\omega + \frac{1}{2}r\omega'\right)_G = I(G)\left(\frac{1}{2}\nabla\times\underset{\sim}{v}\right)_G \tag{19}$$

--the angular momentum is given by assigning Δ an angular velocity equal to the vorticity in the original flow. This result can be linked with a general theorem in hydrodynamics, found in Jeffreys and Jeffreys (1956). Consider a small element of fluid Δ of mass-center G, and suppose a small rigid body with the same shape and density distribution to have the same angular momentum about G. Provided Δ has principal axes of inertia which coincide with those of the rate-of-strain tensor e_{ij} of the fluid motion, then the angular velocity of the rigid body is $\frac{1}{2}\nabla\times\underset{\sim}{v}$. In our problem, with the fluid flow one of simple non-uniform rotation, the components of e_{ij} referred to OG,GQ are

$$e_{11} = e_{22} = 0, \qquad e_{12} = \frac{1}{2}r\omega'. \tag{20}$$

For this tensor the principal axes are indeed inclined at $\pi/4$ to the radius vector, so that substitution of $\lambda = \mu = 1/\sqrt{2}$ in (18) should indeed reduce $\underset{\sim}{H}(G)$ to $I(G)(\frac{1}{2}\nabla\times\underset{\sim}{v})_G$. If Δ has kinetic symmetry about the direction $\hat{\underset{\sim}{z}}$--e.g., if it is circular and uniform--then again (19) holds. If the angular velocity ω is locally uniform, the tensor e_{ij} vanishes, and (19) holds without any restriction on the shape and density distribution, with $\frac{1}{2}\nabla\times\underset{\sim}{v} = \underset{\sim}{\omega}$, in agreement with the elementary kinematics of rigid rotation.

If we represent the rotation law locally by $\omega \propto r^n$, then (18) becomes

$$\underset{\sim}{H}(G) \equiv H(G)\hat{\underset{\sim}{z}} = \hat{\underset{\sim}{z}}\sum mR^2\left\{\frac{(1+n\lambda^2)\sum mX^2 + (1+n\mu^2)\sum mY^2}{\sum m(X^2+Y^2)}\right\}. \tag{21}$$

In the realistic case of a disk with centrifugal force balancing self-gravitation, $n \geq -3/2$, equality applying to the Keplerian case with all the mass effectively concentrated at the center:

$$\omega^2 r = \frac{GM}{r^2} \rightarrow \omega \propto r^{-3/2}. \tag{22}$$

In the galaxy and M31, 21-cm line measurements reveal that $n \simeq -1$ from $r = 3$ to $r = 10$ kpc (Schmidt 1967; Davies and Gottesman 1970 a,b). If $n > -1$, $\underset{\sim}{H}(G)$ has the same sense as $\underset{\sim}{\omega}$ for all elements Δ, but if $n < -1$, $H(G)$ can have either sign, and so can be made to vanish by suitable choice of shape. For example, needle-like regions with $\sum mY^2 \simeq 0$ have $H(G) = 0$ if $\lambda = \pm(1/-n)^{\frac{1}{2}}$: as seen from the point G, the relative velocities at neighboring points in these two directions are either towards or away from G, so that the needle-like region acquires zero angular momentum about G, an observation first made by Edgeworth (1946). If n = -1, a needle-shape directed along the radius vector from 0 likewise does not acquire any spin, because of the absence of linear shear in the rotational velocity.

The above study is pedagogically useful in that it brings out the very limited conditions under which the vorticity of a fluid flow is a precise measure of the angular momentum of an element. However, it is clear that only for very special elements will the angular momentum be much below the estimate used earlier, and so its astronomical relevance is probably slight: it is difficult to find convincing reasons for the preferential formation of needle-shaped clouds directed along the galactic radius. In fact, it can be argued (Hoyle 1945; Ostriker 1970) that the figures used above are conservative, in that the turbulent motions of the normal gas clouds will in general contribute more to the initial angular momentum of a gravitationally-bound cloud than the galactic rotation. If so, this only reinforces the conclusion that an adequate theory of star formation must include as an essential feature an angular momentum transfer process.

<u>Non-uniformly rotating clouds, and the origin of the solar system</u>

We have noted in the last section that the rotation of the galaxy is highly non-uniform. Although we are not primarily concerned with problems of the formation and structure of the galaxy, it is worth noting that the model of a rotating cloud collapsing into a disk-like structure is applicable on a galactic as well as on a star-cluster scale. In fact, the quasi-uniform galactic rotational velocity can be explained as resulting from a large-scale gravitational instability to which the uniformly rotating disk is subject (Mestel 1963; Hunter 1963). In the infinitely-thin disk limit, the centrifugal forces associated with a uniform rotational velocity V are balanced by the gravitational field of the mass distribution per unit area

$$\sigma(r) = \frac{V^2}{2\pi G r}\left[1 - \frac{2}{\pi}\sin^{-1}\left(\frac{r}{R}\right)\right], \qquad (23)$$

where R is the radius; and the distribution of mass with respect to angular momentum is almost the same as that in the uniformly rotating sphere or spheroid, implying that the centrally condensed non-uniformly rotating disk (23) can result from an axially-symmetric perturbation, in which each element conserves its angular momentum. The same linear stability analysis (Hunter 1965) shows that the disk (23) is now stable against such a disturbance [though not against smaller-scale modes (Toomre 1964)]. It is therefore not difficult to understand why the distribution of mass in the galaxy (most of it in the form of the stellar "disk population") is such as to enforce a non-uniform rotation.

The same mathematical theory applies to a highly flattened interstellar gas cloud: there is a spontaneous tendency towards axial condensation and a highly non-uniform rotation law. Cameron (1970) has extended the above results, showing that for each reasonable distribution of angular momentum with respect to cylindrical mass shells, there are again two equilibrium models: one in quasi-uniform rotation and the other with a more nearly flat rotational velocity curve.

The fragmentation problem is more difficult in the presence of a large zero-order shear that tends to disrupt any small blob that would otherwise form by gravitational instability. Prima facie, it is only axially-symmetric disturbances--a sub-disk and surrounding rings--that are unaffected by the shear. During early phases of star formation the galactic magnetic field trapped within the cloud probably plays a dominant role in the dynamics, not least in its effect on the shear, but at later stages the field may very well be negligible, and we must appeal to turbulent friction to transfer angular momentum and to reduce shear. In a study of the later phases of star formation, Cameron (1969) has suggested that both the quasi-uniformly rotating disk and the strongly-sheared, axially-condensed model are cosmogonically relevant: the first breaks up into a proto-binary star system, while the other is the precursor of a single star surrounded by a nebula which ultimately becomes a planetary system. In order to give the gas the extra central condensation necessary for it to emulate the proto-sun surrounded by a Keplerian envelope (with $\omega \propto 1/r^{3/2}$ rather than $1/r$), Cameron has revived the von Weizsäcker picture, with thermal convection--due to the high opacity--replacing dynamical turbulence as the cause of angular momentum transport. It is not yet clear that the model will survive the criticism brought against the von Weizsäcker theory (ter Haar 1950; Lüst 1952) that there is insufficient energy available to maintain the turbulence against its decay for long enough for it to be able to transport a significant amount of angular momentum. Ebert (1964) has made a similar study, again taking the V = constant disk as the zero-order state, but including the effect of a trapped magnetic field. His model has the interesting feature that the ring-like condensations forming by gravitational instability are spaced at intervals close to those of the Titius-Bode law. He appeals to magnetic transport of angular momentum to give the disk the extra central condensation.

Whatever one's view of the details of Cameron's or Ebert's model, one can hardly help feeling that the rotating disk is a very plausible initial assumption for a theory of the origin of the solar system, simply because of the high angular momentum normally to be expected in condensations from the interstellar gas. And we should also note that our newly-acquired knowledge of the braking of the rotation of the sun and solar-type stars during their main-sequence lifetime (cf. lectures III and IV) alters considerably one of the constraints on the theory. It is no longer plausible to argue (e.g., Hoyle 1960) that the low spin of the sun is due to its having supplied the high angular momentum of the planets at their formation. Again, in a classical study Jeans (1919) remarked that even if all the angular momentum in the planets were put back into the sun, the ratio of centrifugal force to gravity would be well below unity; he therefore argued against a rotating nebula model for the primeval solar system, and for the famous tidal theory involving an accidental close encounter of two stars. We need not summarize the objections to this now discarded theory, but merely emphasize that we have no reason to doubt that the sun in fact formed rotating rapidly, and that the proto-planetary material had high angular momentum right from the start.

REFERENCES

Cameron, A.G.W. 1969, Symposium on Low Luminosity Stars, Charlottesville (New York: Gordon and Breach).

______. 1970, private communication.

Davies, R.D., and Gottesman, S.T. 1970a, Mon. Not. R. Astr. Soc., 149, 237.

______. 1970b, Mon. Not. R. Astr. Soc., 149, 263.

Ebert, R. 1964, Zur Theorie der Entstehung von Planetensystemen, Habilitationschrift, Un. of Frankfurt-am-Main.

Edgeworth, K.E. 1946, Mon. Not. R. Astr. Soc., 106, 484.

Haar, D.ter 1950, Astrophys. J., 110, 32.

Hoyle, F. 1945, Mon. Not. R. Astr. Soc., 105, 302.

______. 1960, Quart. J. R. Astr. Soc., 1, 28.

Hunter, C. 1963, Mon. Not. R. Astr. Soc., 126, 299.

______. 1965, Mon. Not. R. Astr. Soc., 129, 321.

Jeans, J.H. 1919, Problems of Cosmogony and Stellar Dynamics (Cambridge: Cambridge Univ. Press).

Jeffreys, H., and Jeffreys, B.S. 1956, Methods of Mathematical Physics (Cambridge: Cambridge Univ. Press).

Lin, C.C., Mestel, L., and Shu, F.H. 1965, Astrophys. J., 142, 1431.

Lüst, R. 1952, Zeits. f. Naturforsch, 7a, 87.

Lynden-Bell, D. 1962, Proc. Camb. Phil. Soc., 58, 709.

______. 1964, Astrophys. J., 139, 1195.

McCrea, W.H. 1961, Proc. Roy. Soc. A., 260, 152.

Mestel, L. 1963, Mon. Not. R. Astr. Soc., 126, 553.

______. 1965, Quart. J. R. Astr. Soc., 6, 161.

______. 1966, Mon. Not. R. Astr. Soc., 131, 307.

Ostriker, J.P. 1970, preprint.

Schmidt, M. 1967, Rotation Parameters and Distribution of Mass in the Galaxy in Galactic Structure, ed. G.P. Kuiper and B.M. Middlehurst (Chicago: Un. of Chicago Press).

Toomre, A. 1964, Astrophys. J., 139, 1217.

von Weizsäcker, C.F. 1944, Zeits. f. Astrophys., 22, 319.

______. 1947, Zeits. f. Astrophys., 24, 181.

Lecture VIII

STAR FORMATION III: THE INTERACTION OF GRAVITATION, MAGNETISM AND ANGULAR MOMENTUM

The galactic magnetic field and the formation of gas clouds

The model of the interstellar medium outlined at the beginning of lecture VI must be modified to take account of the galactic magnetic field (Field 1969). The large-scale component $\underset{\sim}{B}_0$ of the field lies in the galactic plane and has an estimated strength $\simeq 3 \times 10^{-6}$ gauss. This value is consistent with Faraday rotation and plasma dispersion measures for pulsars, theoretical interpretation of interstellar polarization of starlight, and the Zeeman effect on the 21-cm line (cf. below). Earlier arguments for a field stronger by a factor 10 (e.g., Woltjer 1962) depended on the attempt to account for the background synchrotron radiation from the galaxy. This requires knowledge of the electronic component of the cosmic ray gas; more recent measurements (Anand, Daniel and Stephens 1968 a,b) have increased this density and yield a field much closer to the value of B_0 assumed above.

A zero-order, smoothed-out model of the interstellar gas will have the gravitational pull of the stellar background acting normal to the disk balanced by the pressures of the gas, the magnetic field, and of the cosmic ray gas (coupled to the thermal gas via the magnetic field). Parker (1966), Lerche (1967) and Pikel'ner (1967) showed that this configuration is unstable to the formation of gas clouds by motion of gas along field-lines. The relativistic cosmic ray gas stays spread out along the field-lines and helps to distend the field outside the clouds into giant loops. Just as in the non-magnetic Field-Pikel'ner model outlined in lecture VI, the low density inter-cloud gas is heated by sub-cosmic rays and λ-rays to $\simeq 10^4$°, while the efficient cooling mechanisms at higher densities keep the cloud temperatures below 300°K. Thermal pressure balance holds along the field-lines, while across the field the equilibrium is a balance between the gravitational pull of the stellar background and the curvature force exerted by the magnetic field: the cool cloud hangs on the field-lines, rather as in one very plausible model of cool gaseous filaments on the solar surface (Kippenhahn and Schlüter 1957).

Magneto-gravitational equilibrium

We now consider a cool gas cloud, embedded in the hot inter-cloud medium, which has grown by inter-cloud collisions to a mass sufficiently great for it to be in a state of gravitational collapse, if magnetic forces could be ignored. We suppose that the process of formation has not changed the overall structure of the field, which therefore remains a part of the local galactic field, with more-or-less the same direction over the whole cloud. As the cloud attempts to contract under its self-gravitation, the field-lines trapped within it are compressed and so exert a steadily increasing resistance to the contraction. We first study the conditions for magneto-gravitational equilibrium, in which the outward force exerted by the distorted magnetic field is large enough to balance the self-gravitation of the cloud. Such an equilibrium can clearly hold only in directions across the field; the component of self-gravitation along the field must be balanced by the thermal pressure. Construction of a detailed model with the help of an electronic computer is under way. However, the essentials of the problem can be brought out by use of the scalar virial theorem for a magnetic gas cloud (Chandrasekhar and Fermi 1953): a necessary condition for equilibrium of a cloud of volume τ and surface S is

$$2K + 3(\gamma-1)U + \mathcal{M} + \mathcal{V} = \int x_i \left(T_{ik} + p\delta_{ik}\right) dS_k, \qquad (1)$$

where

K = macroscopic kinetic energy (centrifugal or turbulent),

U = the thermal energy,

$\mathcal{M} = \int \frac{B^2}{8\pi}\, d\tau$ is total magnetic energy within τ,

$\mathcal{V}$ = the self-gravitation energy = $\int \rho \underset{\sim}{r} \cdot \underset{\sim}{g}\, d\tau$,

dS_k = the outward vectorial surface element,

$\underset{\sim}{r} = x_i$ is the position vector of an element, referred to the mass-center of the cloud,

p = the thermal pressure at the cloud surface,

and

$$T_{ik} = \frac{\underset{\sim}{B}^2}{8\pi}\,\delta_{ik} - \frac{B_i B_k}{4\pi} \qquad (2)$$

is the Maxwell stress tensor. If we drop the magnetic terms, then (1) describes the equilibrium of the cloud, with the pressure of the Field-Pikel'ner inter-cloud medium assisting self-gravitation against the disruptive effect of the different forms of kinetic energy (cf. lecture VI). The new terms show that the magnetic energy within the cloud does have an overall disruptive effect, although the anisotropy of the magnetic forces is disguised. However, there is now another surface integral, involving the magnetic stresses, so that for a cloud with a magnetic field that is a locally distorted part of the galactic field, it is not sufficient to consider just the volume integral $\mathcal{M}$.

For simplicity, we suppose the cloud is spherical and has formed from a uniform, galactic background, permeated by the uniform field $\underset{\sim}{B}_0$, by the initial motion down the field (which leaves $\underset{\sim}{B}_0$ unchanged) followed by spherical shrinkage from a radius R_0 to the equilibrium radius $\bar{R}$. Detailed analysis of the effect of this type of motion on the field (Mestel 1966) confirms the intuitive picture derived from field-freezing: beyond R_0 the field stays essentially unchanged; within $\bar{R}$ the field remains nearly uniform and parallel to $\underset{\sim}{B}_0$ but is increased to

$$\bar{B} = \frac{B_0 R_0{}^2}{\bar{R}^2}; \tag{3}$$

while in between the field-lines are drawn out into a nearly radial structure, with

$$B_r = B_0 \left(\frac{R_0}{r}\right)^2 \cos\theta, \tag{4}$$

where (r,θ) are spherical polar coordinates based on the mass-center and the direction of $\underset{\sim}{B}_0$. The $1/r^2$ law in (4) ensures zero divergence for $\underset{\sim}{B}$, and the $\cos\theta$ factor ensures continuity of B_r at $r = \bar{R}$ and $r = R_0$. The actual field structure must in fact be more complicated than this (Mestel 1966 and below), but for a virial theorem treatment (4) is adequate.

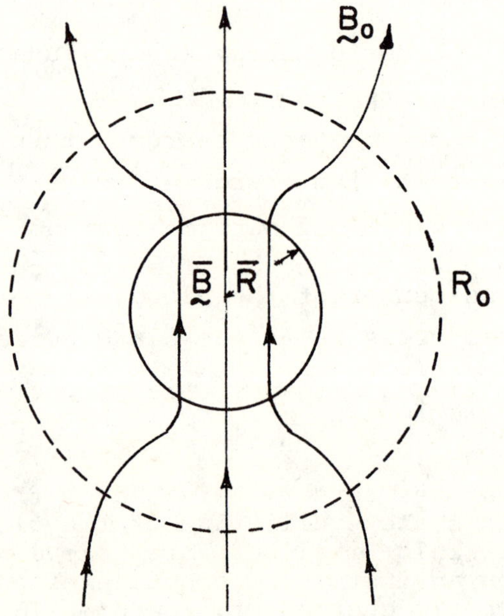

Fig. 1

With most of the mass of the cloud within the radius $\bar{R}$, the gravitational energy term $\simeq -\frac{3}{5}\frac{GM^2}{\bar{R}}$ for a uniformly dense sphere $\bar{R}$, with the numerical factor somewhat greater but still of order unity in a more centrally condensed cloud. The magnetic energy within R_0 is

$$\mathcal{M} = \frac{4\pi}{3}\bar{R}^3\left(\frac{\bar{B}^2}{8\pi}\right) + \frac{1}{8\pi}\int_0^{\pi}\int_{\bar{R}}^{R_0} 2\pi r^2 \sin\theta dr d\theta\left(\frac{B_0{}^2 R_0{}^4}{r^4}\cos^2\theta\right) \tag{5}$$

$$= \frac{1}{6}\bar{B}^2\bar{R}^3 + \frac{2}{3}\left(\frac{1}{\bar{R}} - \frac{1}{R_0}\right)\left(\frac{B_0{}^2R_0{}^4}{4}\right).$$

We must also add in the surface integral of the stresses associated with the uniform field $\underset{\sim}{B}_0$ over a sphere just outside R_0. The simplest way is just to imagine the field within R_0 replaced by the uniform field $\underset{\sim}{B}_0$. The magnetic forces exerted by $\underset{\sim}{B}_0$ are zero (since $\nabla\times\underset{\sim}{B}_0 = 0$) and so can make no net contribution to an integral result such as the virial theorem (which is derived from the equation of motion); hence if we apply the theorem to a sphere R_0 permeated by $\underset{\sim}{B}_0$, the magnetic surface integral on the right of (1) must be equal to

$$\int_{R_0} x_i T_{ik} dS_k = \frac{4\pi}{3} R_0{}^3\left(\frac{B_0{}^2}{8\pi}\right) = \frac{1}{6}\frac{B_0{}^2R_0{}^4}{R_0}. \tag{6}$$

For the configuration of Fig. 1, the total disruptive magnetic effect is from (3), (5) and (6)

$$\mathcal{M} - \int x_i T_{ik} dS_k = \frac{1}{3}\bar{B}^2\bar{R}^3\left(1 - \frac{\bar{R}}{R_0}\right). \tag{7}$$

Thus from (1) and (7) the virial theorem predicts that magnetic forces alone will halt the gravitational collapse if there exists a value of $\bar{R} < R_0$ such that

$$\left(1 - \frac{\bar{R}}{R_0}\right) = \frac{9\pi^2}{5}\frac{GM^2}{F^2}, \tag{8}$$

where F is the (invariant) flux through the cloud

$$F = \pi\bar{B}\bar{R}^2: \tag{9}$$

for a prescribed flux F we arrive at a critical mass M_c, given by

$$\frac{9\pi^2}{5} GM_c{}^2 = F^2. \tag{10}$$

A mass $M \geq M_c$ exerts enough self-gravitation to cause indefinite gravitational collapse. For $M \ll M_c$, only a modest distortion of the magnetic field-lines yields enough magnetic force to halt the contraction at $R \simeq R_0$.

A more detailed treatment of this problem has been given by Strittmatter (1966). Noting that in the realistic problem there would inevitably be some flattening along the direction of the field until thermo-gravitational equilibrium is reached along the field, he gave an analogous treatment for a spheroidal mass distribution in the cloud, finding that the reduction of the critical mass for a given flux--due mainly to the increase in $|v|$ with increasing eccentricity--is by at most a factor 2.

Models of gas clouds in magneto-gravitational equilibrium may very well be relevant to recent observations of Zeeman splitting of the 21-cm line (Verschuur 1969 a,b), indicating a magnetic field strength that is locally as much as 10^{-5} or 2×10^{-5} gauss, as compared with the estimated background field of $\simeq 3 \times 10^{-6}$ gauss. The clouds observed by Verschuur are markedly denser than the background and so should have a stronger (frozen-in) field if there has been any motion perpendicular to the field. If, in addition, the clouds are massive enough, they will exert enough self-gravitation to satisfy the dynamical as well as the kinematical condition (Mestel 1969). Of the first two clouds observed, upper limits for the densities--found from area densities and lower limits for cloud radii--were 120 m_H and 165 m_H, with corresponding lower limits for the masses of 680 $M_\odot$ and 2500 $M_\odot$ (Clark 1965). The second cloud is certainly markedly self-gravitating. Further measurements by Verschuur yield results that are at least not inconsistent with the above picture. However, we urgently need detailed studies of the structure and stability of such clouds, taking account of thermal pressure and of the equilibrium conditions in both directions. A start on this has been made by D. A. Parker (1970).

Clouds of low density--not more than about 10 m_H--will have formed almost entirely by motions along the field (driven by thermal instability--cf. lecture VI) and so the field within them should not be much above the background galactic field. According to Field (1969) the Zeeman measurements seem consistent with this, and do indeed yield a value of $B_0 \simeq 3 \times 10^{-6}$ gauss.

Gravitational collapse and fragmentation

Now suppose that the cloud mass does exceed the value of the critical mass M_c appropriate to its shape, so that the cloud collapses in spite of the resistance of the magnetic forces. The first point to note is that the critical mass is a function of only the prescribed flux F, but not of the density, so that once collapse starts, it can continue indefinitely (assuming no other forces such as thermal pressure or centrifugal force intervene). In contrast to the centrifugal force in a cloud of constant angular momentum (cf. lecture VII), the "magnetic γ" is 4/3; for under isotropic change of scale by a factor λ the conditions of mass and flux conservation, respectively, yield

$$\rho\lambda^3 \propto \text{Mass}, \qquad (11)$$

$$B\lambda^2 \propto \text{Flux}, \qquad (12)$$

so that $B \propto \rho^{2/3}$, and the magnetic pressure

$$\frac{B^2}{8\pi} \propto \rho^{4/3}, \qquad (13)$$

implying that magnetic and gravitational force densities increase proportionately under isotropic contraction. The critical mass for a spheroid of eccentricity e can be written in terms of the mean density and field strength as (Mestel 1965; Strittmatter 1966)

$$M_c = \frac{1}{48\pi^2}\left(\frac{5k}{G}\right)^{3/2}\left(\frac{B}{\rho^{2/3}}\right)^3 \frac{1}{(1-e^2)}, \qquad (14)$$

where the constant k decreases from unity to $\simeq 1/4$ as e increases from zero to unity. It is again apparent that under isotropic contraction--with e and $B/\rho^{2/3}$ constant--M_c is also constant.

Now suppose the cloud is spherical, and that it contracts isotropically, so that its mass M is only slightly greater than the value of M_c for e = 0. Then it is at once clear that no sub-spheres can separate out of the contracting sphere. For consider a sub-sphere of radius $R' = \lambda R$ ($\lambda < 1$), and so containing magnetic flux $F' = \lambda^2 F$. By the mass-flux relation (10), in order for this sub-sphere to be able to contract out of the background, it must have mass $\lambda^2 M_c$, whereas in fact its mass is $M' = \lambda^3 M$. As long as this sub-sphere is essentially part of the cloud, with density and associated field-strength more-or-less the same as the mean cloud values, then the internal and external magnetic stresses balance--or the energy term in the virial theorem (applied to the sub-sphere) is cancelled by the stress tensor integral over the surface R'. But a local density fluctuation of mass M' cannot contract far before the consequently distorted magnetic field exerts local forces too strong for the self-gravitation of the blob, and so the motion will be reversed; and this result is independent of the density of the cloud as a whole, so long as the contraction has been isotropic, so that $B/\rho^{2/3}$ has remained constant (Mestel and Spitzer 1956; Mestel 1965, 1969).

This result would seem at first sight to pose a serious dilemma, for as pointed out by Field (1969) the minimum gravitationally-bound mass that can form by motion down the galactic field-lines under the external pressure of the hot inter-cloud medium is at least of the same order as the Ebert-Bonnor-McCrea critical sphere, which is $\simeq 10^3$ $M_\odot$. However, the assumption of isotropic collapse is certainly questionable; and in fact, a cool, non-rotating spherical cloud with a large-scale magnetic field will tend to flatten spontaneously along the field, because the essentially anisotropic magnetic force acts across the field, and the gravitational field of a spheroid tends to increase the eccentricity (Lin, Mestel and Shu 1965). This preferential flow down the field reduces $B/\rho^{2/3}$, so that by (14) spherical sub-condensations can form (Mestel 1965). For example, suppose the cloud is initially spherical with radius R, and has flattened into a spheroid of semi-minor axis R', so that the initial density

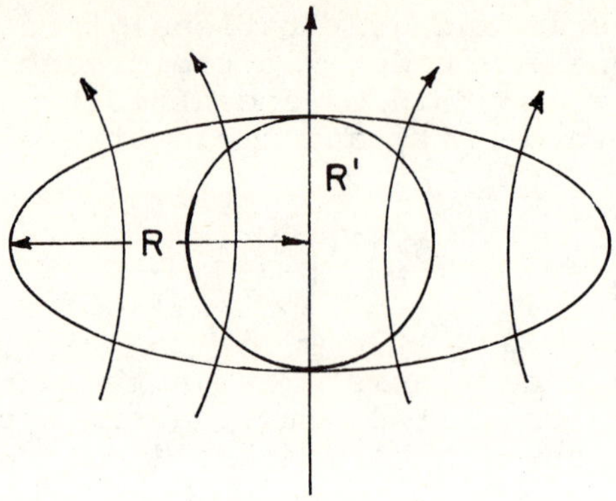

Fig. 2

ρ_i has increased to $\rho = \rho_i R/R'$. A sub-sphere of radius R' again contains flux $F' = (R'/R)^2 F$, but its mass is now $M' = \frac{4\pi}{3}\rho R'^3 = \frac{4\pi}{3}\rho_i R^3 (R'/R)^2 = (R'/R)^2 M$, so that the ratio $F'/M' = F/M$, the same as for the cloud as a whole. Thus, provided the cloud as a whole is able to contract indefinitely, then fragmentation into sub-spheres can take place after flattening along the direction of the field. The condition (10) remains valid; it tells us how much mass is required for the gravitational contraction of a roughly spherical mass containing the prescribed flux F. Earlier we noted that a sub-sphere within the original spherical cloud does not have enough mass to be able to separate out, and that this continues to hold after isotropic contraction. We now see that the necessary mass M' can be accumulated from the cloud itself by motion along the field: the mass initially in a cylinder of height 2R and radius R' is more than adequate when concentrated into a sub-sphere of radius R' and density $\rho_i R/R'$. The modifications to the theory necessary because of finite temperature are easily made (Mestel 1965; Field 1969).

Thus again we see how the spontaneous anisotropic collapse of a strongly magnetic cloud enables the first fragmentation to occur; but there is a crucial difference from the problem of a rotating cloud conserving its angular momentum. Because of the "$\gamma = 4/3$" behavior of a frozen-in magnetic field under isotropic contraction, a sub-mass forming after preferential flow down the field can subsequently contract isotropically to arbitrarily high densities without running into trouble from the magnetic forces, which merely increase proportionately to the gravitational; whereas we saw in lecture VII that an isotropically contracting blob that conserves its angular momentum ultimately will either be forced into further anisotropic motion, or will attain equilibrium with thermal pressure and centrifugal forces jointly balancing gravity. Thus, for example, if the cloud has a mass of $\simeq 10^3\ M_\odot$, flattening by a factor 1/30--either in one stage or several--is sufficient to enable a mass of solar order to form and contract to proto-stellar densities, dragging its trapped flux with it.

However, it must be pointed out that the sub-condensations and proto-stars forming in this way are inevitably "strongly magnetic"; for since $F'/M' = F/M$, the sub-condensations have the same ratio of magnetic to gravitational energy as the cloud as a whole, and so will have much more magnetic flux than we infer for even the strongest observed magnetic stars. The statement of the "fossil" theory of stellar magnetism (cf. lectures I and II) should perhaps be inverted: the problem is prima facie not how to account for stellar magnetic fields, but rather to explain why they are not stronger than they are. Clearly the ratio F/M can be reduced by imagining a correspondingly greater degree of preferential flow down the field-lines; but the anisotropy of flow required to yield F/M in agreement with observation is so great that we are forced to ask whether the assumption of field-freezing remains good all the way to the pre-main sequence stage.

The flux-freezing approximation

According to the Field-Pikel'ner model, even the hot inter-cloud gas is only about 10 per cent ionized, while in normal HI clouds almost all the hydrogen is neutral, the free electrons coming mainly from elements such as carbon and magnesium with low ionization potentials. The familiar magneto-hydrodynamic model of flux-freezing in a fully ionized gas (cf. lecture I) is now replaced to adequate approximation by a model in which the ionized fraction of the gas (electrons-plus-ions) is again coupled by electromagnetic induction to the magnetic field, but the ionized gas plus magnetic field moves relative to the neutral bulk of the gas at a rate given by balancing the ion-neutral gas friction against the magnetic force (Mestel and Spitzer 1956):

$$\frac{\partial \underset{\sim}{B}}{\partial t} = \nabla \times (\underset{\sim}{v}_i \times \underset{\sim}{B}) \qquad (15)$$

and

$$\frac{(\nabla \times \underset{\sim}{B}) \times \underset{\sim}{B}}{4\pi} = \delta(\underset{\sim}{v}_i - \underset{\sim}{v}_H). \qquad (16)$$

Here $\underset{\sim}{v}_i$ is the mean velocity of the ions,

$\underset{\sim}{v}_H$ is the mean velocity of the neutral particles (mainly hydrogen),

and the frictional coupling constant δ is

$$\delta = \sigma_{iH} m_H (v_H)_T \left(\frac{n_i}{n_H}\right) n_H^2 = \alpha\left(\frac{n_i}{n_H}\right)\rho^2, \qquad (17)$$

where

σ_{iH} = ion-neutral collision cross-section,

$(v_H)_T$ = thermal speed of the neutral particles,

n_i, n_H = number density of ions and neutrals, respectively.

The associated rate of energy dissipation per unit volume is

$$\delta(\underset{\sim}{v}_i - \underset{\sim}{v}_H)^2 = \left(\frac{(\nabla\times\underset{\sim}{B})\times\underset{\sim}{B}}{4\pi}\right)^2 \Big/ \delta = \frac{(\nabla\times\underset{\sim}{B})_\perp^{\ 2}}{4\pi\alpha\left(\frac{n_i}{n_H}\right)\rho^2/B^2}, \tag{18}$$

implying that for currents flowing perpendicular to the field the effective resistivity increases with B.

In the star formation problem, the above equations describe the tendency of field-lines, distorted by the gravitational contraction of a cloud or a fragment, to straighten themselves and so to reduce the flux trapped within. (Magnetic <u>energy</u> is destroyed by this process, but not magnetic <u>flux</u>.)

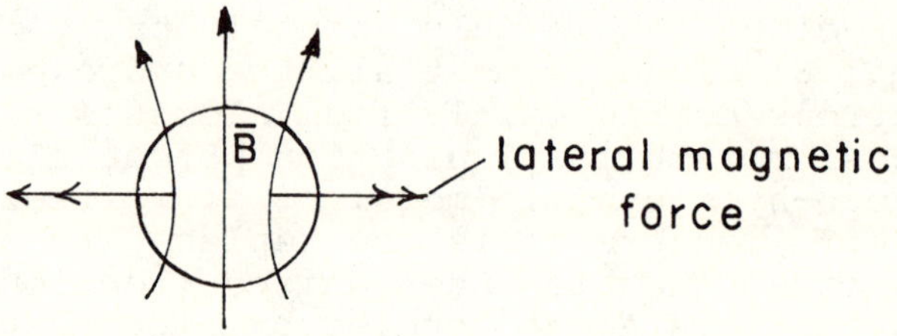

Fig. 3

The characteristic time of this flux-loss is

$$\tau = \frac{\bar{R}}{|\underset{\sim}{v}_i - \underset{\sim}{v}_H|} = \frac{4\pi\delta\bar{R}}{|(\nabla\times\underset{\sim}{B})\times\underset{\sim}{B}|} \simeq \frac{4\pi\alpha\left(\frac{n_i}{n_H}\right)\rho^2\bar{R}^2}{\bar{B}^2} = \frac{9\pi}{4}\,\frac{M^2}{F^2}\,\alpha\left(\frac{n_i}{n_H}\right), \tag{19}$$

where we have written $|\nabla\times\underset{\sim}{B}|$ as $\bar{B}/\bar{R}$. Then if the blob has a flux near the maximum (10) allowed by the virial theorem,

$$\tau \simeq \frac{5}{4\pi}\,\frac{\alpha}{G}\left(\frac{n_i}{n_H}\right). \tag{20}$$

Osterbrock (1961) has estimated $\sigma_{iH} \simeq 10^{-14}$ [a factor 10^2 larger than that used in the first discussion in Mestel and Spitzer (1956)], yielding $\alpha \simeq 6 \times 10^{14}$ in a cloud of temperature of 100°. Then with a standard value of $n_i/n_H \simeq 5 \times 10^{-4}$, τ is longer than the galactic age, so that in normal HI clouds, with the magnetic field scale comparable with the radius of the cloud or the blob, the flux-freezing approximation remains excellent. However, n_i/n_H should certainly fall in dusty HI clouds, firstly because the dust grains absorb the galactic ultraviolet radiation that ionizes the heavy particles with low ionization potentials, and secondly because they provide a locus for the rapid recombination of ions and electrons. As against this, ionization of hydrogen atoms by cosmic rays persists until high densities. In dense clouds the ion density is determined by a balance between cosmic ray ionization, at the rate $\Gamma_{c.r.}$/sec, and attachment to grains:

$$\Gamma_{c.r.} n_H = n_i n_H \left(\frac{\sigma_g n_g}{n_H}\right) (v_i)_T, \qquad (21)$$

where σ_g is the geometrical cross-section of the grains, n_g their number density, and $(v_i)_T$ the thermal speed of the ions. The quantity $\sigma_g n_g / n_H$ is estimated from observed interstellar extinction as $\simeq 4 \times 10^{-22}$ cm^2 (Spitzer 1968). With the cloud temperature near 50° (21) predicts

$$\frac{n_i}{n_H} = \frac{2.6 \times 10^{16} \Gamma_{c.r.}}{n_H}, \qquad (22)$$

so that even for fixed $\Gamma_{c.r.}$ the time (20) will ultimately become less than the free-fall time, which decreases like $1/n_H^{1/2}$ [cf. (16) of lecture VI]. It should be noted that neutralization of the ions by attachment to the grain is crucial, for the rate of radiative recombination of electrons and ions is proportional to n_i^2, so that the appropriate analogue of (21) would yield $n_i/n_H \propto 1/n_H^{1/2}$ and (20) would decrease proportionately to the free-fall time.

With a conservative estimate of $\Gamma_{c.r.} = 7 \times 10^{-18}$ (Spitzer and Tomasko 1968), flux-loss becomes "rapid"--i.e., comparable with the free-fall rate--when $n_H \simeq 3 \times 10^8$. If the low-energy cosmic ray flux is assumed given by linear extrapolation of the observed relativistic range, then $\Gamma_{c.r.}$ would be as high as 6×10^{-15}, and flux-loss would not become rapid until $n_H \simeq 2 \times 10^{13}$. At this density the fragments would in any case have become opaque and built up enough thermal energy for collisional ionization to have restored flux-freezing. However, the absorption of the most energetic cosmic rays is certainly effectively complete by the time the mass per projected area of the blob is $\simeq 80$ gm/cm^2, while smaller values are adequate to extinguish the sub-cosmic rays. Thus we may expect cosmic ray ionization to disappear at some density between 10^8 and 10^{10}/cm^3. The charged grains can act as current carriers and temporarily maintain the field, but the time-constant for dissipation of the field energy by the frictional drag of the neutral gas is short (Spitzer 1963). Ionization by radioactive isotopes (e.g., K^{40}) may, however, take over and keep the ion density high enough for flux-freezing to persist (Cameron 1962).

The upshot is that studies on the early stages of star formation should certainly adopt flux-freezing as an excellent approximation over the bulk of a cloud or fragment, though not excluding local breakdown in singular regions (cf. below). There may be a fairly narrow density range in which the ratio n_i/n_H becomes low enough in dusty clouds or fragments for the flux lines to slip out, before the opaque proto-stellar regime is reached (especially if contraction of fragments is slowed up by centrifugal forces), though this is not absolutely certain. If a proto-star does reach the pre-main sequence phase having retained flux comparable with the virial theorem maximum, it may very well find itself subject to large-scale hydromagnetic instabilities, leading to accelerated shrinkage of flux-loops into 0-type neutral points (cf. lecture II). Some such process of flux destruction would seem to be the only alternative, if the slipping of flux-lines from cold fragments does not occur.

Star formation in rotating magnetic clouds

A gas cloud forming in the non-uniformly rotating galaxy is subject to the disruptive effect of the shear. However, this will certainly be negligible if the differential galactic gravitational field acting across the cloud is much less than the self-gravitation: i.e., if

$$\frac{\left|\frac{d}{dr}\left(\frac{V^2}{r}\right) 2R\right|}{GM/R^2} = \frac{2\Omega_c{}^2 R^3}{GM} = \frac{\Omega_c{}^2}{\frac{2\pi}{3} G\rho} << 1, \tag{23}$$

where $V \equiv \Omega_c r$ is the local (roughly constant) galactic rotational velocity and r is the distance from the galactic center. With Ω_c = 250 km/sec/10 kpc. $\simeq 10^{-15}$/sec the inequality (23) requires $\rho >> 10\ m_H$, as is indeed the case for gravitationally-bound clouds of mass $\simeq 10^3$-$10^4\ M_\odot$.

We saw in lecture VI that only very specially shaped clouds will have angular momentum of spin much below that to be expected from the local galactic rotation (and, in any case, that discussion assumed relaxation into a state of pure rotation with the clouds conserving its angular momentum, whereas we can expect the galactic field-lines threading the clouds to exert some torque on the contracting cloud). A realistic discussion must therefore assume a cloud from which a star cluster forms to have both a strong magnetic flux and a significant angular momentum about its mass-center, so that we must attempt a synthesis of the ideas of the last three lectures (Mestel 1965).

Consider first the simplest (but probably not the most plausible) case of a cool cloud with the overall direction of the magnetic field more-or-less parallel to the angular momentum vector. Then there is no obstacle to flattening of the cloud into a spheroid, so that sub-condensations can form, with the crucial ratios $\Omega^2/\frac{4\pi}{3}G\rho$ and F/M both below the upper limits (cf. lecture VII and above). The new feature is again the coupling of the sub-condensations with the surroundings by the magnetic field-lines. The contraction of a blob increases its spin and so also the shear relative to the background gas, and the consequent magnetic torques systematically transport angular momentum outwards. The process continues as long as there is any differential rotation between the blob and the background and as long as the magnetic field-lines emanating from the blob have not detached themselves from the field of the cloud as a whole. This loss of angular momentum of spin is crucial, for otherwise the contraction and further break-up of the blob will be limited by a combination of centrifugal force, magnetic force and thermal pressure (cf. lecture VII).

Detailed studies of magnetic braking in this context are under way. As a first estimate for its likely efficiency, we compare the time of travel of Alfvén waves with the gravitational free-fall time:

$$\frac{\tau_{\text{Alf.}}}{\tau_{\text{fr.f.}}} = \frac{R/\left(B/(4\pi\rho)^{\frac{1}{2}}\right)}{1/(32G\rho/3\pi)^{\frac{1}{2}}} = \left(\frac{24GM^2}{F^2}\right)^{\frac{1}{2}} > \left(\frac{40}{3\pi^2}\right)^{\frac{1}{2}} \simeq \frac{2\sqrt{3}}{3}, \tag{24}$$

where we have used the critical mass (10) on the right of the inequality. This is in fact a lower limit, as the centrifugal and thermal energies must increase somewhat the minimum gravitationally-bound mass. The tentative conclusion from (24) is that the rate of transport of angular momentum is never likely to be so great that the blob's angular velocity is reduced to that of the surroundings. Instead, after the initial flattening of the cloud the sub-condensations separate out and contract rapidly to a state of approximate centrifugal balance. This is maintained during their subsequent contraction, which takes place at a rate determined by the rate of transport of spin angular momentum, necessarily rather less than the rate of free-fall. Further fragmentation will take place if the blobs contract isothermally and so flatten along the joint magnetic and rotation axis.

The following points should be noted. (1) The postponing of possible flux loss to high densities is a great advantage, in that a weak field would be much less efficient in transferring angular momentum. (2) The "flattening" of the cloud should not be taken too literally: all we require is spontaneous preferential flow down the axis, preferably with retention of a good deal of bulk kinetic energy (cf. lectures VI and VII). (3) In a strictly axisymmetric system there can be a mean differential rotation between one field-line and another; however, if the field axis does not coincide exactly with the rotation axis, magnetic pressure gradients in the azimuthal direction will tend to keep the cloud as a whole rotating more-or-less uniformly in the mean, so that the difficulties associated with shear will not arise (cf. lecture VII). (4) The proto-stars entering the pre-main sequence phase are likely to be rapid rotators. Their further contraction is determined by energy loss rather than changes in centrifugal force. Subsequent magnetic braking (lectures III and IV) is therefore no longer immediately off-set by rapid contraction, and once the star has reached the main sequence loss of angular momentum manifests itself as a reduction in spin.

The angular momentum loss will be sharply reduced if the field-lines of the sub-condensation "snap." It can be shown (Mestel 1966; Mestel and Strittmatter 1967) that a field structure such as in Fig. 1 will not in fact persist indefinitely. In the region between $\bar{R}$ and R_0 the magnetic forces build up a high-density equatorial zone with locally large magnetic field gradients, so that the assumption of strict flux-freezing ceases to be valid over the relevant times: the field relaxes to a quasi-dipolar structure, with most of the field-lines of the blob detached from the background. However, this will not occur until after angular momentum transport has caused a considerable contraction of the blob, which will then flatten and fragment once more. We therefore picture the break-up process occurring as a hierarchy. The mean mass of the ultimate proto-stars would be determined by the growth of opacity, and so is likely to be small (cf. lecture VI).

The problem is more difficult if the magnetic and angular momentum vectors are inclined at a large angle (as one might expect to be the case if the angular momentum is due essentially to the galactic rotation). It is not yet clear whether or not such a cloud can fragment, if the flux is strictly frozen-in. The difficulty is that the flow of gas down the field-lines--necessary so as to decrease $B/\rho^{2/3}$--is inhibited by the centrifugal forces. Magnetic transport of angular momentum to the surrounding gas will cause a roughly isotropic contraction of the cloud as a whole rather than flattening along the field. Perhaps the magnetic stresses will simultaneously cause the angular momentum vector to rotate in space until it is more nearly parallel to the direction of the magnetic field. There is again the possibility that once the cloud field-lines have detached themselves from the background and have acquired a structure with 0-type neutral points, hydromagnetic instabilities may lead to rapid destruction of excess flux, so allowing fragmentation to proceed. It is now not obvious that the masses of the final fragments are determined only by the opacity properties of the cloud: the magnetic field may also be a significant factor. Clearly, much more work remains to be done before we have a coherent theory. And it must be remembered that if the magnetic flux is effectively lost at some stage, then any further angular momentum transport is presumably by turbulence, as emphasized by Cameron (cf. lecture VII).

Accretion by a magnetic proto-star

In appealing to the transport of angular momentum by Alfvén waves, we have implicitly ignored the inward transport of angular momentum carried by inflowing gas. In fact, the angular momentum in a gas cloud surrounding a non-magnetic star is a serious obstacle to accretion (cf. lecture VI). However, in a model with magnetic field-lines emanating from a condensation and extending far into the surrounding cloud, as in Fig. 1, the gas between $\bar{R}$ and R_0 will be of much lower density than that within $\bar{R}$, so that the Alfvén speed is likely to be greater than the free-fall speed in this region, though less than it within $\bar{R}$ [cf. (24)]. Thus gas with high angular momentum which in the absence of the magnetic field could not reach the blob can flow down the field-lines, with a small twist in the field exerting sufficient torque to prevent the angular velocity from growing (Mestel 1959). Once the field of the condensation has detached itself, then one cannot speak of gas from the cloud "flowing down the field-lines" into the blob; but instead, gas may be able to reach the star by "elbowing" the field-lines aside. In this case the inflowing gas can reach the star, but its excess angular momentum will not flow to infinity, but instead will be transformed to the star via the distorted magnetic field-lines.

REFERENCES

Anand, K.C., Daniel, R.R., and Stephens, S.A. 1968a, Proc. Ind. Acad. Sci., 67, 267.

______. 1968b, Phys. Rev. Lett., 20, 764.

Cameron, A.G.W. 1962, Icarus, 1, 13.

Chandrasekhar, S., and Fermi, E. 1953, Astrophys. J., 118, 116.

Clark, B.G. 1965, Astrophys. J., 142, 1398.

Field, G.B. 1969, Introductory Report, Liège Symposium.

Kippenhahn, R., and Schluter, A. 1957, Zeits. f. Astrophys., 43, 36.

Lerche, I. 1967, Astrophys. J., 148, 415.

Lin, C.C., Mestel, L., and Shu, F.H. 1965, Astrophys. J., 142, 1431.

Mestel, L. 1959, Mon. Not. R. Astr. Soc., 119, 223.

______. 1965, Quart. J. R. Astr. Soc., 6, 265.

______. 1966, Mon. Not. R. Astr. Soc., 133, 265.

______. 1969, "The Role of the Magnetic Field in Star Formation," Plasma Instabilities in Astrophysics (Asilomar Conference, 1968)(New York: Gordon and Breach).

Mestel, L., and Spitzer, L., Jr. 1956, Mon. Not. R. Astr. Soc. 116, 583.

Mestel, L., and Strittmatter, P.A. 1967, Mon. Not. R. Astr. Soc., 137, 95.

Osterbrock, D. 1961, Astrophys. J., 134, 270.

Parker, D.A. 1970, in preparation.

Parker, E.N. 1966, Astrophys. J., 145, 811.

Pikel'ner, S.B. 1967, Astr. Zh., 44, 1915.

Spitzer, L., Jr. 1963, in Origin of the Solar System (New York: Academic Press), p. 39.

______. 1968, Diffuse Matter in Space (New York: Interscience).

Spitzer, L., Jr., and Tomasko, M.S. 1968, Astrophys. J., 152, 971.

Strittmatter, P.A. 1966, Mon. Not. R. Astr. Soc., 132, 359.

Verschuur, G.L. 1969a, Astrophys. J., 155, L155.

______. 1969b, Astrophys. J., 156, 861.

Woltjer, L. 1962, in Interstellar Matter in Galaxies, ed. L. Woltjer (New York: Benjamin), p. 88.

21
INTENSE MAGNETIC FIELDS IN ASTROPHYSICS

V. CANUTO* and H. Y. CHIU**
Institute for Space Studies, Goddard Space Flight Center, NASA, New York, N.Y., U.S.A.

(Received 6 November, 1970)

Abstract. In this paper we summarize the current knowledge of research on the influence of intense magnetic fields on physical processes. The contents are summarized in the enclosed Table of Contents.

Table of Contents:

1. Introduction

The presence of magnetic fields in nature is a common phenomenon. Our earth possesses an approximate dipole field aligned at about 15° from its rotational axis with a strength of $\frac{1}{2}$ G, and, according to fossil evidence, the field has an ancient history. The Sun, which is an average star, possesses a magnetic field of apparently complicated structure and configuration. The average value of the solar field (on the surface of the Sun) is 1 G, but this average is derived from a very heterogeneous distribution of fields ranging from zero to several thousands of gauss in sunspots. Many stars are also known to possess magnetic fields with strengths in excess of 500 G, which is the present lower limit of detectability of stellar magnetic fields (C67)[†]. In one case the average field strength is in excess of 3.4×10^4 G, about twice the saturation field of iron! Our Galaxy also possesses a magnetic field whose strength is a few times 10^{-6} G.

* NAS-NRC Senior Postdoctoral Resident Research Associate.
** Also with Physics Dept. and Earth and Space Sciences Dept., State University of New York at Stony Brook.
† These numbers refer to the References at the end of this paper.

Reprinted courtesy of Space Science Reviews, vol. 12, 1971, copyright © D. Reidel Publishing Company, Dordrecht-Holland.

The energy density of the field is comparable to the kinetic energy density of gas in our Galaxy, and the galactic field is believed to have a nonnegligible effect on the structure of our Galaxy (W64a, S66).

The role played by magnetic fields in astrophysics has been extensively discussed (S66, C67, W64, Ca67, CC68d, CC68e). Most fields discussed have strengths much less than 10^5 G and such fields will hereafter be referred to as classical fields since for such fields the quantum effect is not likely to play any significant role in astrophysics. Exceptions are: (1) low temperature physics with a cryogenic temperature; (2) Zeeman splitting of atomic lines, which has been discussed elsewhere. Although on occasion fields more intense than 10^5 G can still be regarded as classical, in no case can quantum effects be entirely neglected when the field strength is greater than 10^8 G.

In the following sections we will briefly summarize the properties of classical magnetic fields, in particular, the source of magnetic fields in astronomical objects and the flux conservation law. In the remaining part of this paper, we will be concerned with the quantum effects of magnetic fields in astrophysics.

2. The Source of Magnetic Fields in Astronomical Objects

As far as we know, magnetic fields can only be generated and maintained by one of the following processes:

(1) Moving charges (electric current).

(2) Alignment of spin magnetic moment.

(3) Alignment of magnetic moment due to orbital angular momentum of some types of atoms.

(4) Landau Orbital Magnetized State.

The presence of an electric current in a conductor can generate a magnetic field according to the Maxwell equation:

$$\operatorname{curl} H = \frac{4\pi}{c} j . \tag{2.1}$$

However, the current density is subject to Ohmic dissipation. In conductors of ordinary size (e.g., coils in a small transformer) the Ohmic dissipation will dissipate a current completely in a matter of milliseconds. In the case of astronomical objects such as stars and nebulae, the conductivity is so high and the inductance so large that the time of decay ranges from millions to billions of years.

On the other hand, in order to align the spin or orbital magnetic moment, invariably a solid crystalline structure is needed. Further, the temperature cannot exceed the Curie temperature, which is of the order of 10^3 K. As a result, although (2) and (3) are important processes in solid matter possessing permanent or semipermanent magnetism, they are of negligible importance in astronomical objects. The new process (4) which can give rise to a semipermanent magnetic field in dense bodies such as neutron stars and white dwarfs, will be discussed in a separate section.

A current is made of moving charges. In a system in thermal equilibrium in which the distribution of velocities is random and isotropic there is no net velocity in any direction, and consequently, a system in complete thermodynamic equilibrium does not possess a magnetic field. A magnetic field will be present if one of the charges (usually electrons) in a medium possesses a small net drifting velocity v_d. Let us take a simple configuration, that of a cylinder of charged particles rotating with an angular velocity ω which in turn gives rise to an average linear drift velocity v_d (in cm/sec). Ley ϱ be the density of matter in the cylinder (in g/cm^3) and Z be the average number of free charge per atom of average mass number A. Then the field at the test point is given by Equation (2.1). With a simple calculation one can show that the expression for the field H is given by (R in cm)

$$H/H_q \cong 10^2 \frac{Z}{A}\left(\frac{v}{c}\right)\varrho R\,, \quad H_q = \frac{m^2c^3}{e\hbar} = 4.414 \times 10^{13}\ \mathrm{G}\,. \tag{2.2}$$

The Planck constant $\hbar$ has been purposely introduced only for comparison with quantum regime. Here R is the linear dimension over which the drifting charges extend. If $\varrho=1$, $Z/A=1$, $R\simeq 0.1$, then a field of 10^4 G can be generated by a relatively small drifting velocity of only 1 cm/sec. This is to be compared with the thermal velocity of the electrons v_{th} which is approximately ($T_4 \equiv 10^{-4}\,T$)

$$v_{\mathrm{th}}/c \simeq 10^{-3} T_4^{1/2} \tag{2.3}$$

where T is the temperature. As another example, let us put in (2.2) the conditions appropriate to a neutron star:

$$\varrho = 10^{14}\ \mathrm{g/cm^3}\,, \qquad Z/A \simeq 10^{-3}\,, \qquad v \simeq 1\ \mathrm{cm/sec} \tag{2.4}$$

we find

$$H \cong 10^{15}\ \mathrm{G}\,.$$

Thus, the existence of a relatively strong field implies only a negligible departure from a state of complete thermodynamic equilibrium.

The origin of magnetic fields has been a knotty problem in astrophysics. It is generally believed that turbulence is the cause for macroscopic drifting velocity for one of the two component charges in a plasma, but up-to-date theory of turbulence is still very crude and does not lend any insight to the solutions of the problem (LL60). The condition for turbulence to exist requires a high Reynolds number which is usually satisfied on low density plasma such as interstellar plasma. It is thus believed that magnetic fields observed in stars originated from their prestellar state.

Once a current is established the flux is determined by the subsequent events of development and by the dissipation. The dissipation is usually small, the time for decay being of the order of millions and billions of years. If the subsequent development of the astronomical object carrying a magnetic field follows a simple scaling law, then the magnetic field is roughly proportional to the square of the scaling factor.

This is easily shown as follows. The Maxwell equation is

$$\operatorname{curl} \mathbf{H} = \frac{4\pi}{c}\mathbf{j} \tag{2.5}$$

where **j**, the current density is equal to the density of charge ϱ_e times the drifting velocity. Under a transformation $r \rightarrow \alpha r$ where α is a scaling factor, then the density of charge ϱ_e increases as α^{-3}. Equations (2.5) is invariant if $\mathbf{H} \rightarrow \alpha^{-2}\,\mathbf{H}$. Thus the invariance of the Maxwell equation requires that $|H| \propto r^{-2}$ (this result can also be derived differently fron flux conservation law of a closed loop of current).

3. Classical and Quantizing Fields

In a magnetic field a charged particle suffers a force perpendicular to the field and the direction of motion. The equation of motion is:

$$\dot{\mathbf{p}} = e\mathbf{E} + \frac{e}{c}\mathbf{v} \times \mathbf{H} \tag{3.1}$$

where **E** is the electric field, **p** is the momentum, and $\mathbf{v} \times \mathbf{H}$ is the magnetic force. Consider now the case $E=0$. In this case the force is always perpendicular to the instantaneous direction of the motion, and hence no work is done on the particle. As a consequence the energy of a particle in a magnetic field is invariant.

In a constant uniform magnetic field along the z-axis the solution of (3.1) is readily obtained as (LL62)

$$\begin{aligned} x &= x_0 + R_L \sin(\omega t + \alpha) \\ y &= y_0 + R_L \cos(\omega t + \alpha) \\ z &= z_0 + v_z t \end{aligned} \tag{3.2}$$

where

$$\begin{aligned} R_L &= v_\perp/\omega_L = \frac{v_\perp E}{ecH} = \frac{cp_\perp}{eH} = \gamma v mc/eH = \left(\frac{H}{H_q}\right)^{-1} \gamma\,(v/c)\,(\hbar/mc) \\ R_L &= \left(\frac{H}{H_q}\right)^{-1} \gamma\beta\lambda\!\!\!^{-}_c, \qquad \omega_L = ecH/E, \qquad \gamma^{-2} = 1 - (v/c)^2 \\ H_q &= m^2c^3/e\hbar, \qquad \omega_L = (H/H_q)\,(E/mc^2)^{-1}\,(mc^2/\hbar) \end{aligned} \tag{3.3}$$

Here $\lambda\!\!\!^{-}_c$ is the Compton wavelength for the electron (3.8615×10^{-11} cm). The trajectories are helices with a constant velocity v_z in the z-direction. $v_\perp$ is the velocity of the particle in the plane perpendicular to the field and E is the total energy of the electron:

$$E = \gamma mc^2 \qquad v^2 = v_\perp^2 + v_z^2 = v_x^2 + v_y^2 + v_z^2 \tag{3.4}$$

R_L is the radius of the orbit projected in the x, y-plane and ω_L is the angular frequency of the circular motion in the x, y-plane (the Larmor frequency). (x_0, y_0) is the center of the orbit (the *guiding center*). The presence of a magnetic field therefore confines the motion of the charges particle in the x, y-plane.

If the field is not uniform and if the non-uniformity is small (i.e. $1/H|\text{grad}\, H|^{-1} \gg R_L$) then it has been shown that if grad H is in the x, y-plane, the guiding center slowly moves toward regions of weaker fields. If $|\text{grad}\, H|$ is along the field line, then as the particle moves toward regions of weaker fields the energy in the xy-plane is slowly transferred to that in the parallel direction and as the particle moves toward regions of stronger fields the energy in the parallel direction is transferred into that in the perpendicular plane. Reflections of particles can take place if the magnetic field gradient is sufficiently large.

We shall assume that the gradient of the magnetic field is small and that the circular motion of the electron is unaffected by the non-uniformity of the field.* As the field strength increases R_L decreases and ω_L increases. Consider now the de-Broglie wavelength of a particle

$$\lambda_B = \frac{h}{p} = \frac{h}{m\gamma v} = \frac{2\pi\lambda_c}{\gamma\beta}.$$

Comparing it with the Larmor radius R_L, Equation (3.3) we found that quantum effects are important where

$$R_L \leqslant \lambda_B$$

or

$$\gamma^2\beta^2 \equiv \frac{\beta^2}{1-\beta^2} \lesssim 2\pi \frac{H}{H_q}$$

we have

$$\frac{v}{c} = \left[\frac{2\pi k\,(H/H_q)}{1 + 2\pi k\,(H/H_q)}\right]^{1/2} \qquad k \geqslant 1\,.$$

We see that a particle with a velocity of 10^7 cm/sec ($\simeq 1$ eV) enters in the quantum domain at $H \simeq 10^7$ G. The classical equation of motion (3.1), from which the classical trajectories of a free electron are derived, is therefore expected to be invalid. The solution for the trajectory then calls for the use of the Schrodinger equation or the Dirac equation. According to these solutions the circular orbits in a magnetic field are quantized.

The quantized total electron energy is (R28, P30, H31, JL49, CC68a, b, c)

$$E\,(p_z, n) \equiv E = mc^2\,[1 + (p_z/mc)^2 + 2nH/H_q]^{1/2}\,. \tag{3.5}$$

$n = 0, 1, 2, \ldots \infty$ is the principal quantum number which characterizes the sizes of the orbits, which are referred to as *Landau levels*. Comparing Equation (3.5) with the

* In the case of quantizing fields the Larmor orbit is ~ de-Broglie wavelength $\leqslant 10^{-8}$ cm and the Larmor period is $\leqslant 10^{-8}$ sec. If the field gradient is less than $H/(h/mc)$ and the time rate of change is less than $H/(h/mc^2)$, all fields can be regarded as time constant and homogeneous in space.

usual expression for the energy of an electron

$$E = mc^2[1 + (p_z/mc)^2 + (p_x/mc)^2 + (p_y/mc)^2]^{1/2} \tag{3.6}$$

we find that quantization replaces the x- and y-momenta $p_x^2+p_y^2$ by the quantity $2n(H/H_q)\,m^2c^2$. In other words the momentum of the electron in the plane perpendicular to the field is quantized into a discrete set according to the formula:

$$p_\perp^2 \equiv p_x^2 + p_y^2 = 2n\,(H/H_q)\,m^2c^2\,. \tag{3.7}$$

This discrete set becomes continuous as $n\to\infty$. According to Bohr's correspondence principle, when n is large, classical trajectories of the electrons can be applied. However, for small values of n the discrete behavior of the perpendicular ($\perp$) momentum must be taken into account. If we assume that the parallel ($\|$) momentum is approximately the same as the $\perp$-momentum, then the criteria for application of the classical trajectory is when

$$(p_z/mc)^2 \gg 2\,(H/H_q)\,. \tag{3.8}$$

If we can assign a temperature to the electron, then in the nonrelativistic case $p_z^2/2m \sim kT$, and Equation (3.8) becomes:

$$kT/mc^2 \gg H/H_q \tag{3.9}$$

or, numerically,

$$T\,(\mathrm{K}) \gg 10^{-4}\,H\,(\mathrm{G})\,. \tag{3.10}$$

That is, at $H=10^{11}$ G, classical trajectory analysis is inapplicable even at a temperature of 10^7 K.

In the relativistic limit $p_zc \sim kT$ and Equation (3.8) becomes:

$$(kT/mc^2)^2 \gg 2\,(H/H_q)\,. \tag{3.11}$$

At a field of 10^{13} G even at a temperature of 5.9×10^9 K quantization of orbits have to be taken into account.

In a degenerate medium p_z is replaced by the Fermi momentum p_F. A similar analysis may be made on the applicability of classical trajectory.

A. STRONG COUPLING REGIME

An electron in a magnetic field can make transitions from one orbit to another accompanied by the emission of a photon. Such radiation is called the synchrotron radiation (to be discussed later). In classical regimes the energy loss per orbit is usually negligible, and in computing the trajectories (3.2), the energy loss can be treated as a small perturbation.

In a strong field or when the electron energy is high the energy loss rate will be so large that a substantial amount of energy of the electron will be lost before a complete orbit is described. This is the regime of the strong coupling. Two effects must then be taken into account.

(a) The perturbation theory of computing energy loss rate is not applicable and the complete wave function must be used to evaluate the energy loss rate and classical trajectories cannot be used. This has been worked out by Klepikov (K54) and his result has been extensively discussed by Erber (E66).

(b) The radiation reaction must be taken into account. This has been discussed to some extent by Erber, but no quantitative result is available yet.

This regime is characterized by the condition that the lifetime of electrons against synchrotron radiation loss is smaller or comparable to the Larmor period. From the expression of the Larmor period and the lifetime against synchrotron radiation (to be discussed later), we find that the condition for strong coupling is ($\gamma = E/mc^2$)

$$\Gamma \equiv \frac{e^2}{\hbar c} \gamma^2 \frac{H}{H_q} \gtrsim 1 \quad \text{or} \quad \gamma^2 H \geqslant 6 \times 10^{15} .$$

4. The Impossibility of Spontaneous Pair Creation in a Magnetic Field

It has often been erroneously stated that in a field greater than $2H_q = 8.8 \times 10^{13}$ G, spontaneous pair creation can take place. The basis of this argument is as follows. In a magnetic field the non-relativistic spin interaction hamiltonian is given by $\sim (q\hbar/2mc)\, \boldsymbol{\sigma}\cdot\mathbf{H}$ where $q = -e$ for electron and $q = e$ for positron, with $e > 0$. The non-relativistic magnetic energy of an electron with its spin antiparallel to the field is $-\frac{1}{2}e\hbar/mc\, H$. The corresponding quantity for a positron parallel to the field is $\frac{1}{2}e\hbar/mc\, H$. When H is greater than $2H_q = 2m^2c^3/e\hbar$ the non-relativistic magnetic energy is then greater than $2mc^2$, and it was thus concluded that a pair of electron and positrons would be created with spins opposite to each other properly aligned with the field. However, according to Equation (3.5) the lowest state of energy of an electron in a magnetic field remains unchanged and the separation between the positive and negative energy states is still $2mc^2$. Therefore, spontaneous pair creation cannot take place. When the anomalous magnetic moment of the electron is taken into account (TBZ66, CCFC68), Equation (3.5) is changed to ($s = \pm 1$)

$$E = mc^2 \left\{ x^2 + \left[\sqrt{1 + 2nH/H_q} + s \frac{\alpha}{4\pi} H/H_q \right]^2 \right\}^{1/2}, \qquad x = p_z/mc .$$

When $x = 0, n = 0$ for $\alpha H/4\pi H_q = 1$ or $H \cong 10^{16}$ G, the lowest energy eigenvalue is zero and therefore spontaneous pair creation could occur. This conclusion is however invalidated by the fact that for high magnetic field, the form of the anomalous magnetic moment cannot be taken to be simply $\alpha/2\pi$. This point has been emphasized by Jancovici (J69). An asymptotic expression for the anomalous magnetic moment when $H \gg H_q$ can be found in (TBBD69).

5. Thermodynamic Properties

As is well known in classical plasma physics the presence of a magnetic field introduces an anisotropy only from the magnetic stress (LL62)

$$T_{\alpha\beta} = \frac{1}{4\pi}[-E_\alpha E_\beta - H_\alpha H_\beta + \tfrac{1}{2}\delta_{\alpha\beta}(E^2 + H^2)] \tag{5.1}$$

leaving the equation of state unchanged.* This is because the radius of the classical orbits takes continuous values and all values of ⊥-momentum are available. In the case of strong quantization, the ⊥-momentum must assume the quantized values given in Equation (3.7).

The equation of state for a gas of electrons may be obtained by evaluating the energy momentum tensor (AB65)

$$T_{\mu\nu} = \tfrac{1}{2}\hbar c[\bar{\psi}\gamma_\mu \partial_\nu \psi - \partial_\nu \bar{\psi}\gamma_\mu \psi] \tag{5.2}$$

using the exact wave function solutions of the Dirac (R28, P30, H31, CC68, L149, R52, S60) or Schrodinger (K27, D28, L30, P30, UY30) equation. This has been done in a previous paper (CC68a). In this paper we will present an alternative approach based on a re-interpretation of the velocity and the generalization of the usual procedure in deriving the equation of state (CCFC68a).

Because of the cylindrical symmetry of the problem, $T_{xx} = T_{yy}$. If $P_\perp$ is the pressure in the direction perpendicular to the field and $P_\parallel$ the pressure in the parallel direction, then

$$P_\perp = \langle T_{xx}\rangle = \langle T_{yy}\rangle \qquad P_\parallel = \langle T_{zz}\rangle \tag{5.3}$$

where the symbol ⟨...⟩ stands for a sum of the contribution of all particles according to the distribution function $f(p)$.

The pressure is the force exerted on the wall of a container during the reflection of particles of velocity v from the wall. In a reflection the exchange of momentum is $2p_x$ and the rate of collision is $v_x/2 = \frac{1}{2}\partial E/\partial p_x = \frac{1}{2}c^2 P_x/E$, hence (in the x-direction, for example)

$$P_\perp = \langle 2p_x \tfrac{1}{2}v_x\rangle = \left\langle \frac{c^2 p_x^2}{E} \right\rangle = \left\langle \frac{c^2 p_y^2}{E} \right\rangle \tag{5.4}$$

$$P_\parallel = \left\langle \frac{c^2 p_z^2}{E} \right\rangle. \tag{5.5}$$

In the quantized case we have

$$c^2 p_x^2 + c^2 p_y^2 \to m^2 c^2 (H/H_q)\, 2n \tag{5.6}$$

* As is well known, in the viscosity-free case all diagonal elements vanish and T_{xx}, T_{yy}, T_{zz} are pressure in the x, y, and z direction.

hence we can write Equations (5.4) and (5.5) as

$$P_\perp = \left\langle \frac{c^2 p_x^2}{E} \right\rangle = \left\langle \frac{c^2 p_y^2}{E} \right\rangle = \tfrac{1}{2} m^2 c^4 \frac{H}{H_q} \left\langle \frac{2n}{E} \right\rangle \tag{5.7}$$

$$P_\parallel = \left\langle \frac{c^2 p_z^2}{E} \right\rangle. \tag{5.8}$$

The statistical average, as indicated by the symbol $\langle \ldots \rangle$ is achieved by multiplying the quantity of interest by the Fermi distribution function $f(p_z, n)$

$$f(p_z, n) = \{1 + \exp \tilde{\beta} [E(p_z, n) - \tilde{\mu}]\}^{-1} \tag{5.9}$$

where $\tilde{\beta} = (kT)^{-1}$, T is the temperature and $\tilde{\mu}$ is the chemical potential plus the electron rest mass, mc^2.

The summation over p_z in the average $\langle \ldots \rangle$ is carried out as usual, that is,

$$\sum_{p_z} \to \frac{1}{2\pi\hbar} \int_{-\infty}^{+\infty} \mathrm{d}p_z \tag{5.10}$$

and the summation over p_x and p_y is carried out as follows:

$$(2\pi\hbar)^2 \sum_{p_x} \sum_{p_y} \to \int_{-\infty}^{+\infty} \mathrm{d}p_x \int_{-\infty}^{+\infty} \mathrm{d}p_y = \int_0^\infty p_\perp \, \mathrm{d}p_\perp \int_0^{2\pi} \mathrm{d}\phi = \pi \int_0^\infty \mathrm{d}p_\perp^2 \tag{5.11}$$

where $\phi = \tan^{-1} p_y/p_x$. Quantization requires that

$$p_\perp^2 \to m^2 c^4 (H/H_q)\, 2n \tag{5.12}$$

hence

$$\int_0^\infty \mathrm{d}p_\perp^2 \to \sum_{n=0}^{\infty} \omega_n \tag{5.13}$$

where ω_n is the degeneracy of the level n. ω_n is evaluated as follows: when $H = 0$, the number of levels in $\mathrm{d}p_x$ and $\mathrm{d}p_y$ at p_x and p_y is given by

$$(2\pi\hbar)^{-2} \, \mathrm{d}p_x \, \mathrm{d}p_y. \tag{5.14}$$

In the presence of H these levels coalesce into those of a harmonic oscillator, as shown in Figure 1. The degeneracy of each of these levels is therefore given by integrating Equation (5.14) as follows [K65]

$$\omega_n = (2\pi\hbar)^{-2} \int_{A < p_\perp{}^2 < B} \mathrm{d}p_x \, \mathrm{d}p_y$$

where [see Equation (3.7)]

$$A = m^2c^2 \frac{H}{H_q} 2n \qquad B = m^2c^2 \frac{H}{H_q} 2(n+1).$$

$$\frac{dp_x\, dp_y}{(2\pi\hbar)^2}.$$

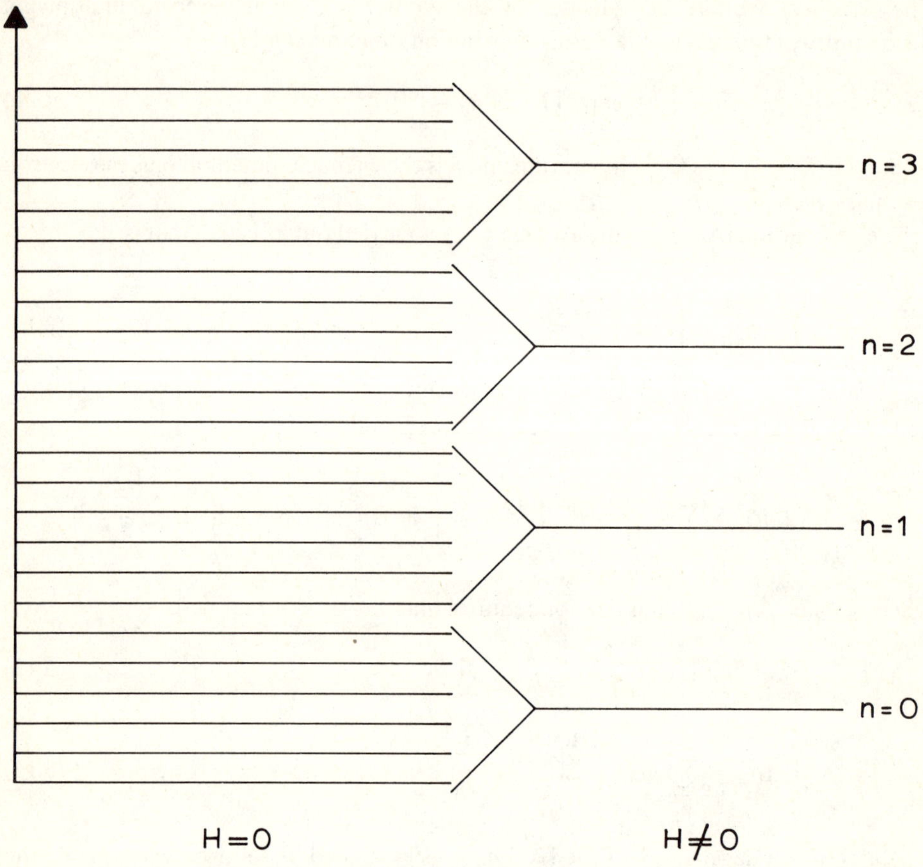

Fig. 1. The coalescence of free particle states into equally spaced harmonic oscillator energy states in the presence of a magnetic field.

Introducing the cylindrical coordinates $(p_\perp, \phi)$, we then obtain:

$$\omega_n = (2\pi\hbar)^{-2} \int_0^{2\pi} d\phi \int_{A<p_\perp^2<B} p_\perp\, dp_\perp = 2\pi(2\pi\hbar)^{-2}\, \tfrac{1}{2}(B-A)$$

$$\omega_n = \frac{1}{2\pi} (\hbar/mc)^{-2} (H/H_q). \tag{5.15}$$

Equations (5.7) and (5.8) therefore become:

$$P_{\perp} = \frac{1}{8\pi^2} \frac{mc^2}{\lambda_c^3} \left(\frac{H}{H_q}\right)^2 \sum_{n=0}^{\infty} \int_{-\infty}^{+\infty} \mathrm{d}\,(p_z/mc) \frac{2n\, mc^2}{E} f\,(p_z, n) \tag{5.16}$$

$$P_{\parallel} = \frac{1}{8\pi^2} \frac{mc^2}{\lambda_c^3} \left(\frac{H}{H_q}\right) \sum_{n=0}^{\infty} \int_{-\infty}^{+\infty} \mathrm{d}\,(p_z/mc) \frac{c^2 p_z^2}{E} f\,(p_z, n). \tag{5.17}$$

On the first look it might be deduced that $P_{\perp} \propto H^2$ and $P_{\parallel} \propto H$. However, since H also appears inside the integrals the functional dependence of P on H is rather complicated and it will be discussed later. The energy density U and the particle density N are obtained in the usual manner as follows:

$$U = \sum_{p_x} \sum_{p_y} \sum_{p_z} (E - mc^2)\, f\,(p_z, n) \tag{5.18}$$

$$U = \frac{1}{4\pi^2} \frac{mc^2}{\lambda_c^3} \frac{H}{H_q} \sum_{n=0}^{\infty} \int_{-\infty}^{+\infty} \frac{E - mc^2}{mc^2} f\,(p_z, n)\, \mathrm{d}\,(p_z/mc) \tag{5.19}$$

$$N = \frac{1}{4\pi^2} \frac{1}{\lambda_c^3} \frac{H}{H_q} \sum_{n=0}^{\infty} \int_{-\infty}^{+\infty} f\,(p_z, n)\, \mathrm{d}\,(p_z/mc). \tag{5.20}$$

These thermodynamic functions appear to be complicated, but they can be simplified considerably by noting that there is a degeneracy between the levels n, $s = +1$ and $n+1$, $s = -1$. This amounts to saying that all the levels are doubly degenerate except the one with $n = 0$.

$$P_{xx} = P_{yy} = P_0 \theta^2 \sum_{n=1}^{\infty} n \int_0^{\infty} \frac{f\,(x, n)\, \mathrm{d}x}{\varepsilon\,(x, n, \theta)} \tag{5.21}$$

$$P_{zz} = P_0 \theta \left[\tfrac{1}{2} \int_0^{\infty} \frac{x^2 f\,(x, 0)\, \mathrm{d}x}{\varepsilon\,(x, 0, \theta)} + \sum_{n=1}^{\infty} \int_0^{\infty} \frac{f\,(x, n)\, x^2\, \mathrm{d}x}{\varepsilon\,(x, n, \theta)} \right] \tag{5.22}$$

$$U = U_0 \theta \left[\tfrac{1}{2} \int_0^{\infty} f\,(x, 0)\, \varepsilon\,(x, 0, \theta)\, \mathrm{d}x + \sum_{n=1}^{\infty} \int_0^{\infty} f\,(x, n)\, \varepsilon\,(x, n, \theta)\, \mathrm{d}x \right] \tag{5.23}$$

$$N = N_0 \theta \left[\tfrac{1}{2} \int_0^{\infty} f\,(x, 0)\, \mathrm{d}x + \sum_{n=1}^{\infty} \int_0^{\infty} f\,(x, n)\, \mathrm{d}x \right] \tag{5.24}$$

where

$$f^{-1}(x, n) = 1 + \exp\{[\varepsilon(x, n, \theta) - \mu]/(T/T_0)\}$$
$$\varepsilon(x, n, \theta) = 1 + x^2 + 2n\theta, \qquad \theta = H/H_q, \qquad \mu = \tilde{\mu}/mc^2$$
$$P_0 = U_0 = \frac{1}{\pi^2}\frac{mc^2}{\lambda_c^3} = 1.4407 \times 10^{24}\ \mathrm{erg/cm^3}$$
$$= 1.4407 \times 10^{24}\ \mathrm{dyn/cm^2}$$
$$N_0 = \frac{1}{\pi^2}\frac{1}{\lambda_c^3} = 1.7598 \times 10^{30}\ \mathrm{cm^{-3}},$$
$$T_0 = mc^2/k = 5.903 \times 10^9\ \mathrm{K}. \tag{5.25}$$

By means of the following transformations (CC68b)

$$v = x/a_n \qquad a_n^2 = 1 + 2n\theta$$
$$\varepsilon(x, n, \theta) = a_n(1 + v^2)^{1/2} \tag{5.26}$$

the equations of state can be expressed in terms of the following functions:

$$C_1(T, \mu) = \int_0^\infty \frac{f(\mu, T, v)}{\sqrt{1 + v^2}}\, dv \tag{5.27}$$

$$C_2(T, \mu) = \int_0^\infty \frac{v^2}{\sqrt{1 + v^2}} f(\mu, T, v)\, dv \tag{5.28}$$

$$C_3(T, \mu) = \int_0^\infty (1 + v^2)^{1/2} f(\mu, T, v)\, dv = C_1(T, \mu) + C_2(T, \mu) \tag{5.29}$$

$$C_4(T, \mu) = \int_0^\infty f(\mu, T, v)\, dv \tag{5.30}$$

where

$$f(\mu, T, v) = \left[1 + \exp\left(\frac{\sqrt{1 + v^2} - \mu}{T/T_0}\right)\right]^{-1}. \tag{5.31}$$

The results are (CC68b)

$$P_{xx} = P_{yy} = P_0\theta^2 \sum_{n=1}^{\infty} n\, C_1(T/a_n, \mu/a_n) \tag{5.32}$$

$$P_{zz} = P_0\theta \left[\tfrac{1}{2} C_2(T, \mu) + \sum_{n=1}^{\infty} a_n^2\, C_2(T/a_n, \mu/a_n)\right] \tag{5.33}$$

$$U = U_0\theta \left[\tfrac{1}{2} C_3(T, \mu) + \sum_{n=1}^{\infty} a_n^2\, C_3(T/a_n, \mu/a_n)\right] \tag{5.34}$$

$$N = N_0\theta \left[\tfrac{1}{2} C_4(T, \mu) + \sum_{n=1}^{\infty} a_n\, C_4(T/a_n, \mu/a_n)\right]. \tag{5.35}$$

The C_k functions are average values of dynamic variables of a one-dimensional gas. $C_1(T, \mu)$ is the average value of E^{-1} where $E^{-1}=(1+v^2)^{1/2}$ is the total energy of a one-dimensional particle of unit mass and momentum v. $C_2(T, \mu)$ is the average value of $v\,\mathrm{d}v/\mathrm{d}E$ whose statistical average gives the pressure of a one-dimensional gas. C_3 is the average of E and C_4 is the average particle density (in appropriate units). The properties of a magnetized Fermi gas therefore are closely related to those of a one-dimensional gas. This is intuitively clear since an electron in a magnetic field is quantized in energy in the $\perp$-direction and moves freely only in the $\|$-direction.

The properties of the C_k functions have been studied extensively in the general case of a Fermi gas [CC68b]. No simple inversion formula expressing T and μ in terms of C_k's is known. In the following we will study 2 cases, the non-degenerate and the degenerate case.

In the nondegenerate case the factor 1 can be neglected in the denominator of the integrand of the C_k's. Since the relativistic case is always marked by some degree of degeneracy (because of pair creation at relativistic temperatures) we will consider the non-relativistic case, $v \ll 1$. In this case ($\mu' = \mu - 1$)

$$f(\mu, T, v) = \exp[\mu'/(T/T_0)] \exp[-v^2/(2T/T_0)] \quad (5.36)$$

and

$$C_1(T, \mu) = \int_0^\infty (1+v^2)^{1/2} \exp[\mu'/(T/T_0)] \exp[-v^2/(2T/T_0)]\,\mathrm{d}v$$
$$\cong \sqrt{\pi T/2T_0}\, \exp[\mu'/(T/T_0)] \quad (5.37)$$

$$C_2(T, \mu) = \int_0^\infty v^2 \exp[\mu'/(T/T_0)] \exp[-v^2/(2T/T_0)]\,\mathrm{d}v$$
$$= (\pi T/2T_0)^{1/2} (T/T_0) \exp[\mu'/(T/T_0)] \quad (5.38)$$

$$C_3(T, \mu) \simeq \int_0^\infty (1+\tfrac{1}{2}v^2) \exp[\mu'/(T/T_0)] \exp[-v^2/(2T/T_0)]\,\mathrm{d}v$$
$$= (\pi T/2T_0)^{1/2} (1 + T/2T_0) \exp[\mu'/(T/T_0)] \quad (5.39)$$

$$C_4(T, \mu) = C_1(T, \mu) = (\pi T/2T_0)^{1/2} \exp[\mu'/(T/T_0)]. \quad (5.40)$$

We therefore find:

$$P_{xx} = P_{yy} = P_0\theta^2 \sum_{n=1}^{\infty} n(\pi T/2T_0 a_n)^{1/2} \exp[(\mu T_0/T) - (T_0 a_n/T)]. \quad (5.41)$$

In the non-relativistic case $T \ll T_0$ and the requirement of non-degeneracy implies $\mu T_0 \simeq T$. Let us consider the case $n\theta \ll 1$ so that

$$a_n = (1 + 2n\theta)^{1/2} \simeq 1 + n\theta \simeq 1.$$

Then

$$P_{xx} = P_{yy} = P_0\theta^2 (\pi T/2T_0)^{1/2} \exp[\mu'/(T/T_0)] \sum_{n=1}^{\infty} n \exp(-n\theta T_0/T)$$
$$= P_0\theta^2 (\pi T/2T_0)^{1/2} \exp[\mu'/(T/T_0)] \exp(\theta T_0/T) [\exp(\theta T_0/T) - 1]^{-2}. \quad (5.42)$$

Similarly

$$P_{zz} = P_0\theta (\pi/2)^{1/2} (T/T_0)^{3/2} \exp(\mu T_0/T) \left\{\frac{1}{2} + \frac{\exp(\theta T_0/T)}{\exp(\theta T_0/T) - 1}\right\}. \quad (5.43)$$

The internal energy density is a little bit more involved. Note that $a_n^2 = 1 + 2n\theta$ and $a_n^2 \cong 1$, hence

$$U/U_0 = \theta \left\{\tfrac{1}{2}[C_3(T, \mu) - C_4(T, \mu)] + \sum_{n=1}^{\infty} [a_n^2 C_3(T/a_n, \mu/a_n) - a_n C_4(T/a_n, \mu/a_n)]\right\}$$
$$\cong \theta\sqrt{\pi}(T/2T_0)^{3/2} \exp(\mu' T_0/T)$$
$$\times \left\{\tfrac{1}{2} + [\exp(\theta T_0/T) - 1]^{-1} + 2\sqrt{\pi}\frac{T_0\theta}{T}\frac{\exp(\theta T_0/T)}{[\exp(\theta T_0/T) - 1]^2}\right\}. \quad (5.44)$$

Analogously

$$N = N_0\theta\sqrt{\pi}(T/2T_0) \exp[\mu'/(T/T_0)] [\tfrac{1}{2} + \{\exp(\theta T_0/T) - 1\}^{-1}]. \quad (5.45)$$

From quantum statistics the ratio of particles in states separated by an energy ΔE is $\exp-(\Delta E/kT)$. The energy separation between adjacent Landau levels is θmc^2 and $\Delta E/kT$ becomes $\theta T_0/T$. When $\theta T_0/T \gg 1$, most electrons are in the ground state, that is, the state of one-dimensional particle with no $\perp$ motion. Dividing Equation (5.44) by (5.45), we then find that the heat capacity per particle, $c_v = U/N$, approaches $\frac{1}{2}kT$ in the limit $\theta T_0/T \gg 1$, as expected from the law of equipartition which states that each degree of freedom is associated with an energy of $\frac{1}{2}kT$. In this limit P_{xx} and P_{yy} also vanish, to the order $\exp(-\theta T_0/T)$ as expected from the behavior of a one-dimensional gas.

A. DEGENERATE CASE

From the expressions (5.32)–(5.35) the 'equivalent' Fermi energy of the state is μ/a_n. The criteria for degeneracy of the n state is $(\mu/a_n - 1) \gg kT$. As n increases, this inequality becomes weaker and weaker. Therefore, at a given temperature the higher states are always less degenerate.

Similarly (CC68b)

$$C_2(0, \mu) \equiv C_2(\mu) = \tfrac{1}{2}\mu(\mu^2 - 1)^{1/2} - \tfrac{1}{2}\ln[\mu + (\mu^2 - 1)^{1/2}]$$
$$= \tfrac{1}{2}\mu(\mu^2 - 1)^{1/2} - \tfrac{1}{2}C_1(\mu) \quad (5.46)$$

$$C_3(0, \mu) \equiv C_3(\mu) = \tfrac{1}{2}\mu(\mu^2 - 1)^{1/2} + \tfrac{1}{2}\ln[\mu + (\mu^2 - 1)^{1/2}] \quad (5.47)$$

$$C_4(0, \mu) \equiv C_4(\mu) = (\mu^2 - 1)^{1/2}. \quad (5.48)$$

TABLE I

$C_\varkappa(\mu)$.

μ	$C_1(\mu)$	$C_2(\mu)$	$C_3(\mu)$	$C_4(\mu)$
1	0	0	0	0
1.25	0.69315	0.12218	0.81532	0.75000
1.5	0.96242	0.35731	1.31974	1.11803
1.75	1.15881	0.67722	1.83603	1.43614
2.00	1.31696	1.07357	2.39053	1.73215
2.5	1.56680	2.08071	3.64509	2.29129
3.0	1.76275	3.36127	5.12401	2.82843
3.5	1.92485	4.90726	6.83210	3.35410
4.0	2.06344	6.71425	8.77769	3.87298
5.0	2.29243	11.10123	13.39366	4.89898
6.0	2.47789	16.50930	18.98718	5.91608
7.0	2.63392	22.93175	25.56567	6.92820
8.0	2.76866	30.36469	33.13335	7.93725
10.0	2.99322	48.25276	51.24598	9.94987

In Table I the functions C_k are given for $1 \leqslant \mu \leqslant 10$. It is easy to see from the definition of the C_k functions that $C_k(\mu)=0$ if $\mu<1$. Therefore the sum in Equations (5.32)–(5.35) terminates at s such that

$$a_s \leq \mu < a_{s+1}\,. \tag{5.49}$$

Physically this means that energy levels up to $n=s$ are occupied and all levels above $n=s+1$ are vacant. The last level to be occupied is given by the criterion

$$s = \frac{\mu^2 - 1}{2H/H_q} \quad \text{or} \quad \mu = [1 + 2(H/H_q)\,s]^{1/2}\,. \tag{5.50}$$

Because each time when μ exceeds a_s an extra term is added to the sum in the equation of state, discontinuities in the derivatives of thermodynamic variables exist and at $a_s=\mu$, a 'transition' takes place (see Figure 2). These discontinuities are associated with the behavior of the density of states which also shows such discontinuities in the derivatives, as shown in Figure 3.

In particular when $\mu \leqslant (1+2H/H_q)^{1/2}$, only the first term in the sum appears and such a gas behaves as a one-dimensional gas. In this case the $\perp$-stress vanishes. At $H/H_q=1$, the critical density for transition into a one-dimensional gas is approximately 10^6 g/cm^3. Any finite temperature will destroy this one dimensional behavior, however. From the general expressions (5.32)–(5.35), it is easily derived that if degeneracy prevails, the residual $\perp$-pressure is largely due to the state $n=1$ and is given by

$$P_{xx} = P_{yy} = (2\pi^3)^{-1/2}(H/H_q)^2(mc^2/\lambda_c^3)\left[\frac{T/T_0}{1+2H/H_q}\right]^{1/2}\exp(-\Lambda)$$

$$T\Lambda/T_0 \equiv (1+2H/H_q)^{1/2} - \mu\,. \tag{5.51}$$

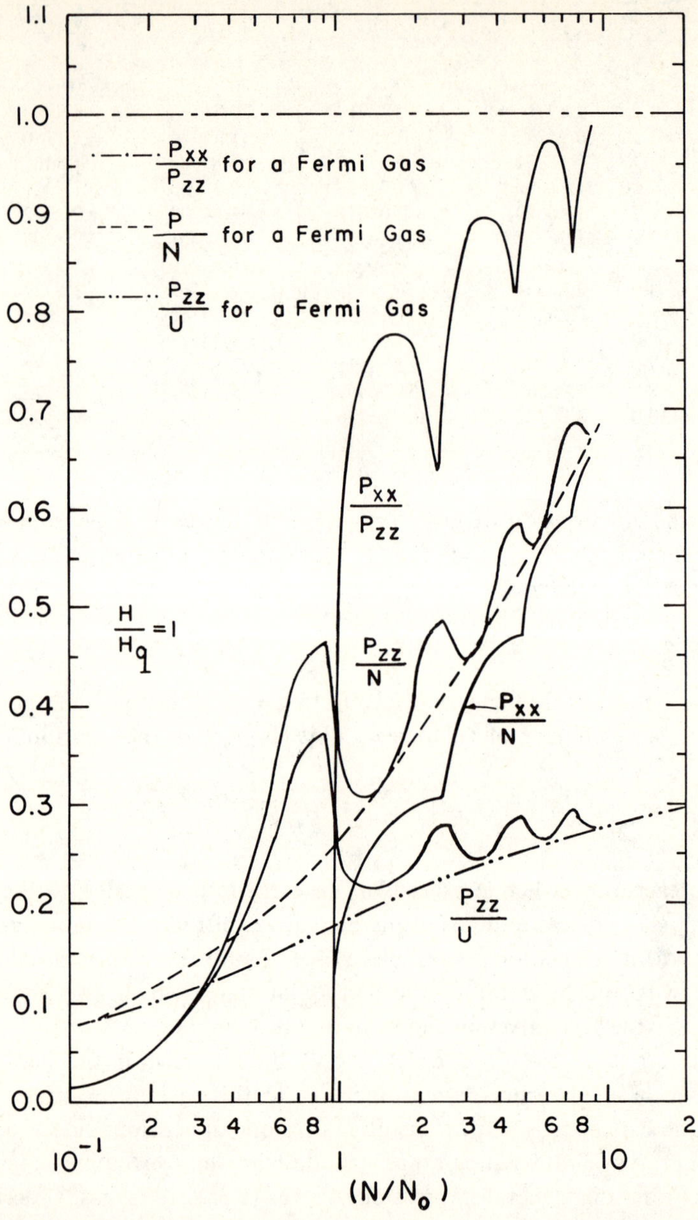

Fig. 2. Functional dependence of P_{xx}/P_{zz}, P_{zz}/N, P_{xx}/N on N/N_0 for the degenerate case at $H/H_q = 1$. The corresponding functions for a Fermi gas are also shown for comparison.

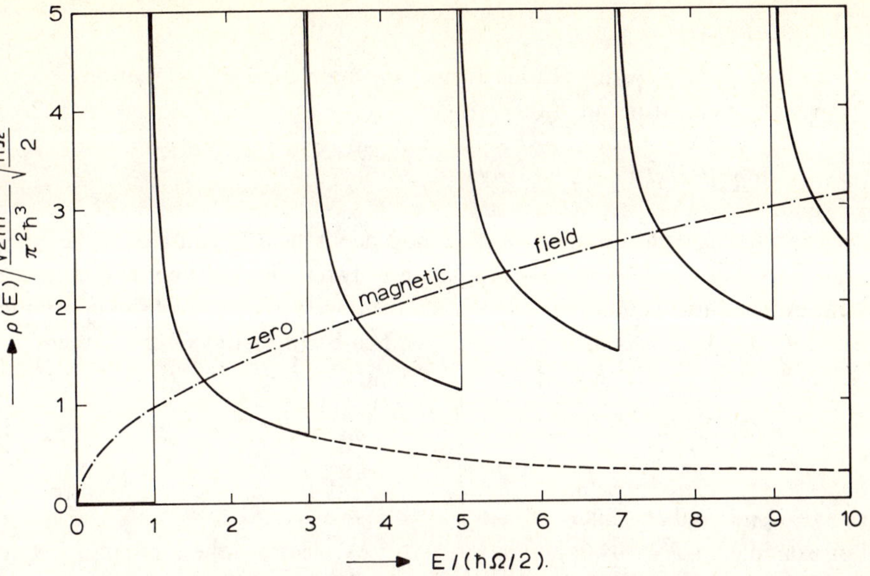

Fig. 3. The oscillating behavior of the density of final states of an electron in a magnetic field (non-relativistic case). [For details on the derivation see KMH65]

6. Radiation Processes in a Magnetic Field

(1) *Synchrotron Radiation.* In a magnetic field an electron can make a transition from one quantized orbit to another, emitting a photon: (if the energy of the photon exceeds $2mc^2$ pairs may be emitted):

$$e^- \rightarrow e^- + \gamma . \tag{6.1}$$

Such a process is strictly forbidden in the classical case. This radiation process is called 'synchrotron radiation' because it was first studied in the design of electron synchrotron accelerators. Equation (6.1) remains the limiting factor in designing circular electron accelerators. In nature, synchrotron radiation from energetic electrons in relatively weak fields is the main source of radio emission (and in some cases also of optical and x-ray emissions).

(2) *Bremsstrahlung Process.* An electron collides with an ion and makes a transition from a state n to another state n' ($n=n'$ and $n \neq n'$ are allowed).

$$e_n^- + (Z, A) \rightarrow e_{n'}^- + (Z, A) + \gamma . \tag{6.2}$$

This radiation process emits a continuum as in the field free case. The bremsstrahlung process takes place in a dense medium and the effect of the medium on the emission process cannot be fully neglected. This problem is discussed in greater detail in Section 12. The rest of this section will then be devoted to the discussion of the synchrotron radiation process.

A. SYNCHROTRON RADIATION

According to the eigenvalues of electrons in a magnetic field (Equation (3.5)), the frequency of synchrotron radiation, ν, is given by the equation:

$$h\nu = E(p_z, n) - E(p_z, n'). \tag{6.3}$$

Therefore synchrotron radiation in a homogeneous field is emitted in the form of discrete lines, but if the field varies by a factor of two in the emitting region these lines are smeared out into a continuum. However, the lowest frequency of this continuum is given by $n = n' + 1$. If we consider the relativistic case with large values of n then

$$\nu = \frac{E(p_z', n) - E(p_z', n')}{h} \simeq \frac{1}{h}\frac{H/H_q}{E/mc^2}\, mc^2 = 2.8 \times 10^{10}\, H\gamma^{-1} \quad \text{(Hz)} \tag{6.4}$$

where $\gamma = E/mc^2$ and H is in G.

For example, at $H = 10^6$ G and $\gamma = 10^6$ (10^{12} eV electron) the minimum frequency ν_m of emission is 3×10^{10} Hz and the corresponding wavelength is 1 cm. ν_m increases with decreasing n, electron parallel energy (p_z), and with increasing H. For example, the energy gap between the ground state $n = 0$ and the first excited state $n = 1$ at a field of 10^{12} G is approximately 50 keV.

As discussed earlier, according to the field strength and the electron energy and the rate of emission, synchrotron radiation is studied in two domains:

a. *The Classical Relativistic Domain.*

This is the regime encountered most frequently in astrophysics: a large electron energy ($\gamma \gtrsim 10^3$) and a low field ($H \lesssim 10^{-2}$ G). Under this circumstance an electron loses a negligible fraction of its perpendicular kinetic energy in each orbit. That is, the mean lifetime of the electron against losing its kinetic energy, τ, must be much larger than the period $P = \omega^{-1}$. As we will show later, this condition is fulfilled if

$$(e^2/\hbar c)\, \gamma^2 (H/H_q) \ll 1$$

or (6.5)

$$\gamma^2 H \ll 6 \times 10^{15}.$$

For example, at $\gamma = 10^6$ the electron no longer radiates according to the classical theory when $H > 10^3$ G. At $\gamma = 10^8$ the limiting field is only 10 G, and at $H = 10^{-6}$ G (galactic field) an electron no longer radiates classically when $\gamma > 10^{11}$ (10^{17} eV).

When (6.5) is satisfied, the 'radiation reaction' is negligible and the rate of emission is given by classical electrodynamics, using the Lienard-Wiechert potential. The rate of emissions has been discussed extensively previously. For completeness we give the results below. We will first give the general quantum mechanical expression for the spectral distribution of the emitted radiation and then some approximate expressions.

Using the exact wave function of a relativistic electron in a magnetic field, Klepikov (K54) first performed the computation of the intensity in the case where the sum over the initial and final quantum numbers can be approximated with an integral. The result turns out to be (per unit distance)

$$I = \frac{3\alpha}{\pi^2}\frac{mc^2}{\lambda\!\!\!^{-}_c}\frac{x}{2+3x}\frac{x^2}{E}\mathcal{M}(x, y) \tag{6.6}$$

$$x = \frac{E}{mc^2}\frac{H}{H_q} \qquad y = \left(\frac{3xE}{2+3x}\right)^{-1} h\nu \qquad \alpha = e^2/\hbar c$$

$$\mathcal{M}(x, y) = \sum_{i=1}^{3} \mathcal{M}_i(h\nu/E)\, J_i(x, y) \tag{6.7}$$

$$\begin{aligned}\mathcal{M}_i(x) &= 1 + (1-x)^{-2} \quad i = 1\\ &= 2(1-x)^{-1} \quad i = 2\\ &= [x(1-x)^{-1}]^2 \quad i = 3\end{aligned} \tag{6.8}$$

$$J_1(xy) = \int_0^\infty ds \cosh^5 s K_{2/3}^2(t)$$

$$J_2(xy) = \int_0^\infty ds \cosh^3 s \sinh^2 s K_{1/3}^2(t) \tag{6.9}$$

$$J_3(xy) = \int_0^\infty ds \cosh^5 s K_{1/3}^2(t), \qquad t = y\cosh^3 s[2 + 3x(1-y)]^{-1}.$$

The structure of Equation (6.6) is quite complex and it is hard to compare it with the classical expression

$$I = \frac{\sqrt{3}\alpha}{2\pi}\frac{mc^2}{\lambda\!\!\!^{-}_c}\frac{x}{E}\, k\left(\frac{2}{3x}\frac{h\nu}{E}\right) \tag{6.10}$$

where the function $k(s)$ is defined as

$$k(z) = z\int_z^\infty dx K_{5/3}(x) \rightarrow \begin{cases} 2.14\, z^{1/3} & z \ll 1 \\ 1.25\, z^{1/2} e^{-z} & z \gg 1 \end{cases}. \tag{6.11}$$

($K_n(x)$ is the McDonald function.) Its behavior can be seen in Figure 4.

There is however, one regime, i.e., when

$$h\nu \ll E \tag{6.12}$$

in which the complex mathematical nature of Equation (6.6) can be reduced to the

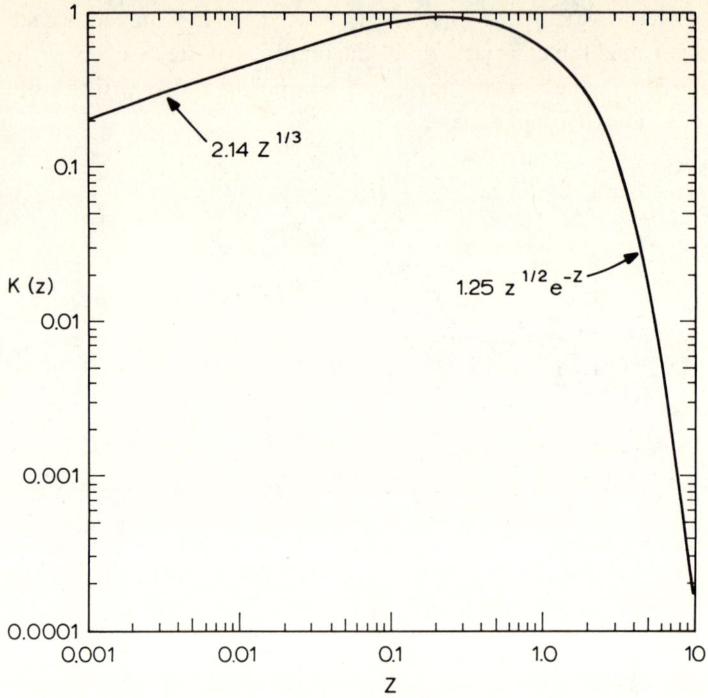

Fig. 4. The bremsstrahlung function $\kappa(z)$, Equation (6.11) of the text.

following simple formula

$$I = \frac{\sqrt{3}\alpha}{2\pi} \frac{mc^2}{\lambda\!\!\!^{-}_c} \frac{x}{E} (1 - h\nu/E)\, k(2s)$$
$$s \equiv \frac{y}{2 + 3x(1-y)} \simeq (3x)^{-1} \frac{h\nu}{E}\left(1 + \frac{h\nu}{E}\right) \tag{6.13}$$

which is almost identical with the classical expression Equation (6.10). The explicit reduction of Equation (6.6) to Equation (6.13) is done in Erber article (E66). One final remark about Equation (6.6) is that its validity is related to the following inequalities being satisfied:

$$\gamma \equiv E/mc^2 > 1 \qquad \gamma - h\nu/mc^2 \gg 1\,.$$

The total energy dissipated per unit distance Δl was obtained by Klepikov by using the general expressions (6.6). The result is

$$\frac{\Delta E}{\Delta l} = \tfrac{2}{3}\alpha \frac{mc^2}{\lambda\!\!\!^{-}_c}\, g(x) \tag{6.14}$$

$$g(x) = \begin{cases} x^2(1 - 5.95x) & x \ll 1 \\ 0.556\, x^{2/3} & x \gg 1\,. \end{cases} \tag{6.15}$$

The total power of emission is given by

$$-\frac{dE}{dt} = \tfrac{2}{3}\alpha \frac{mc^2}{\hbar}\left(\frac{H}{H_q}\right)^2 mc^2 \beta^2\gamma^2 \equiv \Gamma \frac{\beta^2}{1-\beta^2} = \Gamma \frac{E^2 - m^2c^4}{m^2c^4} \tag{6.16}$$

with

$$\Gamma = \tfrac{2}{3}\alpha \frac{mc^2}{\hbar}\left(\frac{H}{H_q}\right)^2 mc^2 .$$

Integration of Equation (6.16) gives the variation of the energy with time, i.e.,

$$E/mc^2 = \coth(\Gamma t/mc^2 + k) . \tag{6.17}$$

As $t \to \infty$, $E \to mc^2$, as it should because the particle is completely stopped. The value of the constant k is easily found to be

$$k = \tfrac{1}{2} \ln \frac{\gamma_0 + 1}{\gamma_0 - 1} \tag{6.18}$$

with

$$\gamma_0 = (E/mc^2)_{t=0} . \tag{6.19}$$

The half-life of the electron, τ, is defined by the equation

$$\coth(\Gamma\tau/mc^2 + k) = \tfrac{1}{2}\gamma_0$$

and it is given by

$$\tau = \tfrac{1}{2} t_s \ln\left(\frac{\gamma_0 + 2}{\gamma_0 - 2}\,\frac{\gamma_0 - 1}{\gamma_0 + 1}\right), \qquad t_s = \frac{mc^2}{\Gamma} \tag{6.20}$$

For $\gamma_0 \gg 1$

$$\tau \simeq t_s/\gamma_0 \tag{6.21}$$

i.e.,

$$\tau = \left[\frac{2}{3}\alpha \frac{mc^2}{\hbar}\left(\frac{H}{H_q}\right)^2 \gamma_0\right]^{-1} . \tag{6.22}$$

At $\gamma_0 \simeq 10^6$ and $H = 10^3$ G, the lifetime is only 10^{-5} sec, corresponding to a mean free path of 3×10^{-5} cm. If we now require that τ should be greater than the period ω_L where ω_L is given by Equation (3.3), we easily obtain

$$\frac{e^2}{\hbar c}\gamma^2 \frac{H}{H_q} < 1 \tag{6.23}$$

or

$$\gamma^2 H < 6.10^{15}$$

as discussed previously [see Equation (6.5)].

b. *The Low Quantum Number Region* (CFC69)

When the energy of the electron is small, only the lower quantum states are occupied. The transition between neighboring states are then important. The nature of the synchrotron radiation is very different from that of the case of large quantum number n.

The radiation rate has been computed in analogy with the general case. The transition between two states n and n' gives rise to a photon of energy

$$\hbar\omega = mc^2 \frac{\varepsilon - \cos\theta p_z/mc}{\sin^2\theta}\left[1 - \left\{1 - \frac{2\sin^2\theta\,(n-n')}{(\varepsilon - \cos\theta p_z/mc)^2}\right\}^{1/2}\right] \tag{6.24}$$

where ε is the energy of the initial state:

$$\varepsilon = E/mc^2 = [1 + (p_z/mc)^2 + 2nH/H_q]^{1/2} \tag{6.25}$$

n' is the quantum number of the final state, and θ is the angle of the emitted photon with respect to the axis of the magnetic field.

It may appear that the emission gives rise to a continuum on account of the θ-dependence. In the limit $H/H_q \ll 1$ we can expand Equation (6.24) with the following result

$$\hbar\omega/mc^2 = (n-n')\,H/H_q + \theta\left[(\hbar\omega/mc^2)^2\cos^2\theta\right] \tag{6.26}$$

which corresponds to a narrow line emission of width $\simeq (\hbar\omega/mc^2)^2 mc^2$. In the case $H/H_q \gg 1$ the line is smeared into a band and the wavelength of the emitted radiation depends on the angle of emission.

The radiation rate (radiation energy per unit volume per time) is

$$I(n) = \frac{e^2 c}{\lambda_c^2}\left[\frac{1}{4\pi^2}\frac{H}{H_q}\frac{1}{\lambda_c^3}\int_{-\infty}^{+\infty} \mathrm{d}x f(x)\right]\sum_{n'}\int_0^{\pi}\sin\theta\,\mathrm{d}\theta\,[1 - f(x')] \times \frac{w^2\varepsilon' F(n, n', w, \theta)}{\varepsilon' - (x-w)\cos\theta} \tag{6.27}$$

with

$$x' = x - w\cos\theta, \qquad x = p_z/mc, \qquad \varepsilon^2 = 1 + x^2 + 2nH/H_q. \tag{6.28}$$

$w = \hbar\omega/mc^2$ is given by Equation (6.24). $f(x)$ is the usual Fermi distribution. The function $F(n, n', w, \theta)$ has the following form

$$\begin{aligned} F(n, n', w, \theta) = {} & [\omega_1 I_{n'-1,n}(y) - \omega_2 I_{n',n-1}(y)]^2 \\ & + \cos^2\theta\,[\omega_1 I_{n'-1,n}(y) + \omega_2 I_{n',n-1}(y)]^2 \\ & + \sin^2\theta\,[\omega_3 I_{n'-1,n-1}(y) - \omega_4 I_{n',n}(y)]^2 \\ & - 2\sin\theta\cos\theta\,[\omega_1 I_{n'-1,n}(y) + \omega_2 I_{n',n-1}(y)] \\ & \times [\omega_3 I_{n'-1,n-1}(y) - \omega_4 I_{n',n}(y)] \end{aligned} \tag{6.29}$$

$$y = w^2\sin^2\theta/[4(H/H_q)]$$

where

$$\left(\eta^2 \equiv 1 + 2n \frac{H}{H_q}\right)$$

$$\begin{aligned}
\omega_1 &= \tfrac{1}{4}[(1 + s'/\eta')(1 - s/\eta)]^{1/2} \\
&\qquad \times \{(1 - w\cos\theta/\varepsilon')^{1/2} + ss'(1 + \cos\theta/\varepsilon')^{1/2}\} \\
\omega_2 &= \tfrac{1}{4}[(1 - s'/\eta')(1 + s/\eta)]^{1/2} \\
&\qquad \times \{ss'(1 - w\cos\theta/\varepsilon')^{1/2} + (1 + w\cos\theta/\varepsilon')^{1/2}\} \\
\omega_3 &= \tfrac{1}{4}[(1 + s'/\eta')(1 + s/\eta)]^{1/2} \\
&\qquad \times \{(1 - w\cos\theta/\varepsilon')^{1/2} - ss'(1 + w\cos\theta/\varepsilon')^{1/2}\} \\
\omega_4 &= \tfrac{1}{4}[(1 - s'/\eta')(1 - s/\eta)]^{1/2} \\
&\qquad \times \{(1 + w\cos\theta/\varepsilon')^{1/2} - ss'(1 - w\cos\theta/\varepsilon')^{1/2}\}.
\end{aligned} \tag{6.30}$$

The $I_{\alpha\beta}(x)$ functions are defined in the Appendix I.

Equation (6.30) is too complicated to be analyzed in full generality. We will confine ourselves to the non-relativistic case and to the transition $n=1$ to $n'=0$.

In this case the various ω_k simply become

$$\omega_1 = \omega_3 = \omega_4 = 0, \qquad \omega_2^2 = \tfrac{1}{2}H/H_q. \tag{6.31}$$

The function $F(1, 0, w, \theta)$ reduces to

$$F(1, 0, w, \theta) = \tfrac{1}{2}\frac{H}{H_q}(1 + \cos^2\theta)\, I_{0,0}^2 = \tfrac{1}{2}\frac{H}{H_q}(1 + \cos^2\theta). \tag{6.32}$$

For the non-degenerate case we can take $f(x') < 1$. The first parenthesis in Equation (6.27) is therefore seen to be the particle density N_e. (see Equation (5.20)). The final expression is simply

$$I(1, 0, \theta) = \tfrac{1}{2}\alpha\frac{mc^2}{\hbar} mc^2 N_e w^2 \left(\frac{H}{H_q}\right)(1 + \cos^2\theta). \tag{6.33}$$

c. *Other Related Synchrotron Radiation Processes*

In a magnetic field, there are a number of processes involving free photons and free electrons which are normally forbidden. Most of these processes are of theoretical interest but may be important in astrophysics. These processes have been discussed by Erber (E66). They are:

(1) Pair production by a free photon (of energy MeV) in a magnetic field:

$$\gamma \rightarrow e^- + e^+.$$

This process is normally forbidden in the field free case but is allowed here because electrons in Landau states behave kinetically as one-dimensional particles. The lifetime of the process is given by

$$\tau = l/c$$

where (when $H \ll H_q$)

$$l^{-1} = \frac{1}{2} \frac{\alpha}{\lambda_c} \frac{H}{H_q} T(x), \qquad x \equiv \frac{1}{2} \frac{h\nu}{mc^2} \frac{H}{H_q}$$

with

$$\begin{aligned} T(x) &= 0.60\, x^{-1/3} && x \gg 1 \\ &= 0.46 \exp[-(4/3x)] && x \ll 1 . \end{aligned}$$

The maximum of the function $T(x)$ is at $x \simeq 6$ at which $T(6) = 0.1$. At $H = 10^{-8} H_q$ an $h\nu = 6.10^6$ eV, the mean free path l of the photon is $\simeq 1$ cm or $\tau \simeq 10^{-10}$ sec.

(2) *Photon splitting*

$$\gamma \rightarrow \gamma + \gamma .$$

This process has been computed by Skobov (S58) using Schwinger's Green's function which is valid only for $H \ll H_q$. The exact form of the Green's function is discussed in Appendix II. In this case the attenuation coefficient is computed to be

$$l^{-1} = \frac{5}{3} \frac{1}{(144\pi)^2} \frac{\alpha^3}{\lambda_c} \frac{h\nu}{mc^2} \left(\frac{H}{H_q}\right)^2 .$$

(See however Adler *et al.*, 1970, A70.)

7. Neutrino Processes in Magnetic Fields

It is known that neutrinos can strongly dissipate thermal energy of stars at later stages of stellar evolution (Ch66a, CCFC69, R65). Most neutrino processes in the field free case also operate in the presence of a magnetic field, but in addition, those photon emission processes which are allowed in the presence of a field, can also emit neutrinos via the $(e\nu)(e\nu)$ interaction.

As in all other cases neutrinos are emitted in pairs. This is due to the nature of the $(e\nu)(e\nu)$ interaction. In the following we list a number of examples of neutrino processes in a magnetic field. For a summary see Canuto (C71).

A. Examples of processes allowed in the absence of field and in a field:
   (1) Bremsstrahlung process $e^- + (Z, A) \rightarrow e^- + (Z, A) + \nu + \bar{\nu}$
   (2) Photo neutrino process $e^- + \gamma \rightarrow e^- + \nu + \bar{\nu}$
   (3) Electron pair annihilation process $e^- + e^+ \rightarrow \nu + \bar{\nu}$
   (4) Plasma neutrino process $\gamma \rightarrow \nu + \bar{\nu}$
   (although this process in a vacuum without a field is forbidden, it is still allowed in a field. It is analogous to the photon splitting case (see B (2) below).

B. Examples of processes only allowed in the presence of a field.
   (1) Synchrotron process $e^- \rightarrow e^- + \nu + \bar{\nu}$
   (2) Photon splitting process $\gamma \rightarrow \nu + \bar{\nu}$, $\gamma \rightarrow \gamma + \nu + \bar{\nu}$

One thing worth noting in the precence of a field is that cross-section loses its meaning. In a field free case, particles can move freely in any direction and a cross-section can be defined such that a beam of particles of density N travelling with velocity v striking a group of stationary particles of density N_2 has a reaction rate proportional to $N, N_2 v$, where the constant of proportionality is the cross-section σ. In the presence of a field the motions of particles are confined and there is only one velocity component in the usual sense. Therefore cross-section has no physical meaning. On the other hand, the transition probability (which in the field free case is σv) is well defined. In any case, in evaluating the energy loss rate (or other quantities of physical interest) one only needs σv which is equivalent to the transition probability. We will give the results for the process (B.1). The result can be easily modified to compute (A.3). We start with the usual V-A type of interaction

$$S = \int d^4x \mathscr{L}(x) = \sum_k \int [\bar{\psi}_e(x)\, O_k \psi_e(x)]\, [\bar{\psi}_\nu(x)\, F_k \psi_\nu(x)]\, d^4x$$

with

$$F_k = 2^{-1/2} g_k O_k (1 + \gamma_5).$$

The index k runs only for vector $O_k = \gamma_\mu$ and axial $O_k = i\gamma_\mu\gamma_5$. The $\psi_e(x)$ are the exact electron wave functions in a magnetic field given in Appendix I. Using the standard field-theoretical method one can then compute the neutrino energy loss

$$-\frac{du}{dt} = \sum_i \sum_f (E_i - E_f)\, f(E_i)\, [1 - f(E_f)]\, W$$

where the transition probability per unit time and volume W is defined as

$$W \equiv \frac{|S|^2}{\Omega T}.$$

A. NEUTRINO SYNCHROTRON ENERGY LOSS

The neutrino luminosity for this process turns out to be (CCCFC70)

$$l = l_0 \frac{H}{H_q} \sum_n \sum_{n'} \int_{-\infty}^{+\infty} dx \int_{-\infty}^{+\infty} dx' \,[\varepsilon_n(x) - \varepsilon_{n'}^*(x')]$$
$$\times f_n(x)\, [1 - f_{n'}(x')] \int_0^1 d\varrho\, I(x, x', \varrho) \qquad (7.1)$$

where

$$l_0 = \frac{1}{6} \frac{1}{(2\pi)^8} \frac{g^2}{c\hbar^2} \frac{mc^2}{\lambdabar_c^8} = 1.7475 \times 10^{18} \text{ erg/cc sec} \qquad (7.2)$$

and

$$\begin{aligned} I(x, x', \varrho) &= q_M^4 (A + \varrho B) + q_M^2 (q_3 C + q_0 G)^2 \\ q_M^2 &= [\varepsilon_n(x) - \varepsilon_{n'}(x)]^2 - (x - x')^2 \end{aligned} \tag{7.3}$$

$$\begin{aligned} A &\equiv \omega_1^2 \phi_1^2 + \omega_2^2 \phi_2^2 - 4\omega_3 \omega_4 \phi_3 \phi_4 \\ B &\equiv \omega_1^2 \phi_1^2 + \omega_2^2 \phi_2^2 + 4\omega_3 \omega_4 \phi_3 \phi_4 \\ C &\equiv \omega_3 \phi_3 - \omega_4 \phi_4 \\ D &\equiv \omega_3 \phi_3 + \omega_4 \phi_4 \end{aligned} \tag{7.4}$$

$$\phi_1 = \Phi(n|n' - 1) \qquad \phi_2 = \Phi(n - 1|n')$$
$$\phi_3 = \Phi(n|n') \qquad \phi_4 = \Phi(n - 1|n' - 1)$$

$$\begin{aligned} \Phi(n|n') &= (n!\, n'!)^{-1/2} e^{-t/2} t^{n+n'/2} {}_2F_0(-n', -n; -t^{-1}) \\ t &= (2H/H_q)^{-1} q_M^2 (1 - \varrho). \end{aligned} \tag{7.5}$$

A numerical integration of Equation (7.1) was performed for $H/H_q = 1$ and four different temperatures. The results are reported in Table II and III. They are usually larger by a factor of $\sim 10^2$ from the Landstreet results (L66) which were obtained

TABLE II

ϱ_6(g/cm³)		I(erg/cm³ sec)
	$T = 5.9303 \times 10^7$ K	
3.7146		0.26037×10^{-1}
5.598		0.4998
5.8417×10		4.054×10^5
2.0736×10^2		1.9397×10^7
	$T = 3.7418 \times 10^8$ K	
4.802×10^{-1}		1.2063×10^{12}
1.44		5.092×10^{12}
3.7146		2.107×10^{13}
5.598		3.7116×10^{13}
5.8417×10		1.2984×10^{14}
2.0736×10^2		2.5338×10^{13}

TABLE III

ϱ_6(gr/cm³)		I(erg/cm³ sec)
	$T = 5 \times 10^8$ K	
6.7082×10^{-2}		1.64071×10^{13}
3.7146		1.7649×10^{14}
5.598		2.675×10^{14}
5.8417×10		7.37×10^{14}
	$T = 9.3988 \times 10^8$ K	
6.7082×10^{-2}		1.659×10^{15}
3.7146		8.51×10^{15}
5.598		1.0586×10^{16}
5.8417×10		3.1804×10^{16}

after integrating over all the quantum numbers n and n'; this procedure is valid only when the problem is quasi-classical, i.e., when the density is much higher than ϱ_6; this is surely not the case considered in the numerical analysis performed in (CCCFC69). Therefore the comparison with Landstreet results cannot be taken too seriously. In Figure 5 we report the region in the ϱ–T plane where the neutrino synchrotron process is important.

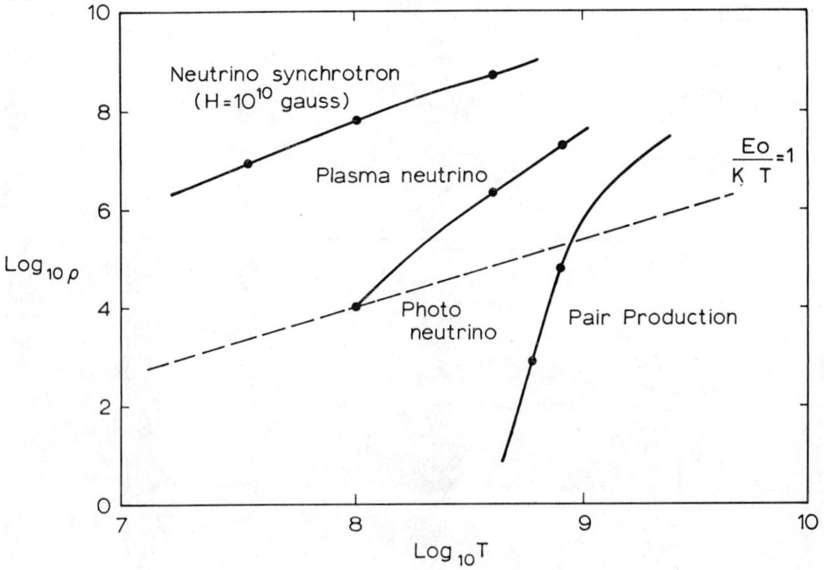

Fig. 5. Region describing the relative importance of the neutrino synchroton, plasma neutrinos, photoneutrinos, and neutrino pair processes. (L66)

B. PLASMAN NEUTRINOS (CCC70)

As seen in Figure 5 the plasmon neutrino process

$$\gamma \to e^- + e^+ \to \nu + \bar{\nu}$$

is of primary importance at relatively low temperature and high density. This process has been repeated in the presence of a strong magnetic field (CCC70). For the transverse decaying photon the neutrinos energy loss is given by

$$Q(\theta = 0) = \bar{Q}\omega_p^4 \int_{\omega_0}^{\infty} \mathrm{d}\omega \, \frac{\omega^6}{(\omega \pm \omega_c)^2} N_l [1 - N_l^2] f(\omega) \tag{7.6}$$

where

$$N_l^2 = 1 - \frac{\omega_p^2}{\omega^2} \frac{\omega}{\omega \pm \omega_c}, \quad \omega_p^2 = \frac{4\pi e^2 N_e}{m}, \quad \omega_c = \frac{eH}{mc} \tag{7.7}$$

$$f(\omega) = \{\exp(\hbar\omega/kT) - 1\}^{-1} \tag{7.8}$$

$$\bar{Q} = \frac{1}{12\alpha}\frac{1}{(2\pi)^5}\frac{g^2}{\hbar c^q}\left(\frac{\text{erg}}{\text{cc sec ster}}\right). \tag{7.9}$$

The lower limit ω_0 has to be chosen such that $N_e^2 < 1$. Equation (7.6) is for propagation along the magnetic field, $\theta = 0$. The plus or minus ($l = 1, 2$) in the refractive index refer to ordinary (0) and extraordinary (X) waves.

An analogous computation gives for $\theta = \pi/2$

$$Q_0(\theta = \pi/2) = \bar{Q}\omega_p^4 \int_{\omega_p}^{\infty} d\omega\, \omega^4 N_0 (1 - N_0^2) f(\omega) \tag{7.10}$$

$$N_0^2 = 1 - \omega_p^2/\omega^2 \tag{7.11}$$

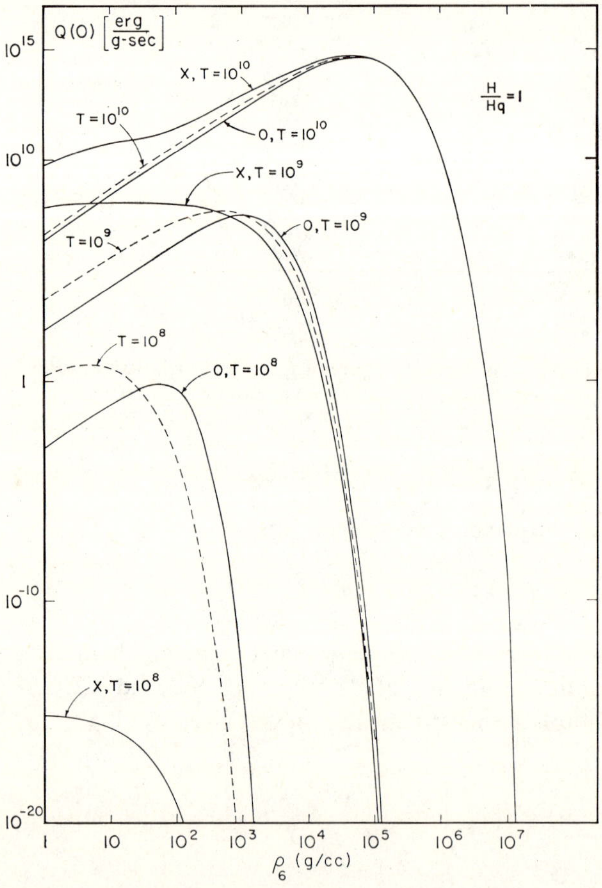

Fig. 6. Energy loss per unit mass and unit solid angle vs ϱ_6 and different temperatures for $H = H_q$ and $\theta = 0$. As explained in the text the symbols O and X stand for ordinary and extra-ordinary modes. The dashed lines correspond to $H = O$.

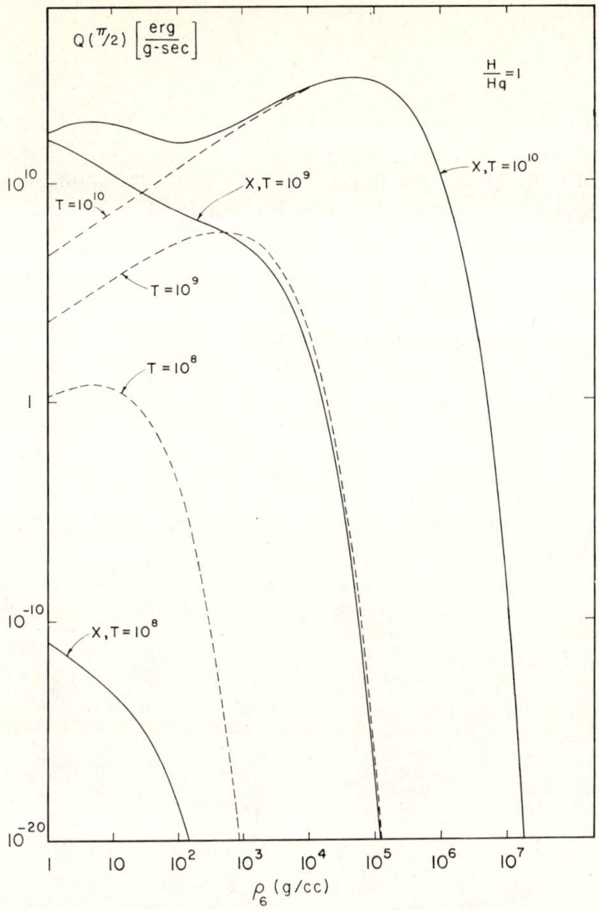

Fig. 7. The same as in Figure 6 for $\theta = \pi/2$. The ordinary mode coincides with the $H = O$-case.

$$Q_x(\theta = \pi/2) = \bar{Q}\omega_p^4 \int_{\omega_0}^{\infty} d\omega \frac{\omega^8}{(\omega^2 - \omega_c^2)^2} N_x[1 - N_x^2] \times \left[1 + (1 - N_x^2)\left(\frac{\omega_c}{\omega}\right)^2\right] f(\omega) \qquad (7.12)$$

$$N_x^2 = 1 - \left(\frac{\omega_p}{\omega}\right)^2 \frac{\omega^2 - \omega_p^2}{\omega^2 - \omega_p^2 - \omega_c^2}. \qquad (7.13)$$

Equations (7.6), (7.10) and (7.12) were solved numerically for $H = H_q$ for various densities and temperatures. The results are shown in Figures 6 and 7. The general conclusion is that a sizable effect can be found only for low densities and very high field, a situation not met in white dwarf or neutron star interiors. A different situation is encountered when one considers the longitudinal plasmon.

At $\theta = 0$ of the two possible dispersion relations

$$\begin{aligned} \omega^2 &= \omega_p^2 \\ \omega^2 &= \omega_c^2 \end{aligned} \tag{7.14}$$

the first does not depend on the magnetic field while the second is purely magnetic field dependent and correspondent neutrino luminosity is given by

$$Q(\theta = 0) = \bar{Q}\omega_p^4 (\omega_c/\mu)^5 f(\omega_c/\mu). \tag{7.15}$$

At $\theta = \pi/2$ only one dispersion relation is important

$$\omega^2 = \omega_h^2 = \omega_c^2 + \omega_p^2 \tag{7.16}$$

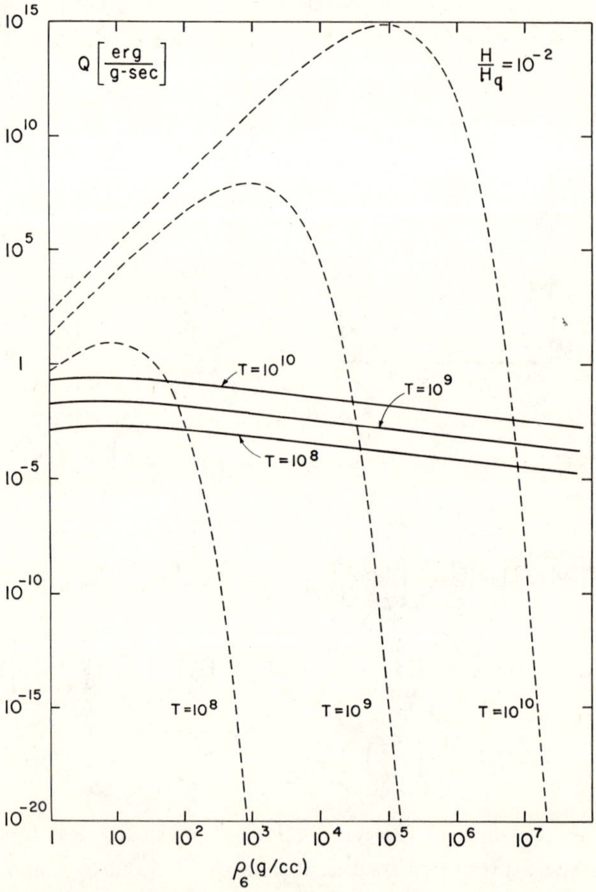

Fig. 8. Energy loss per unit mass and unit solid angle at $\theta = 0$ as a function of ϱ_6 and different temperatures. The solid curve refers to the mode $\omega = \omega_c$ and $H/H_q = 1\bar{0}^2$. The dashed curve refers to the mode $\omega = \omega_p$ which is independent of the field.

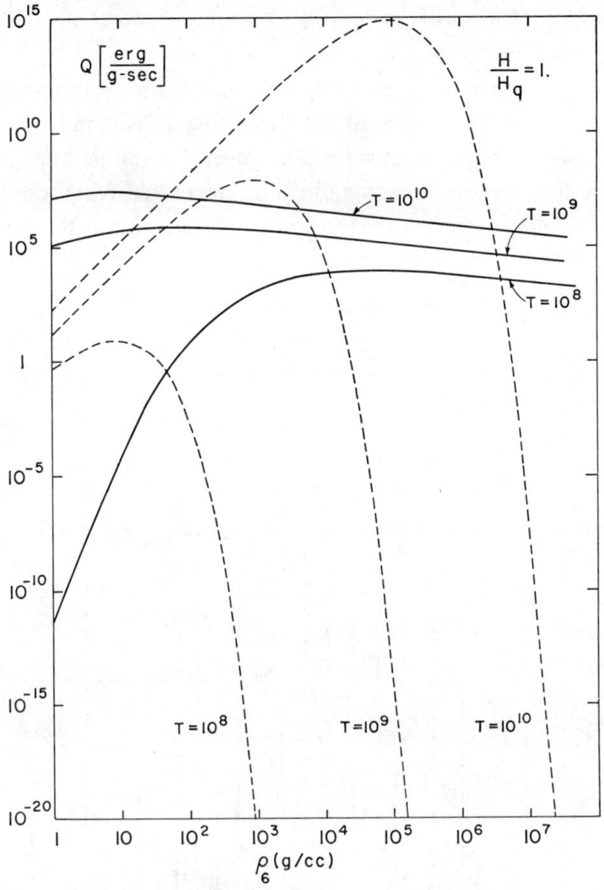

Fig. 9. Same as in Figure 8 for $H = H_q$.

since the other gives

$$\omega^2 = 0 .$$

The neutrinos luminosity turns out to be

$$Q(\theta = \pi/2) = \bar{Q}\omega_p^2\omega_h^5(\omega_p^2 + \tfrac{7}{4}\omega_c^2) f(\omega_h) . \tag{7.17}$$

Since $\omega_p \gg \omega_c$, $\omega_h \approx \omega_p$ and therefore no great difference is expected from the case with $H = 0$, at least at $\theta = \pi/2$. In Figures 8 and 9 we reproduce the neutrinos luminosities (Equation 7.14) for $H = 10^{-2} H_q$ and $H = H_q$. It is important to note that this magnetic field dependent mode survives at those densities at which the free field case is exceedingly small.

8. Neutron Beta Decay

Because of the Landau levels the electron final states are strongly modified. Strictly speaking, all states of the neutron and the proton are also affected by a field. However, the effect of the field is proportional to m^2. For fields $H \ll 10^{19}$ G, the states of the proton and neutron are not affected.

The neutron mean life τ in a magnetic field has been considered, but the most complete work is due to Fassio-Canuto (FC69). She considered two cases: vacuum and a highly degenerate magnetized state. The general result is that the neutron mean life in a magnetic field begins to be significantly decreased when the field strength is greater than 10^{10} G. The physical reason for the decreasing of the lifetime lies in the fact that the phase space for a one-dimensional particle is dp_z (instead of $p^2\,dp$) and

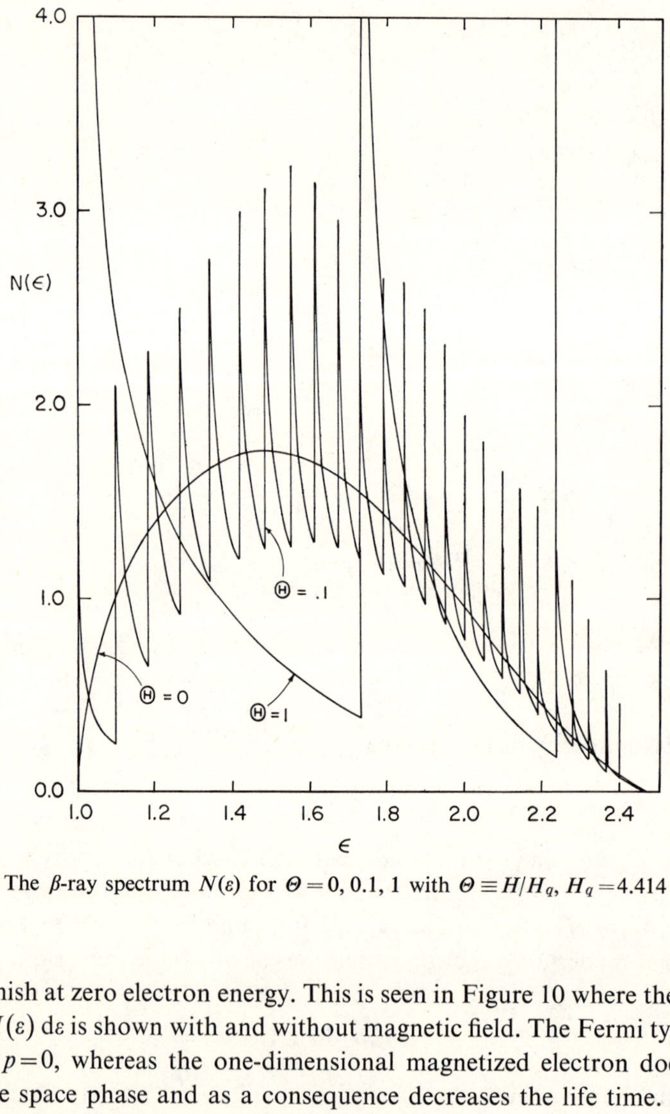

Fig. 10. The β-ray spectrum $N(\varepsilon)$ for $\Theta = 0, 0.1, 1$ with $\Theta \equiv H/H_q$, $H_q = 4.414 \times 10^{13}$ G.

does not vanish at zero electron energy. This is seen in Figure 10 where the differential spectrum, $N(\varepsilon)\,d\varepsilon$ is shown with and without magnetic field. The Fermi type spectrum vanishes at $p=0$, whereas the one-dimensional magnetized electron does not. This increases the space phase and as a consequence decreases the life time. The discontinuities are due to the density of final states (Figure 3).

The expression for the neutron decay half-life τ is

$$\tau^{-1} = (\bar{\tau})^{-1} \frac{H}{H_q} \sum_{n=0}^{N} \int_{a_n}^{\Delta} (1 - \tfrac{1}{2}\delta_{n0})\, d\varepsilon \left[1 - f(\varepsilon)\right] \frac{\varepsilon(\Delta - \varepsilon)^2}{(\varepsilon^2 - a_n^2)^{1/2}} \tag{8.1}$$

where

$$a_n^2 = 1 + 2nH/H_q \qquad \Delta = m^{-1}(M_n - M_p)$$
$$(\bar{\tau})^{-1} = g_V^2 (1 + 3\lambda^2) \frac{m^5 c^4}{4\pi^3 \hbar^7} \tag{8.2}$$

where $\lambda \equiv g_A/g_V$ is the ratio of the axial and vector coupling constant and N must be chosen in such a way that $\varepsilon^2 - a_N^2 > 0$. In the nondegenerate case $f(\varepsilon) \ll 1$. The integral can be exactly integrated and we have, with $x \equiv \Delta/a_n$

$$\tau^{-1} = (\bar{\tau})^{-1} \frac{H}{H_q} \sum_{n=0}^{N} (1 - \tfrac{1}{2}\delta_{n0})\, a_n^3 \times \left[\tfrac{1}{3}(x^2 - 1)^{1/2}(2 + x^2) - x \ln\left(x + \sqrt{x^2 - 1}\right)\right] \tag{8.3}$$

$$N = (\Delta^2 - 1)/(2H/H_q).$$

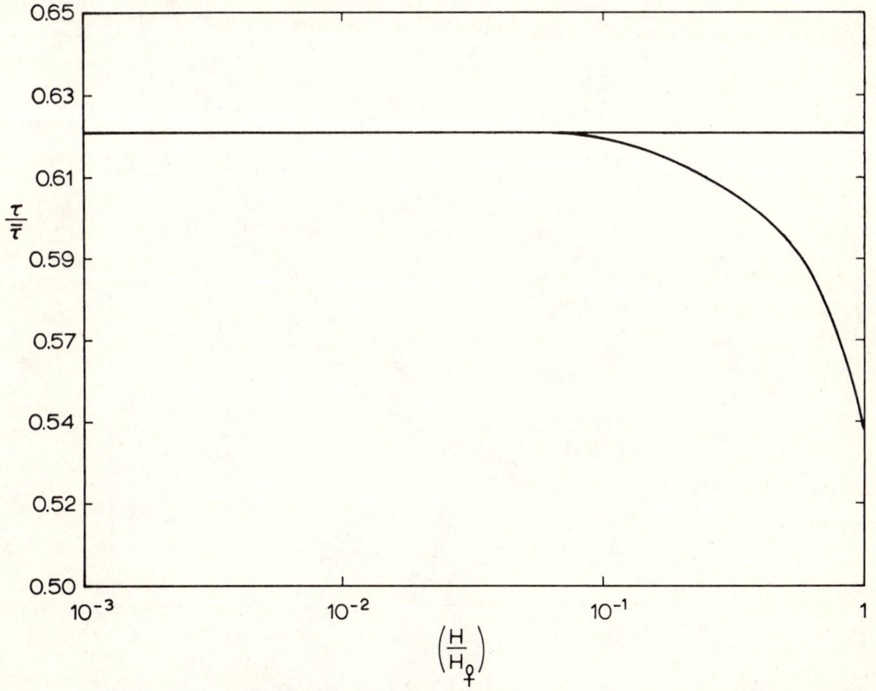

Fig. 11. The neutron mean life in vacuum normalized to a constant factor $\bar{\tau}$, Equation (8.3), is plotted as a function of the magnetic field. The straight line is the neutron mean life when $H = 0$.

When

$$\Delta^2 < 1 + 2H/H_q$$

only the first term contributes and the result is easily seen to be

$$\tau^{-1} = 1.3\,(\bar{\tau})^{-1}\,\frac{H}{H_q}. \tag{8.4}$$

At higher fields $\tau \sim H^{-1}$. The lifetime τ (Equation 8.3) has been computed numerically as a function of H/H_q. Its behavior is shown in Figure 11 where we also give the free neutron lifetime. As said before, the effect of the magnetic field becomes appreciable only at fields greater than 10^{10} G. As the density increases degeneracy appears. In the completely degenerate case we have

$$1 - f(\varepsilon) = H(\varepsilon - \mu) \tag{8.5}$$

where H is the Heaviside step function. As a consequence the lower limit in Equation

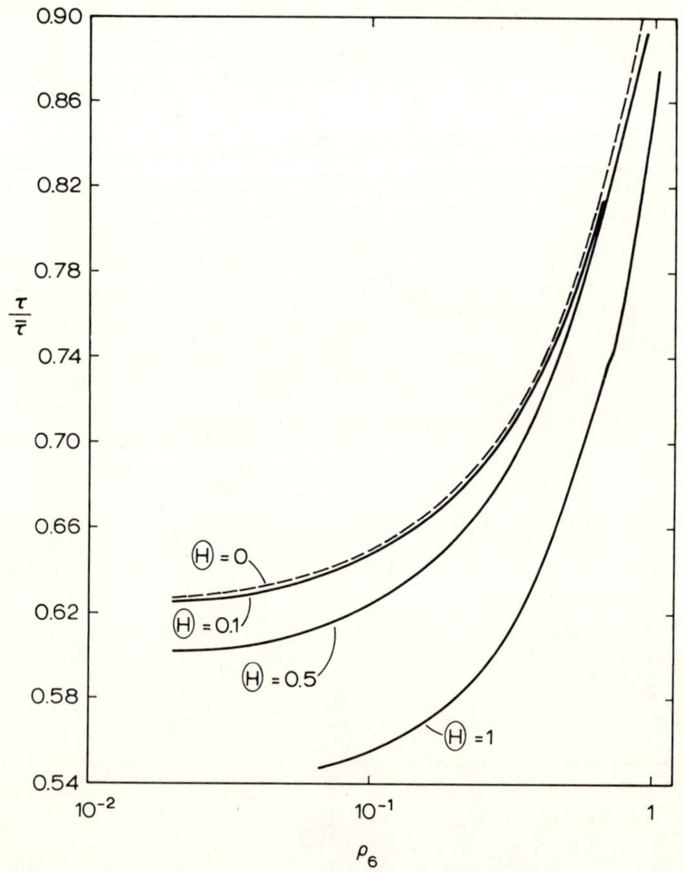

Fig. 12. The neutron mean life normalized as in Figure 11 is plotted vs $\varrho_6 \equiv 10^{-6}\,\varrho/\mu_e$ (ϱ in g/cc) for different values of $\Theta = H/H_q$.

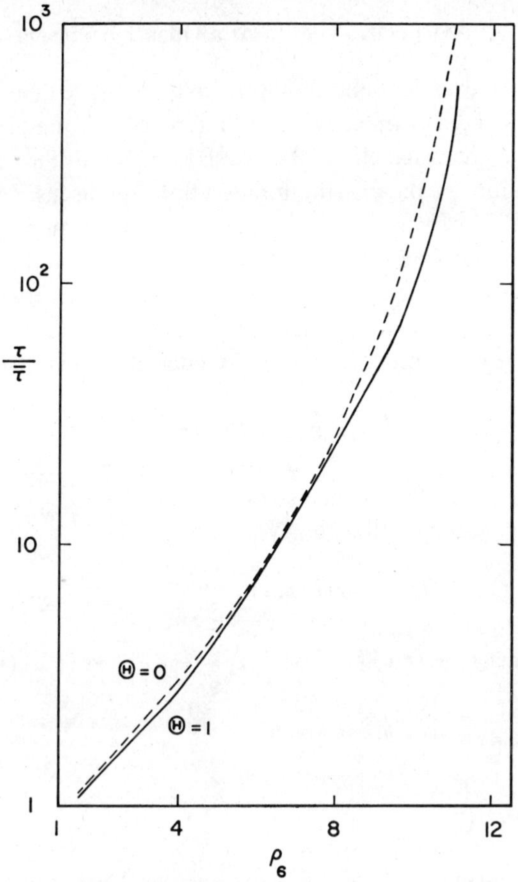

Fig. 13. The same as in Figure 12. At $\varrho \simeq \varrho^* \simeq 10^7$, the neutron mean life becomes infinite, i.e., the neutron becomes a stable particle.

(8.1) is changed to μ. The result is

$$\frac{\bar{\tau}}{(\tau)_{\mathrm{Deg}}} = \frac{\bar{\tau}}{\tau} - \frac{H}{H_q} \sum_{n=0}^{N} \left(1 - \tfrac{1}{2}\delta_{n0}\right) a_n^3 \{(y^2 - 1)^{1/2}$$

$$\times \left[x(x - y) + \tfrac{1}{3}(2 + y^2)\right] - x \ln\left[y + (y^2 - 1)^{1/2}\right]\} \qquad (8.6)$$

with

$$y \equiv \mu/a_n \qquad N \equiv \frac{\mu^2 - 1}{2H/H_q}. \qquad (8.7)$$

In Figure 12, and 13 the lifetime is shown as a function of the density for some values of H/H_q. As the density approaches the values ϱ^* at which the Fermi energy is equal to Δ, the neutrons become stable and the lifetime is infinite. As in a vacuum, the effect of the magnetic field is that of lowering the neutron lifetime. Some astrophysical consequences of this shortening have recently been considered (G69).

9. Dielectric Tensor for a Quantum Plasma

The dispersive character of a quantum plasma in a magnetic field has been the subject of numerous papers and an exhaustive list of references is found in a recent article by Green *et al.* (Gr69). The usual classical analysis is based on the solution of the Boltzmann-Vlasov equation for the distribution function. As discussed at length in Kelly's paper (Ke64) there is no unique way to treat the problem on a quantum mechanical basis. The reason is simply because no precise quantum counterpart of the classical distribution exists. An improved fully quantum-mechanical derivation has recently been achieved and studied in detail (Canuto and Ventura; CV71). The method employed by Kelly is based on the use of the Wigner function [BC62]

$$f(r, p, t) = (\pi\hbar)^{-3} \int \exp(2i\mathbf{p}\cdot\boldsymbol{\lambda})\, \varrho(\mathbf{r}-\boldsymbol{\lambda}, \mathbf{r}+\boldsymbol{\lambda})\, d^3\lambda \tag{9.1}$$

where $\varrho(r, r', t)$ denotes the single particle density matrix. The Wigner function $f(r, p, t)$ is shown to satisfy Boltzmann-Vlasov equation

$$f(r, p, t) = f_0(p) + f_1(r, p, t) \tag{9.2}$$

$$\frac{\partial f_1}{\partial t} + \mathbf{v}\cdot\boldsymbol{\nabla} f_1 + (e/c)(\mathbf{v}\times\mathbf{B})\nabla_p f_1 = -e[E_1 + (1/c)(\mathbf{v}\times\mathbf{B}_1)]\nabla_p f_0 \tag{9.3}$$

with

$$\begin{aligned} \mathbf{E}_1 &= -\frac{1}{c}\frac{\partial \mathbf{A}_1}{\partial t} - \boldsymbol{\nabla}\phi \\ \mathbf{B}_1 &= \boldsymbol{\nabla}\times\mathbf{A}_1 \end{aligned} \tag{9.4}$$

where $\phi(r, t)$ and $A_1(r, t)$ are the scalar and vector components of the self-consistent electromagnetic field felt by each electron. Introducing the Fourier transform of $f_1(r, t)$ and $E_1(r, t)$ and using standard plasma physics techniques the general expression of the dielectric tensor turns out to be

$$\varepsilon_{\alpha\beta} = (1 - \omega_p^2/\omega^2)\,\delta_{\alpha\beta} - (\omega_p^2/\omega^2)\, n_{\alpha\beta} \tag{9.5}$$

with (we only quote the components of $n_{\alpha\beta}$ we shall need)

$$\begin{aligned} n_{11} &= S[(nJ_n/b)^2\, p_\perp \{v_\perp q_\| f_\| + n\omega_c f_\perp\}] \\ n_{12} &= -n_{21} = S[i(nJ_n J'_{n'}/b)\, p_\perp \{v_\perp q_\| f_\| + n\omega_c f_\perp\}] \\ n_{22} &= S[(J'_n)^2\, p_\perp \{v_\perp q_\| f_\| + n\omega_c f_\perp\}] \\ n_{33} &= S[J_n^2 p_\| \{v_\| q_\| f_\| + v_\perp^{-1} v_\| n\omega_c f_\perp\}] \end{aligned} \tag{9.6}$$

where

$$\begin{aligned} S[x] &\equiv \sum_{n=-\infty}^{+\infty} \int \frac{d^3p\,[x]}{\omega + q_\| v_\| + n\omega_c}, \quad \omega_c = eH/mc, \quad \omega_p^2 = 4\pi e n_e/m \\ f_\| &= \partial f_0/\partial p_\| \quad . \ f_\perp = \partial f_0/\partial p_\perp \qquad b = q_\perp p_\perp/m\omega_c \end{aligned} \tag{9.7}$$

J_k is the Bessel function of order i. As customary in plasma physics, the wave vector q is taken as $\mathbf{q} = (q_\perp, 0, q_\parallel)$; on the other hand, the momentum variable is taken as $\mathbf{p} = (p_\perp \cos\phi, p_\perp \sin\phi, p_\parallel)$. Having obtained the tensor $\varepsilon_{\alpha\beta}$ the dispersion relation is found by solving the usual equation

$$\det\left[c^2 (q^2 \delta_{\alpha\beta} - q_\alpha q_\beta) - \omega^2 \varepsilon_{\alpha\beta}\right] = 0. \tag{9.8}$$

In order to evaluate the tensor $n_{\alpha\beta}$ one still needs the equilibrium distribution $f_0(p)$ given by Equation (9.1) where the equilibrium density matrix

$$\varrho(\mathbf{r}, \mathbf{r}') = \sum_j w_j \phi_j^* (\mathbf{r}') \phi_j(\mathbf{r}) \tag{9.9}$$

with

$$w_j = \{1 + \exp[\tilde{\beta}(E_j - \tilde{\mu})]\}^{-1}$$

requires the use of single particle wave-function of an electron in a uniform magnetic field. For the Maxwell-Boltzmann case, w_j is simply given by $\exp[-\tilde{\beta}(E_j - \tilde{\mu})]$. Using the wave functions given in Appendix I, Kelly's result is the following

$$f_0(p_\perp, p_\parallel) \equiv \frac{\operatorname{tgh}\theta}{m\hbar\omega_c}\left(\frac{\beta}{2m\pi^3}\right)^{1/2} \exp\left[-\frac{\beta p_\parallel^2}{2m} - \operatorname{tgh}\theta \frac{p_\perp^2}{m\hbar\omega_c}\right] \tag{9.10}$$
$$\theta \equiv \tfrac{1}{2}\beta\hbar\omega_c$$

for the Maxwell-Boltzmann distribution and $[w^2 \equiv (p_\parallel^2 + p_\perp^2)/m\hbar w_c]$

$$f_0(p_\perp, p_\parallel) = \frac{2e^{-w^2}}{N_e (2\pi\hbar)^3} \sum_{n=0}^{\infty} \sum_{s=\pm 1/2} \times \frac{(-)^n L_n(2w^2)}{1 + \exp[\beta p_\parallel^2/2M + \beta\hbar\omega_c(n + s + \frac{1}{2}) - \beta\mu]} \tag{9.11}$$

for Fermi-Dirac statistics. Both $f_0(p_\parallel, p_\perp)$ are normalized in such a way that

$$\int f_0(p_\parallel, p_\perp)\, \mathrm{d}^3 p = 1.$$

With these formulae we are now able to study the propagation parallel and perpendicular to the magnetic field. For the first case $q_\perp = 0$, $q_\parallel = q$ and the result turns out to be

$$n_{11} = n_{22} = n_+ + n_- \tag{9.12}$$
$$n_{12} = -n_{21} = i(n_+ - n_-)$$

$$2n_{\pm}^{MB} = \frac{\mp\omega_c}{\omega \pm \omega_c} + \frac{q^2}{m\beta(\omega \pm \omega_c)^2}\left\{\theta\coth\theta \mp \frac{\omega_c}{\omega \pm \omega_c}\right\}$$

$$2n_{\pm}^{FD} = \left(\frac{q^2\hbar\omega_c}{2m}\sum_{n=0}^{n_F}\frac{\alpha_n V_n}{(\omega \pm \omega_c)^2 - q^2V_n^2}\right.$$

$$\left. \mp \frac{\omega_c}{q}\sum_{n=0}^{n_F}\alpha_n \ln\left|\frac{\omega \pm \omega_c + qV_n}{\omega \pm \omega_c - qV_n}\right|\right)\Big/\sum_{n=0}^{n_F}\alpha_n V_n \tag{9.13}$$

$$n_{33}^{MB} = 3q^2/m\beta\omega^2$$

$$n_{33}^{FD} = (q^2\mu/m\omega^2)\,\chi(\hbar\omega_c/\mu)$$

with

$$\chi(s) = \sum_{n=0}^{n_F}\alpha_n(1 - sn)^{3/2}/\sum_{n=0}^{n_F}\alpha_n(1 - sn)^{1/2}$$

$$V_n^2 = \frac{2}{m}(\mu - n\hbar\omega_c), \qquad \alpha_n = 2(\perp - \tfrac{1}{2}\delta_{n0}). \tag{9.14}$$

The dispersion relation (9.5) becomes

$$\omega^2 = \omega_p^2(1 + n_{33})$$

$$\omega^2 = q^2c^2 + \omega_p^2(1 + 2n_{\pm}) \tag{9.15}$$

for longitudinal and circularly polarized waves. For waves propagating perpendicular to the magnetic field $q_{\perp} = q$, $q_{\parallel} = 0$ the results are

$$\varepsilon_{11} = 1 - \frac{\omega_p^2}{\omega^2 - \omega_c^2} + \frac{q^2\hbar\omega_c}{2m}\frac{\omega_p^2}{\omega^2}(\Delta_1 - 4\Delta_2)\coth\theta$$

$$\varepsilon_{12} = -\varepsilon_{21} = \frac{i\omega\omega_c}{\omega^2 - \omega_c^2}\frac{\omega_p^2}{\omega^2} - \frac{iq^2\hbar\omega_c}{m}\frac{\omega_p^2}{\omega^2}(\Delta_1 - \Delta_2)\coth\theta$$

$$\varepsilon_{22} = 1 - \frac{\omega_p^2}{\omega^2 - \omega_c^2} + \frac{q^2\hbar\omega_c}{2m}\frac{\omega_p^2}{\omega^2}(3\Delta_1 - 4\Delta_2)\coth\theta \tag{9.16}$$

$$\varepsilon_{33} = 1 - \frac{\omega_p^2}{\omega^2}\left(1 + \frac{q^2}{m\beta[\omega^2 - \omega_c^2]}\right), \qquad \Delta_n = (\omega^2 - n^2\omega_c^2)^{-1}$$

for a Boltzmann gas. For a Fermi gas at $T = 0\,\mathrm{K}$, the corresponding dielectric tensor is

$$\varepsilon_{11} = 1 - \frac{\omega_p^2}{\omega^2 - \omega_c^2} + \frac{q^2\hbar\omega_c}{2m}\frac{\omega_p^2}{\omega^2}(\Delta_1 - 4\Delta_2)\,\Psi(\hbar\omega_c/\mu)$$

$$\varepsilon_{12} = -\varepsilon_{21} = \frac{i\omega\omega_c}{\omega^2 - \omega_c^2}\frac{\omega_p^2}{\omega^2} - \frac{iq^2\hbar\omega_c}{m}\frac{\omega_p^2}{\omega^2}(\Delta_1 - \Delta_2)\,\Psi(\hbar\omega_c/\mu) \tag{9.17}$$

$$\varepsilon_{22} = 1 - \frac{\omega_p^2}{\omega^2 - \omega_c^2} + \frac{q^2\hbar\omega_c}{2m}\frac{\omega_p^2}{\omega^2}(3\Delta_1 - 4\Delta_2)\,\Psi(\hbar\omega_c/\mu)$$

$$\varepsilon_{33} = 1 - \frac{\omega_p^2}{\omega^2}\left[1 + \frac{2q^2\mu}{3m(\omega^2 - \omega_c^2)}\chi(\hbar\omega_c/\mu)\right] \tag{9.17}$$

$$\Psi(s) = 2\sum_{n=0}^{n_F}\alpha_n n(1 - sn)^{1/2} / \sum_{n=0}^{n_F}\alpha_n(1 - sn)^{1/2}.$$

10. Transport Processes – Electron Conduction

In general, transport processes may be classified into 2 categories: energy transport and material transport. Generally speaking, material transport often involves energy transport. The former includes radiative transport of energy and electron conduction and the latter includes convection and diffusion.

While all these processes have been studied in the field free case, with certain exceptions (in the non-relativistic low temperature case for ordinary solids) few of these processes have been studied in the case of strong fields. The problem of radiative transport has not been studied at all in the case of strong fields.

The computation of the electrical conductivity is extremely important by itself because problems related to the decay in time of a magnetic field superimposed on a plasma depends on it. Using Maxwell's equation and Ohm's law, one can easily prove that the diffusion equation for a magnetic field H is (J62)

$$\frac{\partial H}{\partial t} = -\left(\frac{c^2}{4\pi\sigma}\right)\nabla^2 H \tag{10.1}$$

where σ is the electrical conductivity. Approximating $\nabla^2 H$ by HL^{-2} where H is the field and L is the dimension of the field. Equation (10.1) yields an exponential solution for the field H, i.e.,

$$H(t) = H(0)\exp\left(-\frac{c^2}{4\pi\sigma L^2}t\right)$$

which shows that the initial configuration of H will decay in a diffusion time τ given by

$$\tau = \frac{4\pi\sigma L^2}{c^2}.$$

The problem of electron conduction in the absence of a field has been studied since 1932, and the most recent computation has been given by Hubbard and Lampe (HL69) and by Canuto (C69) for the non-relativistic and relativistic case respectively. These works include the ion-ion correlation to eliminate the forward divergence problem of the Coulomb scattering cross-section. The results of Hubbard-Lampe and Canuto are believed to be, up to the time of this article, the best electron conductivity in the absence of a field. The results reported below will be compared to the values of these authors.

In classical theory of conductivity it has been shown that the ratio of thermal conductivity to electrical conductivity is given by the Wiedeman-Franz law, which states that

$$\sigma_{\text{th}}/\sigma_0 = \frac{\pi^2}{3}\frac{k^2}{e^2}T$$

where σth is the thermal conductivity and σ is the electrical conductivity defined (C69) ($\varrho_6 \equiv \varrho \times 10^{-6}$, ϱ in g/cc)

$$\begin{aligned} \sigma_0 &= \bar{\sigma}_0 G(\varrho_6^{2/3}) \\ \bar{\sigma}_0 &= \tfrac{4}{3}[(2\pi)^3 \lambda_c^3 Z \mathcal{N}_i]^{-1} (\alpha Z\hbar/mc^2)^{-1} \end{aligned} \tag{10.2}$$

$$G(x) = x(1+x)^{-1/2}\left\{\int_0^\pi d\theta \sin\theta(1-\cos\theta)[2 + x(1+\cos\theta)]\right.$$

$$\left.\times [x(1-\cos\theta) + \frac{2\alpha}{\pi}x^{1/2}(1+x)^{1/2}]^{-2}\phi(x,\theta)\right\}^{-1} \tag{10.3}$$

$$\phi(x,\theta) = 1 + 3\int_0^\infty ds(s\xi)^{-1}\sin(s\xi)\,g(s) \tag{10.4}$$

$$\begin{aligned} \xi &= 2.69\,\mu_i^{1/3}\mu_e^{-1/3}(1-\cos\theta)^{1/2}, \qquad \mathcal{N}_i = \frac{N_{\text{ions}}}{\Omega} \\ \mu_i &= A, \qquad \mu_e^{-1} = Z/A. \end{aligned} \tag{10.5}$$

The function $g(x)$ is the pair-correlation function which takes into account the ion-ion interactions. Its value depends on a parameter $\Gamma^{-1} = (z^2e^2/KT)[(4\pi/3)(\mathcal{N}_i/\Omega)]^{1/3}$ which measures the strength of the ion-ion interaction. For high values of $\Gamma\,(\Gamma \gg 1)$ the function $g(x)$ is only known numerically. Tables of the function $G(x)$ are given in (C69) for different values of Z and Γ.

Since in a strong field σ is modified by the field itself, it is necessary to know σ in order to estimate τ. Canuto's conductivities have a limited range of validities. When the proton gas becomes degenerate the collision term of the Boltzmann equation has to be treated in a different way to include degeneracy. The modification to Canuto's conductivities due to this phenomenon has been studied recently by Baym *et al.* (BPP69), for the nonrelativistic case and by Kelly (Ke70) and Gentile *et al.* (CG70) for the relativistic case. The general effect is an increase of an amount of the order of $\sim 10^6$ over the previous conductivities ($\approx 10^{23}$ sec^{-1}) for the interior of neutron stars. This increases the magnetic field decay time by the same amount and therefore the pulsar lifetime with respect to the one computed by Gunn and Ostriker (GO69). A different scattering mechanism, i.e., phonon scattering has been recently shown to give the same results as the impurity mechanism used by Canuto, at least for values of $\Gamma \lesssim 75$. A simple analytic expression for σ_0, Equation (10.2) which also reproduces the phonon mechanism, is simply

$$\sigma_0 = 1.08\, T_6^{-1} \varrho_6^{1/3}\, 10^{22} \quad (\sec^{-1})$$
$$T_6 = 10^{-6}\, T\,, \qquad \varrho_6 = 10^{-6}\, \varrho/\mu\,. \tag{10.5}$$

As discussed earlier, this formula is valid for the neutron star crust (CS70).

A. LONGITUDINAL CASE (ALONG THE MAGNETIC FIELD) (CC69)

In this case the electric field is parallel to the magnetic field. The electric field induces a motion of the electron which will be in the direction of the field. Since the motion of the electron along the field is a free motion, this problem can be treated using the correct wave functions for the electron in a magnetic field. In other words, the Hamiltonian of an electron in an electric field parallel to the magnetic is diagonal in the representation where only a pure magnetic field exists.

The most important scattering process is the Coulomb scattering process

$$e^-_{np_z} + (Z, A) \to e^-_{n'p_z'} + (Z, A)\,. \tag{10.6}$$

The nucleus can be assumed to be infinitely heavy. In this process n, p_z, and n', p_z' are related by the energy conservation relation,

$$2nH/H_q + (p_z/mc)^2 = 2n'H/H_q + (p_z'/mc)^2\,. \tag{10.7}$$

In the absence of an electric field the motion of the electron is linear in either direction with a zero average motion. When an electric field is applied in the z-axis, the z-motion of the electron will become accelerated in the direction of the field, but this acceleration is dissipated by scattering processes into a constant macroscopic drift velocity v_d. We are therefore interested in the rate that z-momentum of the electron is lost by collision.

For a given set of n and n' (10.7) gives two solutions for p_z:

$$\begin{array}{ll} p_z > 0 & p_z' > 0 \\ p_z > 0 & p_z' < 0\,. \end{array}$$

There are two additional sets of solutions with the reverse sign for p_z. These 2 solutions will be included in (10.6) when p_z is integrated from 0 to ∞. The time scale over which the z-momentum is lost is (AA56, AH59, A58):

$$\tau_n^{-1} = \sum_f (1 - p_z'/p_z)\, w(i \to f) \tag{10.8}$$

where i is the initial state and f is the final state. This gives:

$$\begin{aligned} \tau_0/\tau_n = & \sum_{n'=0}^{\infty} E[E^2 - a_{n'}^2]^{-1/2} \left[1 + \sqrt{\frac{E^2 - a_{n'}^2}{E^2 - a_n^2}}\right] R_+ \\ & + \sum_{n'=0}^{\infty} E[E^2 - a_{n'}^2]^{-1/2} \left[1 - \sqrt{\frac{E^2 - a_{n'}^2}{E^2 - a_n^2}}\right] R_- \end{aligned} \tag{10.9}$$

$$R_\pm \equiv R(u, \mp u') \quad u' = (2\theta)^{-1/2}(E^2 - a_n^2)^{1/2} \quad a_n^2 = 1 + 2n\theta$$

$$R(u, u') = \int_0^\infty \mathrm{d}t \frac{|A_{n,n'}(t, u, u')|^2}{[t + (u - u')^2]^2} \Phi(\xi)$$

$$|A_{n,n'}|^2 = [\omega_1 \Psi(n|n') - \omega_2 \Psi(n-1|n'-1)]^2$$

$$\Psi(n|n') \equiv (n!\, n'!)^{-1/2} e^{-t/2} t^{(n+n')/2}\, {}_2F_0\left(-n', -n; -\frac{1}{t}\right)$$

$$\Phi(\xi) = 1 + 3\int_0^\infty (x\xi)^{-1} \sin(x\xi)\, \mathrm{d}x \qquad (10.9)$$

$$\xi_\pm = 2.69\, \varrho^{-1/3} z^{1/3} \theta^{1/2} [t + \{u \mp u'\}^2]^{1/2}$$

$$2\omega_1^2 = 2\omega_2^2 = 1 + E^{-2} \pm E^{-2}(E^2 - a_n^2)^{1/2}(E^2 - a_{n'}^2)^{1/2};$$

$$\omega_1\omega_2 = \theta E^{-2}(nn')^{1/2}.$$

As explained in a review article by Kahn and Frederiske (KF59), the Boltzmann equation (which is the master equation for the computation of conductivity in the field free case) for the longitudinal case is the same as that in the classical case because the electron motion in the z-direction is still free (see, however, KMH65). If f_n is the occupational number for the state n, then the first order solution for the Boltzmann equation is:

$$f_n = f_n^{(0)} + eE\tau_n \partial f_n^{(0)}/\partial p_z$$

where $f_n^{(0)}$ is the equilibrium occupational number for the nth state:

$$f_n^{(0)} = \left\{1 + \exp\left[\frac{\varepsilon_n(x) - \mu}{kT/mc^2}\right]\right\}^{-1}. \qquad (10.10)$$

By definition the current J is in the z direction. Since the current associated with a single electron is:

$$j_z = eV_z = ecE^{-1}(E^2 - a_n^2)^{1/2} \qquad (10.11)$$

we find

$$J \equiv J_z = -\frac{1}{\hbar}\sum_n \int j_z f_n N(p_z)\, \mathrm{d}p_z \qquad (10.12)$$

where $N(p_z)\, \mathrm{d}p_z$ is the density of states between $p_z + \mathrm{d}p_z$ and is given by:

$$N(p_z)\, \mathrm{d}p_z = g(2\pi)^{-2} \lambda_c^{-2}(H/H_q)\, \mathrm{d}p_z. \qquad (10.13)$$

The electrical conductivity $\sigma_\parallel$ is obtained from Equations (10.9) and (10.12) and the Ohm's law $J = \sigma_\parallel E$; we obtain

$$\sigma_\parallel = -\hbar^{-1} eg \sum_n \int_{-\infty}^{+\infty} j_z \tau_n (\partial f_n^{(0)}/\partial p_z)\, N(p_z)\, \mathrm{d}p_z. \qquad (10.14)$$

Substituting into (10.14) the expression of τ_n (10.8), we find that

$$\sigma_{\parallel} = \bar{\sigma}_{\parallel} (H/H_q)^2 f(\phi, \mu, \theta) \tag{10.15}$$

where

$$\begin{aligned} \bar{\sigma}_{\parallel} &= 4[(2\pi)^3 \lambda_c^3 Z \mathcal{N}_i]^{-1} (\alpha Z \hbar/mc^2)^{-1} \\ \bar{\sigma} &= 3\bar{\sigma}_0, \quad \text{(Equation 10.2)} \\ \phi &\equiv kT/mc^2, \qquad \theta \equiv H/H_q \end{aligned} \tag{10.16}$$

and

$$f(\phi, \mu, \theta) = \int_1^\infty \frac{dE}{E^2} \times \sum_{n=0}^{\infty} \frac{(E^2 - a_n^2)^{1/2} (\partial f_0^{(n)}/\partial E)}{\sum_{n'=0}^{\infty} (E^2 - a_{n'}^2)^{1/2} [1 + g(E)] R_- + \sum_{n'=n}^{\infty} (E^2 - a_{n'}^2)^{1/2} [1 + g(E)] R_+}$$

$$g^2(E) = (E^2 - a_{n'}^2)/(E^2 - a_n^2), \quad a_n^2 = 1 + 2n\theta \tag{10.17}$$

where R_+ and R_- have been defined earlier, Equation (10.9).

The thermal conductivity coefficient λ_H is defined through the relation (M41, M50)

$$Q = -\lambda_H \frac{dT}{dx} \tag{10.18}$$

where Q has the dimensions of erg cm^{-2} sec^{-1} and λ_H of erg cm^{-1} deg^{-1}. The Wiedemann Franz law relates the parameter λ_H with the electrical conductivity σ via the relations

$$\lambda_H = \frac{\pi^2}{3} \frac{k^2}{e^2} \sigma T. \tag{10.19}$$

As mentioned earlier, Equation (10.19) has been shown to be valid in a strong magnetic field by Zyrianov (Z64). The conductivity opacity coefficient K_c^H defined as (Ch68):

$$K_c^H = \left(\frac{4ac}{3\varrho}\right) \lambda_H^{-1} T^3 \tag{10.20}$$

becomes:

$$\begin{aligned} K_c^H &= \bar{K}_c^H (H/H_q)^2 f(\phi, \mu, \theta) \\ \bar{K}_c^H &= \tfrac{3}{5} (2\pi)^3 m_H^{-1} \alpha^2 \phi^2 \lambda_c^2 Z^2 \mu_i^{-1} \\ &= 20.26 \times 10^{-8}\ T_6^2 Z^2/A \quad (\text{cm}^2\ \text{g}^{-1}). \end{aligned} \tag{10.21}$$

Equation (10.17) contains a summation in the denominator and even if an approximate expression exists, this approximate expression will be very hard to use. Equation (10.17) has been solved numerically for degenerate and non-degenerate case for $H/H_q \equiv \theta = 1, 0.1 - 1.2 \leqslant \log \phi \leqslant 0.2$, where $\phi = kT/mc^2$ and $1 \leqslant \mu \leqslant 15$. Extensive numerical tables are given in (CC69). In Figs. 14, 15, 16 we report the behavior of $f(\phi, \mu, \theta)$ for $\phi=0$ and $\theta=1$, 0.1 and for $\phi=1$, $\theta=1$. The discontinuities are due to the density of final states. They almost disappear at high temperature. In Table IV we compare $\sigma_{\parallel}$ with σ_0, Equations (10.2) and (10.15) at $\theta=1$ for densities ranging

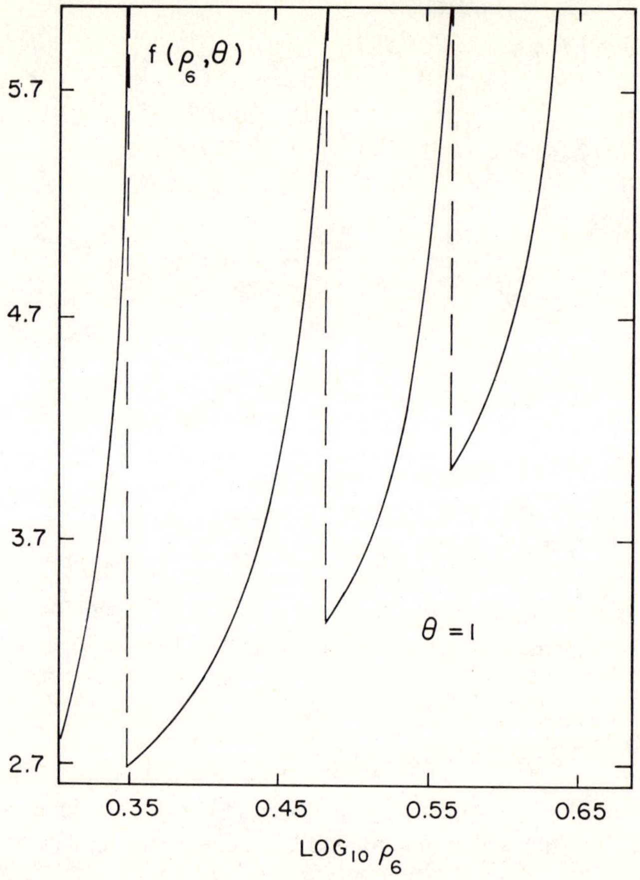

Fig. 14. The function $f(\mu, \theta)$, see Equation (10.15), for a degenerate electron gas as a function of the density ϱ_6 for $\theta \equiv H/H_q = 1$. The relation between ϱ_6 and μ is taken to be $\varrho_6^{2/3} = \pi^2 - 1$. The undulating behavior is related to the density of final states (Figure 3).

TABLE IV

ϱ_6	$\sigma_{\parallel} \times 10^{-21}$ (sec^{-1})	$\sigma_0 \times 10^{-21}$ (sec^{-1})	$\sigma_{\perp} \times 10^{-21}$ (sec^{-1})
8.0	69.37	3.14	1.68
14.7	74.08	3.12	1.13
22.6	71.93	3.32	1.00
31.6	72.61	3.49	0.97
41.6	76.48	3.59	0.98
52.4	81.56	3.45	1.01

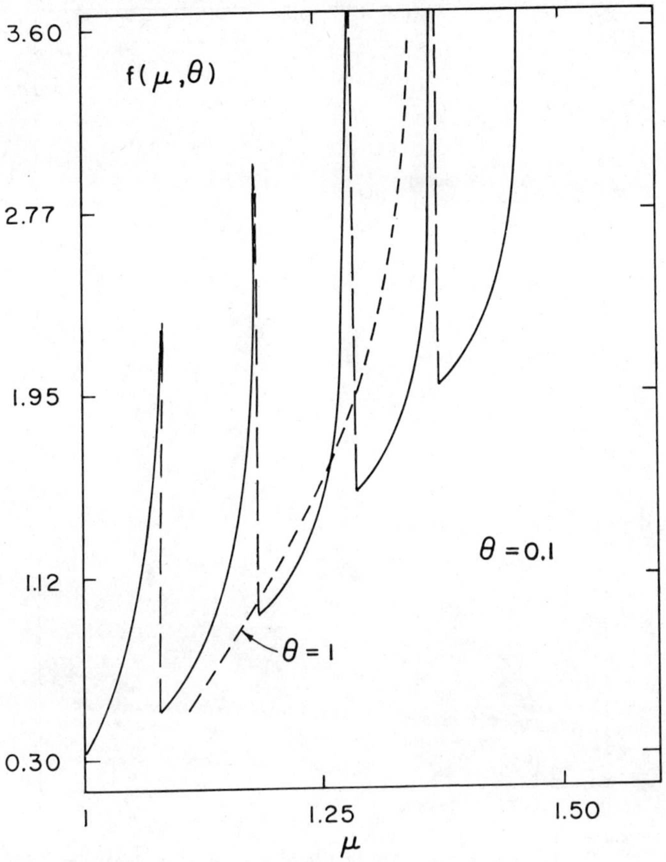

Fig. 15. The same as in Figure 14 for $\theta = 0.1$. In this case the relation between ϱ and μ is more complicated than in the case of Figure 14. The number of discontinuities is noticeable increased with respect to the case $H = H_q$.

between $0.4 \leqslant \log p_6 \leqslant 1.6$. In general it is seen that $\sigma_{\parallel} \gg \sigma_0$. This phenomenon is known as negative magnetoresistence. At higher densities $\sigma_{\parallel}/\sigma_0$ tend to decrease and eventually the density effect will dominate even at $H \simeq H_q$. We therefore expect $\sigma_{\parallel}/\sigma_0 \to 1$ at $\varrho_6 \gg 10^6$ g/cc.

It is seen that the conductivity is increased by the presence of the magnetic field by a factor of 70 at $\theta = 1$. As the density is increased, this difference gradually diminishes and at very high density the effect of the field becomes small and the theory with no magnetic field can be applied.

B. TRANSVERSE CASE ($E_{\perp}$ TO THE MAGNETIC FIELD)

In this direction the motion of the electron is quantized and the ordinary Boltzmann formulation fails. Instead, the Wiedemann-Franz law may be applied to obtain the thermal conductivity from the electrical conductivity. The method of computing

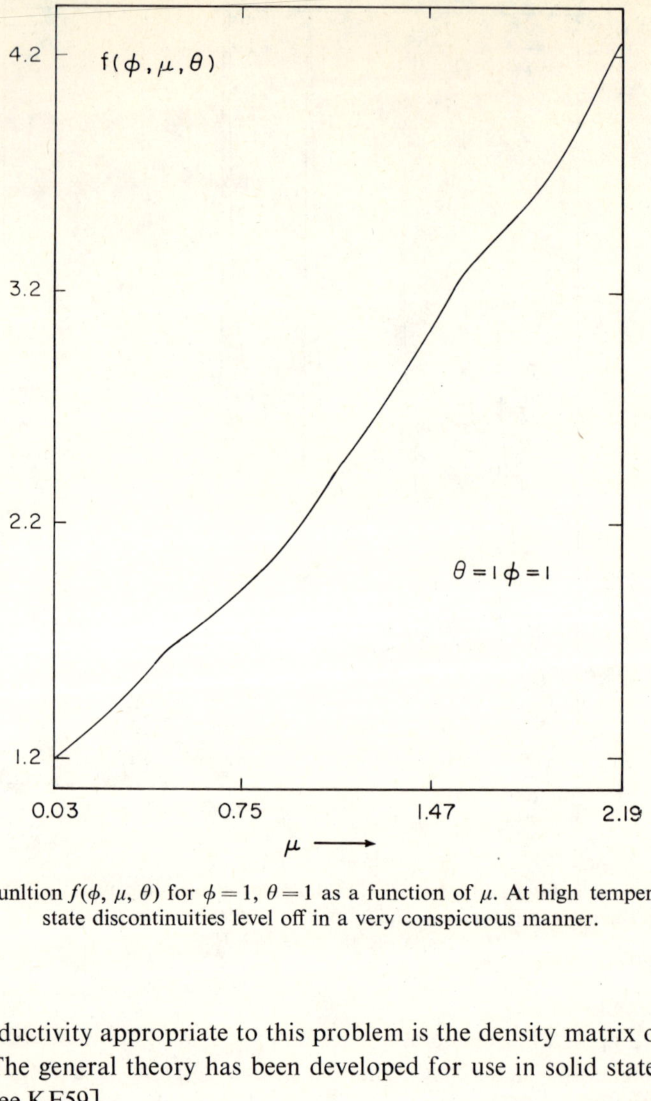

Fig. 16. The funltion $f(\phi, \mu, \theta)$ for $\phi = 1$, $\theta = 1$ as a function of μ. At high temperature the final state discontinuities level off in a very conspicuous manner.

electrical conductivity appropriate to this problem is the density matrix or the 'Kubo formalism'. The general theory has been developed for use in solid state physics for a long time [see KF59].

The average velocities of v_x and v_y are zero for an electron in a magnetic field. An impressed electric field $\perp$ to H will cause particles to move. The non-relativistic solution for an electron in a crossed electric field and magnetic field has been known for some time. The energy eigenvalues and the wave function are given by ($\hbar\omega_0 = mc^2 H/H_q$) (CCH69, KF59)

$$E_{np_zp_y} = mc^2 (n + \tfrac{1}{2}) \frac{H}{H_q} + \frac{1}{2m} p_z^2 - \frac{eEp_y}{m\omega_0} - \frac{1}{2} \frac{e^2E^2}{m\omega_0^2} \tag{10.23}$$

$$\Psi_{np_zp_y} = \frac{1}{L} \phi_n \left(x + \frac{p_y}{m\omega_0} + \frac{eE}{m\omega_0^2} \right) e^{ik_z z} e^{ik_y y} . \tag{10.24}$$

The wave function ϕ differs from those in the absence of a magnetic field by a displacement from the center by the following given amount:

$$\frac{eE}{m\omega_0^2} = \left(\frac{E}{H}\right)^2 \frac{\hbar}{mc} \tag{10.25}$$

but otherwise have the same analytic form, i.e.,

$$\phi_n(x) = e^{-x^2} H_n(x).$$

The effect of an impressed electric field in the x-direction is therefore to remove the degeneracy with respect to motion in the y-direction, and to shift the center of the orbit.

The conductivity in the $\perp$ direction is no longer a scalar as in the $\parallel$ case. We can define the conductivity tensor as (Z65)

$$\sigma_{\alpha\beta} = \begin{pmatrix} H^{-2} A_{xx} & -H^{-1} A_{yx} & -H^{-1} A_{zx} \\ H^{-1} A_{yx} & H^{-2} A_{yy} & -H^{-1} A_{zy} \\ H^{-1} A_{zx} & H^{-1} A_{zy} & A_{zz} \end{pmatrix} \tag{10.26}$$

where the coefficient A is independent on the magnetic field. The transverse resistance is defined as

$$\varrho_t = \frac{1}{\sigma_{xx}} = \frac{1}{\Delta H^2} (A_{yy}A_{zz} + A_{zy}^2)$$

where Δ is the determinant of the matrix $\sigma_{\alpha\beta}$. At the lowest power in $1/H$

$$\frac{1}{\sigma_{xx}} \to \frac{A_{yy}A_{zz} + A_{zy}^2}{A_{zz}A_{yx}^2}$$

i.e., the transverse magnetoresistance tends to a constant at high field.

In Figure 17 we show the experimental behavior of $(\Delta\varrho/\varrho_0)_{\parallel,\perp}$ in the case of mercury telluride at 4.2 K as a function of H (G67).

A complete quantum-mechanical computation of the transverse conductivity was performed by using the density matrix approach (CC70). The expression for σ_{yy} and σ_{xy} are given by

$$\sigma_{yy} = \frac{4Z^2\alpha^3}{\sqrt{2\pi}} \frac{mc^2}{\hbar} \lambda_c^3 \mathcal{N}_i \left(\frac{H}{H_q}\right)^{-1/2} f(\mu, \theta) \tag{10.27}$$

$$\sigma_{xy} = \alpha \frac{mc^2}{\hbar} \lambda_c^3 \mathcal{N}_e \left(\frac{H}{H_q}\right)^{-1} \tag{10.28}$$

with

$$f(\mu, \theta) = \sum_{n=0}^{\infty} \sum_{n'=0}^{\infty} (1 - \tfrac{1}{2}\delta_{n0})(1 - \tfrac{1}{2}\delta_{n'0}) F(n, n') \times \left[\int_0^{\infty} \mathrm{d}s\, \Lambda_- G(s, n, n') + \int_0^{\infty} \mathrm{d}s\, \Lambda_+ G(s, n, n')\right] \tag{10.29}$$

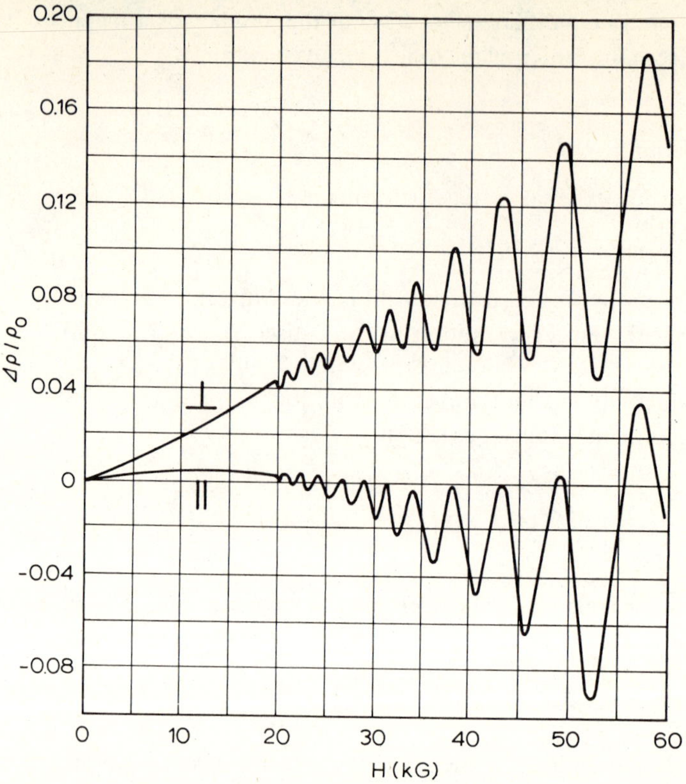

Fig. 17. SdH oscillations in mercury telluride in ∥ and ⊥ magnetic field, at 4.2K; the electron density is $1.18 \times 10t^8$/cm^3. The oscillations are in phase.

$$F(n, n') = \frac{\mu^2}{(\mu^2 - a_n^2)^{1/2}(\mu^2 - a_{n'}^2)^{1/2}} \quad a_n^2 = 1 + 2nH/H_q \equiv 1 + 2n\theta \tag{10.30}$$

$$G_\pm = \frac{s^{1/2}}{\left\{s + \frac{1}{2\theta}\left[\sqrt{\mu^2 - a_n^2} \pm \sqrt{\mu^2 - a_{n'}^2}\right]^2\right\}^2}. \tag{10.31}$$

The function $\Lambda_\pm$ depends on a certain combination of spinor coefficients as reported in the appendix of the original paper (CC70). In Figure 18 the function $f(\mu, \theta)$ is given *vs.* $\mu = (1 + \varrho_6^{2/3})^{1/2}$ for $H = 10^{-1} H_q$. Compared with the equivalent function for the longitudinal case (Figure 14), we see that in each jump the behaviour is just the opposite. The $\theta = 1$ case is perfectly analogous although the number of jumps per density interval is reduced. Once we know the function $f(\mu, \theta)$ the transverse conductivity can be easily computed from

$$\sigma_\perp = \frac{\sigma_{xy}^2 + \sigma_{yy}^2}{\sigma_{yy}}. \tag{10.32}$$

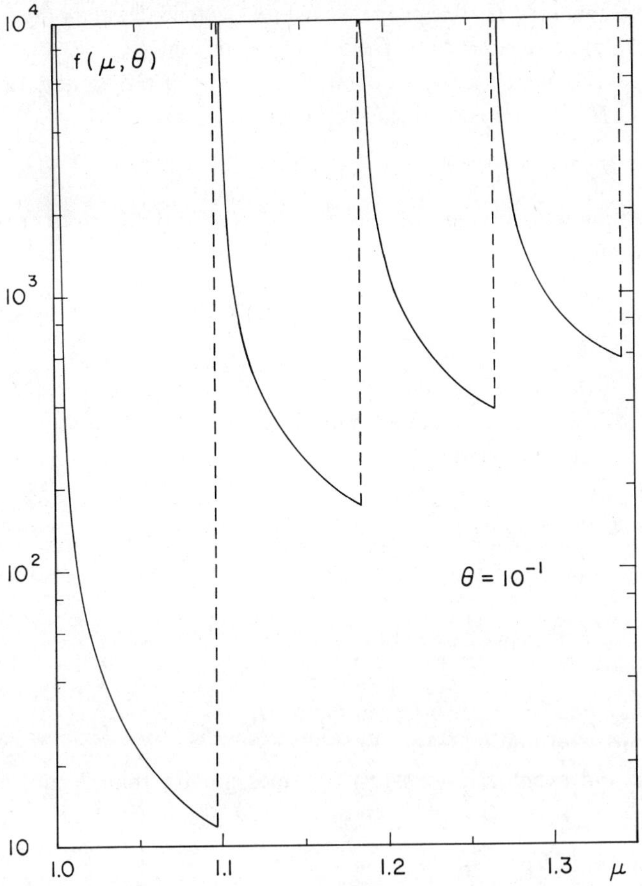

Fig. 18. The function $f(\mu, \theta)$ vs μ (in units of mc^2) for $\theta = H/H_q = 10^{-1}$.

A comparison of the transverse, zero-field and longitudinal conductivities, Equations (10.32), (10.2) and (10.15) is given in Table IV, *vs.* density at $H = H_q$. A sizable effect can be found only if the density is low enough. At higher density the three conductivities will coincide and no magnetic effect would be left whatsoever. This situation is analogous to the one encountered in the neutron beta decay (see Figure 13).

11. Magnetization of an Electron Gas. Semi-Permanent Magnetism (LOFER)

An electron possesses a magnetic moment $\mu_\beta = e\hbar/2mc$ (Bohr magneton) and macroscopic magnetization can result from this magnetic moment and from the orbital motion of the electrons. As is well known in plasma physics, the magnetic moment associated with a classical electron is diamagnetic. However, under certain circumstances an electron gas can possess a net magnetic moment.

In previous sections we have made no distinctions between B, the magnetic induction, and the external field H. The properties of an electron depend only on the strength of field near the electron and this is the field we talk about. However, as has been shown previously, (Ra44, Wa47), an orbiting electron in a magnetic field senses not the external field H but the magnetic induction B such that

$$B = H + 4\pi M \tag{11.1}$$

where M is the magnetic moment of the gas. As a result the Larmor frequency of the electron ω_L is given by:

$$\omega_L = \frac{eB}{mc} \quad \text{and not} \quad \frac{eH}{mc}. \tag{11.2}$$

If there is a net magnetic moment of an electron gas, M, then $B \neq H$. However, as M depends on ω_L which in turn depends on B, M must be a function of B. Hence the magnetic moment of an electron gas must be given by the following non-linear equation:

$$B = H + 4\pi M(B). \tag{11.3}$$

Under ordinary conditions (i.e., high temperature and low density) $4\pi M \ll H$, i.e.,

$$\frac{B}{H} = \mu = 1 + 4\pi \frac{M}{H} \simeq 1 \tag{11.4}$$

and no distinction need be made between B and H.

At high density and low temperature, however, as we will show below, the value of M increases and eventually $4\pi M(B)$ becomes greater than B and solutions can exist such that

$$B = 4\pi M(B), \qquad H = 0. \tag{11.5}$$

This means that an electron gas can become self-magnetized. Such a state of permanent magnetization is distinctively different from the ordinary ferromagnetism. The magnetization associated with Equation (11.5) will be referred to as 'Landau Orbital Ferromagnetism' or LOFER (LCCC69, CCCL69).

The magnetic moment M of an electron gas has been shown to be (CC68c):

$$M = \frac{1}{\Omega}\frac{1}{kT}\frac{\partial}{\partial H} \ln Z(T, H) \tag{11.6}$$

where Ω is the volume of the system and Z is the grand partition function, which is given by

$$\frac{1}{\Omega} \ln Z = \frac{eB}{\pi\hbar}\left\{\frac{1}{2}\int_{-\infty}^{+\infty} dx \ln\left[1 + \exp\left\{-\frac{\varepsilon(x, 0) - \mu}{T/T_0}\right\}\right.\right.$$

$$\left.\left. + \sum_{n=1}^{\infty}\int_{-\infty}^{+\infty} dx \ln\left[1 + \exp\left\{-\frac{\varepsilon(x_1 n) - \mu}{T/T_0}\right\}\right]\right\} \tag{11.7}$$

where

$$\varepsilon^2(x, n) = 1 + x^2 + 2nB/B_q \quad x \equiv p_z/mc, \quad B_q \equiv H_q. \tag{11.8}$$

After substituting (11.7) into (11.6) and carrying out some reductions as in reference (CC68c) we find:

$$M \equiv \frac{2}{\pi^2}\frac{\mu_B}{\lambda_c^3}\left\{\frac{1}{2}C_2(T, \mu) + \sum_{n=1}^{\infty} a_n^2 C_2(T/a_n, \mu/a_n) - \frac{B}{B_q}\sum_{n=1}^{\infty} n\, C_1(T/a_n, \mu/a_n)\right\} \tag{11.9}$$

or

$$M = \frac{B_q}{B}\frac{\mu_B}{mc^2}(P_{\|} - P_{\perp})$$

where $P_{\|}$ is the parallel stress and $P_{\perp}$ is the normal stress. C_n are the functions defined earlier (see Equations 5.27–5.30).

A. CLASSICAL LIMIT

We will show that at high temperature and low density M does reduce to the Curie-Langevin law for the magnetic susceptibility. In the non-degenerate and non-relativistic case:

$$\varepsilon(x, n) \to 1 + \tfrac{1}{2}x^2 + nB/B_q \quad \exp\left(\frac{\varepsilon - \mu}{T/T_0}\right) \gg 1. \tag{11.10}$$

In the definition of the C_k function when 1 is neglected against $\exp[(\varepsilon-\mu)/(T/T_0)]$, Equation (11.9) becomes:

$$\left[M = \frac{2}{\pi^2}\frac{\mu_B}{\lambda_c^3}\frac{1}{2}\left(\frac{T}{T_0}\right)\exp\frac{\mu - 1}{T/T_0} + \frac{T}{T_0}\sum_{n=1}^{\infty}\exp\left(\frac{\mu - 1}{T/T_0} - \frac{nB/B_q}{T/T_0}\right) - \frac{B}{B_q}\sum_{n=1}^{\infty} n\exp\left(\frac{\mu - 1}{T/T_0} - \frac{nB/B_q}{T/T_0}\right)\right]\int_0^{\infty}\exp\left(-\frac{x^2}{T/T_0}\right)dx. \tag{11.11}$$

Equation (11.11) can be rearranged into a geometrical series and its derivatives. After carrying out the sum we find:

$$M = \frac{(2\pi T/T_0)^{1/2}\,\mu_B}{\pi^2\lambda_c^3}\exp\frac{\mu - 1}{T/T_0}\frac{\partial}{\partial\eta}(\eta\coth\eta) \tag{11.12}$$

$$\eta \equiv \frac{B\mu_B}{kT}. \tag{11.13}$$

In the case $\eta \ll 1$ we then find:

$$\frac{\partial}{\partial\eta}(\eta\coth\eta) \to \tfrac{2}{3}\eta. \tag{11.14}$$

As we will show later $B \sim H$ and $M \ll B$ so that in (11.12) B may be replaced by H. We therefore find for the magnetic susceptibility

$$\chi \equiv \left(\frac{M}{H}\right)_{H=0} = \frac{2}{3}\frac{\mu_B^2}{kT}\exp\frac{\mu-1}{T/T_0}\left[\frac{1}{4\pi^3}\left(\frac{2\pi kT}{mc^2}\right)^{3/2}\frac{1}{\lambdabar_c^3}\right]. \tag{11.15}$$

From the definition of Z we find that in the limit (11.10)

$$\ln Z = \frac{\Omega}{4\pi^3}\frac{1}{\lambdabar_c^3}\left(\frac{2\pi kT}{mc^2}\right)^{3/2}. \tag{11.16}$$

The particle density N is given by

$$N = \frac{1}{\Omega}\lambda\frac{\partial}{\partial\lambda}\ln Z\,, \qquad \lambda \equiv \exp\frac{\mu-1}{T/T_0} \tag{11.17}$$

and therefore

$$\chi = \tfrac{2}{3}\left(\frac{\mu_B^2}{kT}\right)N \propto \frac{1}{T} \tag{11.18}$$

which is the Curie-Langevin law.

This classical result includes both the diamagnetic part (due to particle orbital motions) and the paramagnetic part (due to spin magnetic moment). The paramagnetic part is $N\,\mu_B/kT$ and the diamagnetic part is minus $\frac{1}{3}$ that of the paramagnetic part, and the sum of the two yields a factor of $\frac{2}{3}$ in (11.18).

It is difficult to separate the general expression (11.9) into the corresponding diamagnetic and paramagnetic parts. Nevertheless we can still decompose (Gordon decomposition) the 4 current j_μ (Sa67)

$$j_\mu = ie\bar{\psi}\gamma_\mu\psi = j_\mu^{(1)} + j_\mu^{(2)} \tag{11.19}$$

such that

$$j_\mu^{(1)}A_\mu = \frac{ie\hbar}{2m}\left[\partial_\mu\bar{\psi}\psi - \psi\partial_\mu\bar{\psi}\right]A_\mu - \frac{e^2}{mc}A_\mu^2\bar{\psi}\psi \tag{11.20}$$

becomes the Landau Hamiltonian giving rise to diamagnetism in the non-relativistic limit and the other part

$$j_\mu^{(2)}A_\mu = -\frac{e\hbar}{2mc}\left[\frac{1}{2}\frac{\partial A_\mu}{\partial x_\nu}(\bar{\psi}\sigma_{\mu\nu}\psi) + \frac{1}{2}\frac{\partial A_\nu}{\partial x_\mu}(\bar{\psi}\sigma_{\mu\nu}\psi)\right]$$
$$\sigma_{\mu\nu} = -i\gamma_\mu\gamma_\nu \quad \mu \neq \nu \tag{11.21}$$

becomes the Pauli Hamiltonian giving rise to spin paramagnetism in the non-relativistic limit.

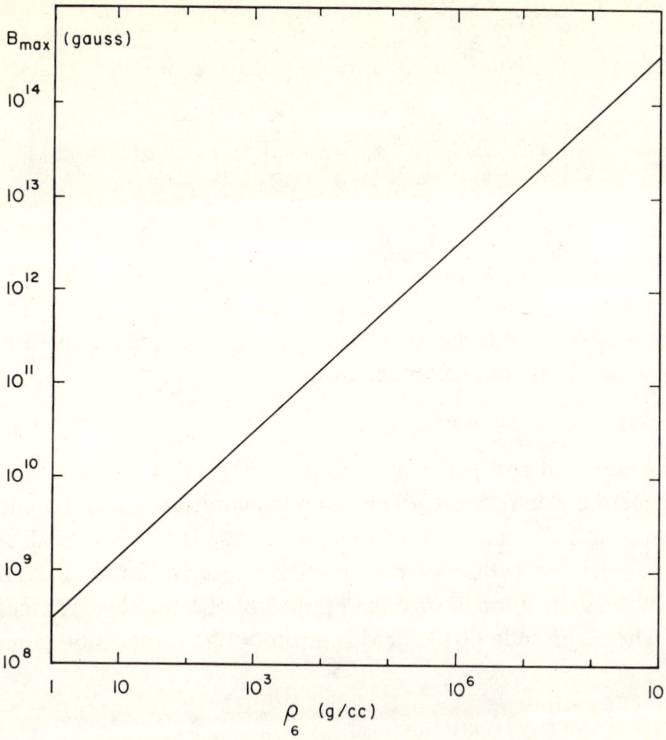

Fig. 19. As explained in the text at a fixed density (or μ) Equation (11.22) admits many solutions because of the oscillatory character of $M(B)$. In this figure only the maximum values of B are plotted vs the corresponding densities. It can be checked that $B_{max} \simeq \varrho_6{}^{2/3}$ which is what the flux conservation law would predict.

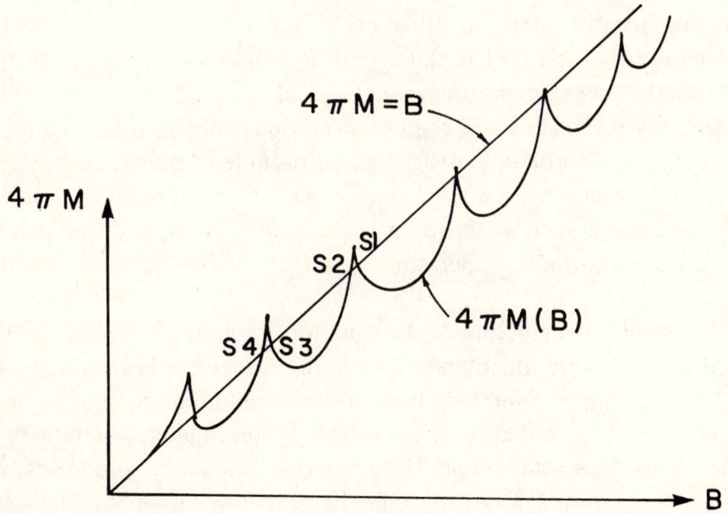

Fig. 20. Behaviour of the two functions $4\pi M$ and $4\pi M(B)$ as given by Equation (11.22). The last point of intersection corresponding to B_{ax} is plotted in Figure 19.

B. DEGENERATIVE EXPRESSIONS FOR M

This limit is analogous to that in the equation of state. We find

$$M = \frac{2}{\pi^2} \frac{\mu_B}{\lambda_c^3} \left[\frac{1}{2} C_2(\mu) + \sum_{n=1}^{s} a_n^2 C_2(\mu/a_n) - \frac{B}{B_q} \sum_{n=1}^{s} n C_1(\mu/a_n) \right]$$
$$= \frac{B}{B_q} \frac{\mu_B}{mc^2} [P_{\|}(B) - P_{\perp}(B)]. \tag{11.22}$$

$P_{\|}$ and $P_{\perp}$ show oscillatory behavior as a function of the density N, and so will M. This oscillatory behavior is entirely due to orbital quantization and is responsible for the existence of a state of permanent magnetism, as we shall now show.

C. LOFER (LANDAU ORBITAL FERROMAGNETISM)

This is a quasi-stable self-consistent macroscopic magnetism associated with a non-interacting degenerate electron gas. (This state is characterized by an energy higher than the ground state.) This macroscopic magnetization is the sum of all microscopic magnetizations associated with electrons in their respective Landau levels while the Landau levels of the system are in turn maintained by the macroscopic magnetization of the system. The magnitude of the field is given by the solution of Equation (11.3) with $H=0$, i.e.,

$$B = 4\pi M(B) \quad \text{or} \quad M = M(4\pi M). \tag{11.23}$$

$M(B)$ or $M(4\pi M)$ is derived by using in the interaction terms of the Hamiltonian-$\mathbf{J}\cdot\mathbf{A}$, the induced current $\mathbf{J}$ and the vector potential A due to the orbital motions of all electrons. There are a number of solutions of the LOFER state for a given density of electrons and the maximum value of LOFER magnetism is shown in Figure 14 for densities ranging from 10^6 g/cm^3 to 10^{15} g/cm^3.

Thermodynamically the LOFER state is not the minimum energy state and hence not the most stable state in the thermodynamical sense. But a system that is not thermodynamically stable may still require a very long time to reach the most stable state. (An example is a bottle of hydrogen gas at room temperature; the thermodynamically most stable state is when all hydrogen nuclei are catalyzed into a piece of iron, but the lifetime against such a transition is over 10^{100} years, and one can regard a bottle of hydrogen at room temperature as a truly stable state for practically all purposes.)

The LOFER state is the solution of the equation $B=4\pi M(B)$ and in the $4\pi M$ *vs* B plot, the LOFER states are the intersections of the curves $B=4\pi M$ and $B=4\pi M(B)$. Since $4\pi M(B)$ is a spiked function, there are two solutions S_1 and S_2 associated with each spike as shown in Figure 15. As will be discussed later, a state with a higher magnetization is the more stable one. If the systems attains the solution S_2 then it is certainly unstable against a transition into the next lower level S_3. The stability of this system is therefore the same as the stability against a transition from the S_1 state into the S_2 state (or S_3 into S_4, and so on).

In general $M(B)$ is also a function of the temperature and at high temperature $M(B)$ reduces to the classical expression, for which no intersection exists between the curve $4\pi M(B)$ and $4\pi M$, therefore there must exist a critical temperature T_c above which no LOFER state exists. This is analogous to the classical case of Ferromagnetism in which above the Curie temperature only paramagnetism exists. The value of the critical temperature has not been ascertained for the non-relativistic case. It is found that the condition for LOFER to exist is:

$$\mu/\kappa T > 10^7 (c/v_F)^2 \tag{11.24}$$

where μ is the Fermi energy and v_F the Fermi velocity. For white dwarfs (for which (11.24) is barely applicable) $v_F \simeq c$. The condition for LOFER magnetism to exist is therefore $T \lesssim 10^3$ K ($\mu \simeq mc^2$ in the center of most white dwarfs). However, (11.24) expresses a much more stringent conditions for the existence of LOFER state. The actual transition temperature is probably $\simeq 10^4$ K. This temperature is achievable in white dwarfs after a few times 10^9 years if crystallization does take place (AM59, S61, MR67).

Below this temperature the LOFER state is stable. In order for the gas to make a transition form S_1 to S_2 it is necessary to cross an energy barrier of the amount $\beta \Delta\Omega_B \cdot V$ where V is the characteristic volume of the system, and $\Delta\Omega_B$ is the free energy difference at the peak of the $M(B)$ curve and at S_1. If we choose $V = 4\pi R_L^3/3$ then it is shown that below T_c

$$V\Delta\Omega_B \gg \kappa T$$

so that the probability for the system to make a transition from S_1 to S_2 which is proportional to $\exp(-V\Delta\Omega_B/\kappa T)$, is negligibly small.

12. Coulomb Bremsstrahlung in a Magnetized Plasma

In addition to the synchrotron radiation process (Section 6) the bremsstrahlung process is an important radiation process. Further, it is the most important continuum emission process.

Unlike the synchrotron radiation the bremsstrahlung process must take place in the presence of a plasma. The presence of a plasma strongly affects the propagation of electromagnetic radiation. In this section the bremsstrahlung process in the presence of a plasma and a magnetic field will be discussed in detail.

Consider a test particle moving in a magnetized plasma. All possible effects can be understood in terms of the equation of motion whose solution gives us the current $j_\alpha(r, t)$ ($\alpha = 1, 2, 3$). Quantum mechanically this current is given by

$$j_\alpha(r, t) = \frac{ie\hbar}{2m}[\Psi^* \tilde{\nabla}_\alpha \Psi - \Psi \tilde{\nabla}_\alpha \Psi^*] \qquad \tilde{\nabla} \equiv \nabla - \frac{ie}{\hbar c} A_{\text{ext}} \tag{12.1}$$

where Ψ describes the state of the particle. If the particle is thought to be acted upon only by an external magnetic field the Ψ to be used is the solution of Schrodinger or Dirac equation with a magnetic field [see Appendix I]. This particle, moving in a medium produces an electromagnetic field, which in turn can be determined, using the Maxwell equation, as a function of $j_\alpha(r, t)$ itself. In a magnetic medium the electric field produced by such a current is given by [Sh66]

$$\mathbf{E}(r, t) = \int \mathbf{E}(\kappa, \omega)\, e^{i\mathbf{k}\cdot\mathbf{r} - i\omega t}\, \mathrm{d}^3\kappa\, \mathrm{d}\omega$$

$$\mathbf{E}(\kappa, \omega) \equiv \sum_{l=1}^{2} \frac{4\pi i}{\omega} \frac{\mathbf{a}(l)\,\mathbf{j}\cdot\mathbf{a}^*(l)}{\kappa^2 c^2/\omega^2 - n_l^2}. \tag{12.2}$$

The quantity n_l^2, the refractive index for the ordinary ($l=1$) and extraordinary wave ($l=2$) respectively, is given by

$$n_l^2 = -\frac{B}{2A} \pm \frac{1}{2A}(B^2 - 4AC)^{1/2} \tag{12.3}$$

$$A = \varepsilon_{11} \sin^2\theta + \varepsilon_{33} \cos^2\theta + \varepsilon_{13} \sin^2\theta$$

$$B = (\varepsilon_{12} \sin\theta - \varepsilon_{33} \cos\theta)^2 + \varepsilon_{13}^2 (\cos^2 2\theta + \sin^4\theta)$$
$$- \varepsilon_{11}\varepsilon_{33} - \varepsilon_{22}[\varepsilon_{11} \sin^2\theta + \varepsilon_{33} \cos^2\theta + \varepsilon_{13} \sin 2\theta]$$

$$C = \varepsilon_{11}\varepsilon_{22}\varepsilon_{33} - \varepsilon_{33}\varepsilon_{12}^2 - \varepsilon_{11}\varepsilon_{23}^2 - \varepsilon_{22}\varepsilon_{13}^2 - 2\varepsilon_{12}\varepsilon_{23}\varepsilon_{13}.$$

Here θ is the angle between κ and H. The dielectric tensor of the medium, $\varepsilon_{\alpha\beta}$, is usually given in a system (x_0, y_0, z_0) in which H is along the z_0-axis and κ lies in the plane (x_0, z_0).

The vector **a** has been shown by Shafranov to be

$$\mathbf{a} = a_y \{i\alpha_x^0, 1, i\alpha_z^0\} \tag{12.5}$$

where $[a_y^2 = (1 + \alpha_x^2)^{-1}]$ the quantities α_x^0 and α_z^0 given by

$$\alpha_x^0 = \frac{\varepsilon_{12}\varepsilon_{33} + \varepsilon_{23}\varepsilon_{13} - n_l^2[\varepsilon_{12} \sin\theta - \varepsilon_{23} \cos\theta] \sin\theta}{n_l^2(\varepsilon_{11} \sin^2\theta + \varepsilon_{33} \cos^2\theta + \varepsilon_{13} \sin 2\theta) - \varepsilon_{11}\varepsilon_{33} + \varepsilon_{13}^2} \tag{12.6}$$

$$\alpha_z^0 = -\frac{\varepsilon_{12}\varepsilon_{13} + \varepsilon_{11}\varepsilon_{23} + n_l^2[\varepsilon_{12} \sin\theta - \varepsilon_{23} \cos\theta] \cos\theta}{n_l^2(\varepsilon_{11} \sin^2\theta + \varepsilon_{33} \cos^2\theta + \varepsilon_{13} \sin 2\theta) - \varepsilon_{11}\varepsilon_{33} + \varepsilon_{13}^2} \tag{12.7}$$

are the ratio of the component of the electric field and therefore determine the polarization of the wave. In fact $\alpha_x^0 = \alpha_x \cos\theta + \alpha_z \sin\theta$, $\alpha_z^0 = a_z \cos\theta - \alpha_x \sin\theta$ satisfy the following equation

$$\alpha_x^2 + i\frac{\eta_{xx} - \eta_{yy}}{\eta_{yx}}\alpha_x - 1 = 0 \tag{12.8}$$

with

$$\eta_{xx} = N[\varepsilon_{11}\varepsilon_{33} - \varepsilon_{13}^2]$$
$$\eta_{yy} = N\{\varepsilon_{22}[\varepsilon_{11}\sin^2\theta + \varepsilon_{33}\cos^2\theta] - \varepsilon_{12}^2\sin^2\theta + [\varepsilon_{22}\varepsilon_{13} + \varepsilon_{23}\varepsilon_{12}]\sin 2\theta - \varepsilon_{23}^2\cos^2\theta\} \quad (12.9)$$
$$\eta_{yx} = -iN\{[\varepsilon_{12}\varepsilon_{33} + \varepsilon_{23}\varepsilon_{13}]\cos\theta + [\varepsilon_{12}\varepsilon_{13} + \varepsilon_{23}\varepsilon_{11}]\sin\theta\}$$
$$N^{-1} = \varepsilon_{11}\sin^2\theta + \varepsilon_{33}\cos^2\theta + \varepsilon_{13}\sin 2\theta\,.$$

Depending on the angle θ, Equation (12.8) gives the various types of polarizations. Once the electric field is known as a function of j_α, the Maxwell equation gives the definition of the intensity per second or power emitted by the particle, as

$$I = \frac{1}{T}\int \frac{d^3\kappa\, d\omega}{(2\pi)^4}[j_\alpha(k,\omega)E_\alpha^*(k,\omega) + E_\alpha(k,\omega)j_\alpha^*(k,\omega)] \quad (12.10)$$

Substituting Equation (12.2) into Equation (12.10) we obtain

$$I = \int_0^\infty d\omega \int_{4\mu} d\Omega\, I(\omega,\Omega) \quad (12.11)$$

where the emissivity $I(\omega,\Omega)$ is defined as

$$I(\omega,\Omega) = 2\,\mathrm{Re}\int d^3\kappa\, G_{\alpha\beta}(k,\omega)\, L_{\alpha\beta}(k,\omega) \quad (12.12)$$

with

$$G_{\alpha\beta} = j_\alpha(k,\omega)\, j_\beta^*(k,\omega)$$
$$L_{\alpha\beta} = \frac{4\pi i}{\omega}\sum_{l=1}^{2}\frac{a_\alpha(l)\,a_\beta^*(l)}{k^2c^2/\omega^2 - n_l^2}\,. \quad (12.13)$$

In Shafranov's paper many examples with different $j_\alpha(k,\omega)$ are discussed in detail. To compute Equation (12.11) in the case of Coulomb bremsstrahlung in a magnetic field, i.e.,

$$e^- + (Z,A) \rightarrow e^- + (Z,A) + \gamma$$

we must compute the current given in Equation (12.1). From perturbation theory we know that

$$\Psi_\alpha(r) = \psi_\alpha(r) - \sum_\beta \frac{\langle\beta|\,A_0\,|\alpha\rangle}{E_\beta - E_\alpha}\psi_\alpha(r) \quad (12.14)$$

where

$$A_0 = Ze/r\,.$$

The Fourier transform of the electric current, $j(k,\omega)$, is simply given by

$$j(k,\omega) = 2\pi\hbar\, j(k)\,\delta(E_i - E_f - \hbar\omega) \quad (12.15)$$

with ($\alpha = x, y, z$)

$$j_\alpha(k) = \frac{e}{2m}\sum_I \left\{\frac{\langle f|\, e^{-ik\cdot r}\Pi_\alpha\, |I\rangle\, \langle I|\, A_0\, |i\rangle}{E_i - E_I - i\Gamma} + \frac{\langle f|\, A_0\, |I\rangle\, \langle I|\, e^{-ik\cdot r}\Pi_\alpha\, |i\rangle}{E_i - E_I - \hbar\omega}\right\} \tag{12.16}$$

where i, f and I stand for initial, final and intermediate state.

A. DIELECTRIC TENSOR FOR A COLD PLASMA

With Equation (12.16) substituted in Equation (12.2) and then in Equation (12.10) one obtains in principle the energy loss at any angle and for any form of dielectric tensor. The problem is quite involved and the final form too complicated to analyze. We will therefore study separately the propagation $\parallel$ and $\perp$ to H as usually done in magnetized plasma, and will specify the type of plasma to be worked with. We will use the tensor $\varepsilon_{\alpha\beta}$ as given in the magneto-ionic theory, i.e., (S62)

$$\varepsilon_{\alpha\beta} = \begin{pmatrix} S & -iD & 0 \\ iD & S & 0 \\ 0 & 0 & P \end{pmatrix} \tag{12.17}$$

where

$$\begin{aligned} R &= 1 - \frac{\omega_p^2}{\omega^2}\frac{\omega}{\omega - \omega_H} - \frac{\Omega_p^2}{\omega^2}\frac{\omega}{\omega + \Omega_H} \\ L &= 1 - \frac{\omega_p^2}{\omega^2}\frac{\omega}{\omega + \omega_H} - \frac{\Omega_p^2}{\omega^2}\frac{\omega}{\omega - \Omega_H} \\ P &= 1 - \frac{\omega_p^2}{\omega^2} - \frac{\Omega_p^2}{\omega^2} \qquad 2S = R + L \qquad 2D = R - L\,. \end{aligned} \tag{12.18}$$

The various symbols are defined in the following way

$$\begin{aligned} \omega_p^2 &= \frac{4\pi N_e e^2}{m} \qquad & \omega_H &= \frac{eH}{mc} \\ \Omega_p^2 &= \omega_p^2\frac{m}{M_i}Z \qquad & \Omega_H &= \omega_H\frac{m}{M_i}Z_i \end{aligned} \tag{12.19}$$

where M_i is the mass of the ion, and Z_i its charge, $n_i Z_i = n_e$. With this notation, Equation (12.3) for H_l^2 reduces to

$$N_l^2 = \frac{1}{2A}(B \pm F) \tag{12.20}$$

with

$$\begin{aligned} A &= S\sin^2\theta + P\cos^2\theta \qquad & B &= RL\sin^2\theta + S(1 + \cos^2\theta) \\ C &= PRL \qquad & F^2 &= (RL - PS)^2\sin^4\theta + 4P^2D^2\cos^2\theta \end{aligned} \tag{12.21}$$

Equation (12.20) can also be written in a more transparent form (S62)

$$\mathrm{tg}^2\,\theta = -\frac{P(N_x^2 - R)(N_0^2 - L)}{(SN_x^2 - RL)(N_0^2 - P)} \tag{12.22}$$

from which the dispersion relation at $\theta=0$ and $\theta=\pi/2$ are easily obtained as

$$\begin{array}{lll} \theta = \pi/2 & N_0^2 = P\ (\text{ordinary}) & N_x^2 = RL/S\ (\text{extraordinary}) \\ \theta = 0 & N_0^2 = L & N_x^2 = R\,. \end{array} \tag{12.23}$$

The terminology, ordinary and extraordinary, is well established at $\theta=\pi/2$ where it is seen that the minus sign in Equation (12.20) gives rise to $N_-^2=P$, independent of the magnetic field: i.e., this mode propagates as it would in the absence of H (ordinary). The plus sign gives rise to $N_+^2=RL/S$, i.e., it does depend on H (extraordinary). The same terminology is retained for $\theta=0$. The parameters α, α_x and α_z are easily transformed to ($l=\pm$)

$$\begin{aligned} \alpha(l) &= -PD\cos\theta\,[AN_l^2 - PS]^{-1} \\ \alpha_x(l) &= D(N_l^2\sin^2\theta - P)(AN_l^2 - PS)^{-1} \\ \alpha_z(l) &= N_l^2 D\sin\theta\cos\theta\,(AN_l^2 - PS)^{-1}\,. \end{aligned} \tag{12.24}$$

B. PROPAGATION AT $\theta=0$.

From the previous equations one can easily see that at $\theta=0$, one obtains

$$\begin{array}{lll} A = P & B = 2PS & C = PRL \\ N_0^2 = L & N_x^2 = R & \alpha(0) = -\alpha(X) = 1 \\ \alpha_z(0) = \alpha_z(X) = 0 & & \alpha_x(0) = -\alpha_x(X) = 1 \end{array} \tag{12.25}$$

i.e.,

$$\begin{aligned} \mathbf{a}(0) &= \frac{1}{\sqrt{2}}[i, 1, 0] \\ \mathbf{a}(x) &= \frac{1}{\sqrt{2}}[-i, 1, 0] \end{aligned} \tag{12.26}$$

From here it follows that since $iE_x/E_y=-\alpha_x$, the ordinary wave is left-hand circularly polarized. This means that at $\theta=0$, only circularly polarized waves can propagate in a plasma. However, the emerged radiation may still be unpolarized or linearly polarized as a result of a combination of both R- and L-polarization states.

Substituting the polarization vector a into the expression for the electric field a little algebra gives ($\alpha=1, 2, 3$)

$$E_\alpha j_\alpha^* + E_\alpha^* j_\alpha = \frac{4\pi^2}{\omega}\,\delta(\Lambda)\,|j_x - ij_y|^2 \tag{12.27}$$

for the extraordinary wave and

$$E_\alpha j_\alpha^* + E_\alpha^* j_\alpha = \frac{4\pi^2}{\omega}\,\delta(\Lambda)\,|j_x + ij_y|^2 \tag{12.28}$$

for the ordinary wave. The delta function comes from the fact that

$$\begin{aligned} \frac{1}{\Lambda} &= P(\Lambda) - i\pi\delta(\Lambda) \\ \Lambda &= \kappa^2 c^2/\omega^2 - N_l^2 \end{aligned} \tag{12.29}$$

where P stands for the principal value.

Remembering that (ST68, AR69)

$$\begin{aligned} (\Pi_x - i\Pi_y)\,|n\rangle &= mc\,(2nH/H_q)^{1/2}\,|n-1\rangle \\ (\Pi_x + i\Pi_y)\,|n\rangle &= mc\,[2(n+1)\,H/H_q]^{1/2}\,|n+1\rangle \end{aligned} \tag{12.30}$$

we obtain for the extraordinary wave

$$\begin{aligned} \frac{dI(\omega,\Omega)}{d\omega\,d\Omega} &= I_0 \frac{\omega^2 N}{\sqrt{\varepsilon - \omega - n'\omega_H}} \int_0^\infty \frac{dt}{(t+\lambda)^2} \\ &\times \left\{\frac{1}{E_1}\sqrt{1+n'}\,\mathscr{I}(n, n'+1, t) + \frac{1}{E_2}\sqrt{n}\,\mathscr{I}(n-1, n')\right\}^2 \end{aligned} \tag{12.31}$$

$$I_0 = \frac{Z^2\alpha^3}{\sqrt{2}\,8\pi}\,\frac{mc^2}{\hbar/mc^2}\,N_i\lambda_c^3$$

$$N^2 = R, \qquad \varepsilon = E_i/mc^2, \qquad \omega \equiv \omega/(mc^2/\hbar), \qquad \omega_H \equiv H/H_q$$

$$\lambda = (2\omega_H)^{-1}\{\sqrt{2(\varepsilon - n\omega_H)} - \omega N - \gamma\sqrt{2(\varepsilon - \omega - n'\omega_H)}\}^2$$

$$\gamma = \pm 1$$

$$\mathscr{I}(n, n', t) = (n!\,n'!)^{-1/2}\,e^{-t/2}\,t^{(n+n')/2}\,{}_2F_0\left(-n, -n'; -\frac{1}{t}\right) \tag{12.32}$$

$$\begin{aligned} E_1 &= \omega - \omega_H - \tfrac{1}{2}\omega^2 N^2 - \gamma\omega N\sqrt{2(\varepsilon - \omega - n'\omega_H)} \\ E_2 &= \omega - \omega_H + \tfrac{1}{2}\omega^2 N^2 - \gamma\omega N\sqrt{2(\varepsilon - n\omega_H)}. \end{aligned}$$

A perfectly analogous computation gives the following result for the ordinary wave

$$\begin{aligned} \frac{dI(\omega,\Omega)}{d\omega\,d\Omega} &= I_0 \frac{\omega^2 N}{\sqrt{\varepsilon - \omega - n'\omega_H}} \int_0^\infty \frac{dt}{(t+\lambda)^2} \\ &\times \left\{\frac{1}{E_3}\sqrt{n'}\,\mathscr{I}(n'-1, n, t) + \frac{1}{E_4}\sqrt{n+1}\,\mathscr{I}(n+1, n', t)\right\}^2 \end{aligned} \tag{12.33}$$

$$E_3 = \omega + \omega_H - \tfrac{1}{2}\omega^2 N^2 - \gamma\omega N\sqrt{2(\varepsilon - \omega - n'\omega_H)}$$
$$E_4 = \omega + \omega_H + \tfrac{1}{2}\omega^2 N^2 - \gamma\omega N\sqrt{2(\varepsilon - n\omega_H)} \qquad (12.34)$$
$$N^2 = L.$$

Simple cases of n, and n' will be discussed later.

C. PROPAGATION AT $\theta = \pi/2$.

In this case the only wave we consider is the extraordinary one since the ordinary will not propagate if

$$N_0^2 = P = 1 - \frac{\omega_p^2}{\omega^2} - \frac{\Omega_p^2}{\omega^2} < 0. \qquad (12.35)$$

Equation (12.44) is satisfied in the radio wave region and at the surface of a neutron star where the density is of the order of $N_e \simeq 10^{20}/\text{cc} \simeq 10^{-4}$ g/cc The extraordinary mode has the following polarization vector

$$a(x) = (0, 1, 0). \qquad (12.36)$$

This form has been deduced from Equations (12.6) and (12.7), omitting the longitudinal component since our unit module's normalization is valid for pure transverse waves. In this case the quantity of interest is simply given by

$$E_\alpha j_\alpha^* + j_\alpha E_\alpha^* = \frac{8\pi^2}{\omega} |j_y(k, \omega)|^2 \, \delta(\Lambda).$$

Using Equation (12.39) we obtain

$$\frac{dI(\omega, \Omega)}{d\omega\, d\Omega} = \frac{1}{2} I_0 \frac{\omega^2 N}{\sqrt{\varepsilon - \omega - n'\omega_H}} \int_0^\infty \frac{dt}{(t + \tilde{\lambda})^2}$$
$$\times \left\{ \frac{1}{(\omega - \omega_H)^2} \left[\sqrt{1 + n'}\, \mathscr{I}(n, n'+1) + \sqrt{n}\, \mathscr{I}(n-1, n)\right]^2 \right. \qquad (12.37)$$
$$\left. + \frac{1}{(\omega + \omega_H)^2} \left[\mathscr{I}(n+1, n')\sqrt{n+1} + \sqrt{n'}\, \mathscr{I}(n'-1, n)\right]^2 \right\}$$

$$\tilde{\lambda} = \frac{1}{2\omega_H} \left\{\sqrt{2(\varepsilon - n\omega_H)} - \gamma\sqrt{2(\varepsilon - \omega - n'\omega_H)}\right\}^2 \qquad (12.38)$$
$$N^2 = RL/S.$$

D. BREMSSTRAHLUNG EMISSION IN THE LOW QUANTUM NUMBER REGION

In this section we will give the explicit form of the radiation intensity for a few quantum numbers, namely, $n=0$, $n'=0$, $n'=1$, $n=1$, $n'=0$. Using Equations (12.31) and

(12.38), the following notation

$$\tilde{I}(n, n'; O, X) = \frac{1}{I_0}\frac{\mathrm{d}I(\omega, \Omega)}{\mathrm{d}\omega\,\mathrm{d}\Omega} \tag{12.39}$$

we obtain the emissivity along the direction of the field, $\theta = 0$:

$$\begin{aligned}
\tilde{I}(0, 0; O) &= \sqrt{L}\,\frac{\omega^2}{\sqrt{\varepsilon - \omega}}\,\frac{1}{E_4^2}\,C_1(\lambda)\\
\tilde{I}(0, 0; X) &= \sqrt{R}\,\frac{\omega^2}{\sqrt{\varepsilon - E}}\,\frac{1}{\omega_1^2}\,C_1(\lambda)\\
\lambda &= (2\omega_H)^{-1}\,[\sqrt{2\varepsilon} - \omega\sqrt{R, L} - \gamma\sqrt{2(\varepsilon - \omega)}]^2\\
E_1 &= \omega - \omega_H - \tfrac{1}{2}\omega^2 R - \gamma\omega\sqrt{2R(\varepsilon - \omega)}\\
E_4 &= \omega + \omega_H + \tfrac{1}{2}\omega^2 L - \gamma\omega\sqrt{2L\varepsilon}
\end{aligned} \tag{12.40}$$

$$\begin{aligned}
I(1, 0; O) &= \sqrt{L}\,\frac{\omega^2}{\sqrt{\varepsilon - \omega}}\,\frac{1}{E_4^2}\,C_3(\lambda)\\
E_4 &= \omega + \omega_H + \tfrac{1}{2}\omega^2 L - \gamma\omega\sqrt{2L(\varepsilon - \omega_H)}\\
\lambda &\equiv (2\omega_H)^{-1}\,\{\sqrt{2(\varepsilon - \omega_H)} - \omega\sqrt{L} - \gamma\sqrt{2(\varepsilon - \omega)}\}^2\\
I(1, 0; X) &= \sqrt{R}\,\frac{\omega^2}{\sqrt{\varepsilon - \omega}}\left\{\frac{1}{E_1^2}[C_2 + C_3 - 2C_1]\right.\\
&\quad \left. + \frac{1}{E_2^2}C_2 + \frac{1}{E_1E_2}[2C_1 - 2C_2]\right\}\\
E_1 &= \omega - \omega_H - \tfrac{1}{2}\omega^2 R - \gamma\omega\sqrt{2R(\varepsilon - \omega)}\\
E_2 &= \omega - \omega_H + \tfrac{1}{2}\omega^2 R - \gamma\omega\sqrt{2R(\varepsilon - \omega_H)}\\
\lambda &= (2\omega_H)^{-1}\,[\sqrt{2(\varepsilon - \omega_H)} - \omega\sqrt{R} - \gamma\sqrt{2(\varepsilon - \omega)}]^2
\end{aligned} \tag{12.41}$$

Analogously the emissivity perpendicular to the magnetic field, $\theta = \pi/2$ is given by:

$$\begin{aligned}
\tilde{I}(0, 0; X) &= \frac{1}{2}\sqrt{\frac{RL}{S}}\,\frac{\omega^2}{\sqrt{\varepsilon - \omega}}\left[\frac{1}{(\omega - \omega_H)^2} + \frac{1}{(\omega + \omega_H)^2}\right]C_1(\lambda)\\
\lambda &= (2\omega_H)^{-1}\,[\sqrt{2\varepsilon} - \gamma\sqrt{2(\varepsilon - \omega)}]^2\\
\tilde{I}(1, 0; X) &= \frac{1}{2}\sqrt{\frac{RL}{S}}\,\frac{\omega^2}{\sqrt{\varepsilon - \omega}}\left[\frac{1}{(\omega - \omega_H)^2} + \frac{1}{(\omega + \omega_H)^2}\right]C_3(\lambda)\\
\lambda &= (2\omega_H)^{-1}\,[\sqrt{2(\varepsilon - \omega_H)} - \gamma\sqrt{2(\varepsilon - \omega)}]^2\\
\tilde{I}(0, 1; X) &= \frac{1}{2}\sqrt{\frac{RL}{S}}\,\frac{\omega^2}{\sqrt{\varepsilon - \omega - \omega_H}}\left[\frac{1}{(\omega - \omega_H)^2} + \frac{1}{(\omega + \omega_H)^2}\right]C_3(\lambda)\\
\lambda &= (2\omega_H)^{-1}\,[\sqrt{2\varepsilon} - \gamma\sqrt{2(\varepsilon - \omega - \omega_H)}]^2
\end{aligned} \tag{12.42}$$

$$C_1(x) = e^x(1+x)E(x) - 1 \qquad C_2(x) = \frac{1}{x} - e^x E(x)$$
$$C_3(x) = 1 + x - (2x + x^2)\,e^x E(x) \qquad E(x) = \int_x^\infty \frac{e^{-s}}{s}\,ds. \tag{12.43}$$

The present calculation differs from the previous one (CCFC69) in several significant ways. First, the Green's function used here is an exact one, while in the previous case the free particle Green's function has been used. Second, the plasma effect has been incorporated into our calculation. The final result, for example, Equation 12.40, can be interpreted easily. The factor $C_1(\lambda)$ arises from the Coulomb field of the nucleus, the factor ω^2 arises from the density of state of the photon, and finally, the factor E_4^{-2} comes from the Green's function and the factor $1/\sqrt{(\varepsilon-\omega)}$ comes from the density of the final state of the electron. As discussed earlier in an intense magnetic field an electron exhibits one dimensional behavior; instead of the usual expression $p^2\,dp/dE$, the density of the final state for a one dimensional particle is just $dp/dE \simeq 1/p$. The effect of the refractive medium on the photon is to alter the relation between ω and k, and this is taken into account throughout the calculation. This process has recently been applied to pulsar emission radiation (CHC70).

13. Astrophysical Applications

In the laboratory the generation of a steady state magnetic field up to a strength of 10^5 G is relatively easy. By using a generator and a capacitor (Kapitza magnet) an oscillatory field of strength of 10^6 G can be achieved. By using shaped charge and proper configurations a field of 10^7 G can be achieved. At a field of 10^7 G the Landau level spacing is about 1 eV, matter will disintegrate because of the enormous Zeeman level splitting the outer shell electrons suffer. As Regge pointed out, the chemistry of matter under intense fields is quite different from those with normal fields. In fact, effects of strong magnetic fields on biological systems have been detected.

Steady fields of the order of 10^7 G or greater can only be found in astrophysical bodies. According to the flux conservation law, the field of a current carrying plasma can increase as the square of the contraction ratio, C. In the case of white dwarfs, C is of the order of 100, and fields of the order of 10^7 can exist in white dwarfs. More drastically, in the case of neutron stars, C is of the order of 10^5 and fields up to 10^{14} G can exist. Such fields have been speculated in the past and have been ridiculed for being unrealistic. However, the discovery of pulsars (pulsed radio sourches) left no doubt that, not only neutron stars exist, but fields of the order of 10^{13} G or greater also exist.

However, much of the treatment of strong fields in the literature still follows the pattern of classical electrodynamics which we, among others including groups working actively in the Soviet Union, and Erber, have shown must fail under conditions currently associated with neutron stars. The most severe effect of strong magnetic fields on a plasma is the synchrotron radiation rate loss, which in a field of 10^{13} G and an electron energy of 1 MeV, is already as short as 10^{-19} sec. As the application of the physics of strong fields to neutron stars is still in progress, we will not go on into details here. (Ch69)

Appendix 1. Wave Function of an Electron in a Uniform, Constant Magnetic Field

A detailed derivation of the wave function is given in (JL49). We will report here only the result. The general feature is that the wave function contains or depends on a parameter which represents the degeneracy related with the location of the electron orbit in a magnetic field. The cylindrical coordinates ($x=r\cos\phi$, $y=r\sin\phi$, $z=z$) the wave function ψ has the form [S60].

$$\psi = \begin{pmatrix} \psi_1 \\ \psi_2 \\ \psi_3 \\ \psi_4 \end{pmatrix} \tag{A.1}$$

$$\psi_{1,3} = e^{-iEt/\hbar} \frac{e^{ik_z \cdot z}}{\sqrt{L}} \frac{e^{i(l-1)\phi}}{\sqrt{2\pi}} f_{1,3}(\varrho) \tag{A.2}$$

$$\psi_{2,4} = e^{-iEt/\hbar} \frac{e^{ik_z \cdot z}}{\sqrt{L}} \frac{e^{il\phi}}{\sqrt{2\pi}} f_{2,4}(\varrho) \tag{A.3}$$

$$f_{1,2,3,4} = \sqrt{2\gamma} \begin{pmatrix} C_1 & I^s_{n-1}(\varrho) \\ iC_2 & I^s_n \ (\varrho) \\ C_3 & I^s_{n-1}(\varrho) \\ iC_4 & I^s_n \ (\varrho) \end{pmatrix} \tag{A.4}$$

with

$$E = \eta mc^2 [1 + x^2 + 2nH/H_q]^{1/2} \qquad \eta = \pm 1 \tag{A.5}$$

$$\begin{aligned} \gamma &= \tfrac{1}{2}\lambda_c^{-2} H/H_q \qquad \varrho = \gamma r^2 \\ I^s_n(x) &= (n!\, s!)^{-1/2} e^{-x/2} x^{(n-s)/2} Q^{n-s}_s(x) \end{aligned} \tag{A.6}$$

$$Q^l_s(x) = (-)^s \sum_{j=0}^{s} (-)^j \frac{s!\, n!\, x^{s-j}}{j!\,(s-j)!\,(n-j)!}. \tag{A.7}$$

$n=1+s=0, 1, 2\ldots$ principal quantum number. $s=0, 1, 2\ldots$ the radial one and $1=0, \pm 1, \pm 2 \ldots (-\infty < l < \infty)$ the azimuthal ones. The 4 coefficients C_k in (A.4) at this stage are completely arbitrary except for the normalization condition

$$\sum_{i=1}^{4} |C_i|^2 = 1. \tag{A.8}$$

Their determination requires the introduction of a new operator $\hat{O}$ which commutes with the initial Hamiltonian. Two choices are usually made (TBZ66).

(1) $\hat{O}=\boldsymbol{\sigma}\cdot\boldsymbol{\pi}$, $\boldsymbol{\pi}=\mathbf{p}-e/(c)\,\mathbf{A}$: this operator describes the projection of the spin on the direction of motion. The requirement that

$$\hat{O}\psi = \hbar\tilde{k}\zeta\psi \tag{A.9}$$

with $(\hbar c\tilde{k})^2=E^2-m^2c^4$ determine uniquely the four C_k. The eigenvalue ζ can take the value ± 1. The C_k comes out to be

$$\begin{aligned} &C_1 = \zeta\tilde{\alpha}\tilde{A} \qquad C_2 = \tilde{\alpha}\tilde{B} \qquad C_3 = \tilde{\beta}\tilde{A} \qquad C_4 = \zeta\tilde{\beta}\tilde{B} \\ &2\tilde{A}^2 = 1+\zeta k_3/\tilde{k} \qquad 2\tilde{B}^2 = 1-\zeta k_3/\tilde{k} \\ &2\tilde{\alpha}^2 = 1+mc^2/E \qquad 2\tilde{\beta}^2 = 1-mc^2/E \end{aligned} \tag{A.10}$$

(2) $\hat{O}=\Pi_{12}$

$$\begin{aligned} \Pi_{12} &= mc^2\,\sigma_3 + c\varrho_2\,(\boldsymbol{\sigma}\times\boldsymbol{\pi})_3 \\ \Pi_{12}\psi &= \hbar c k\zeta\psi\,. \end{aligned} \tag{A.11}$$

In this case $\zeta=\pm 1$ characterizes the state of the spin polarization relative to the direction of the magnetic field: $\zeta=1$ along the field and $\zeta=-1$ against the field. In this case we found

$$\begin{aligned} &C_1 = aA \qquad C_2 = -\zeta bB \qquad C_3 = bA \qquad C_4 = \zeta aB \\ &2A^2 = 1+\zeta\lambda_c^{-1}/k \qquad 2B^2 = 1-\zeta\lambda_c^{-1}/k \\ &2a = (1+p_3c/E)^{1/2} + \zeta(1-p_3c/E)^{1/2} \\ &2b = (1+p_3c/E)^{1/2} - \zeta(1-p_3c/E)^{1/2}\,. \end{aligned} \tag{A.12}$$

The wave function used in References (K54) and (FC69) correspond to the first choice. In (K54) the wave function is given Cartesian coordinates and in this form it was taken in (CC68a).

Appendix II. Green Function for an Electron in a Constant Magnetic Field

We shall quote here 3 forms for the Green function.

1. NONRELATIVISTIC CASE *(Zero temperature)* (KMH65)

$$\begin{aligned} G(r,r'\,|\,E) \equiv \lim_{\varepsilon\to 0} \sum_{n,\,p_z,\,s} \frac{\psi_{n,\,p_z,\,s}(r)\,\psi^{\dagger}_{np_zs}(r)}{E(np_z)-(E\pm i\varepsilon)} \equiv \frac{m}{2\pi_l{}^2\hbar^2} \\ \times\Bigg\{\pm\sum_{n=0}^{N-1} J_n(x,y-y',x')\,\lambda_n(E)\exp[\pm i\,|z-z'|\,\lambda_n^{-1}(E)] \\ -\sum_{n=N}^{\infty} J_n(x,y-y',x')\,\lambda_n(E)\exp[-|z-z'|\,\lambda_n^{-1}(E)]\Bigg\} \end{aligned} \tag{A.13}$$

with

$$J_n = \exp\left\{\left(\frac{H/H_q}{2\lambda_c^2}\right)\right.$$

$$\left.\times\left[i(x+x')(y-y') - \tfrac{1}{2}(x-x')^2 - \tfrac{1}{2}(y-y')^2\right]\right\}\mathscr{L}_n(\alpha) \qquad \text{(A.14)}$$

$$\alpha \equiv \frac{H/H_q}{2\lambda_c^2}\left[(x-x')^2 + (y-y')^2\right]. \qquad \text{(A.15)}$$

The quantity N is the smallest integer, not smaller than $E/(mc^2\theta)-1$. When $|r-r' \ll \ll (H/H_q)^{1/2}\,\lambda_c^{-1}$. Equation (A.13) can be simplified to ($\hbar\omega_c \equiv mc^2 H/H_q$)

$$G(r-r'\,|\,E) = \frac{m}{2\pi\hbar^2}\left\{|r-r'|^{-1} + \frac{\hbar/l^2}{\{2m[(n+\tfrac{1}{2})\hbar\omega_c - E]\}^{1/2}}\right.$$

$$\left.\pm i\sum_{n=0}^{N}\frac{\hbar/l^2}{\{2m[E-(n+\tfrac{1}{2})\hbar\omega_c]\}^{1/2}}\right\}$$

2. Nonrelativistic case *(Finite temperature)* (L68)

The usual definition of the finite temperature Green function is

$$G(x,x') = \beta^{-1}\sum_n \exp[-i\omega_n(\tau-\tau')]\,G_n(x,x')$$

$$\omega_n = (2n+1)\pi\beta^{-1}, \qquad \beta^{-1} = \kappa T, \qquad n = 0, \pm 1, \pm 2, \ldots.$$

The function $G_n(x,x')$ can be shown to have the following general form

$$G_n(x,x') = \exp\left[-\frac{i}{2}m\omega_c(y+y')(x-x')\right]\bar{G}_n(\mathbf{x}-\mathbf{x}')$$

where the vector $\mathbf{x}-\mathbf{x}'$ has the following components

$$\mathbf{x}-\mathbf{x}' = [(\mathbf{x}-\mathbf{x}')_\perp, z-z'].$$

The function $\bar{G}_n(\mathbf{x}-\mathbf{x}')$ is now Fourier transformed

$$\bar{G}_n(\mathbf{x}-\mathbf{x}') = \int\frac{d^3p}{(2\pi)^3}\exp[i\mathbf{p}(\mathbf{x}-\mathbf{x}')]\,G_n(p)$$

and $G_n(p)$ is finally computed to be

$$G_n(p) = 2e^{-\eta/2}\sum_{N=0}^{\infty}\frac{(-)^N L_N(\eta)}{i\omega_n - E(Np_z)}$$

with

$$\eta \equiv \frac{2}{m\omega_c}\left(p_x^2 + p_y^2\right)$$

$$E(N, p_z) = \frac{1}{2m} p_z^2 + (N + \tfrac{1}{2})\,\hbar\omega_c - \mu$$

L_N is the Laguerre polynomial of order N and μ is the chemical potential.

3. RELATIVISTIC CASE *(Zero temperature)* (KU64)

The first computation of the Green function in a constant magnetic field was performed by Schwinger in 1949 (Sc49). We will report here a more recent computation by Kaitna and Urban (KU64). The relativistic Green function satisfies the following differenticl equation

$$\left(i\partial_\nu\gamma^\nu + \frac{e}{\hbar}A_\nu\gamma^\nu - \lambda_c^{-1}\right)G(xy) = \delta^4(x - y).$$

With the ansatz

$$G(xy) = \left(i\partial_\mu\gamma^\mu + \frac{e}{\hbar}A_\mu\gamma^\mu + \lambda_c^{-1}\right)\bar{G}(xy)$$

the Fourier transform of $\bar{G}(x, y)$ i.e.,

$$\bar{G}(p, q) = \frac{1}{(2\pi)^2}\int \exp(ipx)\exp(-iqy)\,\bar{G}(x, y)\,\mathrm{d}^4x\,\mathrm{d}^4y$$

is shown to be given by

$$G(p, q) = (32\omega_c^2\pi^4)^{-1}\,\delta(p_0 - q_0)\exp\left[\frac{i}{\omega_c}(p_1q_2 - q_1p_2)\right]Z^{-1/2-a/2b}$$

$$\times\begin{pmatrix}[\Gamma(1 + a/b)]^{-1/2} & w_{-(1/2)-(a/b),\,0}(Z) & 0 \\ 0 & & \Gamma(a/b)\,w_{-a/b,\,0}(Z)\end{pmatrix}$$

with

$$Z \equiv \omega_c^{-1}[(p_1 - q_1)^2 + (p_2 - q_2)^2],\quad a \equiv p_0^2 - p_3^2 - \lambda_c^{-2},\quad b \equiv -4\omega_c.$$

List of Symbols

α	Sommerfeld's fine structure constant
$\alpha_x, \alpha_y, \alpha_z$	components of polarization vector
$\mathbf{a}$	polarization vector
A	mass number of a nucleus
$\tilde{A}$	vector potential operator
A_0	scalar Coulomb's potential
A	coefficient of plasma dispersion relation
$\mathbf{A}_1(\mathbf{r}, t)$	vector potential of the self consistent electromagnetic field in a plasma

A_μ	four vector potential
B	magnetic induction
B_q	critical magnetic field strength
C	coefficient of plasma dispersion relation
c	speed of light
c_v	specific heat at constant volume
D	plasma parameter as defined by Stix
$\delta_{\alpha\beta}$	Kronecker delta
$E \equiv E(p_z, n)$	total electron energy
$\mathbf{E}$	electric field vector
E_α, E_β	electric field components
$\varepsilon_{\alpha\beta}$	plasma dielectric tensor
$\varepsilon_{11}, \varepsilon_{12}, \varepsilon_{23}$ etc.	are components of the plasma dielectric tensor
e^-	electron
e	charge of the electron
e_n^-	electron in the state n
ε	total electron energy (in a strong magnetic field) in units of rest energy of the electron mc^2

$$f_{\parallel} \equiv \frac{\partial}{\partial p_{\parallel}} f_0, \; f_{\perp} \equiv \frac{\partial}{\partial p_{\perp}} f_0$$

F	plasma parameter as introduced by Stix
$f(\mathbf{r}, \mathbf{p}, t)$	Wigner distribution function
$f_0(p)$	equilibrium distribution
$f_1(rp\lambda)$	perturbed first order Wigner function
$f(x)$	Fermi distribution, $x \equiv p/mc$
g_A	axial vector coupling constant
g_V	vector coupling constant
γ	electron total energy in units of mc^2
γ_μ, γ_5	Dirac matrix $\mu = 1, 2, 3, 4$
$\frac{\partial}{\partial x_\mu} \equiv \partial_\mu$	covariant derivature
$\mathbf{H}$	magnetic field vector
H_q	strength of the critical magnetic field
h	Planck's Constant
$\hbar$	rationalized Planck's Constant
$\tilde{I}$	radiation intensity
$\mathbf{j}$	electric current vector
$\tilde{j}_\alpha$	quantum mechanical current operatc
K	Boltzmann's constant
K_H	conductive opacity coefficient
k	propagation vector
$\mathscr{L}$	Lagrangian

L	plasma Parameter introduced by Stix
l	attenuation coefficient
λ_H	thermal conductivity coefficient in magnetic field H
M	magnetization
$\lambda \equiv g_A/g_V$	magnetic moment of electron gas
M	
m	rest mass of the electron
M_i	rest mass of ion
μ_β	Bohr magnetron
μ	chemical potential
N	particle density
n_l	refractive index of magnetoactive plasma
n	quantum number
N_0	electron number density
n_+, n_-	refractive index for circularly polarized waves
N	index of refraction
$n_{\alpha\beta}$	tensor related to the dielectric tensor $\varepsilon_{\alpha\beta}$
$\eta_{xx}\ \eta_{xy}$ etc.	plasma parameters related to plasma dielectric tensor
ω	circular frequency
ω_n	level degeneracy factor in magnetic field
ω_C	cyclotron frequence for electron ($\omega_C \equiv \omega_H$)
ω_L	relativistic Larmor frequency
ω_p	plasma frequency for electron
Ω_p	plasma frequency for ion
Ω_H	cyclotron frequency for ion
Ω	volume
w_j	Fermi distribution function
w	electron energy in units of mc^2
$P_{xx}, P_{yy} \equiv P_\perp$	
$P_{zz} \equiv P_\parallel$,	
$P_{zz} \equiv P_\parallel$	Pressure along the magnetic field
$P_\perp$	Pressure perpendicular to magnetic field
$\mathbf{p}$	momentum vector of electron
p_x, p_y, p_z	components of electron momentum
$\omega_1, \omega_2, \omega_3, \omega_4$	spin dependent functions
Q	Heat flow
q	wave vector
$\varrho(\mathbf{r}, \mathbf{r}')$	equilibrium density matrix
$\varrho_6 \equiv 10^{-6}\varrho$	density in g/cm^3
ϱ_t	transverse electrical resistance
R_L	Larmor radius
S	s matrix
S	Stix's plasma parameter

s	spin quantum number
$T_{\alpha\beta}$	electromagnetic stress tensor
$T_{\mu\nu}$	energy momentum tensor
T	temperature (in K)
$T_0 \equiv \frac{mc^2}{\mathrm{K}}$	relativistic temp
τ	half life of electron
U	energy density
v	electron velocity
$v_{\parallel}$	electron velocity parallel to magnetic field
$v_{\perp}$	electron velocity perpendicular to magnetic field
$\sigma_{\alpha\beta}$	electrical conductivity tensor
$\sigma_{\parallel}$	electrical conductivity along magnetic field
$\sigma_{\perp}$	electrical conductivity perpendicular to magnetic field
$\sigma_{\mu\nu}$	product of Dirac matrix ($= -i\gamma_\mu\gamma_\nu$)
z	atomic number
Z	grand partition function
χ	magnetic susceptibility
x	electron momentum in units of mc
ψ	wave function of electron
ϕ	scalar potential
θ	angle
$\theta \equiv H/H_q$	field strength parameter ($H_q \equiv m^2c^3/e\hbar = 4.414 \times 10^{13}$ G)
λ_B	de-Broglie wave length
$\tilde{\lambda}_C$	Compton wave length of the electron divided by 2π
ν	frequency
ν_m	minimum frequency of emission
ν	neutrino
$\bar{\nu}$	antineutrino

References

A58 Argyres, P. N.: 1958, *Phys. Rev.* **109**, 115.

A70 Adler, S. L., Bahcall, J. N., Callan, C. G. and Rosenbluth, M. N.: 1970, *Phys. Rev. Letters* **25**, 1061.

AA56 Argyres, P. N. and Adams, E. N.: 1956, *Phys. Rev.* **104**, 900.

AB65 Akhiezer, A. I. and Berestetsky, V. B.: 1965, *Quantum Electrodynamics*, Interscience Publ. Inc., New York, Ch. IV.

AH59 Adams, E. N. and Holstein, T. D.: 1959, *Phys. Chem. Solids* **10**, 254.

AM59 Auluck, F. C. and Mathur, V. S.: 1959, *Z. Astrophys.* **48**, 28.

AR69 Arunalsalam, V.: 1969, *Math. Phys.* **10**, 1305.

BC62 Brittin, W. E. and Chappel, W. R.: 1962, *Rev. Mod. Phys.* **34**, 620.

BPP69 Baym, G., Pethick, C., and Pines, D.: *Nature* **224**, 674.

C67 Cameron, R. C.: 1967, *The Magnetic and Related Stars*, Mono Book Corp.

Ca67 Cameron, A. G. W.: 1967, *Astrophys. J. Letters* **1**, 9.

C69 Canuto, V.: 1970, *Astrophys. J.* **159**, 641.

C71 Canuto, V.: 1971, *Astrophys. J.*, to be published.

CC68a Canuto, V. and Chiu, H. Y.: 1968, *Phys. Rev.* **173**, 1210.
CC68b Canuto, V. and Chiu, H. Y.: 1968, *Phys. Rev.* **173**, 1220.
CC68c Canuto, V. and Chiu, H. Y.: 1968, *Phys. Rev.* **173**, 1229.
CC68d Canuto, V. and Chiu, H. Y.: 1968, *Astrophys. J. Letters.* **153**, 157.
CC68e Chiu, H. Y. and Canuto, V.: 1968, *Phys. Rev. Letters* **21**, 2, 110.
CC69 Canuto, V. and Chiu, H. Y.: 1969, *Phys. Rev.* **188**, 2446.
CC70 Canuto, V. and Chiuderi, C.: 1970, *Phys. Rev.* **D1**, 8, 2219.
CCC70 Canuto, V., Chiuderi, C., and Chou, C. K.: 1970, *Astrophys. Space Sci.* **7**, 407.
CCCFC70 Canuto, V., Chiu, H. Y., Chou, C. K., and Fassio-Canuto, L.: 1970, *Phys. Rev.* **D2**, 2, 281.
CCCL69 Canuto, V., Chiu, H. Y., Chiuderi, C., and Lee, H. J.: *Nature* **225**, 47.
CCFC68 Chiu, H. Y., Canuto, V., and Fassio-Canuto, L.: 1968, *Phys. Rev.* **176**, 1438.
CCFC68a Canuto, V., Chiu, H. Y., and Fassio-Canuto, L.: 1969, *Astrophys. Space Sci.* **3**, 258.
CCFC69 Chiu, H. Y., Canuto, V., and Fassio-Canuto, L.: 1969, *Phys. Rev.* **185**, 1607.
CCH69 Canuto, V. and Chiuderi, C.: 1969, *Nuovo Cimento* **2**, 223.
CFC69 Chiu, H. Y. and Fassio-Canuto, L.: 1969. *Phys. Rev.* **185**, 1614.
CG70 Cameron, A.G.W. and Gentile, A.: private communication.
Ch66 Chiu, H. Y.: 1966, *Ann. Rev. Nucl. Sci.* **16**, 591.
Ch66a Chiu, H. Y.: 1966, in *Stellar Evolution* (ed. by R. F. Stein and A. G. W. Cameron), Plenum Press, New York.
Ch68 Chiu, H. Y.: 1968, *Stellar Physics*, Blaisdell Publ. Co., Waltham, Mass., Ch. 5.
CHC70 Chiu, H. Y. and Canuto, V.: 1971, *Astrophys. J.* **163**, 577.
CS70 Canuto, V. and Solinger, A.: 1970, *Astrophys. J. Letters* **6**, 141.
CV71 Canuto, V., Ventura, J.: 1971, *Nuovo Cimento*, to be published.
D28 Darwin, B. C.: 1928, *Proc. Roy. Soc.* **117**, 258.
E66 Erber, T.: 1966, *Rev. Mod. Phys.* **38**, 626.
FC69 Fassio-Canuto, L.: 1969, *Phys. Rev.* **187**, 2141.
G67 Giriat, W.: 1967, *Phys. Letters* **24A**, 515.
G69 Greenstein, G.: 1969, *Nature* **223**, 938.
GO69 Gunn, J. E. and Ostriker, J. P.: 1969, *Nature* **223**, 813.
Gr69 Greene, M. P., Lee, H. J., Quinn, J. J., and Rodriquez, S.: 1969, *Phys. Rev.* **177**, 1019.
H31 Huff, I. D.: 1931, *Phys. Rev.* **38**, 510.
HL69 Hubbard, W. B. and Lampe, M.: 1969, *Astrophys. J., Suppl.* **163**, 18.
HS61 Hamada, T. and Salpeter, E. E.: 1961, *Astrophys. J.* **134**, 683.
J62 Jackson, J. D.: 1962, *Classical Electrodynamics*, John Wiley & Sons, New York, p. 313.
J69 Jancovici, B.: 1969, *Phys. Rev.* **187**, 2275.
JL49 Johnson, M. H. and Lippmann, B. A.: 1949, *Phys. Rev.* **76**, 828.
K27 Kennard, E. H.: 1927, *Z. Physik* **44**, 263.
K54 Klepikov, N. P.: 1954, *JETP* **26**, 19.
K65 Kubo, R.: 1965, *Statistical Mechanics*, North-Holland Publ. Co., Amsterdam, pp. 278–280.
Ka64 Kaitna, R. and Urban, P.: 1964, *Nucl. Phys.* **56**, 518.
Ke64 Kelly, D. C.: 1964, *Phys. Rev.* **134**, A641.
Ke70 Kelly, D. C.: 1970, *Am. Phys. Soc. Bull.* **15** (11), No. 1, 124.
KF59 Kahn, A. H. and Frederiske, H. P. R.: 1959, *Solid State Phys.* **9**, 257.
KMH65 Kubo, R., Miyake, S. J., and Hashitsume, N.: 1965, *Solid State Phys.* **17**, 270.
L30 Landau, L.: 1930, *Z. Physik* **64**, 629.
L66 Landstreet, J. D.: 1967, *Phys. Rev.* **153**, 1372.
L68 Lee, H. J.: 1968, *Phys. Rev.* **166**, 459.
LCCC69 Lee, H. J., Canuto, V., Chiu, H. Y., and Chiuderi, C.: 1969, *Phys. Rev. Letters* **23**, 390.
LL60 Landau, L. D. and Lifshitz, E. M.: 1960, *Electrodynamics of Continuous Media*, Pergamon Press, Ch. VIII.
LL62 Landau, L. D. and Lifshitz, E. M.: 1962, *Classical Theory of Fields*, Pergamon Press, Ch. 3.
M41 Marshak, R. E.: 1941, *Ann. Acad. Sci.* **41**, 49.
M50 Mestel, L.: 1950, *Proc. Cambridge Phil. Soc.* **46**, 337.
MR67 Mestel, L. and Ruderman, M. A.: 1967, *Monthly Notices Roy. Astron. Soc.* **136**, 27.
P30 Page, L.: 1930, *Phys. Rev.* **36**, 444.

P130 Plesset, M. S.: 1930, *Phys. Rev.* **36**, 1728.
R28 Rabi, I. I.: 1928, *Z. Physik* **49**, 507.
R52 Robl, B.: 1952, *Acta Phys. Austriaca* **6**, 105.
R65 Ruderman, M. A.: 1965, *Rep. Progr. Phys.* **28**, 411.
Ra44 Rasetti, R.: 1944, *Phys. Rev.* **66**, 1.
S58 Skobov, V. G.: 1958, *JETP* **35**, 1315.
S60 Sokolov, A. A.: 1960, *Introduction to Quantum Electrodynamics*, U.S. Atomic Energy Commission: AEC-tr-4322, pp. 174–223.
S61 Salpeter, E. E.: 1961, *Astrophys. J.* **134**, 669.
S61 Stix, T. H.: 1962, *The Theory of Plasma Waves*, McGraw-Hill Book Co., New York.
S66 Shklovsky, I. S.: 1966, *Soviet Astron. AJ* **10**, 6.
Sa67 Sakurai, J. J.: 1967, *Advanced Quantum Mechanics*, Addison-Wesley Publ. Co., Reading, Mass., pp. 107–110.
Sc49 Schwinger, J.: 1949, *Phys. Rev.* **75**, 898.
Sh66 Shafranov, V. D.: 1966, in *Reviews of Plasma Physics* **3**, Consultants Bureau, New York.
ST68 Sokolov, A. A. and Ternov, I. M.: 1968, *Synchrotron Radiation*, Pergamon Press, New York.
TBBD69 Ternov, I. M., Bagrov, V. G., Bordovitsyn, V. A., and Dorofeev, O. F.: 1969, *Soviet Phys. Doklady*, No. 13, 1219.
TBR64 Ternov, I. M., Bagrov, V. G., and Rzaev, R. A.: 1964, *JETP* **19**, 225.
TBZ66 Ternov, I. M., Bagrov, V. G., and Zhukovskii, V.: 1966, *Moscow Univ. Phys. Bull.* **1**, 21.
UY30 Uhlembeck, G. E. and Young, L. A.: 1930, *Phys. Rev.* **36**, 1721.
W64 Woltjer, L.: 1964, *Astrophys. J.* **140**, 1309.
Wa47 Wannier, G. H.: 1947, *Phys. Rev.* **72**, 304.
Z64 Zyrianov, P. S.: 1964, *Fiz. Metal. Metalloved.* **18**, 161.
Z65 Ziman, J. M.: 1965, *Principles of the Theory of Solids*, Cambridge Univ. Press.

22

STELLAR COALESCENCE

A.G.W. Cameron

Yeshiva University and
Institute for Space Studies
New York, N.Y.

The subject of stellar collisions is of little interest in our own galaxy. Such collisions cannot be expected with a frequency greater than one per billion years, or so. Recently it has become evident that the frequency may be much higher in some massive elliptical galaxies with dense centers. Furthermore, it has been observed that the light energy emanating from the nuclei of some of these galaxies is of variable intensity, the radiation presumably coming largely from the synchrotron process. It has been suggested that quasars may be dense stellar systems, and that most of the energy release occurs as an indirect result of stellar collisions. In such a system stellar velocities would be quite high, perhaps up to thousands of kilometers per second.

Colgate has discussed the development of such a system of stars. There would be a period of soft collisions when the main result of collision would be the coalescence of stars into larger bodies. For the moment, assume that there is no mass loss in stellar coalescence. If one began with, say, stars of one solar mass, then the result of an amalgamative collision would be a star of 2 $M_\odot$. This star will have a radius about double the 1 $M_\odot$ stars, an area four times as great, and correspondingly greater collision cross section. Subsequent collisions will occur more rapidly, resulting in more and more massive stars. This process will eventually reach a limit, estimated to be about 50 $M_\odot$, at which point the star will be very large, but of such low density that further collisions with stars of a few solar masses will not lead to further coalescence. The evolutionary lifetime to the supernova stage for such a star will be about 3×10^6 years. The amount of energy released by the supernova will be so much greater than the energy released in the stellar collisions that the collisions should be thought of primarily as a mechanism for producing frequent and massive supernovae.

Studies of the development of star clusters have shown that some stars will be lost by evaporation from the cluster, and that a fraction of the cluster will be concentrated more deeply into a sort of cusp at the bottom of the cluster's potential well. In a steady state situation, stellar encounters throughout the body of the cluster would continually feed in stars to the dense center. In a qualitative way, we can think of a relatively small number of stars at the bottom of the potential well sweeping up other stars as they are brought in, going to supernovae, and new amalgamating centers being formed in further collisions. Such a process could be fairly steady, repetitive, or quite sporadic. This is an area for further theoretical study. There is some observational evidence, from radio galaxies which do not have the extremely high densities we would associate with quasars, that the process may go in spurts. On the other hand, in a quasar one can imagine energy pumped in from several supernovae per year and a fairly steady energy output.

Consider a general collision between two stars. Even if the stars start from rest at infinity, they will have a finite relative velocity at collision corresponding to their escape velocity from each other. We know from the virial theorem that the gravitational binding energy of a star will be twice its thermal energy. Now the central regions of a star have high density and high temperature, and the surface layers have low density and generally low temperature. If two stars collide at escape velocity, the collisional velocity will be supersonic for the outer layers of the stars, but subsonic for the central regions. That means that during the collisions strong shocks will develop in the outer layers and pressure irregularities will be transmitted as sound waves in the inner regions. Of course, if the collisional velocity is increased the fraction of the mass which participates subsonically goes down. We wish to know what fraction of the colliding mass will remain bound for various velocities equal to, greater, and less than the velocity which gives the system zero total energy, i.e., which could just disperse the particles composing the stars to infinity with zero energy. It is evident that at that critical energy the system sould not be completely disrupted, since some of the mass will be sent to infinity with non-zero energy, and the rest will remain bound. We want to know the breakeven point at which at least half of the total colliding mass remains.

The computations which will be presented shortly are for two polytropes of index n = 3.0 of 1 $M_\odot$. The critical velocity for this pair is 1512 km./sec. relative velocity.

In Colgate's discussion, a glancing collision was considered. Since the collision would normally be supersonic in the surface region there would be dissipation of energy in shock waves. This would reduce the relative velocity. Thus if the relative velocity were initially equal to the escape velocity, then after the collision the stars would be in a bound system which would result in further collisions. Colgate estimated that the breakeven point occurs when the total kinetic energy of the system is less than 1/3 of the critical energy required to disperse the system to infinity. For 1 $M_\odot$ this corresponds to a relative velocity of about 900 km./sec.

There have been a few attempts to study stellar collisions by numerical hydrodynamics in the past. Slow collisions of the kind that interest us here have been studied by Mathis. He used one-dimensional hydrodynamics, but in a very clever way. He considered head-on collisions of identical stars, and took the tangent plane at the point of initial contact to be a plane of mirror symmetry. He divided the star into a series of planes parallel to the mirror plane. In addition to considering the hydrodynamics of the forward motion, he also considered what would happen as the material in the successive slabs was heated up and proceeded to expand perpendicular to the direction of collision. His results indicate that stars colliding with mutual escape velocity lose approximately 7% of their total mass to infinity.

The work described below was done primarily by Fred Seidl, who used two dimensional hydrodynamics. With a two-dimensional calculation we are still limited to head-on collisions, and for accuracy it is still easier to consider a collision with a mirror image. (For a discussion of the technique of one-dimensional hydrodynamic calculations, see S. Colgate, this volume.) In the two-dimensional calculations, the difference equations are much more involved, and to obtain second-order accuracy it was necessary to take second order differences in the primary variables and third order differences in the artificial viscosity. In hydrodynamic calculations of spherically symmetric systems it is convenient to use Lagrangian coordinates. For the collision it is necessary to use Eulerian "space-fixed" coordinates. Near the surface of a stellar model, where the scale height is small, the density can change by orders of magnitude between two points on the coordinate grid. This explains the need for higher order differences. At low surface densities (less than about 10^{-6} gm/cm^3) the hydrodynamical calculations are not accurate. A good test for the computations is to simply try to move the star across the grid and observe whether deformation occurs.

In the figures below, two tones are used. The darker shading represents nonvolatile material (i.e. with negative total energy), and the lighter shading represents volatile material, which can presumably escape from the configuration. Also plotted are density contours for 1.0, 0.5, 0.1, 0.01 and 0.001 of the initial central density, and the shading extends down to 10^{-6} of the initial central density, which is taken as the surface of the star. Only one quadrant of the collision is shown, since the quadrants are mirror images. Velocity vectors are shown superimposed on the drawings.

Figures 1 through 8 show various stages in the collisions of the two polytropes corresponding to the initial condition that the relative velocity at infinity is 1000 km./sec. The following features should especially be noted:

1. When the stars first come into contact, the collision velocity is supersonic in the outer layers. A strong shock develops which ejects material in the central plane of

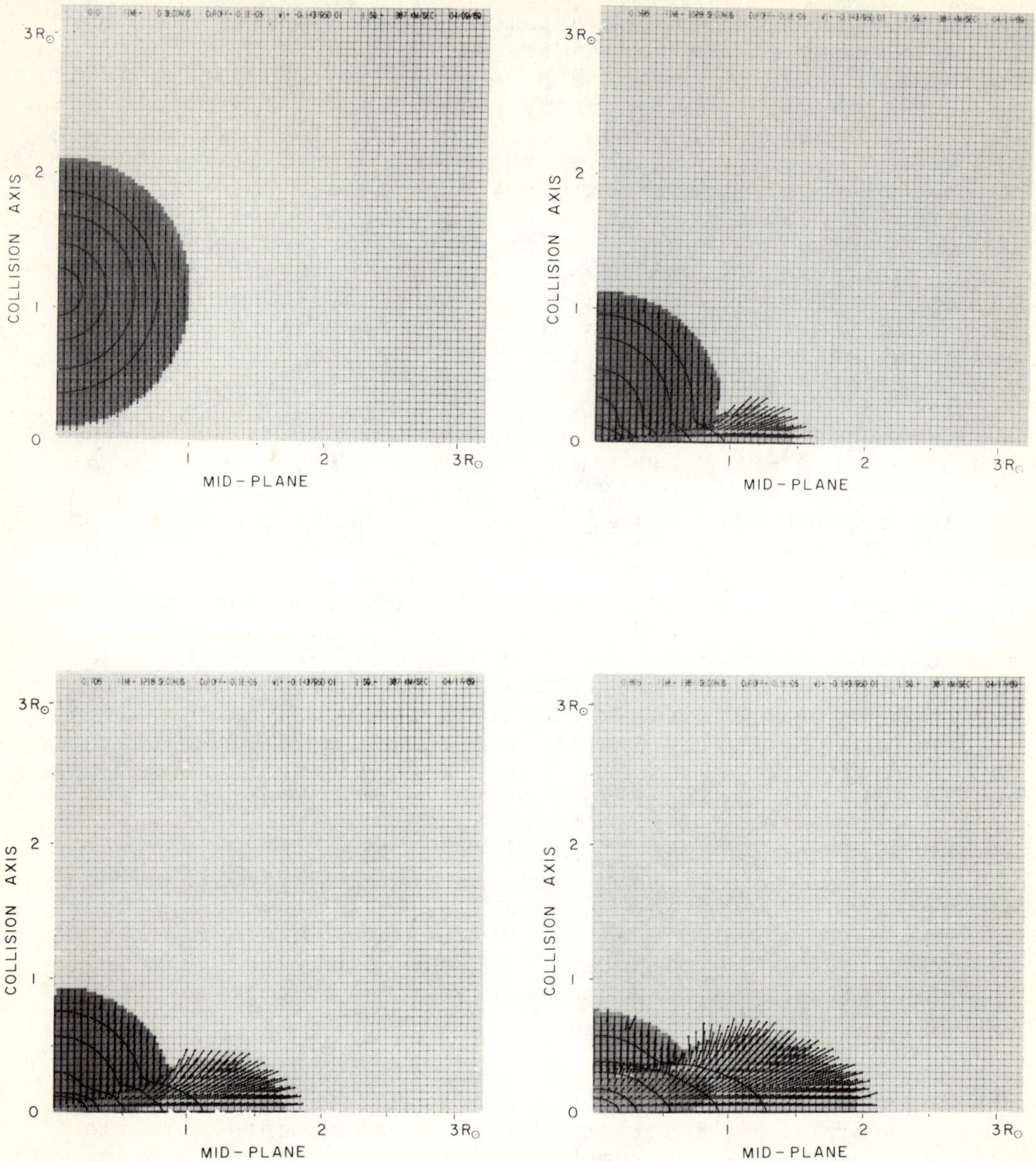
3R⊙
2
1
0
COLLISION AXIS
1
2
3R⊙
MID-PLANE
3R⊙
2
1
0
COLLISION AXIS
1
2
3R⊙
MID-PLANE
3R⊙
2
1
0
COLLISION AXIS
1
2
3R⊙
MID-PLANE
3R⊙
2
1
0
COLLISION AXIS
1
2
3R⊙
MID-PLANE

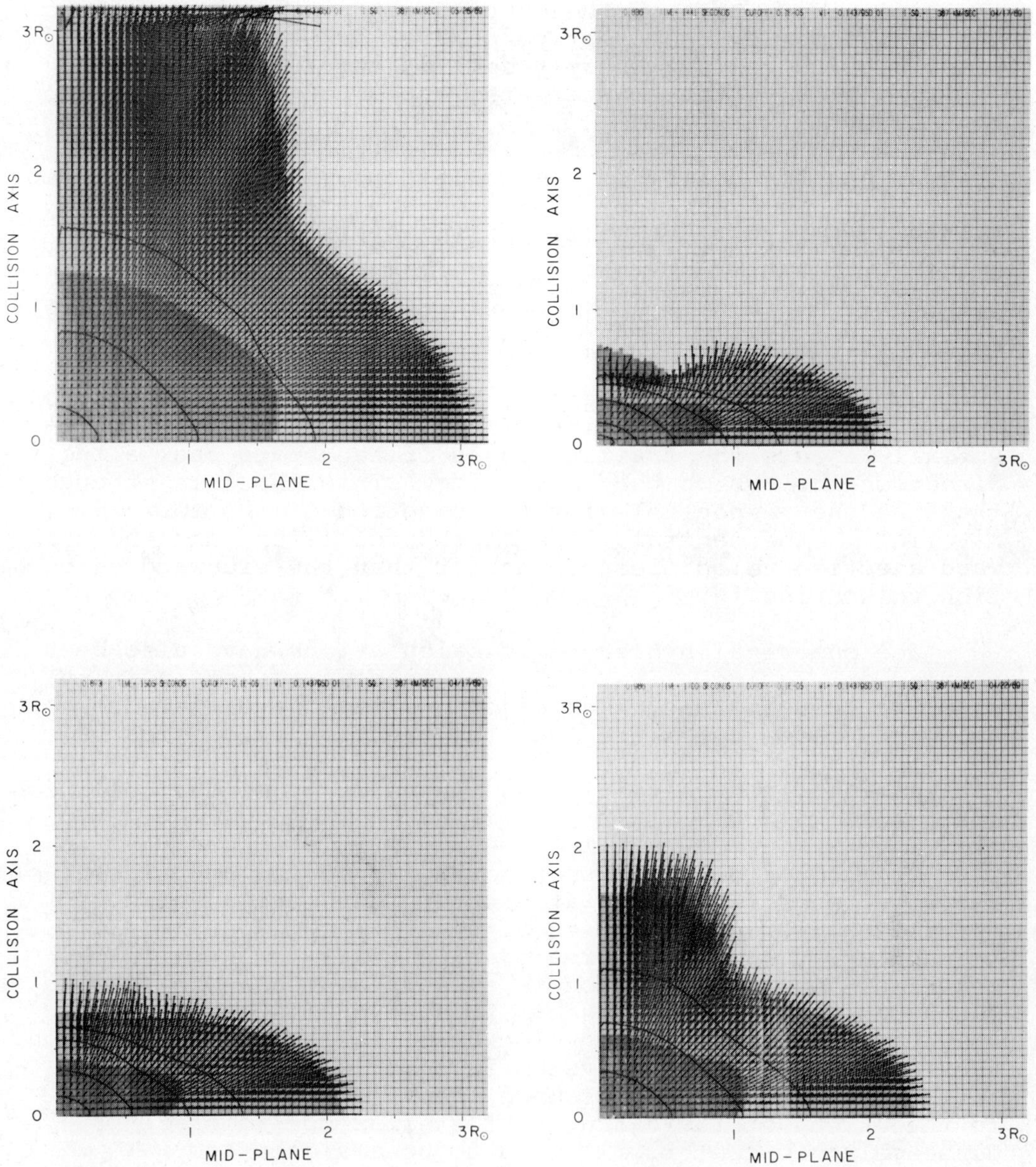
COLLISION AXIS
MID-PLANE
3R⊙
0
1
2
3R⊙
COLLISION AXIS
MID-PLANE
3R⊙
0
1
2
3R⊙
COLLISION AXIS
MID-PLANE
3R⊙
0
1
2
3R⊙
COLLISION AXIS
MID-PLANE
3R⊙
0
1
2
3R⊙

the collision. Some of the ejected material acquires a significant velocity component perpendicular to the central plane, and there is a sharply-defined curved shock surface in the stellar interior where the motion of the material is bent abruptly outward.

2. At the height of the collisions the amalgamated central region is moderately compressed relative to the initial central density of the participating stars. However, the duration of the collision is less than one hour, and maximum compression lasts for only about a minute, so that no significant number of nuclear reactions occurs during the collision.

3. In the later stages of the collision a strong shock develops along the direction of the collision which preferentially ejects the trailing edges of the stars back along the direction from which they have come. This shock is stronger than the one responsible for the ejection of the forward material in the central plane of the collision, and the backward ejection velocities are larger than the sideward ejection velocities.

The numerical hydrodynamic calculations have also been carried out for relative collision velocities (at infinity) of 0 and 2000 km./sec. The major qualitative differences among the three cases are that the ejection velocities of the sideward-ejected material increase rapidly relative to the backward ejection velocities as the velocity of the collision increases.

It was found that approximately 4, 17 and 60 per cent of the total mass is ejected at greater than escape velocity for initial collision velocities of 0, 1000, and 2000 km./sec., respectively. Thus the breakeven velocity below which the net result of the collision is stellar amalgamation lies in the vicinity of 1700 km./sec. This is of the order of but slightly greater than the velocity of 1512 km./sec. at which the total energy of the system is zero. These calculations support the stellar coalescence theory of Colgate, at least to the extent of indicating that stellar collisions in a dense galactic center will tend to be amalgamative.

This research has been supported in part by the National science Foundation and by the National Aeronautics and Space Administration. I am indebted to Mr. S. Ridgway for preparing a preliminary draft of these notes.

Bibliography

Colgate, S.A. 1967, Astrophys. J., 150, 163.

Mathis, J.S. 1967, Astrophys. J., 147, 1050.

OHIO UNIVERSITY LIBR